D1505031

Foundations of Neurobiology

Foundations of Neurobiology

Fred Delcomyn

University of Illinois at Urbana-Champaign

W. H. Freeman and Company

New York

Acquisitions Editors: Deborah Allen and Sara Tenney
Project Editors: Christine Hastings
Jane Judge Bonassar, Editorial Services of New England
Cover Designer: Diana Blume
Text Designer: Editorial Services of New England
Illustration Coordinator: Susan Wein
Illustration: HRS Electronic Text Management; Carlyn Iverson
Production Coordinator: Paul W. Rohloff
Compositon: Black Dot
Manufacturing: R R Donnelley & Sons Company

cover: Section through the hippocampus of a rat immunostained for paravalbumin. Superimposed on the section are axonal and dendritic arbors of intracellularly stained hippocampal interneurons. Image from T.F. Freund and G. Buzsáki (1996) "Interneurons of the hippocampus" *Hippocampus* 6(4):347–470, copyright © 1996 Wiley-Liss, Inc. Reprinted by permission of Wiley-Liss, Inc., a division of John Wiley & Sons, Inc.

Backcover: Intracellularly stained interneuron from the third thoracic ganglion of an American cockroach, *Periplaneta americana.*

The recording trace that runs across both covers is an intracellular record, also from an interneuron in *Periplaneta.*

Library of Congress Cataloging-in-Publication Data

Delcomyn, Fred.
Foundations of Neurobiology / Fred Delcomyn.
p. cm.
Includes bibliographical references and index.
ISBN 0-7167-2627-0 (EAN 9780716726272)
1. Neurobiology. I. Title.
QP355.2.D45 1997
612.8—dc21 97-10864
CIP

Printed in the United States of America

Second printing

To the memory of Graham Hoyle,
who saw this coming long before I did

Brief Contents

Contents

List of Boxes

Preface

I never wanted to write a textbook. In discussions with colleagues about the need for a neuroscience text that was not too narrow in its focus or oriented too strongly toward premedical students, I was loud and clear in disavowing any interest in providing the text that many of us were searching for. Yet here I am, preparing a preface for the book I said I would never write. I include this little bit of personal history because the reasons that finally persuaded me to put fingers to keyboard are really a summary of what you can expect to find in this book.

As an instructor of an introductory neurobiology course for undergraduates, I sought four important features in a text:

1. I wanted a book written for students who had solid grounding in biology, but who had not necessarily taken any advanced courses. I think it is important for students to be able to study neuroscience early in their careers, and I always felt that a good writer should be able to make difficult concepts understandable even for students who did not have much background.
2. I wanted a book that put facts in context. There is never a shortage of information—beginning students often drown in it. The trick is to present the specifics in a broad context. This not only makes the facts easier to remember, it makes them easier to understand as students begin to make the connection between individual experimental observations and the generalizations that are based on them.
3. I wanted a book that presented neurobiology in all its wondrous diversity. It should treat insects and frogs, cats and mollusks with equal ease, weaving a tapestry of general principles that drive home the important point that there is a unity of neural function shared by virtually all animals. There are principles of neural function that can be discovered by studying a diversity of animals, principles that would be difficult to uncover if only a few major groups were investigated. Even students headed for medical school benefit from (and often appreciate) such an approach.
4. I wanted a book that covered the gamut of modern neurobiological research, from molecules to behavior, from development to cognitive function. Students should come away from having studied such a book ready to begin further study in any area of neuroscience, to understand the scientific basis of current neurobiologically related social issues discussed in the pages or *Newsweek* or *Time,* and even to understand the importance and basic nature of the scientific questions being asked in neuroscience research papers.

There are several fine neuroscience texts on the market. Some of the best known are listed at the end of Chapter 1. However, none provide what I believe students need. Some are beautifully and clearly written but narrow in focus. Some are broad in coverage but difficult to understand. Some are too advanced. Some are simply overwhelming. In the end—well, here I am, writing my preface. I have tried to provide what I want for my students, on the grounds that what will benefit my own students will benefit others. That is, a clearly written book for undergraduates that assumes no advanced knowledge, a book that puts facts in the context of general principles of neural function, a book that is comparative in its coverage, treating invertebrate as well as vertebrate animals, and a book that describes neuroscience at all levels from molecules to behavior. You can think of these as the guiding principles around which the book is organized.

There is more to this book than these guiding principles, however. Pedagogical aids have been included to make the book easier to read and to learn from. For instance, no scientific idea about neural function can be understood fully without some knowledge of the experimental findings that led to the idea and support it. Thus, the book describes experiments as well as experimental results and the generalizations drawn from them. It is easy enough to digress to describe a method during a lecture, but I found that doing so in the text was disruptive to the flow of the discussion. For this reason, methods are described in boxes distributed throughout the book. This allows the reader to skip the experimental detail for a time to finish a section, then to go back and read about the method later. The boxes are also a valuable resource, providing concise and easy-to-find summaries of the main methods used in neuroscience research.

Each part of the book is preceded by a short introduction that gives an overview of and sets the context for the chapters in that part, not unlike how the overture sets the mood for each act of an opera. Each chapter is itself headed by a brief paragraph that sets out the central issue or theme dealt with in that chapter. This, too, helps to provide an important context for the details presented in the chapter. A summary follows each section within each chapter to serve as both a self-check for the reader and a quick review of key points.

Each chapter is followed by a list of additional readings—books, reviews, and research articles—that cover the topic of the chapter. In addition to the specialized works cited after individual chapters, there are a number of fine, general neurobiology texts. Rather than list them after each chapter, I have listed these general texts together at the end of Chapter 1. They provide other points of view or additional information on some topics.

In terms of subjects represented, the book is divided into six parts, each organized around a single theme. The first part is an introduction to neurobiology, from a description of the cellular elements of nervous systems to an overview of how those elements are assembled. The second part treats communication between neurons, from the membrane potentials that allow communication to occur to the synaptic transmission that is the means. The third, fourth, and fifth parts deal with the three basic functions of nervous systems: processing sensory input, organizing motor output, and integrating and making decisions about information. The final part considers the nervous system's changeability, starting with neural development and ending with the effects of hormones on neural function. Throughout the text, I have tried to tie in physiological detail to the ultimate behavior of an animal wherever possible. Further, although I believe a book that is too heavily oriented toward vertebrate neurobiology shortchanges students, students (and others) have a natural interest in how the principles of neuroscience apply to themselves, so I have provided as many human examples as possible.

No book like this could be completed without the help of many people. First and foremost, I wish to thank the students in the introductory neurobiology class at the University of Illinois, whose interest in and enthusiasm for the subject have served as a constant source of motivation for providing them with the best textbook possible. I also thank the neuroscience students taught by Carl Thurman at the University of Northern Iowa and Jack Kinnamon at the University of Denver for providing valuable feedback on early versions of the text.

Further, I wish to thank the reviewers and colleagues who gave me the benefit of their reactions to all or part of the text. I thank the following people, who read one or more chapters of the manuscript:

Ruth R. Bennett, University of Massachusetts at Boston
Phillip Best, University of Illinois at Urbana-Champaign
Peter Brodfuehrer, Bryn Mawr College
James Buchanan, Marquette University

Ansgar Büschges, University of Kaiserslautern
Susan Fahrbach, University of Illinois at Urbana-Champaign
Michael Fine, Virginia Commonwealth University
Barbara Finlay, Cornell University
Cole Gilbert, Cornell University
Ronald Hoy, Cornell University
Janice Juraska, University of Illinois at Urbana-Champaign
David Jensen, Tomball College
Gregory Lnenicka, State University of New York at Albany
Jim Nardi, University of Illinois at Urbana-Champaign
Peter Narins, University of California at Los Angeles
Kiisa Nishikawa, Northern Arizona University
Donata Oertel, University of Wisconsin
Marc Roy, Beloit College
Andrea Simmons, Brown University
Harold Zakon, University of Texas at Austin

I thank the following people, who read the entire text:

Tom Anastasio, University of Illinois at Urbana-Champaign
Joseph Bastian, University of Oklahoma
Douglas Eder, Southern Illinois University at Edwardsville
Steve George, Amherst College
Mark Nelson, University of Illinois at Urbana-Champaign
Dennison Smith, Oberlin College
Carl Thurman, University of Northern Iowa

I also thank the following colleagues who contributed suggestions, advice and encouragement, all of which have helped make this a better book: David Bentley, Akira Chiba, Tom Christensen, David Clayton, Neal Cohen, Lars-Gösta Elfvin, Eve Gallman, Tomas Hökfelt, Eric Horn, Darcy Kelley, Sue Kinnamon, David Lampe, Ellis Macleod, Joe Malpeli, Ed Roy and Bruce Wheeler.

Original photos or micrographs, many previously unpublished, were kindly provided by Deborah Allen, Art Arnold, Rod Bates, Dennis Baylor, Bill Bosking, Dennis Bray, Peter Brink, György Buzsáki, Rene Couteaux, Chris Comer, John Conner, Darryl Daley, Hannah Damasio, Bob Doyle, Marla Feller, Bill Greenough, Felix Gribaken, Tove Heller, Akimichi Kaneko, John Kauer, Chen Liu, Margaret Livingston, Joe Malpelli, Venkata Mattay, Bruce McNaughton, James Pickles, Marcus Raichle, Cedric Raine, Aloicia Schmid, Constantino Sotelo, Jim Truman, Bruce Ware, and Steve Zotolli. Original photos or micrographs are credited in the figure caption. Others are credited at the end of the book. Uncredited micrographs are my own. Space limitations preclude mention here of the many colleagues who answered my e-mail questions or gave me permission to use their data.

No compendium of those who contributed to making this book what it is would be complete without mention of the people at W. H. Freeman and Company who contributed to this book: my editor, Deborah Allen, who from the first shared with me a vision of what the book should be; Tina Hastings, who expertly coordinated the book's production; Trimette Roberts, who dealt with the many details of a project of this magnitude; and Liz Widdicombe, who saw the potential of the book and committed the resources necessary to make it a reality. Further, the book could not have been done without the expert work of Susan Ecklund and Jane Judge Bonassar in copyediting the manuscript and overseeing its production, or the incredible efforts of the artists at Hudson River Studios, who did the illustrations.

Finally, I want to thank my family for their forbearance during the course of this project. My wife, Nancy, and my children Julia, Michael, and Erik gave up vacations and family time so that this project could be finished. Fortunately, they are all old enough to understand why dad was rarely home in the evenings. Those now in college may even have a better understanding of the lives of their own professors!

As should be clear from these acknowledgments, this book, although written by a single author, is in many respects a collaborative effort. Not only did many people contribute to its production, many specialists and teachers in the neuroscience community contributed their expertise to make the book accurate, up-to-date, and clear. In many respects the book can be considered a collective effort of the entire neuroscience community, without whose research there would be nothing to write about. If this book conveys even a small part of the excitement and potential of contemporary neuroscience research, it will have served its purpose.

Urbana, Illinois
October, 1997

Foundations of Neurobiology

PART 1

Elements of Neurobiology

Neurobiology is the study of the nervous system, the system of the body that confers on animals the ability to sense the environment, to process information, and to move around. From the detection of a featherlight touch to the perception of a Monet painting, from the decision to eat a morsel of food to the contemplation of Newton's law of gravity, and from the slow churning of a stomach to the exquisite movements of a ballerina, the nervous system regulates and controls almost every aspect of the lives of animals and humans alike. The nervous system is able to carry out its functions because of the properties of its constituent cells and the way they are organized. The chapters in this part lay the foundation for your study of neurobiology by introducing the nervous system and what it does, the characteristics of the cells found in it, and the anatomical arrangements that make their action possible.

CHAPTER 1

Introduction to the Nervous System

The science of neurobiology encompasses the study of the nervous system in animals as diverse as jellyfish and humans, and stretches from the roles of individual molecules in neural function to the expression of neural activity as behavior. The ultimate objective of neurobiological research is to understand how the nervous system carries out its functions. The goal of this book is to describe how far researchers have come toward reaching this objective and to set out the fundamental principles that are the foundation of neurobiology.

"They move!" This is certainly one answer you will likely hear if you ask a child what sets animals apart from other living things. The ability of animals to move has fascinated people for millennia. During most of human history, people could only speculate as to how movement was possible. From a study of primitive cultures and early historic records, we know that for thousands of years, people believed there was some nonmaterial spirit that inhabited both animals and humans, a spirit that made movement possible. Many ancient Egyptians, and later Greeks, gradually came to recognize an association between the brain and behavior, but even Aristotle, who made so many fundamental contributions to our understanding of the natural world, did not make the connection, believing instead that the function of the brain was to cool the blood. Today, of course, even children in elementary school are taught that the nervous system is responsible for the behavior of all animals, including humans. However, this biological viewpoint has become firmly established only in the past few hundred years.

The Brain and Behavior

People throughout history certainly had ample evidence to support the view that the brain is responsible for what a person or an animal does. After all, the profound effects of head wounds acquired during war were well known to ancient physicians. Accidents also provide many clues to the role of the brain in behavior.

Perhaps one of the most dramatic examples of such an accident occurred in 1848 to a young railway foreman named Phineas Gage. Before his accident, Phineas was described as a likable, calm individual, a person who did not swear much and who was generally quite responsible in his behavior. An accident with a tamping rod changed his life. A tamping rod is a metal rod, more than a meter long and pointed at one end, that was used to set up controlled explosions to break up rock for railroad rights-of-way. Normally, gunpowder was placed into a hole drilled in the rock, sand was poured over the powder and tamped in with the rod, and the powder was ignited with a fuse. Distracted during the course of his work, Phineas dropped the rod into a hole that contained the explosive but no protective layer of sand. The rod produced a spark when it struck the side of the hole, and the resulting explosion shot the tamping rod, point first, into Phineas's face just under his left cheekbone, up into his brain, and out through the top of his skull (Figure 1-1). It was found later, still covered with evidence of its journey.

Amazingly, the passage of an iron rod completely through his head did not kill Phineas. Physically, he recovered fully after a period of some months. However, he was a changed man. He tried to regain his job as railroad foreman but was refused. In place of the calm, responsible individual he had been before the accident, he was now irritable, impulsive, and given to fits of temper if his desires were thwarted. He swore frequently and was quite childlike in many ways. The physician J. M. Harlow, who studied Phineas, described changes in his intellectual capacity as well, but what most caught the attention not only of the medical community but also of Phineas's friends was the complete change in his personality. This change seemed obviously related to the massive injury sustained by the frontal part of his brain, the main part damaged in the accident. Phineas lived to be 37, dying nearly 13 years after the accident that changed not only his life but the man himself. He never regained his former personality.

This famous case clearly illustrates that there is far more to behavior than simply the control of movement. Indeed, all the aspects of a person's behavior that we think of as personality, as well as intellectual abilities and talents, stem from the actions of the nervous system. It is the objective of scientists who study the nervous system to explain every aspect of behavior in terms of neural function.

Historical awareness of the relationship between the brain and behavior developed from study of injuries to the head. One of the most dramatic demonstrations of the relationship between brain and behavior is the striking personality change of a man who had most of the frontal part of his brain destroyed in an accident.

A

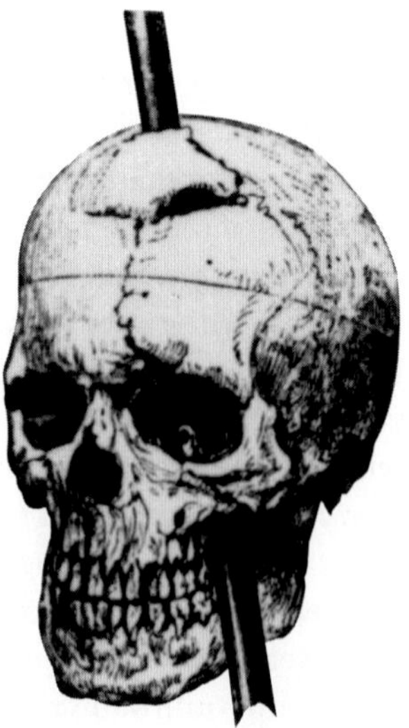

B

FIGURE 1-1. (A) The skull of Phineas Gage, showing the route of the tamping rod through his skull. The angle of entry of the rod shot it behind the left eye and through the front part of the brain, sparing regions at the base of the brain that control vital functions like breathing and the beating of the heart. (B) A computer reconstruction of Phineas's skull and the rod, made to determine as accurately as possible which areas of his brain were destroyed.

What Does the Nervous System Do?

In spite of the many instances in which wars and accidents have shown that damage to the brain can irrevocably affect a person's actions or behavior, it is not all that easy to grasp just what the nervous system actually does, to say nothing of how it does it. Take the simple example of a cat walking along the floor. For just this action, the cat must coordinate the actions of the 15 or so muscles in each leg, as well as seeing to it that the four legs act together so that she maintains her balance while also moving forward. Now suppose that having smelled the remnants of your dinner, she jumps onto the table to snatch a piece of chicken you had left there. The requirements for this simple action—identifying the smell of the chicken, deciding that she wants to have the chicken more than she wants to avoid the swat of the newspaper that may follow, as well as the coordination of gauging her leap, grabbing the chicken, and clearing the edge of the table as she jumps down—are truly staggering.

Part of the problem students and scientists alike face in understanding neural function is that there is no ready analogy to help picture the nervous system in action. Consider the heart as an example of the usefulness of analogy. When the heart is compared to a pump forcing blood through the blood vessels of the body, its operation is immediately apparent even to the most biologically naive person. All of us are likely to be familiar with pumps as machines, so we can easily picture the heart working as a pump, and we can relate the mechanical structure of a pump to the organic structure of the heart. But to what can we compare the nervous system? Until the advent of computers, there was nothing at all that could serve as a reasonable analogy. Unfortunately, even a computer analogy is not satisfactory because computers actually work quite differently from the way nervous systems do. Furthermore, most people find it hard to imagine how anything machinelike could possibly be the basis of their own unique mental experience and individuality.

A way around this problem is to think of the function of the nervous system in terms of specific tasks instead of the rather nebulous function of generating behavior. Put simply, the nervous system can be considered to have three main tasks: to receive and interpret information about the internal and external environment of the body, to make decisions about this information, and to organize and carry out action. These functions are obviously interrelated, but it is useful to consider them separately.

To receive and interpret information about the internal and external environment is the task of what we call the **sensory system**. The sensory system is usually defined to include all the sense organs of the body, the nerves that carry information from them, and the areas of the nervous system that are responsible for interpreting sensory information. Your cat's ability to identify an odor as chicken and to determine that it comes from the table are both consequences of the operation of her sensory system (Figure 1-2A). Internal cues, such as the amount of force being generated by the muscles in her legs, are also processed by the sensory system.

To organize and carry out action is the task of what we call the **motor system**. The motor system is usually defined to include all of the muscles and ducted glands of the body, the nerves that control them, and the parts of the nervous system that are necessary for their activation. The ability of your cat to walk to the table, jump onto it, grab the chicken, turn, and then jump down are all manifestations of the motor system (Figure 1-2C). Each of these actions has its own requirement for balance and coordination.

To make decisions about information is the task of what we can call the **integrating system**. The decision to eat some chicken and the direction of her movements toward and onto the table are examples of the integrating system in action (Figure 1-2B). Actually, the tasks of integrating information from sense organs, calling on the stored record of previous experiences (memory), and making decisions about both the new and the stored information are all part of what most people

A

B

C

FIGURE 1-2. The nervous system has three main functions. (A) It receives sensory input, like this cat smelling chicken on the table. (B) It integrates (makes decisions about) this input, like the cat deciding to snatch a piece of the chicken. (C) It organizes motor action, like the cat leaping to the tabletop, grabbing the chicken, and jumping back down.

consider the central role of the nervous system, and usually no special name is given to the parts that carry them out. However, just to make it easier to think about how the nervous system works as a whole, the term integrating system can be used to refer to the parts that carry out these operations.

The brain's integrating systems are at the core of any animal's behavior, including that of humans. Phineas Gage's personality, both before and after his accident, was an expression of action of the integrating systems of his brain. Even in primitive animals, the brain does more than simply serve as a relay between input and output. Your cat probably will not be interested in your dinner if she has just eaten. If you have tried to keep her off the table on previous occasions, she will probably not risk a swat if you are standing nearby, even if she is a little hungry. She must assess her bodily needs and balance these needs against the likely consequences of her actions, based on her unique experience. This is a complex process that requires many different parts of her nervous system and involves much more than merely selecting and coordinating a sequence of movements.

The nervous system is responsible for generating and coordinating an animal's behavior. It does this by carrying out three interrelated functions. The sensory system receives and interprets information from the internal and external environment, the motor system controls the actions of muscles and glands, and the integrating system makes decisions about the actions to take, based in part on past experience and current conditions.

The Field of Neurobiology

The study of the nervous system is the purview of **neurobiology.** Some researchers prefer to use the term **neuroscience** for the field, including under this term medical disciplines such as neurology. Neurobiology is a relatively new field, but one that has seen astonishing growth over the past several decades. For example, the Society for Neuroscience, the North American organization devoted to neurobiology, has grown from an initial membership of 500 in 1969 to a membership of over 26,000 in 1996. Similar growth has been experienced by neuroscience organizations in other parts of the world.

One reason for the great scientific interest in the study of the nervous system is that recent technical advances have made it possible to start answering some of the most fundamental questions about neural function. How do genes regulate the types of proteins embedded in nerve cell membranes, and how does the presence of these proteins influence what a nerve cell may do? How do nerve cells convey information to one another? How is information about sensory stimuli encoded and interpreted by the brain? How are complex movements planned and coordinated? How does the intricate and highly specific organization of the nervous system come about during embryonic development? What are learning and memory, and how do they occur? What parts of the brain are used in human cognitive tasks such as thinking, and how do these parts carry out their functions? These are only a small sample of the fundamental questions that are successfully being addressed at the molecular, cellular, and behavioral levels by modern neuroscience research. In this book you will learn the answers to some of these questions. You will also find that many questions do not yet have satisfactory answers.

The excitement in neuroscience is not restricted to the scientists who do the research; it has reached the general public as well. Public interest stems from two sources. First, there is tremendous intellectual interest in how the brain works, both from the standpoint of simple curiosity and for the practical implications such knowledge has for social policy. The very idea that it might be possible to describe a thought strikes a responsive chord in many people. For this reason, television specials and newspaper and magazine articles that deal with the brain and its functions appear quite regularly. People are also intensely interested in how the brains of exceptional individuals work—on the one hand to promote the development of genius, such as for music or mathematics, and on the other hand to control extreme antisocial or aggressive behavior.

The second source of public interest is in the practical applications of neuroscience research. The only hope for treatment of diseases such as Alzheimer's or Huntington's, and for the effects of events such as a stroke or a spinal cord injury, is a thorough understanding of the way in which the brain normally carries out the functions that are affected by these ailments. In some cases, scientists are still far from providing much useful information for physicians. In other cases, research has provided new knowledge that is helping the afflicted. In the mid-1990s, for example, new treatments for stroke and spinal cord injury have proved extremely effective in limiting the effects of these events on the nervous system when the treatments are administered quickly, whereas even in the late 1980s very little could be done for a person who suffered from either.

In addition to the impact of new knowledge on treatment of disease or injury to the brain or spinal cord, the marriage of medicine and engineering has opened possibilities that until recently were confined to the realm of science fiction: the development of a "bionic" person. As depicted in a popular American television series of the mid-1970s, such a person could have surgically implanted artificial organs such as muscles and eyes that could be used just like the natural organs but would be an improvement over the original. Although biomedical engineers cannot yet offer an artificial nose, much less one that is more sensitive than the original, work is currently being carried out on artificial eyes and ears that will send their signals to the proper processing centers in the brain for interpretation just like signals from the natural sense organs. An artificial limb controlled directly by the central nervous system through a microchip interface implanted in the stump of the natural limb is yet another possibility that is currently being explored. All these practical applications are possibilities not only because of advances in microelectronics but also because of new knowledge about how the nervous system works.

Neurobiology is the study of the nervous system and its function. It is a subject that has attracted considerable interest recently from both researchers and the general public. Some of this interest stems from curiosity about how the brain works, and some from the intriguing possibilities that now exist for the practical use of neurobiological knowledge in medicine.

Studying the Nervous System

Much of the fascination with neuroscience among the general public and students alike naturally centers around the human brain. However, to characterize neuroscience as the study of the brain alone, and just the human brain at that, does an injustice to the extraordinary richness and diversity of the field. Neurobiology as a discipline encompasses not just the brain, not even just the nervous systems of higher animals, but every aspect of neural function in every animal. Not only can you best gain a full appreciation of the richness of neurobiology by studying the nervous systems of all animals, but such an approach will actually enhance your understanding of the human brain.

The Comparative Approach

Even the simplest animals have nervous systems and can exhibit complex and interesting behavior. Consider, for example, the cnidarian *Stomphia coccinea* (Figure 1-3A), a sea anemone that lives off the northern Atlantic and Pacific coasts of North America. If you touch this anemone with an inanimate object, it will either ignore the touch or withdraw its tentacles and contract its body, depending on how hard you touch it (Figure 1-3B). Drop a piece of fish on it, however, and the anemone reacts quite differently. It discharges its stinging nematocysts into the fish, curls its tentacles up around it, and brings the morsel into its mouth (Figure 1-3C). A third type of stimulus elicits yet another response. If *Stomphia* is touched by the predatory starfish known as the Pacific leather star, *Dermasterias imbricata*, the anemone will elongate its column, spread its tentacles, detach itself from the place where it has been anchored, and "swim" away by vigorously bending from side to side (Figure 1-3D).

These reactions illustrate the three basic tasks of a nervous system: sensory processing, motor organization, and decision making. *Stomphia* obviously discriminates between the different sensory stimuli. Light touch by itself causes no reaction, but strong touch will cause the anemone to withdraw. One particular chemical stimulus causes feeding, and another elicits swimming. Each of these reactions requires very different types of motor activity. Furthermore, the reactions are not just reflexive; they require a kind of decision making. Is the stimulus a touch? Then how hard is it? Is a chemical substance present as well? Then which chemical? Even an animal as simple as an anemone is obviously ableto analyze the combination of stimuli thatit receives in order to determine the appro-priate response.

Nevertheless, the question remains as to whether studying the neurobiology of animals other than mammals is of any real value to a researcher wanting to understand human neural function, or to you as a student whose interest may be primarily in human medicine. If the objective is to understand the fundamental principles by which nervous systems operate, there is no doubt as to the answer. Studies of animals from nearly every major phylum have contributed in important ways to our understanding of neural function.

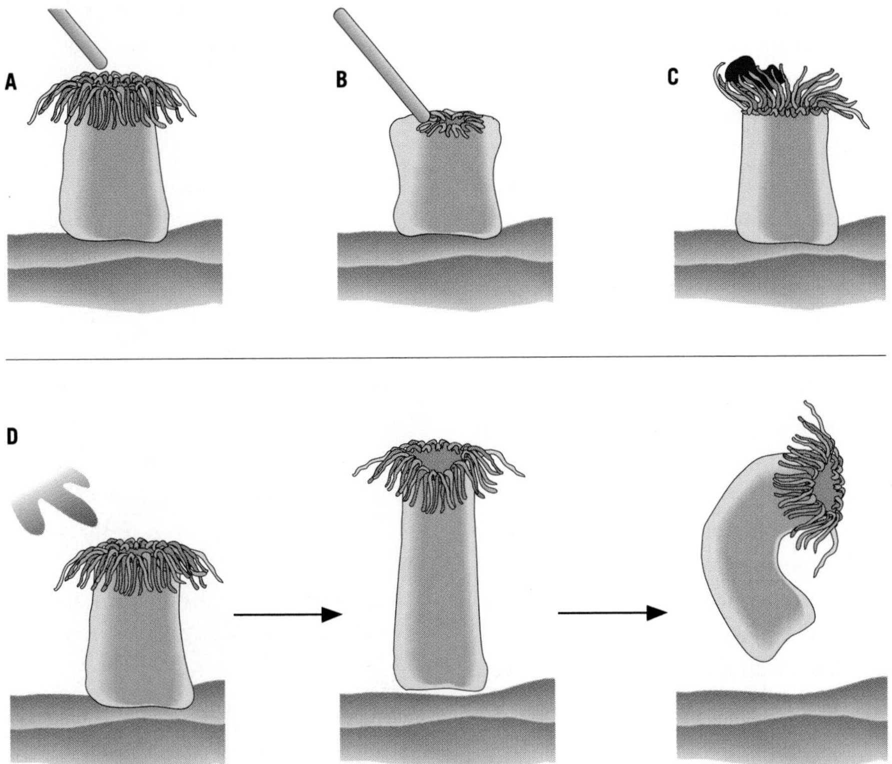

FIGURE 1-3. The sea anemone *Stomphia* will react differently to a mechanical stimulus depending on how strong it is and on chemical stimuli that may accompany it. (A) A gentle touch has no effect on the anemone. (B) A vigorous touch will cause it to withdraw. (C) The touch of a piece of fish, which is accompanied by many chemicals, causes the tentacles to curl up and around the fish before ingestion. (D) The touch of a predatory starfish brings about a complex response involving detachment and swimming.

Consider a few examples. At the cellular level, study of the nerve cells of squid have revealed the ionic basis of signal conduction in neurons (described in Chapter 5), and studies of frogs have revealed the ionic basis of chemical communication between nerve cells (see Chapter 7). Much of our knowledge of sensory systems also comes from studying various animals. Many of the properties of sensory nerves as they react to stimuli were first worked out in a horseshoe crab, a type of marine arthropod (see Chapter 9), and the ways in which sense organs on the tongue respond to different tastes was determined by studying salamanders (see Chapter 13). Even some of the complex actions of nervous systems have been revealed in so-called simple animals. Study of flying in locusts has helped to reveal the neural basis of many simple types of behavior (see Chapter 16), and a possible molecular basis of learning, a phenomenon typically associated with human behavior, has come from study of a simple sea slug (see Chapter 24).

Even from such an abbreviated list, you may recognize that major discoveries have been made from investigations of many types of animals. The question remains as to why a researcher should even consider studying the nervous system of a squid, locust, or frog. There are three good reasons. First, a particular animal may be selected as a subject for study because that animal is suitable for particular types of experiments. In the parlance of the field, the animal presents a favorable "preparation" for the study of a particular neurobiological process of interest. Second, a researcher may simply be curious about and interested in the diversity of solutions that have been devised by nervous systems to deal with a particular problem (Figure 1-4). Third, by studying the nervous systems of many different animals, it may be possible for researchers to identify general principles of neural organization that cross phyletic lines. These general principles can then be tested in other animals and, if valid, can suggest avenues for further research and increase researchers' understanding of all nervous systems, including those of humans.

It is because of the importance of studying many kinds of animals in neurobiological research that a comparative approach has been adopted in this book. Even if you are especially interested in human neurobiology, you would fail to learn about significant neurobiological findings if you were presented only with information gained from a study of primates, or even mammals. Furthermore, a major emphasis of the book is that there are common principles of neural organization that cut across phyletic lines. Hence, only a description of studies of many different kinds of animals can support these principles. Finally, a comparative point of view will enable you to appreciate better that the best approach to a fundamentally significant problem, even one relating primarily to human medicine, may require the study of a lowly worm.

FIGURE 1-4. A neurobiologist at work. Many important scientific discoveries have been made as a result of experiments carried out to satisfy simple curiosity. (Courtesy of Tove Heller.)

The nervous systems of even the simplest animals show the same three functions that are exhibited by vertebrates: processing sensory information, organizing motor output, and making decisions. Hence, comparative study of a diversity of animals can lead to new insights into general principles of neural organization.

Levels of Organization

Neurobiology not only involves the study of all types of nervous systems but also encompasses study of every aspect or level of neural function, from molecules to behavior. Specific molecules affect the activities of individual cells, the activities of cells affect the operation of the networks of which they are a part, and the operation of the networks affects the ultimate behavior of the animal (Figure 1-5). Thus, for you to understand how the nervous system operates as a unified system, it is important to think about the consequences that particular processes may have for neural function at many levels.

Consider one dramatic behavioral consequence of molecular events. In Japan, the puffer fish—a scaleless fish, named for its habit of inflating itself with air when it is caught—is considered a great delicacy. Unfortunately, like many other tropical scaleless fish, puffer fish are poisonous, and eating the fish can be fatal if it is not prepared properly. The internal organs of puffer fish contain tetrodotoxin, a neurotoxin that interacts with proteins in the membranes of nerve cells. As you will learn in Chapter 5, these molecules are essential for the cells to carry messages. The presence of tetrodotoxin blocks the membrane proteins, leaving the nerve cells unable to carry out their usual function. Proper preparation of the fish can make it safe to eat, but a poorly trained chef may cause the death of the unfortunate gourmet; ingestion of tetrodotoxin causes the nerves supplying the muscles of the diaphragm to cease working, leading to suffocation.

The precise structure of the membrane protein allows it to help in the conduction of neural messages. That same molecular structure also allows tetrodotoxin to bind with it and to interfere with its normal function. Hence, events at the molecular level have profound effects at higher levels; the binding of tetrodotoxin starts a deadly cascade by binding with the membrane proteins, causing malfunction of individual nerve cells, the failure of those cells in networks, and, finally, the death of the individual. Although certainly not every molecular event has a readily identifiable and specific behavioral consequence, and not every behavioral event has a single molecular cause, there are many instances like this in which a direct cause-and-effect relationship exists.

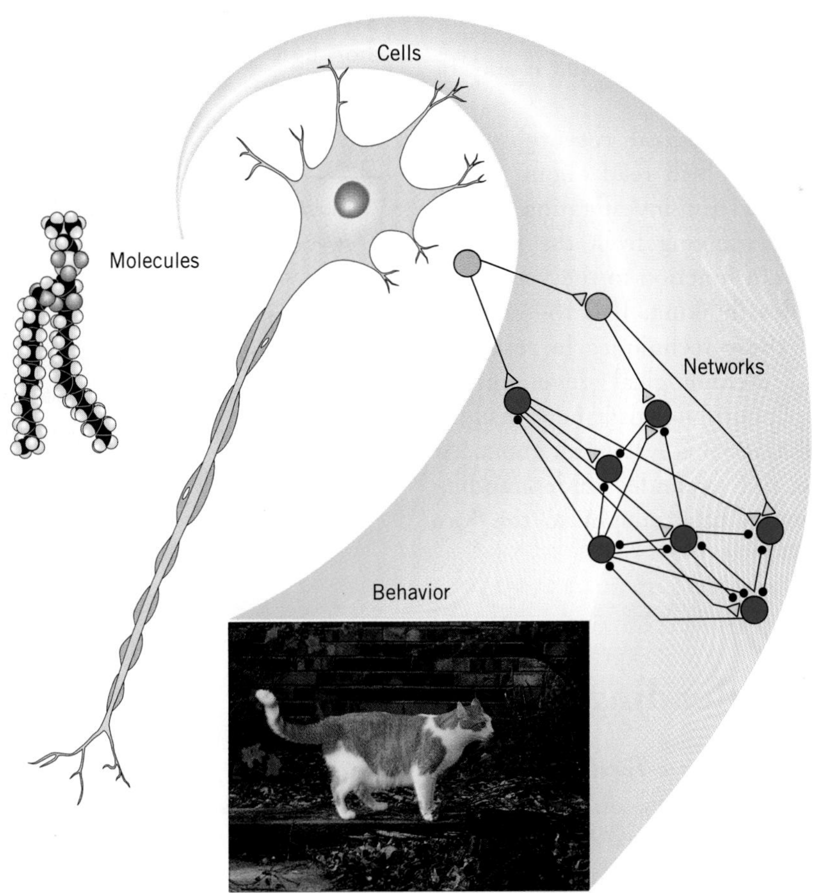

FIGURE 1-5. Levels of organization in nervous systems, from molecules to behavior. Events at one level may have clear and identifiable effects at the next higher level. (Photo courtesy of Deborah Allen.)

Neurobiology encompasses the study of the nervous system at many levels, from individual molecules to complex behavior.

The Future of Neurobiology

Neurobiology is an exciting and vibrant field of biology in part because it deals with some of the most riveting intellectual questions about the nature of our own existence. What determines each person's capabilities and talents? What determines personality? In short, what makes us human? These questions focus on issues of compelling importance to many people. Yet, these kinds of questions have been asked for centuries. What excites scientists now is the real possibility that it may be possible to answer them.

Modern neurobiological research techniques, such as you will read about in this book, have opened for investigation fundamental questions ranging from the roles of molecules in brain function to the way brain activity underlies thinking. It is the creative application of these techniques in research, and continual technological development, that has given neurobiologists new insights into the way the nervous system functions. At the same time, this expanded understanding of neural function has brought to the fore important moral issues. If it can be shown that certain types of early experience have a profound impact on a person's capabilities in later life, how should we as a society respond? If it becomes possible to identify children who are at risk for extreme aggressive behavior, how should we treat them? These and other social issues that arise from the knowledge that research brings have no easy resolution, but without doubt, factual information about these matters is a better basis than ignorance for making decisions.

The study of neurobiology, whether it is the first step in a lifelong career or a diversion on the way to other subjects, will equip you to understand the scientific facts that underlie discussion of these issues. It may also leave you with a sense of wonder at the capabilities of the three pounds of tissue that sits in your head, quietly allowing you to understand these words. And there is still considerable room for wonder. Neurobiologists still do not have a thorough understanding of the reason the destruction of part of the brain of Phineas Gage had the consequences for him that it did.

Much of the current interest in neurobiology stems from new knowledge and insights into brain function that modern techniques have brought. This new knowledge presents modern society with difficult ethical choices. Study of neurobiology can provide you with information on which to base your own choices.

Additional Reading

Neurobiology/Neuroscience Textbooks

Bear, M. F., B. W. Connors, and M. A. Paradiso. 1996. *Neuroscience: Exploring the Brain*. Baltimore: Williams and Wilkins.

Dowling, J. E. 1992. *Neurons and Networks: An Introduction to Neuroscience*. Cambridge, Mass.: Belknap Press.

Kandel, E. R., J. H. Schwartz, and T. M. Jessell, eds. 1991. *Principles of Neural Science*. 3d ed. New York: Elsevier.

Nicholls, J. G., A. R. Martin, and B. G. Wallace. 1992. *From Neuron to Brain*. 3d ed. Sunderland, Mass.: Sinauer.

Purves, D., G. J. Augustine, D. Fitzpatrick, L. C. Katz, A. S. LaMantia, and J. O. McNamara, eds. 1997. *Neuroscience.* Sunderland, Mass.: Sinauer.

Shepherd, G. M. 1994. *Neurobiology*. 3d ed. New York: Oxford University Press.

Books on the History or Philosophy of Neuroscience

Changeux, J.-P. 1985. *Neuronal Man: The Biology of Mind*. New York: Oxford University Press.

Finger, S. 1994. *Origins of Neuroscience: A History of Explorations into Brain Function*. New York: Oxford University Press.

Jacobson, M. 1993. *Foundations of Neuroscience*. New York: Plenum Press.

Research Article

Damasio, H., T. Grabowski, R. Frank, A. M. Galabarda, and A. R. Damasio. 1994. The return of Phineas Gage: Clues about the brain from the skull of a famous patient. *Science* 264:1102–5.

CHAPTER 2

Cellular and Molecular Building Blocks

The ability of nerve cells to receive and transmit information is the functional basis of all nervous system activity. It would be easy to think that this signaling ability derives from special and unique properties of nerve cells. Yet it actually derives from molecules and organelles that are present in nearly all cells. In this chapter, you will learn about the cells that make up nervous systems, and about the molecules and subcellular structures that determine the special properties of nerve cells.

The nervous system is a complex organ system that is uniquely challenging to study. In most organs of the body, the cells found in a particular type of tissue are relatively uniform in structure. Not only are muscle cells in your legs similar to those in your neck, but they also are similar to the muscle cells in the legs and neck of a grasshopper. But in the nervous system, diversity is the rule. Nerve cells, also known as **neurons**, can be so different that it is hard to believe they all belong to a single organ system. Although they do typically have a characteristic branching structure that allows them to be recognized, this structure can be quite different from cell to cell. Some neurons are small, some large; some are highly branched, some have little branching; some have long processes, some have only short ones. Neurons can be different even in the same part of an animal's nervous system. For example, Purkinje cells and granule cells are two types of neurons found together in a part of the vertebrate brain known as the cerebellum. Despite this proximity, they do not look at all alike (Figure 2-1).

The Neuron Doctrine

The history of scientific investigations of the nervous system reflects the difficulty that the complexity of neural structure has posed for scientists. One of the most significant achievements of nineteenth-century biology was the formulation of the **cell theory**. In 1838, the botanist Matthias Schleiden suggested that all plants consisted of small individual units called cells. Theodor Schwann, noting in 1839 that animal tissue also consisted of similar units, elaborated on Schleiden's suggestion and proposed that cells were the smallest living units of all life, the postulate we call the cell theory. In subsequent years, biologists examined every type of tissue imaginable to test Schleiden and Schwann's theory. By the last two decades of the nineteenth century, the universality of the cell theory had been accepted for every tissue in the body—except nerve tissue.

As late as the mid-1880s, many neuroanatomists still supported the **reticular theory**, the idea that the nervous system is a

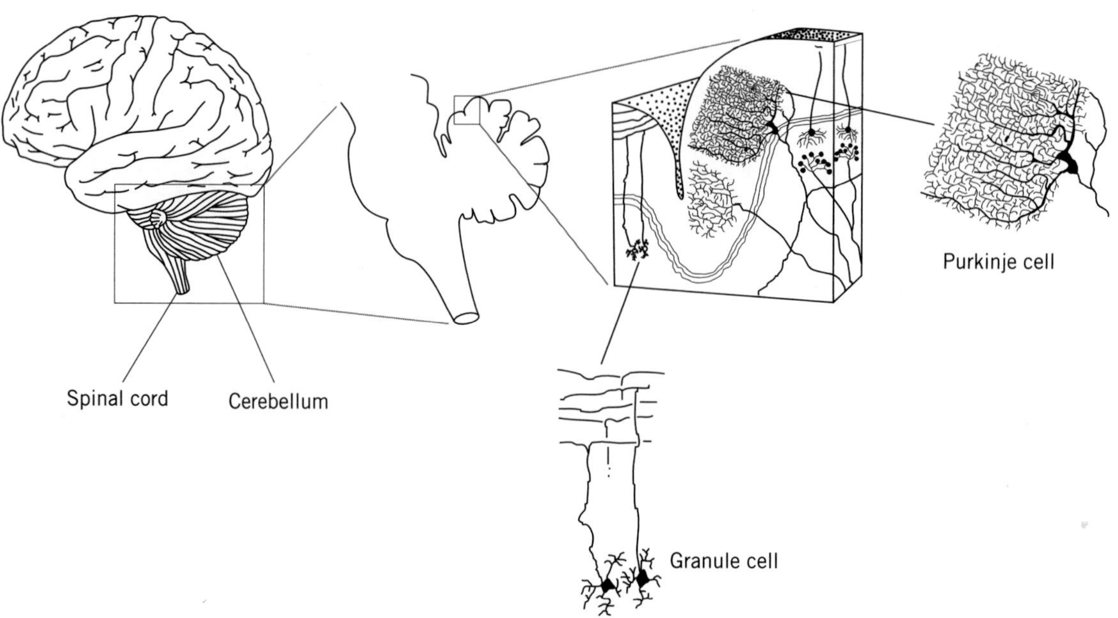

FIGURE 2-1. The diversity of neuronal types. Even neurons in a single region of the human brain, such as the cerebellum shown here, may contain cells that are quite different in size and shape from one another. At the left is a whole brain. Subsequent enlargements are of sections cut through the brain to show single cells in the cerebellum.

syncytium—a network of living material having multiple nuclei and cytoplasmic continuity from one place in the network to another. Other anatomists supported the **neuron theory** (also called the **neuron doctrine**), the idea that the nervous system is cellular. That is, they thought that the nervous system, like other bodily systems, was composed of discrete cellular elements, each having just one nucleus and being entirely surrounded by its own membrane.

The argument raged on because of the lack of sufficient optical power to resolve individual cell membranes at the points of contact between cells. The resulting uncertainty meant that proponents of the neuron theory could use only indirect arguments to support their position. They used three main lines of argument. First, neurons are clearly seen to be individual cells during embryonic development; therefore, it was argued that they ought to remain separate in adult animals. Second, making small lesions in the nervous system causes discrete, localized degeneration, as if parts of individual cells have been severed. Third, certain stains highlighted what appeared to be individual cells, a result argued to be unlikely if the nervous system were a syncytium. At the same time, proponents of the reticular theory continued to point out quite correctly that these arguments could all be wrong, and that whatever arguments were used, it was still true that individual cells could not be distinguished in adult animals.

The most prominent proponent of the neuron theory, and the person most directly responsible for its eventual acceptance, was Santiago Ramón y Cajal, a Spanish neuroanatomist. Using staining techniques developed by the Italian anatomist Camillo Golgi (Box 2-1), Ramón y Cajal was able to describe the shapes and distributions of individual neurons in many different parts of the nervous system. In spite of arguments that the staining effects were artifacts, the sheer volume of careful, detailed anatomical studies conducted by Ramón y Cajal eventually convinced all but the most die-hard reticulist. In 1906, Golgi and Ramón y Cajal shared the Nobel prize for Physiology or Medicine for their contributions to our understanding of nervous system structure and organization. It

BOX 2-1

Making Neurons Visible: Staining Techniques for Nervous Systems

Being able to see the minute structure of neurons in the brain and elsewhere has been an objective of scientists interested in studying the nervous system since well before the microscope was invented. However, even after the development of adequate optical instruments, the structure of neurons could not have been studied without the use of dyes that stained specific features of the tissue under investigation. Many early neuroanatomists, such as the inventor of the **Golgi stain**, the Italian Camillo Golgi, devoted their lives to finding new dyes that would color specific neurons or parts of neurons, and thereby make visible some specific component of neural tissue.

Finding these dyes was often more a matter of art than of science. Investigators first "fixed" the tissue, that is, treated it chemically so that

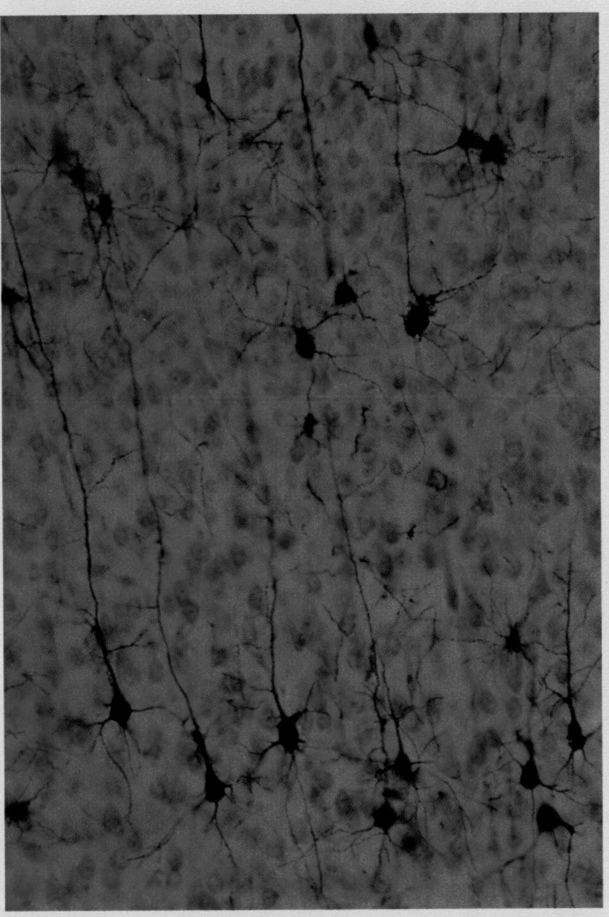

FIGURE 2-A. A section of the outer part of the brain (the cortex) of a rat, stained with Nissl stain to show all neurons (blue background), and with Golgi stain to show individual neurons (black). Notice how each Golgi-stained neuron is stained fully, even the small branches. (Courtesy of Mario Saltarelli and William T. Greenough.)

(continued)

Making Neurons Visible: Staining Techniques for Nervous Systems *(continued)*

proteins and other labile constituents were structurally stabilized, and hence would not fall apart as dead tissue normally did. After embedding the fixed tissue in wax for stability, they then carefully sliced thin pieces off the tissue, mounted these pieces on a microscope slide, and applied a series of chemicals, one after another, in the hope that these chemicals would bind with and make visible some component of the cells. Because some dyes required specific chemical conditions in order to stain the tissue, such as just the right amount of acidity, the most successful neuroanatomists were those who developed a feel for the effect that specific chemicals would have on neural tissue. In this way, stains for muscle and for nerve tissue, for nuclei and for mitochondria, and for many other cell types and cell components were discovered.

The disadvantage of the early stains was that they colored equally every cell or cellular component with which they would react. This was acceptable if the objective was to gain an impression of the general appearance of some particular part of the nervous system. It did not, however, reveal the structure of or connections between individual neurons because the sheer mass of stained neural processes completely obscured such detail. Here is where Golgi made his contribution. Without intending to, Golgi discovered a procedure that stained only about 5% of the cells exposed to it. However, these cells stained completely, revealing the marvelous detail of neural cell structure (Figure 2-A). It is still not understood why this stain has this capricious effect, but for over 100 years the result has allowed neuroanatomists to study the structure of individual neurons in the central nervous system.

is one of the ironies of science that Golgi supplied the method for Ramón y Cajal to establish the neuron doctrine, even though Golgi himself was, even until his death in 1926, a strong supporter of the reticular theory.

The neuron doctrine is the idea that the nervous system, like other bodily systems, is composed of individual and discrete cells. The lack of adequate technology to study the structural complexity of the nervous system delayed the acceptance of this idea until late in the nineteenth century.

Neurons and Their Connections

A neuron usually consists of a cell body and one or more slender branches that grow out from it. The cell body, which contains the cell nucleus, is often called the **soma**, and the branches are termed **neurites**. In most neurons, there is one long neurite called an axon and several shorter neurites called dendrites. An **axon** is functionally defined as a neurite that conveys information *away* from the cell body. A **dendrite** is a neurite that conveys information *toward* the cell body. Some neurons do not have readily identifiable axons or dendrites. Because of the diversity of neuronal types, it is difficult to depict just one as a representative kind, so two types are illustrated in Figure 2-2.

Communication between neurons is the foundation on which every action of the nervous system is built. It takes place at specialized sites called **synapses**, where one neuron comes close to or touches another neuron. The process of communication occurs via a mechanism known as **synaptic transmission**. You will learn more about it later in this chapter and in Chapter 6.

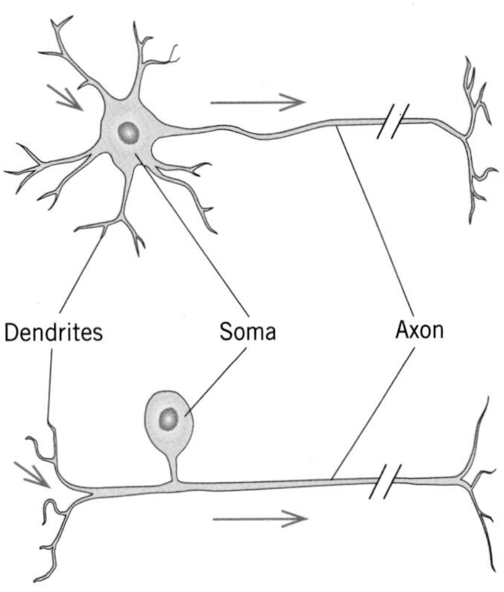

FIGURE 2-2. Two typical neurons, showing the soma and neurites. The arrows indicate the direction of information flow.

Types of Neurons

Because there are so many different kinds of neurons, it is often convenient to think about them in related groups. Neurons have been classified in a number of ways, and how to classify them is as much a matter of personal preference as anything else. One scheme of classification is based on what a neuron does. In this scheme, a kind of functional classification, neurobiologists identify three main types of neurons: sensory neurons, motor neurons, and interneurons, each roughly corresponding to one of the three main functions of the nervous system (see Chapter 1). **Sensory neurons** respond to sensory stimuli and hence are part of the sensory system that receives and interprets sensory information. **Motor neurons** deliver output to muscles or glands and thus are part of the motor system for producing action. **Interneurons** act largely in a decision-making capacity, corresponding to the integrative functions of the nervous system. Sensory and motor neurons have branches that extend into the body of the animal, conveying information in from sense organs or out to muscles or glands, respectively. Interneurons are confined to the main mass of neural tissue, such as the brain or spinal cord.

The classifications of sensory and motor neurons are universally accepted. Yet in articles that concern vertebrate animals, you may see the term interneuron used in a more restricted sense, to mean a neuron that is confined to one region of the brain or spinal cord. Neurons that send branches from one place to another more distant location within these regions are then called **projection neurons**. In this book, the term interneuron will be used in its more general sense, to refer to any neuron that is not a sensory or a motor neuron.

Another way to classify neurons is by their shape or general appearance, particularly by the number of neurites that branch from the soma (Table 2-1). Using this scheme, four main types of neurons are identified. **Anaxonal neurons** have no neurites at all. **Monopolar neurons** have a single neurite extending from the cell body, although this neurite typically branches into an axon and many dendrites shortly after it leaves the soma. **Bipolar neurons** have two neurites, one an axon and the other a dendrite, extending roughly from opposite sides of the cell body, and **multipolar neurons** have a single long axon and many dendrites extending from the cell body. Although it is difficult to make generalizations that do not have exceptions regarding the locations of particular types of neurons, researchers do expect to find certain structural types of neurons associated with specific functions in different animals.

TABLE 2-1. The Main Morphological Types of Neurons and Their Locations

Type	Appearance	Where Found
Anaxonal		Sensory systems, such as in the vertebrate eye or ear
Monopolar		Invertebrate interneurons and motor neurons; many vertebrate sensory neurons
Bipolar		Invertebrate sensory neurons
Multipolar		Vertebrate interneurons and motor neurons; some invertebrate sensory neurons

Neurons are classified by function into sensory neurons, motor neurons, or interneurons. A neuron may also be classified by its appearance based on the number of neurites that arise from its cell body.

Synapses

Neurons can communicate in two ways. In the process called **chemical (synaptic) transmission**, communication takes place via a chemical intermediary called a **neurotransmitter**, which is released by one neuron and influences another. This type of synapse is referred to as a **chemical synapse**. In the process known as electrical (synaptic) transmission, communication takes place by the flow of electrical current directly from one neuron to another. A synapse at which this occurs is called an **electrical synapse**. Electrical synapses are also known as **gap junctions**. Specific terms are applied to the sending and receiving neurons at synapses. The neuron that sends the message is called the **presynaptic neuron**, and the receiving neuron is called the **postsynaptic neuron**.

Chemical synapses differ from electrical synapses both structurally and functionally. Structurally, chemical synapses have a small gap, the **synaptic cleft**, between the communicating neurons. The neurotransmitter must diffuse across this gap in order to influence the postsynaptic neuron. There is no synaptic cleft in electrical synapses, since the communicating neurons touch one another. Chemical synapses are also characterized by the presence of small spherical or oval organelles called **synaptic vesicles** in the presynaptic terminals (Figure 2-3A). These vesicles, which contain the chemical neurotransmitter used in transmission, are described in more detail in Chapter 7. There are no synaptic vesicles at electrical synapses (Figure 2-3B).

Chemical and electrical synapses differ functionally as well as structurally. The main functional difference is the defining one that a chemical intermediary is used at chemical synapses whereas none is used at electrical

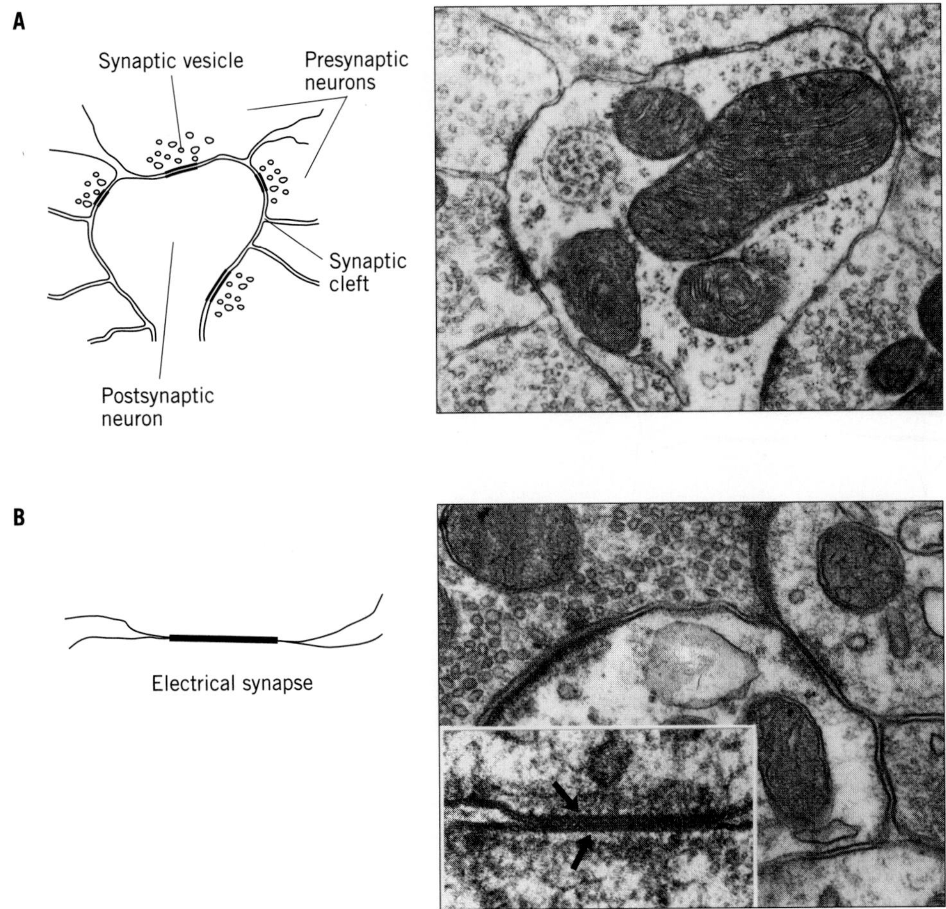

FIGURE 2-3. Synapses, the sites of communication between neurons in the nervous system. (A) A chemical synapse, showing part of one neuron receiving inputs from several other neurons. In electron micrographs (right), chemical synapses are typically identifiable by the presence of synaptic vesicles on the presynaptic side of the membrane. (B) An electrical synapse. Note the absence of synaptic vesicles in the electron micrograph (at arrows).

synapses. In addition, chemical synapses are polarized, meaning that communication can take place in only one direction. The direction in which information is transferred can usually be determined by the presence of synaptic vesicles, since these are concentrated in the region of the synapse in presynaptic neurons. Electrical synapses are not polarized. Communication across electrical synapses takes place by the direct flow of electrical current from one neuron to another, and the current may flow in either direction.

In some cases, it is possible to find two chemical synapses immediately adjacent to one another, one transmitting information in one direction, the other transmitting in the other direction. Such synapses are referred to as **reciprocal synapses** (Figure 2-4A). In this case, the designation of the presynaptic and postsynaptic neuron depends on which synapse is being considered. Since *presynaptic* always refers to the neuron sending a message, it is perfectly possible for the neuron to be presynaptic to its partner at one moment and yet be

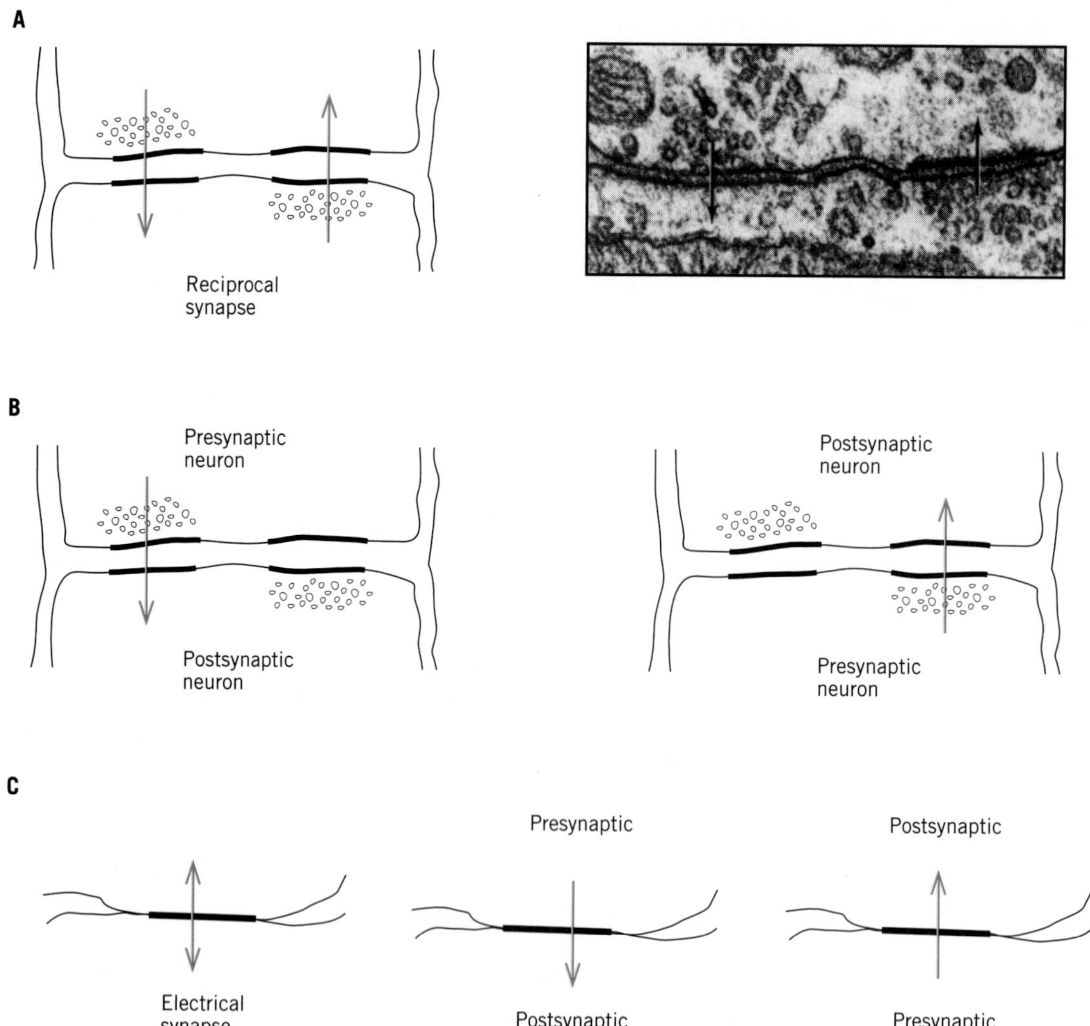

FIGURE 2-4. (A) A reciprocal synapse, in which two cells make mutual synaptic contact with one other. (B) At a reciprocal synapse, the presynaptic and postsynaptic neurons are identified based on the direction in which the message is being passed at any one time; one cell may be presynaptic at one instant and postsynaptic at another. (C) An electrical synapse may send messages in either direction. Here, too, the terms presynaptic and postsynaptic are applied to the two cells depending on the direction in which information is being sent at any given moment.

postsynaptic to that same neuron at another moment (Figure 2-4B). A neuron that forms an electrical synapse with another may also be either presynaptic or postsynaptic to its partner, depending on the direction in which information is being passed (Figure 2-4C).

Information flow along a neuron and from one neuron to another is typically depicted as moving from dendrites, considered the input zone, to axons, considered the output zone, and from an axon of one neuron to a dendrite of another. It is certainly true that synapses in which information is transferred from an axon to a dendrite are common, but this arrangement is by no means the only one that can be found. Actually, almost every part of any neuron may form a synapse with any other part. In addition to the typical synapses formed by axons on dendrites, synapses may be formed by axons on the cell soma, by dendrites on other dendrites, by axons on other axons, and even by dendrites on axons. These different types of synapses may play a variety of specific functional roles in neural interactions.

A synapse may be chemical or electrical. The two types differ functionally and structurally. The term presynaptic is applied to the neuron sending a message across a synapse, and the term postsynaptic to the neuron receiving the message. Synapses may form between different parts of neurons.

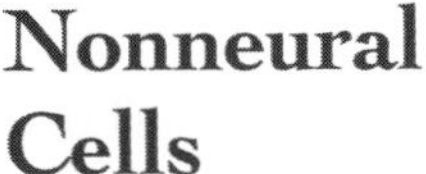

Nonneural Cells

When you think of the nervous system, you probably think of neurons. Yet the majority of the cells of any nervous system are the nonneural cells collectively known as **glia** or **neuroglia**. In the human brain, for example, it is estimated that there may be two to five times as many glial cells as the 10 to 50 billion estimated neurons. Glial cells actually have many similarities to neurons. Like neurons, glial cells have an electrical potential difference across their membranes and are sensitive to changes in this potential as well as to the presence of certain chemicals in the fluids surrounding the cells. Some even look rather like neurons, sporting neuritelike branches from the cell body. But glial cells differ from neurons in important ways. For example, they do not show an active electrical response like that which will be described for neurons in Chapter 5; their cell bodies are rather small; and what branches they have are usually symmetrical, unlike the dendrite-axon asymmetry typical of neurons. Their most important characteristic, however, is the functional relationship they form with many neurons.

Types of Glial Cells

All nervous systems contain glial cells, but they have not been studied as intensively in invertebrate animals as they have in the vertebrates. In vertebrate nervous systems, four main types of glial cells are recognized: **astroglia, microglia, oligodendroglia**, and **Schwann cells** (Table 2-2). The first three of these are usually referred to by the names of their individual cells, **astrocytes** (astroglial cells), microglial cells, and **oligodendrocytes** (oligodendroglial cells). These three are found only in the brain and spinal cord, whereas Schwann cells are found only in peripheral nerves. A variety of types of glia have been identified in invertebrates as well, but these are often lumped together and simply referred to as glial cells.

Different kinds of glial cells serve different functions in the nervous system. Astrocytes provide physical support for neurons and help regulate the extracellular concentration of potassium ions. They might also help supply nutrients to neurons, but the evidence in support of this possibility is not yet conclusive. One of the most intriguing recent suggestions about astrocyte function is that they may play a role in neuronal communication. This suggestion comes from two kinds of experiments. First, astrocytes seem able to influence directly the level of free Ca^{2+} in neurons, which in

TABLE 2-2. Types of Vertebrate Glial Cells

Type	Appearance	Features and Functions
Astroglia		Star-shaped, symmetrical; nutritive and support function
Microglia		Small, mesodermally derived; defensive function
Oligodendroglia		Asymmetrical; form myelin around axons in brain and spinal cord
Schwann cell		Asymmetrical; wraps around peripheral nerves to form myelin

turn can influence communication between the neurons themselves, as described in Chapter 6. Second, recent experiments have shown that astrocytes can physically cover or uncover synaptic regions. By baring synaptic regions between neurons, astrocytes may enhance synaptic transmission, or the formation of new synapses. The evidence also suggests that in some situations this synaptic enhancement may play a role in learning, thereby perhaps giving astrocytes a hitherto unsuspected role in this phenomenon.

Microglial cells are unusual in that they develop from mesodermal rather than ectodermal tissue in the embryo. All other glial types are ectodermal in origin, as are all neurons. Based on the observation that microglial cells can become quite mobile and are able to consume cellular debris stemming from injury or disease, they are thought to have a defensive function.

Recent work has emphasized the importance of interactions between glia and neurons. Like neurons, glial cells are now known to be sensitive to changes in the electrical potential difference across their membranes, as well as to the presence of neurotransmitters and other chemicals in the fluids surrounding the cells. This sensitivity allows for communication in both directions, from a neuron to a glial cell, and from a glial cell to a neuron. For example, glial cells are responsive not only to the neurotransmitters used in chemical synaptic transmission but also to a variety of compounds known as growth factors (see Chapter 22). These compounds can strongly influence the growth and differentiation of neurons during development of the embryo and during regeneration after an injury, helping to guide sprouting in axon ends. Conversely, neurons can also have a strong influence on glial cells. During development and regeneration, the proliferation of Schwann cells is strongly stimulated by the growing axon.

Glial cells are nonneural cells in nervous systems. They provide support and perhaps serve other functions for neurons. Glial cells can interact with neurons via chemicals that they and the neurons release.

Myelin

The most thoroughly studied role of some types of glial cells is their formation of **myelin**, a membrane that is wrapped tightly several times around most axons and long dendrites in the nervous system. Glial cells surround the neurons of invertebrates as well, but in a loose association that does not form true myelin (Figure 2-5A, B). Myelin acts as an electrical insulator, helping to increase the speed of conduction of nerve impulses in neurons, as discussed in Chapter 5. The importance of myelin for normal nerve function in vertebrates can be appreciated from consideration of the devastating effects of demyelinating diseases like multiple sclerosis, in which paralysis is caused by failure of nerve impulses in the demyelinated motor neurons to reach the muscle.

Myelin is formed when a glial cell wraps all or part of itself so tightly around a neurite (Figure 2-5C) that nearly all the cytoplasm of the glial cell is squeezed into the outer loop, leaving only the glial cell membrane in the inner layers. It does not encase the entire

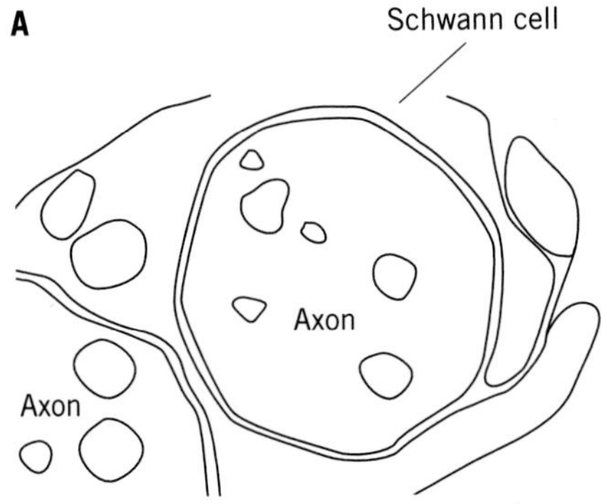

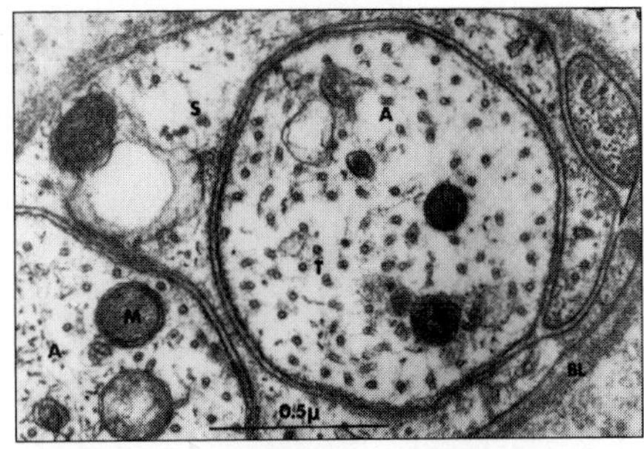

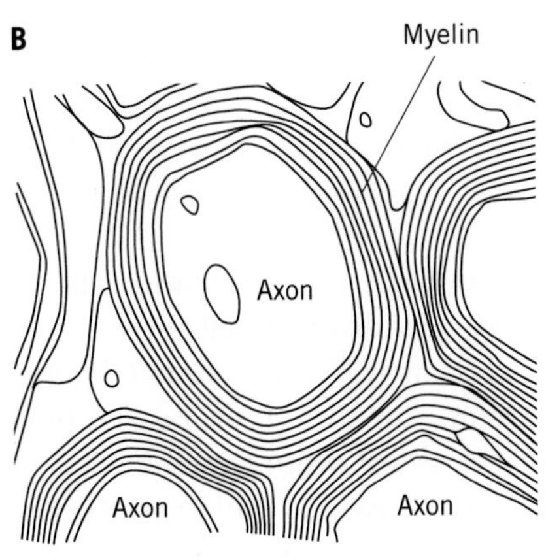

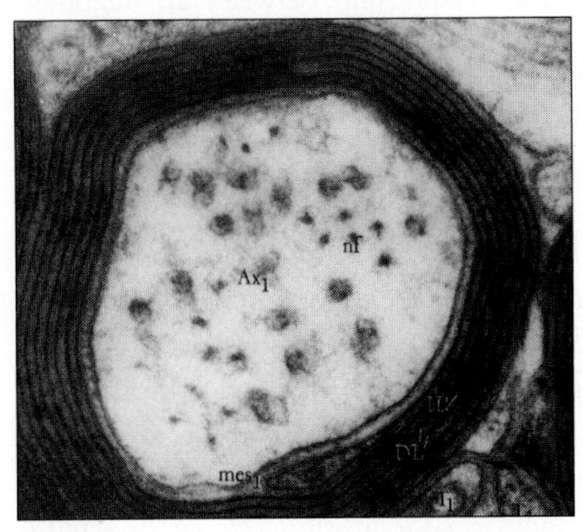

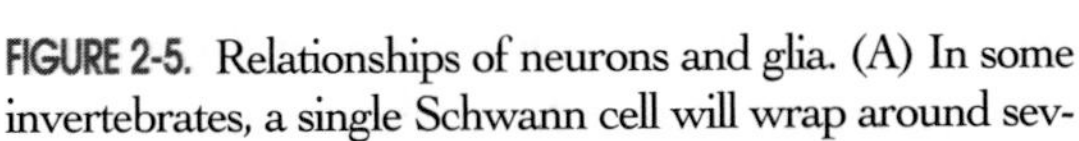

FIGURE 2-5. Relationships of neurons and glia. (A) In some invertebrates, a single Schwann cell will wrap around several neurons. (B) In the vertebrate nervous system, glial cells form the specialized tight wrapping known as myelin.

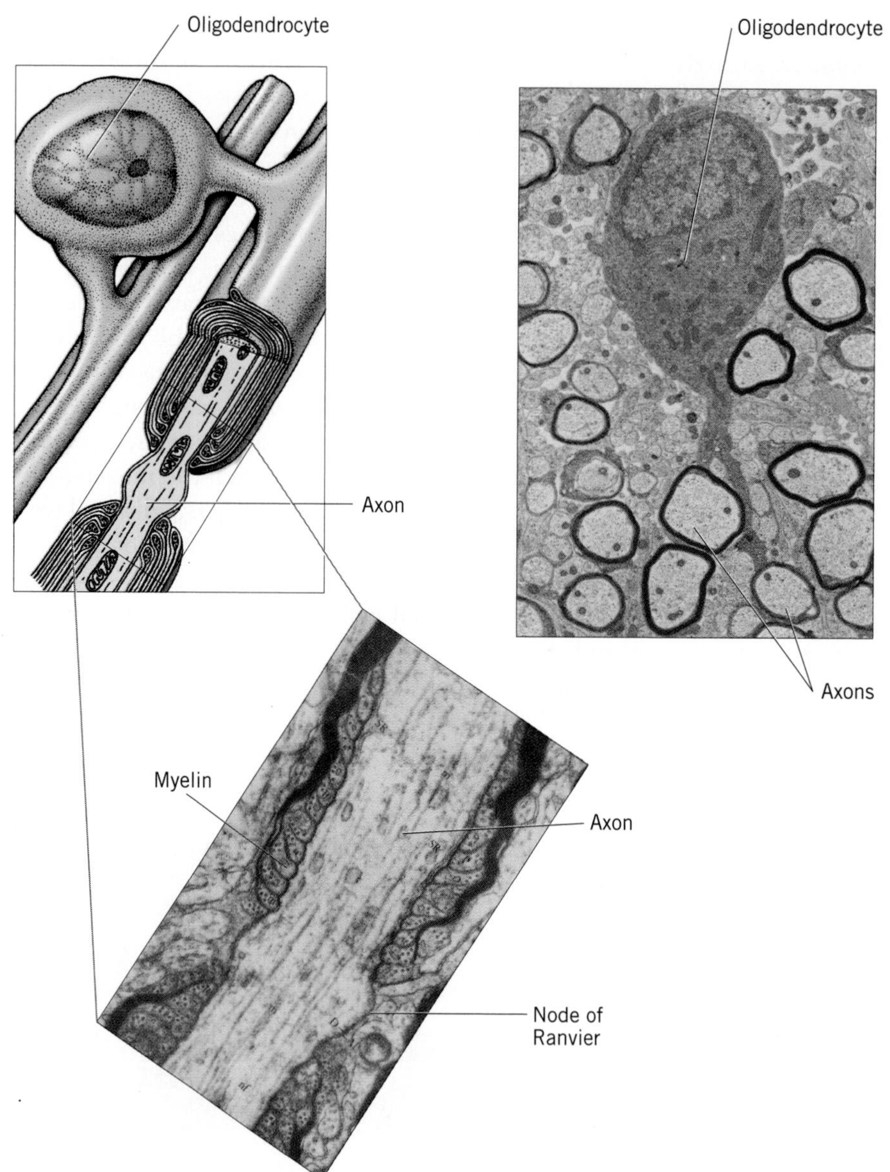

FIGURE 2-6. The relationship between an oligodendroglial cell and the neurites it wraps around. Each oligodendroglial cell sends processes to several neurites. The processes form cylinders of myelin around the neurite, each section of myelin separated from the next by a small gap, the node of Ranvier. O, oligodendroglial cell; ax, axon; m, myelin; n, node of Ranvier; N, nucleus of oligodendroglial cell.

length of an axon or dendrite. Instead, it forms discrete bundles, each pair of bundles separated by a gap called a **node of Ranvier**. In nerves outside the brain and spinal cord, myelin is formed by Schwann cells, with each Schwann cell forming a single myelin bundle around a section of a single axon or dendrite, separated from adjacent bundles by one node of Ranvier on each side. Within the brain and spinal cord, myelin is formed by oligodendrocytes. In contrast to Schwann cells, however, each oligodendroglial cell forms myelin bundles on many different axons (Figure 2-6).

Myelin, consisting of nearly pure membrane, is shiny white. Unmyelinated neural tissue, in contrast, is dull gray. This color difference is obvious even in living tissue and allows researchers to distinguish regions of the brain and spinal cord that contain bundles of myelinated axons from those regions that contain unmyelinated cell bodies and synapses. The two regions are designated **white matter** and **gray matter**, respectively.

Myelin is formed by the tight wrapping of the membranes of oligodendrocytes or Schwann cells around short sections of axons or dendrites in the nervous systems of vertebrates. It aids in the rapid conduction of nerve signals.

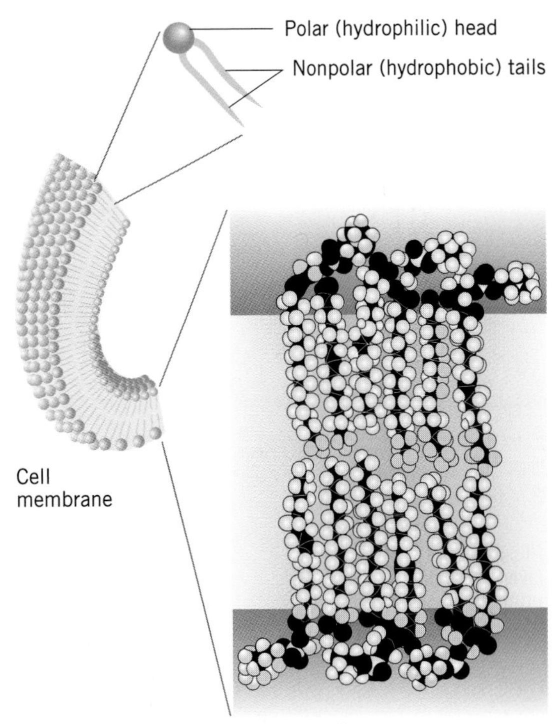

FIGURE 2-7. The molecular structure of a lipid bilayer membrane. The membrane consists of a double layer of phospholipid molecules aligned so that the lipid (fatty acid) hydrophobic tails point in toward each other and the phosphorus-containing hydrophilic heads point out into the liquid medium.

The Cell Membrane

Like all other cells, neurons are enclosed by a cell membrane, which is composed mainly of phospholipids and proteins. The chemical composition of membrane confers to every cell a number of specific properties.

Membrane Structure

The phospholipid components of cell membranes consist of a hydrophilic ("water-seeking") head and two hydrophobic ("water-avoiding") fatty acid tails. Hydrophilic molecules or parts of molecules mix freely with water, whereas hydrophobic molecules like oil do not. The phospholipids in cell membranes are arranged in two layers back-to-back so that the hydrophobic lipid tails of one layer face the hydrophobic tails of the other, and the hydrophilic heads point outward away from the middle of the membrane. Because of this double-layered structure, a membrane is referred to as a **lipid bilayer** (Figure 2-7).

The protein components of membranes are embedded in the lipid bilayer. Some proteins extend all the way through the membrane and hence come in contact with both the intracellular and extracellular fluid that bathes it. Other proteins are embedded only partly in the membrane and hence come in contact with either intracellular or extracellular fluid but not both (Figure 2-8). Membranes are typically 4 nm (nanometers, 10^{-9} meters) thick.

The lipid composition and bilayer structure of the cell membrane have two important consequences. First, the hydrophobic tails of the lipid molecules tend to be chemically incompatible with water-soluble substances such as inorganic ions. As a result, the membrane acts as a barrier to such molecules, which cannot simply diffuse through it. The only way that water-soluble ions and other charged molecules can enter or leave a cell is via the proteins that span the membrane. This is especially important for neurons because it is the controlled flow of ions through the neuron's membrane that is the basis for the cell's ability to send and receive messages, as explained in Chapters 4 and 5.

Second, the presence of lipids makes membranes fluidlike. This is because lipid molecules, rather than binding tightly to one

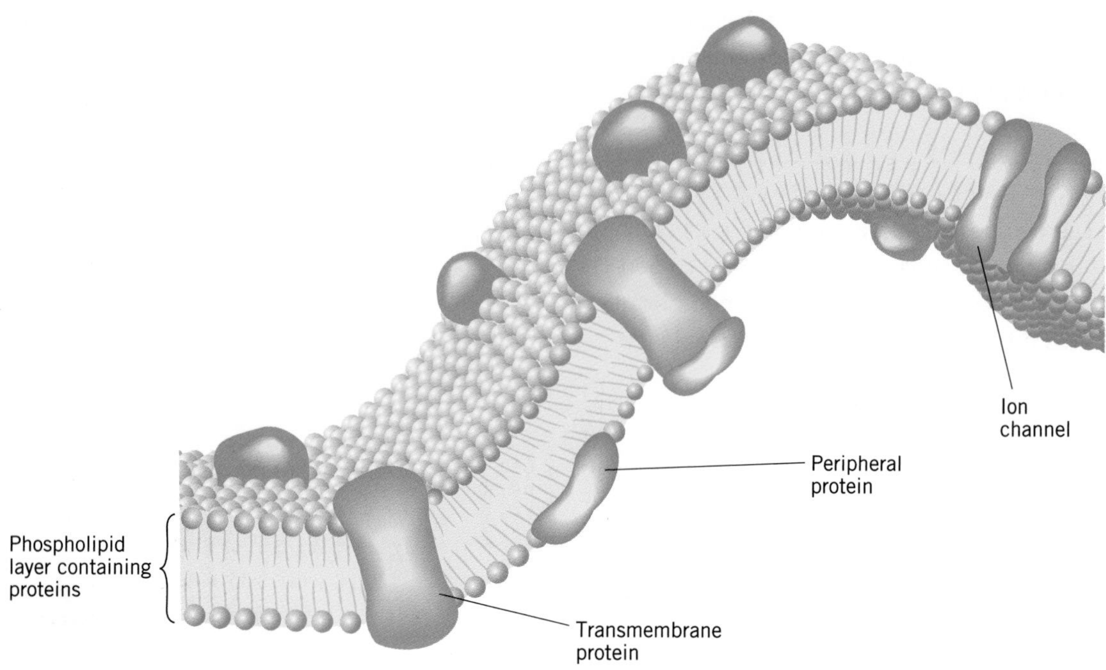

FIGURE 2-8. Proteins in a cell membrane. Some proteins are embedded in or attached to one face but do not extend all the way through the membrane. Transmembrane proteins extend from one side of the membrane to the other. Some transmembrane proteins have hollow openings in the center that allow ions to pass through, and hence are ion channels.

another as amino acids do in a protein, are only loosely associated, leaving them free to slide by one another. As a result of this fluid property, many embedded proteins are able to move around within the membrane, floating about somewhat like icebergs floating in the ocean. Movement is not completely free, since some proteins are anchored to the cell's internal skeleton and others are restricted to movement within a confined area, but the fact that proteins can move at all is important because in some cases the movement of a protein is an essential element in the response of a neuron to an external signal. This is because proteins are often spaced some distance apart. It is only when they come together, an action triggered by an external event, that they can interact to initiate a biochemical cascade that leads to the cell's response. Hence, if proteins could not move around in the cell membrane, many important forms of communication between neurons would not be possible.

The cell membranes of neurons are fluid lipid bilayers, consisting of phospholipids and proteins. The phospholipids make the membrane impermeable to ions and allow the proteins to move around. Movement of membrane proteins is important in allowing the cell to respond to external chemical signals.

Membrane Proteins

Protein is the main and most important nonlipid component of all cell membranes. It constitutes more than 20% of the membrane of a typical neuron and confers on the neuron many of the characteristics that distinguish it from other types of cells, and indeed that distinguish one type of neuron from another. Both membrane-spanning and partly embedded proteins contribute to the properties that give each type of neuron its unique characteristics. Different proteins can have different properties, and it is the presence of particular types of proteins with specific characteristics that confers on each neuron a particular set of functional features. Neurons can also change the types of proteins present in their cell membranes, hence changing their own functional characteristics.

Proteins, in spite of their bewildering array of types, each with different characteristics, capabilities, and functions, can be grouped into several categories. Three categories will be discussed in this and subsequent sections of this chapter: transport proteins, signaling proteins, and a miscellaneous group consisting mainly of proteins of attachment (binding proteins). **Transport proteins** allow or facilitate the movement of water, water-soluble ions, and other substances into or out of the cell. **Signaling proteins** receive or respond to chemical messages coming from outside the neuron. **Binding proteins** help bind cells to one another or to an extracellular matrix of molecules that is present around some cells. These three categories, and the principal types of proteins they include, are summarized in Table 2-3 and discussed in more detail in the sections that follow. The structure of these proteins, which is critical to their function, can be investigated using molecular techniques (Box 2-2).

Specific types of protein confer specific properties on neurons. Three important groups of protein are transport proteins, which facilitate ion and molecular movement across the cell membrane, signaling proteins, which aid signaling between cells, and binding proteins, which bind cells to one another and to other parts of the body.

TABLE 2-3. The Important Types of Neuronal Membrane Proteins

Type of Protein	Function
Transport proteins	
Ion channel	Allows the passive flow of ions through the membrane
Ion pump	Expends energy to move one or more ions across the membrane against a concentration gradient
Other transporters	Facilitate the transfer of molecules across the membrane
Signaling proteins	
Receptor	Binds with a signaling molecule, and initiates the neuron's response
G protein	Initiates a cascade of biochemical reactions that leads to a neuron's response to a signaling molecule
Other enzymes	Catalyze biochemical reactions that are part of a neuron's response to a signaling molecule
Binding proteins	
Adhesion proteins	Anchor the neuron to other cells
Cytoskeletal binding proteins	Anchor the cell membrane to the internal cytoskeleton

BOX 2-2

Unraveling the Structure of Proteins: Recombinant DNA Techniques

Proteins are critical to the function of every neuron. Receptors, ion channels, and enzymes all play essential roles in receiving and passing on signals from one neuron to another. Understanding how proteins carry out their functions, how they can be activated and inactivated, and how pharmacological agents interact with them requires a thorough knowledge of the sequence of amino acids that make up the protein and the three-dimensional structure into which they are arranged. This information is beyond the reach of ordinary analytic techniques.

The advent of recombinant DNA techniques has made it possible to investigate protein structure and function at the molecular level. In the method of complementary DNA (cDNA) cloning, the sequence of amino acids that make up a protein is determined by decoding the DNA that directed its synthesis. The key to the method is to introduce into bacteria the DNA that codes for a specific protein, and then to isolate genetically pure cultures, called clones, of these bacteria. Each bacterial clone will produce large quantities of a single protein, under the control of the inserted DNA (Figure 2-B).

To isolate a particular protein, such as an ion channel, a researcher will first extract messenger RNA (mRNA) from neural tissue that is rich in the channel of interest. The tissue contains mRNA for all the proteins that are synthesized in the tissue, but each mRNA codes for just one protein. Since mRNA contains a sequence of nucleic acids that was determined by the nucleic acids in the DNA strand from which it was originally assembled, the mRNA can be used to reconstruct the sequence of the original DNA. The isolated mRNA is first processed to remove the noncoding introns. Next, under the proper chemical conditions, each molecule of mRNA will serve as a template for the assembly of DNA that complements (hence the name, complementary DNA) that specific mRNA molecule in a process called reverse transcription. Each strand of cDNA is a compact form of the original gene; the cDNA codes for the specific protein whose synthesis is directed by the mRNA transcribed from the original gene.

These procedures generate many strands of cDNA, each of which codes for a different protein in the original tissue and which collectively form a gene library from the neural tissue. Next, researchers insert each strand into a single plasmid, a form of bacterial extrachromosomal DNA, and insert the plasmids into bacteria. This process usually leaves just one plasmid in each bacterial cell. Hence, as each cell grows, it will produce a colony, or clone, of identical bacteria that makes only one of the proteins

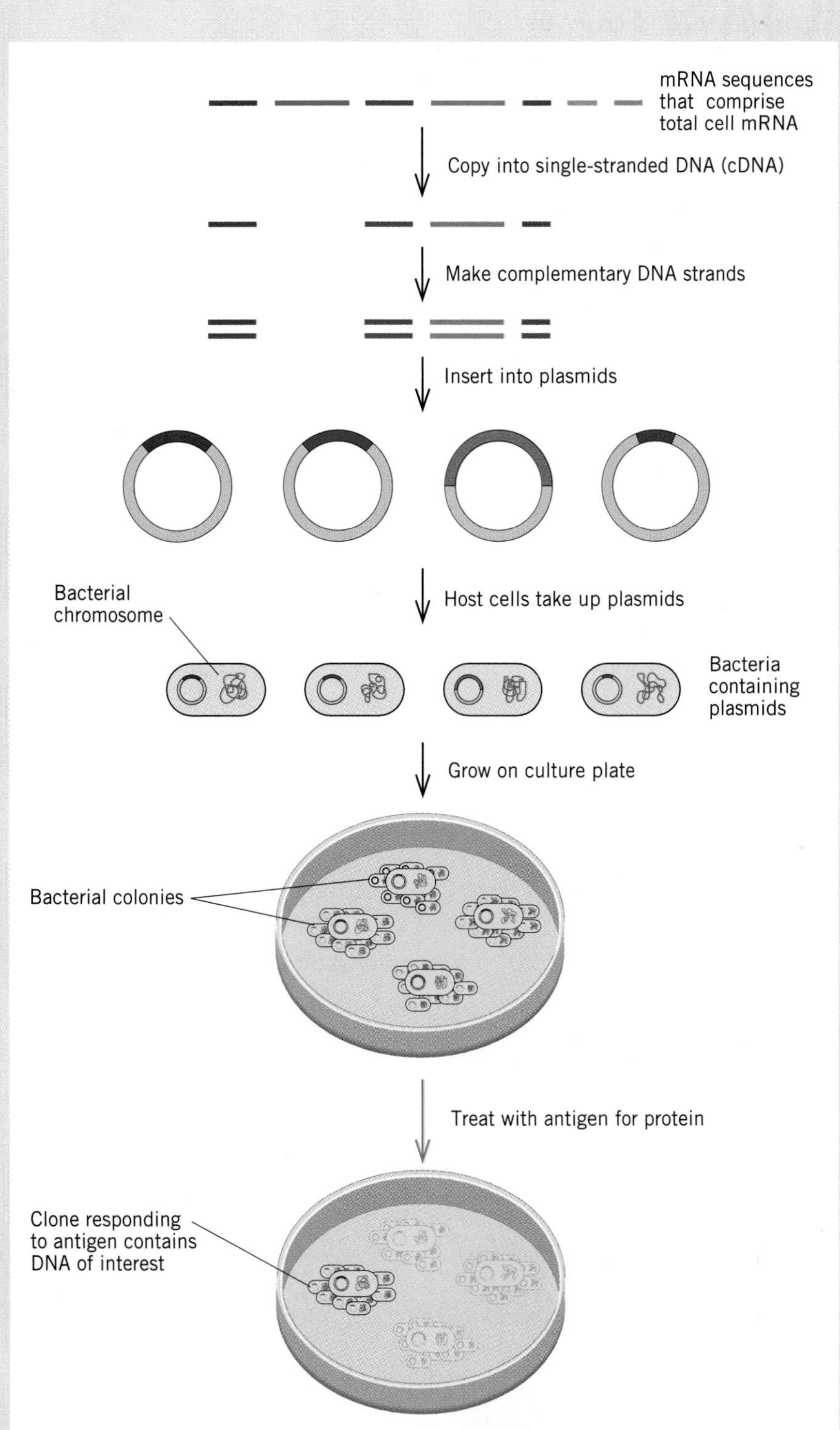

FIGURE 2-B. The procedure for making cDNA clones for specific proteins. The procedure involves extracting the mRNA from tissue rich in the proteins of interest, making cDNA from it, inserting individual segments of cDNA into host bacteria, and growing colonies from each individual bacterium.

(continued)

Unraveling the Structure of Proteins: Recombinant DNA Techniques *(continued)*

from the original tissue. The trick then is to identify the clone that is producing the protein of interest. This can be done by exposing all the bacterial colonies to an antibody for the protein and noting which colonies show a reaction with the antibody.

After researchers have determined the amino acid sequence of the protein from an analysis of the cDNA, they may use this information to determine the likely way that the molecule is folded in three-dimensional space and oriented in the membrane. For example, analysis of the sequence of amino acids of the protein can yield information about which parts of the molecule are hydrophobic and which are hydrophilic. The hydrophobic parts are likely to be embedded in the membrane, whereas the hydrophilic parts are likely to be outside the membrane, extending into the intracellular or extracellular space. This information can then be used to direct the search for parts of the protein that allow the channel to be open or closed, or activated or inactivated, leading to greater understanding of how the channel works.

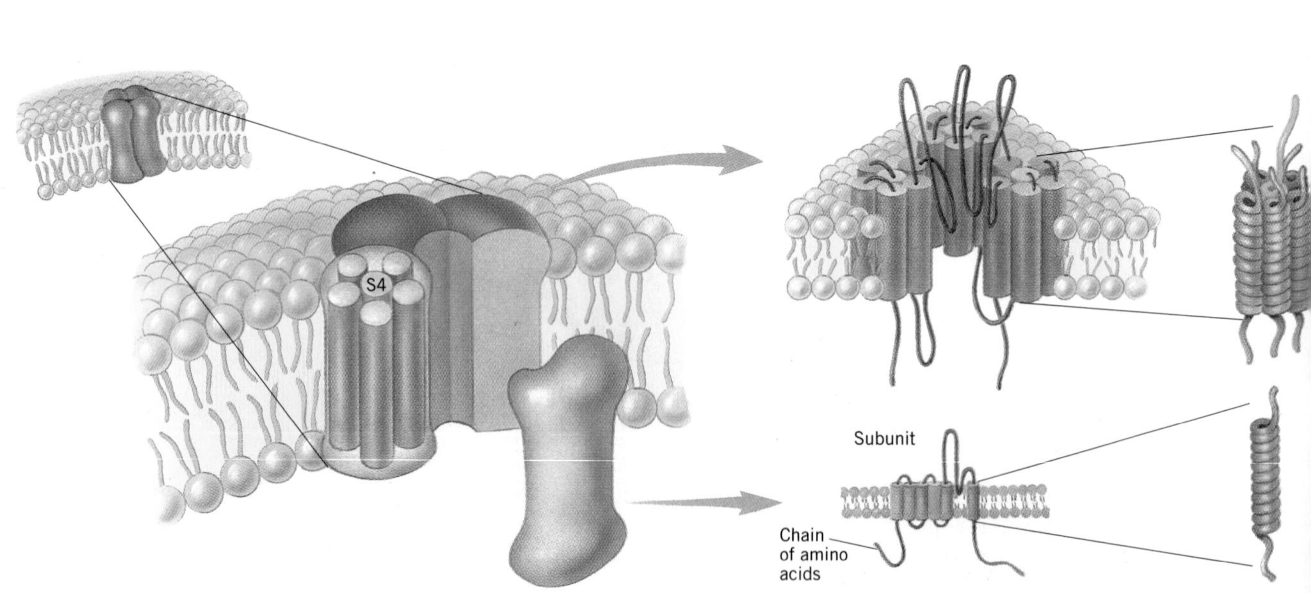

Transport Proteins

Transport proteins mediate the transfer of ions or molecules across the membrane in a wide variety of ways. Although the term transport suggests an active process, current usage encompasses proteins that operate both actively and passively. At one extreme are the proteins known as **ion channels**, which are cylindrically shaped, hollow-cored proteins that span the membrane and allow ions passively to move across the membrane by diffusion. At the other extreme are the **ion pumps**, which use the energy stored in adenosine triphosphate (ATP) to transfer ions or molecules across the membrane in the process known as **active transport**. Some proteins are highly selective about which ions or molecules they allow to cross the membrane, whereas others are not.

Transport proteins are especially important in shaping how a neuron responds to stimulation from other neurons or from the outside world. Whether a neuron responds well to weak stimulation or only to strong stimulation depends in large part on the types and numbers of transport proteins found in its cell membrane. Furthermore, modulation of the properties of transport proteins is an essential component of much of the adaptability of neurons and neuronal circuits described in Part 6.

Ion Channels

Ion channels are proteins that allow the passive diffusion of ions through them. They share a similar structure, being cylindrical and having a central core through which the ions flow. The hollow core of a typical channel is about 0.3 to 0.6 nm in diameter. An ion channel ordinarily consists of several chemically similar subunits, each of which is composed of helical strands that cross the membrane of the cell, called transmembrane domains (Figure 2-9). These transmembrane domains are connected by chains of amino acids (shown as looping solid lines in the figure). The amino acid chains help confer to the channel its specific properties. Some ion channels allow almost any small ion to pass through. Most, however, are **ion-selective**; that is, they allow the passage of only a specific type of ion, such as K^+ or Na^+. The structural basis of this selectivity is the size of the pore in the center of the protein. In some cases, the electrical charge of the chains of amino acids that dip down into the channel is also a factor.

Although all ion channels share a similar structure, they can differ in function. **Leakage channels** are ion channels that are open to the flow of ions all the time. Their presence in a cell ensures that there will be a continual movement of ions across the mem-

Two types of transport proteins are ion channels, which allow the passive movement of ions from one side of the membrane to the other, and ion pumps, which use the energy of ATP to move molecules across actively.

Transport proteins shape how a neuron responds to external input.

FIGURE 2-9. *(left)* The molecular structure of an ion channel, the sodium channel of axonal membranes. (A) Three of the four channel subunits, shown in position in a membrane. Each subunit is made of six transmembrane domains, represented here as cylinders in the membrane. The central pore is formed at the junction of the four subunits. The transmembrane domains and the subunits themselves are connected by amino acid chains, represented here by lines. (B) The channel opened like the wedges of an orange and spread out along the membrane. The solid loops are the chains of amino acids that connect the transmembrane components to one another and to the adjacent subunits.

brane. Most ion channels are regulated, however, meaning they are open to ion flow at some times and closed to ion flow at others. Channels that can open and close are called **gated channels**. They open or close through a change in the structure of the channel protein, called a **conformational change**. The structural change may involve an increase or decrease in the diameter of the central pore, or the movement of an obstructing part of the protein to open or block the pore to the flow of ions. The part of the molecule that moves to occlude or open the channel is referred to as a **gate** (Figure 2-10). Some channels have more than one gate, a factor that is important for their role in neuronal function.

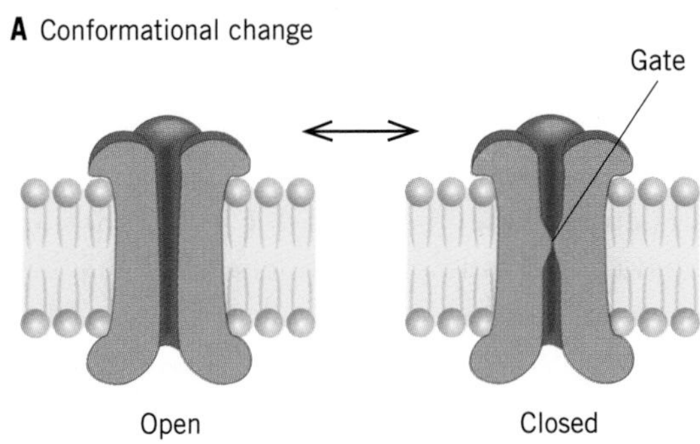

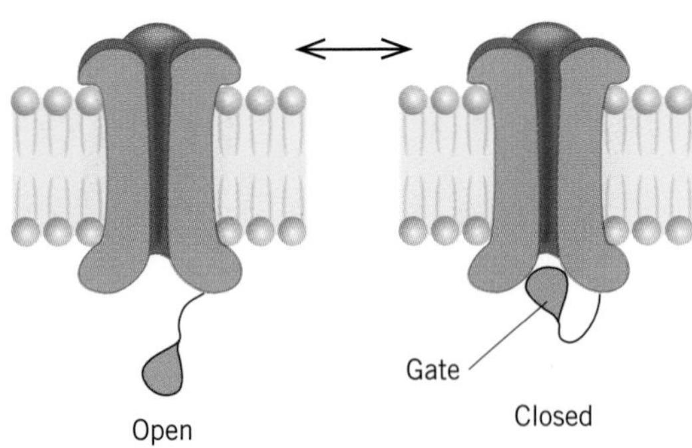

FIGURE 2-10. The gating of ion channels. (A) Some channels undergo a conformational change that widens or narrows the central pore. (B) Other channels open and close by the movement of a small flap or other projection that either uncovers or occludes the central pore.

The mechanism by which a gated channel opens or closes depends on which of the three types of channels it is. One type, called a **ligand-gated channel**, or **ligand-sensitive channel**, is regulated by the presence of a particular kind of signaling molecule and opens (or closes) only in the presence of that molecule. The second, called a **voltage-gated channel**, or **voltage-sensitive channel**, is regulated by the difference in electrical potential across the neuronal membrane. The channel opens as the potential difference increases (or decreases) and closes when the voltage changes in the opposite direction. The last type, called a **stretch-sensitive channel**, opens or closes when a mechanical force is applied to it.

Specialized channels called **connexons** form gap junctions (electrical synapses). Each synapse typically consists of as many as 100 tightly packed pairs of connexons that connect the interiors of two adjacent neurons (Figure 2-11A). The channels in the two neurons are aligned so that the central pore of a connexon in one neuron is directly in line with the pore of a connexon in another neuron. The pore is relatively large, around 2 to 3 nm, allowing ions and many molecules to pass easily and directly from one neuron to the other. Connexons can also be gated. Each connexon consists of six parts, called **connexins**, which can rotate and twist slightly within the membrane. When the connexins twist in one direction, the central pore opens. When they twist in the other direction, the central pore closes (Figure 2-11B). This twisting can be influenced by the presence or absence of specific chemical signals.

The rich diversity of gated channels in neuronal membranes imparts to neurons their unique signal processing capabilities. Voltage-sensitive channels are the basis of the active electrical responses of neurons that allow signals to travel along the length of long neurites. Ligand-sensitive channels are the basis of other important functions, including reception of information from other neurons, modulation of a neuron's responses over time, and even certain types of sensory reception. Stretch-sensitive channels mediate all mechanical sensitivity.

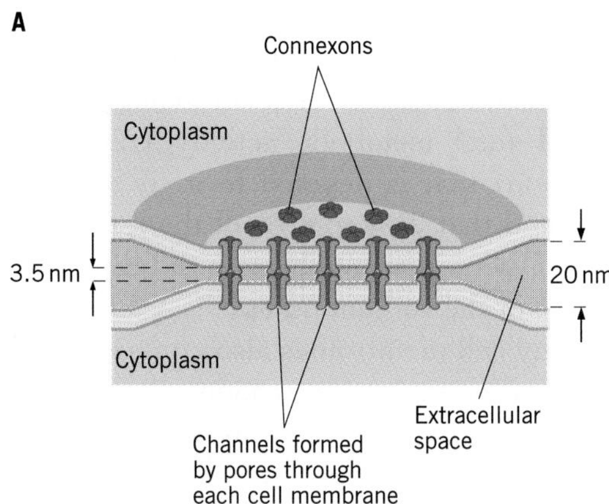

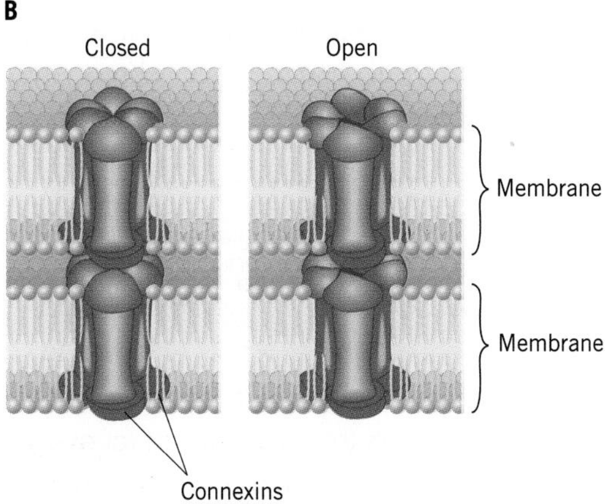

FIGURE 2-11. The structure of a gap junction (electrical synapse). (A) Each synapse consists of many connexons, with the channels in the membrane of one cell lined up with the channels in the membrane of the other cell. (B) Individual connexons open or close by the lateral twisting of the connexins that make up the channel, opening or occluding the central pore.

Ion channels are transport proteins that allow ions to diffuse across the membrane through a central pore. If the pore can open and close, the channels are called gated. Ligand-sensitive channels are opened or closed by the presence of a particular chemical; voltage-sensitive channels open or close by changes in the potential difference across the neuronal membrane; stretch-sensitive channels open or close by mechanical contact. Connexons are ion channels that directly connect one nerve cell with another at electrical synapses.

Ion Pumps and Other Transport Proteins

An ion pump is a second type of transport protein. Ion pumps require a source of energy in the form of ATP in order to transfer ions across the membrane. The energy is used to move ions from a region of lower concentration to a region of higher concentration. By pumping ions against a concentration gradient, an ion pump can build up a high concentration of ions on one side of a membrane. This is possible because an ion pump has no central pore through which ions diffuse freely. Instead, ions enter one side of the pump molecule and are actively (i.e., with the aid of ATP) moved to the other side, where they are released. Because ions move through a pump in a multistep process, whereas they flow straight through open ion channels, an ion can cross the membrane through an open channel as much as 100 times faster than it can through an ion pump. The main differences between ion channels and ion pumps are summarized in Table 2-4.

Of the several types of ion pumps found in neurons, one is particularly important: the **sodium pump**, also called the **sodium-potassium exchange pump** (Figure 2-12A). As this latter name implies, the important feature of this pump is that two types of ions, Na^+ and K^+, are moved across the membrane but in opposite directions. They do not move in a simple one-for-one exchange, however. In every pump cycle, three Na^+ are moved from inside the neuron out to the extracellular fluid, and two K^+ are moved from the extracellular fluid into the neuron. One molecule of ATP is hydrolyzed to adenosine diphosphate (ADP) to power this movement.

The mechanism by which the hydrolysis of a molecule of ATP can trigger the movement of these ions is still not fully understood, but the main steps involved in the ion transfer are known. When the pump is open to the interior, three Na^+ ions enter the open chamber, promoting the splitting of ATP. This brings about a reconfiguration of the pump, so that the interior opening closes and the exterior one opens. The Na^+ ions then leave, to be replaced by two K^+ ions. The presence of the

A

B

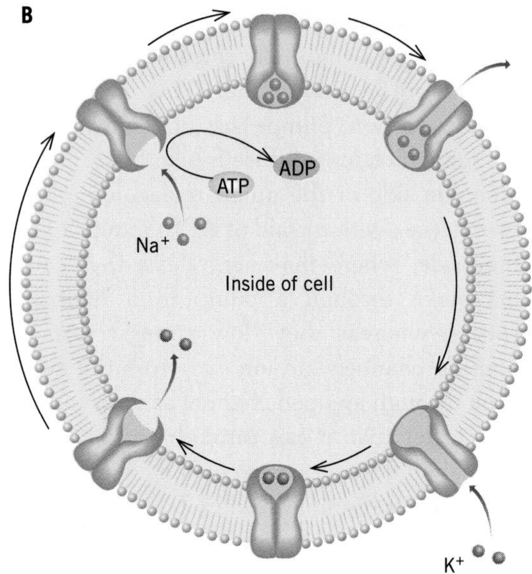

FIGURE 2-12. The sodium-potassium exchange pump. (A) A model of the pump, showing its orientation in the cell membrane. The model shows the contours of the entire molecule at a resolution of about 0.5 nm, as well as revealing the deep cleft in the center through which ions are believed to be exchanged. (B) The operation of the pump, showing the expulsion of three atoms of Na^+ and the intake of two atoms of K^+ during one cycle.

K^+ ions destabilizes the external opening, and the pump closes, causing a subsequent opening to the interior again and the release of the K^+ ions, thereby readying the pump for another cycle. These steps are summarized in Figure 2-12B.

Activity of the sodium-potassium exchange pump is crucial for normal nerve function because it allows a neuron to maintain the proper concentrations of Na^+ and K^+. A proper distribution of ions is essential to the generation and maintenance of the small difference in voltage that is present across the membranes of all nerve cells, a difference that is the basis of communication between neurons. The importance of the sodium pump in nerve metabolism is reflected in the proportion of total metabolic activity (ATP consumption) that is devoted to it. It has been estimated that 25% to 40% of the total metabolic activity in a vertebrate brain represents activity of the sodium pump.

Many cell membranes also contain a variety of other transport proteins that require ATP. In neurons, the most important of these bring back into the cell the chemical intermediaries used during chemical synaptic transmission. They also concentrate these molecules in the storage sites within each neuron. You will learn more about these actions in Chapter 7.

An ion pump is a type of transport protein that moves one or more ions across the cell membrane against a concentration gradient, using energy in the form of ATP. The sodium-potassium exchange pump is an ion pump that expels three ions of Na^+ for every two ions of K^+ that it brings into the neuron. Other types of transport proteins help to bring the chemical intermediaries released during chemical synaptic transmission back into the neuron.

Signaling Proteins

Signaling proteins, necessary for information to pass from one cell to another, are a second major category of membrane proteins. Some signaling proteins are important for the release of neurotransmitter from the presynaptic terminal of the sending cell, as you will learn in Chapter 6. Most, however, are necessary for receiving and responding to the message in the postsynaptic neuron. Of these, some are important mainly in message reception. Signaling proteins of this type are known as **receptor proteins**, or just **receptors** for short. Broadly

TABLE 2-4. **Differences Between Ion Channels and Ion Pumps**

Ion Channel	Ion Pump
Ions move unaided through a central pore	Ions actively transferred through the protein
No external energy required to move ions	Energy from ATP required to move ions
Ions move down a concentration gradient	Ions move against a concentration gradient
Ions move through rapidly	Ions move through slowly

defined, a receptor is a protein that has a high binding affinity for some other molecule. In the context of membrane-bound signaling proteins, the term is defined in a more specific sense to mean a protein that responds to the presence of a signaling molecule by initiating the chain of events that brings about the response of a neuron. Hence, the receptors make it possible for a neuron to respond to the presence of a specific signaling molecule. There is no specific term for those signaling proteins that are important in responding to a message. Proteins of this type are all enzymes, catalyzing one or more chemical reactions to orchestrate the response of the neuron to the message.

Receptors

Receptors serve as sites for the recognition and binding of chemicals that impinge on the cell from the outside. For this reason, they are usually highly concentrated in the membrane of postsynaptic cells at the synapse. Because the shape of the signaling molecule must complement that of the site at which it binds with the receptor, each receptor is specific for only one type of signaling molecule and its close chemical analogs.

Receptors are classified according to the effect they produce in the postsynaptic cell. Those that cause the opening or closing of ion channels, a strong flow of ions, and a quick, short-lasting electrical response in the neuron are called **ionotropic receptors** (Figure 2-13A). In some cases, the receptor protein itself is the ion channel. These receptors contain a central pore, which may open or close

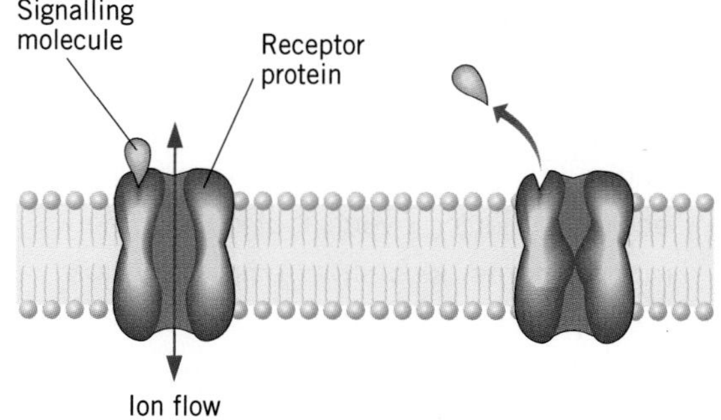

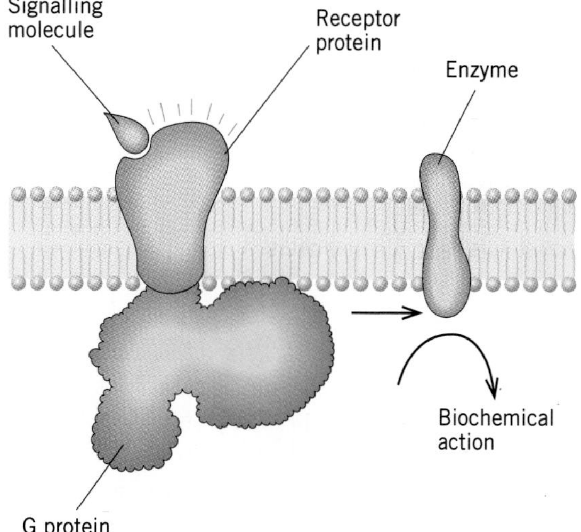

FIGURE 2-13. (A) An ionotropic receptor, which itself is often an ion channel as shown here. The channel opens when it is activated by a signaling molecule and closes when the molecule detaches. (B) A metabotropic receptor interacts with G proteins, thereby initiating a cascade of biochemical activity within the neuron.

in the presence of the appropriate signaling molecule. (For this reason, these receptors are also properly called ligand-sensitive channels.) In other cases, the receptor may be chemically coupled to an adjacent ion channel. The effect, however, is the same.

The other type of receptor initiates a biochemical cascade that not only takes longer to develop but also lasts longer. Because the biochemical reactions that are facilitated are considered a part of the cell's normal metabolic machinery, these are called **metabotropic receptors** (Figure 2-13B). Metabotropic receptors exert their effects via G proteins (see the next section). You will learn more about their effects on postsynaptic cells in Chapters 6 and 7.

Receptors are membrane proteins that respond to the presence of a signaling molecule by initiating events that cause a neuronal response. Ionotropic receptors initiate the rapid opening of ion channels and a strong electrical response in the postsynaptic cell. Metabotropic receptors initiate a biochemical cascade that has slower but longer-lasting effects on the cell.

G Proteins and Other Enzymes

Metabotropic receptors exert their effects by initiating a chain reaction, or cascade, of biochemical events. They usually do this by activating a series of proteins, some of which are membrane-bound enzymes. The first protein in the cascade, the one activated directly by the metabotropic receptor, is known as **guanine nucleotide binding protein**, or **G protein** for short. G proteins are capable of binding the two guanine nucleotides, guanosine triphosphate (GTP) and guanosine diphosphate (GDP); G proteins are classified as enzymes because they catalyze the conversion of GTP to GDP. The importance of G proteins, however, stems not from their enzymatic activity as such but from their influence on other membrane-bound proteins.

When G proteins are activated by a receptor that has bound with a specific molecule, they initiate a biochemical sequence that produces a response in the neuron. They do this by activating some other membrane-bound protein, such as adenylyl cyclase, phospholipase C, phosphodiesterase (PDE), or another enzyme. Activation of one of these enzymes induces an increase or decrease in the availability of a specific ion or molecule, which in turn controls some critical event in the neuron. Hence, the G proteins and the other membrane-bound enzymes are critical participants in the metabotropic response of a neuron to a specific chemical message received via chemical synaptic transmission. This entire chain of events is discussed in greater detail in Chapter 6.

The guanine nucleotide binding proteins, or G proteins, represent the first step in the cellular response to activation of a metabotropic receptor. Activation of a G protein initiates a cascade of linked biochemical reactions that either increases or decreases the availability of a critical molecule or ion, which then causes the neuron's response.

Binding Proteins

Binding proteins, such as adhesion and anchor proteins, are a third group of proteins associated with cell membranes. Adhesion proteins help bind one cell to another. They project from the external surface of a cell and help fasten the cell to other cells or to the extracellular matrix that surrounds many cells. Adhesion proteins play a crucial role during the development of the nervous system. As you will learn in Chapter 22, adhesion molecules not only act to fasten cells to one other, but they also help a developing neuron identify the proper cells along which to grow or with which to associate. In this role they act as receptors, responding to the presence of specific signaling molecules.

Anchor proteins face into the cell rather than out. They fasten the membrane to the internal network of proteins that constitutes an internal skeleton for the neuron. This binding of the membrane to the interior of the cell gives the cell structural stability as well as its specific form.

Adhesion molecules help anchor a cell to other cells and play a role during development. Anchor proteins fasten the membrane to the internal skeleton of the cell.

Subcellular Elements

Neurons are cells and hence contain in the cell body the organelles typical of all living cells. These include a nucleus, mitochondria, ribosomes, the Golgi complex, lysosomes, vesicles, and the system of internal membranes called the endoplasmic reticulum. These organelles serve the same function in neurons as they do in other cells, directing protein synthesis, synthesizing ATP, and so on. Some of these functions are summarized in Table 2-5. However, neurons also have unique needs that can be met by an elaboration of one or more of the organelles to serve special functions.

Two organelles are especially important in neurons. One, rough endoplasmic reticulum, is so prominent in neurons that early cytologists were easily able to recognize it in stained neural tissue and gave it a name of its own: **Nissl substance.** This organelle is abundant in neurons because it synthesizes new membrane and the proteins that are embedded in it. Neurons need membrane for their long axons and dendrites; they replace the membrane continually. It has been estimated that small neurons may synthesize an area of membrane equal to their own surface area every hour. Nissl substance often has a distinctly different distribution in different types of neurons, which sometimes allows researchers to distinguish one type of neuron from another in electron micrographs.

The other important neuronal organelle is the small spherical or ovoid inclusion known as a vesicle. Most eukaryotic cells contain vesicles, which usually serve as containers for the transfer of materials from one site to another within a cell. In neurons, however, vesicles also store the neurotransmitters that are used for chemical synaptic transmission (see Figure 2-3A). Vesicles are typically hollow, spherical organelles that range from about 30 to 160 nm in diameter. Vesicles that are clear in electron micrographs are concentrated in the presynaptic terminals of synapses. This is one reason they are thought to contain the molecules used in synaptic transmission. Other vesicles, called **dense-core vesicles,** appear solid in electron micrographs. They are present in neurons that contain certain types of signaling molecules (peptides) and in those that secrete hormones.

In addition to organelles, neurons contain a number of fine, filament-like proteins that form the neuron's cytoskeleton, which is present in all cells and allows the cell to maintain a particular shape. It can be especially important in neurons since many of these cells maintain a shape that is far from spherical. Three kinds of filament-like proteins—microfilaments, neurofilaments, and microtubules—play especially important roles in neurons.

Microfilaments, with a diameter of about 5 nm, are the smallest filament-like type of protein. These short, ubiquitous filaments consist primarily of actin. They often form a loose network just under the cell membrane and allow a cell to move about. Hence, they are especially important during development, when the growing axon is continually probing its environment to find the correct path along which to grow.

Neurofilaments, called **intermediate filaments** in nonneural cells, have a diameter of about 10 nm. They consist of twisted coils of

TABLE 2-5. Subcellular Organelles

Organelle	Features	Function
Endoplasmic reticulum	A system of internal membranes in cells	Protein and lipid synthesis
Golgi complex	An organelle of stacked membranes	Separation and sorting of protein components of the cell membrane and proteins for secretion
Lysosome	A roughly spherical, membrane-bound organelle	Enzymatic breakdown of cell products and materials taken up by the cell
Mitochondrion	Ovoid, membrane-bound organelle with a complex interior membrane scaffold	Oxidative metabolism, production of ATP
Nucleus	A relatively large region of the cell body surrounded by a membrane	Contains DNA, directs development and activity of the cell
Ribosome	A complex protein usually associated with the endoplasmic reticulum	Site of protein synthesis
Vesicle	A small, membrane-bound inclusion in the cell	Serves as a container for material being moved within or out of the cell

rodlike strands of protein, with a minute opening in the center. The composition and length of neurofilaments can vary considerably. They can be found in both axons and dendrites but are more common in the former. Neurofilaments are structural elements that help to stabilize and strengthen long neurites. They are also the elements stained by the Golgi stain. They form dense tangles in patients with Alzheimer's disease, but the significance of these tangles is not known.

Microtubules, which in neurons are sometimes called **neurotubules,** are hollow, with an outer diameter of about 23 to 25 nm. They are composed of the globular protein **tubulin,** arranged in a spiral to form a hollow cylinder. Tubulin is a 120-kdalton dimer, meaning that it consists of two subunits, α and ß. Microtubules are used in the elaborate system of long-distance internal transport characteristic of neurons (see the next section) and for structural support. In living cells, they exist in a state of dynamic equilibrium, continually being assembled and disassembled. Drugs like colchicine that prevent or reduce the growth of microtubules upset this equilibrium and consequently cause a reduction in their number and length. Microtubules are polarized; that is, their two ends are functionally and morphologically different. The end nearest the cell body, arbitrarily called the minus (–) end, is generally capped with microtubule-associated proteins. The other end is open. Growth of a microtubule occurs when tubulin dimers are added to the open, or plus (+) end. Disassembly occurs when dimers disassociate from the tubule. The distinguishing features of the three types of filament-like proteins are summarized in Table 2-6.

Neurons contain the same subcellular elements found in nonneural cells. Most of these play the same role in neurons as in nonneural cells. Some, however, such as Nissl substance, have expanded roles, and others, such as vesicles, have additional roles. Some of the filamentous proteins of neurons are especially adapted to help the neuron maintain its unique structure or to carry out neuron-specific tasks like long-distance transport.

TABLE 2-6. Filamentous Proteins

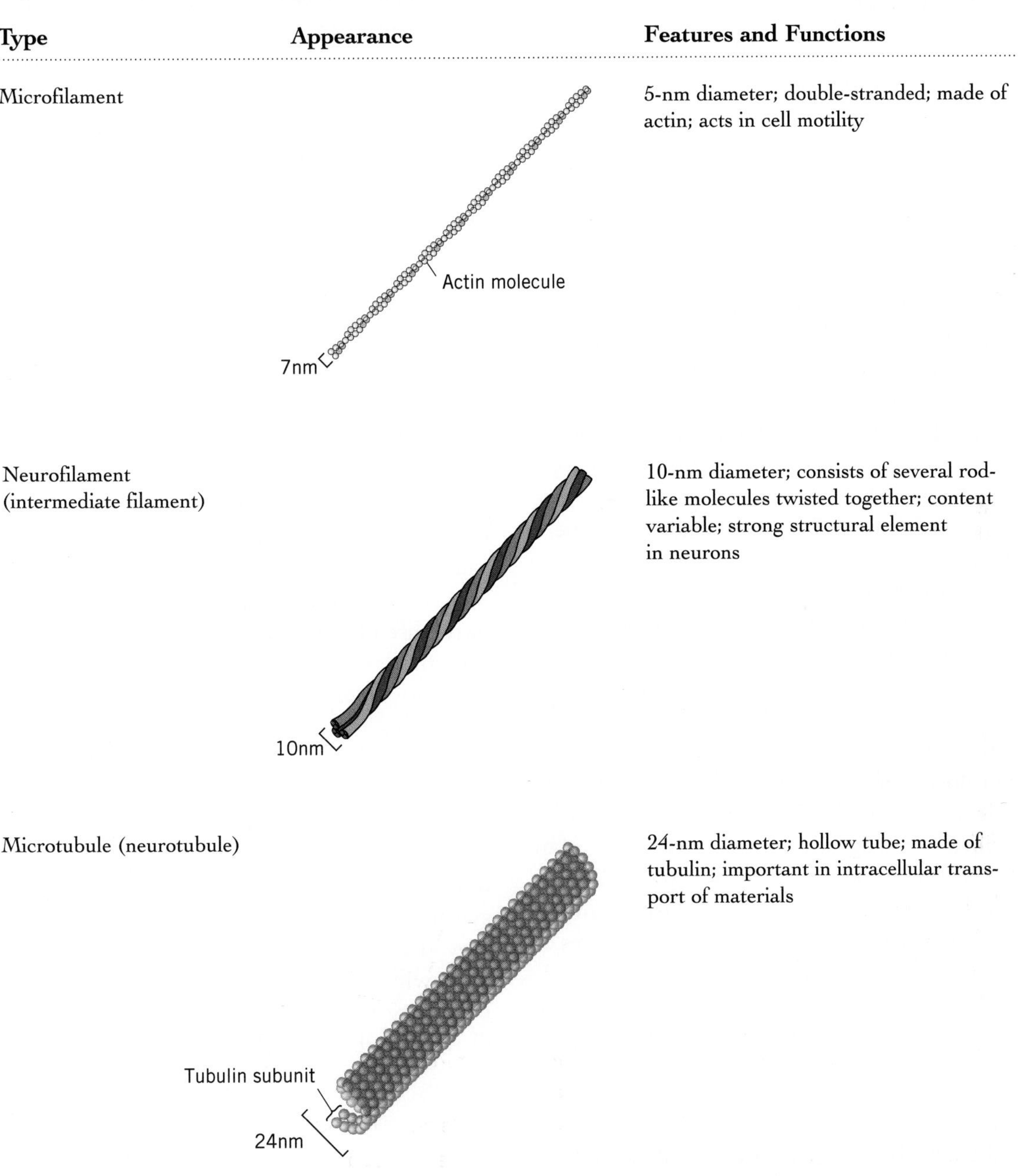

Type	Appearance	Features and Functions
Microfilament		5-nm diameter; double-stranded; made of actin; acts in cell motility
Neurofilament (intermediate filament)		10-nm diameter; consists of several rod-like molecules twisted together; content variable; strong structural element in neurons
Microtubule (neurotubule)		24-nm diameter; hollow tube; made of tubulin; important in intracellular transport of materials

Axonal Transport

Like other cells, neurons must ensure that organelles and complex molecules such as membrane channel proteins are properly distributed throughout the cell. Most cells use a system of internal transport for moving such elements from one place to another. In neurons, this system has been elaborated to cope with the long distances substances must travel to get from the cell body to the end of the axon. Although this transport system is not confined to axons, it is known as **axonal transport**, or **axoplasmic transport**.

Types of Transport

Even in the early part of the twentieth century, Ramón y Cajal and others had argued that a mechanism for moving materials from the cell body to the tip of an axon had to be present. The earliest evidence that such a mechanism actually existed, however, was an observation reported by Paul Weiss and H. B. Hiscoe in 1948. They found that tying a fine thread tightly around an axon, a process called ligation, resulted in swelling of the axon on the side of the ligature toward the cell body but no swelling on the side away from it, as if material were being pushed up against the ligature. Many subsequent studies of axonal flow, in which a variety of techniques have been used (Box 2-3), have revealed that there are actually several types of transport.

Anterograde transport is the movement of materials from the cell body toward the periphery. The rate of anterograde transport ranges from a low of 0.5 mm/day to a high of 400 mm/day. The extreme rates are often designated with their own names, **fast axonal transport** and **slow axonal transport**. Movement of materials from the periphery toward the cell body is termed **retrograde transport.** So far, researchers have found just one speed of retrograde transport, about half the fastest anterograde rate, or 200 mm/day. The main features of axonal transport are summarized in Table 2-7.

Axonal transport moves virtually everything in the cell that is not fastened down. Complex molecules like enzymes, structural elements like tubulin, and cell organelles like vesicles and mitochondria are all transported. However, the different transport mechanisms have affinities for specific items. Fast anterograde transport is specific for membrane-bound organelles or materials. It can actually transport a variety of molecules, including many complex proteins, but substances such as enzymes that do not already have a covering membrane are packaged into vesicles before being transported. Retrograde transport also moves only molecules packaged inside vesicles. Slow anterograde transport moves mainly soluble proteins. However, the different types of slow transport carry different molecules. One type seems to carry mainly α and ß tubulin and the proteins that make neurofilaments. Another transports molecules such as actin, the Ca^{2+} regulator calmodulin, and various enzymes.

Axonal transport plays a fundamental role in the life of a neuron. Any function, such as growth, development, or repair, that requires the movement of molecules or cellular organelles from one part of the neuron to another will depend on axonal transport. Anterograde transport is necessary because all synthesis of proteins takes place in the cell body, so receptors, signaling proteins, and binding proteins, as well as enzymes for the synthesis of neurotransmitters, must all be moved from the soma to the sites at which they will be used. Retrograde transport is necessary for communication from the ends of neurites back to the cell body, such as to

TABLE 2-7. Types of Axonal Transport

Anterograde	Axonal transport away from the cell body of the neuron
Fast transport	Anterograde transport at the rate of up to 410 mm/day
Intermediate transport	Anterograde transport at the rate of about 200 mm/day
Slow transport	Anterograde transport at the rate of less than 6 mm/day
SCa	The "slow" component of slow transport, about 0.5–3 mm/day
SCb	The "fast" component of slow transport, about 5–6 mm/day
Retrograde	Axonal transport toward the cell body of the neuron

BOX 2-3

Seeing Molecules Move: Video Techniques for Axonal Transport

Simple ligation experiments such as those conducted by Weiss and Hiscoe are sufficient to demonstrate that axonal transport occurs, but understanding the mechanisms underlying the phenomenon requires quantitative data. One method of acquiring such data involves introducing radioactively labeled substances into individual neurons. As the labeled material moves along the length of a neurite, its course can be followed by the movement of the "hot spot" representing the radioactivity. By introducing different labeled materials into the cell, it has been possible to study the rate of transport of different substances and to demonstrate that they are not all transported at the same rate.

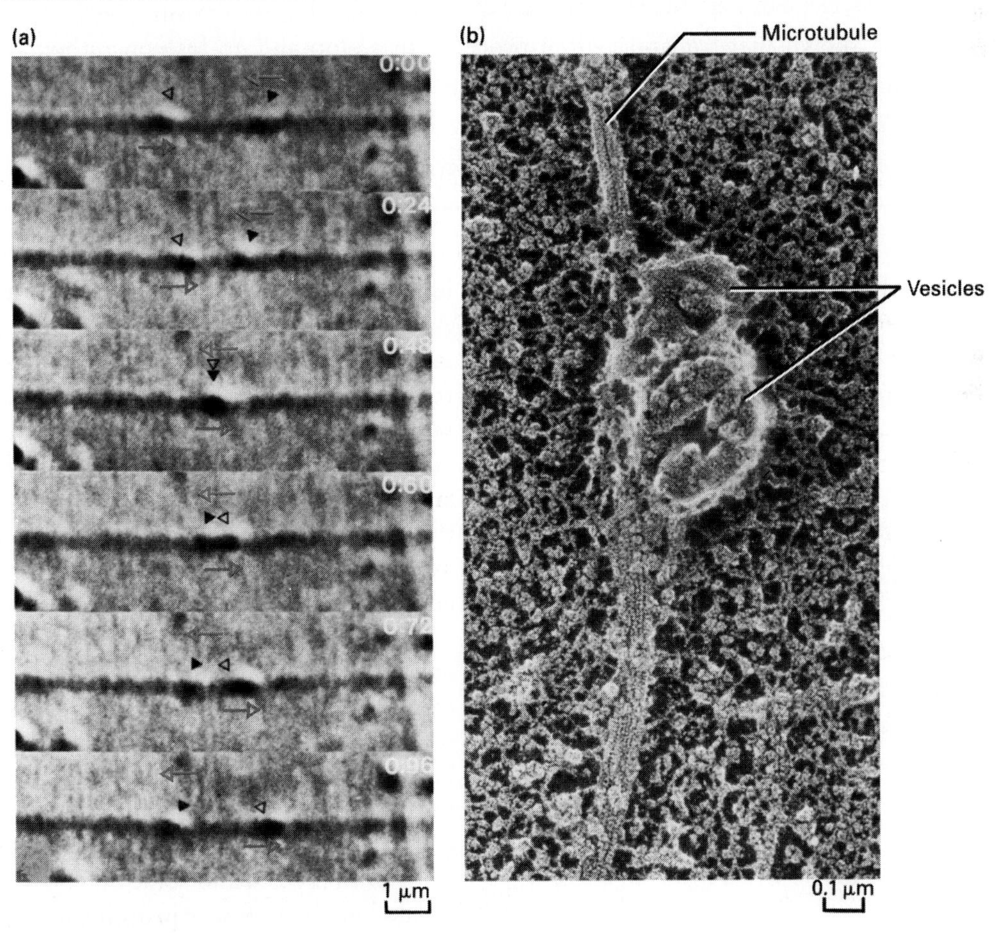

FIGURE 2-C. A sequence of frames from an enhanced video microscopal sequence showing the movement of two individual particles along a microtubule from squid giant axon. The numbers in the upper right of each frame give the elapsed time in seconds. The two particles can be distinguished by a slight difference in size. The video sequence shows that the particles pass one another without apparent difficulty.

(continued)

Seeing Molecules Move: Video Techniques for Axonal Transport (continued)

A newer technique, video-enhanced microscopy, gives researchers a direct view of the movement of organelles in living neurons. This method, developed by R. D. Allen and colleagues in the early 1980s, combines computer-enhanced video techniques with an advanced microscopal technique called differential interference contrast microscopy, which allows the resolution of small particles in living tissues. This combination of methods allows researchers to resolve organelles as small as 25 nm in diameter, an order of magnitude below the size of particles that can normally be resolved with the light microscope.

Video-enhanced microscopy has yielded several important findings. First, it has demonstrated conclusively that anterograde and retrograde transport can use the same track simultaneously (Figure 2-C). Second, it has shown that transport often proceeds in a stop-and-go fashion rather than at a steady, even rate. This has forced researchers to look for mechanisms that engage and disengage materials with the transport system as well as for the mechanism of propulsion itself.

tell the cell body during development that a neurite has reached a target cell and can stop growing, or to report that a distant neurite has been damaged and needs repair.

Within neurons, materials are transported away from the cell body by anterograde axonal transport and toward it by retrograde transport. Different types of transport have affinities for specific substances. Most types of transport move materials enclosed by a membrane, but slow transport will move unbound soluble proteins. Axonal transport is necessary for the movement of molecules to and from the cell body.

Mechanisms of Transport

Axonal transport needs two elements. First, it requires energy, obtained from the hydrolysis of ATP. If the cell is poisoned by a metabolic inhibitor (which shuts down production of ATP), all axonal transport ceases. Second, axonal transport requires microtubules, which serve as the structural grid along which most transport takes place. This is shown by the application of pharmacological agents that dissociate tubulin, which stops transport, and by the finding that transport can take place along microtubules that have been extracted from cells as long as ATP is available.

The molecular basis of transport has been most thoroughly investigated for retrograde transport and fast anterograde transport, which work on similar principles. The membrane of the vesicle or other organelle being transported has attached to it receptors that have a high affinity for a type of protein that, along with myosin and some others, belongs to a class known as **motor proteins** for their ability to transform energy into physical movement through the hydrolysis of ATP. **Kinesin**, a motor protein that powers anterograde movement, and **dynein**, which powers retrograde movement, were the first motor proteins discovered for axonal transport. By the mid-

1990s, however, it had become clear that there are at least seven families of motor proteins that power movement of materials along microtubules. These families are members of a single superfamily known as the kinesin family (KIF) motor proteins. Different proteins move different materials along the microtubules, as indicated by experiments in which the anterograde transport of vesicles, for example, can be inhibited without affecting anterograde transport of mitochondria.

Transport occurs when one end of the motor protein binds to a receptor on the organelle that is to be transported and the other end binds to and releases a microtubule in such a way as to move the organelle along. How movement is actually accomplished has been investigated most thoroughly for kinesin. A kinesin molecule consists of a head and two tails. The head binds with the kinesin receptor attached to the organelle membrane. The tails bind with a microtubule. The hydrolysis of one molecule of ATP by kinesin results in a complex conformational change in the tail, causing one part to detach from the microtubule, step in the anterograde direction, and then reattach. The molecule then repositions itself in the new location, moving the organelle along with it. Experiments conducted in 1993 have shown that the kinesin tail moves about 8 nm per step, which is just about the size of one tubulin dimer (Figure 2-14). Retrograde transport works in a similar way, except that dynein is the motor protein that binds to the vesicles and the microtubules. One of the satisfying features of this scheme is that it allows simultaneous bidirectional transport along a single stretch of microtubule, a phenomenon noted in microscopic videos taken during transport in living axons (Box 2-3).

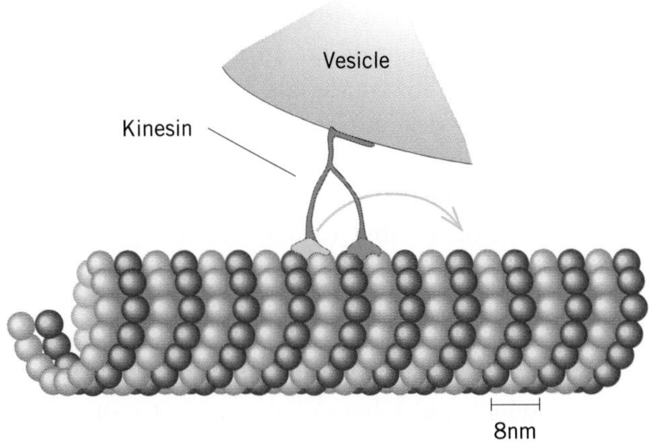

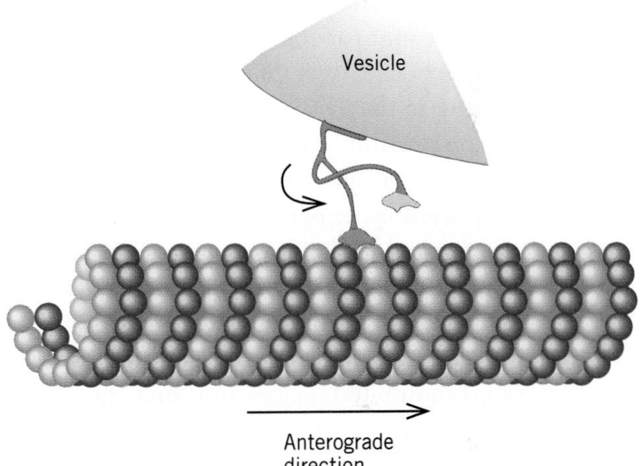

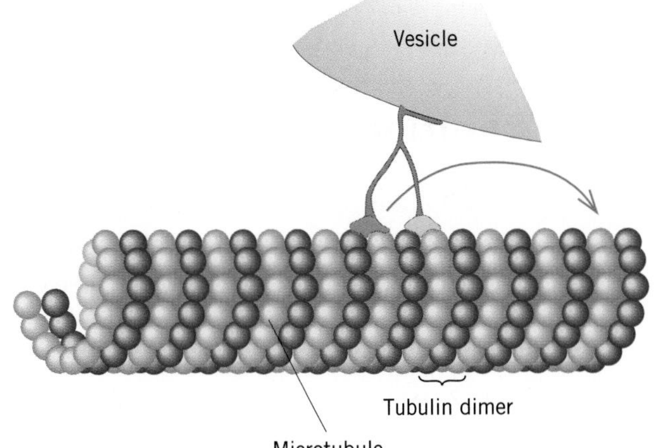

FIGURE 2-14. The mechanism of anterograde axonal transport, showing how vesicle-bound kinesin steps along a microtubule. The feet of the kinesin molecule alternately bind with the microtubule and release from it. The release is accompanied by a conformational change in kinesin that straightens the "leg" still attached to the microtubule, thereby moving the attached vesicle along. Meanwhile, the released "leg" swings around to attach farther along the microtubule.

It is still not clear how the neuron regulates which materials are to move in which direction, or how materials are translocated to the proper destination. One possibility is that all the protein motors bind with every vesicle or organelle that can be transported, but that the neuron is able to activate or inactivate the motor molecules individually. Vesicles on which only kinesin or another anterograde motor protein is activated will then move in the anterograde direction, and vesicles on which only dynein is activated will move in the retrograde direction. "Errors" in the process of activation and inactivation could then account for the small fraction of vesicles of a particular type that can be observed to be moving in the "wrong" direction.

The possibility that the neuron can activate or inactivate specific protein motors has been supported by experiments. Protein kinase A, which can bind a phosphate group to protein in a process known as phosphorylation, can phosphorylate kinesin. When it does so, the transport of vesicle-enclosed materials is transiently inhibited, suggesting that the cell may regulate some transport by regulating the state of phosphorylation of kinesin. Phosphorylation is an important regulatory mechanism in many neuronal cellular processes (see Chapter 6). Ensuring that vesicles are directed to the proper location is also complex and may be done by molecules that identify and catch an organelle as it moves by a specific site in the neuron.

Slow transport is both similar to and different from fast and retrograde transport. It is similar to fast and retrograde transport in being energy-dependent and in requiring fully functional microtubules; it differs in transporting cytoskeletal elements and other molecules that are not enclosed in any type of membrane, and in using a different motor protein than does fast or retrograde transport.

Axonal transport is an energy-requiring process that involves the transient binding of the transported material to the network of microtubules. Kinesin, a member of a family of motor proteins, is the binding agent for anterograde transport, and dynein is the agent for retrograde transport. A third, unidentified protein is necessary for slow anterograde transport.

Additional Reading

General

Alberts, B., D. Bray, J. Lewis, M. Raff, K. Roberts, and J. D. Watson. 1994. *Molecular Biology of the Cell*. 3d ed. New York: Garland.

Hall, Z. W. 1992. *An Introduction to Molecular Neurobiology*. Chapters 1, 8, 9. Sunderland, Mass.: Sinauer.

Lodish, H., D. Baltimore, A. Berk, S. L. Zipursky, P. Matsudaira, and J. Darnell. 1995. *Molecular Cell Biology*. 3d ed. New York: Scientific American Books.

Siegal, G., B. Agranoff, R. W. Albers, and P. Molinoff, eds. 1994. *Basic Neurochemistry: Molecular, Cellular, and Medical Aspects*. 5th ed. Chapters 1–3, 27. New York: Raven Press.

Research Articles and Reviews

Barres, B. A. 1991. New roles for glia. *Journal of Neuroscience*. 11:3685–694.

Catterall, W. A. 1991. Structure and function of voltage-gated sodium and calcium channels. *Current Opinion in Neurobiology*. 1:5–13.

Hirokawa, N. 1996. Organelle transport along microtubules—the role of KIFs. *Trends in Cell Biology*. 6:135–41.

Okada, Y., R. Sato-Yoshitake, and N. Hirokawa. 1995. The activation of protein kinase A pathway selectively inhibits anterograde axonal transport of vesicles but not mitochondria transport or retrograde transport *in vivo*. *Journal of Neuroscience*. 14:3053–64.

Schroer, T., and M. Sheetz. 1990. Functions of microtubule-based motors. *Annual Review of Physiology*. 53:629–52.

Vallee, R. B., and G. S. Bloom. 1991. Mechanisms of fast and slow axonal tranport. *Annual Review of Neuroscience*. 14:59–92.

CHAPTER 3

The Structure of Nervous Systems

The nervous system of every animal has a distinctive structural organization, a precise arrangement of the axons and dendrites of its constituent neurons that allows it to carry out its functions. Neuroanatomy, the description of this organization, is like a map, indicating the main structural features of the nervous system and the routes by which neurons in one place communicate with neurons in another. This chapter introduces this neural map, outlining the organization of nervous systems and pointing out some of the main connecting pathways.

Neuroanatomy is an essential element of neurobiology. Not only must you have some knowledge of anatomical terms in order to understand any discussion of brain function, you also need a knowledge of neuroanatomy in order to appreciate the functional connections between different parts of the nervous system. To use a map analogy, just as it would be difficult for you to take the freeway from St. Louis to New Orleans without knowing that you should follow the signs toward Memphis, so it would be difficult for you to understand the organization of motor systems in a cat without knowing that some motor control circuits pass through the thalamus and basal ganglia before going on to the motor cortex.

The Parts of the Nervous System

The nervous systems of all animals are divided into central and peripheral parts. The **central nervous system (CNS)** consists of the brain and the nerve cord (spinal cord in vertebrates), which constitute the main aggregation of neural tissue in the body. The brain and nerve cord are a mixture of processing regions and **nerve tracts** (sometimes just called **tracts**), bundles of axons that connect the different processing regions to one another. The processing regions receive sensory input, organize motor output, and make decisions about the animal's current situation based on its past experience. One of the main challenges of neuroanatomy is to determine the precise connections between the different processing centers (Box 3-1).

The **peripheral nervous system** consists of all the nerves that connect to the CNS, as well as the small groups of nerve cell bodies that lie outside the CNS. **Nerves** consist almost exclusively of the dendrites or axons of nerve cells. Some nerves carry information toward the CNS. These are referred to as **afferent nerves**, and the individual neurites in them are called **afferent fibers**. Information that enters the CNS originates from sense organs, so afferent nerves are also **sensory nerves**. Other nerves carry information away from the CNS.

BOX 3-1

Where Do They Go? Neuroanatomical Tracing of Nerve Pathways

The prominent tracts in the brains and spinal cords of vertebrates represent the main routes of communication within the central nervous system. But how do neuroscientists trace the path of a bundle of axons? You cannot simply tug at one end and see what moves at the other end as you can with a fistful of string because everything is tightly bound together. Early investigators devised methods for attacking two questions they needed to answer: where do the axons go that leave a particular region, and where do the axons come from that enter a particular region?

Early methods for finding the pathways and origins of nerve tracts involved making small lesions in the central nervous system and then making a careful examination of sections of fixed and stained nerve tissue to find evidence of the lesion. When an axon is cut, the part that is separated from the cell body will degenerate. The chemical changes that accompany this degeneration can be made visible by applying chemicals that stain the degenerating axons. This method, although useful only for tracts that are large and easy to find, allows a tract to be traced to its end.

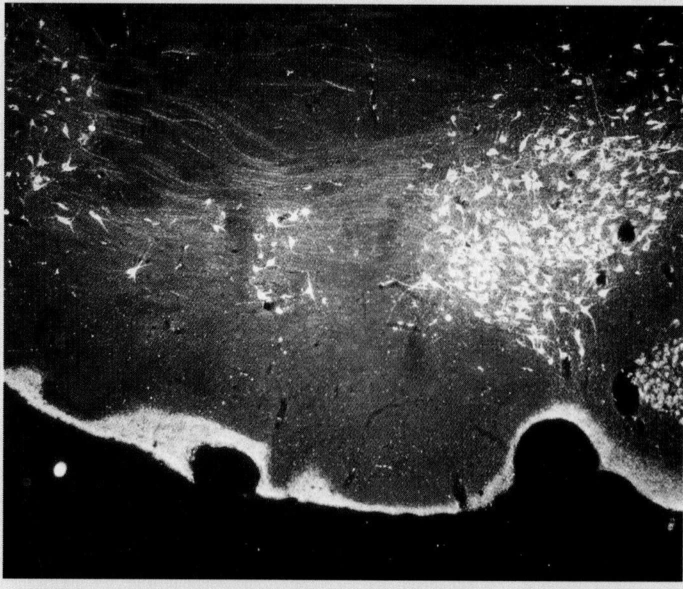

FIGURE 3-A. Horseradish peroxidase staining of cell bodies (light spots) in the basal forebrain of a rat. The arrow shows retrogradely labeled axons from an olfactory center. The photomicrograph is viewed under dark-field illumination, in which unstained areas appear dark and stained areas appear white.

The lesion method can also be used to identify the origin of a nerve tract. When an axon is cut, the loss of chemical signals from its terminals stimulates the cell body to synthesize materials needed for regeneration of the cut end. Appropriate stains can be used to detect the characteristic chemical changes that accompany regeneration in the cell body, which can then be located in the brain. The changes are subtle and variable, however, and the method requires some idea of where to look.

More powerful techniques for tracing nerve tracts have been developed since the early days of neuroanatomy. One method is to inject a specific region of the nervous system with a molecule like **horseradish peroxidase (HRP)**, an enzyme extracted from the horseradish plant. HRP is taken up by individual neurons in the region where it was injected and is moved by axonal transport along the length of their axons to the parent cell bodies. By cutting sections of the tissue and treating them chemically to make the HRP visible (see Box 2-1), anatomists can locate not only the tracts through which the axons run but also the cell bodies of origin (Figure 3-A) and, in favorable instances, the terminals of the axons as well.

These are called **efferent nerves**, and the individual neurites in them are called **efferent fibers**. Information leaving the CNS is destined for muscles or glands, so efferent nerves are also **motor nerves**. A third category of nerve is the **mixed nerve**, which contains both afferent and efferent fibers. In addition to nerves, the peripheral nervous system includes ganglia. A **ganglion** (singular) is an aggregation of cell bodies and neurites surrounded by a connective tissue sheath.

The peripheral nervous system of a typical vertebrate has traditionally been divided into two main functional parts. The **autonomic nervous system** consists of the nerves and ganglia that regulate autonomic functions, such as maintaining blood pressure and controlling the beating of the heart, over which an animal has little or no voluntary control. The remaining nerves and ganglia constitute the **somatic nervous system**, which processes input from sense organs and coordinates voluntary activity. Although the brain is certainly involved in both autonomic and somatic functions, the two terms have traditionally been applied mainly to peripheral nerves. There is some physical separation of somatic and autonomic nerves, but most peripheral nerves carry both somatic and autonomic nerve fibers. The relationships between the different anatomical and functional divisions of the vertebrate nervous system are shown schematically in Figure 3-1.

Most invertebrates also have an autonomic nervous system, but much less is known about it because it has not been extensively studied. The CNS is somewhat different in invertebrate and vertebrate animals, and therefore will be described in more detail separately for the two groups of animals.

The nervous system is divided morphologically into a central component, which consists of the brain and nerve cord, and a peripheral component, which consists of all the outlying nerves and ganglia. The peripheral nerves and ganglia are divided functionally into autonomic and somatic nervous systems. The former controls the involuntary functions of the body; the latter controls the voluntary functions.

Afferent fibers

Efferent fibers

Brain

Spinal cord

Central nervous system

Peripheral nervous system

Ganglion

Somatic nerve

Autonomic nerve

FIGURE 3-1. Structural and functional divisions of the vertebrate nervous system. Everything outside the CNS is part of the peripheral nervous system. Efferent (motor) fibers are shown on the right side only, and afferent (sensory) fibers on the left. Functionally, the peripheral system is divided into somatic (dashed lines) and autonomic (solid lines) parts.

Invertebrate Nervous Systems

The plan of nervous systems in invertebrate animals can vary considerably from species to species. Sponges actually have no nervous systems at all. The cnidarians, which include the sea anemones, corals, and jellyfish, have a diffuse nerve net but no centralized mass of neural tissue (Figure 3-2A). Echinoderms, such as starfish and sand dollars, have a nerve ring and an extensive network of peripheral nerves. In mollusks like clams, snails, and squid, the ganglia that constitute the CNS may be located close together in one restricted area rather than being distributed throughout the body (Figure 3-2B).

The most common morphological plan of the nervous system in invertebrate animals, however, is the arrangement that is typical of many annelids and arthropods. In this plan, a series of ganglia are arranged in the body from the head to the tail, joined to one another by a pair of large nerves called **connectives** and connected to the sense organs and muscles of the body by smaller peripheral nerves. An evolutionarily

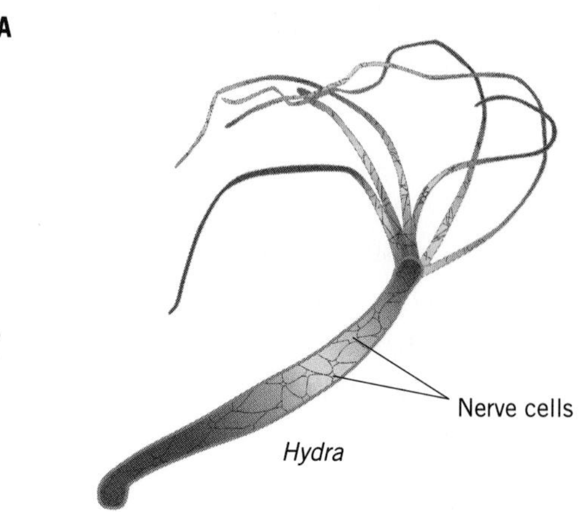

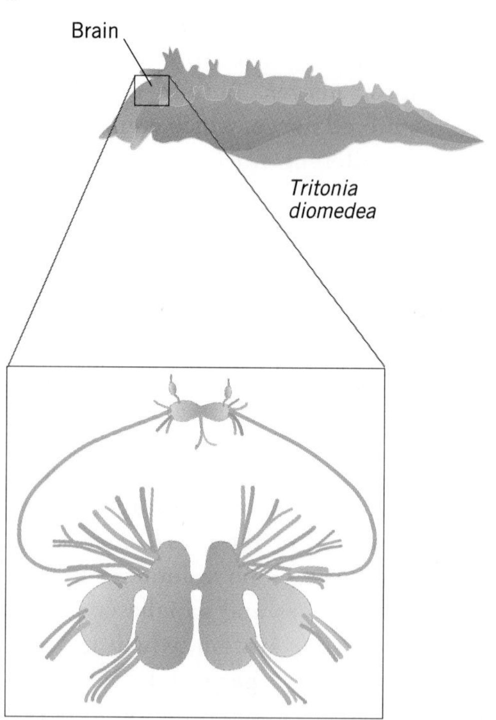

FIGURE 3-2. The diversity of invertebrate nervous systems. (A) The nerve net of the cnidarian *Hydra*, which has no CNS. (B) The concentrated nervous system of the shell-less marine nudibranch *Tritonia diomedea*, a molluscan relative of snails.

early form of this arrangement can be seen even in primitive flatworms (Figure 3-3A). In the annelids and in many arthropods, there is generally one ganglion for each body segment (Figure 3-3B). In other arthropods, the ganglia for several body segments may fuse together during embryological development, so that in the adult the number of ganglia is less than the number of body segments. In some arthropods, such as crabs and flies, for example, this embryonic fusion can be extensive, leaving only the brain and a single large ganglion in the body of an adult (Figure 3-3C).

The nervous systems of invertebrates can vary from a diffuse nerve net that contains no centralized part to two large ganglia. The most common arrangement is a series of ganglia, about one per body segment, joined by paired connectives.

The Central Nervous System

The CNS of a typical invertebrate consists of one or two large ganglia that constitute the brain, the chain of ganglia that lies in the body, and the connectives that join the ganglia to one another and to the brain. The ganglia that represent the bulk of the CNS have a distinctive structural organization. The neurons in them are typically monopolar. Furthermore, synapses do not form on the cell body of any invertebrate neuron. Presumably as a consequence of this, all the cell bodies are arrayed around the outer rim of the ganglion, just under the connective tissue sheath that encloses it (Figure 3-4). The interior of each ganglion is called the **neuropil**, which consists mostly of dendrites and synapses. However, nerve tracts, both those that join two processing regions within a ganglion and those passing through one ganglion on their way to another, also travel through

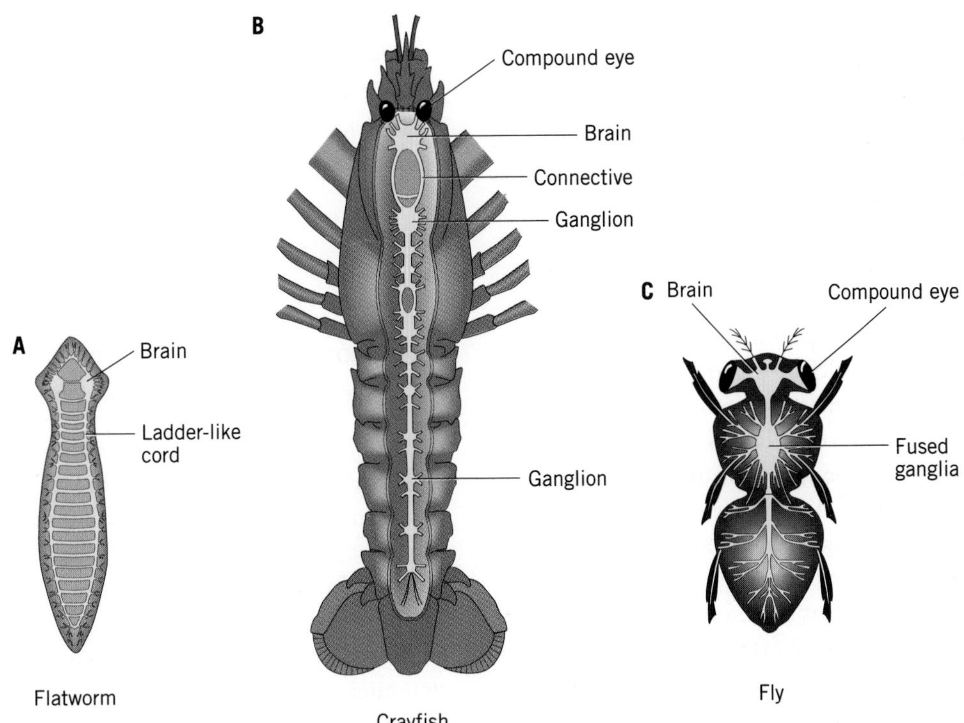

FIGURE 3-3. Typical ladderlike and compact invertebrate nervous systems. (A) Flatworm. The right and left connectives are quite distinct, and ganglia are rudimentary. (B) Crayfish. The right and left connectives run together through much of the nervous system, with distinct ganglia in each body segment. (C) Fly. The right and left connectives are more or less fused, as are all of the body ganglia.

A

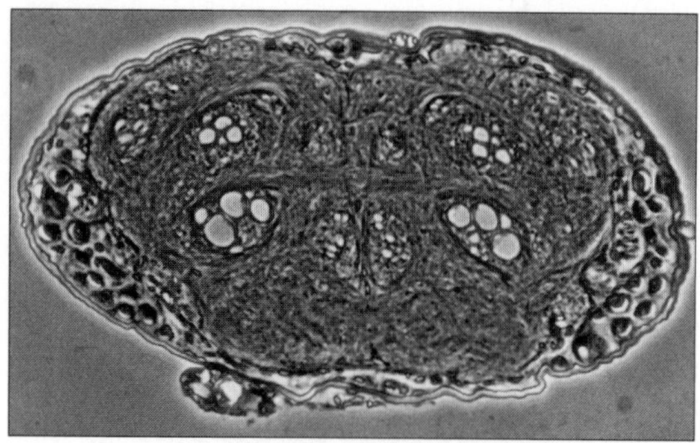

B

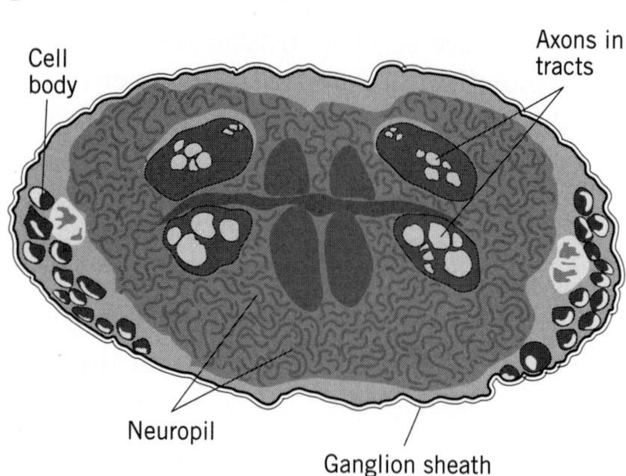

FIGURE 3-4. The structure of an invertebrate ganglion. (A) Photomicrograph of a cross section through a typical insect ganglion, showing the arrangement of nerve cell bodies under the connective tissue sheath of the ganglion. The neuropil is the central region in which synapses between neurites are located. (B) Sketch of the ganglion, showing the different parts.

the neuropil. Peripheral nerves usually enter or leave the CNS at a ganglion rather than branching from a connective. Peripheral nerves are almost always mixed, carrying axons of both sensory and motor neurons.

The connectives that join the ganglia contain tens of thousands of axons (Figure 3-5). Most of the axons are quite small, less than 1 μm in diameter, but some can be more than 50 μm. The axons of invertebrates are not myelinated (see Chapter 2). As you will learn in Chapter 5, the conduction velocity of a nerve impulse increases with the diameter of the axon in which it is traveling. Hence, the large axons are generally used in critically important behaviors like escape. The smallest ones usually carry less important sensory or other information.

The brain is the mass of the nervous system that lies in the head. Most invertebrate animals have a brain, although there are exceptions like the cnidarians and echinoderms. The brain in many an invertebrate is undistinguished, looking only like a particularly large ganglion. However, although an invertebrate brain does not look like the brain of a fish or a dog, it nevertheless shares several features with a vertebrate brain. For one thing, it is larger than any other part of the animal's CNS. More important, however, like any vertebrate brain, it is also morphologically and functionally specialized, meaning that parts of the brain are different from one another in structure and in the functions they carry out.

Invertebrates with a central nervous system usually have a brain and several ganglia joined by connectives. In ganglia, the cell bodies are arranged around the outer rim; dendrites and synapses fill the central neuropil. An invertebrate brain, although quite different in external appearance from the brain of a vertebrate, is nevertheless functionally and structurally specialized.

Functional Organization of Invertebrate Nervous Systems

Although the cellular organization of ganglia and nerve tracts in invertebrates may first strike an observer as a tangled mass lacking any coherent structure, close study has shown that this first impression is wrong. Just as a consistent internal structure in the brain of a cat can be found in all cats, so a consistent internal structure in the brain of a housefly can be found in all houseflies. In many cases, it is even possible to identify the same specific neuron in different individuals of the same species (Box 3-2; see also Chapter 18). This ability to repeatedly find a given neuron in different individual

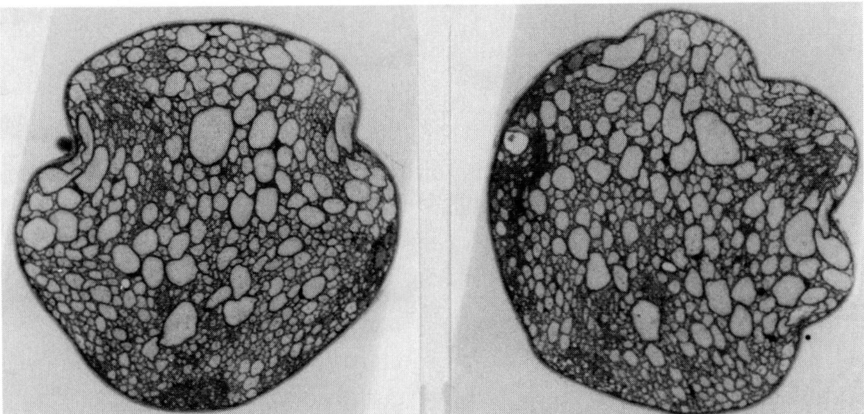

FIGURE 3-5. Photomicrograph of a section through the connectives joining two thoracic ganglia in an insect. Notice the enormous range of axon size in this cross section.

animals has led to important advances in the understanding of the anatomical and functional organization not only of invertebrate nervous systems but of all animals. Much of our current understanding of phenomena such as learning has derived from experiments on individually recognizable neurons in invertebrates.

The brain of an invertebrate plays the same roles in neural function as does the brain of a vertebrate: to process sensory information, to organize motor output, and to make decisions about events. For example, inputs from the eyes or odor receptors on the head are processed in the brain, and a decision to walk, stand still, or eat is made there as well. In some advanced invertebrates, there are even brain regions that are specialized for memory storage, as you will learn in Chapter 24. The ganglia in the body are concerned with local motor control and reflexes. You will learn more about the functional organization of invertebrate CNSs in Part 4 of this book.

The brain of every invertebrate has a distinctive structure and organization that are consistent from individual to individual within a given species. The brain acts as a sensory processing, motor control, and decision-making center, controlling the behavior of the animal.

The Vertebrate Central Nervous System

The CNS of a vertebrate is larger both in absolute terms and relative to body size than is that of an invertebrate. Due in part to its larger size, the vertebrate brain is also more complex structurally. For this reason, the rest of this chapter will be devoted to an outline of vertebrate neuroanatomy. More detail about individual parts of the brain and spinal cord will be introduced as appropriate when specific functional systems are presented in Parts 3 through 6 of this book.

It is assumed throughout this and the remaining chapters that you are familiar with the anatomical terms of orientation used to describe front and back, top and bottom, and so forth. If you are not familiar with them, or if you need to refresh your memory, the diagrams and definitions in the Appendix may be helpful.

The Spinal Cord

The **spinal cord** lies in the dorsally located vertebral column, from the base of the brain to the pelvis. In humans, the spinal cord does not reach the end of the backbone. The posterior few segments are filled with nerves from the end of the cord that supply the pelvis, buttocks, and legs. The spinal cord has

BOX 3-2

Standing Out from the Crowd: Staining Individual Neurons

The ability to stain single neurons in the nervous system has revolutionized neuroscientists' understanding of neural anatomy because it has revealed just how highly organized the nervous system is. The technique depends on the ability of a researcher to impale an individual neuron with a fine, hollow glass tube, called a microelectrode. Substances in the microelectrode can be passed into the impaled neuron, where they diffuse or are transported away from the site of entry and distributed to the whole neuron. If the material is subsequently made visible in the tissue by suitable histological techniques, it becomes possible to see not only the shape of the entire neuron but also where the dendrites and the axon of the neuron lie in relation to other structures.

In general, two types of materials have been used for intracellular staining, those that require chemical treatment to become visible and those that do not. The first type includes metallic salts such as cobaltous chloride and enzymes such as HRP (see Box 3-1). The procedure involves injecting the material, then fixing the tissue and treating it histochemically to render the material visible, thereby revealing the structure of the stained cell (Figure 3-B). In vertebrate animals, the tissue containing the cell is usually first cut into sections about 50 μm to 100 μm thick, from which the researcher must reconstruct the neuron's structure. In invertebrates, the neuron may be

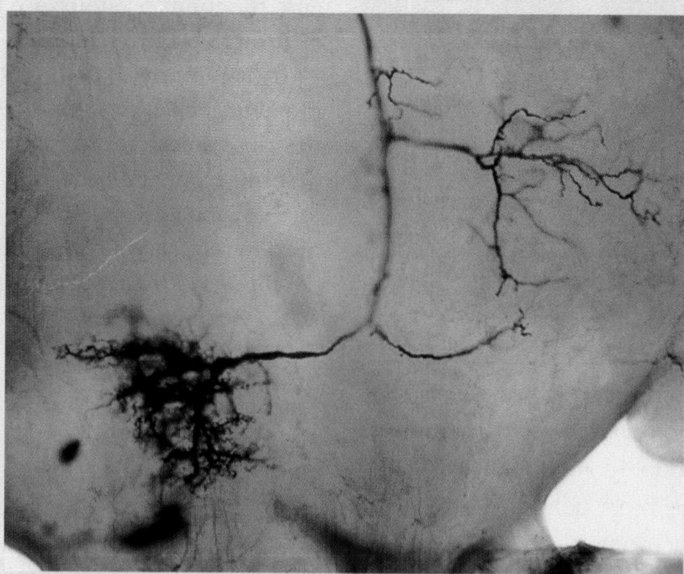

FIGURE 3-B. Interneuron in the nervous system of a cockroach, stained with a metallic salt. The cell body is the dark oval in the bottom left of the photomicrograph. Notice the fine detail of the neurite branching.

viewed in a whole ganglion, as shown in Figure 3-B. One particular advantage of using materials that require chemical treatment is that the histochemical procedures used to make them visible are quite sensitive; thus only relatively small amounts need to be injected in order for the neuron to be seen.

The other type of material used to stain cells fluoresces when exposed to ultraviolet (UV) light. The most popular of these materials is the dye **Lucifer yellow**. The steps in this procedure are to inject the dye, remove and fix the tissue, treat it chemically to make it transparent, then expose it to UV light under a fluorescence microscope. When the tissue is viewed through optical filters that pass only the wavelength of light at which the dye fluoresces, the injected cell and all of the neurites that the dye has filled become visible. An advantage of injected materials is that the electrical activity of a neuron can be recorded before the dye is injected and the cell processed histologically. This characteristic allows neurons to be studied physiologically as well as morphologically, allowing correlations to be made between the structure and location of the cell and its functional characteristics.

two distinct regions, white matter and gray matter, arranged around a small hollow central core called the **central canal**. White matter consists mainly of bundles of myelinated axons that connect different parts of the cord with one another or connect the cord with the brain. Gray matter consists mainly of cell bodies, dendrites, and synapses. The tracts that constitute white matter are grouped around the outside of the cord, and the cell bodies and dendrites that form gray matter are in the interior. Spinal nerves enter and leave the cord at regular intervals, between the bones of the vertebral column. Figure 3-6 shows a cross section of a typical spinal cord.

The spinal cord has a precise organization beyond the simple distribution of white and gray matter. Each tract in the spinal cord, often called a column, carries specific sensory or motor information. In addition, particular areas of the gray matter are also devoted to neurons that serve specific functions. In general, columns in the dorsal area of the cord carry sensory information up to the brain and those in the ventral area carry motor commands down from the brain. Of the lateral columns, about half carry motor signals and half carry sensory signals. This loose dorsal = sensory and ventral = motor organization of the tracts is maintained in the gray matter. The ventrolateral regions of the gray matter, called the ventral horns (Figure 3-6A), contain mainly the cell bodies of motor neurons that innervate the somatic (voluntary skeletal) muscles via the ventral root nerves, whereas the dorsal horns contain neurons that receive sensory information via the peripheral nerves. The cell bodies of efferent autonomic neurons lie in the lateral regions of the gray matter.

Closely associated with the spinal cord are the nerves that carry sensory information into and motor information away from the cord. Most of the sensory information that enters the cord does so through the dorsal roots. Each dorsal root contains a swelling, the **dorsal root ganglion**, which contains the cell bodies of sensory neurons entering the spinal cord from the body (Figure 3-6B). Motor information leaves the nerve cord via the

A

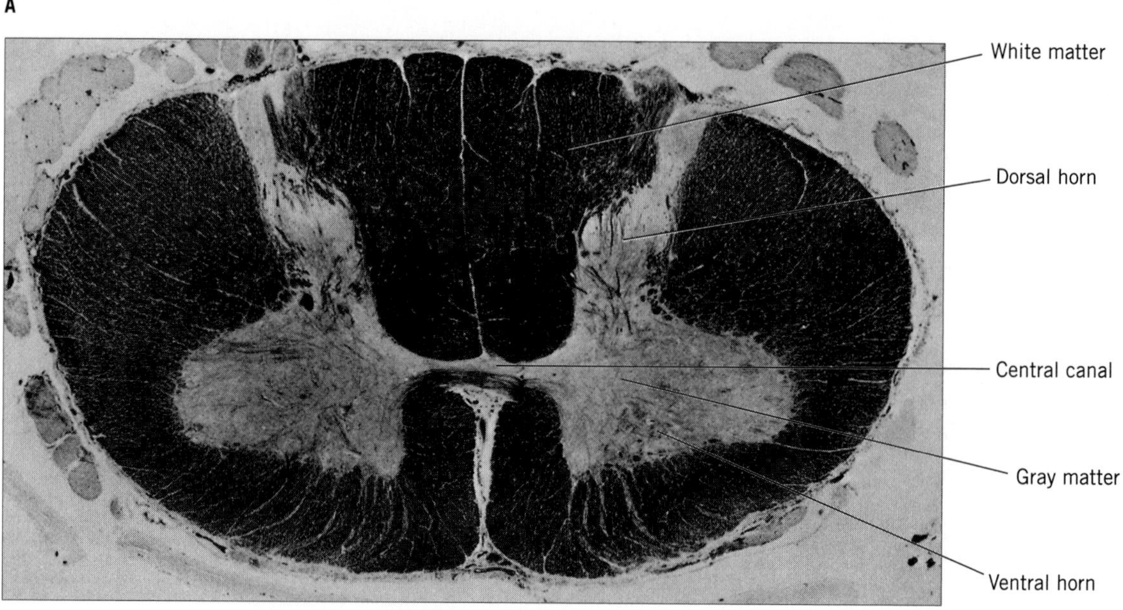

B

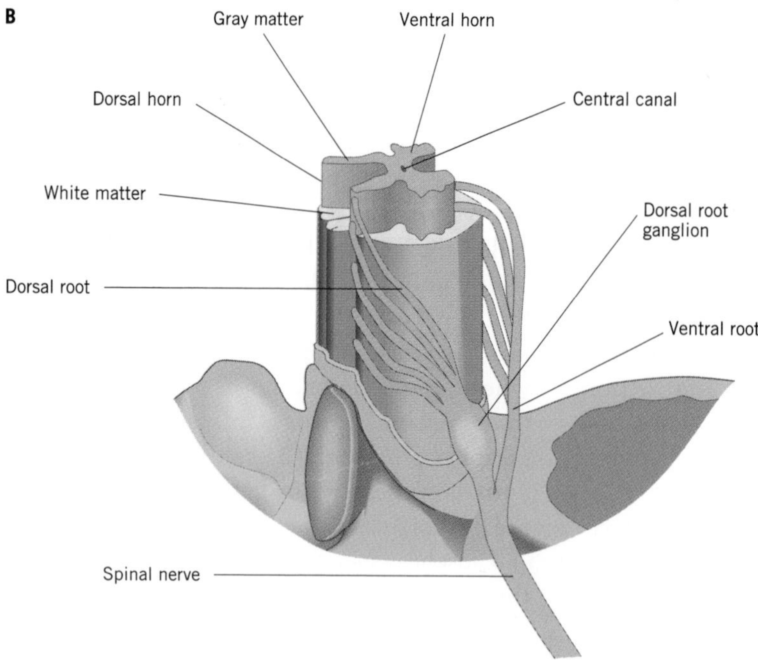

FIGURE 3-6. The spinal cord of a vertebrate. (A) Photomicrograph of a spinal cord, showing the white matter (stained dark in this section) and gray matter. (B) Schematic diagram of a human cord, showing the dorsal and ventral roots.

ventral roots. Because the cell bodies of the motor neurons that send their axons out the ventral root are all in the spinal cord, there is no peripheral ganglion associated with the ventral root. The dorsal and ventral roots fuse close to the spinal cord to form spinal nerves that run out between the vertebrae of the backbone. The spinal nerves are identified and numbered according to the part of the cord they serve and their position in that part, such as the fifth thoracic (T_5) or the second lumbar (L_2) nerve. In humans, there are 8 cervical, 12 thoracic, 5 lumbar, 5 sacral, and 1 coccygeal pairs of spinal nerves (Figure 3-7).

In the spinal cord, nerve tracts are arranged around the periphery, surrounding the cell bodies, dendrites, and synapses. The dorsal part of the cord is mostly devoted to carrying sensory information, whereas the ventral part is devoted to carrying motor information. Sensory information enters the cord through the dorsal root from the dorsal root ganglion, and motor information leaves the cord through the ventral root. Spinal nerves are identified according to the part of the cord from which they originate.

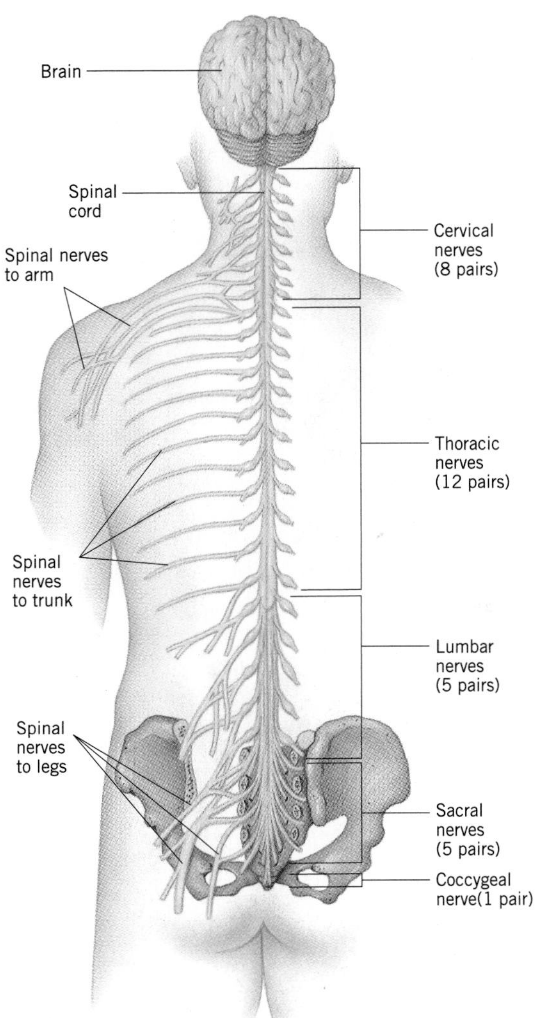

FIGURE 3-7. The human spinal cord, showing its different divisions, and the spinal nerves.

The Vertebrate Brain: An Overview

The brain is located in the head and has the role of supervising much or all of the animal's behavior. In vertebrate as well as invertebrate animals, the size of the brain is roughly scaled to the size of the body; larger animals have larger brains. However, even taking this scaling into account, the brains of vertebrates show an enormous range of size and complexity among the different classes of animals (Figure 3-8). Although it is easy to point out exceptions, there is a trend for the brain to be larger and more complex relative to body mass in mammals and birds than it is in fish and amphibians. Reptiles fall roughly in the middle of the range. The human brain is largest relative to body size (Figure 3-9).

For convenience, the surface of the human brain may be divided into four lobes: the frontal, parietal, occipital, and temporal. Looking at the brain from the side, the **frontal lobe** is the most anterior region, the **parietal lobe** is the middle region, and the **occipital lobe** is the most posterior region. The **temporal lobe** is situated below the parietal lobe, covering the lower parts of it and

Frog

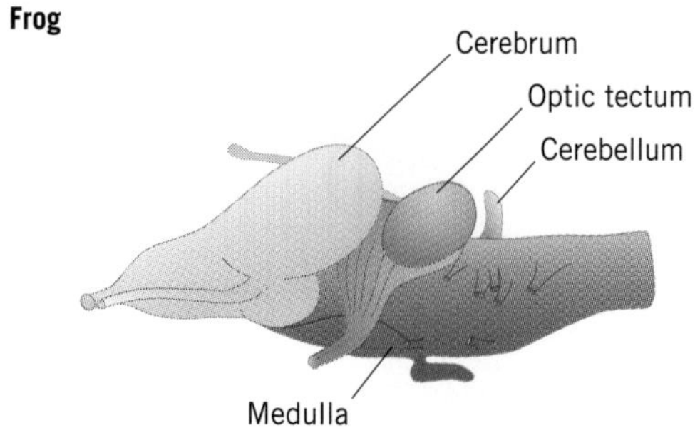

Pigeon

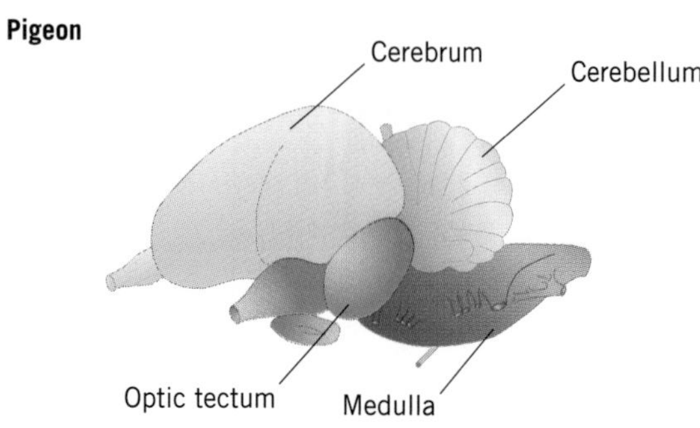

Dog

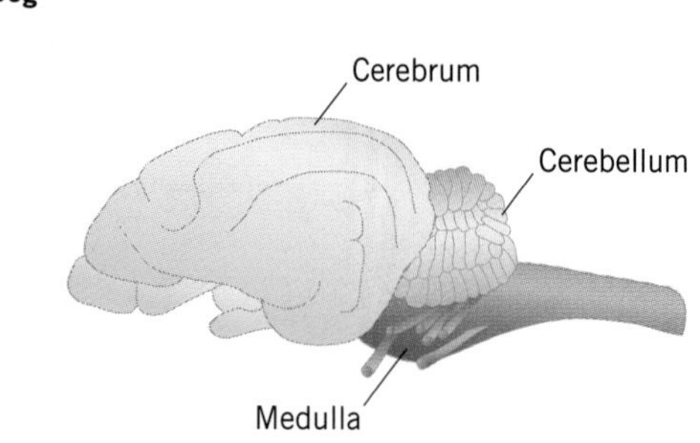

FIGURE 3-8. The brains of three vertebrates—a frog, a pigeon, and a dog—showing the relative sizes of the main parts. Notice that the brain is an expansion of tissue that is more or less in a straight line with the nerve cord. Note also in the dog the substantial physical development of the frontal lobes, which have covered nearly all the other brain structures. The labeled parts are listed in Table 3-1 and are described later in the text.

the frontal lobes. The boundaries between these lobes are not always clearly delineated, so the names of the lobes serve more to indicate a general region rather than a specific location.

The greater size and complexity of the visible parts of the brains of humans and other mammals result from a relatively greater increase in the size and surface area of the front part of the brain compared with the rest. This increase in surface area is accomplished during development by growth that produces folds in the brain surface. The growth and the folding of the surface are even more pronounced in primates, reaching their greatest flowering in the human brain. Each of the folds appears on the surface as a band of tissue, a **gyrus** (plural, gyri), that is bounded by an infolding on each side called a **sulcus** (plural, sulci). The small gyri and sulci may be in different places in different individuals. The larger ones, however, have the same placement in all individuals and hence serve as landmarks for the main lobes of the brain, as shown in Figure 3-9.

Like the spinal cord, the brain also shows a distinctive pattern of cellular organization. However, the clear separation of tracts and cell bodies that is present in the spinal cord is not as apparent in the brain. Some tracts are near the surface, but some are also deeper inside. The regions of gray matter, which in the cord are more or less continuous, tend to break up and form distinct collections of cell bodies called **nuclei** (singular, nucleus). In some regions, tracts and nuclei are so commingled that you cannot easily distinguish between them. The organization of the outer layers of the brain is reversed from that of the spinal cord. These layers, collectively called the **cortex**, consist of gray matter and contain cell bodies, dendrites, and synapses. Under the cortex lies a complex tangle of tracts and, interspersed among these tracts, deep nuclei.

The next four sections of this chapter present a more detailed picture of the organization of the vertebrate brain, using the human brain as an example. The purpose of these sections is to provide an overview of the structural and functional organization of a mammalian brain so that as you learn more about the operation of the nervous system you will be able to relate this knowledge to the neuroanatomy laid out in this chapter.

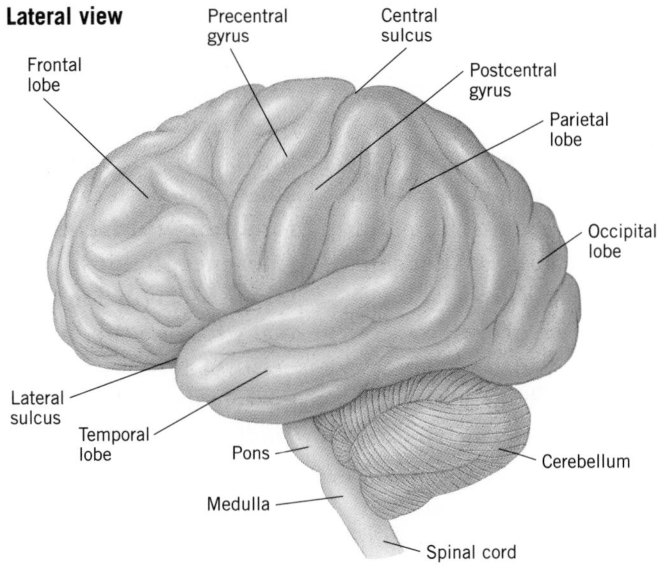

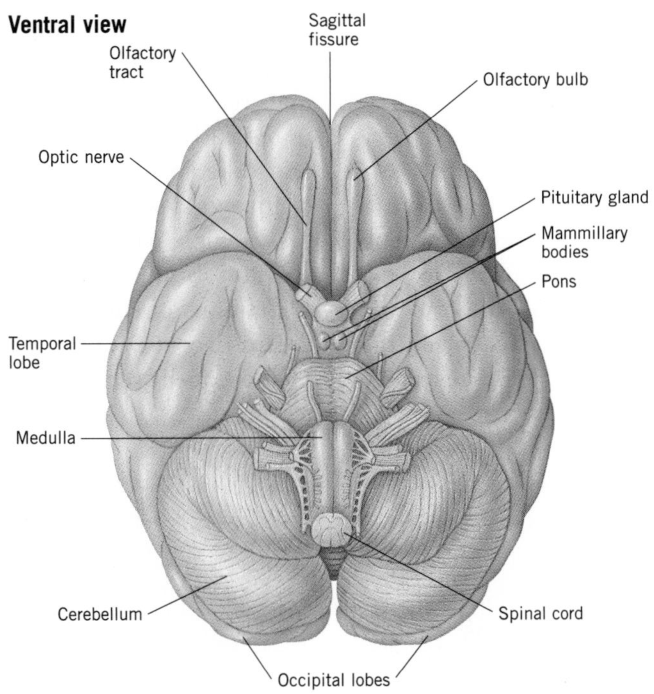

FIGURE 3-9. Views of the human brain from the side (lateral view) and the bottom (ventral view), showing the major gyri and sulci. Notice that the major lobes (frontal, temporal, parietal, and occipital) hide nearly every other part of the brain from view.

Among vertebrates, larger animals have larger brains. However, the brains of mammals and birds are much larger and more complex relative to body size than are the brains of fishes and amphibians. Complexity is especially great in the primates, in which the front part of the brain is much enlarged and shows an increased surface area due to folding of the surface. At the cellular level, cell bodies in the brain are located near the surface and in deeper groups called nuclei. Tracts of axons surround the nuclei.

Brain Development

The enormous growth of some parts of the human brain during development, and the structural complexity that results from this growth, can make anatomical relationships difficult to understand. This is especially true because the front of the brain grows so vigorously during development that it covers and obscures almost all the underlying structures, much like rising bread dough can spill out and cover a pan. Under this mass of tissue, however, the brain retains a linear arrangement of parts. This arrangement is quite apparent when you study the embryological development of the brain.

Emergence of the Brain and Spinal Cord

The brain and spinal cord originate from a tube that runs the length of a developing embryo. As development continues, the posterior part of this tube begins to form the spinal cord and the anterior part forms the brain. The spinal cord arises from a relatively uniform enlargement of the tube. The brain is formed by a more complex pattern of growth. As the anterior end of the neural tube grows, it develops three distinct lobes, which become the three main parts of the adult brain. From anterior to posterior, these parts are called the **forebrain (prosencephalon)**, the **midbrain (mesencephalon)**, and the **hindbrain (rhombencephalon)**. Each of the lobes eventually forms a subdivision of the brain (Figure 3-10). The forebrain grows two main parts, the **telencephalon** and the **diencephalon**. The hindbrain differentiates into the **cerebellum**, the **pons**, and the **medulla oblongata** (also called the **medulla**). Each of these brain regions is discussed in more detail in subsequent sections. The parts of the adult brain in primitive vertebrates like fish are relatively easy to identify and link to their embryological precursors because the brain in these animals retains the linear arrangement that it has in the early embryo. Identification in humans and other primates is more difficult because of the expansion of the forebrain.

The main parts and subdivisions of the vertebrate brain are listed in Table 3-1. Definitions of some of the terms in this table are given as the parts of the brain are discussed in subsequent sections. Table 3-1 is presented

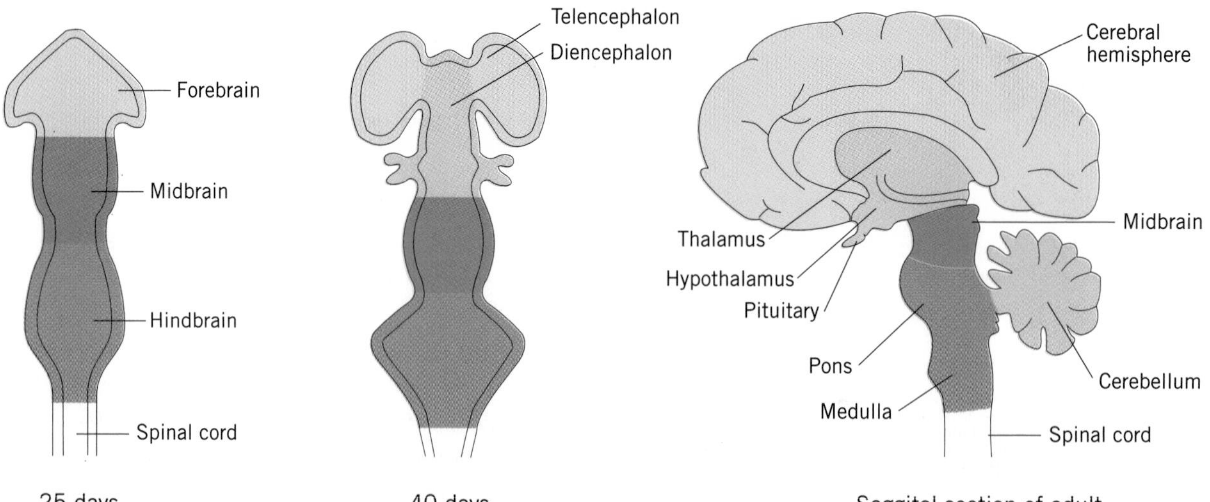

FIGURE 3-10. Differentiation of the anterior part of the neural tube at day 25 into three distinct lobes, which become the forebrain, midbrain, and hindbrain. By day 40, further subdivision of the forebrain has begun. The color-coded section of an adult brain shows the main structures formed by each lobe.

TABLE 3-1. The Main Parts of the Mammalian Brain

Structure	Function
FOREBRAIN (prosencephalon)	
Telencephalon	
Cerebral hemispheres	"Higher" mental functions, sensory processing, motor control
Diencephalon	
Thalamus	Sensory processing center
Lateral geniculate nucleus	Visual relay center
Medial geniculate nucleus	Auditory relay center
Hypothalamus	Homeostatic regulation: thirst, hunger, osmotic balance, control of pituitary gland
Pituitary	Endocrine gland control center
Basal ganglia*	Motor planning and control
Limbic system*	Control of emotions and memory
MIDBRAIN (mesencephalon)	
Tectum	Sensory processing center
Superior colliculus (optic tectum in lower vertebrates)	Oculomotor reflexes (visual processing center in lower vertebrates)
Inferior colliculus	Auditory relay center
Tegmentum	Orientation reflexes and auditory center
Red nucleus	Postural reflexes, motor control
Substantia nigra	Postural reflexes, motor control
HINDBRAIN (rhombencephalon)	
Metencephalon	
Cerebellum	Aids in motor coordination, learning
Pons	Control of respiration
Medulla oblongata	Control of respiration, heart rate, blood pressure, vomiting, coughing

*Groups of nuclei that are distributed in several regions of the brain, but mainly in the forebrain.

here so that you can use it as a summary overview of the relationships between the different parts of the brain before you learn about them in detail. Development of the brain is considered in more detail in Chapter 22.

The brain and spinal cord originate as a straight, tubular cylinder of cells. The anterior part of the tube expands to form the forebrain, midbrain, and hindbrain, whereas the rest forms the spinal cord. In primates, the forebrain expands to cover the midbrain and most of the hindbrain.

The Central Core and External Cover of the Brain

As the brain develops, the hollow central region of the tube from which it originates also changes shape. The hollow core is retained in all parts of the CNS, from the spinal cord to the brain. In the cord it can be seen in tissue sections as the small, hollow central canal that runs through the center of the spinal cord.

In the brain the tube expands to form the system of internal chambers called **ventricles** (Figure 3-11). The chamber in the hindbrain is the fourth ventricle, and that in the forebrain is the third ventricle. These are connected by a narrow channel, the **cerebral aqueduct,** which passes through the midbrain. In the telencephalon, the third ventricle expands to form the right and left ventricles. Both the central canal of the spinal cord and the ventricles are filled with **cerebrospinal fluid (CSF),** a clear, colorless fluid containing ions but nearly devoid of protein and blood cells. The CSF, which also circulates over the surface of the brain, cushions the brain against the shock of impact, as well as providing a suitable chemical environment for neural tissue.

Although the focus in this book is on neural organization and function, the brain does have important constituents other than neurons and glia. Among these are the **meninges,** the layers of tissue that surround the brain, supporting and protecting it. Among other functions, the meninges produce and aid in the circulation of the CSF. The blood vessels that enter the brain are also important. The cells that constitute the lining of these blood vessels are unusual in forming especially tight seals with one another, making it difficult or impossible for many substances that may be in the blood to leave the blood and enter the brain. The barrier formed by the cells of blood vessels, as well as additional protection that lines the ventricles, is called the **blood-brain barrier**. This barrier is critical for keeping the brain isolated from the many metabolites and toxins that may be present in the blood.

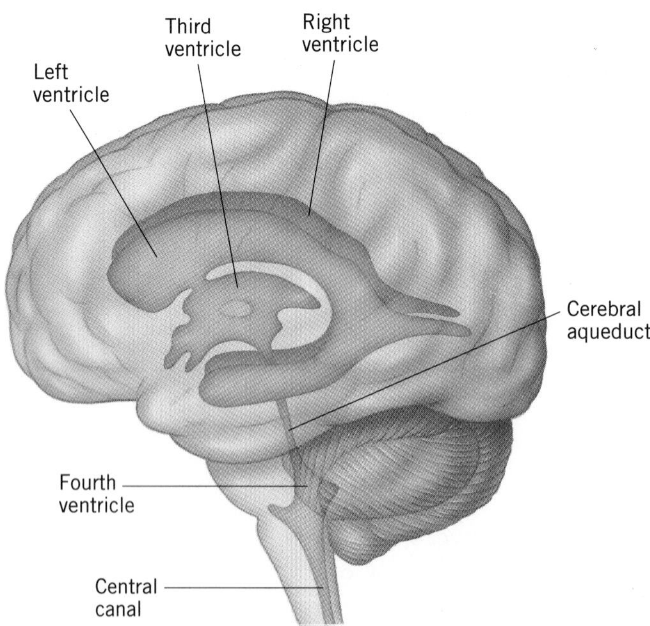

FIGURE 3-11. The ventricles of the human brain, viewed from the side as if looking into the brain. The third ventricle, which lies in the forebrain, expands to form two lobes, called the right and left ventricles. The right ventricle is almost completely obscured in this view.

Within the developing nervous system, the hollow central tube expands into a system of ventricles filled with cerebrospinal fluid. The exterior surface of the central nervous system becomes covered by the meninges. The blood-brain barrier protects the brain from toxins in the blood.

Enlargement of the Brain

As the brain grows, it continues to differentiate (Table 3-2). The regions of the brain transform from a simple, straight tube into a tube with several parts seemingly appended along its length, with an enormous double balloon perched on top. During the second month of development, the central tube, which is originally straight, begins to buckle or flex in several places (Figure 3-12). Imagine holding a long, straight, inflated balloon in your hands in front of you. If you move your hands together, the balloon will buckle and bend. Between 35 and 50 days (Figure 3-12), the medulla and metencephalon (see Table 3-1) grow rapidly, forcing the junction between them into a ventral bend and out of its straight-line position. At the same time, the dorsal part of the metencephalon continues to grow rapidly, bulging up and out as a portion of your balloon would if there were a weak spot and you continued to pump air into it. This bulge forms the cerebellum, which eventually sits astride the pons.

Meanwhile, the telencephalon is also growing rapidly. The most anterior part of the telencephalon, which becomes the cerebral hemispheres, begins to enlarge enormously at about 50 days, growing over and enclosing virtually all the rest of the brain. One consequence of this growth is that the lateral margins of the developing cerebral hemispheres curl in and under the rest (see 100 days, Figure 3-12), and in the adult lie *below* the diencephalon in some places. Nevertheless, because of the clear developmental relationship between the adult structures and their embryological precursors, these regions are still considered part of the telencephalon. Beneath the telencephalon, in

TABLE 3-2. Developmental Origins of the Main Regions of the Vertebrate Nervous System

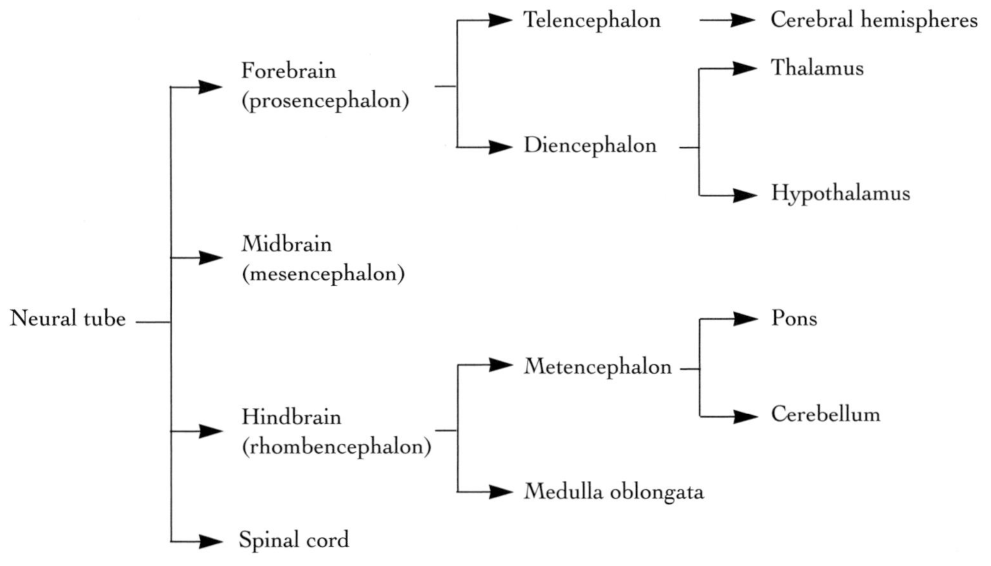

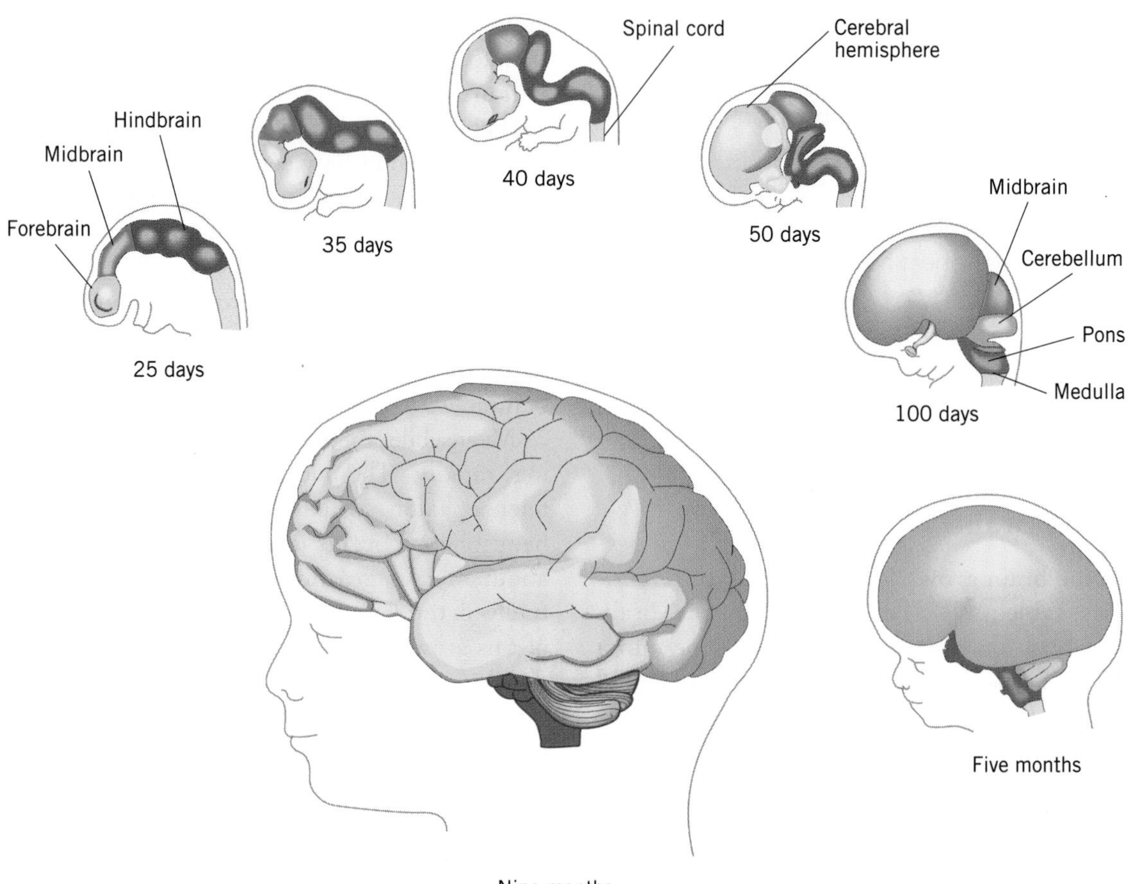

FIGURE 3-12. The development of the human brain. As the neural tube differentiates, it folds and bends on itself. The main lobes of the brain develop last.

the interior of the brain, the diencephalon and its two main constituent structures, the hypothalamus and the thalamus, remain intact, perched atop the midbrain.

Rapid growth of specific regions of the embryonic brain causes formation of the cerebellum on top of the hindbrain and of the cerebral hemispheres, which cover virtually all the brain structures beneath it.

The Lower Brain

When you look at the nervous system in a vertebrate embryo and at the brain of a relatively simple vertebrate like a fish, it is easy to see the fundamentally linear arrangement of the spinal cord and the brain. Even in the highly developed mammalian brain, the medulla, the pons, and the midbrain are arranged in a row, one after the other (Figure 3-13). By analogy with the straight stem of a plant with a flower perched on top, these parts of the brain can be thought of as the stem that supports the flower represented by the cerebral hemispheres. The analogy is particularly apt because these linearly arranged parts of the brain are collectively referred to as the **brain stem**. The cerebellum, because it sits perched above the pons rather than being in line with the other parts of the hindbrain, is not included in the brain stem. Based on its embryological origin, however, it is part of the hindbrain. The parts of the brain stem all contribute to the regulation of basic bodily functions such as respiration and heartbeat.

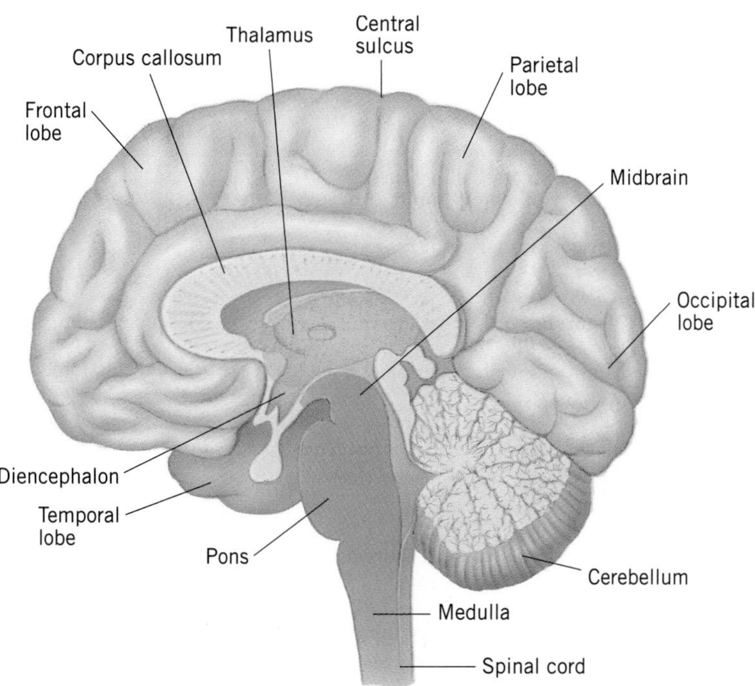

FIGURE 3-13. Section through the middle of the human brain, showing the medulla, pons, and midbrain (collectively the brain stem) in relation to some of the other main structures of the human brain. Some of the labeled parts are discussed in later sections of this chapter.

The Medulla and the Pons

Two components of the hindbrain are part of the brain stem. The medulla arises directly from the spinal cord, and looks somewhat like an enlargement of the cord. Joined to the anterior end of the medulla is the pons, a bulbous part of the brain stem that is distinctive for the fluting of its external surface (see Figure 3-9). From an evolutionary point of view, these parts of the hindbrain constitute the oldest part of the brain. In line with its primitive origins, many parts of the hindbrain control what you might think of as the "primitive" functions in the animal's body, those that are fundamental to life and not under conscious control. Both the medulla and the pons contain important nuclei, control centers containing cell bodies and dendrites of neurons that help to control vital functions like breathing, heart rate, and blood pressure, and that are responsible for initiating the reflexive actions of coughing, gagging, and vomiting.

Some of the medullary and pontine nuclei are associated with **cranial nerves**. These nuclei contain the cell bodies of motor neurons that leave the brain via some of the cranial nerves. Like spinal nerves, cranial nerves serve as the communication links between the CNS, the brain in this case, and the rest of the body. However, whereas spinal nerves are all organized in the same pattern (dorsal roots being sensory, ventral roots being largely motor), not all cranial nerves are the same. Some are efferent, carrying motor information to muscles and glands; some are afferent, bringing sensory information into the brain; and some are mixed, carrying both efferent and afferent fibers. Not all cranial nerves connect with the medulla and pons; some also connect to the mesencephalon and the forebrain. The locations of the 12 pairs of cranial nerves are shown in Figure 3-14, and their functions and the type of information they carry are listed in Table 3-3.

In addition to housing important control nuclei, the brain stem is also a major pathway for communication within the CNS. Some of the nerve tracts found there connect one nucleus in the brain stem to another, some connect nuclei in the brain stem to the

TABLE 3-3. **The Cranial Nerves**

Nerve	Function	Origin of Motor Nerves	Destination of Sensory Nerves
I Olfactory	Sensory: olfactory input		Olfactory bulb
II Optic	Sensory: visual input		Lateral geniculate nucleus (thalamus)
III Oculomotor	Motor: controls most eye muscles	Mesencephalon	
IV Trochlear	Motor: controls superior oblique eye muscles	Mesencephalon	
V Trigeminal	Mixed: carries sensory input from the face; controls muscles that move the jaw	Pons	Mesencephalon, pons, and medulla
VI Abducens	Motor: controls external rectus eye muscles	Pons	
VII Facial	Mixed: carries sensory input from tongue and palate; controls muscles of the face	Pons	Pons and medulla
VIII Vestibulocochlear (vestibular, cochlear)	Sensory: carries auditory input and input concerning balance		Pons and medulla
IX Glossopharyngeal	Mixed: carries sensory input from the tongue and throat; controls throat muscles	Medulla	Pons and medulla
X Vagus	Mixed (autonomic): carries sensory input from the heart, lungs, and viscera; controls movements of the heart, lungs, and viscera	Medulla	Pons and medulla
XI Accessory	Motor: controls neck muscles	Medulla	
XII Hypoglossal	Motor: controls tongue and neck muscles	Medulla	

spinal cord and to nuclei in the rest of the brain, and some go right through the brain stem, connecting spinal centers to nuclei in the midbrain and forebrain. Tracts that interconnect nuclei within the brain stem are important for the proper execution of many vital reflexes.

Among the most prominent of the tracts that course through the brain stem are the two ventrally located **pyramidal tracts**. These tracts, which derive their name from the pyramid-shaped part of the medulla through which they travel (Figure 3-15), carry axons that convey motor information from the motor centers in the forebrain to the motor centers in the spinal cord. Tracts of this type, which carry information from the cortex to the spinal cord, are called **corticospinal tracts**. Tracts conveying information in the opposite direction are called **spinocortical tracts**.

Distributed throughout the brain stem is a loose network of neurons called the **reticular formation**, or **reticular activating system**. Neurons of this network form a number of biochemically and morphologically distinct groups within the brain stem, groups that communicate extensively with one another as well as with other parts of the brain. Researchers have identified four important

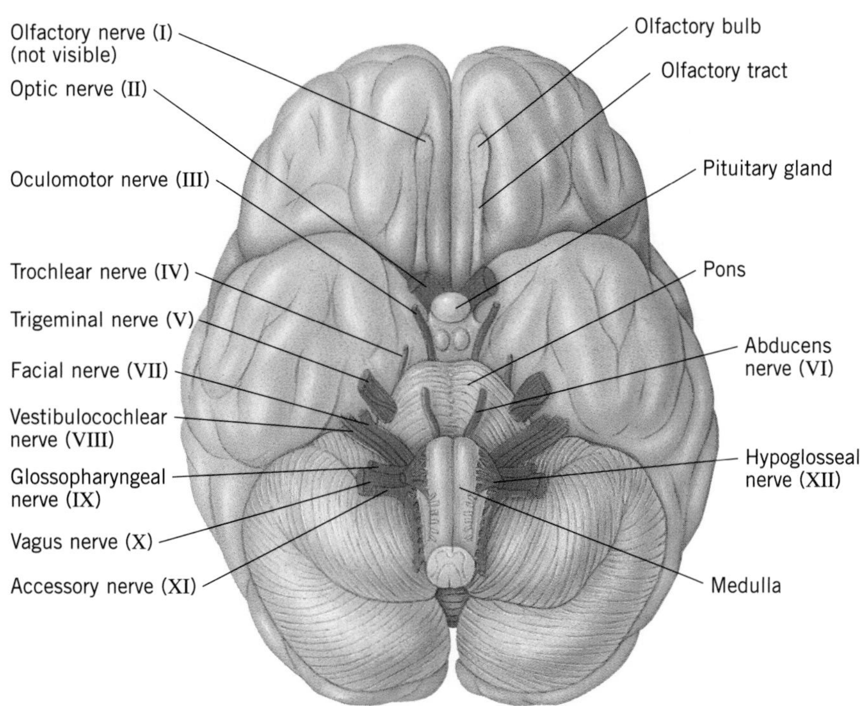

FIGURE 3-14. The cranial nerves of the human brain, shown in ventral view. Several nerves have origins or destinations in more than one part of the brain.

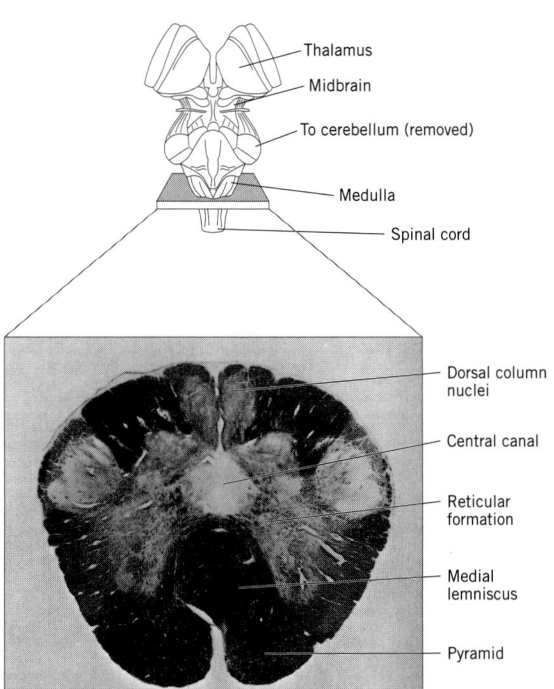

FIGURE 3-15. Photomicrograph of a section through the human medulla. The drawing, a dorsal view of the brain stem, shows the level of the section in a dorsal view.

functions of the system. These are (1) to modulate the sensation of pain; (2) to modulate certain postural reflexes and muscle tone; (3) to help control breathing and heartbeat; and (4) to regulate the level of brain arousal and, in humans, consciousness. The last function is that with which the reticular formation is the most closely associated. For example, the **raphe nuclei** that lie along the midline of the medulla, pons, and midbrain are especially important in maintaining wakefulness (Figure 3-16). Damage to them can result in permanent coma.

All sensory input that enters the brain via the medulla is also sent to neurons of the reticular formation. These may monitor sensory input for importance, and alert higher brain centers when critical input is detected. It has been suggested that the reticular formation is able to do this in part because it receives input from the cortex and uses that input as the basis for its decisions.

The brain stem, consisting of the medulla, the pons, and the midbrain, represents a physical continuation of the spinal cord into the head. It and other parts of the brain communicate with the body via the 12 pairs of cranial nerves. The medulla and pons contain centers for the regulation of vital functions of the body like breathing, swallowing, and heartbeat. They also contain the reticular formation and nerve tracts that carry information through the midbrain.

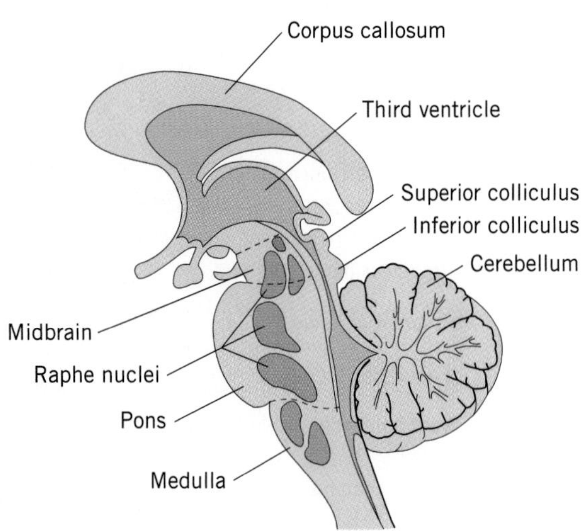

FIGURE 3-16. Midsagittal section of the human brain stem, showing the positions of the raphe nuclei (shaded). The lower three nuclei send axons mainly to the spinal cord, whereas the rest send axons to other places in the brain stem or to the forebrain.

The Cerebellum

The cerebellum is a part of the hindbrain that is not a part of the brain stem. It sits astride the pons, connected to it by a pair of thick **cerebral peduncles**. The peduncles consist of thousands of fibers, the axons of neurons that send information into or out of the cerebellum. The cerebellum is important in coordinating motor action, in learning and memory of motor tasks, and perhaps in certain types of cognitive tasks. Exactly how these functions are carried out is not entirely clear, however.

Many researchers are especially interested in the cerebellum because its cortex (the outer thickness of cell bodies, dendrites, and synapses) consists of three layers containing only a few distinct types of cells and receiving input from just two types of neurons. These neurons and their inputs are arranged in an extraordinarily regular array (Figure 3-17). Input enters the cerebellum via two types of neurons, **climbing fibers**, which originate from the inferior olivary nucleus in the medulla, and **mossy fibers**, which originate from the other brain nuclei that project to the cerebellum. Output from the cerebellar cortex to cerebellar nuclei is exclusively via **purkinje cells**. Neurons in the cerebellar nuclei, in turn, send their axons out of the cerebellum to other parts of the brain. The other types of cells in the cerebellum, such as the granule cells shown in Figure 3-17, are intrinsic to the cortex and form a variety of connections.

Lesion experiments with animals and accidents to humans have provided some insight into four functions of the cerebellum. These

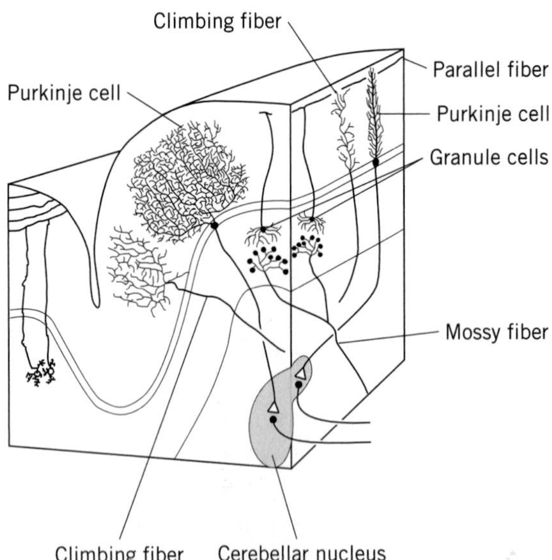

FIGURE 3-17. Cellular organization of the cortex of the cerebellum, showing the arrangement of the main cell types in the cortex. Purkinje cells and the axonal branches of climbing fibers are quite flat and fan-shaped, as shown.

functions are associated with specific regions of the cerebellum, as detailed in Chapter 18. First, some regions of the cerebellum are necessary for maintaining muscle tone and balance during voluntary and reflexive movements, as shown by the reduced muscle tone and difficulty in standing or walking that follow lesions or damage to certain of its parts. Second, some regions are necessary for the planning and execution of simple and complex voluntary movements. This is shown by the shaking associated with any attempt to carry out a movement, or the slowness or other difficulty in executing movements, that is associated with damage to certain of its other parts. Third, parts of the cerebellum are important in learning and memory related to motor tasks. How the animal learns and the involvement of the cerebellum in the process are discussed in more detail in Chapter 24. Fourth, there is some evidence that some regions of the cerebellum are important in certain kinds of cognitive tasks, especially those that require mental manipulation of spatial relationships.

The cerebellum is part of the hindbrain, but not part of the brain stem. The cerebellum helps to maintain balance during reflex movements, is important for the planning and execution of voluntary movements, is important in learning and remembering some types of movements, and may play a role in some cognitive tasks.

The Midbrain

The midbrain sits between the hindbrain and the forebrain (Figure 3-18). Considering its position, it is not surprising that it is a major thruway of axon tracts that connect these two brain regions, but the midbrain also contains several regions that are important in their own right. In lower vertebrates such as fish and amphibians,

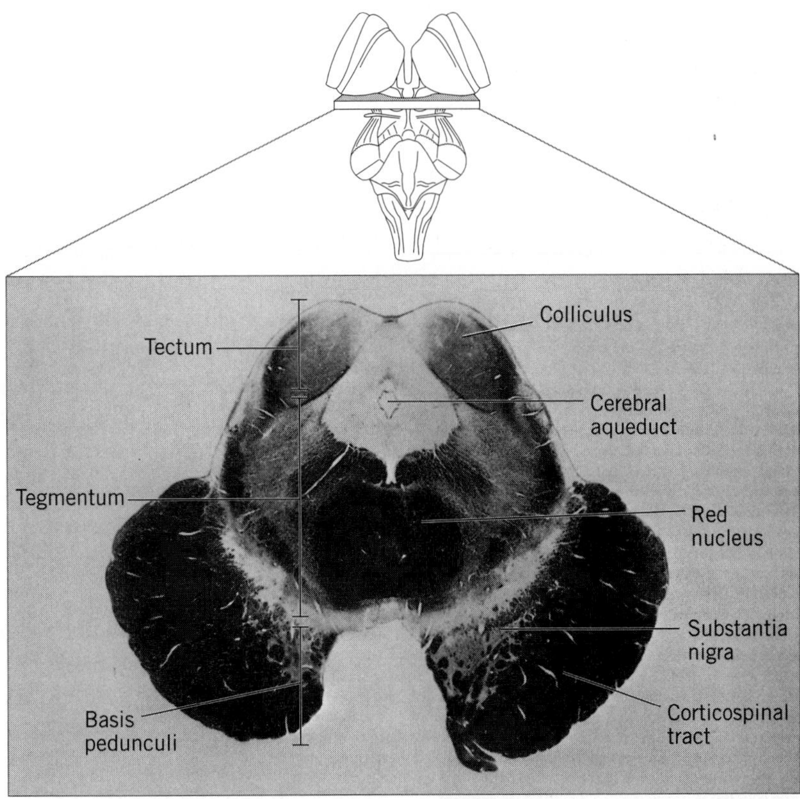

FIGURE 3-18. Photomicrograph of a section through the human midbrain. The drawing shows the level of the section in a dorsal view.

the **tectum**, a layer of brain tissue that lies like a covering (hence its name, "roof") over the cerebral aqueduct, has an important function in processing visual input. The tectum is highly organized, being physically arranged in distinct lamina, or layers. Physiologically, it has a topographical organization, a kind of map of a sensory surface that you will learn about in Chapter 9. In mammals, the corresponding region of the midbrain is occupied by the paired **superior colliculi**, and the term tectum is not used. The superior colliculi also process visual information.

Most of the midbrain lies under the cerebral aqueduct and is called the **tegmentum**. In lower vertebrates, it contains the **torus semicircularis**, which is an important center for processing auditory information. In mammals, the paired **inferior colliculi**, which also process auditory information, constitute the corresponding region. The inferior colliculi are situated posterior to the superior colliculi on the dorsal surface of the midbrain. The tegmentum also contains two prominent regions, the **red nucleus** and the **substantia nigra**, which are especially important in orientation reflexes, in maintaining posture, and in voluntary movements. The substantia nigra is damaged in patients afflicted with the motor disorder Parkinson's disease. You will learn more about Parkinson's disease and motor function in Chapter 18.

The midbrain contains many tracts that connect the forebrain to the hindbrain or to the spinal cord. It consists of the tectum, a center for the processing of visual information, and the tegmentum, for processing auditory information and important in the control of orientation and posture.

The Diencephalon

The most anterior brain region is the forebrain, consisting of the rather small diencephalon, perched on top of or lying in front of the midbrain, and the telencephalon, the brain region whose enormous development in mammals, especially primates, has provided these animals with significantly greater mental abilities.

The diencephalon is the "lower" part of the forebrain, sandwiched between the mesencephalon and the telencephalon. In most vertebrates it is physically covered by parts of the telencephalon, and hence is not visible from the exterior. It contains two structures, the thalamus and the hypothalamus, that play a central role in brain function.

In mammals the **thalamus** is an important relay station for nearly all sensory information on its way to the appropriate cortical center. Specific regions of the thalamus generally process input from specific sense organs. Two sets of paired nuclei are among the most thoroughly studied thalamic nuclei. The **lateral geniculate nuclei** receive input from the optic nerves of the eyes, and the **medial geniculate nuclei** receive auditory information. Other nuclei in the thalamus receive input from neurons of the reticular formation and project to the cerebral cortex; some also participate in motor control.

The **hypothalamus** plays a unique role in the body. Every animal must regulate the internal state of its body in order to maintain the conditions necessary for life. In mammals, maintaining a relatively constant internal environment, or homeostasis, involves regulating parameters like body temperature, blood pressure, and salt and water levels. The hypothalamus plays a central role in homeostasis in three ways. First, it helps to initiate or suppress behaviors such as eating and drinking that affect internal body conditions. Second, through regulation of the pituitary gland, it exerts control over the endocrine system, which in turn adjusts many internal conditions such as salt and water balance in the blood. Third, it is able to regulate many homeostatic conditions by control of the autonomic system, which influences factors like blood pressure and body temperature.

The hypothalamus serves other important functions as well. It is a part of the system that is necessary for the expression of emotions, as described at the end of the next section. Most emotional reactions of mammals involve extensive changes in factors that are controlled by the autonomic system, such as blood pressure, heart rate, and respiratory patterns. These changes happen in humans as well and are manifested by such symptoms as blushing, rapid heartbeat, and sweaty palms. They are induced by stimulation of the appropriate autonomic pathways via the hypothalamus.

The diencephalon consists mainly of the thalamus and the hypothalamus. The thalamus is a major relay and initial processing center for sensory input. The hypothalamus helps to maintain homeostasis, regulates autonomic function, and serves as an important link between "higher" brain centers and the autonomic system.

The Telencephalon

The telencephalon is the "upper" part of the forebrain. In birds and mammals it constitutes more than half of the brain by weight and volume. The most obvious part of the telencephalon is the pair of cerebral hemispheres, but the region also contains a number of functionally related and important nuclei that serve specific functions such as motor control and emotional reactions.

The Cerebral Hemispheres

When most people think of the human "brain," they generally picture the **cerebral hemispheres**, the enormous paired lobes that sit on top of the other structures of the brain. These hemispheres, together called the **cerebrum**, are the main part of the telencephalon. The outer layers, consisting of millions of cell bodies, dendrites, and synapses, and making up the gray matter that covers the interior of the brain, are referred to as the **cerebral cortex**. In fish, amphibians, and reptiles, the cerebral hemispheres are rather small and not especially prominent. They attain a significant size only in birds and mammals, and only in mammals do they expand to become the convoluted structures that are commonly depicted in textbooks. The two cerebral hemispheres are physically and functionally linked to one another by a broad sheet of neural tissue consisting of millions of axons, the **corpus callosum** (see Figure 3-13).

The cerebral hemispheres carry out many different functions. Two of these, interpreting sensory input and organizing motor output, are easy to relate to particular regions because injury to specific brain regions affects specific functions like hearing or control of movements. A major anatomical landmark in the brain is the **central sulcus**, which divides each cerebral hemisphere roughly into anterior and posterior halves, serving as the dividing line between the frontal and parietal lobes. The areas of the cortex devoted to organizing motor output, called the **motor cortex,** lie just ahead of the central sulcus, on the **precentral gyrus** and extending into the regions just anterior to it as well (Figure 3-19). The functions of these parts of the brain are discussed in Chapter 18.

The parts of the brain devoted to processing sensory input are scattered over several regions. Sensory information from the muscles and the body surface is processed in the **postcentral gyrus**, the region just posterior to the central sulcus. The most ventral part of the postcentral gyrus also receives input from taste buds in the mouth. Information from the eyes is processed in the occipital lobe at the back of the head, and that from the ears is handled in parts of the temporal lobes. Information from the nose is processed in the olfactory bulbs, which in humans are quite small and completely covered by the frontal lobes. Figure 3-19 depicts these regions. The processing of information from specific sense organs is discussed in Chapters 11 through 14.

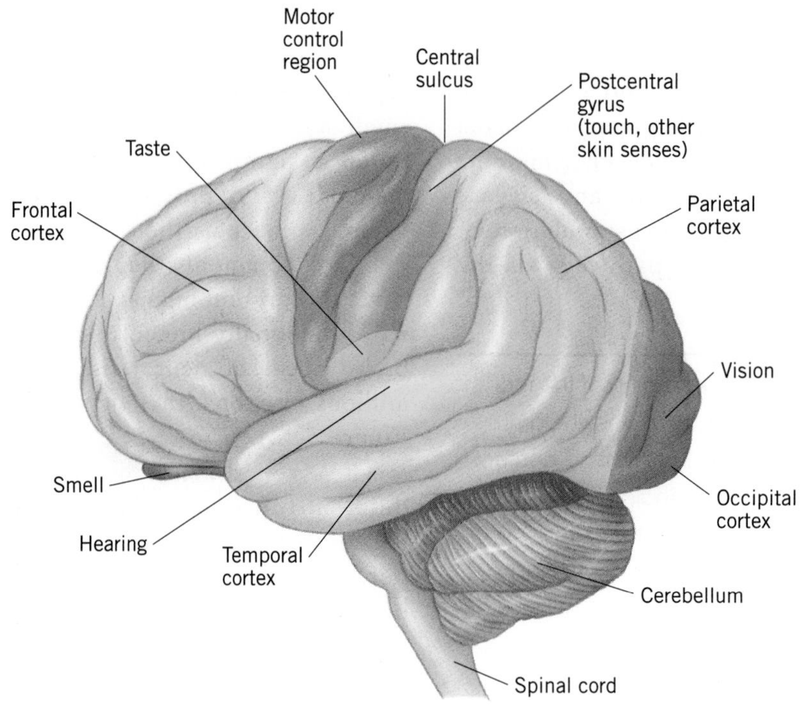

FIGURE 3-19. The parts of the human brain that control motor output and that process input from the major sense organs.

Perhaps the greatest advance in neuroscientists' understanding of brain function in the past few decades has been the recognition that rather specific regions of the cerebrum play quite specific roles in abstract thought. This regional specialization often does not seem to follow any logical plan. For example, everyone can name an object like a fork or a rose and describe what it is. You might think that if you know the meaning of the word *fork* you would automatically be able to identify a fork when you saw one. However, some people who have suffered a mild stroke, a disruption of the blood supply to some region of the brain, may be completely unable to identify a fork or say what it is used for when they see one, even though in a conversation about dinner they would be able to use the word quite properly. The region of the brain deprived of blood is damaged by lack of oxygen, and the absence of its normal contribution to the completion of specific tasks is manifested by a specific deficit. You will learn more about the functions of specific brain regions in Chapter 21.

The cerebral hemispheres are the paired lobes that sit astride other brain regions in birds and mammals. Known collectively as the cerebrum, they are part of the telencephalon. The two lobes are linked by the corpus callosum. Specific areas of the cerebrum are devoted to organizing motor output and processing sensory input. Even abstract thought is carried out by specific brain regions, not by collective action of the entire brain.

The Basal Ganglia

The **basal ganglia** are a group of nuclei that are clustered around the thalamus (Figure 3-20). They are considered as a group because they have a common functional role. The term *ganglia* as applied to them is an outdated designation for

CNS structures now called nuclei; the basal ganglia are the only group of nuclei inside the brain to which this old term is still applied, although concentrations of nerve cell bodies outside the CNS are still called ganglia.

The main components of the basal ganglia are the **putamen**, the **globus pallidus**, and the **caudate nucleus**. The putamen and the globus pallidus lie adjacent to one another and form a lens-shaped structure. For this reason, they are together referred to as the **lentiform nucleus**. The putamen and the caudate nucleus share some structural and functional features and hence have been given a common name, the **striatum**, or **neostriatum**. The globus pallidus alone is sometimes called the **paleostriatum**, and it and the striatum are collectively called the **corpus striatum**. The relationship between these terms is shown in Figure 3-21. As if this terminology is not confusing enough, some neuroanatomists include other structures when the functional role of the basal ganglia is under discussion. For example, the substantia nigra, a part of the mesencephalon, and the **subthalamic nucleus**, which lies in the diencephalon at its junction with the midbrain, are usually included when the function of the basal ganglia is under consideration.

The basal ganglia play an essential role in the control of voluntary behavior. Because of this, there is a trend today to refer collectively to the basal ganglia, including nuclei like the substantia nigra and the subthalamic nucleus, as the **extrapyramidal motor system**. The extrapyramidal system consists of all the axons that carry motor control information but do not pass through the pyramidal system. The pyramidal system consists of the motor cortex and the axons it sends to the spinal cord via the pyramids of the medulla. The components of the extrapyramidal system seem to be necessary for proper planning of movements, as well as for control of tone in the "antigravity" muscles, the large muscles used to maintain a normal posture against the force of gravity. These functions are considered in more detail in Chapter 18.

The basal ganglia are a set of nuclei in the telencephalon and nearby brain structures that are located near the thalamus. They are important for control of posture and voluntary movement.

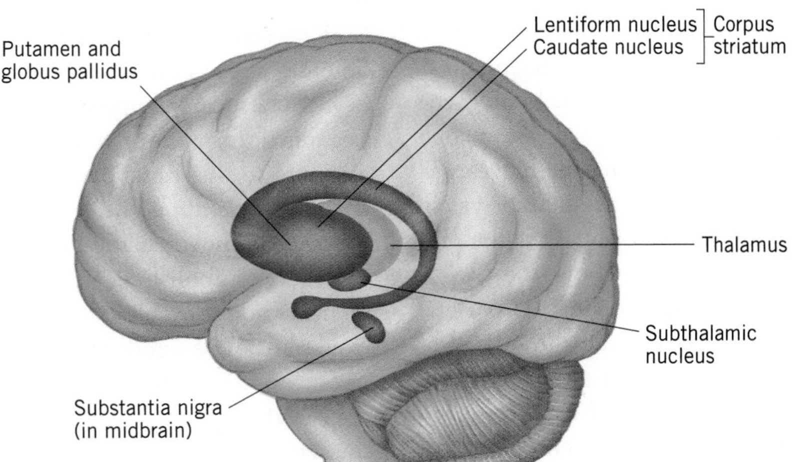

FIGURE 3-20. The basal ganglia of the human brain, shown in the whole brain in a see-through view.

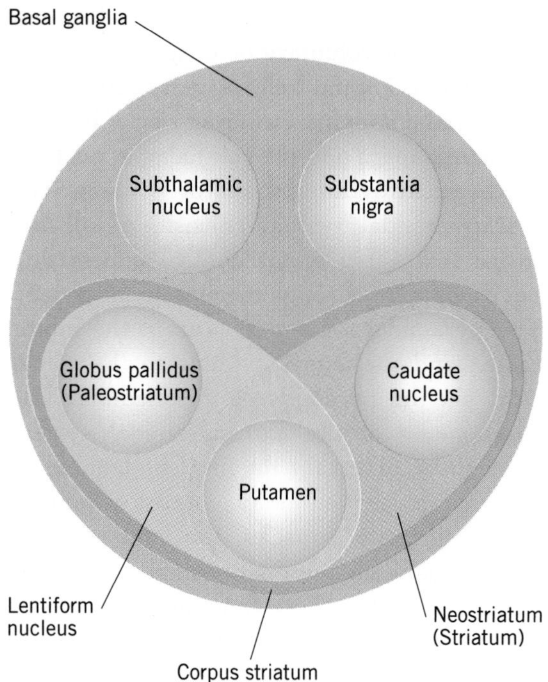

FIGURE 3-21. The basal ganglia, and the relationships among the terms for the different parts.

The Limbic System

The **limbic system** is another group of nuclei in the telencephalon that are unified more by a common function than by being located in a single part of the brain. These nuclei and the fiber tracts that connect them are wrapped around the diencephalon and the basal ganglia, much as your fingers might wrap around the knob at the end of a walking stick. Slicing a brain in half along its midline and separating the halves will expose many components of the limbic system (Figure 3-22). The most dorsal part is the **cingulate gyrus**, which lies just above the corpus callosum along the medial part of the parietal and frontal lobes. Lying ventral to the level of the diencephalon and brain stem are the **parahippocampal gyrus** and the **uncus**, both part of the medial surface of the temporal lobe. The **fornix** lies just above the third ventricle.

The limbic system also has important internal components that are located more laterally (Figure 3-23). The principal internal components are the **amygdala**, the **hippocampal formation**, the **septal nuclei**, and the **mammillary bodies**, which are part of the hypothalamus. Some authorities include other hypothalamic and even thalamic nuclei in the system as well. Whether or not these other nuclei are formally included as part of the limbic system, it is clear from anatomical studies that there are extensive connections between the first four structures named here.

The limbic system, or parts thereof, has two important functions. Its main function is to control emotional behavior and the changes in bodily state that accompany such behavior. This can be shown by electrical stimulation of parts of the limbic system, via electrodes implanted painlessly in the brain. Such stimulation can elicit behavior that represents extreme aggressiveness (rage) or extreme fear in an animal. Judging from the symptoms Phineas Gage showed after his accident (see Chapter 1), this unfortunate man probably sustained damage to neural circuits that allowed him to modulate his emotional responses, leaving the limbic system freer than usual to exert its influence. The limbic system also exerts considerable influence over the hypothalamus, which orchestrates autonomic responses. Electrical stimulation of other limbic regions can induce many of the changes in the autonomic system normally associated with changes in emotional state, such as an increase in heart rate.

A second important function of part of the limbic system is the storage of long-term memory. The **hippocampus**, a part of the hippocampal formation, has been shown to be essential to this function. In humans in which the hippocampus has been damaged or destroyed, the formation of such memory is completely abolished. One famous patient, identified only by his initials, H. M., tragically exemplifies this condition. In 1953, before the importance of the hippocampus was fully appreciated, a lesion was made in a portion of the temporal lobe of each hippocampus of this patient as a treatment for severe and intractable epilepsy. Afterward, H. M. could remember only for a few hours almost anything he experienced. He could not even remember from one day to the next people with whom he had met

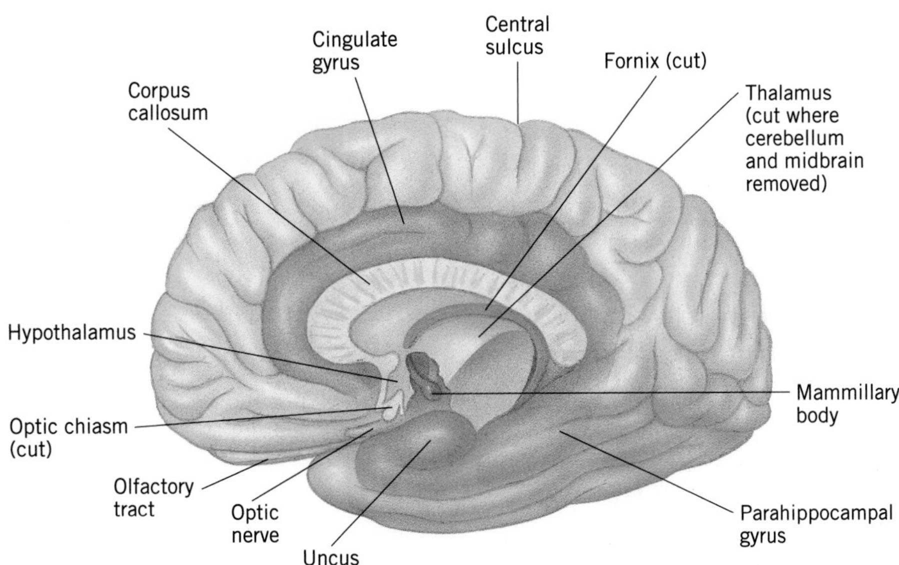

FIGURE 3-22. The limbic system. The brain has been bisected along the midsagittal line to expose the medial surface of one cerebral hemisphere. Only the parts of the limbic system visible on this surface are shown here.

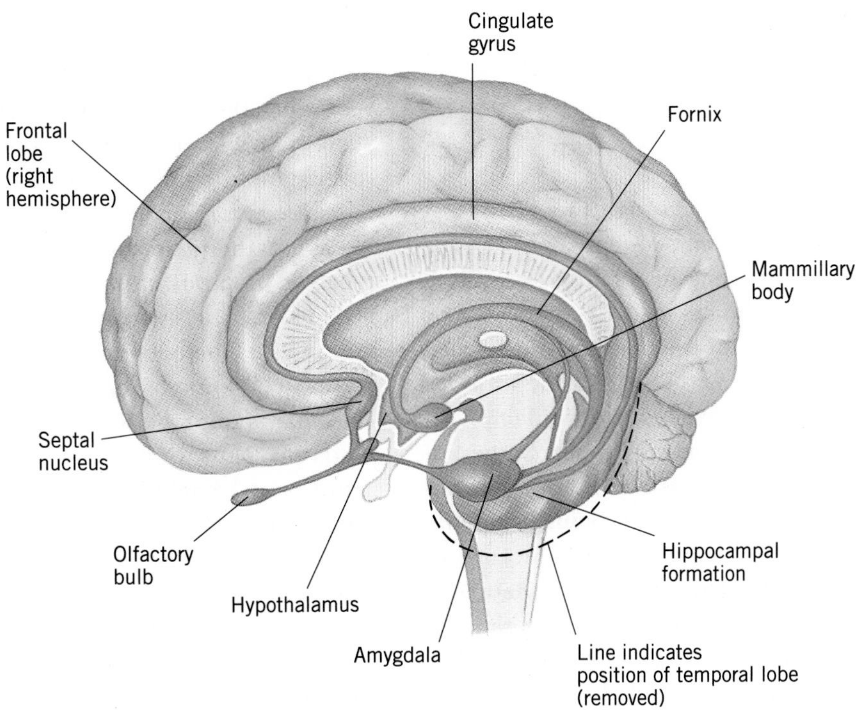

FIGURE 3-23. The limbic system, showing the parts of the system in relation to the whole brain.

almost daily over a period of 30 years, although his memory of people and events he encountered before his surgery remained intact.

The limbic system consists of several regions of the cerebral hemispheres surrounding the diencephalon and mesencephalon. The system controls emotional behavior and, through the hypothalamus, the bodily changes such as an increase in heart rate that accompany such behavior. The hippocampus helps to form long-term memory.

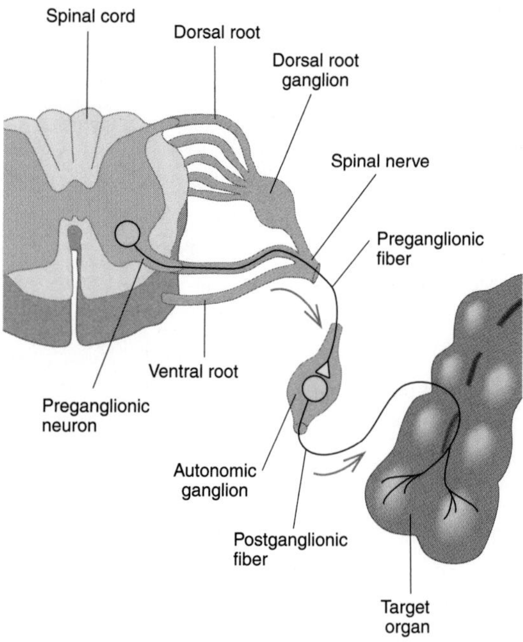

FIGURE 3-24. The interactions of autonomic neurons from the spinal cord with body organs. Rather than traveling directly to each organ, autonomic preganglionic neurons synapse in a peripherally located ganglion, from where postganglionic fibers continue on to the organ.

The Autonomic Nervous System

The term "autonomic nervous system" is typically applied only to the autonomic nerves and ganglia of the peripheral nervous system that are associated with and control internal organs of the body such as the heart and the viscera. However, parts of the brain play a key role in regulating the actions of the autonomic system.

Organization of the Autonomic Nervous System

The route of neurons through the peripheral nerves of the autonomic system is not as simple as that of neurons in somatic peripheral nerves. The cell bodies of all somatic motor neurons lie in the brain stem or spinal cord and send their axons directly to the muscles or glands with which they make contact, with no intervening synapses. The cell bodies of many autonomic neurons also lie in the brain stem or spinal cord, but their axons terminate in peripheral autonomic ganglia. Because of this, they are called **preganglionic neurons**, or **preganglionic fibers**. Inside the ganglia, the preganglionic fibers synapse with another set of neurons, typically called **postganglionic neurons**, or **postganglionic fibers**, which in turn make contact with the various organs of the body (Figure 3-24).

The autonomic system has long been considered to consist of two main functional divisions, the **sympathetic autonomic system** and the **parasympathetic autonomic system**. Most internal organs of the body receive input from both systems, but the systems are antagonistic to one another. That is, if a sympathetic neuron excites an organ that receives its input, a parasympathetic neuron inhibits it. The sympathetic autonomic system arouses the body for action, as in the case of the well-known "fight-or-flight" reaction. The parasympathetic system, in contrast, tends to shut down the body, inhibiting action and bringing about a state of rest or relaxation. It also enhances vegetative processes like digestion.

Researchers also recognize a third functional division of the autonomic system, the **enteric nervous system**, which consists of two arrays of ganglia and nerves distributed along the gut. One array, called the **myenteric plexus**, consists of ganglia and nerves that are located between the longitudinal and circular muscle of the intestines. The other, the **submucosal plexus**, consists of ganglia and nerves within the submucosa. These neural elements were originally considered part of the parasympathetic system. However, they differ from typical parasympathetic ganglia in receiving input from the sympathetic as well as the parasympathetic nervous system, as well as input from sensory neurons in the gut. Furthermore, the ganglia also contain many local neurons that allow the enteric system to function semiautonomously.

The autonomic nervous system consists of preganglionic neurons, whose cell bodies lie in the brain stem or spinal cord, and postganglionic neurons, whose cell bodies lie in autonomic ganglia. The sympathetic division of the autonomic system arouses the body, whereas the parasympathetic division generally calms it down. Two sets of ganglia and nerves distributed throughout the gut constitute the enteric nervous system.

The Sympathetic Nervous System

The sympathetic system has several distinctive structural features. First, preganglionic fibers are usually relatively short, whereas postganglionic fibers are relatively long. Most of the sympathetic ganglia in which the preganglionic fibers terminate lie close to the nerve cord, in a pair of ganglionic chains called **paravertebral chains**, which lie next to the vertebral column. Some of the sympathetic ganglia, such as the **superior cervical**, the **celiac**, and the **superior** and **inferior mesenteric ganglia**, lie in the body cavity, a bit farther from the spinal cord. In either case, the postganglionic neurons whose cell bodies lie in these ganglia all have relatively long axons because all of the ganglia are still some distance from the part of the body with which the postganglionic fibers make contact. A second distinct structural feature of the sympathetic system is that all sympathetic preganglionic neurons leave the spinal cord through one of the thoracic or lumbar spinal nerves. None originates from cranial, cervical, sacral, or coccygeal nerves. Figure 3-25 illustrates the main sympathetic ganglia and their innervation from the spinal cord.

The path of a preganglionic neuron from the spinal cord to the ganglion in which it terminates can be complex. The preganglionic fiber leaves the spinal nerve via a short branch called the **white communicating ramus** (because of the color of its myelinated axons), which joins the paravertebral chain of ganglia. Once in the paravertebral chain, the fiber may take one of three routes (Figure 3-26). First, it may synapse in the ganglion with a postganglionic neuron, which travels through the **gray communicating ramus** back to the spinal nerve and out to a body organ (Figure 3-26A). As its name suggests, the gray ramus contains unmyelinated axons. Second, the neuron may pass through the sympathetic ganglion without synapsing, exit the ganglion via an autonomic nerve, and terminate in a more distant ganglion such as the celiac or superior mesenteric ganglion in the body cavity (Figure 3-26B). Third, if the neuron leaves the cord via an upper thoracic or lower lumbar spinal nerve, it may ascend or descend to a sympathetic ganglion above T_1 or below L_2 in the chain, respectively (Figure 3-26C).

Nearly all parts of the body receive fibers from the sympathetic nervous system. In addition to internal organs like the viscera, heart, lungs, and so on, the skin and all the arteries throughout the body receive sympathetic fibers. The effects of the sympathetic fibers on the structures with which they make contact vary from organ to organ. In general, the sympathetic system prepares the body for action. When it is activated, heart rate increases, blood pressure rises, and blood is directed to the skeletal muscles and away from the viscera. These effects can be brought about quite rapidly, as when you are suddenly frightened. In these cases, nuclei in the

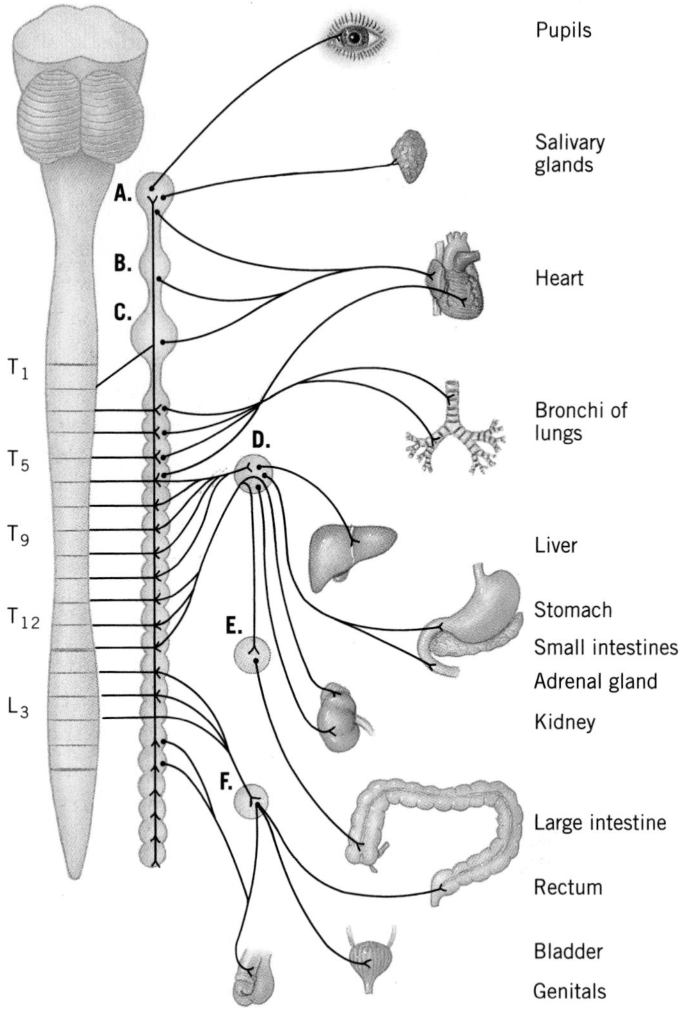

FIGURE 3-25. The ganglia and pattern of contact between efferent sympathetic nerves and the body organs they influence. Sympathetic neurons leave the CNS only via the spinal nerves between the first thoracic (T_1) and the second lumbar (L_2) segments of the spinal cord. For the sake of simplicity, not all the sympathetic nerves are depicted here.

hypothalamus and brain stem send signals to the adrenal glands, as well as to internal organs like the heart. The release of adrenaline (also called epinephrine) from the adrenal glands helps mobilize the body for action in the so-called fight-or-flight response.

Although the action of the sympathetic autonomic system is commonly associated with periods of stress or great excitement, the system is actually active at all times, not just in times of great fear or great joy. Under the control of the hypothalamus, it is continually making adjustments that help maintain the homeostatic balance of the body.

The sympathetic system originates in the thoracic and lumbar regions of the spinal cord. It contains relatively short preganglionic fibers that terminate in a ganglion of the paravertebral chain adjacent to the vertebral column, or in some other ganglion nearby in the body cavity. Postganglionic fibers, which are relatively long, make synaptic contacts throughout the body. The main function of the sympathetic system is to prepare the body for action, but it also maintains homeostasis.

A

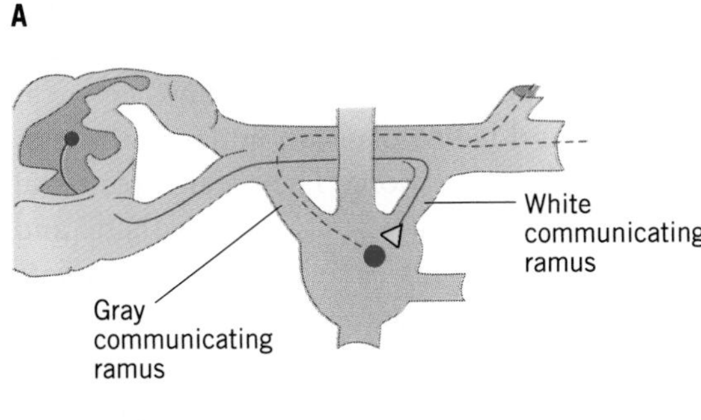

B

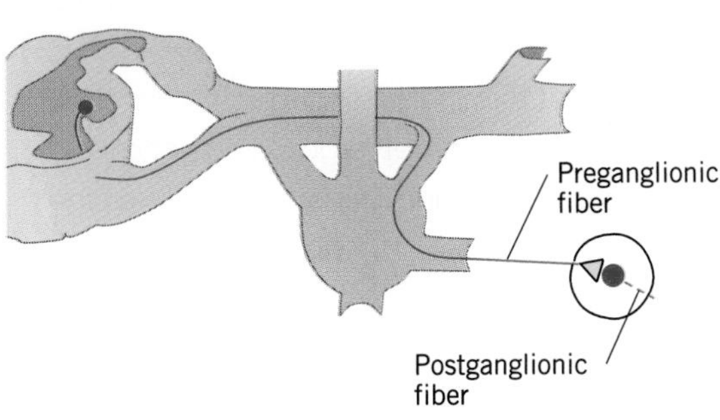

C

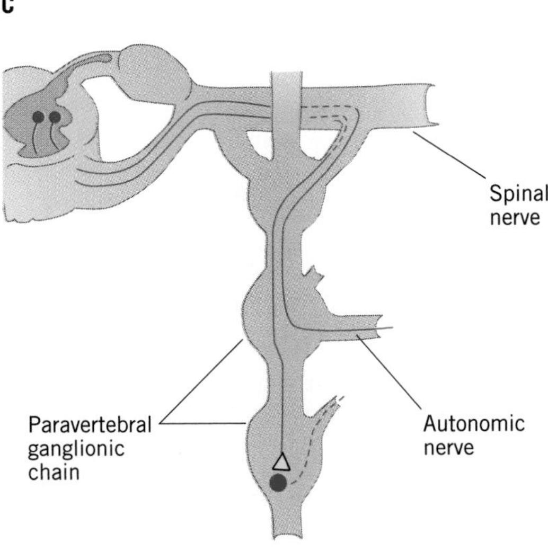

FIGURE 3-26. The various routes of sympathetic information from the spinal cord to its target organ. The sympathetic neuron that leaves the spinal cord may: (A) synapse with a postganglionic neuron in the nearest sympathetic ganglion. The postganglionic fiber loops through the gray communicating ramus and rejoins the spinal nerve before traveling to the target organ. (B) Leave the sympathetic ganglion via an autonomic nerve and travel to a ganglion in the body, such as the celiac ganglion. (C) Ascend or descend to a different ganglion before synapsing or exiting the sympathetic ganglionic chain if it exits from an upper thoracic nerve or a lumbar nerve. Here, only a descending path from a lumbar nerve is shown.

The Parasympathetic Nervous System

The parasympathetic division of the autonomic nervous system differs from the sympathetic system in several ways. First, parasympathetic preganglionic fibers have two sources, the brain stem, which they leave via cranial nerves III, VII, IX, and X (see Table 3-3), and the sacral part of the spinal cord, which they leave via spinal nerves S_2 through S_4. No parasympathetic fibers leave the cord by cervical, thoracic, or lumbar spinal nerves. Second, the ganglia in which parasympathetic preganglionic fibers terminate typically lie in the body close to the organ or body part with which synaptic contact is made. Consequently, parasympathetic preganglionic fibers are quite long, whereas postganglionic fibers are short. Figure 3-27 illustrates the distribution of fibers of the parasympathetic system.

The path of a preganglionic neuron from the spinal cord to the ganglion in which it terminates is straightforward. Because none of the parasympathetic neurons have any communication with the paraventricular chain of sympathetic ganglia, they do not travel through either the white or gray communicating rami. Instead, they stay in the same cranial or sacral nerve from which they exit the CNS, leaving it only when they reach the parasympathetic ganglion that is their destination. There they synapse with short postganglionic neurons, which then innervate the target organ.

All the organs of the body except the liver are contacted by neurons of the parasympathetic nervous system. However, parasympathetic fibers generally do not innervate the skin or blood vessels, which receive almost exclusively sympathetic innervation. Just as with the sympathetic system, the effects of the parasympathetic system on the organs with which it makes synaptic contact depend on the organ. In general, the parasympathetic system tends to inhibit activity in the innervated organ and generally relaxes the body, although it increases activity in the digestive system and a few other organs. Under parasympathetic stimulation, heart rate decreases, blood pressure falls, and

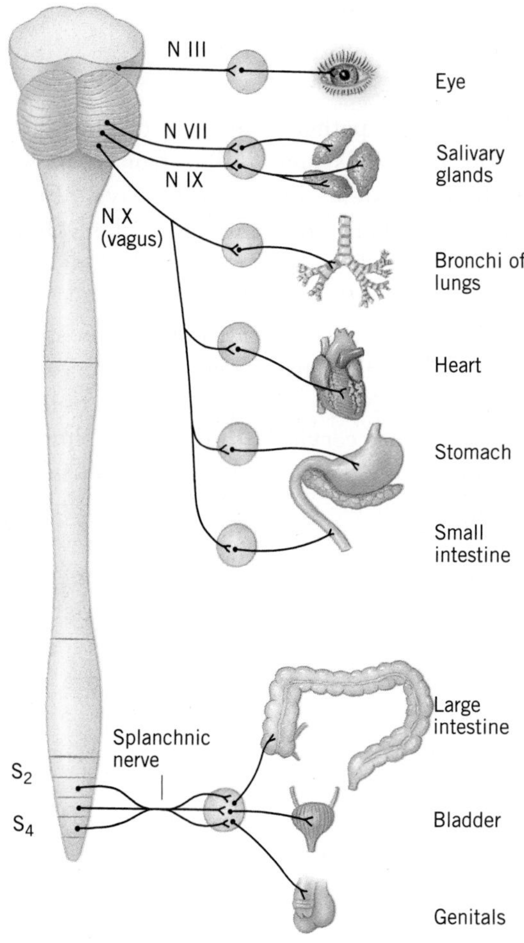

FIGURE 3-27. The ganglia and pattern of innervation of body organs of the parasympathetic autonomic system. Parasympathetic neurons leave the CNS only via cranial nerves III, VII, IX, and X, and the second through fourth spinal nerves of the sacral segments (S_2–S_4) of the spinal cord.

blood is directed away from the skeletal muscles and to the viscera. Like the sympathetic system, the parasympathetic system is active at all times, and hence plays an important role in homeostatic regulation. The main structural and functional features of the sympathetic and parasympathetic divisions of the autonomic nervous system are summarized in Table 3-4.

The parasympathetic system originates in the brain stem and the sacral region of the spinal cord. It contains long preganglionic fibers that terminate in ganglia that lie close to or on the target organ. Postganglionic fibers are short and innervate nearly all the internal organs of the body. An important function of the parasympathetic system is to calm the body, but it also increases digestive activity and plays a role in homeostasis.

The Enteric Nervous System

Although the enteric nervous system was described more than 70 years ago, it was not until the 1990s that most researchers began to appreciate its full structural and functional complexity. By current estimates, the human enteric nervous system contains as many neurons as the spinal cord, about 100 million. The

TABLE 3-4. Differences Between the Sympathetic and Parasympathetic Autonomic Nervous Systems

Sympathetic System	Parasympathetic System
Nerve fibers leave CNS via the thoracic and lumbar spinal nerves	Nerve fibers leave CNS via the cranial nerves and the sacral spinal nerves
Peripheral ganglia near the spinal cord	Peripheral ganglia near the target organ
Postganglionic neurons release norepinephrine as neurotransmitter	Postganglionic neurons release acetylcholine as neurotransmitter
Prepares the body for action	Calms the body

system forms a complex network of interconnected ganglia that lie in two regions of the gut (Figure 3-28A). The myenteric plexus lies between the layers of longitudinal and circular muscles of the alimentary canal, from the pharynx to the anus. The submucosal plexus lies within the submucosal layer of the intestines, from the junction with the stomach to the anus. Not only are the ganglia within each plexus interconnected by an extensive array of nerves (Figure 3-28B), but there also are extensive connections between one plexus and another.

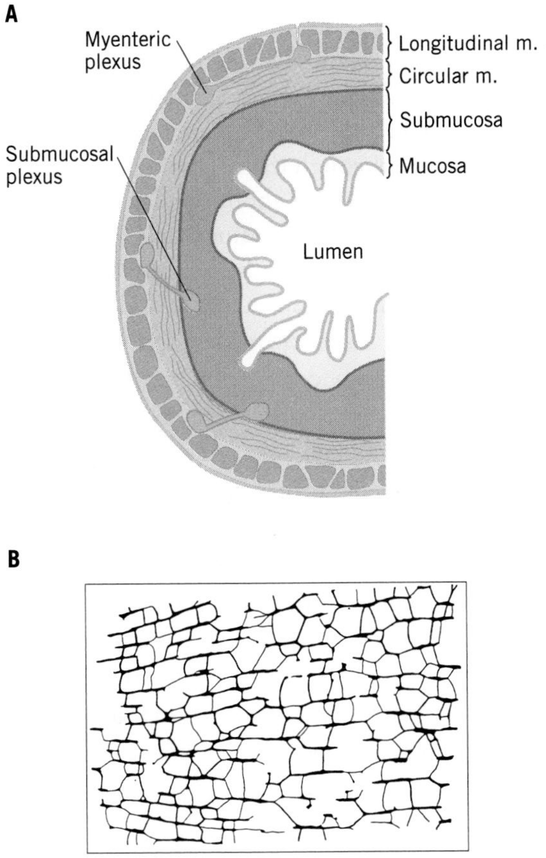

FIGURE 3-28. The enteric nervous system. (A) Cross section through the intestine of a mammal, showing several ganglia of the myenteric and submucosal plexuses. Note that the ganglia are interconnected. (B) The myenteric plexus of a guinea pig small intestine. The intestine has been cut lengthwise and laid flat in this view.

The main function of the enteric nervous system is to control activity of the gastrointestinal tract. It carries out this function in three ways: by controlling intestinal peristalsis, by modulating blood flow through the gut, and by regulating the release of secretions from gastrointestinal glands. Each of these activities can be influenced by signals from sympathetic and parasympathetic nerves that provide input to the enteric system. However, the availability of considerable sensory information from the intestines and the complex interactions between enteric ganglia also make it possible for the enteric system to act semiautonomously. That is, the system can control local activity in the gastrointestinal tract without having to wait for input from the brain. Input that does arrive from sympathetic and parasympathetic fibers serves to adjust the activity of the gut as a whole to the moment-to-moment needs of the animal.

The enteric nervous system is a complex array of ganglia that lie between the muscle layers of the gut and in the submucosal layer of the intestines. It controls peristalsis, modulates blood flow, and regulates the release of gastrointestinal secretions. Its activity can be influenced by the sympathetic and parasympathetic autonomic systems, but it also acts semiautonomously.

The Autonomic Brain

Traditionally, the autonomic nervous system is considered to include only the peripheral autonomic nerves and ganglia. Nevertheless, it is obvious that the activity of the axons that run through peripheral nerves must be regulated by neurons that lie in the brain or spinal cord. In fact, several areas of the CNS are primarily or exclusively devoted to the regulation of autonomic function. When you read about the brain stem, you learned that this region of the brain contains several centers that are necessary for the control of body homeostasis. Heart rate, blood pressure, and body temperature, all of which are physical parameters whose regulation is important for the maintenance of a constant internal environment, are regulated mainly from centers in the brain stem. Control of these parameters is exerted through the autonomic system.

It is also clear, however, that the regions in the brain stem that control these autonomic functions do not work alone. Severing the brain stem between the diencephalon and the mesencephalon is sometimes done as an experimental procedure in mammals to study functions like motor control (see Chapter 18). An animal on which this operation has been performed will survive for some time, but blood pressure and body temperature are no longer well regulated, even though all the brain stem centers that control these functions are intact.

The main control of the brain stem regions is exerted by the hypothalamus, which because of its influence over the autonomic system is often referred to as the homeostatic regulatory center of the body. The hypothalamus not only serves as the seat of homeostatic and hence autonomic control, but it also serves as the site at which the other parts of the nervous system interact with the autonomic system. The close physical relationship between the hypothalamus and the rest of the limbic system is reflected in a close functional relationship as well. Most emotional reactions are accompanied by extensive, and sometimes quite dramatic, changes in autonomic function. If you have ever seen the famous scene in the film *Psycho* in which a woman is murdered in a shower, you may well remember that the fear you felt was accompanied by many manifestations of sympathetic action— a racing heart, sweaty palms, a dry mouth, and more. Strong emotions such as rage, grief, and joy are also accompanied by many physiological changes in the body, all brought about by interactions between the limbic system in the brain and the autonomic nerves in the body, mediated by the hypothalamus.

The hypothalamus, and therefore the autonomic system, can also be affected by other mental processes. Although you cannot directly command your heart to speed up or slow down, you can certainly increase your heart rate by thinking of something frightening or exciting, or decrease it by thinking about a calm pastoral scene. With training, it is even possible to exert a small amount of control over autonomic functions without having to go through the trick of conjuring up scenes that will bring about the desired effect. All these methods of influencing autonomic function work through the hypothalamus.

Some parts of the brain stem and the forebrain are important for the regulation of autonomic function. Centers in the brain stem that regulate heart rate, blood pressure, and body temperature do so through the autonomic nervous system. In the forebrain, the hypothalamus acts as the master control center of autonomic functions, coordinating homeostatic functions and exerting control over the operation of the autonomic system. The hypothalamus is also the site at which higher brain centers can influence autonomic function.

Additional Reading

General

Brodal, P. 1992. *The Central Nervous System: Structure and Function*. New York: Oxford University Press.

Bullock, T. H. 1977. *Introduction to Nervous Systems*. New York: W. H. Freeman.

Carpenter, M. B. 1991. *Core Text of Neuroanatomy*. 4th ed. Baltimore: Williams and Wilkins.

Cramer, G. D., and S. A. Darby. 1995. *Basic and Clinical Anatomy of the Spine, Spinal Cord, and ANS*. New York: Mosby.

Martin, J. H. 1996. *Neuroanatomy: Text and Atlas*, 2d ed. New York: Appleton and Lange.

Nauta, W. J. H., and M. Feirtag. 1986. *Fundamental Neuroanatomy*. New York: W. H. Freeman.

Research Articles and Reviews

Jansen, A. S. P., X. V. Nguyen, V. Karpitskiy, T. C. Mettenleiter, and A. D. Loewy. 1995. Central command neurons of the sympathetic nervous system: Basis of the fight-or-flight response. *Science* 270:644–46.

Pitman, R. M., C. D. Tweedle, and M. J. Cohen. 1972. Branching of central neurons: Intracellular cobalt injection for light and electron microscopy. *Science* 176:412–14.

Stewart, W. W. 1981. Lucifer dyes: Highly fluorescent dyes for biological tracing. *Nature* 292:17–21.

Warr, W. B., J. S. de Olmos, and L. Heimer. 1981. Horseradish peroxidase: The basic procedure. In *Neuroanatomical Tract-Tracing Methods*, ed. H. Lennart and M. J. Robards, New York: Plenum Press. 207–62.

PART 2

Cellular Communication

In a sense, the nervous system is a vast communications network, shuttling information from one neuron to another. To understand even the simplest aspects of neural function, you must understand how neurons communicate.

A neuron transmits information either by causing a brief change in the difference in electrical potential across the target cell's membrane or by altering the way in which that membrane responds to input from other neurons. Hence, every aspect of neural function depends on the electrical properties of nerve cells and how those properties may be modified. This part lays the foundation for your study of the nervous system by describing the electrical properties of neurons, how they enable neurons to transmit information over long distances, and how neurons use their electrical properties to send messages to one another.

Part 2 Cellular Communication *88*

CHAPTER 4

The Electrical Potential of a Resting Neuron

Communication is at the heart of all neural function. It is also heavily dependent on the electrical potential across neuronal membranes. This chapter lays the foundation for a discussion of communication by describing this electrical potential and how it is generated.

Animal cells are like minute, living batteries. Across the membrane of each cell is a small difference in electrical potential, the inside of the cell being electrically negative with respect to the outside. This difference is called the **membrane potential**. Neurons and muscle cells are unusual in having relatively large membrane potentials that can change rapidly under certain circumstances. Nearly every aspect of neural and muscle function depends in some way on the membrane potential. For example, as you will learn in this chapter, the concentration of potassium ions (K^+) outside a cell is important in determining the magnitude of the membrane potential. In a disease known as familial periodic paralysis, foods containing high concentrations of K^+ cause abnormally high concentrations of K^+ in body fluids and a consequent reduction in the membrane potentials of muscle cells. That is, the membrane potential becomes less negative, or "depolarized." Because muscle cells cannot generate normal levels of force with a reduced membrane potential, individuals with this disease exhibit muscle weakness that persists until the concentration of K^+ returns to normal.

Membranes and Ion Movement

The membrane potential results from a separation of positive and negative charges across a cell membrane, generated by the movement of ions through the membrane. Three factors can induce an ion to cross a membrane: a difference in concentration of the ion on the two sides of the membrane, an electrical potential difference across the membrane, or the action of an ion pump. Movement of an ion caused by a concentration difference or a difference in electrical potential is passive, meaning that it requires no metabolic energy. An ion pump requires energy in the form of adenosine triphosphate (ATP) (see Chapter 2).

A concentration difference results in diffusion, a net movement of ions (or molecules) away from a region of greater concentration toward a region of lesser concentration. An ion that is diffusing is said to be moving down its concentration gradient. The presence of a membrane may affect the time required for diffusion of a given number of ions from the inside to the outside of a cell or vice versa, but it does not affect the process itself as long as the membrane contains open channels through which the ions can pass (see Chapter 2). Ions will flow in both directions through open channels. However, if there are more ions of a particular type per unit volume on one side of a cell membrane than on the other, there will be a net transfer of ions toward the side of lower concentration, meaning that more ions will move toward the lower concentration than will move away from it (Figure 4-1A). The greater the concentration difference, the greater the net transfer. Mathematically, it is possible to treat the net movement as the result of a force propelling the ions down the concentration gradient, a force that is proportional to the concentration difference across the membrane.

The electrical potential difference represented by the membrane potential of a cell is another passive factor that influences ion movement. The potential difference is due to a thin layer of anions (negative ions) distributed just inside the cell membrane and a similar layer of cations (positive ions) along the outside. This separation of charge generates an electrical field that spans the membrane and influences the movements of ions in the channels. Once an ion enters an open channel it will be propelled toward the side of the membrane that has a potential opposite to the charge it carries (e.g., a positive ion will move toward the negative interior of the cell) (Figure 4-1B). Ions that are already on the side with a polarity opposite to their own will be impeded from moving through a channel. The force acting on the ion is proportional to the strength of the electrical field, and hence to the magnitude of the potential difference across the membrane.

In addition to these passive factors, ions may also be moved by ion pumps. Such pumps use the energy stored in ATP to move ions across the membrane against their concentration gradients, thereby building up a higher concentration of the ions on one side. Although moving only a few hundred ions of a particular type across a membrane in a second will have no significant effect on the concentration of that ion inside or outside the cell over the course of even a few minutes, by

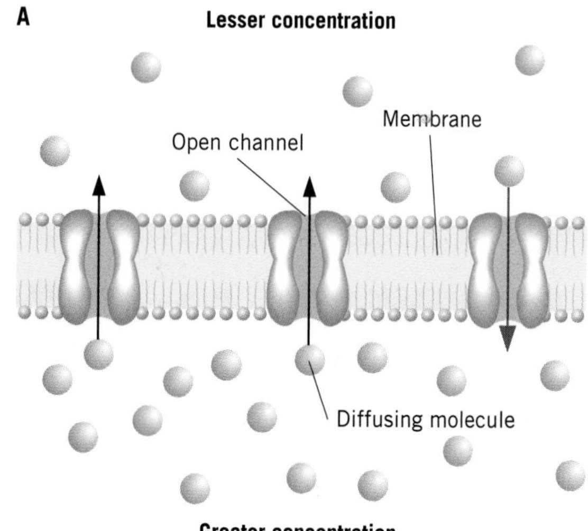

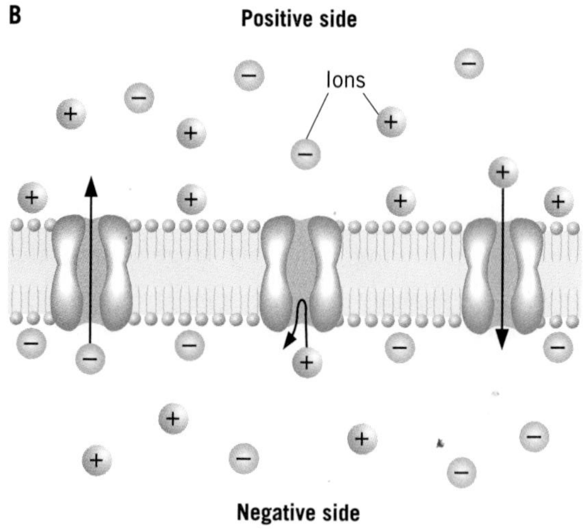

FIGURE 4-1. The passive factors that can influence the movements of ions across cell membranes. (A) Diffusion. A difference in concentration of an ion across a membrane will result in a net movement away from the region of higher concentration toward the region of lower concentration. The concentration gradient can be treated as imparting a force that gives the ion in question its net tendency to move. The greater the concentration difference, the greater the force. (B) Electrical field. An electrical potential difference across the membrane will impart a force to ions near an open channel to move the ion toward the side oppositely charged and to repel it from the side similarly charged.

operating continuously hour after hour, it is possible for an ion pump to have a significant effect on the concentration of the ion. In addition to the sodium-potassium exchange pump that transports sodium and potassium (see Chapter 2), some cell membranes contain pumps for transporting calcium (Ca^{2+}) and chloride (Cl^-) ions as well.

Under normal conditions, neurons are in a **steady state** with regard to the distribution of ions, meaning that except for brief periods during which a neuron is electrically active (see Chapter 5), the number of ions of each type that enters the neuron equals the number that leaves it. Hence, the movement of each type of ion across the membrane in one direction due to one or more of the factors already mentioned is balanced by the movement of the same number of ions of that type across the membrane in the opposite direction, under the influence of the other factor(s), as depicted for K^+ in Figure 4-2. Similar diagrams can be drawn for other ions.

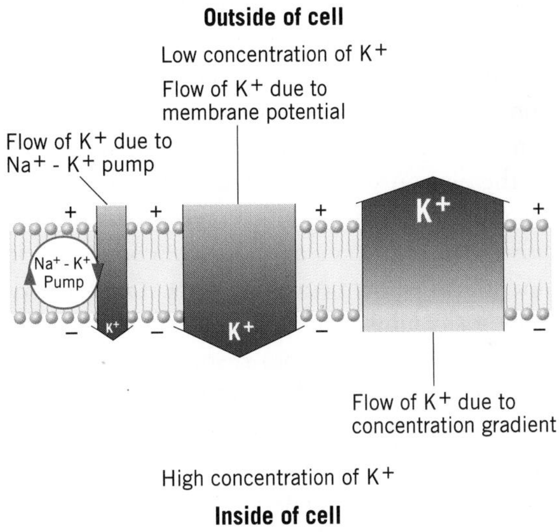

Outside of cell
Low concentration of K+
Flow of K+ due to membrane potential
Flow of K+ due to Na+ - K+ pump
Na+ - K+ Pump
K+
Flow of K+ due to concentration gradient
High concentration of K+
Inside of cell

FIGURE 4-2. The steady-state conditions of a resting neuron for the movement of K^+ across the membrane. The outward flow of K^+ under the influence of its concentration gradient is countered by the K^+ that enters the neuron due to the electrical gradient and the K^+ that is pumped into the neuron by the Na^+-K^+ pump. (The expulsion of Na^+ by the pump is not shown in this diagram.) In steady-state conditions, the total movement of K^+ inward equals the total movement of K^+ outward, and hence there is no net gain or loss of K^+.

The movements of ions across a cell membrane are passively influenced by the concentration gradient of an ion across the membrane and by the membrane potential. Energy-requiring ion pumps also affect the distribution and movements of some ions. Except when electrically active, neurons are in a steady state, neither gaining nor losing ions of any particular type.

Ionic Basis of the Resting Potential

Every neuron has a membrane potential, but this potential can change, and specific terms are used to describe it in specific circumstances. Since it is clear that the membrane potential of a neuron is not the same when the neuron is actively engaged in communication as when it is not, the name **resting potential** is used to refer to the membrane potential in electrically inactive neurons. (Of course the neuron is still metabolically active. It is "resting" only in the sense that its membrane potential is not changing.)

Ions and the Resting Potential

The nineteenth-century physiologist Julius Bernstein was among the first to suggest that a resting potential exists across the membrane of every neuron. Although the source of this potential was not clear, three simple observations provided important clues. First, physically tearing the membrane of a neuron with a sharp needle will cause the membrane potential to quickly decline to zero. This result suggests that an intact membrane is essential for the resting potential to be maintained. Second, changing the relative concentrations of an ion across the membrane can affect the resting potential. Neurons normally contain a relatively high internal concentration of K^+, especially compared with the low concentration of K^+ outside the cells. If you substantially increase the external concentration of K^+ by adding something like potassium acetate to the medium bathing a neuron, the membrane potential of the neuron will rapidly decline toward zero (depolarize). This finding suggests that the relative concentration of potassium ions inside and outside the neuron is important. Third, chemically inhibiting the metabolic activity of a neuron will cause the resting potential to decline to zero over the course of some hours. This response indicates that some energy-requiring process is also necessary to maintain a membrane potential over the long term.

These and other findings have led researchers to the current view that three factors are responsible for the generation and maintenance of the resting potential: (1) the selective permeability of the neuronal membrane; (2) the unequal distribution of certain ions across it; and (3) the action of energy-requiring ion exchange pumps that are located in the cell membrane.

Selective permeability, or **semipermeability**, refers to the ability of a membrane to allow some ions or small molecules to pass into or out of the cell relatively easily, while greatly restricting or preventing the passage of others. Selective permeability arises from the selectivity of ion channels, which allow passage of some ions but not others. Ions or molecules for which there are few or no open channels cannot readily pass through the membrane. The ability of gated channels to allow or not allow ions to pass is essential for the expression of the active electrical responses of a neuron.

As for an unequal distribution of ions, many different ions and small, electrically charged molecules are present in animals, but their concentrations are different inside and outside cells. In general, neurons contain relatively low internal concentrations of Na^+ and Cl^-, and high concentrations of K^+, as well as an abundance of small, electrically charged organic molecules like isethionate and aspartate. The extracellular fluid, in contrast, usually contains few small charged molecules and little K^+, but relatively large amounts of Na^+ and Cl^-. Table 4-1 lists typical values for the concentrations of some important ions and molecules in one representative invertebrate and one vertebrate animal.

Ion exchange pumps are the third factor that influences the generation and maintenance of the membrane potential. The sodium-potassium exchange pump (Na^+–K^+ pump) maintains a high concentration of K^+ inside a neuron and a high concentration of Na^+ outside a neuron. These ion concentrations are necessary for maintaining the resting membrane potential.

The resting potential of every neuron is influenced by three factors: the selective permeability of the membrane, an unequal distribution of ions across the cell membrane, and metabolically driven pumps for sodium and potassium ions.

Generation of a Resting Potential

In a typical axon, the ion selectivity of the membrane interacts with the unequal distribution of ions to generate a resting potential of about -60 mV, meaning that the inside of the cell is 60 mV negative relative to the outside. To understand how this negative potential can come about, imagine a hypothetical cell containing a compound such as potassium aspartate, placed in a solution containing common salt, which dissociates into Na^+ and Cl^-. The inside of the cell is electrically neutral because there are equal numbers of positive potassium ions and negative aspartate ions inside it. The outside of the cell is also neutral, having equal numbers of positive sodium ions and negative chloride ions (Figure 4-3A). Now imagine that this cell contains open channels that are selective for K^+, allowing only K^+ to cross the membrane; Na^+, Cl^-, and aspartate cannot enter or leave the cell.

TABLE 4-1. The Distribution of Some Important Ions Inside and Outside Representative Neurons

Squid Giant Axon

Ion	[Inside]	[Outside]
K^+	400 mM	20 mM
Na^+	50 mM	440 mM
Cl^-	51 mM	560 mM
Ca^{2+}	0.4 mM	10 mM
Mg^{2+}	10 mM	54 mM
organic anions	360 mM	—

Cat Motor Neuron

Ion	[Inside]	[Outside]
K^+	150 mM	5.5 mM
Na^+	15 mM	150 mM
Cl^-	9 mM	125 mM

Given these conditions, what will happen? At first, K^+ will move through the potassium-selective channels in the membrane, down the potassium concentration gradient (Figure 4-3B). If K^+ were uncharged, the potassium concentrations would eventually equalize on the two sides of the membrane because there would be a net transfer of K^+ away from the side with the higher concentration until the concentrations were equal. This equalization of concentration would occur because the probability of K^+ moving out of the cell will be greater than the probability of K^+ moving back into it, as long as there is more K^+ inside than outside the cell. However, each potassium ion carries a unit of positive charge, so the diffusion of each ion out of the cell leaves the inside slightly more negative relative to the outside. This separation of charge causes the generation of a potential difference and an electrical field across the cell membrane.

The presence of the electrical field increases the probability that potassium ions accumulating outside the cell will move back inside when they encounter an open channel because they are attracted by the negative charge. For the same reason, the presence of the field reduces the probability that K^+ will leave the cell. At first, the **efflux** (outward flow) of K^+ will be greater than its **influx** (inward flow) because initially the net amount of K^+ moving out of the cell due to the concentration gradient will be greater than the net amount moving into the cell due to the electrical field. However, every potassium ion that leaves the cell increases the strength of the field. Soon, sufficient K^+ will have moved from the cell to make the electrical field strong enough to counter the effects of the concentration difference. That is, K^+ will have reached an equilibrium at which the amount that *enters* the cell under the influence of the electrical gradient will just be balanced by the amount that *leaves* due to the concentration gradient.

At equilibrium, there is no further net movement of K^+ into or out of the cell. Hence, neither the potential difference across the membrane nor the concentrations of

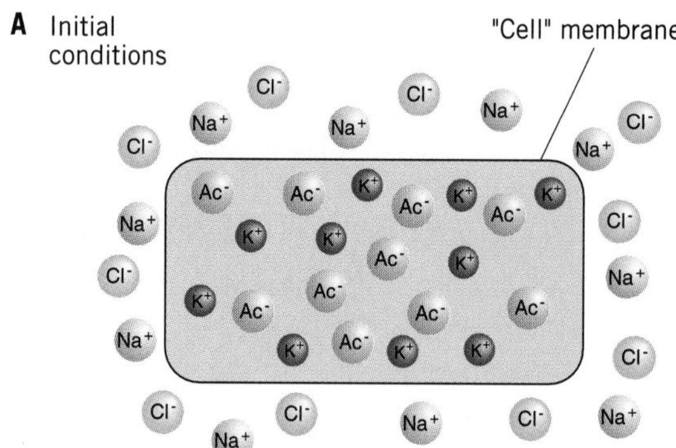

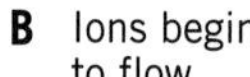

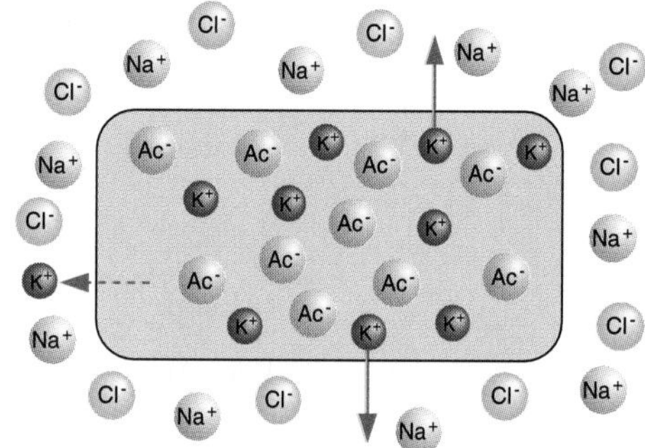

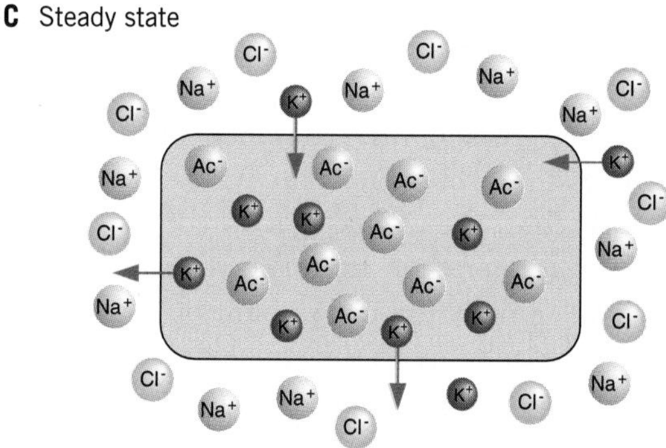

FIGURE 4-3. The generation of a potential difference across the membrane of a hypothetical cell in which there are open channels for only one ion (K^+). (A) Under the initial conditions, the cell is filled with potassium acetate (KAc), and surrounded by salt (NaCl). (B) Potassium ions are small enough to pass through the open channels; diffusion of potassium from the inside to the outside of the cell leaves the interior of the cell negative relative to the outside because neither acetate nor chloride can cross the membrane. (C) In the steady state, just as much potassium is drawn into the cell by the electrical gradient that has been generated as diffuses out as a result of the high internal concentration of potassium.

potassium inside or outside the membrane will change further (Figure 4-3C). In this hypothetical case, then, a membrane potential has been established solely by the passive movements of a single ion and the selective permeability of the membrane that keeps other ions from entering or leaving the cell.

In a hypothetical cell containing a high concentration of potassium and whose membrane contains channels selective only for this ion, K^+ will diffuse out of the cell. This generates a membrane potential by leaving the interior of the cell negative relative to the outside.

Ion Movements in Cells

In real neurons, more than one type of ion can cross the membrane. Nevertheless, a mechanism like that just described operates in a living neuron as well as in a hypothetical one because most of the open channels in a resting neuron are selective for just one ion, potassium. In living neurons, K^+ will leak out of the cell and a potential difference will be generated across the cell membrane. Because relatively few channels are open for the passage of ions such as sodium through the membrane, these other ions play only a minor role in the generation of the resting potential.

The main difference between a living neuron and a hypothetical one is that the membrane of a living cell is not completely impermeable to ions such as Na^+. The number of sodium ions that enter a neuron in a second is quite small compared with the number of potassium ions that leave. Nevertheless, over a period of hours, the slight inward leak of positive charge that these ions represent would neutralize the negative charge built up by the efflux of K^+. Expelling the Na^+ that leaks into the neuron is the task of the Na^+-K^+ pump, which removes three atoms of sodium while bringing two atoms of potassium back into the cell. The importance of this pump can be demonstrated by blocking its action. Metabolic inhibitors (drugs that inhibit production of ATP) will shut down the pump (as well as other energy-requiring processes), resulting in a slow decline of the membrane potential toward zero over the course of several hours. Removing the metabolic inhibitor restores ATP production, allows the Na^+-K^+ pump to restart, and restores the membrane potential.

Sodium is not the only other ion that can cross the membrane of a living neuron. One that crosses relatively easily in most neurons is Cl^-. Chloride ions do not contribute to the membrane potential in most neurons, however; they merely distribute themselves according to the magnitude of the resting potential, with the highest concentration on the positive side of the membrane.

In living neurons, the membrane potential is established by the diffusion of potassium ions out of the cell. The potential is maintained by the action of the sodium-potassium exchange pump, which expels the small number of sodium ions that leak into the neuron while taking up the potassium ions that have leaked out.

The Equilibrium Potential

In the hypothetical situation described in the previous section in which a membrane is exclusively permeable to potassium, K^+ moves out of the cell until the potential difference reaches a value great enough so that the electrical force inducing K^+ to move into the cell is just balanced by the chemical (diffusion) force inducing it to move out. Under these circumstances, the influx and efflux of potassium ions will be equal. Although ions continue to cross the membrane, just as many potassium ions move out due to the concentration gradient as move in due to the electrical gradient. Hence, there is no net transfer of ions in either direction. The potential difference at which this equal movement occurs is called the **equilibrium potential** for potassium.

The equilibrium potential for any ion represents the balance point between net ion flux in one direction due to the concentration difference and net ion flux in the other direction due to the potential difference. Think for a moment about what would happen if you changed the concentration of potassium inside or outside the hypothetical cell. Since the equilibrium conditions would no longer be met, there would be a net flow of ions in one direction or the other until a new equilibrium is reached.

Suppose you increased the external concentration of potassium by adding potassium acetate to the external medium of the cell. The solution you add is electrically neutral (it contains just as many negative acetate ions as positive potassium ions), but adding extra K^+ outside the cell causes a decreased outward flow of K^+ due to the reduced difference in concentration of K^+ across the membrane. The electrical potential difference that had already been built up, however, would at first continue to drive K^+ into the cell at the old rate, resulting in a net influx of K^+ and hence of positive charge. This transfer of positive charge into the cell would make the interior of the cell less negative and bring the potential difference closer to zero. The potential difference would continue to move toward zero until it reached a value that would balance the new concentration difference, and the cell would once again be in equilibrium.

The equilibrium potential for an ion is the potential difference that must be present across the membrane of a cell in order to balance the concentration gradient for that ion. When the membrane potential equals the equilibrium potential for an ion, there will be no net movement of that ion across the membrane.

Calculating the Equilibrium Potential

It is important to be able to calculate the equilibrium potential for an ion because its value relative to the membrane potential determines the direction in which that ion will flow through open ion channels, as you will see in the next section. This is important because ion flow often determines the response of a neuron to synaptic input.

The equilibrium potential is calculated from the concentrations of an ion inside and outside the cell by the **Nernst equation**, named for Walther Nernst, the physical chemist who formulated it in the late 1880s. The general form of the equation is shown here, and the constants and symbols in it are defined in Table 4-2.

$$E_{ion} = \frac{RT}{FZ} \ln \frac{(\text{ion})_o}{(\text{ion})_i}$$

Interpreted in words, the Nernst equation states that the equilibrium potential of an ion is equal to a constant times the natural logarithm (log to the base *e*) of the ratio of the external to internal concentrations of the ion. The constants (*RT* and *FZ*) take account of the thermal energy of the cell and its surroundings, plus the current carried by the ion under consideration.

An important feature of the Nernst equation is that it applies to only one type of ion, the ion whose concentrations are used in it. It is possible to calculate E_{ion} for just one ion because the movement of any one ion due to its concentration gradient is independent of the presence or movement of any other ion. Hence, the equilibrium potential for any one ion can be calculated without regard to the presence or absence of other ions.

Another important feature of the Nernst equation is that the equilibrium potential that it provides is independent of the membrane potential. The equilibrium potential is the potential difference that will just balance a given concentration difference. Alone, it neither predicts nor is dependent on the actual membrane potential in any way. In fact, as you will see in the next section, the value of the equilibrium potential relative to that of the membrane potential gives important information about the factors that influence the distribution of the ion for which the equilibrium potential is calculated.

Consider some specific examples. First, suppose that you want to calculate the equilibrium potential for potassium in the giant axon of a squid, based on the concentrations of ions given in Table 4-1. The concentration of potassium inside the axon is 400 mM, and the concentration outside it is 20 mM. At 20°C, and with $Z = +1$, $RT/FZ = +25$ mV, and the equilibrium potential for potassium becomes:

$$E_K = 25 \text{ mV} \times \ln \frac{20 \text{ mM}}{400 \text{ mM}} = 25 \text{ mV} \times \ln (.05) = -75 \text{ mM}$$

Some people prefer to work with logarithms to the base 10 rather than with natural logarithms. Natural logs can be converted to log to the base 10 by multiplying by 2.3. In this case, the Nernst equation for the equilibrium potential for potassium becomes:

$$E_K = 2.3 \times 25 \text{ mV} \times \log \frac{20 \text{ mM}}{400 \text{ mM}}$$
$$= 57.5 \text{ mV} \times \log (.05) = -75 \text{ mM}$$

The value of E_K (-75 mV) represents the potential difference that will produce a net rate of K^+ movement into the cell that is equal to the net rate at which K^+ diffuses out of the cell due to its high internal concentration. Because of the way the Nernst equation is derived, the sign of the equilibrium potential refers to the polarity of the inside of the cell. Thus, in order to balance exactly the tendency for K^+ to diffuse out of the squid giant axon due to the concentration gradient, the cell must have a membrane potential of -75 mV, inside negative. The situation at equilibrium can be depicted in a diagram such as that shown in Figure 4-4. The fact that the actual membrane potential of a squid giant axon is about -60 mV rather than -75 mV will be considered further in the next section.

TABLE 4-2. The Nernst Equation, Its Symbols, and Its Constants

$$E_{ion} = \frac{RT}{FZ} \ln \frac{(ion)_o}{(ion)_i}$$

E_{ion}	= Equilibrium potential for "ion"
R	= Universal gas constant (8.31 joules/mole/°K)
T	= Temperature, in degrees Kelvin (273 + °C)
F	= Faraday constant (charge per mole: 96,500 coulombs/mole)
Z	= Valence (electrical charge) of the ion
ln	= Natural log (log to the base e)
$[ion]_o$	= Outside concentration of the ion under consideration
$[ion]_i$	= Inside concentration of the ion under consideration

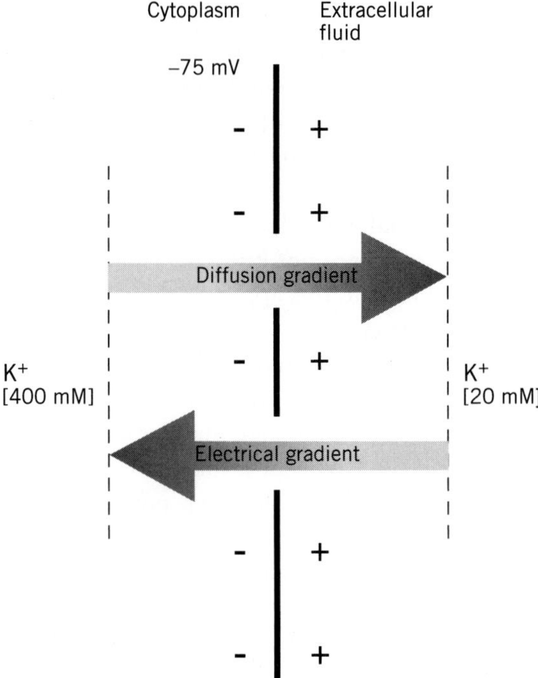

FIGURE 4-4. The equilibrium condition for potassium ions. The force of the diffusion gradient, established by high internal and low external concentrations of potassium, is exactly countered by the equal and opposite force imparted by the electrical gradient. Under these conditions, the membrane potential reaches -75 mV, and there is no net movement of K^+ into or out of the cell.

What is the equilibrium potential for sodium in the squid giant axon? Before calculating it, imagine that the membrane of a cell is permeable to Na^+ and not at all permeable to other ions. If there is a high concentration of Na^+ outside the cell and a low concentration inside, Na^+ will diffuse down its concentration gradient and into the cell. The accumulation of positively charged ions inside the cell will make the inside positive relative to the outside, and the developing electrical field will then begin to move some of the sodium ions that had entered the cell back out again. If the concentrations of Na^+ outside and inside the cell are 440 mM and 50 mM, respectively, as they are in the squid axon, then what must the potential difference be in order for the influx of Na^+ due to diffusion to be balanced by the efflux due to the electrical field? The Nernst equation gives the answer. Assuming that the cell is at 20°C, the equilibrium potential for sodium is:

$$E_{Na} = 57.5 \text{ mV} \times \log \frac{440 \text{ mM}}{50 \text{ mM}}$$
$$= 57.5 \text{ mV} \times \log (8.8) = +54 \text{ mM}$$

This potential represents the value of the electrical gradient that will produce a net tendency for Na^+ to move out of the cell

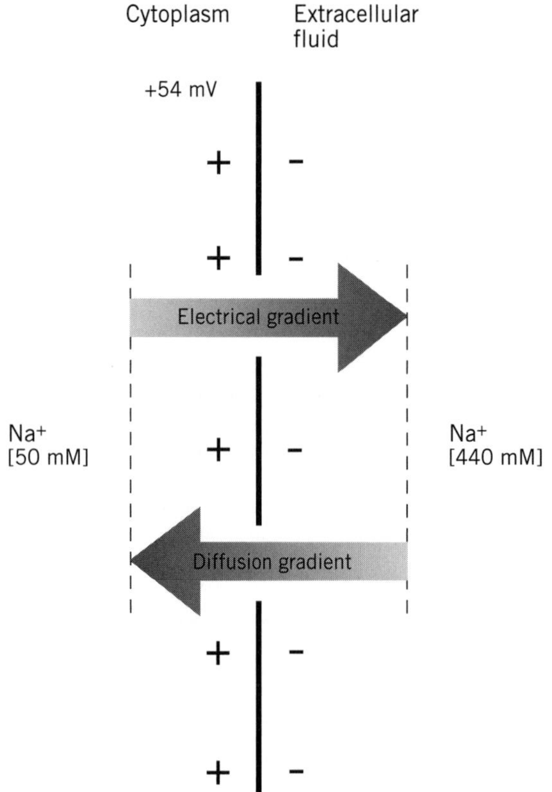

FIGURE 4-5. The equilibrium condition for sodium ions. The diffusion gradient is inward because there is a higher concentration of sodium outside the cell than inside. The electrical gradient that just balances the force of diffusion (+54 mV) must then be directed outward, as indicated by the direction of the arrow. (Note that the interior of the cell becomes *positive* in this case.)

that is equal to the net tendency for Na^+ to diffuse into the cell due to its high external concentration. The absolute value of E_{Na} is different from that of E_K because the internal and external concentrations of sodium are different from those of potassium. The sign of the potential is different because the high concentration of Na^+ is outside the cell rather than inside. But as in the case of E_K, E_{Na} is entirely independent of the distribution of other ions and of the membrane potential. This is indicated by the different values for E_{Na} and E_K in the same cell. The situation at equilibrium is shown in Figure 4-5.

Using the Nernst equation, you should work out the equilibrium potential for chloride at 20°C in the squid giant axon, given the distributions for Cl^- inside and outside that neuron as listed in Table 4-1. Remember that the charge on Cl^- is -1. The equilibrium potential for chloride is given in the next section.

The equilibrium potential for a given ion can be calculated with the Nernst equation. The equilibrium potential is calculated independently for each ion and is independent of the resting membrane potential of a neuron.

Equilibrium and Ion Concentrations

Equilibrium for any particular ion is established when the ion moves across the cell membrane and generates an electrical gradient that is strong enough to counter the diffusion gradient for that ion. But if ions move from one side of the membrane to the other, does this not change the internal and external concentrations of these ions? And if the concentrations are changed, how can an equilibrium potential be calculated when the new concentrations are not known?

The solution to this apparent dilemma lies in a consideration of just how many ions must move in order to generate a potential difference of a particular magnitude, and how that number compares with the total number of ions present in the cell. Consider a spherical cell with a diameter of 100 μm, containing potassium at a concentration of 0.40 mol/l, immersed in a solution containing 0.020 mol/l potassium (Figure 4-6). Such a cell would contain a total 1.26×10^{14} potassium ions. Assuming that only potassium can move through the membrane, the total number of potassium ions that must leave the cell to generate a membrane potential of -75 mV, the equilibrium potential for potassium, is only 1.47×10^8, or about 147 million. This may seem like a considerable number, but it is only 1 of about every 860,000 ions in the cell. The effect on the concentration of potassium of moving this number of ions is far too small to be measured or to have any effect on ion movement. Even in a neuron whose cell body is only 25 μm in diameter, a typical size for a small neuron, the number of ions that must move out of the cell is only 1 of about 214,000, which is still too small to have any measurable effect on the concentration of potassium in the cell. The external volume is usually much larger than the volume of the cell, so the external concentration of potassium will not be affected either.

Although ions must move across the cell membrane to establish a membrane potential that opposes the concentration gradient, ion concentrations are not likely to be affected because the number of ions that must move is far too small relative to the number present.

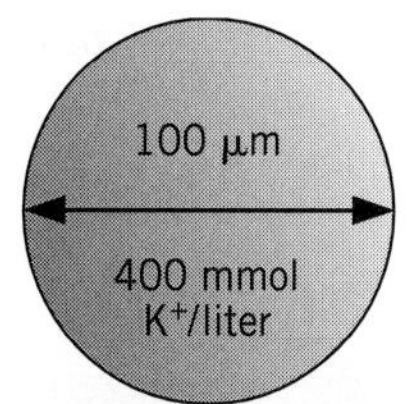

Volume: 5.24×10^{-10} liter

Number $K^+ = 5.24 \times 10^{-10}$ liter x 400 mmol/liter x 6.02×10^{23} ions/mol
$= 2.09$ mol $K^+ \times 6.02 \times 10^{23}$ ions/mol
$= 1.26 \times 10^{14}\ K^+$

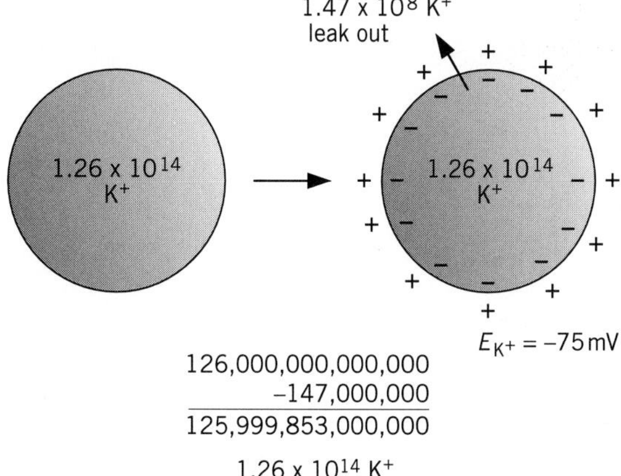

126,000,000,000,000
−147,000,000
125,999,853,000,000

$1.26 \times 10^{14}\ K^+$

FIGURE 4-6. The effects of diffusion on the concentration of an ion. A spherical cell 100 μm in diameter containing 400 mmol potassium will contain 1.26×10^{14} potassium ions. At equilibrium, 1.47×10^{8} ions will have diffused out of the cell, a number so small relative to the number of ions in the cell that it has no practical effect on the concentration of ions in the cell.

Ion Movements in Resting Neurons

If the equilibrium potential for an ion is not dependent on or related to the membrane potential of a cell, what is its importance? When the membrane potential of a neuron is different from the equilibrium potential for an ion, the ion is by definition not at equilibrium. If the cell is in a steady state (which electrically inactive neurons are), there must be some additional factor at work to influence ion flow. This factor counteracts the ion movement brought about by the unbalanced passive forces and thereby keeps the total ion influx equal to efflux. Its magnitude and direction can be determined by comparing the membrane potential with the equilibrium potential for the ion.

Consider the case of sodium in the squid giant axon. The equilibrium potential for sodium for this neuron is +54 mV, as calculated from the Nernst equation. Yet the membrane potential in the cell is -60 mV, which indicates that the membrane potential does *not* impart a rate of outward movement equaling that generated by the concentration gradient. Far from it. Because the inside of the cell is negative, the membrane potential actually propels Na^+ into the cell, just as the concentration gradient does (Figure 4-7A). The amount of Na^+ that actually enters, however, is relatively small, because there are few open channels through which it can enter. The small amount of Na^+ that does enter is balanced by the action of the Na^+-K^+ pump, which transports Na^+ out of the cell as fast as it leaks in, thereby maintaining the concentration gradient that is typical of neurons.

A similar analysis can be carried out for any ion. If the equilibrium potential does not equal the membrane potential, some active

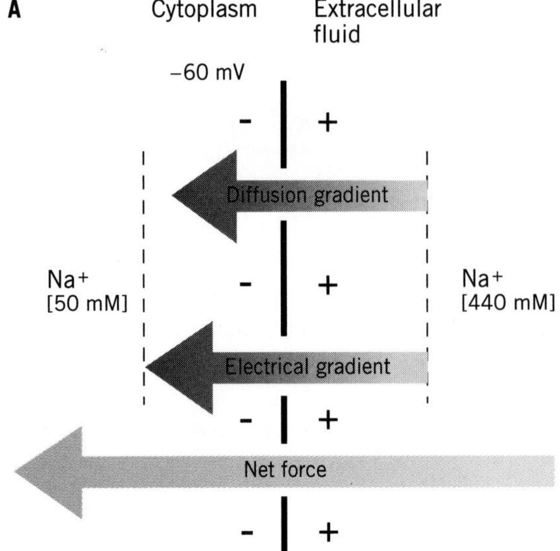

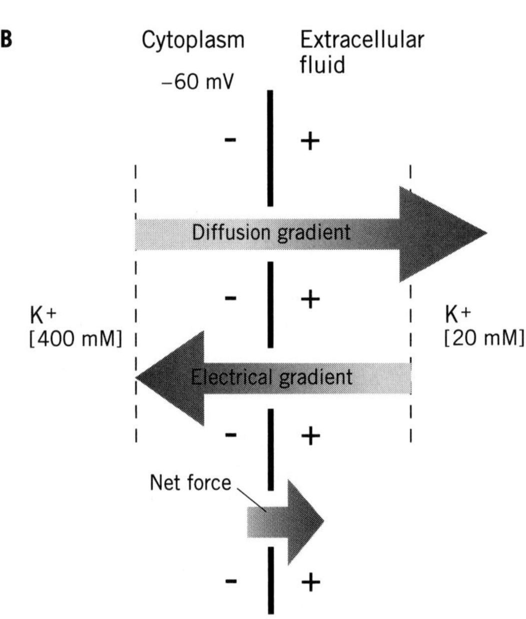

FIGURE 4-7. The direction and magnitude of the passive forces acting on (A) sodium and (B) potassium ions in a squid giant axon. The forces are generated by the concentration differences (diffusion gradient) of the ions and the membrane potential (electrical gradient) of the cell. The algebraic difference between the vectors for the two forces gives the direction and magnitude of the net movement of ions in the resting axon, as shown by the lower arrows in each case.

factor in addition to the passive factors of diffusion and the electrical field across the membrane must be at work. For potassium, the equilibrium potential as calculated by the Nernst equation is -75 mV. Since the membrane potential in the squid giant axon is only -60 mV, K^+ is not at equilibrium. The actual membrane potential is not large enough to fully counter the force of diffusion pushing K^+ out of the cell, so there is a small but constant K^+ efflux (Figure 4-7B). This passive efflux is compensated for by the active influx of K^+ due to the operation of the Na^+-K^+ pump, which brings two potassium ions into the cell for every three sodium ions that are expelled.

If the calculated potential for an ion does equal the membrane potential, then the ion is passively distributed, with no energy-requiring processes acting on it. In the squid giant axon, Cl^- falls into this category. The equilibrium potential for chloride in the squid is -60 mV, which is equal to the membrane potential. Chloride ions, therefore, are passively distributed and are not transported by any pump.

For an ion that is not at equilibrium, what is the net force propelling it in one direction or another across the membrane? Clearly there is more force moving Na^+ into the cell than there is moving K^+ out, even though neither ion is in equilibrium. It is possible to calculate precisely the strength of the net force, but initially it may be easier to visualize the meaning of the calculation with a graphic. Consider the diagrams in Figure 4-7. The forces exerted on each ion by the concentration and potential differences can be represented as arrow vectors. The direction and length of each arrow represent the direction and magnitude of each force, and the vector sum of the arrows gives the net force acting on the ion.

To calculate the net force, assign numerical values to the vector arrows. Consider K^+ first (Figure 4-7B). The electrical gradient of the membrane potential imparts a force that moves K^+ into the cell. This force is proportional to the electrical potential difference of -60 mV due to the membrane potential. The diffusion gradient established by the concentration difference for K^+ requires a potential difference of -75 mV to balance it. Hence, diffusion imparts a force equivalent to that imparted by a potential difference of +75 mV driving K^+ out of the cell. The algebraic sum of the diffusion and electrical forces is a net driving force of +15mV directed out of the cell.

A similar analysis for Na^+ yields a net force equivalent to -114 mV driving Na^+ into the cell (Figure 4-7A). There is a resting potential of -60 mV driving Na^+ into the cell, as well as a diffusion force equivalent to a potential difference of -54 mV driving it into the cell. In spite of the strong resulting net force, relatively little Na^+ actually does enter, because in the resting neuron the membrane has a very low permeability to this ion. The Na^+ that does enter the cell is transferred back out again by the Na^+-K^+ pump.

In the case of passively distributed Cl^- ions, the force imparted on the ions by the membrane potential (-60 mV) is equal and opposite to the electrical equivalent (+60 mV) of the diffusion force as obtained from the Nernst-calculated chloride equilibrium potential, yielding a net force of zero.

If the diffusion and electrical forces on an ion are not equal and opposite in a neuron when the neuron is at rest, the distribution of that ion is not at equilibrium. The Nernst equation can be used to help determine the net force on an ion for which the diffusion and electrical forces are not balanced.

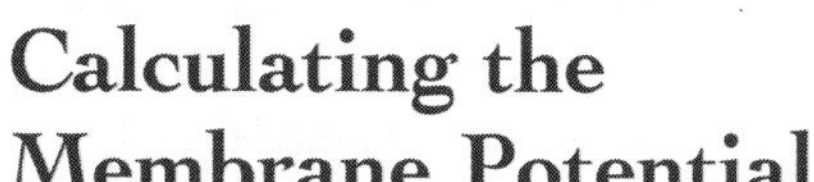

Calculating the Membrane Potential

If the membrane of a neuron were permeable to only one ion, the resting potential of the neuron would be equal to the equilibrium potential for that ion, as you learned in the preceding section. However, no cell is permeable to only one ion. Hence, to determine the resting potential, it is necessary to take into account the contribution that each ion makes to it. The Nernst equation cannot be used for this purpose because it deals with one ion only and cannot deal with the separate contributions of other ions. Furthermore, the contribution of a particular ion to the resting potential will depend in part on the permeability of the membrane to that ion. This factor is also missing from the Nernst equation.

To determine the resting potential of a neuron, the concentration difference and membrane permeability of each ion that can cross the membrane must be taken into account. You can see how this works qualitatively by considering, for example, K^+ and Na^+ in the squid giant axon. If the membrane were permeable only to K^+, then the resting potential would be -75 mV, as described previously. But if the membrane were just slightly permeable to Na^+ as well as being permeable to K^+, a small amount of Na^+ would be able to leak into the cell. Over the short term this would make the interior of the cell slightly less negative, due to the presence of additional positive charges inside the cell. If the permeability of the membrane to Na^+ were increased further, the cell would become even less negative as more Na^+ entered the cell. If the permeability of the membrane to K^+ were equal to that of Na^+, the resting potential would be about halfway between the equilibrium potentials of the individual ions, or about -10 mV.

You can imagine a similar situation in a hypothetical cell permeable only to Na^+, and hence having a resting potential of +54 mV. If such a cell were made slightly permeable to K^+, the resting potential would become slightly less positive, because of the loss of positively charged K^+ ions from the inside of the cell. The more permeable the membrane to K^+, the less positive the membrane potential, reaching a level of about -10 mV when the permeabilities of the two ions are equal. As this analysis indicates, the membrane potential will be dominated by the distribution of the ion to which the membrane is most permeable, assuming, of course, that some sort of ion pump is present to keep the concentration gradients stable over the long term.

Although the Nernst equation is not appropriate for the task, it is nevertheless possible to calculate the membrane potential of a cell from the permeabilities and concentrations of the ions that can cross the membrane. The equation for making this calculation is known as the **Goldman equation**.

Taking into account the main ions that can cross the membrane in neurons (K^+, Na^+, and Cl^-), the form of the equation is:

$$E_m = \frac{RT}{F} \ln \frac{P_K(K^+)_o + P_{Na}(Na^+)_o + P_{Cl}P_K(Cl^-)^i}{P_K(K^+)_i + P_{Na}(Na^+)_i + P_{Cl}(Cl^-)_o}$$

where P_K, P_{Na}, and P_{Cl} are the permeabilities of K^+, Na^+, and Cl^-, respectively, and E_m is the membrane potential. R, T, and F are the same constants used in the Nernst equation, and are defined in Table 4-2. The Goldman equation is also known as the **constant field equation** because one of its assumptions is that the electrical field of the membrane potential is constant across the span of the cell membrane.

Notice that if P_{Na} and P_{Cl} are zero, leaving the membrane permeable only to K^+, the equation reduces to:

$$E_m = \frac{RT}{F} \ln \frac{(K^+)_o}{(K^+)_i}$$

the equilibrium potential for potassium. (The value of Z from the Nernst equation, which is 1, is implicit.) If P_K and P_{Cl} are zero, leaving the membrane permeable only to Na^+, the equation becomes:

$$E_m = \frac{RT}{F} \ln \frac{(Na^+)_o}{(Na^+)_i}$$

the equilibrium potential for sodium. This again shows the dependence of the membrane potential on the distribution of the ion that can most readily pass through the membrane.

Absolute permeabilities are often difficult to measure, so the Goldman equation is frequently recast in terms of relative permeabilities. The equation then becomes:

$$E_m = \frac{RT}{F} \ln \frac{(K^+)_o + (P_{Na}/P_K)(Na^+)_o + (P_{Cl}/P_K)(Cl^-)_i}{(K^+)_i + (P_{Na}/P_K)(Na^+)_i + (P_{Cl}/P_K)(Cl^-)_o}$$

Using experimentally derived permeability ratios for P_K:P_{Na}:P_{Cl} of 1:0.04:0.45 and the values for internal and external ion concentrations given in Table 4-1 for the squid, you should calculate E_m, the membrane potential, for the squid giant axon, and compare it to the measured membrane potential. (Remember that you can convert natural logarithms, ln, to log to the base 10 by multiplying by 2.3.) You should find that the calculated value equals the measured value, -60 mV.

The Goldman equation can be used to calculate the membrane potential of a neuron from information about ion permeabilities and about concentrations inside and outside the cell. The membrane potential of a neuron is most strongly influenced by the concentration difference of the ion to which the membrane is most permeable.

Experimental Verification

In this chapter, the description of the ionic basis of the resting potential has been largely theoretical. In summary, the resting membrane potential is established mainly by an outward diffusion of K^+ ions that leaves negatively charged anions trapped inside the cell. A part of the potential established by this diffusion is negated by the small inward leak of Na^+, which leaves the cell's resting potential a bit closer to zero than the equilibrium potential for potassium, although the distribution of potassium ions is still the most important influence on the membrane potential.

Several important consequences follow if the membrane potential is actually dependent mainly on the distribution of K^+ as suggested by the theoretical discussion presented here. These consequences have been tested experimentally (Box 4-1) and support the theoretical picture. First, the membrane potential should be almost entirely independent of the external concentration of sodium. This is because the P_{Na}:P_K ratio is so low that the effect of small changes in $[Na]_o$ will be negligible in the Goldman equation. Experiments show this to be true. Until very low concentrations of external sodium are reached, changing the external sodium concentration has essentially no effect

BOX 4-1

Neurons at Rest: Recording the Resting Potential

It is possible to measure the resting potential of a neuron by using a device like a voltmeter, just as one uses a voltmeter to measure the potential difference across the terminals of a battery. It is easy to bring a wire from the measuring device to the exterior of the cell but more difficult to establish an electrical connection with the inside of the cell.

The process of detecting and displaying an electrical potential from a living cell is called **recording**. It is done by using probes called **electrodes** to pick up the electrical signal. Since recording a resting potential requires that one electrode be placed inside the neuron, this method is called **intracellular recording**. It is necessary to amplify the potential difference in order to display it conveniently, just as you must amplify the signal from your FM antenna so it can drive the speakers on your radio.

In practice, making an intracellular recording is not easy. Neurons are usually very small, making it difficult to insert anything into the cell without destroying it. The first record of intracellular recording was published by A. L. Hodgkin and A. F. Huxley in 1939 on the squid giant axon, whose diameter of 1 mm is large enough for insertion of a recording electrode through a small nick in the membrane (Figure 4-Aa). By connecting

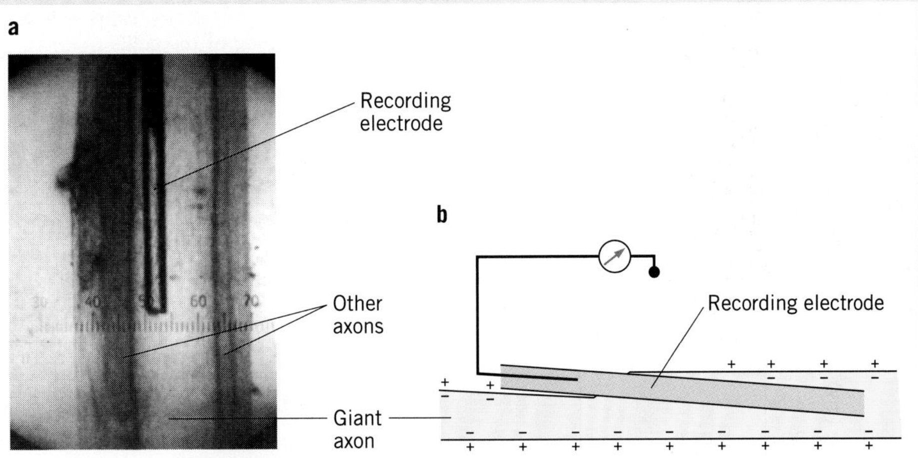

FIGURE 4-A. (a) The first intracellular recording arrangement, an electrode inserted inside a single squid giant axon. The giant axon, the clear cylinder in the center, is flanked by smaller, darker axons on each side. The electrode was inserted longitudinally so that the tip could be positioned away from the injured point of entry, which was tied off to prevent loss of ions from the interior of the axon. The scale is 33 μm per division, making this giant axon about 0.5 mm (500 μm) in diameter. (b) The arrangement for recording intracellularly with this electrode. The circle containing an arrow represents a combined amplifier and voltage measuring device. The small dot at the end of the external wire represents the bare end of the external recording electrode.

(continued)

Neurons at Rest: Recording the Resting Potential *(continued)*

the other end of the electrode to an amplifier, along with a second wire that was positioned outside the axon, Hodgkin and Huxley were able to record the potential difference across the cell membrane (Figure 4-Ab).

A

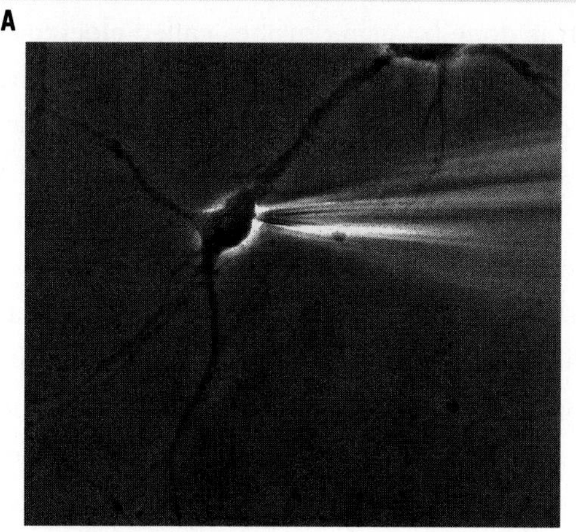

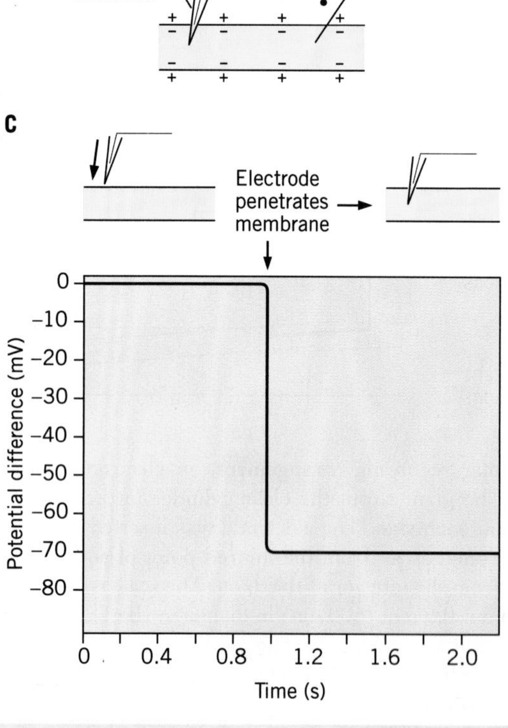

FIGURE 4-B. Modern intracellular recording with a glass microelectrode. (A) A photomicrograph showing an electrode impaling a neuron. (B) The arrangement for recording using a fine glass microelectrode. (C) An intracellular recording of the membrane potential. The sudden drop in the recorded potential at the arrow signifies that the electrode has penetrated the cell membrane. The membrane quickly seals itself around the electrode, allowing a recording of the potential difference between the inside and outside of the cell.

The electrode in the cell can be made of wire, but it does not have to be; anything that will provide a conducting path from the interior of the cell to the amplifier will do. Modern intracellular recording is done using a small, hollow, glass tube (Figure 4-Ba). One end of this tube is heated and drawn into a fine tip less than 1 μm in diameter, and the interior of the tube is filled with an electrolytic (conducting) solution, generally containing potassium. A wire connecting the solution to an amplifier is inserted into the large open end of the tube, and the tip is carefully driven through the cell membrane, which seals itself around the glass. The circuit is completed with a second wire positioned outside the cell to provide an electrical reference (Figure 4-Bb). Because of the small size of its tip, an electrode such as this is referred to as a **microelectrode**. An advantage of using a hollow tube as a microelectrode is that the tube can be filled with dye or other materials that can subsequently be injected into the cell for intracellular staining (see Box 3-2).

on the membrane potential. Changing the internal sodium concentration can affect the membrane potential transiently, primarily because of the effect such changes have on the operation of the Na^+-K^+ pump.

Second, changes in the external concentration of potassium should have significant effects on the membrane potential. Experiments show that, over much of its range, the membrane potential is almost a linear function of the logarithm of external potassium concentration (Figure 4-8). The relationship between $[K]_o$ and E_m can be deduced from an examination of the Goldman equation, but it is also instructive to think about how this comes about from a consideration of what forces are working on potassium ions when the external concentration of potassium is changed. To take a specific example, suppose you were to add enough potassium acetate to the medium bathing a squid giant axon to increase $[K]_o$ from 20 mM to 54 mM. The acetate ions are too large to pass through the membrane and therefore do not contribute to the potential changes that ensue. However, increasing the external potassium concentration to 54 mM results in a change in E_K from -75 mV to -50 mV. Because this value is more positive than the membrane potential of -60 mV, there will now be a net *inward* flow of K^+. This net transfer of K^+ into the neuron, in turn, will make the inside of the cell less negative (more positive) than it was before, as shown in Figure 4-8.

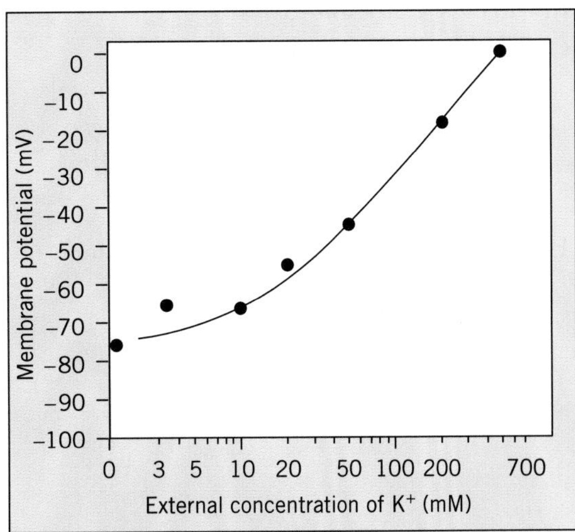

FIGURE 4-8. The relationship between membrane potential and the log of the external concentration of potassium in a squid giant axon. The relationship is nearly linear except at low levels of external K^+; thus, over most of its range, the membrane potential is influenced mainly by the concentration of potassium outside the cell.

The most important point here is that, over the short term, the resting membrane potential of a neuron is due entirely to movements of ions down their concentration gradients and to the selective permeability of the cell membrane. If either of these factors changes, as they do when a neuron conducts a nerve impulse, there are significant changes in membrane potential, as you will learn in Chapter 5.

Studies in which ion concentrations inside and outside a neuron are changed have shown that the membrane potential is strongly dependent on the external concentration of potassium and independent of the external concentration of sodium, except when the external concentration of sodium is very low.

Additional Reading

General

Junge, D. 1992. *Nerve and Muscle Excitation.* 3d ed. Sunderland, Mass.: Sinauer.

Katz, B. 1966. *Nerve, Muscle, and Synapse.* New York: McGraw-Hill.

Matthews, G. G. 1991. *Cellular Physiology of Nerve and Muscle.* 2d ed. Boston: Blackwell Scientific.

Miles, F. A. 1969. *Excitable Cells.* London: Heinemann.

Research Articles and Reviews

Curtis, H. J., and K. S. Cole. 1942. Membrane resting and action potentials from the squid giant axon. *Journal of Cellular and Comparative Physiology* 19: 135–44.

Hille, B. 1977. Ionic basis of resting and action potentials. In *Handbook of Physiology.* Section 1: *The Nervous System.* Vol. 1, *Cellular Biology of Neurons,* ed. E. R. Kandel, 99–136. Bethesda, Md.: American Physiological Society.

Hodgkin, A. L., and R. D. Keynes. 1955a. Active transport of cations in giant axons from *Sepia* and *Loligo. Journal of Physiology (London)* 128:28–60.

—— 1955b. The potassium permeability of a giant nerve fibre. *Journal of Physiology (London)* 128:61–88.

Huxley, A. F., and R. Stämpfli. 1951. Effect of potassium and sodium on resting and action potentials of single myelinated nerve fibres. *Journal of Physiology (London)* 112:496–508.

Katz, B. 1948. The electrical properties of the muscle fibre membrane. *Proceedings of the Royal Society of London. Series B—Biological Sciences.* 135:506–34.

CHAPTER 5

The Nerve Impulse

Communication within the nervous system often requires some way to pass messages long distances along neurites without loss of information. Neurons use the nerve impulse, a brief reversal of membrane potential that can sweep along the length of a neurite, to accomplish this task. How ion channels allow nerve impulses to be generated, and how these impulses can move along neurites, are the subject of this chapter.

As you learned in Chapter 4, the resting potential of a neuron develops in part because of the properties of the ion channels that pierce the cell membrane. Similarly, the ability of a neuron to convey information for some distance also relies on the properties of specific ion channels in the cell membrane. The resting potential is generated by the ion selectivity of channels that are open when the neuron is electrically inactive, whereas a nerve impulse is generated by the ion selectivity of channels that open when the neuron is active. The ability of certain ion channels to open and close gives rise to resting and active states, and allows a neuron to propagate nerve impulses without loss of amplitude along axons or dendrites that may be several feet long.

The Nerve Impulse: An Overview

A nerve impulse, more technically called an **action potential**, is a brief, transient reversal of the membrane potential that sweeps along the membrane of a neuron (Figure 5-1). It starts with a rapid **depolarization** of the membrane—a change of the membrane potential from its resting value toward zero, making the interior of the cell less negative. As an action potential develops, depolarization continues until the potential difference across the cell membrane has reversed and the interior of the cell is positive relative to the outside. The membrane then repolarizes, regaining its normal, interior negative, condition. Typically,

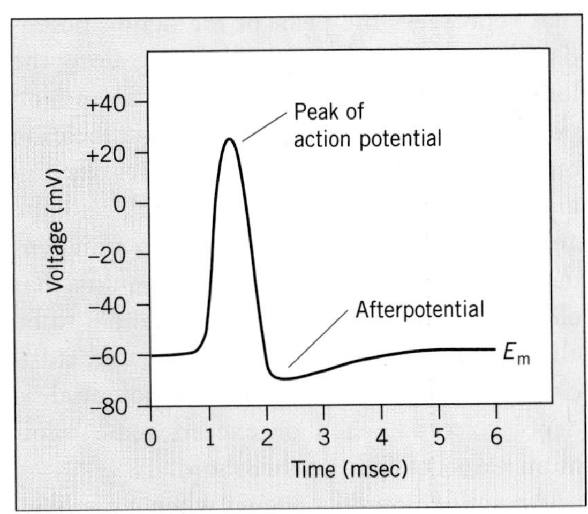

FIGURE 5-1. A typical action potential, showing how the potential difference across the membrane changes with time. The potential rises from a resting level (E_m) of about -60 mV up to a peak of as much as +40 mV, then falls again. After every action potential the membrane potential temporarily becomes more negative than E_m, in what is called an afterpotential.

a small afterpotential, a hyperpolarization, is present for a short period after the membrane has regained its normal resting potential. **Hyperpolarization** refers to any change of the membrane potential to a value more negative than the resting potential. Hence, the membrane potential actually becomes slightly more negative than it is at rest during this period (Figure 5-1). These changes in membrane potential during the action potential are brought about by movements of ions across the cell membrane and can be measured under the appropriate recording conditions (Box 5-1).

The Characteristics of Action Potentials

An action potential has several important features. First, it is temporary, usually lasting only about a millisecond at any one place on the membrane. However, through mechanisms that are explained in the next section, the generation of an action potential at one location on a membrane induces an action potential in the adjacent membrane as well. Hence, the reversal in membrane potential that represents the peak of the action potential gives the appearance of moving along the length of the neurite. Second, an action potential is **all or none**. At any one location on the membrane, it either reaches its full amplitude or does not occur at all, and the amplitude of the action potential is independent of the magnitude of the stimulus that elicited it. Third, for an action potential to be elicited, the neuron must be stimulated sufficiently so that the membrane potential is depolarized to reach or exceed some minimum value, called the **threshold**.

An action potential occurs when a depolarizing stimulus changes the states of ion channels in the membrane of a neuron. These changes typically increase the ability of Na^+, then K^+ to cross the cell membrane. As a result, the membrane depolarizes as positively charged Na^+ enters the cell, then shortly thereafter repolarizes as Na^+ stops entering and K^+ flows out.

An action potential is a brief, all-or-none reversal of membrane potential, elicited by depolarizing stimulation above a threshold, that sweeps along the membrane of a neurite. It is caused by the movement of ions across the membrane, which takes place because ion channels have opened.

Action Potentials and Gated Channels

As you learned in Chapter 2, ion channels that regulate, or gate, movement of ions through them are called gated channels. The gate is a movable portion of the protein molecule that comprises the channel. When the gate occludes the hollow core of the channel, the channel is closed. When the core opens by a shift in position of the gate due to a conformational change in the structure of the channel protein, the channel is open. The channels that are involved in generation of an action potential are voltage-sensitive (voltage-gated); their gates open or close in response to the magnitude and polarity of the electrical potential across the membrane. Only cells whose membranes contain voltage-sensitive channels are capable of generating and conducting action potentials because only voltage-sensitive channels will respond to changes in the electrical potential across the membrane of the cell by opening or closing, thereby increasing or decreasing the flow of ions across the membrane.

Opening ion channels increases the permeability of the membrane to the ion that can pass through those channels. A full understanding of the action potential requires that this change in permeability be described quantitatively. For convenience, neurophysiologists use a readily measured electrical equivalent of permeability, **conductance**, symbolized by the letter g, instead of permeability. In electrical terms, it is the inverse of resistance, a measure of how difficult it is for a charged particle to move through a conductor.

BOX 5-1

Neurons in Action: Recording and Displaying Neural Activity

Recording a resting potential does not require much in the way of display. A sensitive voltmeter will adequately show its value. Action potentials, on the other hand, consist of large, rapid swings in membrane potential, and hence require a means of displaying these changes in potential over time. There is a trend toward the use of computers to display electrical signals, but the display instrument of choice in neurobiology is still an **oscilloscope** because of its ease of use. The main parts of this instrument are a cathode ray tube (the screen on which the "picture" appears) whose inner face is coated with a material that emits light when it is struck by a beam of electrons, and an electron "gun" that generates the beam (Figure 5-A). The beam can be made to move horizontally and vertically. Horizontal movements, which represent time, are generated by the internal circuitry of the

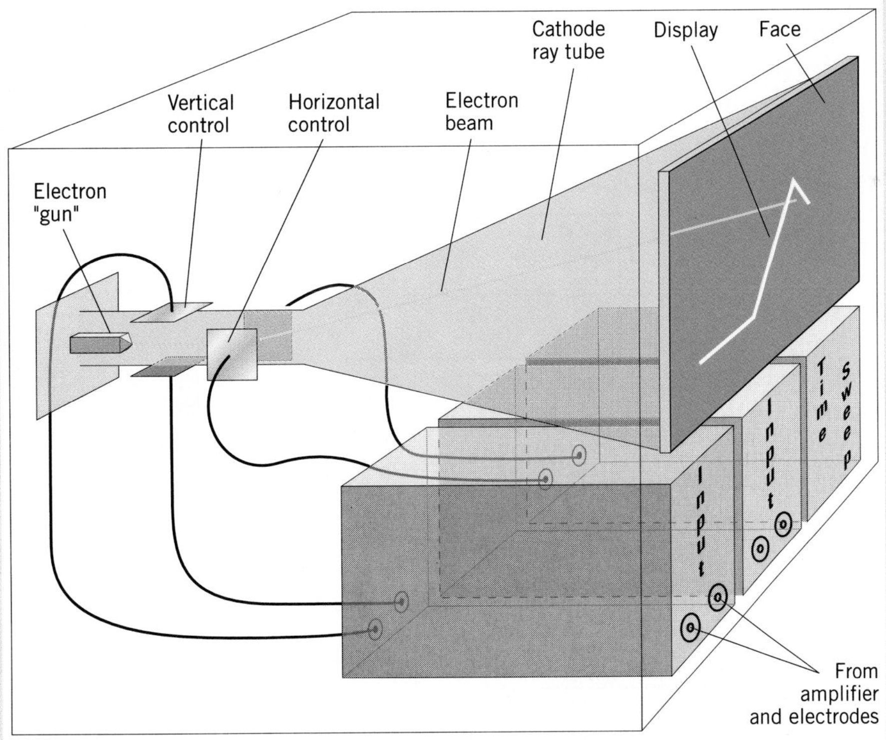

FIGURE 5-A. A simplified schematic of an oscilloscope. A picture of the variation of an input voltage over time is drawn on the face of the cathode ray tube by a beam of electrons. The position of the beam is controlled by parallel pairs of vertical and horizontal plates. The vertical plates control horizontal movement, representing time. The horizontal plates control vertical movement, representing the amplitude of the voltage applied to the input of the oscilloscope.

(continued)

Neurons in Action: Recording and Displaying Neural Activity *(continued)*

instrument. Vertical deflections are caused by an external voltage source connected to the input of the oscilloscope. If the external voltage changes, the beam "draws" a picture of the voltage as it changes with time (Figure 5-B).

Recording a resting potential requires that one electrode be placed within the neuron (an intracellular electrode) and the other outside it, because the potential difference *across* the membrane of the cell must be measured, as described in Box 4-1. Action potentials, however, can be recorded without inserting any electrode into the axon, because the currents that generate an action potential produce local but temporary differences in potential between one place and another on the outer surface of an active axon. Recordings in which both electrodes are outside the neuron are called **extracellular recordings**.

Extracellular recordings cannot show the resting potential; when the neuron is at rest, there is no potential difference between the two extracellular electrodes (Figure 5-Ca). When an action potential sweeps by, however, first one and then the other electrode will record a positive

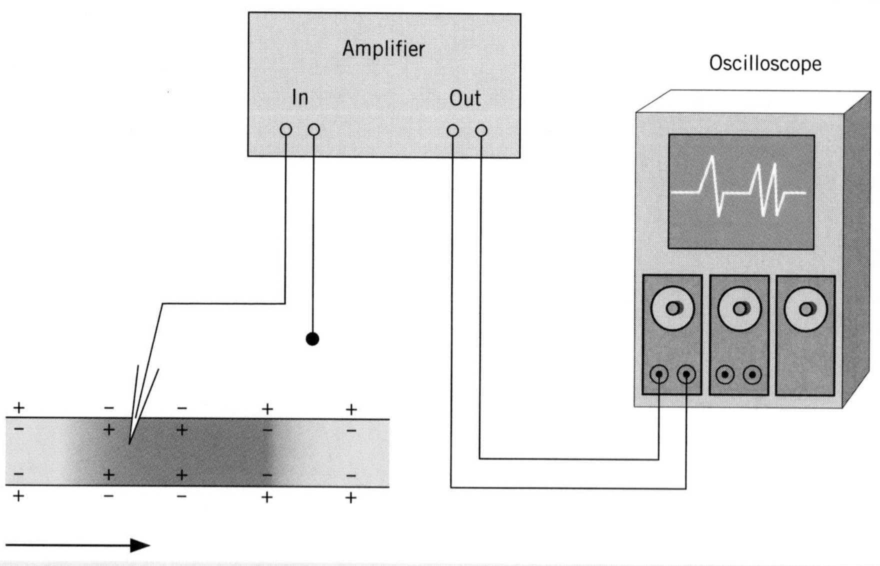

FIGURE 5-B. Recording an action potential with an intracellular electrode. As the peak of the action potential sweeps by the recording electrode, the membrane potential at that point changes from resting to as much as +40 mV and back to resting again. The time course of this potential change can be displayed on the screen of an oscilloscope.

potential relative to its partner (Figure 5-Cb). The action potential recorded by extracellular electrodes looks different from an intracellularly recorded one because the electrodes see the difference in potential between two points along the axon, not across the cell membrane.

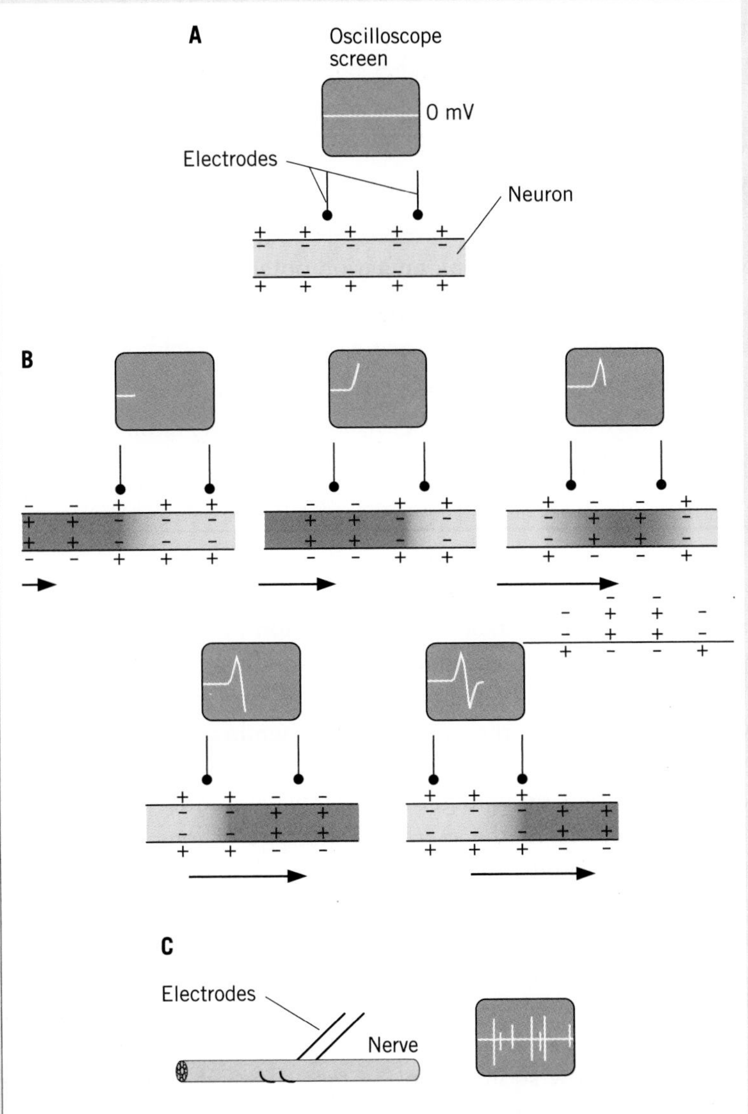

FIGURE 5-C. Extracellular recording. Compare these oscilloscope displays with the one shown in Figure 5-B. (A) When the neuron is at rest, there is no potential difference between the electrodes, so no potential difference is recorded. (B) When an action potential sweeps by, a potential difference is recorded between the two electrodes, which is shown as an upward, then downward, deflection of the beam of the oscilloscope. (C) Extracellular recording from a nerve containing several active axons will generate spikes of several different heights, the largest spikes coming from the largest-diameter axons.

(continued)

Neurons in Action: Recording and Displaying Neural Activity (continued)

Action potentials from extracellular recordings are frequently displayed on an oscilloscope using a slow time scale, so they appear as vertical lines only (Figure 5-Cc). It is from this appearance that an action potential obtains the name **spike**. In recordings from nerves containing several active axons, spikes of different heights will be recorded. The potential difference measured along the length of an axon during an action potential will be greater for axons with a larger diameter because larger axons generate more current flow during an action potential.

One advantage of extracellular recording is the technical ease of carrying it out. Another advantage is that one electrode, by picking up signals from several cells at once, can show coherence of firing among different neurons. The disadvantage is that when a nerve contains many axons, an investigator can rarely be certain which potentials belong to which neurons.

Because voltage is the product of resistance and current, resistance can readily be calculated by placing an electrode inside the neuron, passing a known current across the membrane, measuring the voltage difference that is produced, and dividing current by voltage. If there are many open channels in the membrane for an ion to pass through, the resistance of the membrane to passage of the ion is low, and the conductance of the membrane to the ion is high. Conversely, if there are few open channels for an ion, the membrane has a high resistance to that ion and a low conductance. Resting neurons have low sodium conductance (g_{Na}) because few sodium ions are able to enter the cell. The conductance to potassium (g_K) is higher than for sodium, but still not large, because at rest only some potassium channels are open in addition to the leakage channels that allow small amounts of both Na^+ and K^+ to cross the membrane.

The conductance of a particular patch of membrane to an ion depends on three factors: the availability of ions (measured by the ion's concentration), the density of channels for that ion in the membrane, and the number of those channels that are open. The concentration of an ion is important because, obviously, the membrane's conductance cannot be high if only a few ions are available to cross it. However, under normal conditions there is no shortage of ions; furthermore, the concentrations of these ions inside and outside the neuron do not change significantly. Hence, ion concentration has a constant effect and does not contribute to changes in conductance.

The density of channels that will allow the passage of a particular type of ion is also important. Even if plenty of ions are available, the conductance will clearly be small if channels that allow the ion to cross the membrane are rare. In the giant axon of a squid, so called because it has an unusually large diameter (up to 1 mm), there are about 300 voltage-sensitive sodium channels and about 20 to 30 voltage-sensitive potassium channels per square micrometer. Perhaps you can gain a better impression of the density of channels if you imagine 1 μm^2 scaled up to the area of an American football field (about 91 × 49 m).

Distributing 300 channels evenly on the field would leave about 4.1 yd (3.9 m) between the centers of adjacent sodium channels. The potassium channels would be 13.3 to 16.3 yd (12.2 to 14.9 m) apart from center to center. Leakage channels that allow both Na^+ and K^+ to pass across the membrane are also present, but the density of these channels in the membrane has not been determined. The number of channels in a patch of membrane may change over a period of days or weeks, but it is constant for the duration of an action potential.

The most significant factor to influence conductance is the number of open channels; clearly, a large number of channels for an ion will have no effect on conductance if those channels are not open. In a resting neuron, nearly all the sodium channels and many of the potassium channels are closed. Except for those ions that are transported across the membrane by an ion pump, most ions cross the membrane through leakage channels when a neuron is at rest. This situation changes dramatically when the neuron becomes active.

Only cells whose membranes contain gated, voltage-sensitive channels are capable of generating and conducting action potentials. Action potentials can be described by quantitative changes of conductance of the membrane to specific ions. Conductance depends on ion concentration, channel density, and the number of open channels. Although both ion concentration and channel density are important, under normal conditions only the number of open channels has any significant short-term influence on conductance.

Ionic Basis of the Action Potential

In intact nervous systems, action potentials are initiated when one neuron stimulates another or when a sensory stimulus impinges on a sensory neuron, as you will learn in the next few chapters. For now, however, consider what happens when a local depolarization is induced in an axonal membrane by a brief electrical stimulation. If the stimulus is weak, there will be a small local depolarization of the membrane but no action potential (Figure 5-2A). A somewhat stronger stimulus will produce a somewhat larger local depolarization but still no action potential (Figure 5-2B). A sufficiently strong stimulus will produce first a large local depolarization and then an action potential (Figure 5-2C).

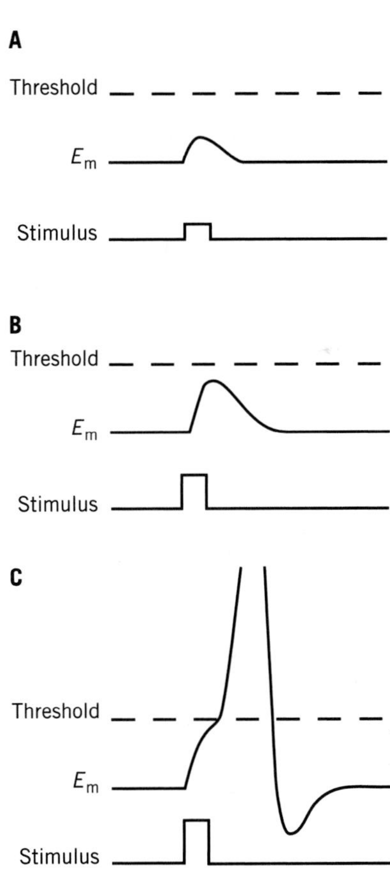

FIGURE 5-2. Stimuli must depolarize the neuron at least to the threshold in order for an action potential to be generated. (A) A weak stimulus will generate only a weak local depolarization of the membrane. (B) A stronger stimulus will produce a somewhat larger local depolarization but still no action potential. (C) A strong stimulus will produce first a large local depolarization and then an action potential.

The threshold is the trigger point for an action potential. A weak local depolarization that does not reach threshold causes some of the voltage-sensitive sodium channels in that region of the membrane to open, allowing Na^+ to flow into the cell under the influence of the concentration gradient for sodium and the electrical gradient. However, relatively few sodium channels open, and the sodium ions that enter the cell move away from the site of entry rapidly enough due to diffusion and their mutual electrical repulsion that they have minimal effect on the membrane potential. As a result, the membrane potential will quickly return to normal, the open sodium channels will close again, and no action potential will be generated.

An extremely strong depolarization, on the other hand, will cause so many sodium channels to open that Na^+ enters the cell more rapidly than it can move away from the site of entry. By accumulating even for a short while in the stimulated region, these ions cause further depolarization, which causes additional sodium channels to open, allowing the entry of more Na^+, further depolarization, and an action potential. The threshold is the transition point. Below threshold, the sodium ions move away from the open channels inside the neuron as rapidly as they cross the membrane, and hence have minimal local effects on the membrane potential. At and above the threshold, sodium ions accumulate in the region of the open channels faster than they disperse into the surrounding cytoplasm, causing further depolarization and the entry of more ions, resulting in the rapid depolarization that characterizes the rising phase of the action potential. During this depolarizing phase, so much positive charge enters the neuron through the open sodium channels that the polarity of the membrane reverses, making the inside positive for a short time (Figure 5-3). In the squid giant axon, the membrane potential may reach as much as +40 mV.

Even as the sodium gates are opening, however, the stage is being set for repolarization of the membrane. For reasons that will be discussed shortly, sodium channels do not remain in an open state. That is, each open sodium channel closes again a short time after it opens, even though the membrane may still be depolarized, and remains closed for a brief

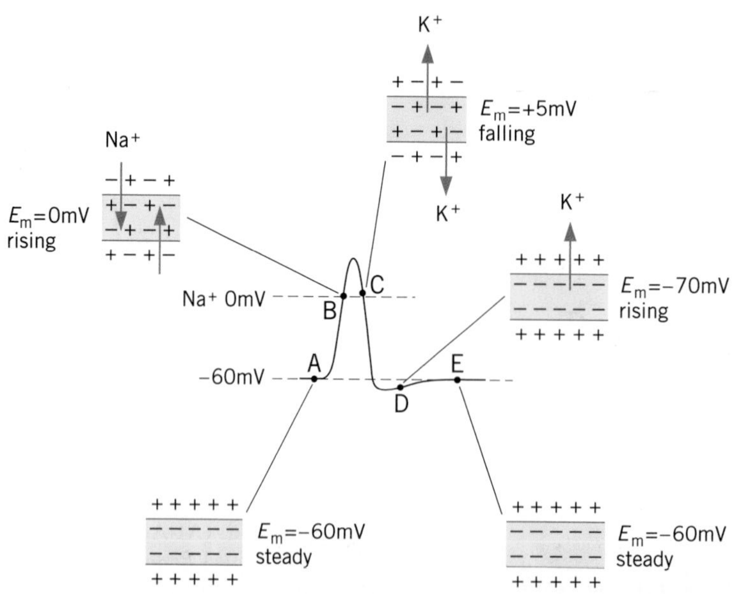

FIGURE 5-3. Ion flow during an action potential. The potential difference across the membrane and the dominant ion flow are shown for each marked position on the action potential (A-E). The small amount of ion flow at rest has been omitted for clarity.

period. This phenomenon, called **sodium inactivation**, stops the influx of Na^+ and allows the cell membrane to recover from the influx of positive charge.

Depolarization of the cell membrane also causes the gates of potassium channels to open. These begin to open slightly later than the gates of the sodium channels; consequently, the peak flow of K^+ out of the cell occurs later than the peak flow of Na^+ into it (Figure 5-3C). As the potassium gates open, K^+ moves out of the cell under the influence of the concentration gradient for K^+ and the membrane potential, which by this time is positive inside. Since K^+ efflux is delayed relative to Na^+ influx, the flow of K^+ ions becomes significant just as the depolarization due to the influx of Na^+ is reaching its peak and sodium channels have already begun to close. This allows the efflux of K^+ to repolarize the membrane to resting levels. As the membrane repolarizes, the potassium channels close once more, but again slowly. Potassium channels in squid giant axons do not inactivate in the same way as sodium channels; they will remain open for several seconds if the membrane is kept depolarized. However, potassium channels in neurons or muscle cells of other animals may inactivate, staying open for a relatively short time even if the membrane is held in a depolarized state.

Because the potassium channels do not respond instantly to changes in membrane potential, there are still a significant number of open channels when the membrane potential has reached its original resting level. Thus, at the instant the membrane potential has returned to the level of the original resting potential, the membrane is more permeable to K^+ than it was before the action potential began. Hence, potassium will diffuse out of the neuron, driven by the difference between the resting potential at that instant (-60 mV) and the equilibrium potential for potassium, (-75 mV) (see Chapter 4). This efflux of K^+ causes the small afterpotential that characteristically follows an action potential (Figure 5-3D). It disappears as the potassium channels close and the resting condition is restored.

Two aspects of these events are especially important. First, the ions move *entirely* under the influence of the factors that bring about passive ion flux: diffusion down an ion's concentration gradient and the influence of an electrical field on charged particles. The hyperpolarization that follows an action potential will never be more negative than E_K because once the membrane potential reaches the equilibrium potential for potassium, the net force on potassium ions will be zero, preventing any further net movement and further hyperpolarization. In the squid giant axon, the peak of the action potential will always be less than +54 mV because that is E_{Na}, and at that potential the net force on sodium ions will be zero, preventing any further depolarization. In fact, the peak of the action potential does not reach E_{Na} because by the time the sodium channels are fully open, some potassium channels are already open, negating some of the Na^+ current.

Experimental results substantiate these thoretical considerations, strengthening researchers' belief that the passive movements of ions are the sole basis of the action potential. For example, poisoning a neuron with a metabolic inhibitor that prevents synthesis of adenosine triphosphate (ATP), and hence stops operation of the sodium-potassium exchange pump (Na^+-K^+ pump), does not prevent the generation of action potentials until the concentrations of the ions in and around the poisoned neuron begin to be affected.

The second point is that, although it might seem that an exchange of ions across the cell membrane would change the concentration gradients enough to influence the diffusion forces acting on the ions, especially if several action potentials are generated in quick succession, this does not occur. Both calculations and experimental tests have indicated that in the squid the net number of ions that enter (Na^+) or leave (K^+) the neuron during one action potential is about 4 pmol/cm^2 of membrane. However, the number of sodium ions in a volume of an axon covered by 1.0 cm^2 of membrane (1.25×10^{-6} mol) is about 3 million times this amount (Figure 5-4). The effect on the internal concentration of sodium of adding a few picomoles of the ion is roughly like the effect on the weight of a diesel locomotive of having a mouse climb aboard. Even if a neuron were only 5 μm in diameter, the amount of sodium that would enter during an action potential would be only 0.032% of the amount in the volume of the axon enclosed by 1.0 cm^2 of membrane. This is analogous to the weight of a mouse relative to that of a small elephant. Hence, thousands of action potentials can be generated even in a relatively small neuron before the ion balance begins to be affected.

An action potential is initiated by a local depolarization that triggers the opening of voltage-sensitive sodium channels. The inward rush of Na^+ further depolarizes the membrane. At the same time but more slowly, voltage-sensitive potassium channels open, an event that allows an efflux of K^+. This efflux of K^+, along with inactivation of the sodium channels, repolarizes the membrane.

Measuring the Flow of Ions

Action potentials were described in neurons more than a century and a half ago by the German physiologist Emil Du Bois-Reymond, but the question of just what ions were involved in producing them and the quantitative relationship between the movements of the ions and the size and shape of the resulting potential remained a contentious issue until the middle of the twentieth century. How was the issue resolved?

Even before the advent of intracellular recording, the importance of different ions in generating an action potential could be demonstrated by simple substitution experiments. If the influx of Na^+ is the main driving force in neurons for the rising phase of the action potential, then changing the external concentration of Na^+ should affect the amplitude of the action potential, just as changing external K^+ concentration in a resting neuron changes the value of the resting potential. Experiments bear out this prediction. For example, increasing the external Na^+ concentration, which can be done by adding NaCl to the external medium, causes an increase in

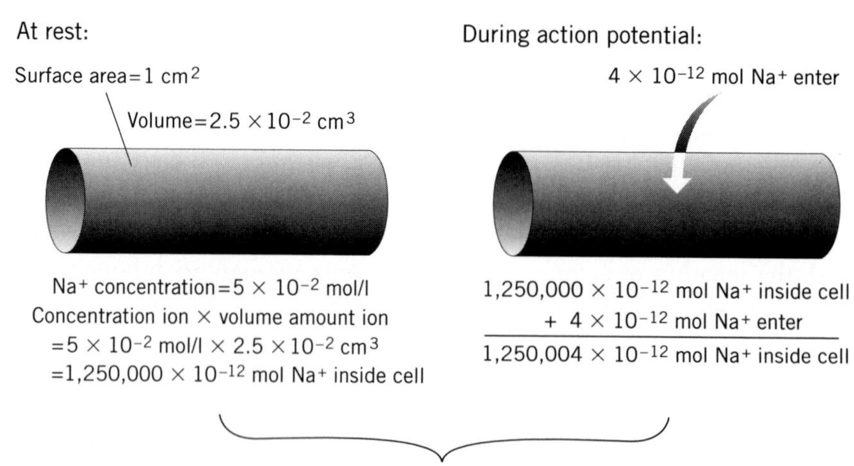

FIGURE 5-4. The effects of an action potential on the concentration of Na^+ ions inside an axon. A section of axon enclosed by 1 cm^2 of membrane will contain 1.25×10^{-6} mol of Na^+. During an action potential, about 1/3,000,000 of this amount will enter the axon section, a quantity so small that it has no practical effect on the concentration of sodium in the cell.

the size of the action potential. Reducing the external Na^+ concentration by placing the neuron in a low-salt medium reduces the size of the action potential. If a neuron is placed in a Na^+-free medium, the action potential is abolished altogether.

Qualitative experiments that measure the types of changes in the action potential following various kinds of ion-replacement procedures provide useful clues to the underlying ionic mechanisms. A full understanding of the action potential, however, requires a quantitative description of the current flow contributed by each ion and how that current flow leads to its final shape and amplitude. Such a quantitative description was provided in the 1950s by Alan Hodgkin and Andrew Huxley in a series of elegant experiments that in 1963 earned them the Nobel Prize for Physiology or Medicine.

Hodgkin and Huxley were able to make measurements of current flow by using a device called a **voltage clamp** (Box 5-2), an electronic instrument that allows researchers to select a particular membrane potential (the "clamp" potential) and to hold ("clamp") the membrane of a cell at that potential. The membrane potential is kept constant even if there are ion movements across the membrane that would change it. The voltage clamp device itself detects any movement of the membrane potential away from the clamp potential and generates a current that will bring the membrane potential back to its clamped value. By measuring the current produced by the device to keep the membrane potential constant under various experimental conditions, researchers can measure quantitatively the ion currents produced by the neuron under those conditions.

The most useful voltage clamp experiments are those in which as few variables as possible are allowed to influence ion movements. For example, clamping the membrane potential to zero will eliminate any movement of ions due to the electrical field, leaving only the concentration gradients of the ions as a motive force. Changing the external concentration of an ion to equal the internal concentration of that ion will eliminate any movement of the ion due to concentration differences.

Hodgkin and Huxley combined these conditions to study ion flux in squid giant axons (Figure 5-5). In their first experiments, they placed the axon in a saline solution that contained the normal concentrations of ions found in squid extracellular fluid; they also

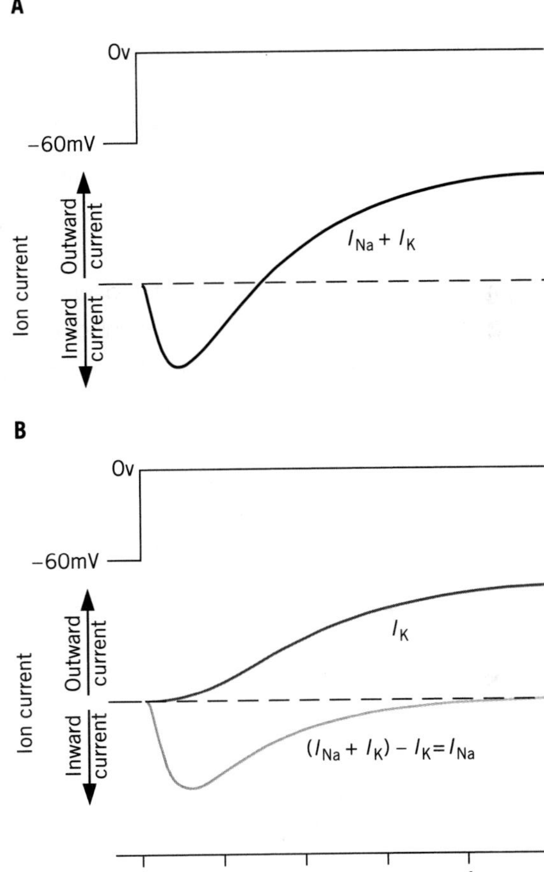

FIGURE 5-5. Current flow during a voltage clamp experiment. (A) When the membrane potential of a squid giant axon is depolarized to 0 V in normal saline (seawater), the current generated by the voltage clamp device to keep the membrane potential at 0 V is that necessary to oppose the sodium (I_{Na}) and potassium (I_K) currents generated by the axon as it responds to the stimulus. (B) When the membrane potential is clamped at 0 V in saline containing only 50 mM NaCl, which equals the concentration of Na^+ inside the neuron, the current generated by the voltage clamp device is that necessary to oppose only the potassium current. This occurs because there will be no flow of Na^+ in the absence of a voltage gradient and a sodium concentration gradient. Subtracting I_K from the summed $I_K + I_{Na}$ gives the value of the I_{Na} by itself (lower line).

BOX 5-2

Measuring Current Flow-I: Voltage Clamping the Membrane

The technique of voltage clamping neurons, developed by the neurophysiologist Kenneth Cole in the late 1930s, was crucial for Hodgkin and Huxley's experiments on the action potential. The method is conceptually simple but requires complex circuitry. As described in the text, the idea is to set the membrane potential at some specified level, then measure the current flow necessary to maintain ("clamp") the membrane potential at that level. The current flow will represent the net flow of ions across the membrane under the clamp conditions and therefore can be used to determine which ions move across the membrane under various experimental conditions.

The method requires three electrodes, two inserted inside the neuron, the third outside (Figure 5-D). These electrodes are connected to the voltage clamp device. When the experimenter has selected a voltage and activated the device, one of the electrodes inside the neuron will be used to measure the potential difference across the membrane. If this potential is different from the voltage set by the user (the clamp voltage), the other electrode will be used to generate sufficient current to bring the voltage across the membrane to the desired level and keep it there. This is done by generating a current that flows between the external (ground) and second internal electrodes until the membrane potential equals the clamp potential selected by the user. The device also allows a record of the current that it generates to be displayed on an oscilloscope.

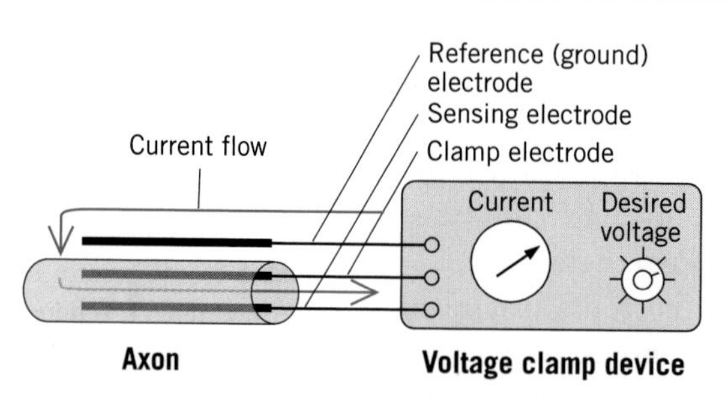

FIGURE 5-D. Simplified schematic of a voltage clamp setup. The electrodes inside and outside the axon whose membrane potential is being clamped are connected to the voltage clamp device, which has a current monitor. One internal electrode (sensing electrode) is used to monitor the membrane potential of the axon. The other (clamp electrode) is used to generate a current to keep the voltage at the specified level, as represented by the arrows. The external electrode completes the circuit through which the current travels. The current that flows between the internal and external electrodes represents the current necessary to counter the ion current flowing across the cell membrane under clamp conditions.

The objective of using a voltage clamp is to measure the ion currents that are flowing across a specific region of membrane of a neuron under various experimental conditions. Difficulties can arise, however, in trying to ensure that the only currents that are measured are those due to ion flow through open channels. If only a small region of membrane is voltage clamped, current flow from the clamp region to adjacent unclamped regions of the cell will also be measured and hence will influence the results. Hodgkin and Huxley were able to minimize such unwanted currents by inserting rather long electrodes into the squid axon and clamping a sufficiently long region of the axon to allow them to measure current flow through membrane channels quite accurately.

set the voltage clamp to zero. When the clamp was activated, the axon was immediately depolarized to zero. As the gates of the sodium and then the potassium channels opened in response to this depolarization and allowed Na^+ and then K^+ to flow through the membrane, the voltage clamp device generated currents that opposed those caused by the ion flow. The summed current, $I_{Na} + I_K$ (I is the electrical symbol for current), shown in Figure 5-5A, gave a picture of the inward flow of Na^+, represented by the negative current necessary to oppose it, and then the outward flow of K^+, represented by the positive current necessary to oppose it. (Positive currents are defined as the movements of positive charges *into* a cell. Hence, the inward flux of Na^+ is a positive current, and the current to oppose it that is generated by the voltage clamp is a negative current.)

From a number of other experiments, Hodgkin and Huxley knew that only Na^+ and K^+ contributed to the summed current they measured in the voltage clamp experiment in normal saline. What they still needed to know was the individual contribution that each type of ion made to the summed current, which they could find by completely eliminating the movement of one ion. For this reason, their next experiment was carried out on an axon immersed in saline solution containing 50 mM sodium, a concentration equal to that of sodium inside the axon. By once again clamping the membrane potential of the axon to zero volts, Hodgkin and Huxley were able to measure I_K, the current generated by the flow of K^+ only (Figure 5-5B). Since the concentration of sodium was the same inside the axon as outside, there was no movement of Na^+ due to a concentration gradient; since the membrane potential was zero, there was no movement of Na^+ due to an electrical field, either. Hence, under these clamp conditions, only K^+ would flow across the membrane in response to the opening of the potassium channels. Therefore, the only current generated by the voltage clamp device was the current necessary to oppose this movement of potassium ions.

Determining the contribution made by the flow of sodium ions is not quite so straightforward. As you learned in Chapter 4, setting the external concentration of potassium equal to its internal concentration causes the membrane potential to fall to zero. Under these conditions, the neuron can no longer respond to step changes in voltage such as those applied during a voltage clamp. For this reason, I_{Na}, the sodium current, must be calculated indirectly by subtracting I_K from the summed current, $I_K + I_{Na}$. The summed current is due to both potassium and sodium currents together, so subtracting the potassium current leaves just the sodium current.

From a quantitative study of the currents due to the flow of Na^+ and K^+ during an action potential, Hodgkin and Huxley were

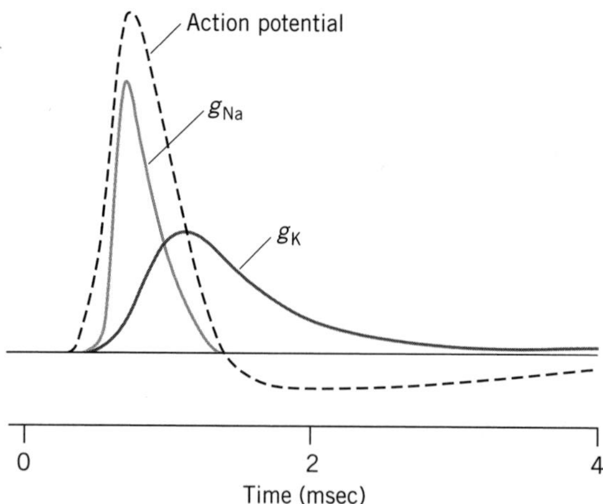

FIGURE 5-6. Changes in conductance of sodium and potassium as calculated from voltage clamp experiments. An action potential is superimposed on the conductance curves to show its temporal relationship to the conductance changes. Sodium conductance rises quickly and rapidly, then declines just as fast, as the sodium channels become inactivated. Potassium conductance rises and falls more slowly, not only repolarizing the membrane but also causing the negative afterpotential.

able to calculate precisely the changes in conductance of the membrane for each of these ions during an action potential (Figure 5-6). The conductance curves show clearly the fast rise and fall of g_{Na} and the slower change in g_K that underlie the rise and fall of the action potential itself. From a knowledge of the conductance changes, Hodgkin and Huxley were then able to calculate the theoretical shape of the action potential, a shape that matched the actual recorded potential remarkably well.

The flow of ions during an action potential has been studied by using a voltage clamp device to hold the membrane potential at a given level. Such studies have allowed the calculation of the precise changes in conductance of the membrane to Na^+ and K^+ flow during an action potential, and a theoretical reconstruction of an action potential.

The Properties of Voltage-Gated Channels

One of the most significant inferences drawn by Hodgkin and Huxley was that the action potential is due to a rapid and independent change in conductance of sodium and potassium ions. From this inference, they proposed that the ions flowed through two types of independent, voltage-gated channels. Whereas Hodgkin and Huxley could only infer the presence of such channels, present-day researchers study them directly.

Activation of Individual Channels

As you look at the rather smooth curves of current flow shown in Figure 5-5, you might easily gain the impression that ion channels open and close gradually, and all in concert. However, experiments using a technique that allows the recording of current through individual ion channels have shown that this is far from the case. The method for recording from single channels was introduced by Erwin Neher and Bert Sakmann in 1976 and immediately had a dramatic impact on physiologists' understanding of cellular electrical phenomena at the molecular level. Neher and Sakmann made such fundamental contributions to the understanding of ion channels by using the method that in 1991 they were awarded the Nobel Prize for Physiology or Medicine. The technique, called the **patch clamp**, involves sealing a fine-tipped, glass micropipette to a small "patch" of cell membrane (Box 5-3). By clamping the voltage of this patch of membrane much as the whole cell was clamped by Hodgkin and Huxley, it is possible to measure the current that flows through individual membrane channels (Figure 5-7).

The results of patch clamp experiments on single sodium channels show that each sodium channel opens only briefly (on average for about 0.7 msec), then shuts again. It is clear from measurements of current that each channel is either open or closed, with no measurable intermediate state. The change in potential recorded in an axon as the action

BOX 5-3

Measuring Current Flow-II: Patch Clamping the Membrane

The voltage clamp used by Hodgkin and Huxley recorded the summed currents that resulted from the opening of all the channels in the region of the membrane being clamped. The development of recording techniques using glass microelectrodes allowed a useful refinement of whole-cell voltage clamping—the patch clamp. In this technique, introduced by Erwin Neher and Bert Sakmann in 1976, a polished glass microelectrode is positioned on the membrane of the cell, and gentle suction is applied (Figure 5-E). When done correctly, this procedure seals the electrode to the membrane and electrically isolates any ion channels in that patch of membrane (hence the name). If the region of the cell where the patch electrode is situated is then voltage clamped, it is possible to record the current generated by the flow of ions through the channels in the patch in response to the applied voltage. With skill, a researcher can seal off and hence record from a single channel, and thereby study channel kinetics under different experimental conditions. Variations of the patch recording method allow experimenters to study the effects of changes in the ion conditions on the inside and outside of the channel (i.e., inside and outside the neuron) as well.

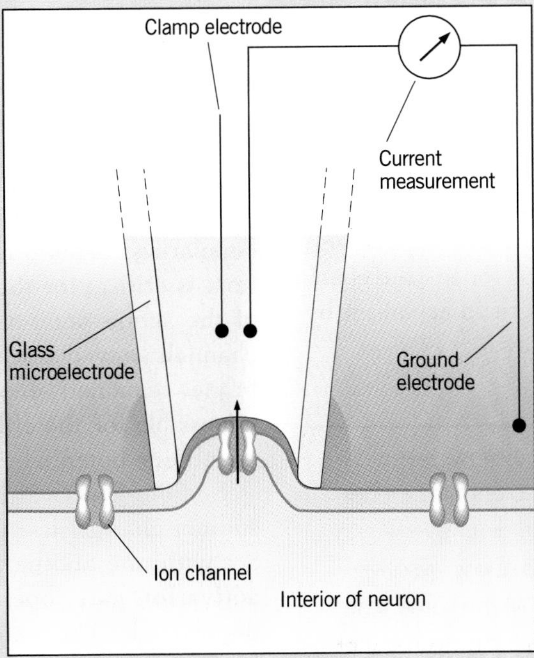

FIGURE 5-E. A patch clamp electrode. A tight seal between the lip of the glass microelectrode and the cell membrane allows the experimenter to record currents from channels in just the small area of membrane encircled by the electrode.

potential sweeps by is the net result of ion flow through all of the individual open channels. When the axon is depolarized, some channels open early, contribute to the early phase of the sodium-caused depolarization, and then close. More channels open a bit later, under the influence of the rapid depolarization of the membrane potential. A few also open much later, actually opposing the repolarization of the membrane due to the efflux of K^+. In general, the number of channels that are open at any one time is proportional to the depolarization of the membrane —the more depolarized the membrane, the more channels are open.

Opening of either a sodium or a potassium channel as a consequence of a membrane depolarization is the result of the physical movement of one of the transmembrane segments of each subunit of the channel protein. As illustrated in Figure 2-9, each sodium channel is made of four subunits, consisting of six transmembrane domains, numbered S1 through S6. The S4 domain, situated in the center of the array, is probably the voltage sensor (Figure 5-8A). When the membrane is at resting levels, S4 is more or less aligned with the others. When the membrane depolarizes, however, S4 is thought to move physically so that it projects out into the extracellular space (Figure 5-8C). This movement is presumably caused by changes in the affinity of the positive charges associated with S4 for the interior face of the membrane as the membrane depolarizes. Movement of S4 presumably opens the channel to the flow of ions, but the mechanism by which this occurs is not known.

The patch clamp technique allows researchers to study the electrical properties of individual ion channels. Such studies have shown that channels open and close relatively quickly but not necessarily at the same time. Channels open because of the movement of one transmembrane segment in response to changes of membrane potential.

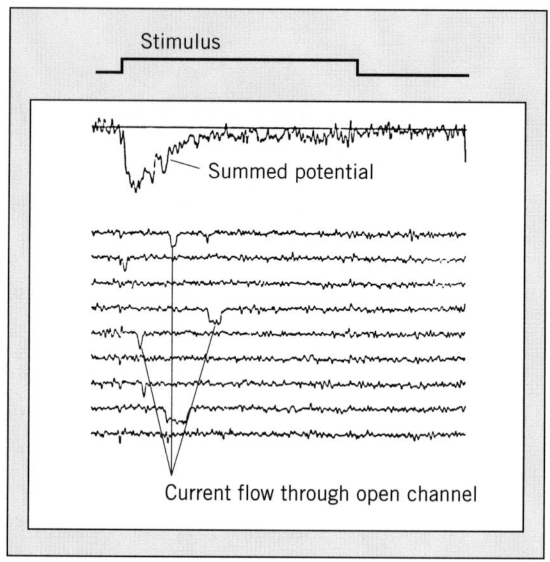

FIGURE 5-7. Current flow through sodium channels in response to an applied step depolarization. The top of the figure shows the averaged response from 300 trials. The individual traces underneath show responses of individual sodium channels. (Downward deflection of the trace represents channel opening.) Note that although the stimulus is the same in each case, individual channels respond differently, some opening for longer periods than others, and some not opening at all.

Channel Gating

Perhaps the most important feature of sodium channels is their inability to stay open for the entire period during which the membrane is depolarized. This property, sodium inactivation, is critical for the successful termination of the action potential. If individual sodium channels stayed open as long as the membrane remained depolarized, it would be impossible for the efflux of K^+ to restore the membrane potential to resting levels.

Sodium inactivation occurs because the sodium channel has two gates, which interact with one another. One gate, termed the **activation gate**, opens when the membrane depolarizes. The nature of the gate is not known, but it has been hypothesized that opening the channel may involve a conformational change in those parts of the protein that line the pore, leading to enough widening of the channel to allow sodium ions to

A

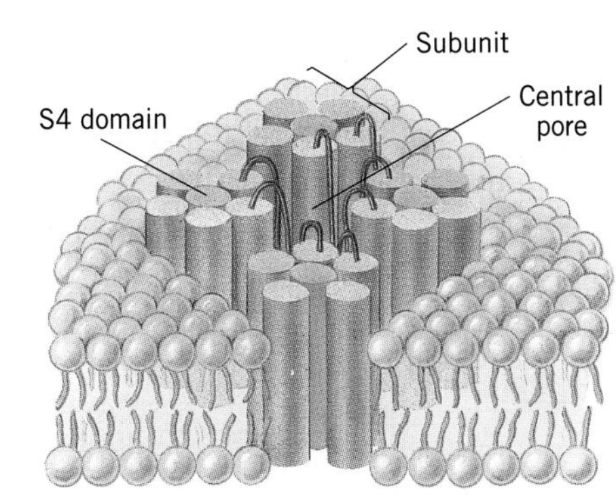

B

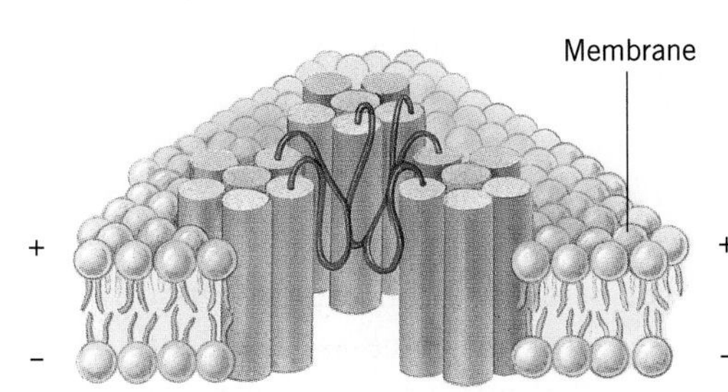

C

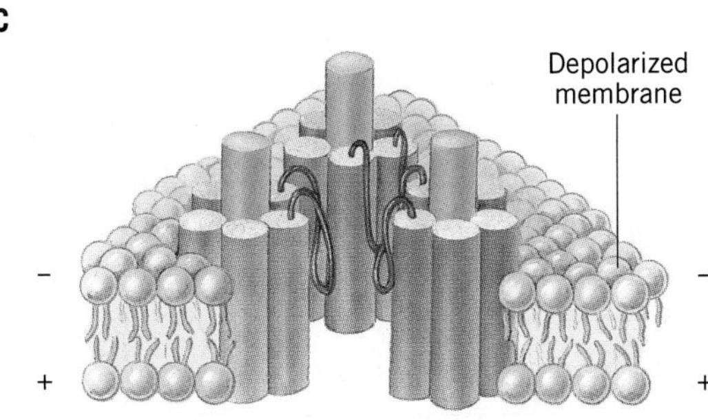

FIGURE 5-8. Voltage sensing in a sodium channel. (A) A single sodium channel consists of four subunits, each with a central voltage-sensing transmembrane domain (S4, light shading). Compare this diagram with the exploded view of a sodium channel in Figure 2-9. (B) At rest, the loops of amino acids that line the channel also occlude it. (C) As the cell membrane depolarizes, the voltage sensor moves out. By means still not understood, this movement opens the channel pore.

pass through (see Figure 5-8C). The conformational change, and hence the opening of the gate, must be triggered by the movement of the S4 domain described in the previous section, but how this might occur is not known.

The **inactivation gate**, is thought to interact with the activation gate. As long as the activation gate is closed, the inactivation gate is held open. When the activation gate opens, the inactivation gate will, after a short delay, swing shut, thereby closing the channel. The nature of the interaction between the two gates is also not known, but the inactivation gate itself consists of the amino acid loop connecting two of the subunits of the sodium channel. This loop swings over and occludes the channel pore after the activation gate has opened. As long as the membrane is depolarized, the activation gate will remain open, and the inactivation gate will remain closed. Only after the membrane has repolarized will the activation gate close and the inactivation gate open, readying the channel for a response to a subsequent depolarization.

Inactivation of voltage-sensitive sodium channels has an important effect on the threshold of the neuron. Following every action potential is a brief period during which it is more difficult to excite a neuron to generate another action potential. This time, called the **refractory period** of the neuron, typically lasts for several milliseconds, and consists of two parts. The **absolute refractory period** occurs while the membrane is repolarizing and immediately afterward. During this time no new action potential can be generated because the threshold is effectively infinite. Following the absolute refractory period is the **relative refractory period**, in which a new action potential can be generated if the stimulus is sufficiently strong. During the relative refractory period, the threshold returns to its normal level (Figure 5-9). The refractory period is caused by inactivation of sodium channels and the elevated g_K that immediately follow an action potential. During the absolute refractory period, so many sodium channels are inactivated that no amount of stimulation will cause enough of them to open to elicit another action potential. As

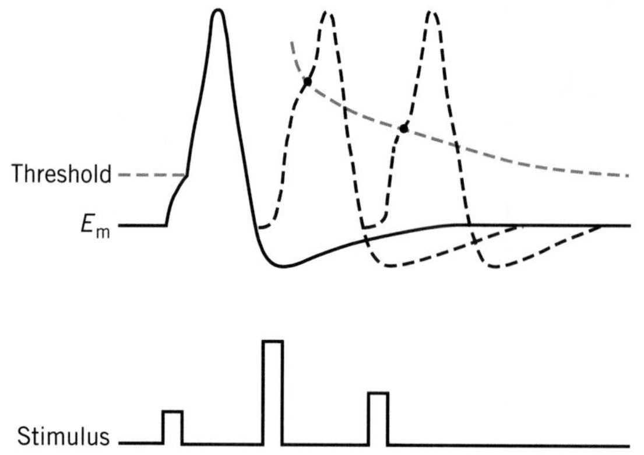

FIGURE 5-9. The effect of the refractory period on the threshold for generation of an action potential. Immediately after an action potential, the threshold for elicitation of another action potential is extremely high. The threshold quickly falls back to resting levels.

Sodium channels have two gates whose interaction leads to the phenomenon of sodium inactivation and the refractory period of the neuron. Potassium channels have only a single gate, and in the squid inactivate quite slowly.

more and more sodium channels recover from their inactivated state and g_K returns to resting levels, the strength of a stimulus required to elicit another action potential gradually returns to normal; the neuron's threshold recovers.

Potassium channels are also voltage-sensitive, but they have different properties. First, they have only a single gate. In the squid giant axon, this gate opens when the membrane depolarizes and closes when it repolarizes. If the membrane is held in a depolarized state, these potassium channels will remain open for quite a long time before they eventually inactivate. Second, it takes the potassium gates significantly longer to open than it does the sodium gates, as a comparison of the conductance curves for potassium and sodium in Figure 5-6 reveals. This longer response time is what allows the action potential to develop because it lets the explosive depolarization of the membrane potential as Na^+ rushes into the neuron proceed without any significant countercurrent due to the efflux of K^+. Third, potassium channels, in the squid at least, inactivate quite slowly. They will stay open for some seconds if the membrane depolarization is sustained experimentally.

Channel Diversity

One of the surprising findings that has emerged from studies of single channels is the diversity of channel types. That is, the voltage-sensitive sodium or potassium channels that are found in neurons in different animals, or in different neurons in the same animal, or even in different parts of a single neuron, are not necessarily identical. This channel diversity can be demonstrated both physiologically and structurally. Physiologically, channels may differ in the kinetics of the currents they carry. Some may open or close more quickly than others or may remain open for longer or shorter periods, leading to different profiles of current flow when they are activated. DNA cloning can sometimes show that the structure of important parts of the channel molecule may vary between channels that exhibit different kinetic current properties.

Voltage-sensitive sodium channels show the least diversity of structure and function. Wherever these channels are present, they exhibit relatively rapid opening, a short open period, and inactivation. The main differences between different sodium channels are in the exact time required for them to open, the length of time they stay open, and their inactivation characteristics. In general, however, they are remarkably uniform in their overall characteristics.

Potassium channels, on the other hand, occur in bewildering variety. Well over 50 types of voltage-gated potassium channels have already been described. Table 5-1 lists seven common types and their main characteristics.

Some of these channels open when the membrane depolarizes, some open when it hyperpolarizes. Some inactivate slowly, some quickly. Some are strongly influenced by the presence or absence of ions such as calcium or molecules such as serotonin. Nevertheless, by regulating the flow of potassium ions, all of these potassium channels have a strong influence on the excitability of the cell, on the duration of the action potential, or on the rate at which action potentials are generated when a neuron is stimulated. Although not all of the various types of potassium channels are found in any one neuron, many neurons do contain several types of these channels. The presence of particular types of potassium channels confers to a neuron a particular constellation of electrical properties, determining such things as how quickly the cell will recover from an action potential. Furthermore, specific chemical substances may change the properties of some potassium channels, leading to a change in the way the cell responds to stimuli from a neighboring cell. You will learn more about such changes in the next chapter.

Not all channels that are selective for a particular ion have the same structure or the same properties. Voltage-sensitive sodium channels vary only slightly in their physiology. Potassium channels, on the other hand, vary considerably in their physiological characteristics.

Channels and Neuronal Properties

The properties of the ion-selective channels that are present in a neuron determine that neuron's particular set of electrical properties. These properties, in turn, determine how the neuron will respond when it is stimulated, as well as other aspects of its activity. Consequently, a good deal of research effort has been put into an identification of channels and channel properties. Consider three

TABLE 5-1. A Selection of Potassium Channels

Channel Type	Common Designation	Function/Action
Delayed rectifier	K channel	Opens when membrane is depolarized; slow inactivation; repolarizes membrane after action potential
A channel	K_A channel	Opens when membrane is depolarized; fast inactivation; influences excitability of neuron
Ca^{2+}-activated	K_{Ca} channel	Opens in presence of Ca^{2+}; influences excitability of neuron
Inward rectifier	K_{ir} channel	Opens when membrane is hyperpolarized; influences duration of action potential; regulates heartbeat
M channel	K_M channel	Opens when membrane is depolarized; slow inactivation; influenced by neurotransmitters
S channel	K_S channel	Normally open, closed indirectly by presence of serotonin; influences duration of action potential
ACh channel	K_{ACh} channel	Opens when membrane is exposed to acetylcholine (ACh); slows heartbeat

examples of ways in which the presence of specific types of channels with specific properties can affect the characteristics of individual neurons.

As you learned in a previous section, the threshold of a neuron is the level of depolarization that must be reached in order for an action potential to be elicited. The threshold is affected by both sodium and potassium channels. First, it will be affected by the density of voltage-gated sodium channels. Other factors being equal, neurons that have a higher density of sodium channels have a lower threshold because when the neuron is stimulated at a certain level there will be more open sodium channels available to carry positive charge inward. For this reason, the balance point between the dissipation of sodium ions away from the point of stimulus and the influx of additional sodium ions that will bring about further depolarization will be reached at lower levels of depolarization in the neuron with a higher density of sodium channels.

The presence of other types of potassium channels can also affect the threshold, although these channels raise the threshold rather than lower it. For example, in the squid, K^+ leaves the neuron through voltage-sensitive potassium channels of the type known as delayed rectifiers (Table 5-1), so called because they respond relatively slowly to changes in membrane potential. If fast-acting potassium channels were added to the membrane, the effect would be to short-circuit a stimulus by opening a path for the loss of positive charge (K^+) at the same time that Na^+ is entering the cell. A large number of fast potassium channels would short-circuit the sodium current enough to prevent the generation of an action potential. However, a few fast potassium channels would affect the excitability of the neuron without entirely preventing the formation of action potentials. They would raise the neuron's threshold by increasing the strength of stimulation necessary to initiate an action potential but would not be present in large enough numbers to prevent the generation of action potentials altogether.

The rate of recovery of a neuron from an action potential, that is, the length of its refractory period, will be affected by the same factors that affect its excitability. More sodium channels will shorten the refractory period, but more of certain kinds of potassium channels will lengthen it by keeping the threshold elevated for longer periods following an action potential.

Some neurons exhibit spontaneous activity, generating one action potential after another without any external stimulus. Such neurons, called **pacemaker neurons**, can be found in organs such as the heart that show continual activity. Pacemaker neurons usually fire continually because they have a high inward sodium current even at rest, a current that continually depolarizes the membrane. When the threshold is reached, the neuron fires an action potential. Hence, the ability of a neuron to act as a pacemaker depends on the presence of particular types of ion channels. In addition, the rate of firing of pacemaker neurons is also strongly affected by the kinds of ion channels that they contain; the presence of certain kinds of potassium channels, for example, will speed up or slow down the rate of pacemaker firing.

Many neuronal properties, such as threshold, refractory period, and pacemaker ability, are determined and influenced by the presence of particular types of ion channels in the neuronal membrane.

Propagation of the Action Potential

One of the essential features of an action potential is that it can move along the length of an axon. This feature allows long-distance communication in the nervous system because action potentials normally maintain the same amplitude along the entire length of a neuron. Conduction takes place by a depolarization and hence excitation of regions of the membrane adjacent to the location of the action potential at any particular instant. The presence of myelin, the membranes of glial cells that wrap tightly around axons as described in Chapter 2, affects the conduction of nerve impulses. The general principles by which an action potential is able to move along a membrane are similar in myelinated and unmyelinated neurons, but the specific events are somewhat different in the two cases because myelin acts as an insulating layer around a neurite and hence affects current flow.

Conduction in Unmyelinated Neurites

Consider first a simplified view of a squid unmyelinated axonal membrane along which an action potential is moving (Figure 5-10). At the peak of the action potential, the interior of the axon is about +40 mV relative to the exterior. Yet in a part of the axon that the action potential has not yet reached, E_m may still be negative. The positive charges in the neuron in the region of the action potential will be repelled by one another and attracted by the nearby negative region. Current, represented by these positive ions, will therefore flow into those adjacent regions, causing depolarization there and the consequent opening of voltage-sensitive sodium channels. As these channels open, Na^+ begins to move into the cell, depolarizing it further, until the adjacent region of membrane also develops an action potential.

This explanation is a simplification. Actually, the spread of current and the depolarization of adjacent membrane happen so quickly that an action potential is usually spread out over several millimeters. In a typical squid axon, for example, a single action potential will be spread out over about 25 mm from beginning to end. For the sake of keeping everything in one drawing, the currents depicted in Figure 5-10 are shown as if the membrane potential of the region of axon just a few micrometers from the

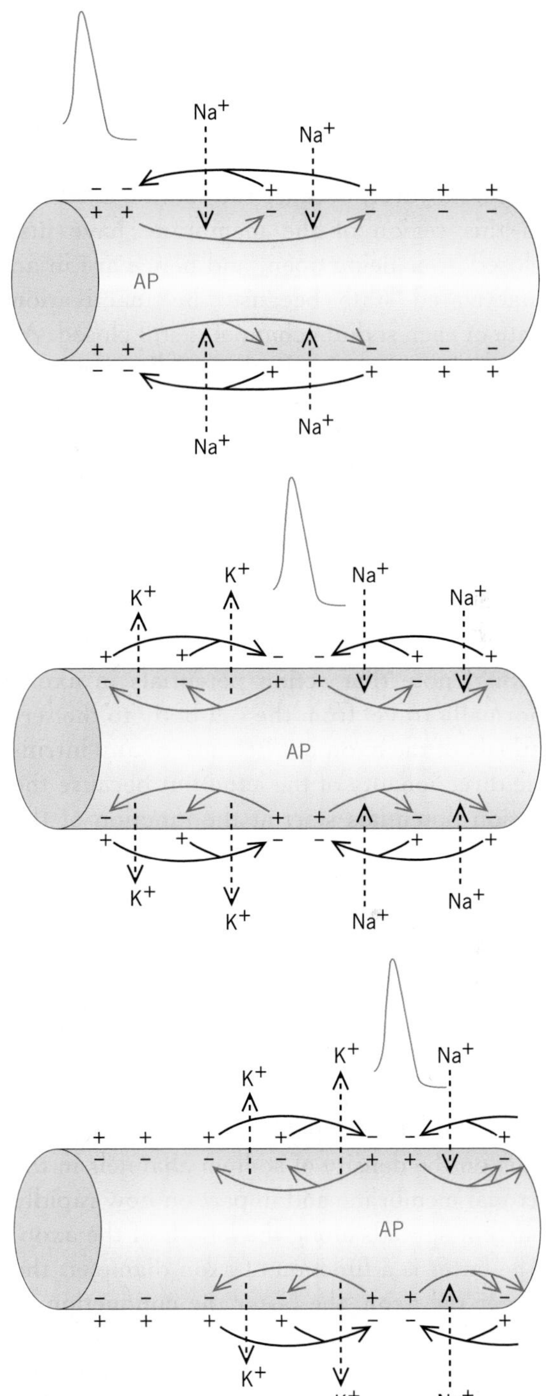

FIGURE 5-10. Diagram of current flow in an axon in advance of an action potential (AP) in an unmyelinated axon. The flow of positive charges from the peak of the AP toward the negatively charged regions in front of it depolarizes those adjacent regions, leading to the opening of sodium channels and the continuance of the AP. Positive charges flow back in the direction from which the AP has just come as well, but the sodium channels do not open because they are still inactivated, so no AP can be sustained there.

peak of an action potential is still negative, which is not the case.

Current flows into the area behind the advancing action potential as well. It does not cause a new action potential to develop there, however, because the sodium channels in this region of the membrane have just closed after being open, and hence are in an inactivated state because the inactivation gate of each sodium channel is still closed. As a consequence, the backward-flowing currents cannot cause a second influx of Na^+ behind the action potential and therefore cannot initiate another action potential there. The refractory period caused by sodium inactivation is hence an important feature that constrains action potentials to travel in only one direction, from the site of initiation to the other end of the neurite. However, you should note that action potentials in axons normally travel from the cell body to the terminals of the axon not because of any intrinsic directionality of the axon but because the action potentials start at the junction of the axon with the cell body. If you were to stimulate an axon at its terminal end, you could elicit a perfectly normal action potential that would travel back to the cell body.

The speed of propagation of an action potential (its conduction velocity) depends on how quickly areas of membrane in front of the moving action potential can be depolarized. The rate of depolarization will depend in part on the density of sodium channels in the axonal membrane and in part on how rapidly current spreads along the length of the axon. The latter is a function of axon diameter: the larger the axon, the faster the conduction of an action potential because larger-diameter axons present less internal resistance to the flow of current to adjacent membrane that has not yet been depolarized. Conduction in a squid axon with a diameter of 350 μm is 25 m/s, whereas conduction in the giant interneurons in a cockroach (diameter 60 μm) is only 7 m/s. In very small (1 μm or less) unmyelinated vertebrate sensory fibers (which convey information about some types of pain; see Chapter 14), conduction velocity is about 1 m/s.

Propagation of an action potential occurs by the depolarizing effect of current spreading out from the locus of the action potential. In unmyelinated axons, conduction velocity is proportional to axon diameter because larger-diameter axons present less internal resistance to the flow of current to undepolarized patches of adjacent membrane.

The Effects of Myelin

The presence of myelin around a neurite strongly affects the velocity of its action potentials by effectively lowering the capacitance of the membrane. **Capacitance**, an electrical term that refers to an ability to store electrical charge, is discussed more fully in Chapter 8, especially in Box 8-1. It is important for conduction velocity because any charge that the cell membrane stores will reduce the charge that can move along an axon to depolarize adjacent regions of membrane and the charge that can be used to depolarize the cell membrane at a particular site. Hence, the higher the capacitance of a membrane (i.e., the greater its ability to store charge), the lower the conduction velocity of action potentials along it. Because the capacitance of patches of membrane covered with myelin is lower than that of unmyelinated membrane, current will spread much faster along myelin-covered stretches of axon, hence boosting conduction velocity.

In myelinated dendrites or axons, action potentials develop only at the nodes of Ranvier, the short gaps in the myelin that is wrapped around them (see Chapter 2). Myelin wraps so tightly around each axon that it effectively prevents any ion flow across the membrane in the wrapped regions; at the same time, it reduces the charge-carrying capacity of the membrane. Hence, current rapidly flows down the axon from one node to the next and causes a new action potential to be generated there, skipping the intervening areas (Figure 5-11). Because action potentials "jump" from node to node in a myelinated neurite, conduction in myelinated neurons is called **saltatory conduction**, from the Latin word for "jump." The speed of conduction in myelinated axons can be considerable. For example, a small myelinated axon in a frog, 12 μm in diameter, will conduct at a velocity of 25 m/s, the same velocity as that of a squid axon that has a diameter about 30 times as great.

Myelinated axons typically have a number of adaptations that exploit the advantages of myelination. One adaptation is that voltage-sensitive sodium channels are present in much higher density at the node of Ranvier than in unmyelinated axons. For example, there are about 300 sodium channels/μm^2 in the squid giant axon, but there may be as many as 3000 channels at a node of Ranvier of a frog myelinated neuron. In terms of the football field analogy presented earlier, sodium channels would only be about 1.3 yd (1.2 m) apart at such a density. The higher density of sodium channels allows action potentials to be generated at the nodes more quickly than they would be at lower channel densities, hence speeding conduction.

A second adaptation is that the neuronal membrane located under the myelin sheath is almost entirely devoid of voltage-sensitive ion channels. This lack of channels is beneficial for the axon because the energy and resources required to synthesize them can be directed to other uses. It is because of these specializations that demyelinating diseases like multiple sclerosis can have such serious consequences; without the insulating presence of myelin, the action potential will move along denuded internodal regions more slowly than normal because of the increased capacitance of the membrane. In some cases it may fail altogether because there are too few voltage-sensitive channels and the action potential fails before it can deliver sufficient depolarizing current to the next nodal region.

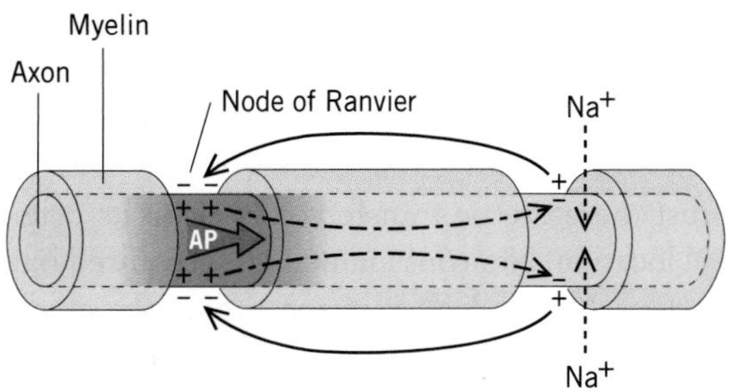

FIGURE 5-11. Diagram of current flow in a myelinated axon in advance of an action potential. In this case, the regions immediately adjacent to the node of Ranvier cannot be depolarized because the myelin sheath prevents the movement of ions away from the outside of the axon. The current hence spreads down to the next node, where ions are free to move and the membrane can be depolarized.

The conduction of action potentials is saltatory in myelinated neurons, meaning that the action potential jumps from node to node. Saltatory conduction is faster than nonsaltatory conduction because myelin reduces the capacitance of the axonal membrane, speeding current flow along the axon. Sodium channels in myelinated axons are concentrated at the nodes of Ranvier.

Energy Requirements of Action Potentials

Neurons do not require any immediate metabolic energy in order to conduct action potentials, as can be demonstrated by temporarily inhibiting the production of ATP in a neuron. A neuron so poisoned can conduct many hundreds of action potentials without difficulty because the energy source for action potentials is the energy stored in the concentration gradients of sodium and potassium across the cell membrane. Hence, the absence of usable ATP will not prevent the generation of action potentials.

Nevertheless, as you have learned, neurons do need ATP over the long term. The Na^+ and

K^+ that leak across the cell membrane at rest, and the small amount of Na^+ and K^+ transferred during the action potential, will accumulate over a period of many hours. Excess Na^+ and K^+ must therefore be transferred back to its original location. Neurons maintain the required balance of Na^+ and K^+ through the action of the Na^+-K^+ pump, as described in Chapters 2 and 4. Because one more sodium ion and its positive charge are removed than potassium ions are brought in during one cycle of activity, the pump keeps the cell slightly more negative inside than it otherwise would be.

The resting potential represents stored electrochemical potential energy. Hence, a neuron in which ATP production has been inhibited can generate action potentials until the ion concentrations are appreciably altered. However, ATP is required for the neuron to maintain the proper ion concentrations inside the neuron over the long term.

Additional Reading

General

Hille, B. 1992. *Ionic Channels of Excitable Membranes*. 2d ed. Sunderland, Mass.: Sinauer.

Hodgkin, A. L. 1964. *The Conduction of the Nervous Impulse*. Liverpool: Liverpool University Press.

Junge, D. 1992. *Nerve and Muscle Excitation*. 3d ed. Sunderland, Mass.: Sinauer.

Katz, B. 1966. *Nerve, Muscle, and Synapse*. New York: McGraw-Hill.

Matthews, G. G. 1991. *Cellular Physiology of Nerve and Muscle*. 2d ed. Palo Alto, Calif.: Blackwell Scientific.

Research Articles and Reviews

Armstrong, C. M. 1992. Voltage-dependent ion channels and their gating. *Physiological Reviews* 72(suppl.): S5–S13.

Catterall, W. A. 1996. Molecular properties of sodium and calcium channels. *Journal of Bioenergetics and Biomembranes* 28:219–30.

Hodgkin, A. L., and A. F. Huxley. 1952a. Currents carried by sodium and potassium ions through the membrane of the giant axon of *Loligo*. *Journal of Physiology* (London) 116:449–72.

—— 1952b. A quantitative description of membrane current and its application to conduction and excitation in nerve. *Journal of Physiology* (London) 116:500–44.

Neher, E. 1992. Ion channels for communication between and within cells. *Science* 256:498–502.

CHAPTER 6

Synaptic Transmission

If neurobiology were to be characterized by any single theme, that theme would be communication between neurons. It is the ability of one neuron to transfer information to another that makes the operation of the nervous system possible. This transfer of information is carried out by synaptic transmission, a two-stage process in which information is sent by one neuron and received by another. In this chapter, you will learn about the different types of synaptic transmission and about the mechanisms that determine the way the receiving neuron responds to a message.

Synaptic transmission is the process by which a signal is passed from one neuron to a target, usually another neuron or a muscle. It takes place at sites known as synapses. Synapses have a unique status in neurobiology because their presence was predicted in the early 1900s by Charles Sherrington on functional grounds, but were not definitively demonstrated morphologically until half a century later (Box 6-1). Both the means by which transmission is accomplished and the effect that transmission has on a postsynaptic neuron can be complex. Nevertheless, although the means may be complex, the purpose of synaptic transmission is relatively simple: to allow one neuron to influence the activity of another.

An Overview of Synaptic Transmission

Synaptic transmission can be classified in at least three ways. First, it can be classified based on the mechanism of transmission. As you learned in Chapter 2, information can be passed between neurons electrically or chemically. In electrical transmission, there is a direct transfer of ions from one neuron to another, and hence a direct influence of electrical current from one on the other. In chemical transmission, a chemical intermediary, the neurotransmitter, is released by the presynaptic neuron, diffuses across the synaptic cleft, and produces an effect on the postsynaptic neuron.

Synaptic transmission can also be classified according to whether it promotes or suppresses an active electrical response of the postsynaptic neuron. **Excitatory transmission**, or just **excitation,** is synaptic transmission that causes depolarization of the postsynaptic neuron. In spiking neurons, this depolarization increases the probability that the postsynaptic neuron will fire an action potential. In neurons that cannot generate action potentials, it increases the amount of neurotransmitter that is released. **Inhibitory transmission**, or just **inhibition**, is the converse—transmission that (usually) causes hyperpolarization of the postsynaptic neuron. This hyperpolarization decreases the probability that the postsynaptic neuron will fire an action potential or, in neurons that do not generate action potentials, decreases the amount of neurotransmitter that is released.

A third, relatively new way of classifying transmission is according to whether its primary effect is to rapidly pass on a message

BOX 6-1

Seeing Is Believing: Electron Microscopy and Neurobiology

One of the great challenges of studying the nervous system is to make visible its component parts and their physical relationships to one another. The Golgi stain and other methods for marking brain nuclei and nerve tracts (see Box 2-1) made it possible to study nervous system morphology at the cellular level. It was not until the advent of the electron microscope, however, that questions about the subcellular organization of neural tissue could be answered.

A prime example of the power of electron microscopy (EM) is the final resolution of the controversy between proponents of the reticular theory and the neuron theory. Although the extraordinary work of Ramón y Cajal in the early decades of the twentieth century had convinced most physiologists that there was a discontinuity between neurons at synapses, even into the 1940s some scientists still thought that neurons in the brain might be syncytial. Early electron micrographs did not resolve the issue, because methods for cutting thin sections of tissue (1 μm or less) and for proper staining were yet not available. Very thin sections of tissue are necessary in order to allow electrons to pass through the specimen to form an image. Proper stains are also essential in order to increase the contrast between different parts of the specimen. Not until the 1950s had techniques improved enough for researchers to obtain clear electron micrographs of synapses that showed the membrane between cells, finally putting the controversy to rest (Figure 6-A).

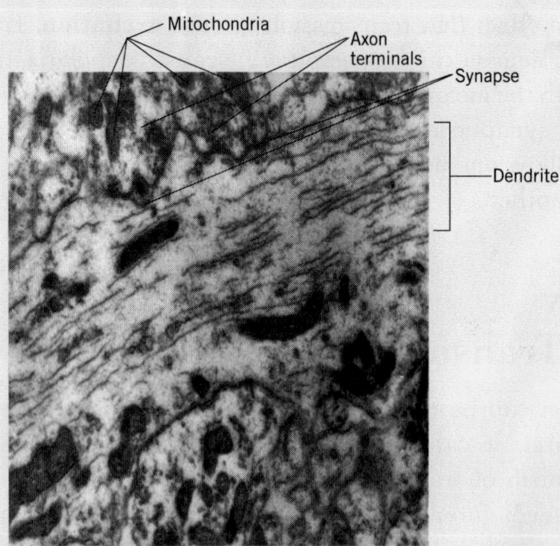

FIGURE 6-A. The earliest transmission electron micrograph (published 1956) that shows an unambiguous separation of the membranes of presynaptic and postsynaptic neurons.

Transmission electron microscopy is the type of EM most commonly used in neurobiological research (Figure 6-B). The specimen to be examined is placed in a vacuum chamber, where a beam of electrons can bombard the specimen without being deflected by atoms of the atmospheric gases. The beam is focused by magnetic lenses. The electrons are scattered to a greater or lesser extent by different parts of the specimen, forming an electron image in the same way an optical image is formed when light passes through a photographic slide. The electrons then strike a phosphor-coated screen much like the face of an oscilloscope, to make the image visible to the user; a photograph of the image can be taken by placing an unexposed negative in the path of the electrons. The dark portions of the image represent parts of the specimen that scattered electrons, whereas the lighter parts are those exposed to the electrons that pass through the specimen.

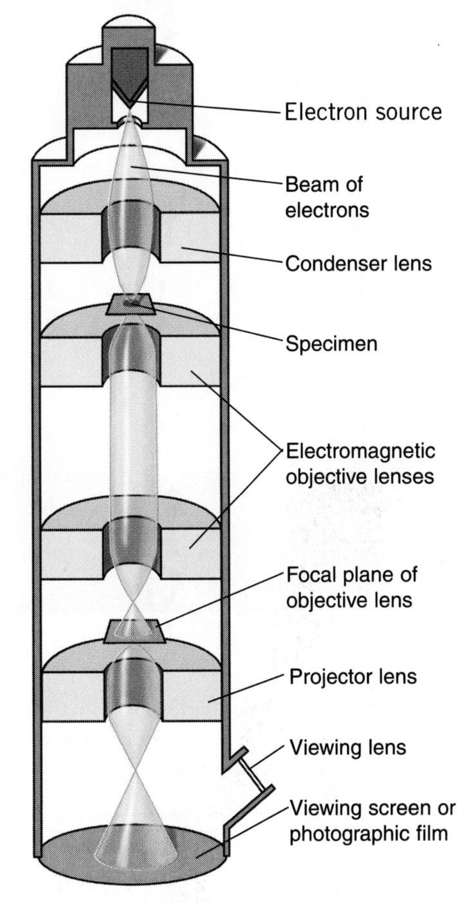

FIGURE 6-B. Principle of operation of a transmission electron microscope. A beam of electrons is emitted at the top and focused by magnetic lenses in the body of the microscope. The electrons pass through the specimen, then onto a screen or photographic plate for viewing or photography.

(continued)

Seeing Is Believing: Electron Microscopy and Neurobiology *(continued)*

The other main kind of EM is scanning electron microscopy. In this method, a specimen is first coated by an extremely fine film of heavy metal atoms. The beam of electrons is then scanned over the surface of the specimen. The electrons of the beam knock other electrons off the metal atoms. These other electrons are counted by a detector, which sends its information to an amplifier that generates an image of the surface of the specimen (Figure 6-C).

Today, EM continues to play a central role in much neuroscience research. By allowing researchers to see the structure of neural elements, EM can validate ideas about neural function and can stimulate further hypotheses about how neural structure relates to function.

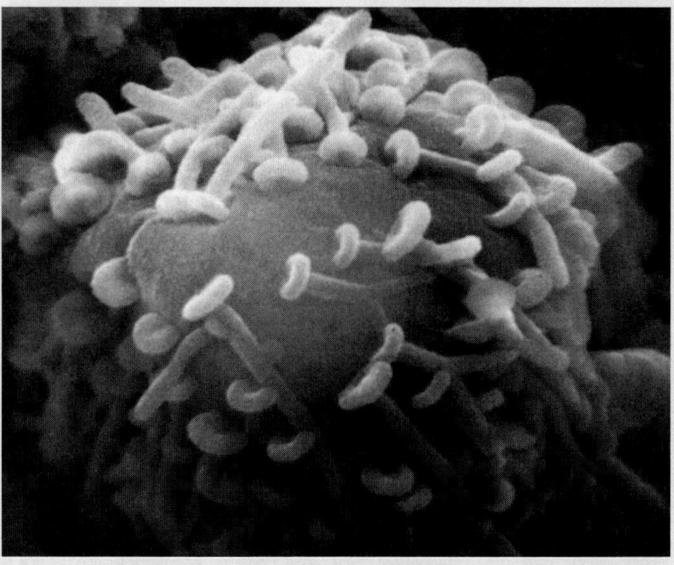

FIGURE 6-C. Scanning electron micrograph of axons forming synapses on another neuron. Scanning EMs give researchers a much better understanding of three-dimensional structural relationships than do transmission EMs.

or to modulate the characteristics of the postsynaptic neuron so that it responds differently to subsequent input than it otherwise would. Synaptic transmission in which a message is passed on is typically referred to as classical transmission, whereas transmission in which a neuron's response to other input is altered is called **neuromodulatory transmission**, or **neuromodulation**. Classical transmission is mediated by ionotropic receptors, which mediate a rapid and strong electrical response in the postsynaptic neuron.

Neuromodulation is mediated by metabotropic receptors, which initiate a biochemical cascade within the postsynaptic neuron (see Chapter 2). Although neuromodulation may have postsynaptic effects that provide some immediate excitation or inhibition to the postsynaptic neuron, the emphasis is usually on the long-term effects of this kind of transmission on the neuron's metabolic activity or on the way in which the neuron will respond to other synaptic inputs. The kinds of synaptic transmission are summarized in Table 6-1.

Synaptic transmission may involve more than one mechanism, or have different kinds of effects on a postsynaptic neuron. The mechanism of transmission may be electrical or chemical. The effects of a chemical neurotransmitter may be classified as excitatory or inhibitory, or classical or neuromodulatory. Classical synaptic transmission has a rapid, strong electrical effect on a postsynaptic neuron, whereas neuromodulatory transmission is slower and changes a postsynaptic neuron's response to other signals.

Electrical Transmission

Electrical transmission occurs only at electrical synapses because at these sites neurons are close enough together for the currents generated by an action potential to depolarize an adjacent neuron to a level above threshold. Current will always follow the path of least electrical resistance. Even though neurons at chemical synapses may typically be only 20 to 40 nm apart (10^9 nanometers [nm] = 1 mm), ions carrying a current diffuse more easily through the low-resistance extracellular fluid than they do across the high-resistance membrane into an adjacent cell (Figure 6-1A). Current flow at electrical synapses occurs through gap junctions, ion channels that provide an opening from one neuron to another, as described in Chapter 2 and shown in Figure 2-11. The channels allow ions to flow readily from one neuron to the next (Figure 6-1B). They are typically clustered in groups containing several hundred channels each, with one group forming a single synapse (Figure 6-2). Hence, the gap junctions provide a low-resistance pathway for the flow of ions that carry current from the presynaptic to the postsynaptic neuron.

Electrical synapses have two features that make them useful in certain circumstances: high-speed transmission of action potentials and faithful transmission of subthreshold

TABLE 6-1. The Main Types of Synaptic Transmission

Type	Description
Electrical transmission	Fast; no chemical intermediary; no receptors required on postsynaptic neuron
Chemical transmission	Slight delay (0.5-1.0 msec); neurotransmitter used as intermediary; neurotransmitter binds with receptors on postsynaptic neuron
Classical	Mediated by ionotropic receptors; main effect is fast and on the membrane potential of the postsynaptic neuron
Excitatory	Promotes depolarization and generation of an action potential in the postsynaptic neuron
Inhibitory	Usually promotes hyperpolarization; suppresses generation of an action potential in the postsynaptic neuron
Neuromodulatory	Mediated by metabotropic receptors; main effect is relatively slow and on the metabolic activity or electrical properties of the postsynaptic neuron

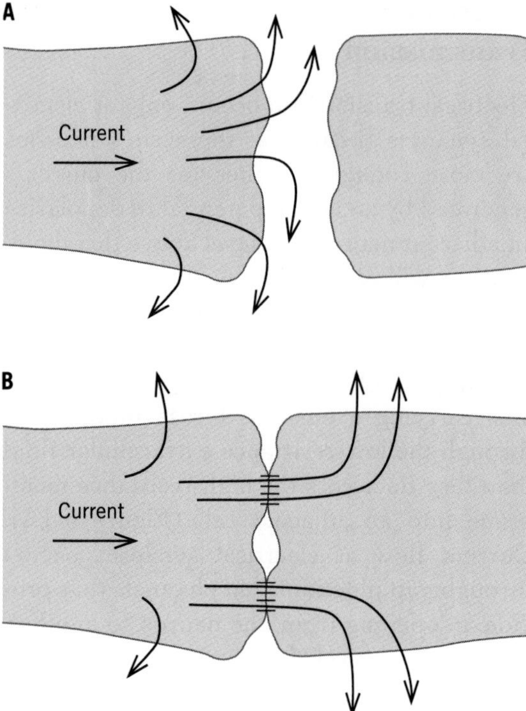

FIGURE 6-1. Current flow at the junction between two neurons. (A) At a typical chemical synapse, the current generated by the action potential flows out between the two neurons because that is the path of least electrical resistance. Hence, there is no appreciable direct electrical effect of these currents on the postsynaptic neuron. (B) At an electrical synapse, current flows from the presynaptic into the postsynaptic neuron through low-resistance gap junctions between them, and the postsynaptic neuron is stimulated.

changes in membrane potential. The first feature is useful in fast-conducting circuits to transfer an action potential from one neuron to another. For example, electrical synapses are often found between neurons that help trigger an escape action, as when an earthworm snaps back into its burrow when it is touched. The rapid transfer of information via electrical synapses shaves about half a millisecond from the transmission time compared with chemical transmission, and hence ensures the most rapid response possible.

The second feature is useful in transferring small changes in membrane potential from one neuron to another within a group of neurons. In such cases, electrical transmission ensures that all the cells act in synchrony. Groups of motor neurons innervating a single muscle in

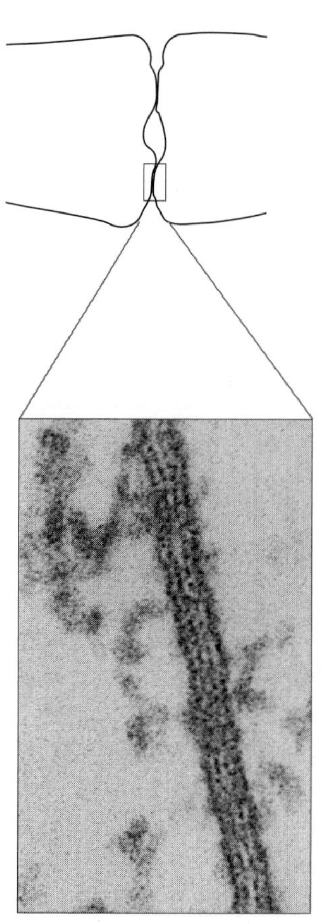

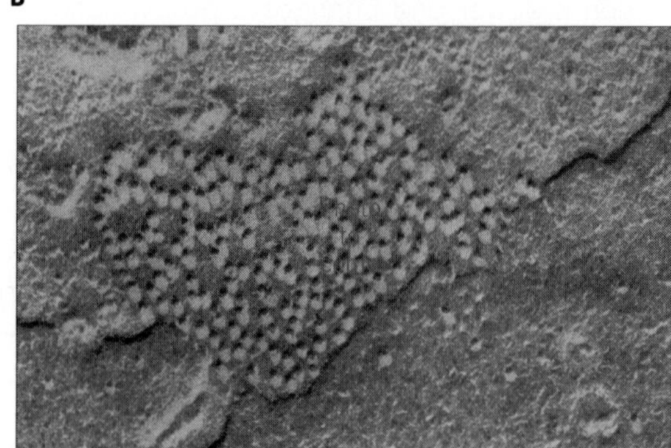

FIGURE 6-2. An electrical synapse in an earthworm. (A) Electron micrographic side view of an electrical synapse showing the close apposition of membranes from the communicating cells. The synapse is the area between the two arrows. (B) A face view of the synapse. To prepare this view, the communicating neurons were split apart, and an electron micrograph of one face of the synapse was made, showing the cluster of gap junctions that constitute the synapse. (Courtesy of Peter Brink.)

invertebrates are often coupled by electrical synapses, ensuring that the motor neurons are excited to fire at about the same time.

Electrical synaptic transmission is the direct transfer of current (ions) from one neuron to another. It requires a low-resistance pathway, formed by gap junctions, between the pre- and postsynaptic neurons. Electrical transmission ensures fast and faithful transmission of an electrical signal from one neuron to another.

Chemical Transmission

One of the very features that makes electrical synapses useful in special circumstances—faithfulness of transmission of an electrical signal—limits their usefulness in nervous systems. In most circumstances, neurons must be able to respond flexibly to synaptic input. For example, it may be necessary for a particular input to produce one response at one time and a different response at another time. Furthermore, sometimes an input must suppress (inhibit) a response altogether or interact with other inputs to produce novel responses.

This flexibility of response is possible in chemical transmission because there is a chemical intermediary, the neurotransmitter, between the presynaptic electrical event and the postsynaptic response. The nature of a postsynaptic neuron's response to a neurotransmitter usually depends on the nature and properties of the receptor protein with which the neurotransmitter binds; by using different receptors or receptors with different properties, a neuron can vary the way it will respond to the neurotransmitter. Furthermore, the modulation of a postsynaptic response on the basis of "experience" enables the nervous system to change its responses to a particular input over time, the basis of learning and memory. Hence, the flexibility of chemical synaptic transmission lies at the very heart of neural function.

The entire process of chemical transmission typically consists of five steps (Figure 6-3). Steps 1 and 2 are the synthesis and storage of the neurotransmitter in vesicles in the vicinity of the synapse. When the presynaptic cell is depolarized, the neurotransmitter is released (step 3) and diffuses across the synaptic cleft to bind with receptors on the postsynaptic membrane. This binding causes a reaction in the postsynaptic cell (step 4). Finally, the neurotransmitter is functionally removed from the synapse, rendering it inoperative and ensuring that it will have only a relatively narrow window of effectiveness (step 5). Because the features of synaptic transmission are easier to appreciate once the responses of postsynaptic neurons to an input are understood, steps 4 and 5 are described in the rest of this chapter. Steps 1 to 3 are described in Chapter 7.

Chemical synaptic transmission allows a postsynaptic response to an input to be modified or modulated, giving the receiving neuron flexibility in its response. Chemical transmission works through the release of a neurotransmitter by one neuron, the diffusion of this intermediary across the minute gap between the communicating cells, and its subsequent effect on the activity of the target cell.

Postsynaptic Channels and the Binding of Neurotransmitter

The effect of synaptic transmission is a change in the activity or responsiveness of a postsynaptic neuron or muscle cell. The process by which a neurotransmitter brings about such a change generally begins with the binding of the transmitter with protein receptor molecules located in the postsynaptic membrane. Each receptor is specific for a particular type of neurotransmitter; that is,

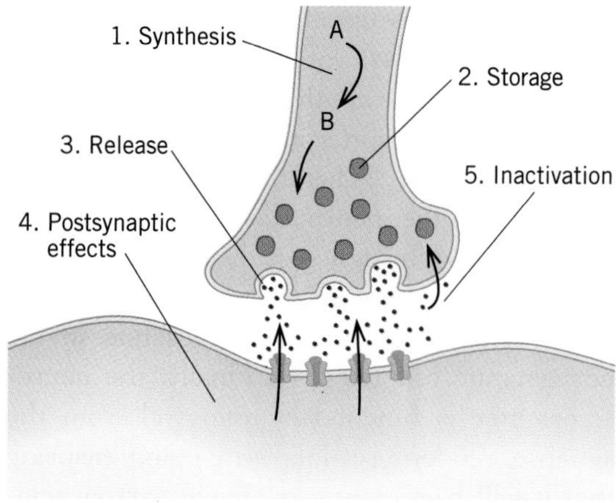

FIGURE 6-3. The five steps in chemical synaptic transmission. For some neurotransmitters, synthesis takes place in the cell body, the neurotransmitter is stored in vesicles, and the filled vesicles are transported down the axon. Inactivation can occur by several mechanisms.

the three-dimensional structure of the receptor allows only one type of molecule, or one of its close chemical analogs, to bind with the receptor.

Neurotransmitters that mediate classical transmission cause rapid and often large electrical responses in a postsynaptic neuron. They do so by binding with specific types of receptor proteins that form ion channels in the membrane of the postsynaptic cell, causing the receptor to undergo a conformational change that opens its central pore so ions can flow through. In the absence of the binding chemical, the structure of the protein is shifted so the central pore is constricted (closed). Because the channels are gated, that is, they open or close according to the presence or absence of a specific chemical substance, they are called ligand-gated. Because they respond to the presence of the chemical, they are also referred to as ligand-sensitive (see Chapter 2).

The mechanism by which ligand-sensitive channels operate has been the subject of considerable research. The channel whose structure and function is best understood is the **nicotinic acetylcholine receptor (nAChR)**, present in various forms at the vertebrate neuromuscular junction, at electric eel electroplaques (the structures that generate the electrical discharges of the fish), at synapses in vertebrate peripheral nervous systems, and at synapses in both vertebrate and invertebrate central nervous systems. Each channel protein consists of five subunits, one of which appears twice (Figure 6-4). The subunits are identified by Greek letters. The nAChRs of eel electroplaque and denervated or developing mammalian muscle consists of subunits α, β, γ, and δ; the α subunit is repeated. A different subunit, ϵ, substitutes for the γ subunit in adult mammalian muscle.

As the name of the receptor suggests, the neurotransmitter with which nAChR binds is acetylcholine (ACh). ACh binds with the nAChR molecule in two places, one molecule each at sites between the α and δ subunits and the α and γ subunits (arrows, Figure 6-4). This binding has two effects on the receptor. First, it causes an immediate conformational change in the protein; this change transiently makes the channel wider, allowing Na^+ and K^+ (and some Ca^{2+} as well) to pass through the central pore. The flow of ions results in a net depolarization of the cell membrane, considered in more detail in the next section. Continued presence of ACh has a second effect,

A

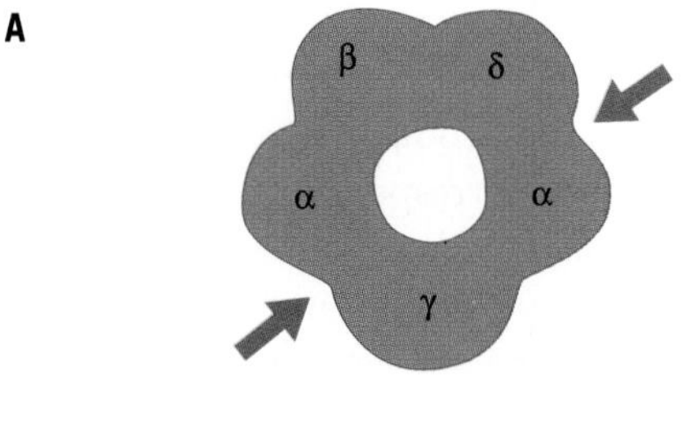

B

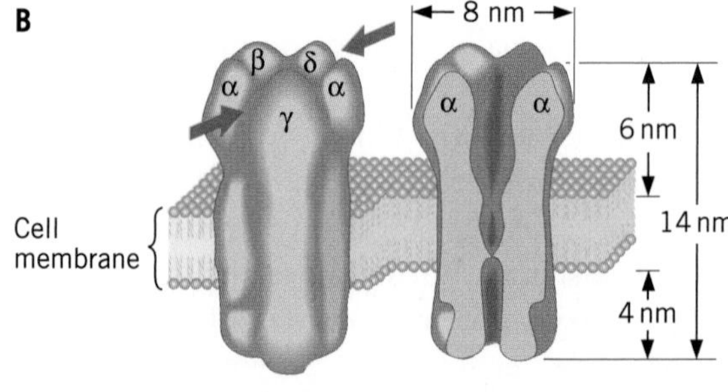

FIGURE 6-4. Diagrammatic (A) top and (B) side views of a nicotinic acetylcholine receptor molecule. The sites at which acetylcholine is thought to bind (two molecules per receptor) are indicated by arrows.

called **desensitization**, which is a slow, further structural change that renders the receptor insensitive to ACh. Recovery from desensitization requires a period of time without the neurotransmitter being present. Many other types of receptors also show desensitization.

The genes responsible for synthesis of the nAChR molecule have been identified in some animals, making it possible for researchers to clone the subunits and determine their amino acid sequences. Considering all the subunits together, the sequence of amino acids in the receptor is 80% conserved between humans and electric fish, suggesting that nearly all of the structure of the molecule is critical for its proper function. It has been found that the genes that code for different subunits may occur on separate chromosomes. In mice, for example, the genes are distributed on three chromosomes. The evolution of the molecular structure of the receptor and the mechanism of assembly of the final structure from the individual subunits are still not known.

There are two important points to recognize about ligand-sensitive channels such as nAChR. First, the nature of the primary postsynaptic response depends entirely on the types of ions that are allowed to flow through the membrane. Since this ion flow depends on the type of receptor-channel complex with which the neurotransmitter interacts and not on the chemical identity of the neurotransmitter, the nature of the postsynaptic response will depend on the receptor. This is illustrated nicely by the fact that the same chemical neurotransmitter can inhibit some neurons and excite others, as will be discussed in a subsequent section of this chapter.

Second, the distribution of ligand-sensitive channels is highly concentrated in the postsynaptic membrane of the postsynaptic cell. For example, nAChR receptors have been estimated to be present at a density of about 20,000/μm^2 on muscle at the synapse, just under the terminus of the motor neuron. Since the receptors are estimated to be about 8 nm in diameter, they must be packed together as tightly as possible, perhaps even being compressed a bit, at the synapse. Away from the synapse, these channels are found at a density of only 20/μm^2 or less. If you scaled 1 μm^2 up to the size of a football field, as suggested in Chapter 5 as a method to help you visualize proportions, there would be about 16.3 yd (14.9 m) between the centers of adjacent nAChR channels at this density. In consequence, a significant electrical response to the neurotransmitter in the postsynaptic cell occurs only at the synapse.

A neurotransmitter initiates a postsynaptic response by binding to a receptor on the postsynaptic neuron. In the case of the acetylcholine receptor, the neurotransmitter ACh binds to the receptor molecule in two places, causing a conformational change that opens the receptor's central channel. The type of response that a neurotransmitter elicits in a postsynaptic neuron depends on the ion selectivity of the channel it binds to, not on the chemical nature of the neurotransmitter.

Excitatory Transmission

The main function of a classical excitatory synapse is to pass on a message from one neuron to another neuron or a muscle. Hence, the typical response of a postsynaptic cell to input at such a synapse is a fast depolarization of the cell membrane that may initiate an action potential.

The Ionic Mechanism of Excitation

The depolarization that is generated by an active excitatory synapse is caused by a flow of ions through ligand-gated channels in the postsynaptic membrane. This depolarization, referred to as an **excitatory postsynaptic potential (epsp)**, can vary in amplitude from a few to more than 40 mV, depending on the neuron involved and its recent history (Figure 6-5A). For historical reasons, the epsp in muscle is referred to as an **end plate potential (epp)**. The name arose because in vertebrate muscle the ends of the terminals of the individual branches of each motor nerve tend to resemble small dinner plates, and hence are called end plates.

How is an epsp or epp generated? One way to answer this question is to eliminate selected ions from the saline that is bathing a synapse from which recordings are being taken, and then analyze the epsp that results when the presynaptic neuron is stimulated. This kind of experiment reveals that at the neuromuscular junction, both K^+ and Na^+ are involved in the generation of a normal postsynaptic response. Because manipulation of ion concentration is rarely possible at synapses in the central nervous system, other experimental approaches must be used there. One approach that has been used with great success is the determination of the **reversal potential** of the epsp—the membrane potential at which a postsynaptic potential is reduced to zero. It represents the balance point for the forces driving the ions that move through the open channels at a synapse and can be used to analyze the ionic basis of the epsp.

To understand how knowing the value of the reversal potential can help an experimenter to analyze the ionic basis of an epsp, consider first what forces act on ions such as Na^+ and K^+ at an active synapse. At synapses where ACh is the neurotransmitter, for example, the presence of ACh causes nicotinic ACh receptor channels to open, allowing Na^+ and K^+ to cross the membrane simultaneously. Each ion will move according to the combined effects of the membrane potential and its own concentration gradient; Na^+ will flow into the neuron, and K^+ will flow out (Figure 6-5B). The equilibrium potential for Na^+, E_{Na} (+54 mV for the squid giant axon, which is fairly typical), will always be considerably farther from the resting membrane potential (-60 mV) than is the equilibrium potential for K^+, E_K (-75 mV). Therefore, there is a considerably greater net force on Na^+ than on K^+ when the nAChR channels first open, and initially there will be a substantially greater influx of Na^+ than an efflux of K^+. The net influx of positive charge that results depolarizes the membrane. As the membrane depolarizes,

A

Membrane potential (mV): +20, 0, -20, -40, -60

Time (msec): 0, 5, 10, 15, 20

B

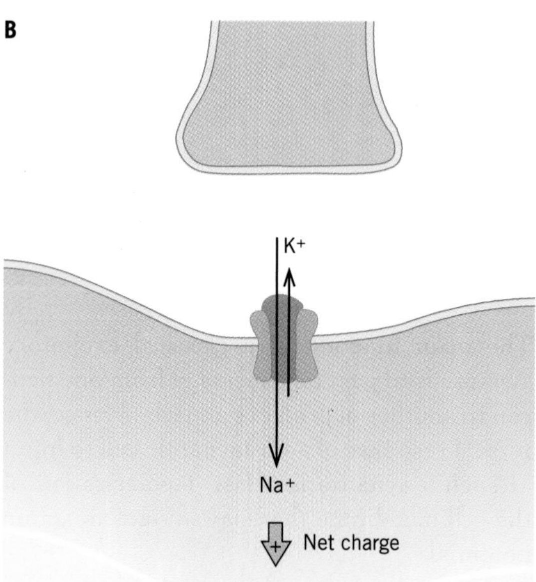

FIGURE 6-5. (A) A recording of an excitatory postsynaptic potential (epsp) in a typical vertebrate neuron in which action potentials have been suppresed. Compare the smaller amplitude and longer duration with the size and duration of the action potential shown in Figure 5-1. (B) Ion movements at a typical excitatory synapse. The channels that open in response to the presence of the neurotransmitter allow both Na^+ and K^+ to pass relatively freely through the membrane. There is a net inward flow of current (positive charge) because more Na+ moves in than K^+ moves out, due to the stronger tendency for the Na^+ to cross the membrane.

however, the membrane potential moves toward E_{Na} and away from E_K, decreasing the driving force on sodium ions and increasing the force on potassium ions. As the membrane continues to depolarize, the forces soon become equal.

How does the reversal potential come into play? Imagine an experiment in which you insert an electrode into a postsynaptic cell and pass current to depolarize it to +31 mV. Suppose also that you have suppressed the generation of action potentials. This is necessary since action potentials are large and all-or-none, and hence their presence makes it impossible to measure the amplitude of the epsps. You can suppress them by adding controlled doses of the neurotoxin **tetrodotoxin (TTX)**, a poison derived from the puffer fish (see Chapter 1). TTX molecules bind with voltage-sensitive sodium channels. When they do so, they block the passage of Na^+ through the channels even when the channels are open. When a small amount of TTX is applied to an axon, the size of the action potentials in the axon is reduced. Adding more TTX causes further reduction in the size of the action potential until, at a sufficiently high concentration, no action potentials will be generated because too many sodium channels are blocked.

If you now stimulate the presynaptic neuron, what happens when channels to Na+ and K^+ open? Sodium ions would still enter the cell, and potassium ions would still leave it, because +31 mV is not as positive as E_{Na} nor as negative as E_K. However, +31 mV is much farther from E_K than it is from E_{Na}, so the greater driving force would now be on K^+ rather than Na^+. As a consequence, more K^+ would leave the neuron than Na^+ would enter, and there would be a net loss of positive charge. Under these circumstances, a recording of the epsp, which is the membrane potential at the synapse when the synapse is active, would show a negative deflection rather than a positive one (Figure 6-6). Since the epsp is positive (depolarizing) if it is evoked when the membrane potential is at the resting level of –60 mV or so, and negative (hyperpolarizing) at +31 mV, there must be some potential between these extremes at which the epsp is zero. This is the reversal potential, which can be determined experimentally by activating the synapse with the membrane potential of the postsynaptic neuron set at different levels, as shown in Figure 6-6.

Knowing the reversal potential of the epsp can help a researcher determine which ions underlie a particular synaptic potential. If the conductance and equilibrium potential for the ion or ions that carry current across an active synapse are known, then the reversal potential for the synaptic potential can be predicted. For example, if the conductances for sodium and potassium, g_{Na} and g_K, are equal at an excitatory synapse, then the reversal potential for the epsp at that synapse will be the average of the equilibrium potentials for the two ions. Using the figures calculated earlier for the squid giant axon, E_{Na} = +54 mV and E_K = –75 mV, the reversal potential would be –10.5 mV. If g_{Na} and g_K are not equal, the reversal potential will be shifted toward the equilibrium potential of the ion

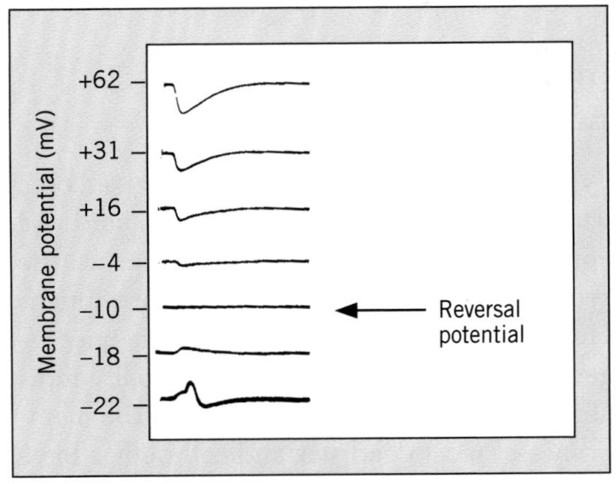

FIGURE 6-6. Experimental determination of the reversal potential at an excitatory synapse. At membrane potentials more negative than the reversal potential (three bottom traces), the epsp is positive because there is more Na+ entering the neuron than there is K^+ leaving. At membrane potentials more positive than the reversal potential (four upper traces), the epsp is negative and hence inverted because there is less Na^+ entering the neuron than there is K^+ leaving. At the reversal potential, the movements of Na^+ and K^+ are equal, leaving the membrane potential unchanged.

with the higherconductance. At the frog neuromuscular junction, E_{Na} = +55 mV and E_K = −100 mV, but the reversal potential, rather than being −22.5 mV, is actually 0 mV because conductance of the membrane to sodium is slightly greater than the conductance of the membrane to potassium through the open nAChr channel. By demonstrating that the observed reversal potential at a synapse is closely matched by the reversal potential calculated by assuming the contribution of particular ion(s), an investigator can strongly support a suggestion that the ion(s) are involved in generation of the epsp.

The excitatory postsynaptic potential (epsp) is the membrane depolarization in a postsynaptic cell excited by an excitatory synapse. It is typically caused by the opening of channels that allow both Na^+ and K^+ to pass. The ions that cause epsps can be identified with the help of the reversal potential of the epsp.

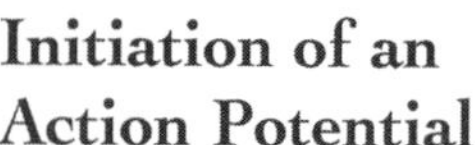

Initiation of an Action Potential

Typically, excitatory synapses are located on a dendrite or cell body of a neuron, and the epsps that are generated develop at the synapses themselves, as described previously. However, any action potentials that may be generated by these epsps are *not* usually initiated at the synapse. This is because the membranes of most dendrites and cell bodies have few, if any, voltage-sensitive ion channels. As described in Chapter 5, it is the presence of voltage-sensitive ion channels in a membrane that allows the explosive depolarization necessary for the generation of an action potential. The absence of such channels in dendritic or soma membrane means that even a substantial depolarization at the synapse will be unable to initiate an action potential there.

The key to understanding where and how spike initiation occurs lies in the behavior of a concentration of positive charge such as represented by the epsp. You have seen how the epsp is the result of a net influx of positively charged ions at the site of the synapse. These positive ions do not stay in one location but instead spread in all directions away from the point of origin, just like water poured onto a tabletop will spread out over the table surface. In a vertebrate motor neuron, the first site at which this spreading current will encounter a substantial number of voltage-sensitive sodium channels is the place where the axon originates, the axon hillock (Figure 6-7). If the depolarization at this point exceeds the cell's threshold, an action potential will be generated. For this reason, the region is called the **spike initiation zone**.

Spike initiation is complicated by the fact that epsps decline in size as they spread along the cell membrane away from the synapse at which they are initiated. The consequences of this phenomenon and interactions between postsynaptic potentials are considered in depth in Chapter 8. For now, the important point is that action potentials are typically generated at the junction of the axon and the cell body, not at the synapse.

Action potentials initiated by active excitatory synapses are not generated at the synapses because the membranes in the region of synapses usually lack sufficient numbers of voltage-sensitive ion channels to allow an action potential to be generated. Instead, action potentials are initiated at the axon hillock, at the region known as the spike initiation zone.

Inhibitory Transmission

Not all synaptic transmission is excitatory. Some is inhibitory, having a hyperpolarizing effect on the postsynaptic neuron and thereby

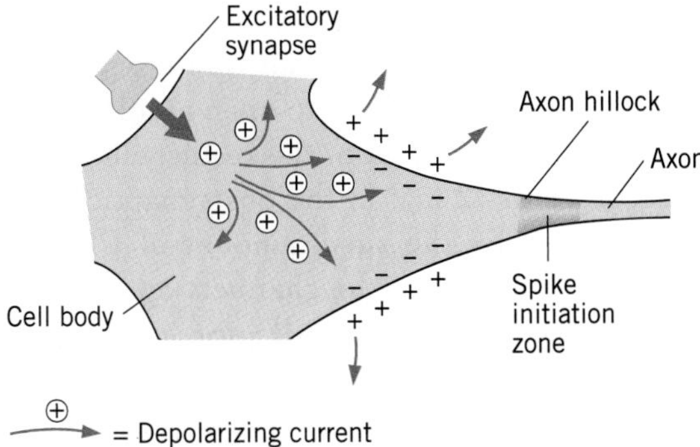

FIGURE 6-7. The flow of current due to the influx of positive charges in a neuron stimulated at an excitatory synapse. In order for an action potential to be generated, there must be sufficient depolarization of the membrane at the spike initiation zone to activate the voltage-sensitive sodium channels there.

reducing the probability that the postsynaptic neuron will fire an action potential. In neurons that cannot generate action potentials, inhibition decreases the amount of neurotransmitter that is released by the neuron. Inhibitory actions normally interact with excitatory ones and do not occur alone. In addition, there are two types of inhibition—postsynaptic and presynaptic—that act by quite different mechanisms.

Postsynaptic Inhibition

Postsynaptic inhibition is characterized by direct synaptic contact between the inhibitory neuron and the neuron whose action is inhibited. It is what is ordinarily meant by the unqualified term inhibition. The typical response of a postsynaptic neuron to inhibitory synaptic input is a hyperpolarization of the cell's membrane potential, making the membrane potential more negative. The hyperpolarization that results from inhibitory synaptic input is called the **inhibitory postsynaptic potential (ipsp)** (Figure 6-8A).

The ipsp is caused by an efflux of positively charged potassium ions or an influx of negatively charged chloride ions, depending on the ion selectivity of the channels activated by the inhibitory neurotransmitter. In either case, the effect is a transient increase of net negative charge inside the cell (Figure 6-8B). One way researchers can determine which of these ionic mechanisms is at work is to determine the reversal potential of the ipsp. Potassium and chloride typically have different equilibrium potentials. Therefore, the reversal potential at synapses that allow K^+ to flow will be different from that at synapses that allow Cl^- to flow.

In neurons capable of generating action potentials, the function of inhibition is to reduce the probability that the postsynaptic cell will generate them. To see how this effect is brought about, consider an excitatory and inhibitory synapse on a typical vertebrate neuron (Figure 6-9). If the inhibitory neuron fires at about the same time as the excitatory one, the channels at the excitatory synapse

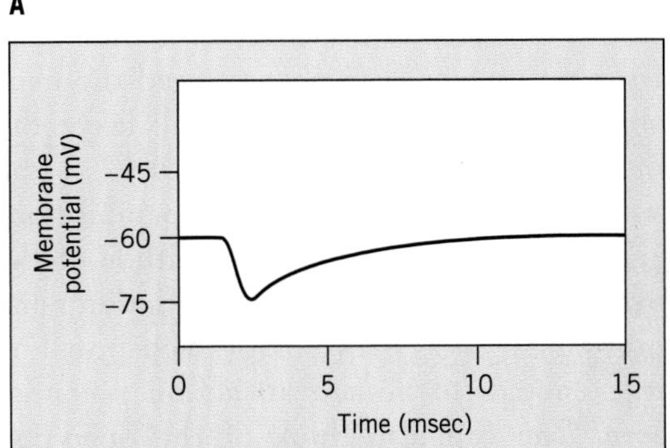

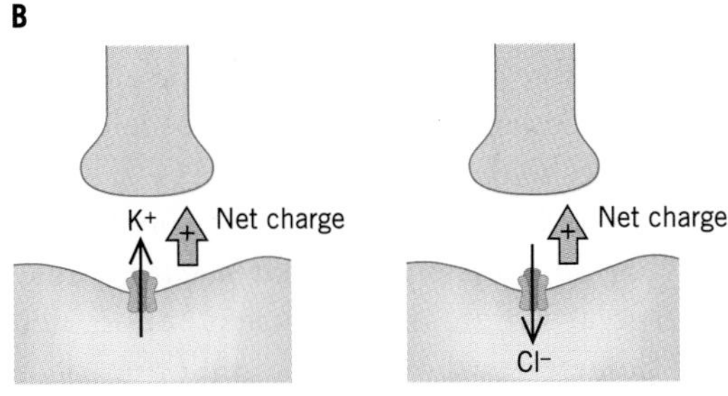

FIGURE 6-8. (A) A typical inhibitory postsynaptic potential (ipsp) in a postsynaptic cell. Note that the ipsp is hyperpolarizing and much smaller than the epsp. Inhibition of the postsynaptic cell results from net reduction of positive charge inside the neuron. (B) This reduction may be brought about by an efflux of (positive) K^+ or an influx of (negative) Cl^-. An influx of negative charge has the same hyperpolarizing effect as does the efflux of positive charge.

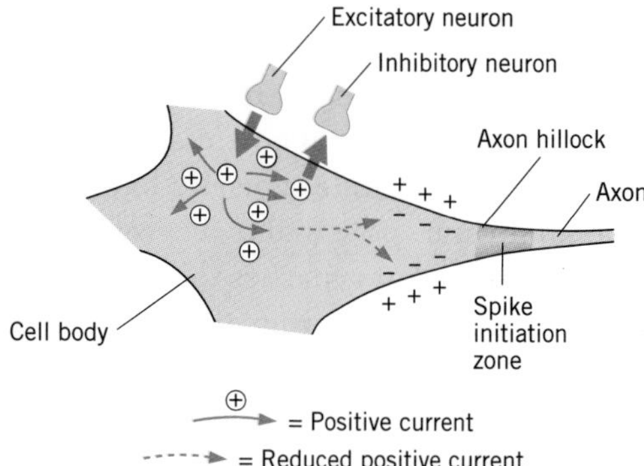

FIGURE 6-9. The flow of current in a neuron when both an excitatory and an inhibitory synapse are active. The loss or neutralization of positive charges at the inhibitory synapse reduces the probability that the epsp will be able to initiate an action potential at the spike initiation zone.

will open to allow a net increase of positive charge at the site of the excitatory synapse at the same time that the channels at the inhibitory synapse open to allow a net loss of positive charge there. The result is that some of the positive charges that entered the neuron at the excitatory synapse will leave the neuron again (or be neutralized, if Cl^- is the ion involved in inhibition) at the inhibitory synapse, thus reducing the magnitude of the current that reaches the spike initiation zone and reducing the chance that the epsp will be large enough to initiate an action potential there. You can also think of the open ion channels at the inhibitory synapse as an electrical short circuit, effectively draining some of the positive charge away from the epsp and possibly reducing its amplitude to below the firing threshold of the neuron.

Inhibitory transmission has been defined as synaptic transmission that *usually* causes hyperpolarization of the postsynaptic neuron. Sometimes, however, stimulation of an inhibitory neuron produces a slight depolarization of the postsynaptic cell. Such an ipsp can nevertheless produce inhibition as long as the reversal potential of the ipsp is below the threshold for action potentials for the neuron.

Consider an example. Suppose the equilibrium potential for Cl^- in a neuron is -62 mV, the resting potential is -65 mV, and the threshold for elicitation of an action potential is -40 mV. If stimulating an inhibitory axon causes chloride channels to open at inhibitory synapses in this neuron, the result will be an ipsp of +1 to +2 mV in the postsynaptic neuron, caused by the small amount of Cl^- that leaves the cell when the channels open. (The membrane potential will not reach E_{Cl} because the chloride channels are not open long enough.) This small depolarizing potential cannot excite the cell because it is far below threshold. If an excitatory synapse is active at the same time as the inhibitory one, the membrane will be much more strongly depolarized. However, as soon as the membrane depolarizes to a value more positive than -62 mV, the reversal potential for the inhibitory synapse, the cell will begin to gain rather than lose Cl^-. This gain of negative charge will reduce the amplitude of the epsp, and hence will reduce the probability that the neuron will fire an action potential.

The inhibitory neuron shown in Figure 6-9 is positioned between the excitatory fiber and the spike initiation zone of the postsynaptic neuron. This is a common arrangement in neurons. However, inhibitory synapses situated at the end of a dendrite can also be effective, since currents spread throughout the cell, and a drain of positive charge or an input of negative charge anywhere will affect the size of the epsp at the initiation zone.

The response of a postsynaptic neuron to direct inhibitory input from another neuron is an inhibitory postsynaptic potential, the ipsp. Typically, ipsps are generated by the efflux of K^+ or the influx of Cl^- through open channels. An action potential is suppressed when the open channels drain sufficient positive charge away from the spike initiation zone to keep excitatory input from depolarizing this region to threshold. Even small depolarizing ipsps can inhibit a neuron as long as the reversal potential of the synapse is below the neuron's threshold.

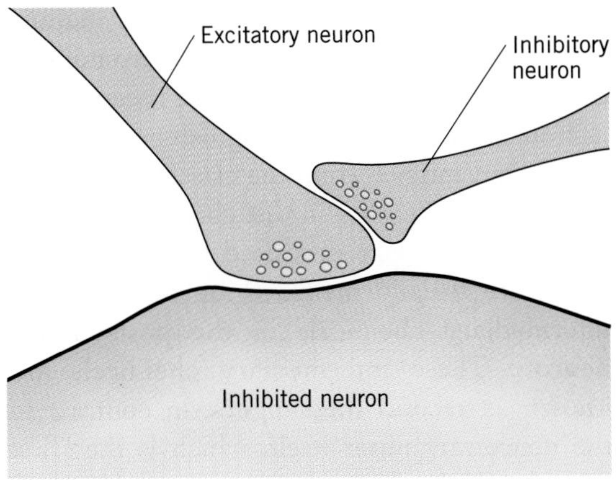

FIGURE 6-10. The physical arrangement of neurons involved in presynaptic inhibition. The excitatory neuron is postsynaptic to the inhibitory neuron but presynaptic to the neuron it excites.

Presynaptic Inhibition

The disadvantage of postsynaptic inhibition is that it momentarily reduces the ability of the postsynaptic cell to respond to any input that it may receive. It may sometimes be useful for an animal to be able to suppress a response to one particular input without suppressing responsiveness to others. For example, stimulation of sensitive hairs at the tail end of an insect normally elicits a reflex escape response. This reflex must be suppressed for mating to occur. However, the insect must remain responsive to other potentially lethal stimuli. Such selective inhibition can be carried out through the mechanism of presynaptic inhibition.

As the name implies, in **presynaptic inhibition** the inhibitory neuron synapses not on the postsynaptic cell that is being inhibited but on the terminal of the excitatory neuron whose input can be reduced, as illustrated in Figure 6-10. The key to presynaptic inhibition is the functional effect of this arrangement on the neuron being inhibited, not just on the excitatory neuron that receives the inhibitory input. Presynaptic inhibition reduces the probability of firing of the neuron whose activity is being regulated, designated the "inhibited neuron" in Figure 6-10. It does so by reducing the amount of neurotransmitter released from an excitatory terminal, thereby reducing the size of the epsp that it generates (see Chapter 7) and making it less likely that the epsp will be able to initiate an action potential, as shown in the figure. Functionally, presynaptic inhibition shuts off the excitatory effects of specific input to a neuron without altering the ability of the postsynaptic cell to respond to other inputs. Postsynaptic inhibition, in contrast, shuts off the output of the cell altogether, reducing the likelihood that any input will cause it to respond.

Presynaptic inhibition can work by any of several mechanisms, which differ not only from the postsynaptic mechanisms described in the previous section but from each other as well. Each mechanism reduces the number of calcium ions that enter the excitatory presynaptic terminal and thereby reduces the amount of neurotransmitter released by the excitatory terminal at which presynaptic inhibition is being applied. Reducing Ca^{2+} availability reduces neurotransmitter release because, as you will learn in Chapter 7, the amount of neurotransmitter released is determined by the level of free Ca^{2+} in the terminal.

Two examples will illustrate presynaptic inhibition. At some synapses, the neurotransmitter released by the presynaptic inhibitory neuron lessens the voltage sensitivity of calcium channels in the excitatory terminal. This reduces the number of channels that open when the action potential reaches the end of the terminal, reduces the amount of calcium that enters the terminal, and hence reduces the amount of neurotransmitter released from the presynaptic terminal. At other synapses, the neurotransmitter increases the conductance of the presynaptic membrane to Cl^- by opening chloride channels. Short-circuiting the currents of the action potential causes a reduction in the height of the action potential, a reduction in the amount of Ca^{2+} entry, and a reduction in the amount of neurotransmitter released from the terminal. In both of these examples, the end result is a reduction in the amplitude of the postsynaptic epsp due to the reduction in the amount of neurotransmitter released from the presynaptic terminal.

The smaller epsp frequently will be too small to generate an action potential in the postsynaptic neuron.

Although postsynaptic and presynaptic inhibition operate by quite different mechanisms, they may be triggered by the same neurotransmitter. Gamma-aminobutyric acid (GABA) has been shown to be the neurotransmitter at many vertebrate presynaptic inhibitory terminals. GABA is also the neurotransmitter at vertebrate and invertebrate postsynaptic inhibitory terminals. The properties of the GABA receptors in the two cases are different, leading to entirely different effects when GABA binds with them and nicely illustrating the point that the effect of a particular neurotransmitter on a postsynaptic neuron is a function of the postsynaptic receptor-channel complex with which it binds, not of the chemical identity of the neurotransmitter itself.

Presynaptic inhibition reduces the probability that one or more specific excitatory neurons will produce an action potential in a postsynaptic neuron. It works by reducing the amount of neurotransmitter released by excitatory neurons, thereby reducing the size of the postsynaptic epsp.

Neuromodulatory Transmission

The phrase "chemical synaptic transmission" generally brings to mind the classical features that you have just read about: rapid opening and closing of postsynaptic ion channels and fast, short-term electrical effects on the membrane of the postsynaptic neuron. The function of this type of transmission is to convey specific information from one neuron to another. However, it is now clear that much, perhaps even most, synaptic transmission has a different function altogether: to regulate a neuron's metabolic activity or the way it responds to other input. As you read earlier in this chapter, this type of transmission, called neuromodulation, is mediated by metabotropic receptors.

Four characteristics distinguish neuromodulatory transmission from the classical transmission described earlier in this chapter. First, the postsynaptic effects produced by neuromodulation are usually mediated by one or more intermediary chemicals in the postsynaptic neuron. These intermediary chemicals are known as **second messengers**, in contrast to the neurotransmitter itself, which is the "first messenger." In classical transmission, the neurotransmitter directly causes the opening of ion channels, and no intermediary molecules are involved. Second, the effects of neuromodulatory transmission are usually long-lasting, from several hundred milliseconds to several hours. Classical postsynaptic effects typically last only 10 to 15 msec. Third, neuromodulation may generate secondary effects by initiating the synthesis of new proteins, effects that may last for many days. There are no such secondary effects in classical transmission. And fourth, the small postsynaptic epsp or ipsp that often accompanies neuromodulatory transmission is usually slow and quite weak, in contrast to the relatively fast, strong classical postsynaptic response. The typical characteristics of neuromodulatory and classical synaptic transmission are summarized and contrasted in Table 6-2.

The second messenger is the key to the response of the postsynaptic neuron because it is this chemical that initiates the cascade of biochemical reactions that direct the neuron's response. For this reason, considerable effort has been expended on the search for second messengers and the chemical events that they trigger. That effort has been amply rewarded, and several second messenger systems are now understood.

The term **neuromodulator** is usually applied to a neurotransmitter that acts via a second messenger system. However, the name does not define a unique set of substances because some neurotransmitters can act both as classical neurotransmitters and as neuromodulators. An example is ACh, which acts as a classical neurotransmitter when it interacts with nicotinic ACh receptors, producing a

TABLE 6-2. Features of Neuromodulation and Classical Transmission

Neuromodulation	Classical Transmission
Effects mediated by second messenger intermediary chemicals	Effects brought about by direct gating of ion channel
Direct postsynaptic effects last from several hundred milliseconds to several hours	Direct postsynaptic effects last for tens of milliseconds
Secondary effects may last days	No secondary effects
Postsynaptic electrical effects are slow and weak	Postsynaptic electrical effects are fast and strong

fast, strong epsp in a postsynaptic neuron or muscle by opening the nAChR channel to K^+ and Na^+. On the other hand, it acts indirectly in some sympathetic and central neurons when it binds with another type of receptor, the **muscarinic acetylcholine receptor**. Activation of the muscarinic ACh receptor inactivates potassium M channels (see Table 5-1) through a second messenger. The M channel ordinarily helps to limit the responsiveness of the postsynaptic neuron to excitatory input by promoting membrane repolarization after an action potential. Inactivation of the M channel by ACh significantly increases the responsiveness of the postsynaptic neuron to excitatory input, even though the ACh by itself generates only a small epsp. Hence, the main effect of ACh acting through muscarinic channels is neuromodulatory.

Neuromodulatory transmission regulates a neuron's metabolic activity or the way it responds to other input. It has slow, weak, and long-lasting direct postsynaptic effects that are mediated by chemical second messengers within the postsynaptic cell. Some neurotransmitters can act as classical neurotransmitters as well as neuromodulators, depending on the specific types of receptors with which they bind.

Mechanisms of Neuromodulation

The details of how neuromodulation works vary considerably from case to case, but there is a common pattern. Binding of a neuromodulator with its receptor causes activation of the membrane-bound proteins known as guanine nucleotide binding proteins, or G proteins (see Chapter 2), that are physically associated with the receptor. In their activated state, the G proteins will move away from the receptor and come into contact either with channel proteins or with membrane-bound enzymes. In the latter case, the enzymes will be stimulated to produce a substance that initiates a cascade of biochemical reactions, as described in the next section. The end result can be a change in the response characteristics of an ion channel, and hence a change in the response of the postsynaptic neuron to subsequent synaptic input.

G Proteins

G proteins are the critical elements that initiate the response of the postsynaptic neuron to the neuromodulatory neurotransmitter. These proteins have a trimeric structure, consisting of a single α subunit (roughly 40-50 kdaltons (kd) in size) and a dimer composed of β (35 kd) and γ (5-10 kd) subunits. The α subunit normally also has bound to it a nucleotide, guanosine diphosphate (GDP). When a G protein is activated by a receptor that is bound with an appropriate neurotransmitter, it releases GDP and picks up guanosine triphosphate (GTP), breaks apart, opens an ion channel or activates another membrane-bound enzyme, catalyzes GTP to GDP and a phosphate group, and then reassembles. This cycle is illustrated in Figure 6-11.

The key to the sequence is the change in affinity that the different subunits have for one another, and for GDP and GTP, in the presence of a neurotransmitter-activated receptor. The binding of a neurotransmitter

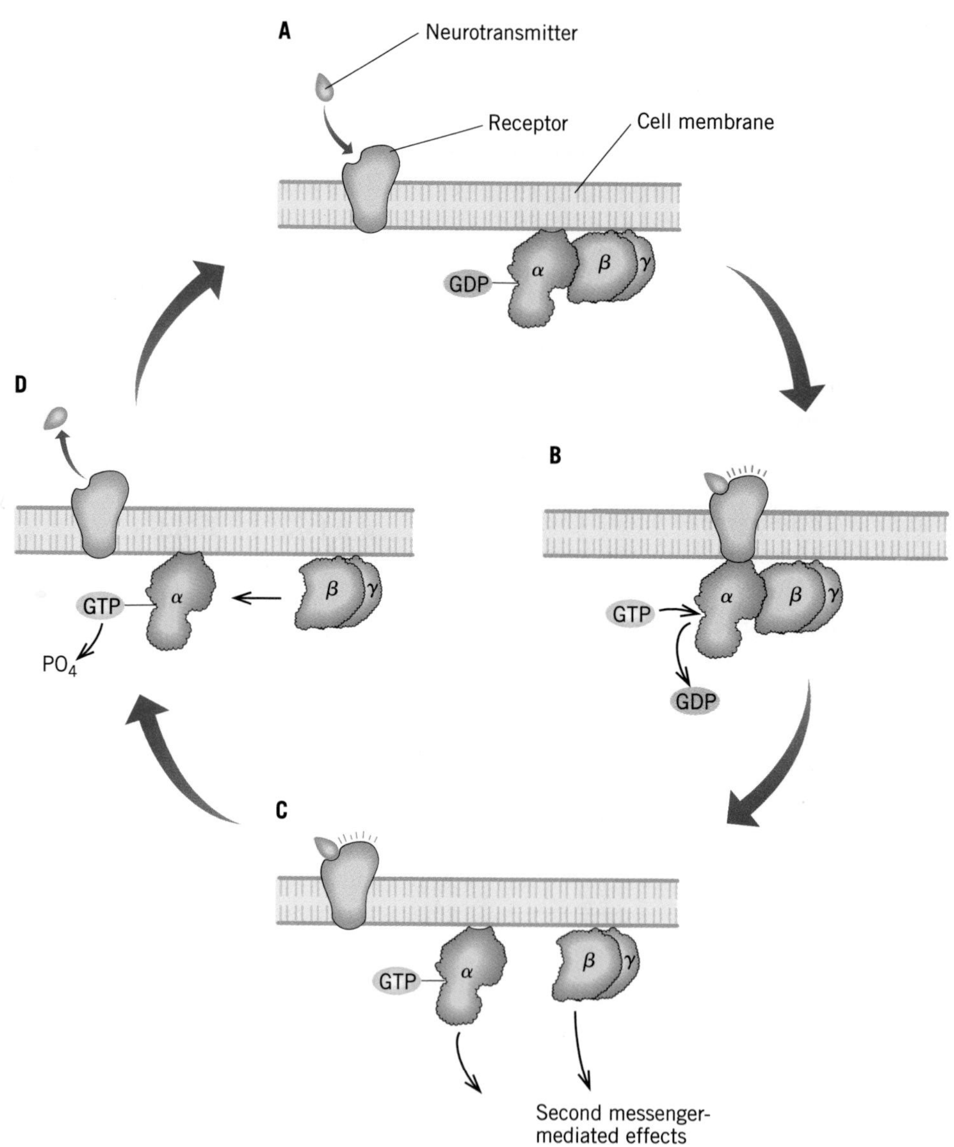

FIGURE 6-11. The cycle of activation of a G protein. (A) In its inactive state, the G protein consists of α and βγ subunits and has a high affinity for GDP. (B) When a neurotransmitter binds with a specific receptor, the G protein is activated; at the same time, GTP displaces GDP. (C) The presence of GTP decreases the affinity of the α subunit for the βγ dimer, and the components separate. Once apart, the α subunit, and often the βγ dimer, initiate an intracellular action that usually causes the production of a second messenger molecule, which triggers a cellular response. (D) The GTPase activity of the α subunit–GTP complex soon splits GTP into GDP and a phosphorus group, leading to inactivation and reassembly of the G protein.

with its receptor causes changes in the distribution of charge within the receptor, and these changes attract nearby G proteins. Because each G protein is free to move within the membrane (see Chapter 2), one will bind with the activated receptor. In this bound (activated) state, the G protein releases GDP and picks up GTP in its place. The GDP normally stabilizes the attachment of the βγ dimer to the α subunit, so its loss causes the G protein temporarily to separate into α and βγ parts. As described in the next section, the α subunit, and often the βγ dimer as well, then produces an intracellular response that makes available a second messenger molecule. Alternatively, the α subunit may directly open an ion channel. At the same time the intracellular effect occurs, the α subunit, which has strong enzymatic activity on GTP, splits one phosphorous group from this molecule, leaving GDP. The presence of GDP allows the βγ dimer to reattach to the α subunit, and the cycle is ready to begin again.

G proteins can vary considerably in the structure of their α, β, or γ subunits. That is, the exact sequence of amino acids that constitute an α, β, or γ subunit varies from one structural type to another. As of 1996, nearly twenty G proteins had been identified, based on differences in the amino acid sequences of their α subunits. These are grouped into four major families based on structural similarities. In addition, five structural types of β subunits and seven structural types of γ subunits have been identified. If all possible combinations of α, β, and γ subunits existed, there would be close to 700 varieties of G proteins. Not every known β or γ subunit can combine with an α subunit to form a functional G protein, but many combinations do exist. Most, if not all, cells contain several different G proteins, each specific for a particular receptor and activating a particular second messenger system.

An important feature of a signaling system that involves multiple molecules is that even a single neurotransmitter molecule can have a substantial intracellular effect. The cycle of G protein dissociation, activation of a second messenger cascade, and reassembly takes very little time compared with the time that the neurotransmitter remains bound to the receptor. As a consequence, each receptor activated by a neurotransmitter will activate several G proteins, one after the other. Because each activated G protein will itself initiate production of many molecules of second messenger, a single neurotransmitter molecule will generate a significant intracellular effect. Hence, G proteins not only direct a particular molecular response in the neuron but also amplify that response relative to the number of stimulating neuromodulatory molecules.

Neuromodulation is typically characterized as a second messenger-mediated system because the effects of neuromodulators on the postsynaptic cell are in most cases indirect, mediated by an intermediate substance. In a few cases, however, the G protein interacts directly with an ion channel. For example, when it binds with a type of muscarinic ACh receptor, ACh has an inhibitory effect on vertebrate heart muscle. This effect is mediated by a direct action of the α subunit of the activated G protein on the K_{ACh} channel of heart muscle (see Table 5-1). The opening of this channel leads to hyperpolarization of the membrane of the heart muscle cell, a prolonged relaxation phase of the muscle, and consequently to a slowing of the heartbeat.

When they are activated by the presence of neurotransmitter, G proteins undergo a cycle of dissociation and reassembly. This cycle initiates a cascade of biochemical events that leads to the neuron's response to the neurotransmitter while at the same time amplifying its response. In a few cases, G proteins may open an ion channel directly.

Second Messenger Systems

In all but the few cases of direct action on an ion channel, G proteins increase the availability of an intracellular second messenger molecule, thereby initiating a biochemical chain reaction. A common end result is a chemical reaction in which a phosphate group is joined to a protein molecule such as an ion channel. The binding of phosphate to a protein, a process called phosphorylation, is promoted by a type of enzyme called a **protein kinase**. Regulation of the activity of protein kinases is an important means by which a cell can regulate the actions of many of its proteins. In the context of neuromodulation, phosphorylation of an ion channel often causes a conformational change in the channel protein, either opening or closing it to ion flow, and thereby changing the way in which the cell will respond to subsequent synaptic input.

Consider a specific example of second messenger-mediated neuromodulation. One of the most widespread and best understood second messenger systems is that in which cyclic adenosine monophosphate (cAMP) is the second messenger and a change in the state of a type of potassium channel is the end result (Figure 6-12). Normally, very little free cAMP is available in neurons. The binding of an appropriate neuromodulatory neurotransmitter, usually serotonin in this particular system, to its receptor, activates nearby G proteins. These G proteins in turn activate the membrane-bound enzyme adenylyl cyclase (AC), which catalyzes the synthesis of cAMP from adenosine triphosphate. The increased level of cAMP in the neuron turns on a cAMP-dependent protein kinase, which either directly or via an intermediate protein causes the potassium channel to close. The effect of this sequence is an increase in the excitability of the neuron to subsequent input from other sources, as described in the next section.

Although there are many forms of G protein, the actions of all of them converge on just a few second messenger systems. Each system tends to use a specific second messenger and, usually, an associated protein kinase. Four of the most important of these systems are listed in Table 6-3, which gives the important second messengers and the protein kinases that they are thought to control. In each case, the release of second messengers is controlled by activation of an enzyme or enzyme system that synthesizes the messenger in question. For example, cAMP is synthesized by AC, cyclic guanosine monophosphate (cGMP) by guanylyl cyclase, and both diacylglycerol (DAG) and inositol triphosphate (IP_3) are synthesized by phosphodiesterase (PDE). Each of these enzymes is turned on by one or more specific types of G proteins that themselves are activated by the binding of a neuromodulator with a receptor.

The cellular effects of neuromodulatory transmission are initiated by membrane-bound G proteins and brought about by a variety of pathways that usually result in the phosphorylation of a membrane channel. This phosphorylation causes a change in the conformation of the channel protein, either opening or closing it.

The Postsynaptic Effects of Neuromodulators

The discovery of neuromodulation and second messenger systems has revolutionized neurobiologists' understanding of neural communication. The new knowledge helps to account for some of the most interesting phenomena in nervous systems: long-term changes in neural function that result from specific experience, or, in common language, learning. You will learn much more about the capability of the nervous system to show functional change, often called neural plasticity, in Chapter 24. Here, just consider these effects in overview.

There are two keys to understanding the function of second messenger systems. First, they usually act on the postsynaptic neuron for longer than just a few seconds. Although the neuromodulatory neurotransmitter molecule may be available at a synapse for only a short while, each biochemical step in the process of activating a second messenger and phosphorylating a protein generally lasts for some time, thereby augmenting the final effect. But further than that, once a protein has been

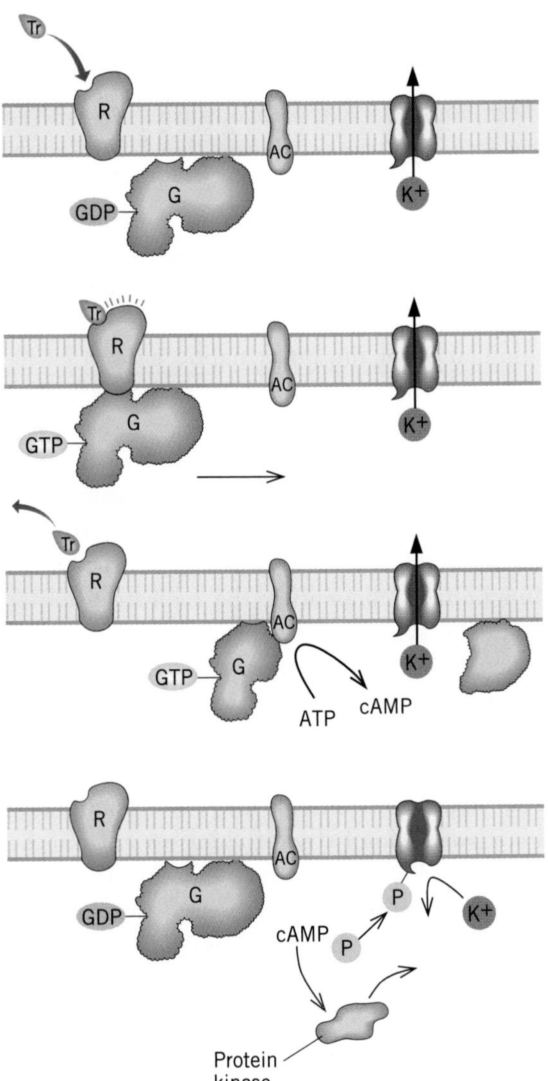

FIGURE 6-12. A typical sequence of events leading to phosphorylation of a membrane channel (resulting in a change in its responsiveness) by the action of a specific neurotransmitter. In this sequence, the receptor (R) becomes activated by binding with the transmitter (Tr) and in turn activates a nearby G protein (G). The G protein splits, and the α subunit moves over in the membrane to activate adenylyl cyclase (AC). The activated AC promotes the synthesis of cAMP, which in turn activates a cAMP-dependent protein kinase. The protein kinase phosphorylates a channel protein, which in this example closes it, shutting off a persistent potassium current and thereby increasing the cell's excitability.

phosphorylated, it can remain phosphorylated for some time, ensuring that the effect of the neuromodulator will last for a period well beyond that necessary to initiate it in the first place.

The second key feature of second messenger systems is that the main functional effect of neuromodulatory neurotransmitters is usually not on the membrane potential of the postsynaptic neuron but on the way the neuron responds to other inputs. For example, in the serotonin-activated cAMP system described earlier, the end result is closure of a specific type of potassium channel. This has an excitatory effect on the neuron, since it shuts off one source of current leak when the neuron is at rest. However, the effects are more subtle than merely increasing the excitability of the cell. Because of the distribution of the affected potassium channels, and because the effect of serotonin can last for many minutes, the

TABLE 6-3. Second Messengers and the Main Protein Kinases They Regulate

Second Messenger	Protein Kinase
Cyclic AMP (cAMP)	cAMP-dependent protein kinase (protein kinase A [PKA])
Cyclic GMP (cGMP)	cGMP-dependent protein kinase
Diacylglycerol (DAG)	Phospholipid-dependent protein kinase (protein kinase C [PKC])
Inositol triphosphate (IP_3)	Calcium-dependent protein kinase

long-term effect of serotonin in this instance is to keep the membrane of the cell depolarized longer when an action potential is generated. This occurs because one source of repolarizing current, the potassium channels, has been shut off. The physiological consequence of this broadening of the action potential is to augment the release of neurotransmitter from the affected neuron when it is excited. Hence, in this case input from the neuron that releases serotonin to the postsynaptic neuron has the effect of influencing the postsynaptic neuron's response to input from other sources.

The same theme is found in case after case. The most profound effect of a neuromodulatory input is rarely the immediate, short-term electrical effect that can be recorded in the postsynaptic neuron. It is instead some other consequence, mediated by a mobilization of specific biochemical pathways that affect the neuron's properties or response characteristics. These effects are frequently brought about by phosphorylation of a channel by protein kinase, as exemplified by the effects of ACh acting through muscarinic receptors in sympathetic neurons.

The emphasis in this section has been on pathways that involve proteins that are already present in the cell. However, modulation of enzymatic activity via second messenger systems can also include activation of specific genes, and hence can influence the synthesis of new proteins. These proteins might be additional channel proteins or channel-regulating proteins; as such, they can then cause changes in the functional characteristics of the target cell that will last not just for a few hours but many days or weeks.

The functional importance of neuromodulation is in its long-term effect on a postsynaptic neuron. By changing the way neurons respond to the input they receive, neuromodulation allows them to adapt to changing circumstances, a hallmark of learning. These adaptations can last from a few hours up to weeks.

Inactivation of Neurotransmitter

The binding of a neurotransmitter with a receptor is a transient event that brings about no permanent change in either the neurotransmitter or the receptor. When the neurotransmitter separates from a receptor, it is free to bind with another and contribute again to the response of the postsynaptic neuron. Perpetual stimulation of the postsynaptic neuron is not desirable because receptors may desensitize (become insensitive to the neurotransmitter) and because the postsynaptic response must be limited to a brief period so that the neuron can respond properly to rapid changes in patterns of input. Hence there must be a process for functionally eliminating the neurotransmitter from the synapse. This process can be called **inactivation**. There are three mechanisms of inactivation: diffusion, enzymatic degradation, and uptake (Figure 6-13).

In cases where precise timing is not critical, neurotransmitters may be functionally inactivated simply by diffusing away from the synapse. Diffusion effectively inactivates a neurotransmitter by reducing the amount available for binding with receptors. However, it is usually supplemented by one of the other methods as well.

Inactivation is the process by which neurotransmitters are functionally eliminated from the synapse at which they work. Diffusion is usually supplemented by enzymatic degradation or uptake.

Inactivation by Enzymatic Degradation

Many neurotransmitters are inactivated by enzymatic degradation. In the case of ACh, for example, the neurotransmitter is disassembled into acetate and choline by the enzyme **acetylcholinesterase (AChE)**. At the neuromuscular junction, this enzyme is

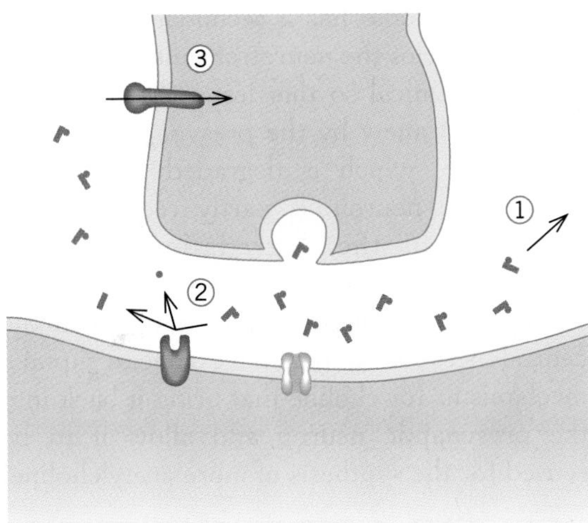

FIGURE 6-13. Mechanisms of neurotransmitter inactivation. 1. Diffusion away from the synaptic cleft. 2. Breakdown by enzymatic degradation. 3. Uptake via neurotransmitter transporters in the surrounding membranes. The illustration is diagrammatic only; the three mechanisms do not necessarily all act at any one synapse.

present on the extracellular matrix formed by the muscle sheath and which encloses the synaptic cleft; in the central nervous system it is found in both the pre- and postsynaptic membranes. The presence of an enzyme that de-grades ACh does not reduce the effectiveness of ACh as a neurotransmitter because typically there are nearly four times as many receptors for ACh at a synapse as there are molecules of AchE. Furthermore, both the receptors and the enzymes are saturated when the synapse is active, which means that there are many more molecules of ACh released at the synapse than there are binding sites for them. Under these circumstances, the AChE cannot break down ACh fast enough to keep the neurotransmitter from producing a strong postsynaptic effect.

Once the presynaptic neuron has stopped releasing acetylcholine, AChE is able to reduce the concentration of ACh to levels too low to exert any significant postsynaptic effect. The importance of AChE in limiting the normal action of ACh is attested to by the powerful effects of poisons like the organophosphorus nerve gases, which block the action of AChE. Death from exposure to these gases usually comes from asphyxiation, brought about by a combination of respiratory failure and bronchial constriction. Acetylcholine is an important neurotransmitter in the respiratory neural centers in the brain stem. Excess ACh leads to overstimulation, which disrupts the networks of neurons that drive the respiratory pattern (see Chapter 16) and leads to reduced activation of the diaphragm. Peripheral effects of the nerve gases are due to their effects on autonomic nerves and on the somatic neuromuscular junction. Persistent stimulation of nACHR by ACh desensitizes the receptor, leading to muscle paralysis.

Enzymatic degradation contributes to the inactivation of many other neurotransmitters, too. For example, diffusion is an effective means of inactivation of peptide neurotransmitters (see Chapter 7) because these are relatively slow-acting neuromodulators and hence do not need to be cycled out of the synapse rapidly. Nevertheless, diffusion is supplemented by enzymatic degradation of the neurotransmitter molecules by enzymes present in and around the synaptic cleft. Further, many neurotransmitters of the type called monoamines (see Chapter 7) are degraded by extracellularly located enzymes such as catechol-O-methyltransferase (COMT) and monoamine oxidase (MAO).

Enzymatic degradation is another method of neurotransmitter inactivation. Degradation enzymes do not interfere with normal neurotransmitter action because much more neurotransmitter is normally released than these enzymes can break down in a short time. Enzymatic degradation may supplement diffusion as a method of inactivation.

Inactivation by Uptake

Diffusion and extracellular enzymatic degradation are the primary means of inactivation of only some neurotransmitters. In most cases, neurotransmitters are functionally inactivated by actively being removed from the synapse. This is done by specialized transport mechanisms in a process known as **reuptake** or, more simply, **uptake**. Uptake is carried out by transporter proteins (see Chapter 2) that are located mainly in the membrane of the presynaptic neuron, but also in the postsynaptic membranes of some neurons, and sometimes in glial membranes as well. Transporter molecules have a strong affinity for one specific type of neurotransmitter. Because of this affinity, they act as a kind of vacuum pump, rapidly clearing the synapse of the neurotransmitter. They can be quite efficient. Some presynaptic neurons can retrieve as much as 50% of the neurotransmitter they have released, and a synapse may effectively be cleared of neurotransmitter within a second. Some drugs, such as Prozac, act by blocking uptake.

The primary role of uptake is to remove neurotransmitter from the synaptic cleft. However, uptake has a secondary benefit as well: it recycles the neurotransmitter, conserving the chemical so that less will have to be synthesized anew by the presynaptic neuron. Even ACh, which is degraded outside the presynaptic neuron, is partly recycled after use. One of the degradation products of this neurotransmitter is choline (see Chapter 7). Both motor neurons and neurons in the central nervous system have specific uptake mechanisms for choline that bring it back into the presynaptic neuron and allow it to be reused for the synthesis of more acetylcholine.

Many neurotransmitters are inactivated by uptake, the process by which they are taken up by cells in the vicinity of the synapse. Uptake is carried out by specific transport molecules in the membranes of surrounding cells, helping to recycle the neurotransmitter chemical as well as to inactivate it.

Additional Reading

General

Eccles, J. C. 1964. *The Physiology of Synapses*. New York: Academic Press.

Kaczmarek, L. K., and I. B. Levitan. 1987. *Neuromodulation: The Biochemical Control of Neuronal Excitability*. New York: Oxford University Press.

Katz, B. 1966. Nerve, Muscle, and Synapse. New York: McGraw-Hill.

Matthews, G. G. 1991. *Cellular Physiology of Nerve and Muscle*. 2d ed. Palo Alto, Calif.: Blackwell Scientific.

Research Articles and Reviews

Jessell, T. M., and E. R. Kandel. 1993. Synaptic transmission: A bidirectional and self-modifiable form of cell-cell communication. *Cell* 72/*Neuron* 10 (review supplement):1–30.

Sakmann, B. 1992. Elementary steps in synaptic transmission revealed by currents through single ion channels. *Science* 256:503–12.

Trimble, W. S., M. Linial, and R. H. Scheller. 1991. Cellular and molecular biology of the presynaptic nerve terminal. *Annual Review of Neuroscience* 14:93–122.

Unwin, N. 1993. Neurotransmitter action: Opening of ligand-gated ion channels. *Cell* 72/*Neuron* 10 (review supplement): 31–41.

CHAPTER 7

Neurotransmitters and Their Release

Chemical communication depends not only on the response of the postsynaptic cell to a neurotransmitter but also on timely synthesis, storage, and release of this substance. In this chapter, you will learn how the release of neurotransmitter is controlled by the presynaptic terminal. You will also learn about the chemical identity of neurotransmitters and how they are synthesized and stored in presynaptic neurons.

The mechanisms by which a neurotransmitter affects the activity of a postsynaptic cell (see Chapter 6) represent only half the process of synaptic transmission. The other half is the set of mechanisms by which neurotransmitter is released and the mechanisms that direct storage and synthesis of neurotransmitter; these processes are discussed in this chapter.

Neurobiologists first began to understand the mechanisms involved in neurotransmitter release when they applied intracellular recording techniques to the study of synapses. The use of molecular biological techniques has now expanded the initial view to provide a much richer and more detailed picture of how the release of neurotransmitter is controlled.

Synaptic Vesicles and the Role of Calcium

Both physiological and morphological approaches were used in early studies of chemical synaptic transmission. Physiological experiments showed that the availability of free calcium ions (Ca^{2+}) in the extracellular fluid was essential for synaptic transmission to take place. Morphological studies using the electron microscope revealed the ubiquitous presence of synaptic vesicles (see Chapter 2) in presynaptic terminals, suggesting that these cellular organelles also played a critical role in the process.

Calcium and the Release of Neurotransmitter

Some of the most important studies of chemical synaptic transmission were carried out by Bernard Katz and Ricardo Miledi in the 1960s. Even at that time it was known that blocking an action potential in a presynaptic nerve terminal eliminated transmission at that terminal. It was also known that reducing the level of external Ca^{2+} available in the vicinity of the synapse had the same effect. From these and other observations, it became clear that both depolarization of the presynaptic terminal and the availability of extracellular free Ca^{2+} are prerequisites for the release of neurotransmitter.

Katz and Miledi were interested in the exact role of these two factors. To investigate the issue, they studied synaptic function in a synapse of the squid giant axon. This is an exceptionally favorable synapse to study because both the pre- and postsynaptic terminals are large and accessible to intracellular microelectrodes. Hence, Katz and Miledi could monitor with great precision electrical events in both the pre- and postsynaptic neurons.

In their experiments, Katz and Miledi first impaled both the pre- and postsynaptic terminals with microelectrodes (Figure 7-1A). They added enough tetrodotoxin (TTX) to reduce the size of the presynaptic action potential without eliminating it (see Chapter 6), stimulated the presynaptic neuron, and measured the amplitude of the resulting excitatory postsynaptic potential (epsp). The epsp is not affected by the presence of TTX because it is generated by the opening of ligand-sensitive channels that allow both Na^+ and K^+ to flow through the membrane (see Chapter 6), and not by the voltage-sensitive sodium channels that are blocked by TTX.

The results of experiments in which the researchers varied the amount of TTX, and therefore the amount of reduction in the amplitude of the presynaptic action potential, showed clearly that neurotransmitter release depends on the level of presynaptic depolarization. Small amounts of depolarization have no effect on the release of neurotransmitter. However, once the presynaptic terminal is depolarized more than about 45 mV above resting level, the amplitude of the post-synaptic epsp is strongly dependent on the amplitude of presynaptic depolarization (Figure 7-1B). Other experiments had shown that the amplitude of a postsynaptic response is proportional to the amount of neurotransmitter present at the synapse; Katz and Miledi's results showed that when the presynaptic cell is partly poisoned with TTX, the amount of neurotransmitter that is released is dependent upon the amplitude of the presynaptic action potential.

In another series of experiments, Katz and Miledi completely suppressed action potentials by applying high concentrations of TTX. They then depolarized the presynaptic terminal directly by passing current into it via the intracellular electrode. By applying different amounts of depolarization, they showed that under these conditions as well, the greater the depolarization, the more neurotransmitter is released.

Katz and Miledi investigated the role of Ca^{2+} in the release of neurotransmitter in a subsequent series of experiments on frog neuromuscular junctions. They reduced the level of external Ca^{2+} at the synapse until synaptic transmission was abolished, then pulsed Ca^{2+} onto the synapse from a nearby electrode. Controlled release of ions from a microelectrode is done by passing current through the electrode, a method called **iontophoresis**, or **ionophoresis** (Box 7-1). Katz and Miledi found that transmission was successful only when Ca^{2+} was made available at the synapse just before the action potential in the presynaptic axon reached the terminal.

A

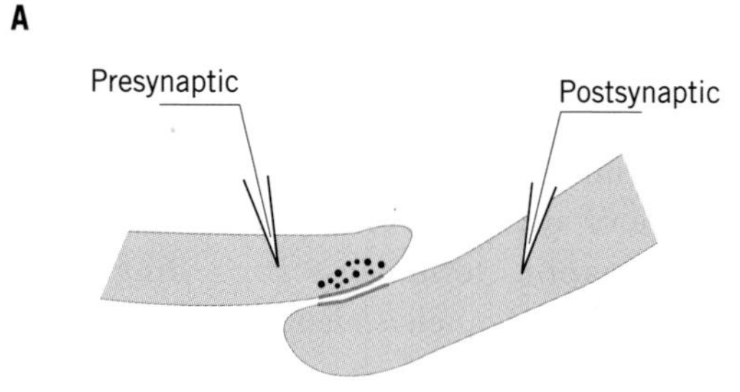

B

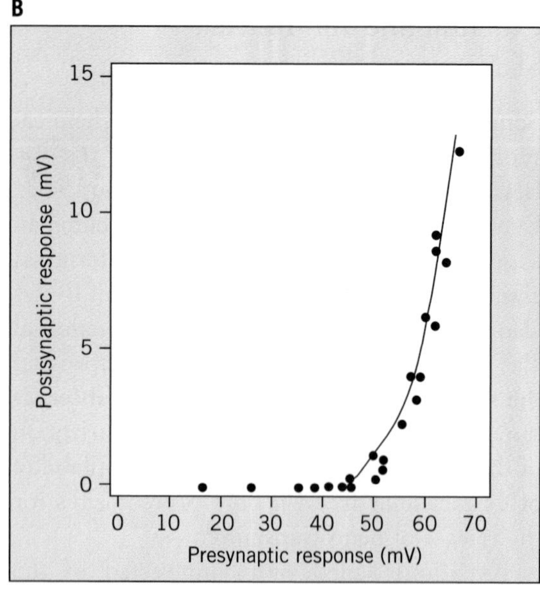

FIGURE 7-1. (A) The experimental arrangement for measuring the relationship between the amplitude of a presynaptic potential and the corresponding amplitude of the postsynaptic response at a squid giant axon synapse. (B) The relationship between pre- and postsynaptic potentials in a squid.

BOX 7-1

Spritzing Ions: The Controlled Release of Charged Molecules

The ability to deliver a precise amount of a chemical at a precise location and time can be critical for the success of many experiments. The technique used to accomplish this, iontophoresis, takes advantage of the electrical charge carried by many molecules that are of interest to researchers. If a solution of $CaCl_2$, for example, is placed into a microelectrode, then by passing (positive) current out of the electrode, a researcher can cause the positively charged Ca^{2+} ions to move out of the tip of the electrode into the surrounding fluid (Figure 7-A). By controlling the time of onset, the duration, and the amplitude of the current, the researcher can regulate precisely the time at which the Ca^{2+} starts to flow out of the tip, how long it flows, and how much leaves the electrode. Katz and Miledi used this method to regulate carefully the timing and amount of Ca^{2+} that was made available at the neuromuscular synapse they were studying.

Inorganic ions are not the only substances that can be applied by iontophoresis. Most neurotransmitters are also electrically charged in solution, and hence can be applied at a synapse through iontophoresis. In fact, an early piece of evidence that acetylcholine is the neurotransmitter at the vertebrate neuromuscular junction was the similarity of the response of the muscle to the application of acetylcholine from a microelectrode to its natural neuromuscular response.

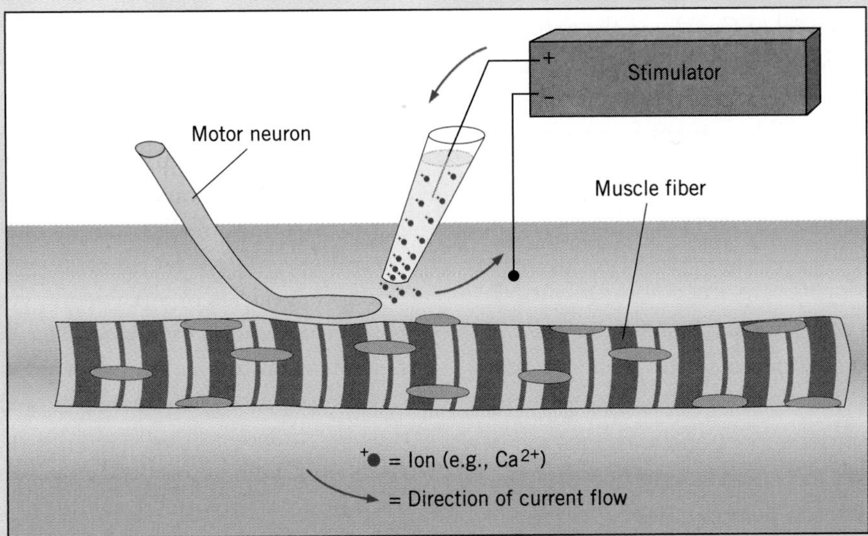

FIGURE 7-A. Schematic showing iontophoresis. A hollow glass microelectrode is filled with a solution containing the charged particles that are to be released, and current of the appropriate polarity is passed through the electrode. Since ions are the carriers of current in solution, passing positive current will cause the migration of cations, Ca^{2+} in the example, out of the electrode.

(continued)

Spritzing Ions:
The Controlled Release of Charged Molecules *(continued)*

Chemicals can be injected into cells by iontophoresis as well. Metallic salts like cobalt chloride and dyes like Lucifer yellow are electrically charged. These substances can be injected into neurons by iontophoresis, and their locations in the nervous system then determined later for identification of individual neurons and neuroanatomical tracing of their processes (see Box 3-2). When the physiology of a neuron is studied first, and the neuron is later marked for anatomical study by iontophoresis of a dye, it becomes possible to correlate the physiological and anatomical characteristics of individual neurons.

Calcium ions exert their effects by entering the presynaptic terminal through voltage-sensitive calcium channels. When the membrane is depolarized because of the invasion of the terminal by the presynaptic action potential, these channels open, allowing Ca^{2+} to flow down its concentration gradient into the neuron. There the ions trigger the fusion of the synaptic vesicles containing neurotransmitter with the neuronal membrane, as detailed later in this chapter. The level of internal free Ca^{2+} in neurons is normally kept extremely low by the active removal of Ca^{2+} from the cell and by its sequestration in the endoplasmic reticulum and mitochondria in the neuron. Sequestered Ca^{2+} is bound, and hence not free to influence any other processes in the neuron.

Neurotransmitter is released from presynaptic terminals when the terminal depolarizes and Ca^{2+} enters. Above a certain level of depolarization, the amount of neurotransmitter released is dependent on the magnitude of presynaptic depolarization. The entry of free Ca^{2+} into the presynaptic terminal through voltage-sensitive calcium channels clustered in the presynaptic terminal membrane is essential for the release of neurotransmitter.

Quantal Release

The ubiquitous presence of synaptic vesicles in early electron micrographs of presynaptic terminals immediately raised the possibility that vesicles are somehow necessary for transmission to take place. When biochemical studies revealed that vesicles contain high concentrations of neurotransmitter chemicals, it was suggested that vesicles are the normal site of storage of neurotransmitter and that they release neurotransmitter into the synaptic cleft. This suggestion is termed the **vesicular hypothesis** of synaptic transmission.

A corollary to the vesicular hypothesis is the hypothesis of **quantal transmission**, the release of neurotransmitter from the presynaptic terminal in discrete packets, each representing the contents of a single vesicle. Synaptic vesicles of a particular type in a neuron are all about the same size. Therefore, if neurotransmitter is stored in these uniformly sized synaptic vesicles and released from them during transmission, synaptic transmission itself ought to be quantal in nature. That is, each postsynaptic potential ought to consist of the sum of a number of smaller elements, just as a tower of wooden blocks consists of a number of individual blocks. Variations in the amplitude of a postsynaptic potential would then be due to variations in the number of elements (the synaptic vesicles) that are released during transmission, just as towers can be built to different heights by using different numbers of blocks. Each unit of height of the postsynaptic potential represents the effects of one **quantum** (packet) of neurotransmitter.

Evidence in support of the quantal nature of synaptic transmission has come from the frog neuromuscular junction. Highly amplified intracellular recordings taken from the muscle in the region of the neuromuscular junction reveal the presence of minute, spontaneous postsynaptic potentials that occur in the absence of any activity in the motor nerve innervating the muscle. These potentials, only a few millivolts in amplitude, are referred to as **miniature end plate potentials** or **mepps** (Figure 7-2). The amplitude of these mepps varies in steps of about 0.5 mV. That is, the smallest mepp is 0.5 mV, the next largest about 1.0 mV, and so on, suggesting that they are generated by the release of one, two, or more quanta of neurotransmitter from the presynaptic terminal, each quantum representing a single synaptic vesicle.

The discovery of mepps was suggestive but did not answer the question of whether full-sized epps were quantal. Unfortunately, carefully measuring the voltage amplitude of epps does not reveal any stepwise variation. The difficulty is that at the neuromuscular junction several hundred synaptic vesicles normally release their contents into the synaptic cleft at one time when the motor nerve fires one action potential. Under these circumstances, minor variations in the amount of neurotransmitter contained in each vesicle make it impossible to observe the effects of adding one or two more vesicles to the number being released. An experimental way around this problem is to reduce the number of quanta that are released when the motor nerve is stimulated. This can be done by partially blocking neurotransmitter release with high levels of magnesium ions (Mg^{2+}) in the external medium. Mg^{2+} blocks calcium channels. When Mg^{2+} levels are sufficiently high, stimulation of the motor neuron causes only very little neurotransmitter to be released, and this release is quantal just as are the mepps, as shown in Figure 7-3.

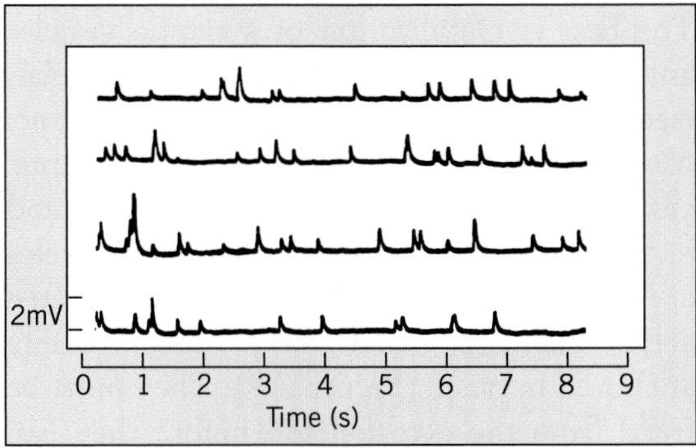

FIGURE 7-2. Miniature end plate potentials (mepps) recorded from a frog neuromuscular junction. The recording is continuous from the top line to the bottom. Note that the potentials vary stepwise in amplitude, demonstrating the quantal nature of neurotransmitter release.

Subsequent studies of a variety of synapses have shown that much chemical synaptic transmission is quantal in nature. At the neuromuscular junction, about 200 quanta are normally released during a full-sized epp. At other synapses, the number may be much lower, even just a single quantum. Neural systems compensate for this low number by having many neurons converge on a single postsynaptic neuron. The number of quanta in autonomic nerves is in the range of 10 to 20, as exemplified by the 20 or so quanta released at synapses in a particular autonomic ganglion, the ciliary ganglion, of the chick. The amount of neurotransmitter in a single vesicle has been measured to be about 10,000 molecules in the case of acetylcholine at the vertebrate neuromuscular junction, with a variance of about 10% from vesicle to vesicle.

According to the vesicular hypothesis of synaptic transmission, neurotransmitter is packaged in synaptic vesicles and released in quanta of roughly equal size. An important line of physiological evidence supporting this idea is the stepwise variation in size of miniature end plate potentials at the neuromuscular junction.

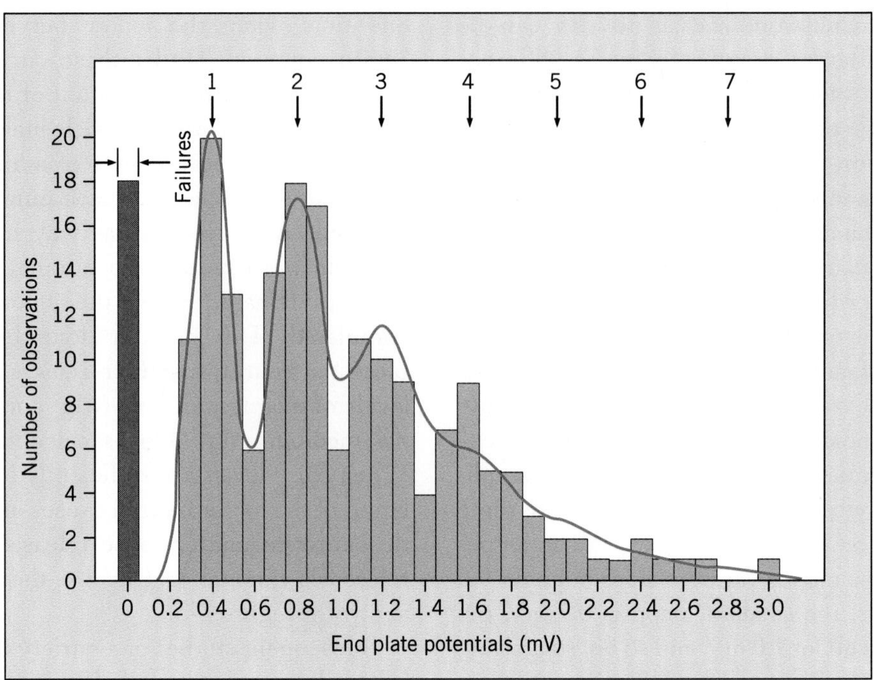

FIGURE 7-3. The distribution of end plate potential (epp) amplitudes at a mammalian neuromuscular junction in the presence of high magnesium, which suppresses synaptic transmission. Each bar represents the number of epps recorded at the indicated amplitude. The solid line represents the theoretical distribution of epp amplitudes, assuming a random distribution of quantal release; the small arrows represent the theoretical peak amplitude of epps due to one, two, three, four, five, six, and seven quanta, as indicated. Note that the peaks of the actual distribution represented by the tallest bars correspond to the peaks of the distribution shown in the theoretical curve. The peaks become less prominent as more quanta contribute to them because there is fluctuation in the amplitude of individual quanta due to variation in the amount of neurotransmitter packaged in each vesicle.

The Molecular Basis of Release

The intracellular studies described in the previous section show the fundamental relationship between presynaptic depolarization and the release of neurotransmitter, as well as the importance of Ca^{2+} in the process. Molecular studies of these events in the 1980s and 1990s have added considerable information about the events that occur in the presynaptic terminal when it is depolarized.

The Distribution and Movement of Vesicles

Since neurotransmitter is stored in synaptic vesicles, there must be specific mechanisms for moving these vesicles to the cell membrane and for releasing the neurotransmitter from the vesicle into the synaptic cleft. Two processes are involved, both requiring the presence of free Ca^{2+} in the nerve terminal. The first is mobilization of synaptic vesicles and their subsequent attachment to specialized sites on the presynaptic membrane. Mobilization is necessary because synaptic vesicles are not normally free to move around in the presynaptic terminal. Instead, vesicles are bound to parts of the cytoskeleton, often actin filaments, that keep them firmly anchored in place (Figure 7-4). They must be freed from the cytoskeleton before they can move to the cell membrane. Furthermore, once they have moved to the membrane, they must be attached to it to prepare for neurotransmitter release. The second process is the physical fusion of the membrane of the synaptic vesicle with the cell membrane and the release of neurotransmitter into the synaptic cleft. Attachment (called docking), fusion, and release are described in the next section.

It seems logical to suppose that the synaptic vesicles mobilized by Ca^{2+} are the vesicles that dock at the membrane, fuse with it, and release their contents into the synaptic cleft during a single transmission event. However, this is not the case. There is a delay between the opening of calcium channels in the terminal and the release of neurotransmitter into the synaptic cleft of only 0.1 to 0.2 msec. It is apparent that this is not nearly enough time for the cascade of biochemical reactions necessary for mobilization to be completed before neurotransmitter release. Based on this observation and other experimental evidence, researchers believe that there are two functional pools of neurotransmitter in the presynaptic terminal.

One group of vesicles—called the **storage pool**—consists of those vesicles that are anchored to the cytoskeleton. The other group—called the **releasable pool**—consists

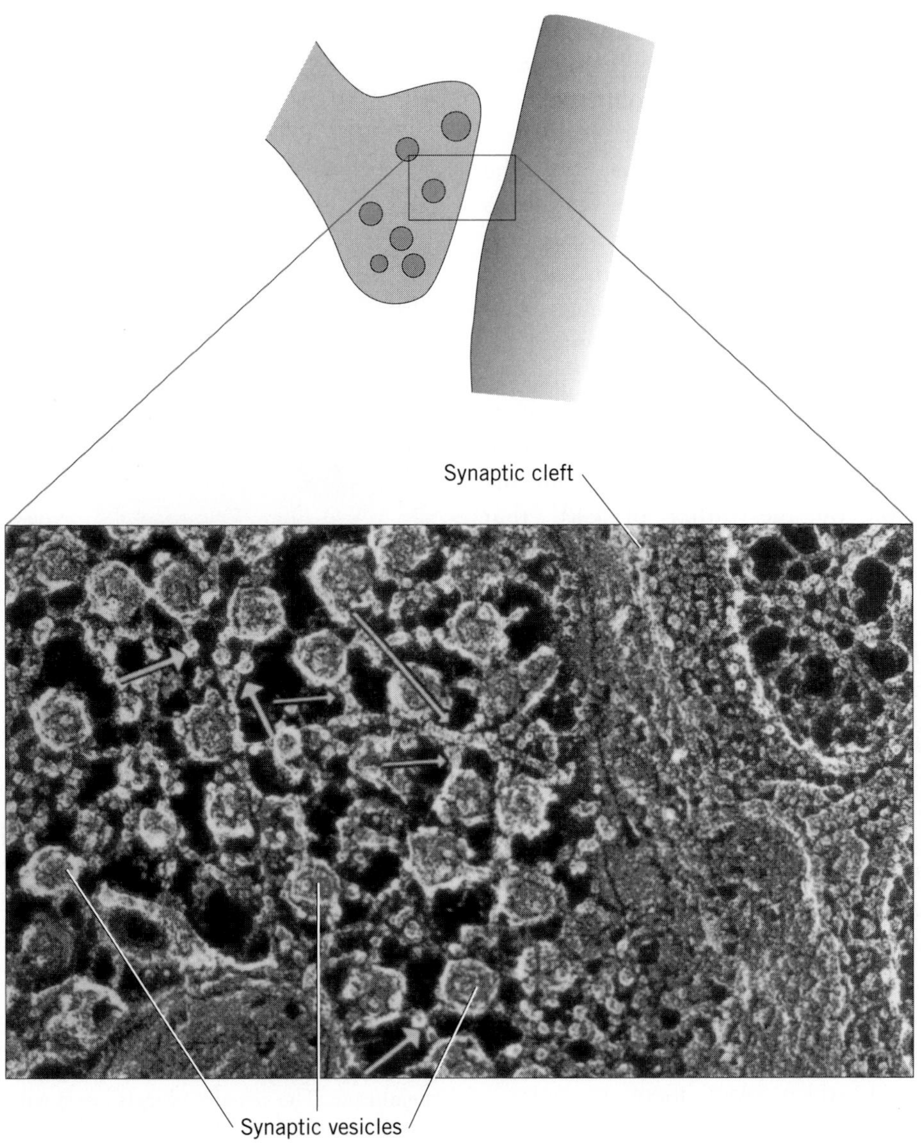

FIGURE 7-4. Electron micrograph of synaptic vesicles in a presynaptic nerve terminal, showing their attachment via short stems (short arrows) to a longer filament (long arrow) of the cytoskeleton.

of the vesicles that are already attached to the presynaptic membrane and are ready to release their contents into the synaptic cleft. Depolarization of the presynaptic terminal, by allowing Ca^{2+} to enter the terminal, has two distinct effects: it causes the release of neurotransmitter from vesicles that are already docked at the presynaptic membrane and, at the same time, mobilizes vesicles from the storage pool and brings them into the releasable pool by promoting their docking with the membrane. Therefore, the vesicles that are mobilized during one event of synaptic transmission are those whose contents are released during a subsequent event.

The role of free Ca^{2+} in the mobilization of synaptic vesicles has been described in some neurons. A protein known as **synapsin I** (a member of a family of proteins that will bind with both vesicular membrane and the cytoskeleton) binds small synaptic vesicles to the cytoskeleton. When synapsin I is phosphorylated, it undergoes a conformational change that causes it to detach from the vesicle membrane. Phosphorylation of synapsin I is promoted by several types of protein kinase, at least some of which can be activated by the presence of Ca^{2+} (see Table 6-3). Hence, vesicles can be released from the cytoskeleton by the influx of Ca^{2+} that accompanies depolarization of the presynaptic terminal. Once mobilized, the synaptic vesicles move toward the cell membrane and dock. The mechanisms that promote movement and docking, however, are not yet understood.

Synaptic vesicles are distributed in a storage pool and a releasable pool. The entry of Ca^{2+} into the presynaptic terminal has two effects. It causes release of neurotransmitter from vesicles in the releasable pool and the mobilization of other vesicles from the storage to the releasable pool. Mobilization occurs when vesicles are detached from synapsin I, which binds them to actin filaments.

Docking and the Release of Neurotransmitter

At every synapse, docking takes place at the **active zone**, a specialized site on the cell membrane of the presynaptic neuron. This region is located on the opposite side of the synaptic cleft from the receptors in the postsynaptic membrane. The membrane in the active zone contains specialized proteins for the attachment of synaptic vesicles, voltage-sensitive calcium channels, and sometimes other channels as well. Electron micrographs of neuromuscular junctions reveal that the synaptic vesicles in the active zone are aligned in double rows (Figure 7-5A). About 20 to 30 nm from these are double rows of particles that are embedded in the presynaptic membrane (Figure 7-5B). Binding studies suggest that these particles are the voltage-sensitive calcium channels, which thus are in the optimal location to allow incoming Ca^{2+} to affect neurotransmitter release.

Docking and fusion are effected by many different proteins acting together. Some of these proteins are free in the cytoplasm, some

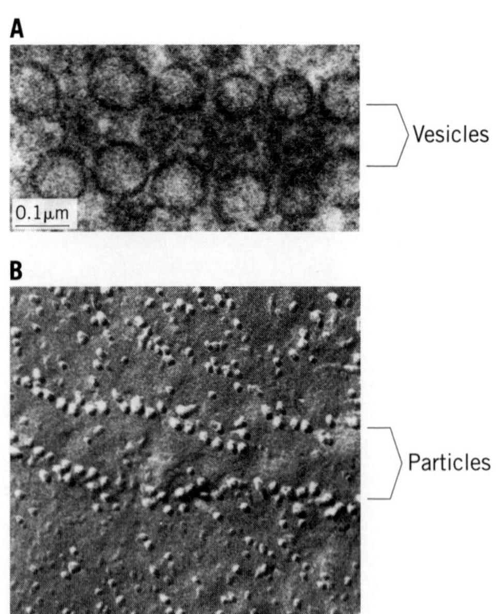

FIGURE 7-5. (A) Electron micrograph of the inside of a presynaptic motor neuron, looking onto the cell membrane. The circular objects are synaptic vesicles. (B) Electron micrograph of "particles" in presynaptic membrane. The view is looking onto the surface of the presynaptic motor neuron that faces into the synaptic cleft, the presynaptic membrane.

are bound to the vesicular membrane, and some are bound to the cell membrane in the active zone of the synapse. The exact role of each of these proteins in these events is not yet clear. The main point to keep in mind as you read about the proteins is that those in the vesicle membrane bind with those in the cell membrane in a lock-and-key fashion to position the vesicle properly. Some also respond to the presence of calcium ions to trigger docking or fusion. The cytoplasmic proteins, mainly *N*-ethylmaleimide-*S*ensitive *F*actor **(NSF)** and the *S*oluble *N*SF *A*ttachment *P*roteins **(SNAPs),** then wrap around the others, helping to stabilize the association between vesicle and membrane, and in the presence of calcium ions and ATP, promoting fusion and release of neurotransmitter.

Of the vesicular membrane proteins that have been identified, several are important in docking and fusion. Those most closely associated with docking are **synaptotagmin** and the *v*esicular *SNA*P *RE*ceptors **(v-SNAREs),** vesicular proteins whose name derives from their ability to bind with the cytoplasmic SNAPs. An important v-SNARE is **synaptobrevin,** which is also known as *V*esicle-*A*ssociated *M*embrane *P*rotein **(VAMP).**

Complementing the v-SNAREs are *t*arget *SNA*P *RE*ceptors **(t-SNAREs),** which are attached to the membrane of the active zone. The membrane at the active zone is often called the target membrane because it is the "target" of the vesicle. Two t-SNAREs are thought to be especially important, **syntaxin** and 25-kdalton *S*y*N*aptosomal-*A*ssociated *P*rotein **(SNAP-25),** which sounds confusingly like a cytoplasmic SNAP but is chemically unrelated. Table 7-1 lists and defines the acronyms for proteins involved in docking and fusion.

The idea that docking occurs by mutual binding among the v-SNAREs, the t-SNAREs, and the cytoplasmic proteins is called the **SNARE** (*SNA*P *RE*ceptor) **hypothesis.** As depicted in Figure 7-6A, the v-SNARE synaptobrevin and the t-SNAREs syntaxin and SNAP-25 align with one another as the vesicle approaches the cell membrane. This alignment positions the vesicle precisely in the active zone. Once in place, the v-SNARE and t-SNARE proteins bind with the SNAPs and NSF that are present in the cytoplasm, forming a multiprotein complex between the vesicle and the synaptic membrane (Figure 7-6B). The entire process of docking is triggered by high levels of free Ca^{2+}

TABLE 7-1. Acronyms for Proteins Involved in Vesicle Docking and Release

Protein location	Acronym	Full name
Cytoplasm		
	NSF	N-ethylmaleimide sensitive factor
	SNAP	Soluble NSF attachment protein
Vesicular membrane		
	v-SNARE	Vesicular SNAP receptor
	VAMP	Vesicle-associated membrane protein (also called synaptobrevin)
Target (presynaptic) membrane		
	t-SNARE	Target membrane SNAP receptor
	SNAP-25	25-kdalton synaptosomal-associated protein

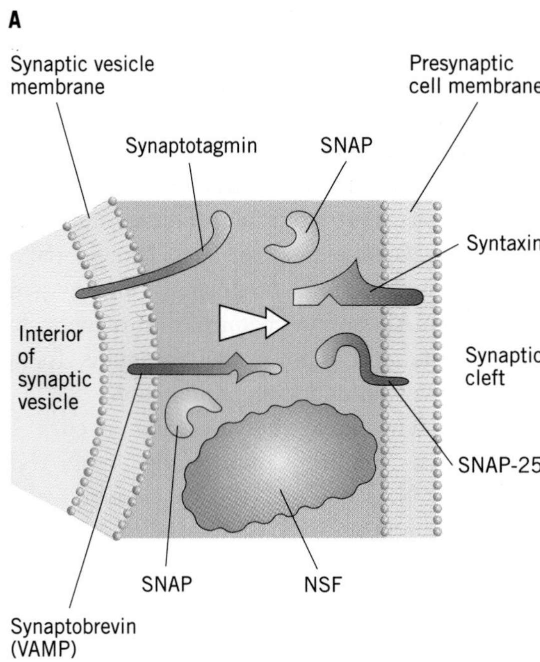

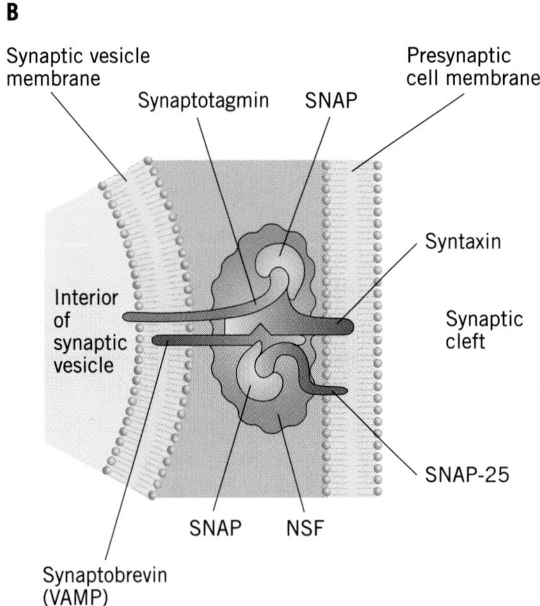

FIGURE 7-6. Docking of the synaptic vesicle at the presynaptic membrane. (A) Proteins associated with the vesicle (v-SNAREs) structurally complement proteins associated with the presynaptic cell (target) membrane (t-SNAREs) allowing the molecules to fit together. NSF and the SNAPs are unattached in the cytoplasm. (B) When the vesicle is securely docked, v-SNAREs and t-SNAREs are bound to one another, and to NSF and SNAP proteins, to form a complex that locks the vesicle in place. Some or all of these proteins are also involved in opening of the fusion pore and release of neurotransmitter into the synaptic cleft. The shapes and positions of the proteins are not intended to represent actual shapes and positions.

in the presynaptic terminal. The process is mediated by synaptotagmin, which binds preferentially with one of the t-SNAREs in the presence of high Ca^{2+}, thereby positioning the vesicle so that the other proteins can make the proper associations.

A vesicle that is docked is ready to fuse with the cell membrane and release its contents into the synaptic cleft. Fusion is not yet understood, but evidence suggests that it starts with the formation of a small channel between the interior of the vesicle and the outside of the cell, triggered by entry of Ca^{2+} into the presynaptic terminal and the presence of ATP. The channel, called a **fusion pore,** is formed by the v-SNAREs, t-SNAREs, NSF, and SNAP-25, probably along with other proteins. The pore lasts only a brief fraction of a millisecond, then usually quickly expands as the contents of the vesicle are dumped into the synaptic cleft. However, it may be possible for a fusion pore to form briefly, allowing some of the neurotransmitter in the vesicle to enter the synaptic cleft, then close again. The pore may flutter open and closed several times. Synaptotagmin appears to be the Ca^{2+} sensor for pore formation, but the exact mechanism by which it works is not yet clear.

It is through interference with the processes of docking and formation of the fusion pore that the deadly tetanus and botulinum neurotoxins exert their effects. By preventing the release of neurotransmitter from inhibitory neurons in the spinal cord, tetanus toxin causes uncontrollable muscle contractions. Botulinum prevents neurotransmitter release from motor neurons, leading to flaccid paralysis.

Not every vesicle that is docked in the active zone will necessarily contribute its neurotransmitter to a synaptic potential. Because one important factor in determining the fate of an individual vesicle is the local concentration of Ca^{2+} around it, researchers have investigated in detail the local changes in Ca^{2+} concentration that follow the depolarization of a presynaptic terminal. An important technique that has been used to good effect in these studies takes advantage of the ability of some dyes to fluoresce at a different wavelength of light when they bind with free Ca^{2+} than they normally do (Box 7-2). By injecting a neuron with such a dye, it is possible to monitor local changes in Ca^{2+} concentration during synaptic transmission. Such studies have indicated that a significantly higher concentration of Ca^{2+}, perhaps as much as 100 to 200 μM (compared with a resting level of about 0.1 μM), may develop in the immediate vicinity of the synapse. This high concentration of Ca^{2+} near the active zone enables Ca^{2+} to exert an extremely rapid effect, since it does not have to diffuse far to reach the docked vesicles. The delay between Ca^{2+} entry and neurotransmitter release is only 0.1 to 0.2 msec, indicating that long-distance diffusion and a complex sequence of biochemical reactions is neither necessary nor possible.

Synaptic vesicles dock at the active zone, a specialized region of the presynaptic membrane that contains docking proteins and calcium channels. According to the SNARE hypothesis, proteins in the vesicular and synaptic membranes, and in the cytoplasm, combine to attach the vesicle to the membrane. High levels of free Ca^{2+} trigger formation of a fusion pore that allows release of neurotransmitter.

The Fate of Synaptic Vesicles

As a synaptic vesicle releases the neurotransmitter that it holds, it fuses with the cell membrane. The membrane of the vesicle actually becomes part of the membrane of the presynaptic neuron (Figure 7-7). Measuring the cell's capacitance (an electrical property of the cell that is proportional to membrane surface area) (see Chapter 8) during and following synaptic transmission shows that this fusion of the membrane of the vesicle with the cell membrane increases the area of the cell membrane at the synapse. It has been suggested that this increased synaptic area could be a basis for the long-term changes that accompany learning. However, theoretical calculations show that the total amount of vesicular membrane that would be incorporated into the presynaptic membrane of a neuron at even a moderately active synapse if every vesicle were incorporated permanently into it is far more than the increase in membrane area actually measured at a synapse. Hence, although every vesicle that releases neurotransmitter fuses with the membrane, there must be some mechanisms for removing some of the cell membrane after such fusion.

BOX 7-2

The Calcium Light Show: Using Indicators of Intracellular Calcium

The concentration of Ca^{2+} in restricted local regions of a neuron can be extremely important in determining what happens in those regions. For this reason, the discovery of molecules that can indicate the level of free Ca^{2+} ions in different parts of living cells has brought significant advances in neuroscientists' understanding of events that are controlled by Ca^{2+} levels, like the release of neurotransmitter.

The first Ca^{2+} indicator to be used was the naturally occurring substance aequorin, a chemical in jellyfish that gives off light in the presence of free calcium ions. Jellyfish use aequorin to generate a luminescent glow as they swim. By the 1980s, neuroscientists were synthesizing designer molecules that exhibited extraordinary sensitivity to small amounts of Ca^{2+}. One of the most popular of these is fura-2.

Fura-2 is fluorescent. That is, it will give off light when it is excited by ultraviolet light of a particular wavelength. In the absence of Ca^{2+} ions, fura-2 will fluoresce when it is excited by ultraviolet light that has a wavelength of about 385 nm. In the presence of high concentrations of free Ca^{2+} ions, on the other hand, the molecule fluoresces when it is excited by ultraviolet light with a wavelength of about 345 nm. The actual concentration of Ca^{2+} ions can be measured quite accurately by measuring the ratio of the amount of light given off when it is excited at the two wavelengths of ultraviolet light. Because fura-2 reacts extremely quickly to changes in Ca^{2+} levels, coupling the optics necessary to measure wavelength ratios to a video system that can monitor living cells allows researchers to measure changes in Ca^{2+} levels over a period of a millisecond or less (Figure 7-B).

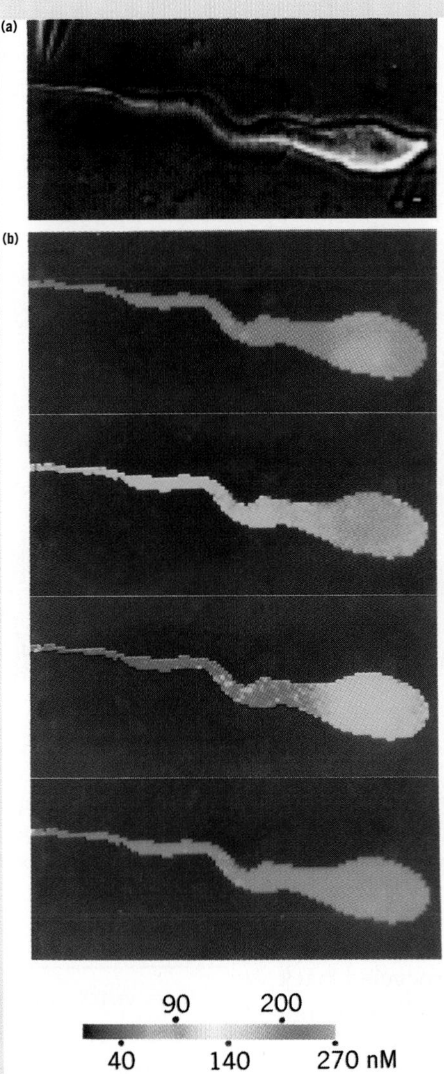

FIGURE 7-B. Imaging of free Ca^{2+} concentration by measurement of the fluorescence emitted by the calcium indicator fura-2. (a) Photomicrograph of an isolated neuron from the hippocampus of a guinea pig showing the cell body, the dendrite, and the tip of a glass microelectrode containing glutamate. (b) A series of false color images showing internal free Ca^{2+} concentration at rest, as the electrode is placed, immediately at the end of a 1-second period during which glutamate is iontophoresed out of the microelectrode, and 1 second after stimulation has ceased. False color images are obtained by sampling the light emitted by fura-2 over a period of 250 msec or 500 msec, and converting the brightness of the light relative to background to a color scale as shown at the bottom of the figure. Red represents the brightest light and therefore the highest concentration of free Ca^{2+}. The application of glutamate leads to a significant increase in the concentration of free Ca^{2+} inside the stimulated dendrite.

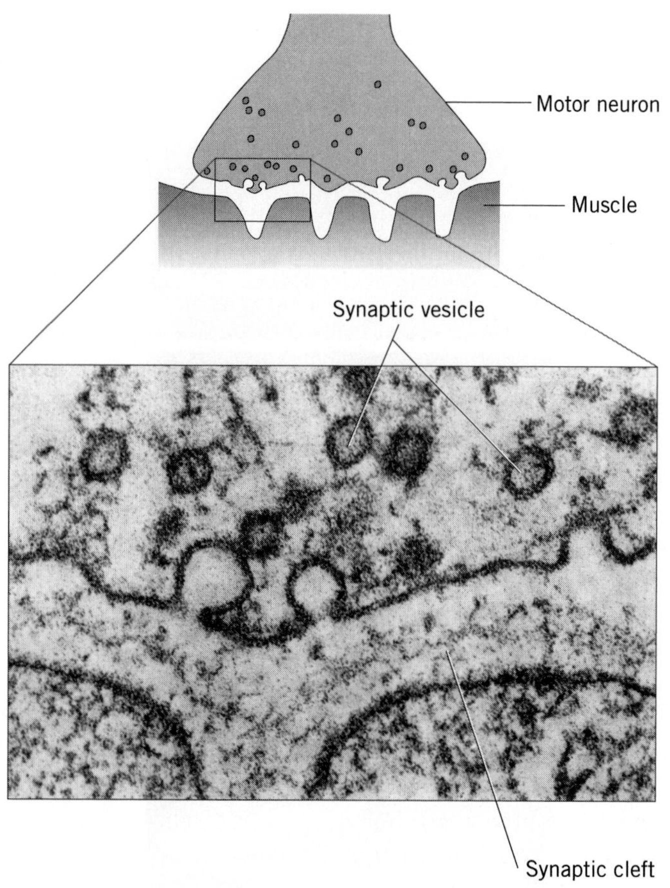

FIGURE 7-7. Electron microscopic evidence for fusion of the membrane of a synaptic vesicle with the membrane of the presynaptic neuron during release of neurotransmitter. The micrograph shows a neuromuscular junction. The muscle membrane has one infolding that extends beyond the bottom margin of the micrograph.

The removed membrane is recycled back to synaptic vesicles. This recycling is not random. As new vesicular membrane is incorporated into the cell membrane, a patch of membrane that came from a particular vesicle is pushed up and away from the synapse to a specialized region where new vesicles are pinched off into the cell. This process leads to the recycling of vesicular membrane, from a vesicle to the cell membrane, and then back into another vesicle. Experiments in which v-SNARE proteins such as synaptotagmin have been tagged with fluorescently labeled antibodies so that specific pieces of vesicular membrane can be identified and followed have provided solid evidence for this concept. The process is remarkably fast, requiring only 45 to 90 seconds to recycle synaptic vesicles, depending on the synapse.

As a vesicle opens and releases its neurotransmitter, its membrane fuses with the cell membrane. The vesicular membrane is then moved up and away from the synapse to a region of the cell where it is reused to form a new vesicle.

Release from Nonspiking Neurons

Some neurons are incapable of generating action potentials, presumably because they lack voltage-sensitive sodium channels. Such cells are called **nonspiking neurons.** In vertebrates, nonspiking neurons are present in the retina (rods, cones, horizontal cells, and bipolar cells), in the inner ear, in the lateral line of fishes, and in the olfactory bulb. Among the invertebrates, some sensory neurons and many interneurons involved in the generation of motor patterns are nonspiking. Nevertheless, even without action potentials it is possible for these neurons to release neurotransmitter and influence neighboring cells.

Action potentials are not necessary for the release of neurotransmitter; they merely facilitate it by supplying the necessary depolarization. For example, Figure 7-2 shows that a motor neuron will release vesicles spontaneously at the neuromuscular junction (mepps) without any depolarization of the presynaptic terminal. In the case of nonspiking neurons, release of neurotransmitter is proportional to the level of depolarization of the neuron. The more depolarized the cell, the more neurotransmitter it releases; the more hyperpolarized it is, the less neurotransmitter it releases. Even minor fluctuations of membrane potential in these cells are sufficient to change significantly the amount of neurotransmitter available to affect a postsynaptic cell. Since the amplitude of the postsynaptic response varies with the amount of neurotransmitter that is released, synaptic transmission in these circumstances is called **graded transmission**.

The realization that graded, spikeless transmission can occur has led researchers to reexamine their fundamental assumptions regarding synaptic transmission. It is now recognized that in a sense all chemical synaptic transmission is graded: The amplitude of a postsynaptic response is always a function of the magnitude of the presynaptic depolarization. This relationship is obscured at most synapses by the magnitude of the depolarization caused by the invading action potential and the resulting large postsynaptic potential. However, the relationship is apparent in the effect of presynaptic inhibition (see Chapter 6), which reduces the amplitude of an action potential and hence reduces the amplitude of the resulting postsynaptic potential. It can also be revealed by experimentally depressing the amplitude of an action potential and observing the resulting drop in the size of the postsynaptic potential. Strictly speaking, however, the term graded transmission is generally reserved for synaptic transmission without spikes, in which even slight fluctuations in membrane potential have substantial effects on the amplitude of a postsynaptic potential.

At some synapses, both graded transmission and release triggered by action potentials can occur. In vertebrate taste cells, for example, low levels of stimulation induce graded release of neurotransmitter that is proportional to stimulus strength. At higher levels, however, the taste cells generate action potentials, which evoke typical strong postsynaptic responses in the neurons on which they synapse.

Nonspiking neurons do not produce action potentials. They communicate via chemical synaptic transmission by releasing neurotransmitter in a graded fashion, in which the amount of neurotransmitter that is released is proportional to the level of membrane depolarization. Some neurons show both graded release and release due to action potentials.

Filling Vesicles

Obviously, synaptic vesicles must be filled with neurotransmitter. However, the mechanism by which filling occurs is not the same for all types of vesicles. Two main types of vesicles have been found in presynaptic terminals (Figure 7-8). Some are small (40–50 nm diameter) and clear in electron micrographs, and are oval or spherical in shape. Others are larger (70–200 nm diameter), have a solid-looking, dense core, and are only spherical in shape. Most of the information currently available about vesicles

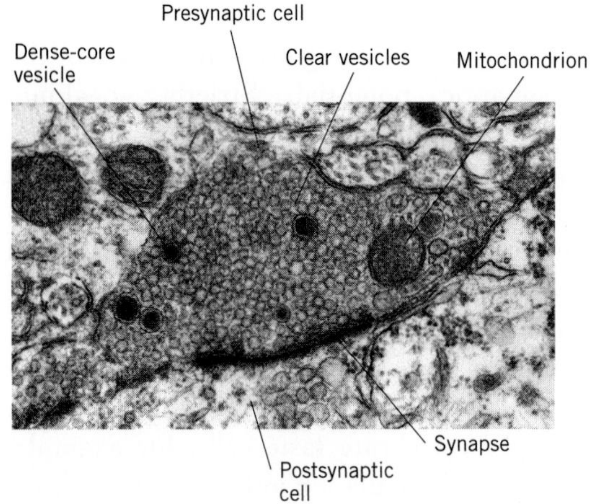

FIGURE 7-8. An axon terminal from a rabbit spinal cord, showing two types of synaptic vesicles: clear vesicles and dense-core vesicles. The latter generally contain peptide neurotransmitter.

described in previous sections of this chapter was gained by a study of clear vesicles. Not as much is known about dense-core vesicles. Some vesicles are filled in the cell body and transported to the site of release; others are filled by neurotransmitter in the axon terminals from which they will be released.

When clear vesicles are reformed from previously used membrane at the synaptic terminal, they are empty of neurotransmitter. New neurotransmitter molecules are synthesized at the terminal, and transported into the empty vesicle by a two-step process. First, energy in the form of adenosine triphosphate (ATP) is expended to transport free protons (H^+) into the interior of the vesicle. The proton gradient established in this way is then used to power the movement of neurotransmitter into the vesicle. Transporter proteins (see Chapter 2) embedded in the vesicle membrane shuttle the neurotransmitter into the vesicle by exchanging neurotransmitter molecules for protons, hence filling the vesicle.

The larger dense-core vesicles are often filled with a peptide neurotransmitter (see the next section), which must be synthesized by the Golgi apparatus in the cell body. For these vesicles, synthesis and filling occur more or less together, and the filled vesicle is moved by axonal transport down the axon to its site of release.

Some synaptic vesicles are reconstituted from cell membrane at the neuronal terminal and are filled with neurotransmitter by transporter proteins. Others are made by the Golgi apparatus in the cell body and are filled with neurotransmitter there.

Chemistry of Neurotransmitters

The discussion of chemical synaptic transmission to this point has covered the process in detail but has given only passing attention to the chemicals that constitute the neurotransmitters themselves. However, considerable research effort has been expended to identify the neurotransmitters at synapses in the central and peripheral nervous systems, in part because researchers are interested in any relationship that might hold between the chemical nature of a neurotransmitter and the role that it plays in the nervous system. Although no compelling evidence for such a relationship is presently available for any single neurotransmitter, certain types do play roles such as neuromodulation more often than others.

Identification of Neurotransmitters

It is rarely easy to decide if a particular chemical acts as a neurotransmitter at any given synapse because most synapses are not readily accessible to experimentation. It is therefore important for researchers to have a clear set of criteria by which to evaluate molecules as neurotransmitter candidates. In general, researchers require that at least four criteria be met in order for unambiguous identification of a chemical as a neurotransmitter to be accepted. The chemical must be synthesized and it must be stored in the neuron from which it is released, it must be released from terminals of these neurons, and, when applied exogenously at the synapse, it must mimic the action of the natural neurotransmitter.

Neuroscientists use a variety of physiological, anatomical, chemical, and pharmacological methods to help establish that these criteria are met. Physiologically, it may be possible to collect and identify the suspected neurotransmitter from the vicinity of active synapses. Furthermore, the technique of iontophoresis (see Box 7-1) can be used to demonstrate that the putative neurotransmitter can mimic the effects of the natural neurotransmitter on the postsynaptic neuron. Anatomical localization techniques may be able to demonstrate that the chemical has a local distribution in the nervous system that matches the known locations of neurons that release it, and, especially, that the chemical is located in the terminals of the presynaptic neurons. Chemical techniques can be used to demonstrate the presence of the enzymatic pathways necessary for the synthesis of the putative neurotransmitter in presynaptic neurons or to show a structural similarity between the neurotransmitter and drugs that mimic its effects. Finally, pharmacological methods may be used to study the effects of drugs at synapses. For example, drugs that are known to block binding of the natural neurotransmitter with its postsynaptic receptors should, at similar doses, also block the effects of iontophoretic application of the chemical.

Although application of a full battery of tests to neurotransmission at a particular synapse is rarely possible, researchers have been able to apply at least some of these tests to many different synapses. As a consequence, the identity of many neurotransmitters is now supported by good evidence.

Four criteria must be met for a chemical to be identified unambiguously as a neurotransmitter. It must be synthesized in, stored in, and released from the presynaptic neuron, and when applied exogenously at the synapse by iontophoresis, it must mimic the effect of the natural neurotransmitter on the postsynaptic neuron. Physiological, anatomical, chemical, and pharmacological methods can help identify neurotransmitters.

The Main Neurotransmitter Chemicals

Most of the known or suspected neurotransmitters fall naturally into three chemically related groups: the amino acids, the amines, and the peptides. Amino acids are the building blocks of proteins, characterized chemically by the presence of acid and amine (NH_2-containing) chemical groups. A surprising number of amino acids have been shown to be neurotransmitters. Glutamate is well known as an excitatory neurotransmitter at most excitatory neuromuscular junctions in insects, as well as in the vertebrate brain. Aspartate is also an excitatory neurotransmitter. Gamma-aminobutyric acid (GABA) and glycine have inhibitory postsynaptic effects.

Amines contain an amine group but lack the acid group that characterizes the amino acids. Amine neurotransmitters, often referred to as biogenic amines because of their biological action, include chemicals like histamine, octopamine, and serotonin. Octopamine is especially important in invertebrate central nervous systems. Some amines contain a catechol group (a six-carbon ring with two OH units) and hence are referred to as catecholamines. The main catecholamine neurotransmitters are dopamine, epinephrine (adrenaline), and norepinephrine (noradrenaline). The amines form one of the most important groups of neurotransmitters. Serotonin and dopamine especially have enormous influence over many brain functions, from mood to simple movement, due to the widespread distribution of neurons that release these neurotransmitters. Norepinephrine is released by sympathetic postganglionic neurons of the autonomic nervous system. Both it and epinephrine are also important neurotransmitters in the central nervous system.

Peptides, sometimes called neuropeptides, which consist of small chains of amino acids, constitute a third group of neurotransmitters. The neurotransmitter action of peptides such as substance P, proctolin, and the endorphins was discovered several decades after that of many other types of neurotransmitters, but researchers now recognize that the peptides play an important role in many neural functions. Substance P, for example, has been identified in vertebrates as the neurotransmitter released by sensory neurons that convey information about pain (see Chapter 14). Some peptide neurotransmitters, such as the endorphins that are important in ameliorating pain (see Chapter 14), were discovered only after receptors for them were identified. Others were discovered to be cotransmitters, that is, substances that were present in a presynaptic terminal and released as a neurotransmitter along with one or more other chemicals. Proctolin, for example, a peptide recognized as a cotransmitter in invertebrates, is released along with glutamate at the neuromuscular junction. Further, a substantial number of peptides have now been identified that are cotransmitters with other neurotransmitters in autonomic neurons (see later section). In fact, one of the few generalizations about neurotransmitters seems to be that peptides are always cotransmitters with other chemicals.

The remaining neurotransmitters constitute a miscellany of substances that have no special chemical relationship to each other or to any other neurotransmitters: acetylcholine (ACh), adenosine triphosphate (ATP), and the gases nitric oxide (NO) and carbon monoxide (CO). ACh is well known as the neurotransmitter at the neuromuscular junctions of vertebrate somatic muscles. It is also released by parasympathetic postganglionic and all preganglionic neurons of the autonomic nervous system, as well as being a neurotransmitter in the central nervous systems of insects. Furthermore, evidence is accumulating that it acts presynaptically in the vertebrate brain. ATP is co-released with catecholamines; it has been suggested that it has a separate neurotransmitter function, especially in the autonomic system.

The gasses NO and CO represent the latest and most unusual neurotransmitters to be identified. Being soluble gases, they are neither stored in nor released from synaptic vesicles but are synthesized as needed. After synthesis, each gas simply diffuses away from the site of synthesis, easily crossing through any cell membrane that it may encounter. Nitric oxide is important in the peripheral autonomic system of vertebrates, and both it and CO have been implicated in the cellular basis of learning in mammals (see Chapter 8). There is evidence that CO is a neurotransmitter at synapses in the vertebrate olfactory bulb. The effects of both neurotransmitters are neuromodulatory; they interact directly with the biochemical machinery of the target neuron instead of interacting with a specific molecular receptor as other neurotransmitters do.

How many chemical substances are actually employed as neurotransmitters in animals? This question has no definitive answer because for some chemicals the evidence that they act as neurotransmitters is still scanty, and because new putative neurotransmitters seem to be identified every year. The best-known and most thoroughly studied neurotransmitters recognized at present are listed in Table 7-2. Many more substances have been proposed to act as neurotransmitters, especially peptides. Including all the peptide molecules that have been shown to have neuromodulatory effects on the nervous system, there is good evidence for the participation of well over 50 compounds as neurotransmitters in one animal or another. Of these, more than half are peptides.

In addition to the terms describing the chemicals that are neurotransmitters, a specialized terminology has also been developed to describe the neurons that release specific neurotransmitters. For example, neurons that release any catecholamine are called **catecholaminergic neurons**, those that release ACh are called **cholinergic neurons**, and those that release serotonin are called **serotonergic neurons**. The terms for neurons that release epinephrine and norepinephrine, **adrenergic neurons** and **nonadrenergic neurons**, refer to the alternative names for these neurotransmitters. In addition, two special terms, nicotinic and muscarinic, are used to describe the two main types of ACh receptors introduced in Chapter 6. The most common terms and their definitions are listed in Table 7-3.

TABLE 7-2. Neurotransmitter Chemicals

Amino Acids:

- Aspartate
- *γ-aminobutyric acid (GABA)
- *Glutamate
- Glycine

Amines (biogenic amines):

- Histamine
- Monoamines:
  - *Octopamine
  - Serotonin (5-hydroxytryptamine; 5-HT)
  - Catecholamines:
    - Dopamine (DA)
    - Epinephrine (adrenaline)
    - Norepinephrine (NE; noradrenaline, NA)

Peptides:

- Substance P
- *Proctolin
- Endogenous opioids (endorphins)
- Enkephalins
- Angiotensin
- Carnosine
- Glucagon
- Neurotensin
- Oxytocin
- Somatostatin
- Vasopressin
- Various hormone releasing factors

Other neurotransmitters:

- Acetylcholine (ACh)
- Adenosine triphosphate (ATP)
- Nitric oxide (NO)
- Carbon monoxide (CO)

*Especially prominent in, or found only in, invertebrates

Most neurotransmitters are amino acids, amines, or peptides. The catecholamines are amines. Other neurotransmitters include ACh, ATP, NO, and CO. More than 50 compounds have been proposed to be neurotransmitters.

Pathways for Synthesis of Neurotransmitters

As discussed in the previous section and shown in Table 7-2, many neurotransmitters have similar chemical properties. Chemically related neurotransmitters also have synthetic pathways in common. As you read about these pathways, keep in mind that many neurons may synthesize and release more than one type of neurotransmitter. Hence, the presence of the enzymes necessary for the synthesis of one neurotransmitter by no means precludes the presence of any other neurotransmitter in that same neuron. In fact, present evidence suggests that some neurotransmitters, especially the peptides, always coexist in a neuron with at least one other neurotransmitter. In some cases, the cotransmitters are released by separate synaptic vesicles. In other cases, two neurotransmitters may be packaged together in a single vesicle. For example, peptide neurotransmitters are always packaged together with some other neurotransmitter in synaptic vesicles.

Neurotransmitters are synthesized in a number of ways. The amino acid neurotransmitters are either obtained from the animal's diet or synthesized in neurons in the same way they are synthesized in other cells of the body. Glutamate and aspartate are synthesized via the Krebs cycle. GABA is formed by decarboxylation of glutamate by the enzyme glutamic acid decarboxylase (GAD); glycine is derived from glucose via the amino acid serine.

A major subgroup of the amines is the group known as catecholamines. These are molecules that include a catechol ring, as illustrated in Figure 7-9. The figure shows the synthetic pathways for the three catecholamines that have been identified as neurotransmitters: dopamine, norepinephrine, and epinephrine. All three are synthesized via a single synthetic pathway, starting from the amino acid tyrosine. The first step in the synthesis is catalyzed by the enzyme tyrosine hydroxylase. This enzyme is somewhat unusual in that it is limited in its action by the presence of the neurotransmitter end products, not DOPA, which is the product of the first step.

TABLE 7-3. Neurotransmitter and Receptor Terminology

Term	Meaning
*Adrenergic	Neurons that release epinephrine
Catecholaminergic	Neurons that release any catecholamine
Cholinergic	Neurons that release acetylcholine
Muscarinic	Receptors for ACh found in the CNS, the autonomic nervous system, and in smooth and cardiac muscle; they show a slow response, either excitatory or inhibitory action, and are blocked by atropine
Nicotinic	Receptors for ACh found at the neuromuscular junction; they have a quick response, are only excitatory, and are blocked by curare
Noradrenergic	Neurons that release norepinephrine
Serotonergic	Neurons that release serotonin

*The term adrenergic arises from adrenaline, the other term for epinephrine.

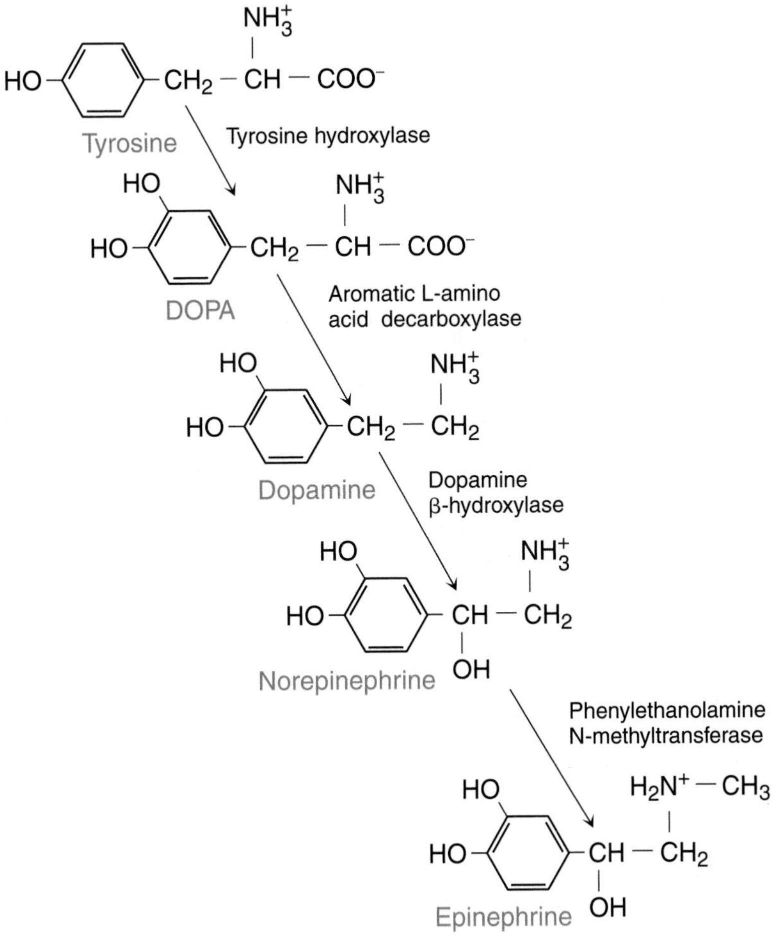

FIGURE 7-9. Pathway of synthesis of the catecholamines. Epinephrine is a derivative of norepinephrine, which in turn is a derivative of dopamine.

Catecholamines are degraded by two enzymes. One, monoamine oxidase (MAO), is an enzyme that will break apart most amines. The other, catechol-O-methyltransferase (COMT), is specific to the catecholamines.

Serotonin, although also an amine, is not a catecholamine but a derivative of the amino acid tryptophan (Figure 7-10). The enzyme tryptophan hydroxylase catalyzes the first step in the synthesis of serotonin. The neurotransmitter end product is sometimes referred to by its chemical name, 5-hydroxytryptamine, and sometimes by the abbreviated form, 5-HT. Its degradation in neurons is carried out mainly by MAO.

The presence of peptide neurotransmitters in the brain has been recognized only since about the 1970s. Because they consist of small chains of amino acids, these neurotransmitters are probably all synthesized in the cell soma, where the chemical machinery for assembling these chains is located. Some peptide neurotransmitters are synthesized directly, but others, such as the endorphins, are cleaved from longer protein chains. They are all degraded by specific and nonspecific aminopeptidases.

Acetylcholine is synthesized from choline and acetyl-CoA, as shown in Figure 7-11. Synthesis of ACh takes place mainly in the terminals of the axon. ACh is typically degraded in the synaptic cleft through the action of acetylcholinesterase

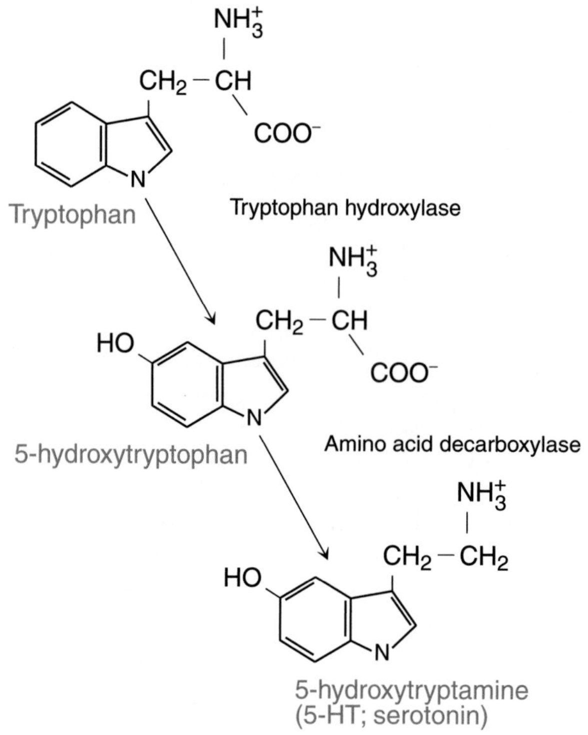

Tryptophan

Tryptophan hydroxylase

5-hydroxytryptophan

Amino acid decarboxylase

5-hydroxytryptamine (5-HT; serotonin)

FIGURE 7-10. Pathway of synthesis of serotonin.

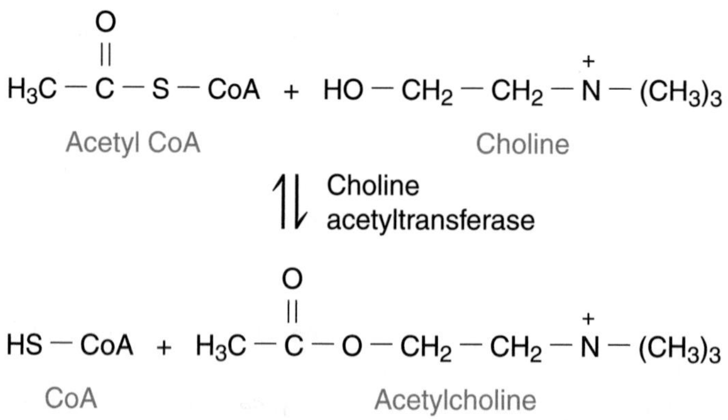

$$H_3C-C(=O)-S-CoA + HO-CH_2-CH_2-N^+-(CH_3)_3$$

Acetyl CoA — Choline

$\rightleftharpoons$ Choline acetyltransferase

$$HS-CoA + H_3C-C(=O)-O-CH_2-CH_2-N^+-(CH_3)_3$$

CoA — Acetylcholine

FIGURE 7-11. Pathway of synthesis of acetylcholine, which is made from acetyl-CoA and choline, catalyzed by choline acetyltransferase.

(AChE), forming choline and acetate. Choline is taken up again by the presynaptic terminal, where it can be reused to form more ACh.

Adenosine triphosphate is synthesized in mitochondria in all cells, including neurons, as a product of the metabolism of glucose. It is formed by phosphorylation of adenosine diphosphate. ATP is most often found packaged in vesicles along with a catecholamine or ACh. It has been shown to act as a neurotransmitter at some synapses but not at all of the synapses at which it is released.

Nitric oxide, which is synthesized by nitric oxide synthase, is a by-product of the conversion of the amino acid arginine to citrulline. As a gas, it can readily pass through any membrane, and therefore is not stored in vesicles. It has an effective life in a cell of only a few seconds, and hence is synthesized as needed. It has been shown to be present in several areas of the brain and in the autonomic nervous system. Carbon monoxide is split from heme by heme oxygenase, forming CO and biliverdin, which is reduced to bilirubin. Carbon monoxide is present in the hippocampus and the olfactory bulb.

Neurons may contain synthetic pathways for more than one type of neurotransmitter, and several neurotransmitters may be packaged together in single synaptic vesicles. Many amino acid neurotransmitters are obtained from an animal's diet. Among the amines, the catecholamines share a single synthetic pathway. The peptide neurotransmitters are synthesized in the cell body, whereas ACh is synthesized in the axon terminal. The gases NO and CO are also synthesized in the axon terminal.

Neurotransmitters in the Central Nervous System

The enormous variety of neurotransmitters presents an interesting problem for neurobiologists—what are they all for? The response of a neuron to a neurotransmitter is dependent on the type and properties of the receptor molecule on the postsynaptic neuron to which that neurotransmitter binds. Why, then, are there so many different neurotransmitters? It would seem that just a handful would be able to serve the needs of the nervous system. It could be a case of evolutionary accident, a matter of mutations in membrane proteins yielding molecules that respond to various chemical substances that nerves happen to release. It is also possible that particular neurotransmitter systems evolved because different chemicals happened to best serve neurons that had specific functions.

At present, investigators are faced with a multitude of neurotransmitter types distributed in various ways in the nervous system. In the periphery, for example, ACh is released by parasympathetic postganglionic neurons and by all preganglionic neurons of the autonomic nervous system. In contrast, norepinephrine is released by sympathetic postganglionic neurons. In the brain, GABA is released mainly by local neurons in the cortex and cerebellum. Dopamine, in contrast, is a neurotransmitter mainly of neurons whose cell bodies lie in the brain stem. Dopaminergic neurons in the substantia nigra project to the striatum, whereas those in the ventral tegmental area of the mesencephalon project to various parts of the limbic system. Other neurotransmitters also have specific distributions in the brain.

A great deal of effort has gone into identifying the neurotransmitters associated with specific brain pathways and functions, in part because of the possible therapeutic value of such knowledge. Continued rapid progress in identifying the specific neural pathways in which particular neurotransmitters are localized can be expected in the near future as a result of the introduction of powerful new techniques such as immunohistochemistry (Box 7-3). In fact, some of the new techniques may even help to identify new neurotransmitter candidates by identifying substances that have localized and highly specific distributions in the central nervous system.

Having the right method for localizing

neurotransmitters is critical. For example, it was in the 1950s that catecholamines were first determined to be present in the brain. Unfortunate-ly, at that time the chemical steps necessary for identification of these substances destroyed brain tissues, so it was not possible to locate the specific neurons with which catecholamines were associated. This difficulty was overcome in 1962 with the introduction of the histofluorescence technique (Box 7-3), in which compounds that fluoresce different colors when bound to amines, catecholamines, and serotonin are formed. Because neurons that release either catecholamines or serotonin contain small amounts of these substances along the length of their axons and in their cell bodies as well as at the synaptic endings, the histofluorescence methods are also excellent for revealing nuclei and tracts of neurons containing any of these neurotransmitters.

Application of the histofluorescence technique to vertebrates has led to useful therapeutic approaches to several diseases that result from malfunctions of specific neurons. For example, **Parkinson's disease** is a debilitating disease affecting motor control that all too often leaves patients frozen into rigid immobility, completely unable to care for themselves. Postmortem examination of individuals afflicted with the disease have revealed considerable degeneration of dopaminergic neurons of the substantia nigra that project to the striatum, a component of the basal ganglia (see Chapter 18). This finding led to treatment of Parkinson's disease with L-dopa, the precursor for dopamine, in the hope that with more precursor, the few dopaminergic neurons that remained would be able to synthesize more dopamine, and hence overcome the behavioral problem. This approach proved quite successful. Administering L-dopa to most parkinsonian patients leads within less than half an hour to movements that look perfectly normal. Treatment has hence allowed many people afflicted with the disease to lead normal lives.

Other neurotransmitters have even more dramatic effects on the brain and brain function, as indicated by the remarkable effects of many drugs that interfere with these neurotransmitter systems. Prozac, for example, is used to treat depression. It blocks uptake of serotonin, leading to an increase in the amount of serotonin available at postsynaptic sites that receive input from serotonergic neurons. A number of other drugs that have specific effects on particular neurotransmitter systems are used to treat medical conditions caused by the failure of one of these systems, but the search for better drugs is ongoing. Knowledge of which neurotransmitters are released by specific neurons and what the functional roles of those neurons are is an important component of this search.

New methods for the localization of neurotransmitters in the central nervous system have been developed to find the specific neurons and pathways associated with particular neurotransmitters. One potential benefit of knowing the distribution of particular neurotransmitters is the development of effective treatments for various brain malfunctions such as Parkinson's disease.

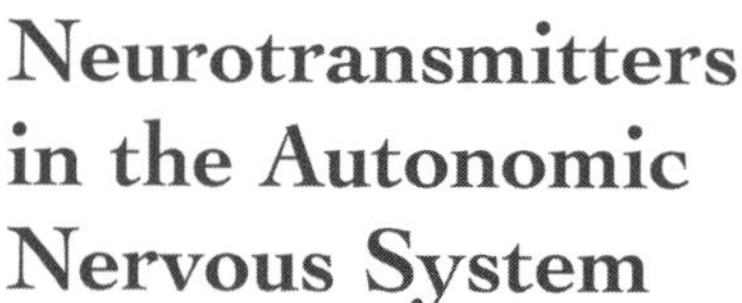

Neurotransmitters in the Autonomic Nervous System

Until the last few decades, the typical view of synaptic transmission in the autonomic nervous system was that ACh and norepinephrine were the only neurotransmitters. The discovery that not all sympathetic function could be blocked by drugs that block the action of norepinephrine suggested that other neurotransmitters must also be released from sympathetic postganglionic neurons. The current view is that, in addition to ACh and norepinephrine, at least three other classical neurotransmitters—dopamine, serotonin, and GABA—are neurotransmitters in the autonomic nervous system. Furthermore, many, if not all, autonomic neurons contain and release a bewildering array of cotransmitters. Many of these are neuropeptides, but ATP and the gas NO are also

BOX 7-3

Where Are Those Molecules?: Histological Localization of Transmission Proteins

Several techniques have been used for the localization of specific neurotransmitters and the proteins associated with them in the nervous system. One of these is **immunohistochemistry**, also called **immunocytochemistry** or **immunostaining**. In this method, a thin slice of the tissue is exposed to a fluorescently labeled molecule that has high affinity for a protein or peptide of interest, thereby revealing the distribution of the protein or peptide.

Immunohistochemistry depends on the ability of mammals to make antibodies to peptides or proteins. Originally, the procedure involved injecting a solution of a protein such as a receptor for a neurotransmitter into a rabbit and inducing it to make an antibody, usually an immunoglobulin such as IgG, to the protein. The antibody, which serves as a probe for localization of the protein of interest, would be extracted from the rabbit, bound to a fluorescent or other kind of marker, and spread over a slice of tissue containing the protein. Viewing the tissue through a fluorescence microscope would then allow a researcher to see the distribution of the protein, since the antibody marker would bind only with it.

This single-antibody procedure has now been replaced by a two-step procedure, in which a secondary antibody is generated that is specific to the primary one. The two-step procedure takes advantage of the molecularly unique tail possessed by all IgGs from one species. Hence, secondary antibodies can be generated to the primary antibody by injecting the primary IgG into another mammal, such as a goat. A marker can then be bound to the secondary antibody to make it visible (Figure 7-C). An advantage of the second step is increased sensitivity because more than one of the secondary antibodies can bind with the primary antibody, hence increasing the intensity of the fluorescence at the sites of the target proteins.

Radioimmunolabeling is another method used to localize proteins or neurotransmitters. In this method, a radioactively labeled secondary antibody is used, and the labeled molecule is made visible by placing unexposed photographic film or emulsion next to a slide containing the tissue being examined, then allowing the radioactive decay to expose the film or emulsion in the region where it is localized. For neurotransmitters like ACh, which are not peptides, researchers can make antibodies to an enzyme, choline acetyltransferase in the case of ACh, that is essential for synthesis of the neurotransmitter. The distribution of the enzyme then gives an indirect indication of the distribution of the neurotransmitter itself.

Histological methods that do not depend on immunological reactions can be used to localize neurotransmitters directly. In the **histofluorescence**

(continued)

Where Are Those Molecules?: Histological Localization of Transmission Proteins (continued)

technique, for example, a tissue slice is exposed to formaldehyde vapors. The formaldehyde reacts chemically with catecholamines like dopamine and norepinephrine to form a green fluorescent compound, whereas it reacts with serotonin to form a yellow-green fluorescent compound. Modern refinements of the method have made it quite sensitive to even small amounts of catecholamines or serotonin, allowing the distributions of these neurotransmitters to be mapped.

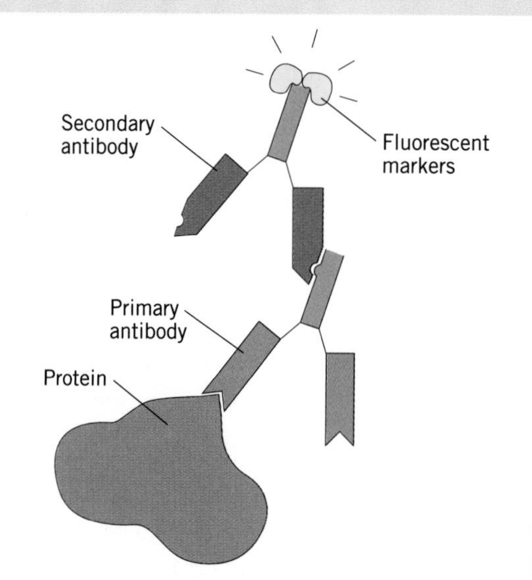

FIGURE 7-C. Antibody-protein binding that is the basis of the immunohistochemical technique for localizing proteins in tissue sections. An antibody is produced that is specific to the protein of interest. A secondary antibody specific to the primary antibody is produced and is bound with a fluorescent marker. The distribution of the marker in tissue sections will match the distribution of the protein of interest.

common. More than a dozen peptide neurotransmitters have already been identified, and many researchers expect additional ones to be discovered as well.

In contrast to the case in somatic neurons, peptide neurotransmitters in autonomic neurons are often packaged in vesicles together with nonpeptide neurotransmitters. Neurons in which this double packaging occurs have two functional types of vesicles. Some contain the nonpeptide neurotransmitter alone, whereas others contain this neurotransmitter along with the peptide. Which vesicles are released when synaptic transmission occurs depends on the frequency at which the presynaptic neuron is activated. Low-frequency stimulation causesrelease only of vesicles containing the nonpeptide neurotransmitter, whereas high-frequency stimulation causes release of both types of vesicles. This dual response will bring about a different postsynaptic effect when the synapse is activated at a low than when it is activated at a high frequency.

In addition to ACh and norepinephrine, neurons of the autonomic nervous system may release serotonin, dopamine, GABA, NO, ATP, and many different peptides as neurotransmitters. In some cases, several neurotransmitters may be packaged together in single vesicles. Since different classes of vesicles will be released at different frequencies of presynaptic activity, the postsynaptic effect of an autonomic neuron depends on the frequency at which it is active.

Additional Reading

General

Cooper, J. R., F. E. Bloom, and R. H. Roth. 1996. *The Biochemical Basis of Neuropharmacology*. 7th ed. New York: Oxford University Press.

Erulkar, S. D. 1994. Chemically mediated synaptic transmission: An overview. In *Basic Neurochemistry*, 5th ed., ed. G. Siegel, B. Argranoff, R. W. Albers, and P. Molinoff, 181–208. New York: Raven Press.

Katz, B. 1966. *Nerve, Muscle, and Synapse*. New York: McGraw-Hill.

——1969. *The Release of Neural Transmitter Substances*. Liverpool: Liverpool University Press.

Research Articles and Reviews

Amara, S. G., and M. J. Kuhar. 1993. Neurotransmitter transporters: Recent progress. *Annual Review of Neuroscience* 16:73–93.

Emson, P. C. 1993. *In-situ* hybridization as a methodological tool for the neuroscientist. *Trends in Neurosciences* 16:9–16.

Hökfeld, T. 1992. Neuropeptides in perspective: The last ten years. *Neuron* 7:867–79.

Kelly, R. B. 1993. Storage and release of neurotransmitters. *Cell 72/Neuron* 10 (review supplement): 43–53.

Littleton, J. T., and H. J. Bellen. 1995. Synaptotagmin controls and modulates synaptic-vesicle fusion in a Ca^{2+}-dependent manner. *Trends in Neurosciences* 18:177–83.

Matthews, G. 1996. Neurotransmitter release. *Annual Review of Neuroscience* 19:219–33.

CHAPTER 8

Integration of Synaptic Action

At one time neural communication was considered to be entirely a matter of yes or no, excitation or inhibition. Researchers now know that communication can actually be modulated and shaped in a wide variety of ways. Not all input to a neuron has the same strength, nor do all neurons respond equally well to a particular input. Furthermore, inputs to a neuron can interact to bring about new types of responses. A given neuron can even respond differently to the same input at different times. These subtleties of synaptic transmission are the subject of this chapter.

Communication between a pair of neurons is relatively simple. However, neurons normally receive input from many neurons at once and are therefore continually bombarded by input from many different cells. These inputs can interact to shape the response of the receiving cell. Furthermore, some types of messages can modify the response of the receiving neuron to subsequent messages. Hence, in the nervous system, communication is strongly influenced by interactions between different input signals. **Integration** is the name given to the process by which signals in the nervous system interact. At the cellular level, you can think of it as the decision-making capability of neurons. Because integration determines the way in which neurons respond to input, it is an integral part of the function of the nervous system.

Integration and Electrical Properties

In Chapter 6 you learned how a postsynaptic potential can be generated in the dendrite of a neuron by the flow of ions through open channels in the postsynaptic membrane. Once it has been generated, this potential spreads largely passively (meaning that there is no action potential) in all directions along the cell membrane, becoming smaller as it spreads. This decrement in the amplitude of any subthreshold potential in a neuron is the reason that voltage-sensitive ion channels and action potentials are necessary to move potentials long distances along an axon. The postsynaptic potential spreads away from its point of origin because there is a difference in electrical potential between the region of the cell membrane where the synaptic potential is at its peak, and nearby parts of the membrane that are still near resting levels. This potential difference causes a flow of current from the region of the synaptic potential to adjacent areas of the membrane.

The loss of amplitude of the postsynaptic potential as it spreads applies equally to excitatory postsynaptic potentials (epsps) and inhibitory postsynaptic potentials (ipsps); for the sake of simplicity, however, this discussion will focus on what happens when an epsp is generated. The amplitude of an epsp at a synapse is determined largely by the net influx of positive ions into the cell through open ion channels, although, as you will learn, the electrical properties of the local membrane contribute as well. As the epsp spreads, its amplitude as recorded at any place on the membrane is dependent on how much of the positive current entering the cell has reached the recording site. Therefore, any factor that can affect the flow of ions will affect the size of the epsp as it spreads away from the synapse.

Three factors are especially important in influencing ion flow: (1) the resistance to ion flow across the cell membrane, (2) the resistance to ion flow inside the cell, and (3) the cell membrane's capacity for storing charge. Each of these factors is an electrical parameter that can be measured or calculated for a neuron. The question is how these factors influence the spread of current, and hence influence the spread of an epsp. In the discussion that follows, the factors will be consiered independently. You should realize, however, that they actually interact with one another, as you will learn in a subsequent section.

The easiest of these factors to understand is the resistance of the cell membrane, which is a measure of how difficult it is for an ion to cross it. Because ions cross through ion channels, a membrane entirely lacking in channels would present a formidable barrier to the flow of ions, and hence would effectively have an infinite resistance. All cell membranes contain ion channels, however, and even at rest there are open channels (see Chapter 4) that allow ions to cross. The higher the density of open channels in the membrane, the lower the membrane resistance; the lower the density, the higher the resistance. Because membranes may differ in the number of leak channels or other channels that are open, the membrane resistance of different cells may not be the same. The resistance may even be different in parts of the same cell. Resistance is important for the epsp because the lower the resistance of the dendritic membrane along which the epsp is flowing (i.e., the more leaky the membrane), the more quickly the epsp will decay.

A second electrical property of neurons that can affect an epsp, and hence the integration of postsynaptic events, is the internal (longitudinal) resistance of the cytoplasm in the cell. Consider the fate of an epsp in a cell. If the internal resistance of the cell is quite high, ions are impeded in their movement along the length of the neurite, and therefore are more likely to leak out close to the origin of an epsp, causing it to dissipate relatively close to its point of origin. On the other hand, if the internal resistance is low, ions will be able to move farther away from the origin of the epsp before they leak out, resulting in a greater distance of travel of the epsp before it dissipates.

The third electrical property that can influence postsynaptic potential is the capacitance of the membrane. Formally, capacitance (C) is the ability of a capacitor to store electrical charge; it is defined as the amount of charge (Q) held by a capacitor per unit of voltage (V), $C = Q/V$, in farads (coulombs/volt). A capacitor consists of two surfaces that can conduct electricity, separated by a nonconducting barrier. In membranes as in electronic devices, capacitance represents a possible reservoir of charge. If the capacitance of the membrane is large, the membrane will be able to store a relatively large number of ions that have not moved away by leaking out of the cell or spreading away from the point at which they entered the cell. If the capacitance of the membrane is small, fewer ions can be stored. The amount of charge stored by the membrane affects the amplitude of an epsp at a particular location and over time.

A water analogy might be helpful for understanding the concept of capacitance. Water will flow in pipes if pressure is applied to it, just as electrons will flow in conductors if a battery or other electromotive force is applied to them, and ions will flow across cell membranes if there is a driving force and an opening for them to pass through. Consider an arrangement of pipes, a source of water pressure, and a rubber diaphragm (Figure 8-1A). If the valve connecting the pipes to the pressure is suddenly opened, water will begin to flow. Initially, some of the water will displace the diaphragm in the center, and some will flow through the constriction on the right (R) (left panel). Only when the diaphragm is stretched to its full extent by the water pressure applied to it will all the water flowing through the valve also flow through the constriction (middle panel). When the valve is closed, water will continue to flow for a short time in the right-hand loop because the rubber diaphragm will now give up the water that had been pushed up against it (right panel).

Now consider the electrical situation. When current from a battery is suddenly allowed to flow through a circuit consisting of resistors and a capacitor (Figure 8-1B), some of the current will flow through the resistor (R), and some will flow to the capacitor (left panel). The current to the capacitor will cause

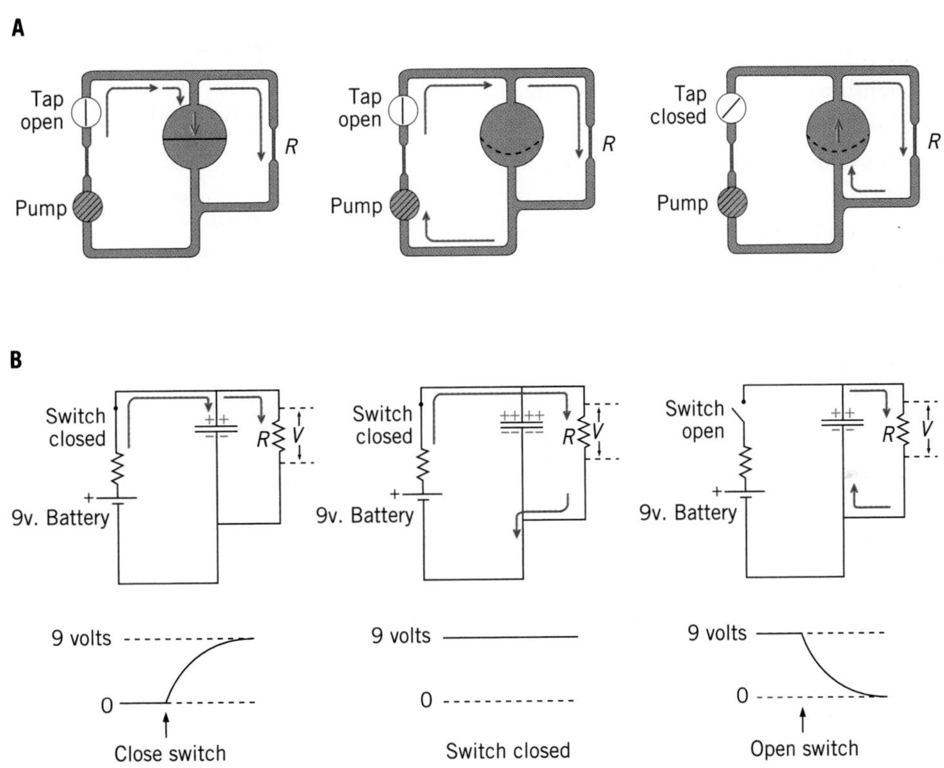

FIGURE 8-1. (A) A water analogy of capacitance. The pump represents an electromotive force like a battery, the tap represents a switch, and the rubber diaphragm in the center pipe represents a capacitor, which can store water pressure just like an electrical capacitor can store electrical charge. Resistance in the circuit is represented by narrowed regions of pipe. Opening the tap (left) causes the diaphragm to stretch like a balloon. Water flows through the pipes until the pressure of water from the pump equals the counterpressure generated by the diaphragm as it stretches (center). When the tap is closed, water will no longer flow in the left loop of pipes. However, the elasticity of the diaphragm will force water along the right loop until the diaphragm regains its original, unstretched position (right). (B) The charging of a capacitor in an electrical circuit. At the instant the switch is closed, current flows from the positive pole of the battery around the loops, charging the capacitor (left). The presence of positive charge on one plate of the capacitor attracts electrons and repels positive charge from the opposite plate. After the capacitor is fully charged, no more current flows through the middle of the circuit (center), even though the switch is still closed. When the switch is opened, current will continue to flow in the right-hand loop for a short while, driven by the accumulated charge on the capacitor (right). The recordings of voltage under the electrical circuits show the voltage that would be recorded across the resistor R when the switch is first closed, when it has been closed for a while, and when it is first opened again. Notice that voltage change is not instantaneous.

a buildup of positive charge on the side of the capacitor connected to the positive pole of the battery. If the second surface is sufficiently close to the first, the positive charge on one surface will attract electrons and repel positive charge from the second surface, making it negative. Only when the capacitor is fully charged by the electromotive force of the battery will all the current flowing through the switch also flow through the resistor R (middle panel). When the switch is opened again, current will continue to flow for a short time in the right-hand loop because the capacitor will now give up the charge that it had stored (right panel).

A cell membrane acts like a capacitor. Positive and negative ions can accumulate on the two sides of the membrane, separated by the nonconducting lipid bilayer membrane itself. Because most membranes are about the same thickness, the main factor that makes the capacitance of one neuron different from that of another is the surface area of the neuron. The membrane will store charge, and this means that the currents generated by an epsp will last slightly longer in cells with high capacitance than in those with low capacitance. Electrical properties of a cell can readily be measured (Box 8-1).

Integration, the process by which signals in the nervous system interact, is mainly influenced by three electrical characteristics of the cell: the resistance of the membrane, the resistance to ion flow inside the cell, and the cell membrane's capacitance. These factors influence the spread of a membrane potential by affecting the flow of ions within the neuron.

Electrical Constants of a Neuron

The electrical characteristics of membranes affect a neuron in two important ways—the distance that a postsynaptic potential can spread along a membrane, and how long a postsynaptic potential will last at any one location. These two factors are crucial determinants of the integrative properties of the membrane and hence of the way in which a neuron can interact with other neurons.

The Length (Space) Constant

The distance that a postsynaptic potential can spread is expressed as the **length** (or **space**) **constant** of the membrane, and is symbolized by the Greek letter lambda (λ). In formal terms, it is defined as the resistance of a unit length of membrane (r_m, in ohm-cm) divided by the resistance of a length of cytoplasm (r_i, in ohm/cm), as shown in equation (1):

$$\lambda = \sqrt{r_m / r_i}$$

It can also be described qualitatively as the distance along a neurite at which a constant applied voltage will decay to about 37% (strictly, $1/e$) of its original value (Figure 8-2).

If the membrane resistance r_m is large, then λ is large, and the postsynaptic potential can spread farther because relatively few ions are lost across the membrane. Conversely, if r_m is low, then there will be a substantial ionic leak, fewer ions will be available to carry current to nearby regions of the neuron, and λ will be smaller. With regard to r_i, a large internal resistance, as found in axons with a small internal diameter, will tend to keep ions from spreading far, and hence will keep λ small. Conversely, a small r_i, as found in large-diameter axons, will not impede the flow of current as much and will allow the epsp to spread farther; λ will therefore be larger. In this way, membrane and internal resistance interact to impart to the cell a characteristic value for the distance that a postsynaptic potential can spread before becoming too small to be

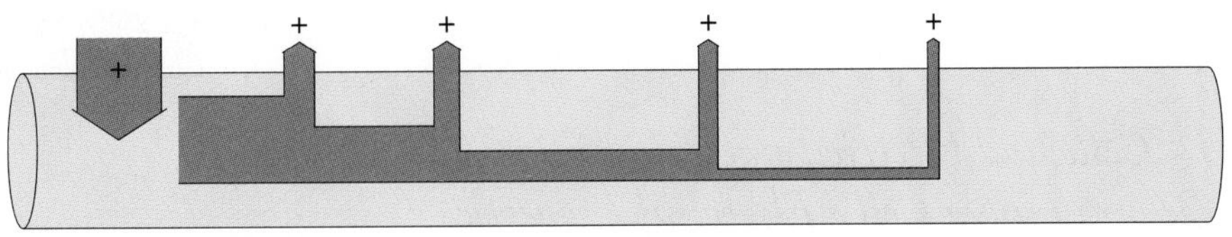

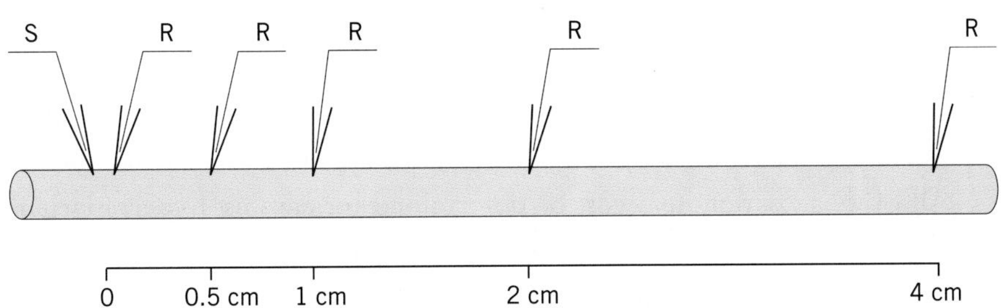

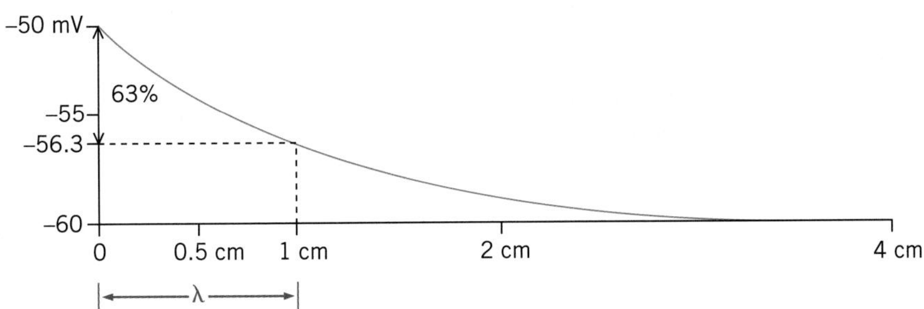

FIGURE 8-2. Graphical depiction of the length (space) constant (λ). A constant voltage stimulus is applied to the electrode labeled S, and the membrane potential recorded at the electrodes labeled R. The voltage drop away from the voltage source is shown in the lower panel. The length constant is the distance over which the voltage drops to about 37% of its value at the stimulus source. The drawing in the top panel depicts the loss of electrical charge along the length of the neurite (in one direction only). Note, however, that the drawing shows current flow out of the cell only at four locations, whereas in actuality current is lost all along the membrane, as shown in the graph in the bottom panel.

effective. Large length constants mean that even synapses some distance from the spike initiation zone can influence the activity of a neuron.

Length constants vary from as little as a few hundred micrometers to as much as several millimeters. The length constant for a dendrite that may be only about 2 mm long can be comparable to the length of the dendrite, 2 mm or longer. Hence, a synapse at the far end of a dendrite with a length constant of this value can still generate an epsp large enough to be measurable at the spike initiation zone (SIZ) of the neuron.

The length constant of a membrane represents the distance along a neurite that an electrical potential such as an epsp can travel before it dissipates. Small-diameter neurites have smaller length constants than do large-diameter neurites.

The Time Constant

Once the ion channels in the postsynaptic membrane have closed, the amount of time

BOX 8-1

Of Cells and Electricity: Determining a Cell's Electrical Properties

The electrical properties of a neuron are critical determinants of the way the neuron responds to synaptic input. Fortunately, they can be estimated from measurements that can easily be made during intracellular recording. The most important electrical properties of interest to neurobiologists are the electrical resistance of the cell membrane, the internal resistance of the cell, and the capacitance of the cell (its ability to store electrical charge).

All of these parameters can be determined by passing hyperpolarizing current into a neuron with a microelectrode and examining the resulting change in the membrane potential (Figure 8-Aa). (Hyperpolarizing

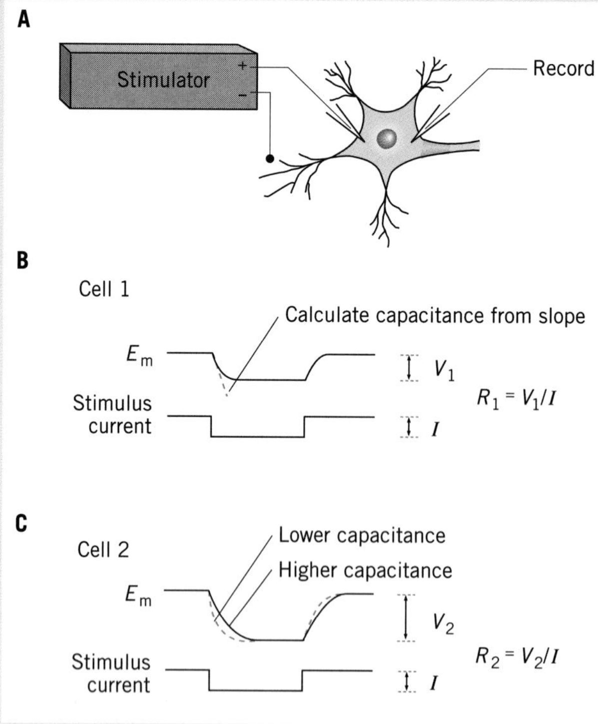

FIGURE 8-A. Computing the electrical properties of a cell membrane. (a) Recording conditions from a neuron with a short axon. The stimulating electrode is used to pass current into the cell and the recording electrode is used to record the resulting changes in membrane potential (E_m). (b) The effect of passing a pulse of hyperpolarizing current into a neuron. The resistance across the membrane is calculated as shown from the change in E_m and the current. The capacitance is calculated from the initial slope of the voltage change. (c) Cells will typically have their own characteristic membrane resistances and capacitances. Cells with higher capacitance (solid line) hyperpolarize more slowly than do cells with lower capacitance (dotted line).

current is used in order to avoid activating voltage-dependent sodium channels.) You may recall from physics that in an electrical circuit, voltage (V) is proportional to current (I) and resistance (R), as described by the equation $V = IR$. The resistance of a cell's membrane can be calculated from this equation by passing a known current into the cell and measuring the resulting change in voltage (Figure 8-Ab). In practice, the R value so calculated must be adjusted to take into account the spread of the current away from the tip of the electrode, but the principle is straightforward. The internal resistance of an axon can be determined in a similar way by inserting two microelectrodes into the axon some distance apart and conducting a similar experiment.

Cell capacitance requires a different approach. Because the capacitance of a membrane reveals itself in the rate at which electrical charge builds up or dissipates across the membrane, the capacitance can best be measured by determining the initial slope of the change in potential across a membrane when current is applied to the cell (Figure 8-Ab). Differences in the cell capacitance will be reflected in differences in the rate at which the cell membrane charges and discharges as current is injected for a short period, as shown by the dotted curves in Figure 8-Ac.

that a postsynaptic potential will last at some location is related to the **time constant** of the membrane, which is symbolized by the Greek letter tau (τ). The time constant is formally defined as the product of the membrane resistance (r_m, in ohm-cm) and the membrane capacitance (c_m, in farads/cm), as shown in equation (2):

$$\tau = r_m \cdot c_m$$

It can also be expressed qualitatively as the time it takes a constant applied voltage to build up to about 63% (strictly, $1 - 1/e$) of its final value (Figure 8-3).

As you can see from equation (2), the time constant is a function of both the membrane resistance and the membrane capacitance of the neuron. Membrane resistance can vary by several orders of magnitude, whereas capacitance varies only by a factor of two to three from cell to cell, so it is clear that variations in resistance contribute more to variations in the time constant than do variations in capacitance. Nevertheless, in order for you to understand the equation, it is useful to consider separately the contribution that these variables make to τ.

If the cell's capacitance is high, then some of the ions that would have flowed across the cell membrane during the epsp are stored instead and therefore do not leave the cell. As the membrane potential begins to return to resting levels, this stored charge keeps it at a depolarized level for a longer time, causing the epsp to last longer than it otherwise would have. Conversely, if the neuron's capacitance is low, there is only a small charge reservoir; thus the epsp decays more rapidly.

The membrane resistance influences the rate at which ions become available for storage in the membrane capacitance. For example, if the membrane resistance is low, ions will leak out of the cell rapidly and the epsp

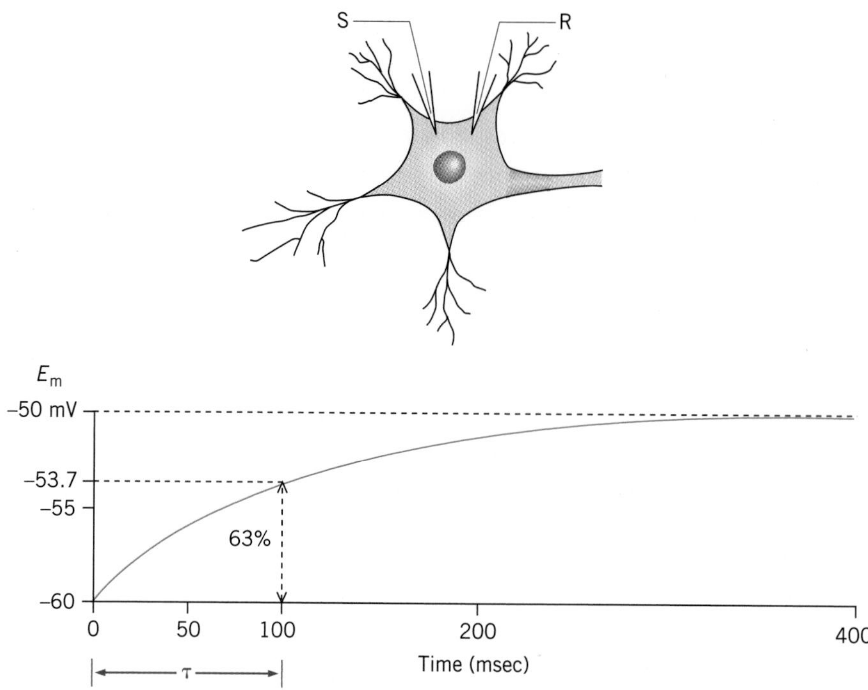

FIGURE 8-3. Graphical depiction of the time constant (τ). A stimulus is applied to S, and the membrane potential is recorded at R. The time course of the buildup of voltage at the place where the stimulus is applied is shown in the lower panel. The time constant is the time it takes for an applied voltage to rise to about 63% of its final value.

will decay quickly, even if the cell has a high capacitance. On the other hand, a high resistance keeps the ions from leaving the cell so quickly, thereby increasing the duration of the epsp. Small time constants for the membrane near a synapse, the product of a relatively small r_m and/or c_m, mean that the epsp will be relatively brief. Large time constants for the postsynaptic membrane, the product of a relatively large r_m and/or c_m, mean that synaptic potentials will be relatively long, and hence will be available for interaction with other potentials longer at a particular site than they would be if the membrane's time constant were small.

The time constant may vary considerably for different excitable cells. A typical value for τ is about 7.5 msec, which is the time constant for a crab axon. The extremes range from 1 msec for the squid giant axon to as much as 35 msec for a frog sartorius muscle.

The time constant of a membrane represents the time it takes for an electrical potential such as an epsp to decay at one location along the membrane. Neurons can have different time constants; dendrites with small time constants will have brief epsps, and those with large time constants will have longer ones.

Summation

The passive electrical properties of neurons as embodied in the length constant and the time constant determine the integrative properties of the cells by influencing how postsynaptic potentials interact with one another. Excitatory potentials can interact with inhibitory ones, as you learned in Chapter 6, but each type of potential also interacts with other potentials of the same kind. It is this type of interaction that will now be considered. For the sake of simplicity, the discussion will focus on the interactions of excitatory potentials. However, inhibitory potentials interact according to the same principles.

Summation is the combined postsynaptic effect of two or more presynaptic inputs of the same type. With regard to the summation of epsps, there are two keys to understanding the phenomenon: (1) In many neurons, perhaps even most, the amplitude of an individual epsp is below the threshold for excitation of the postsynaptic cell, so that by itself it cannot elicit an action potential. (2) Epsps can be added (summed) to one another, which means that two or more subthreshold epsps may sum to produce a potential that is above threshold at the SIZ of the postsynaptic cell. There are two types of summation—spatial and temporal.

Spatial Summation

Spatial summation refers to summation of two or more inputs from different locations that occur at about the same time. You can think of it as summation over the space of the postsynaptic cell membrane. Figure 8-4 illustrates this type of summation for two subthreshold epsps generated at synapses close together and close to the SIZ of the neuron. The potentials, about equal in amplitude, sum in a straightforward fashion to produce a combined postsynaptic potential that is well above threshold at the SIZ.

But consider the summation of epsps if one of them originates at a distant synapse

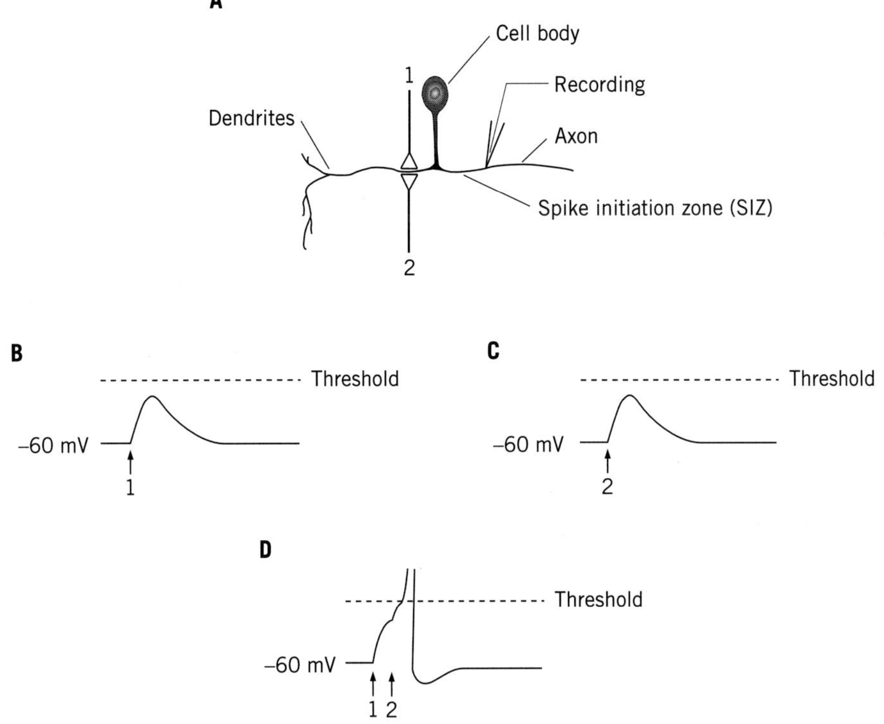

FIGURE 8-4. Spatial summation of inputs near one another. (A) The recording arrangement. An electrode is inserted into the postsynaptic neuron near the spike initiation zone and stimuli are applied to first one (1), then the other (2), then both presynaptic neurons to activate the respective synapses. (B) Activation of synapse 1 alone. The epsp is below threshold. (C) Activation of synapse 2 alone. The epsp is again below threshold. (D) The synapses are activated more or less together. The responses sum to produce a postsynaptic potential that exceeds threshold, and hence generates an action potential.

(Figure 8-5). Because this synapse is farther away, its epsp has decayed more by the time it reaches the SIZ. Therefore, the summed potential may not reach threshold at the SIZ. Thus, whereas two epsps from synapses near the SIZ may exceed threshold when they sum, those from more distant synapses may not. The value of the length constant determines just how effective distant synapses will be; if the length constant is large, an epsp will be able to travel farther without significant loss of amplitude than if the length constant is small.

The functional role of spatial summation can perhaps best be illustrated by an example. Suppose a motor neuron in an insect controls a muscle in a rear leg. It might be useful for the animal to be able to respond to touch of sensory hairs on the leg by contracting the muscle and thereby moving the leg. Perhaps some hairs are in a critical location, so touch on just a few of these should be able to elicit movement. Other hairs should also be able to elicit movement, but only if a stronger stimulus that moves many hairs at once were present. Such a differential response will be possible if the axons from the critical sensory hairs synapse on the motor neuron close to the SIZ, and the sensory neurons from the less important hairs synapse farther away. Few of the critical hairs have to move to elicit a response in the motor neuron because the summation of just a few epsps from their sensory axons is sufficient to bring the membrane potential above threshold. However, many of the less important hairs must move to elicit the same response because even if the epsps generated by input from different hairs are all the

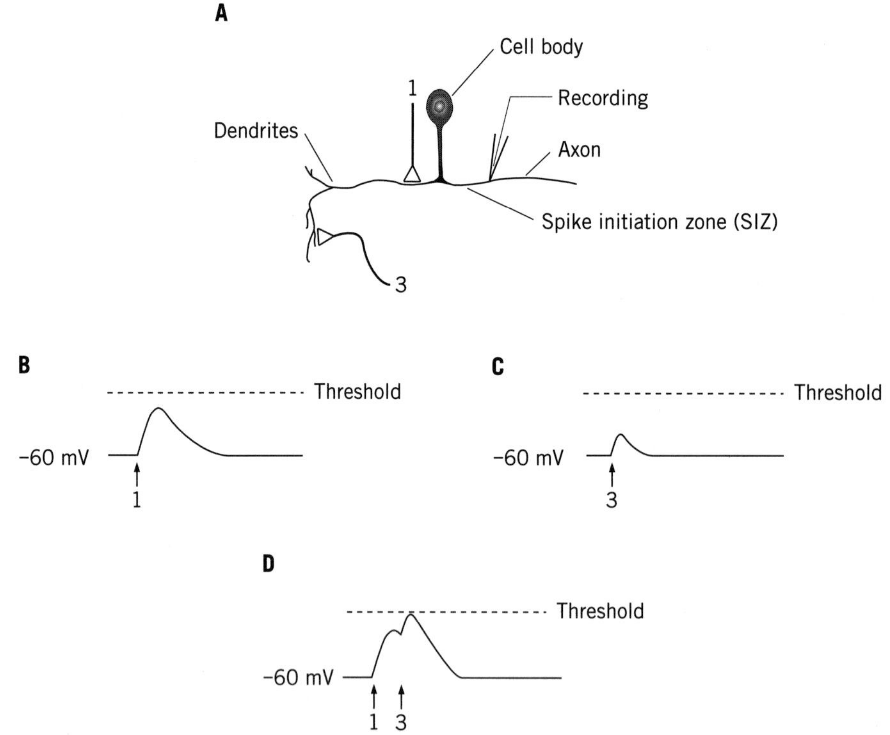

FIGURE 8-5. Spatial summation of inputs from different locations. (A) The recording arrangement is as in Figure 8-4, except that a more distant synapse (3) is present and activated. (B) Activation of synapse 1 alone. The epsp is below threshold. (C) Activation of synapse 3 alone. The epsp, smaller in this instance because the synapse is farther from the recording site, is again below threshold. (D) Synapses 1 and 3 are activated more or less together. Because the epsp from synapse 3 is so small, even the summed potential is too small to reach threshold, and hence no action potential is produced.

same size, the more distant epsps will have become so small by the time they reach the SIZ that many must sum for the membrane potential to exceed threshold.

Summation is the combined postsynaptic effect of two or more presynaptic inputs of the same type. Spatial summation—the summation of two or more inputs from different locations occurring at about the same time—is influenced by the length constant of the membrane. The larger the length constant, the more distant a synapse can be and still be effective in eliciting a postsynaptic response.

Temporal Summation

The discussion so far has concentrated on synaptic input that is more or less synchronous. Summation, however, is not restricted to events that occur at the same time, since epsps do not decay instantly (Figure 8-6). **Temporal summation** refers to summation of two or more inputs that occur at different times. You can think of it as summation over time. This may involve either the activity of one synapse more than once or the activity of different synapses at different times. The effectiveness of temporal summation is a function of the time constant: the larger the time constant, the longer it will take an epsp to decay, and therefore the longer it is available for summation with another epsp that comes along later. For example, a frog sartorius muscle, with a time constant of 35 msec, will respond strongly to inputs that occur at a relatively low frequency because each postsynaptic potential generated in the muscle lasts for a long time and hence can sum with later inputs. The squid giant axon, on the other hand, with a time constant of only 1 msec, is not well adapted for temporal integration because any subthreshold potential produced in it will decay so rapidly that even relatively frequent potentials will decay too quickly to interact.

Many animals take advantage of temporal summation by using neurons that have large time constants to receive important input that may occur at a relatively low frequency. In this way, even low-frequency, subthreshold stimulation of the neuron may result in summed postsynaptic potentials that are large enough to generate action potentials. Conversely, the animal may use neurons with small time constants to receive input to which it only wishes to respond if the input appears at a relatively high frequency. Obviously, there are many ways for differences in length or time constants to be

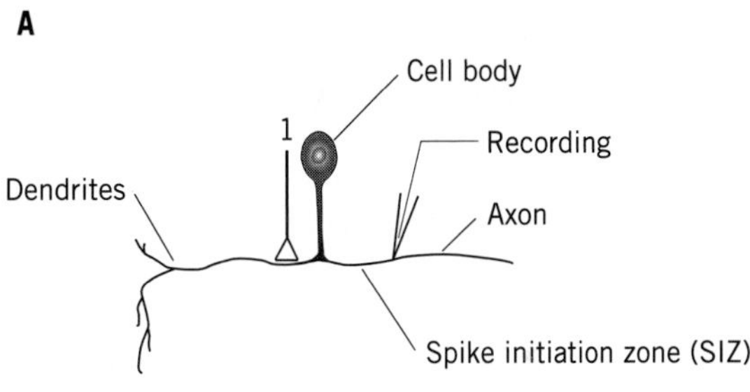

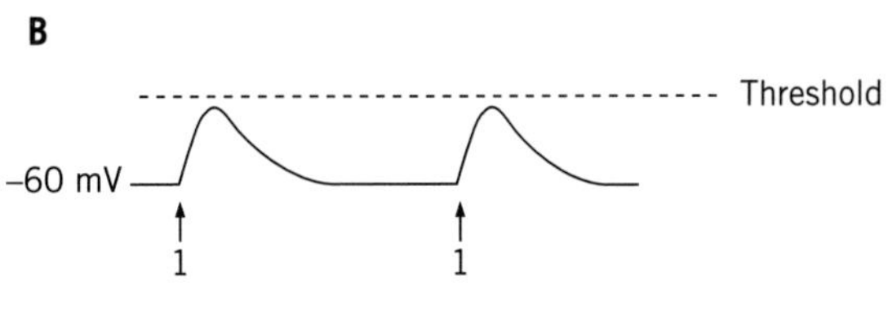

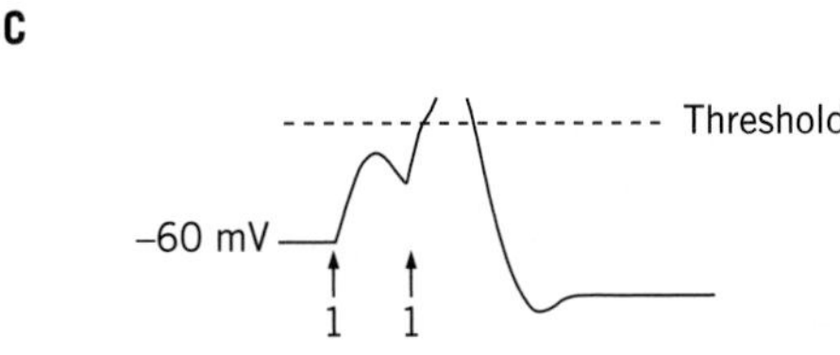

FIGURE 8-6. Temporal summation of a single input at different times. (A) The recording arrangement, showing a single synapse, which is activated twice in succession. (B) The response of the postsynaptic cell to activation of the synapse with some time between activations. The epsps are too far apart to sum. (C) The response of the postsynaptic cell to activation of the synapse twice in close succession. The responses sum to produce a postsynaptic potential that exceeds threshold, and hence generates an action potential.

used by nervous systems. It is important that you recognize the many factors that always work together to achieve a particular result.

Temporal summation is summation of two or more inputs that occur at different times. It may involve the activity of one synapse more than once or the activity of different synapses at different times. The effectiveness of temporal summation is a function of the time constant of the postsynaptic membrane. The larger the time constant, the longer a postsynaptic potential will last.

Integration of Excitatory and Inhibitory Potentials

Historically, summation has referred only to the interaction of postsynaptic potentials of the same type—epsps with epsps, and ipsps with ipsps. However, integration encompasses the interactions of epsps and ipsps as well. These interactions, described in Chapter 6, are also influenced by the electrical properties of the neuron.

Both the time and length constants are important in the integration of excitatory and inhibitory potentials. A relatively large time constant allows ipsps to interact with epsps for a longer period of time. Thus, in neurons with large time constants there is a longer window of opportunity for inhibitory input to suppress the generation of an action potential than in neurons that have smaller time constants. The length constant is also important. In neurons with large length constants, even inhibitory synapses that are distant from the SIZ will be able to exert a significant influence on the generation of spikes.

Neurons may have quite different electrical properties, which arise from two sources: properties of the neuron that arise largely due to the expression of the cell's genome and properties that are readily modified over the life of the cell. For example, cell capacitance and the internal resistance of a neuron are determined in large part by the size and shape of a neurite. Size and shape arise largely from the pattern of growth of the neuron dictated by expression of the cell's genes during development (see Chapter 22), but they can also be influenced by external growth factors that may be present in the neuron's environment during its development (see Chapter 3). Membrane resistance, another electrical characteristic, is determined largely by the number, types, and distribution of ion channels in the cell membrane. These are certainly influenced by the genes expressed during the cell's development, but neurons can actively change the number, type, or distribution of ion channels by synthesizing more or fewer channels and by changing the membrane sites at which they are inserted.

As a consequence, the electrical properties of any neuron can be established and adjusted to suit the needs of the neural population or circuit of which the neuron is a part. If it is advantageous for a particular neuron to integrate input over many neurites and for relatively long times, the neuron will be able to accentuate those features by having large time and length constants. If a neuron will function better by paying selective attention to inputs from just a few neurites or to inputs that arrive within a short time of one another, it will perform better with small time and length constants. Neurons can even change their electrical properties to some extent as a consequence of experience, fine-tuning themselves to better serve the needs of the whole nervous system from hour to hour or month to month.

The description of the passive electrical properties of neurons in the preceding sections of this chapter of necessity dealt with just one aspect of neural integration at a time. You may therefore may have gained the impression that neurons are largely passive elements, sitting quietly and doing nothing until an impulse comes along to which they may have to respond. Such a view is quite misleading. Most neurons are continually bombarded with synaptic input, receiving input from hundreds, even thousands, of other neurons. Hence, the membrane potential of a neuron may continually fluctuate, sometimes exceeding threshold, sometimes dipping well below it as the neuron integrates the input it receives. Although the activity of a single neuron out of millions may seem insignificant, each neuron contributes its voice to the whole; it is the collective activity of many neurons that constitutes the decision-making power of the nervous system, and that is the basis of all behavior. You will learn much more about the collective activity of neurons in subsequent chapters of this book.

The interaction of epsps and ipsps is influenced by the electrical properties of a neuron. The sizes and shapes of the branches of a neuron, by influencing the neuron's electrical properties, will adapt that neuron to serve a particular function in a neural circuit. Electrical properties can also be changed to adapt the neuron to different needs of the nervous system over time. Integration of epsps and ipsps, which occurs continually in most neurons, is the basis of neural decision making.

Activity-Dependent Mechanisms

For any one neuron with a particular set of electrical properties, summation is primarily a passive mechanism because it always gives the same result for the interaction of two epsps irrespective of the past history of the synapse. Some integrative mechanisms, however, are activity-dependent, meaning that they may wax or wane depending on the neuron's pattern of past activity. Some of these activity-dependent mechanisms are relatively simple, but others are quite complex and are thought to underlie learning (see Chapter 24).

Facilitation

Facilitation is an increase in the amplitude of postsynaptic responses as a result of repetitive activation (Figure 8-7). That is, when a single synapse is activated several times in quick succession, the amplitude of the postsynaptic response increases up to some maximum that is characteristic of the synapse and the frequency of stimulation. In summation, in contrast, two or more postsynaptic potentials interact, but the individual postsynaptic potentials themselves do not change in amplitude. Operationally, facilitation may be distinguished from summation by the return of the membrane potential to resting level between the successive epsps (arrows in Figure 8-7). In neurons that show facilitation (not all do), the increase in the amplitude of the postsynaptic potential may range from a low of about 100% (i.e., a doubling of the size of the postsynaptic response) to as much as 700%.

Although the mechanism by which facilitation occurs is still not fully understood, it is

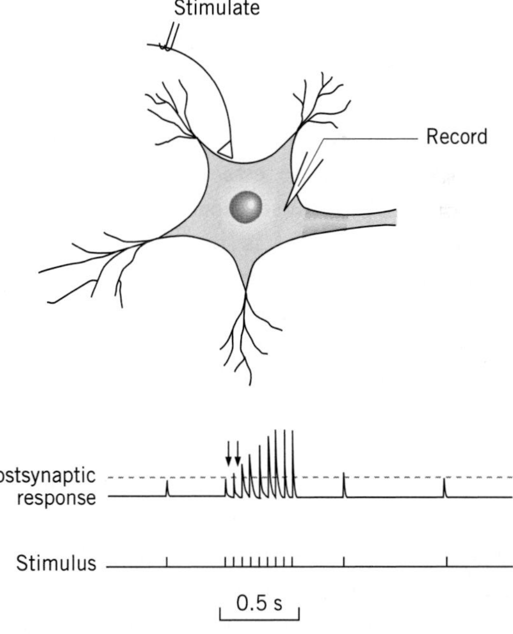

FIGURE 8-7. Facilitation. The amplitude of the postsynaptic response increases when the postsynaptic cell is activated several times in quick succession. Note that the membrane potential in the postsynaptic neuron returns to its resting level between responses (arrows), and that the effects of facilitation last for a short time after the stimulus has been terminated. Compare the peaks of the single responses with the positions of the dashed lines. After a longer period, a single response returns to its unfacilitated size.

clear from experiments by Katz and Miledi in 1968 that calcium ions have an important role. Using a frog neuromuscular junction that was bathed in saline containing little Ca^{2+}, these investigators pulsed Ca^{2+} from a micropipette into the vicinity of the junction at various times relative to the arrival of the motor neuron spike at the presynaptic terminal. They had already shown that it was necessary for external Ca^{2+} to enter the presynaptic terminal in order for transmission to occur (see Chapter 7). In these later experiments, they systematically varied the amount of Ca^{2+} available just before they stimulated the motor neuron; they found that within a certain range of calcium concentration, increasing the amount of Ca^{2+} available at the synapse did not change the amplitude of the first of a pair of end plate potentials (epps) but did increase the amplitude of the second epp. They concluded that a residual level of free Ca^{2+} in the presynaptic terminal was responsible for this facilatory effect.

You will recall from Chapter 7 that Ca^{2+} has the dual effect of promoting transmitter release from docked vesicles and mobilizing more distant vesicles to the docking zone at the synapse. These effects are over in less than 1 msec because most of the Ca^{2+} that enters the terminal is bound or otherwise removed from the cytoplasm within that period after the opening of the calcium channels. This rapid reduction of internal free Ca^{2+} takes place by a combination of diffusion, binding to specialized buffer proteins, and sequestration within calcium-storing vesicles.

In facilitating synapses, it is thought that not all Ca^{2+} is bound or sequestered quickly. Hence, a small, residual amount of Ca^{2+} will still be available after several milliseconds. If the presynaptic terminal is depolarized again before this residual Ca^{2+} has been completely removed, the residual Ca^{2+} will be added to the incoming Ca^{2+}, increasing the internal Ca^{2+} level in the presynaptic terminal above normal for a brief period. This, in turn, will increase the amount of neurotransmitter released from the terminal in response to the next action potential, thereby generating a larger postsynaptic potential in the postsynaptic cell. The increased Ca^{2+} concentration will also increase the number of synaptic vesicles that are mobilized and added to the release pool for the next potential. This hypothesis has received some experimental support from direct measures of residual Ca^{2+} at synapses immediately after synaptic transmission. However, additional factors that also depend on Ca^{2+} apparently are also involved.

One attractive feature of the residual Ca^{2+} hypothesis is that it can explain a complete lack of facilitation, as well as differences in the amount of facilitation at different synapses. If free Ca^{2+} is removed rapidly from the nerve terminal after depolarization, not enough Ca^{2+} will be left to affect subsequent transmitter release; if it is removed relatively slowly, Ca^{2+} will be available for facilitation. The slower Ca^{2+} is removed, the greater will be the facilitation at that synapse.

The facilitation that has just been described can be termed **homosynaptic facilitation** because the activated synapse is the one that facilitates. **Heterosynaptic facilitation** is facilitation of a synapse different than the one that is activated to produce the effect. This type of facilitation was first demonstrated in the marine mollusk *Aplysia* by Eric Kandel and Ladislav Tauc. *Aplysia* is a shell-less, sluglike animal that may grow to the size of a large rat. Its accessible nervous system and the large size of its neurons have made it a favorite experimental animal for many studies of synaptic interaction.

The experimental arrangement that Kandel and Tauc used to demonstrate heterosynaptic facilitation is shown in Figure 8-8. They first placed stimulating electrodes on two of the nerve bundles that carry axons to and from one ganglion of the central nervous system, and then used a microelectrode to impale a specific neuron in the ganglion itself. Some of the incoming axons make synaptic contact with the cell from which the intracellular recording is being taken, and some do not. Stimulating one of the two nerves (labeled "test" in Figure 8-8) evokes an epsp in the monitored cell. Stimulating the other nerve (labeled "priming" in Figure 8-8)

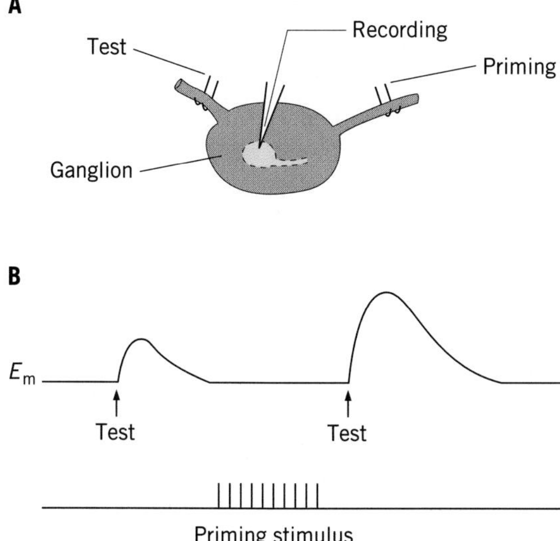

FIGURE 8-8. (A) The recording arrangement and (B) the results of an experiment to demonstrate heterosynaptic facilitation in *Aplysia*. Note that the priming stimulus does not excite the neuron in any way. See text for further explanation.

evokes no response in the monitored cell. However, after a brief period of high-frequency stimulation of the priming nerve, the epsp evoked by stimulation of the test nerve is facilitated, showing a significant increase in amplitude. Like homosynaptic facilitation, this effect lasts for only a short time. This type of facilitation is termed heterosynaptic because it requires that the priming nerve, stimulation of which causes no epsp in the cell at all, be stimulated in order to evoke the response.

Kandel and Tauc suggested that the increase in size of the epsp was caused by a presynaptic facilitation of the synapse (Figure 8-9). Their idea was that neurons excited by the priming stimulus make *presynaptic* contact with the neurons excited by the test stimulus, such that repeated activation of the axons in the priming nerve will lead to an increase in the amount of neurotransmitter released by the test neuron when it is stimulated. This amount then causes an increase in the amplitude of the recorded epsp.

Subsequent experiments suggested how such a presynaptic effect could come about. The heterosynaptic facilatory effect occurs reliably at certain synapses between specific sensory and motor neurons in *Aplysia*. Study of these synapses established that the effect was brought about by action of the presynaptic neurotransmitter, serotonin in this case, on the terminal of the "test" neuron (Figure 8-9). By causing the closure of certain potassium channels, serotonin has the effect of broadening the spike of the excitatory neuron with which it interacts. This increase in the duration (width) of the action potential in the "test" neuron causes an increase in the amount of Ca^{2+} that enters its terminal, and hence an increase in the amount of neurotransmitter that the action potential releases. More neurotransmitter release manifests itself as an increase in the size of the postsynaptic potential of the target neuron. Because it is necessary for the priming and test stimuli to occur close together in time, Kandel suggested that heterosynaptic facilitation might be involved in learning. You will read about the evidence for this view in Chapter 24.

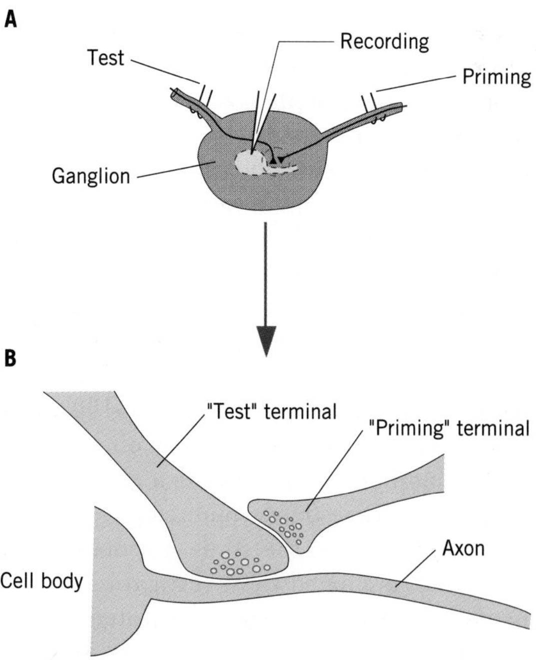

FIGURE 8-9. An arrangement of neurons to explain how heterosynaptic facilitation occurs. The exploded view shows the arrangement of nerve terminals that produces the observed effect.

Facilitation is an increase in the amplitudes of postsynaptic responses as a result of repetitive activation, in which the membrane potential returns to resting levels between the successive potentials. Facilitation can be homosynaptic or heterosynaptic.

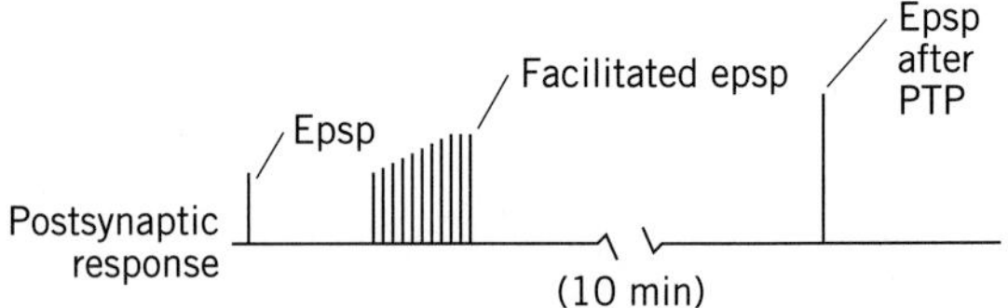

FIGURE 8-10. Posttetanic facilitation. (Recording arrangement as in Figure 8-7.) A single test pulse evokes an epsp of a particular size (first response on the left). After a tetanus (middle), another test pulse evokes a significantly larger epsp. Note that there is facilitation of the response during the tetanus. The potentiated response following the tetanus is much larger than the largest response in the tetanus. Epsps appear as vertical lines because of the compressed time scale.

Potentiation

Potentiation is a form of facilitation in which the increase in amplitude of the postsynaptic potential continues after the stimuli that induced it have ceased. Several types are known, differing mainly in the duration of the effect. One type of short-term potentiation, **posttetanic potentiation (PTP),** is an increase in amplitude of a postsynaptic response in a neuron after that neuron has received a tetanic stimulus (Figure 8-10). The name derives from the concept of a tetanus, which in muscle is a fused contraction caused by a high-frequency burst of impulses in the motor neuron. In the vertebrate literature, tetanus has come to mean any relatively high-frequency burst of impulses in a neuron, even if it has no connection to a muscle. Both homosynaptic and heterosynaptic PTP are known. PTP is similar to ordinary facilitation in the sense that it involves an increase in the size of a postsynaptic response after some previous activity. The main difference is that a fairly long period of stimulation (as long as 20 seconds) at moderately high frequencies (50–100 pulses/sec) is required in order to observe the effect, and that the effect lasts a correspondingly longer time (many minutes).

In recent years, the study of potentiation has been dominated by investigation of one particular form, long-term potentiation, which is a feature of many different neurons in various locations in the vertebrate brain. Although originally referring to potentiation evoked by a long, high-frequency tetanizing stimulus, the phrase **long-term potentiation (LTP)** today refers to any increase in the amplitude of a postsynaptic potential that can be evoked by stimulation and that lasts anywhere from several hours to over a week. At present, two main forms of LTP are recognized: **homosynaptic LTP** requires a tetanizing stimulus across the synapse at which LTP is observed; **heterosynaptic LTP** requires electrical activity both in the neuron in which LTP is observed and in one or more other neurons that synapse on that neuron. Because heterosynaptic LTP requires electrical activity in two different neurons at once, it is also called **associative LTP.** The possible connection of this phenomenon to learning is obvious and has stimulated an enormous amount of work whose objective is to uncover the cellular mechanisms involved. This work has revealed a richly complex synaptic arrangement that allows LTP to occur.

Among the most thoroughly studied synapses that exhibit associative LTP are those formed by neurons in the mammalian hippocampus (see Chapter 2). Glutamate is the most prominent excitatory neurotransmitter at these synapses. Typically, three types of ionotropic receptors for glutamate are located on each postsynaptic neuron. Metabotropic receptors for glutamate are present as well, but these serve a different function. The ionotropic receptors have different pharmacological properties, and their associated channels are not all selective for the same ions. Each type of receptor is named after a pharmacological agent that, like glutamate, acts as an agonist to open the channel.

The three types of ionotropic receptors are K, AMPA, and NMDA receptors. The K receptor is activated by kainate. The **AMPA receptor** is activated by α-amino-3-hydroxy-5-methyl-4-isoxazole-proprionic acid, or AMPA for short. AMPA receptors were formerly classified as Q receptors, since they also respond to quisqualate. The changed terminology resulted from the discovery of two types of "Q" receptors—true Q receptors, which respond to quisqualate but not to AMPA and are metabotropic, and AMPA receptors, which respond to AMPA as well as quisqualate and are ionotropic. The channels formed by K and AMPA receptors allow both K^+ and Na^+ to pass when they are open but usually allow no significant amount of Ca^{2+} to cross the membrane. Synaptic transmission that does not activate LTP involves mainly K and AMPA receptors. Glutamate activates these receptors and opens the associated ion channels to Na^+ and K^+. The ensuing ion flux generates an epsp, as described in Chapter 6.

The third type of glutamate receptor, the **NMDA receptor**, is excited by the glutamate analogue N-methyl-D-aspartate. The NMDA receptor allows significant numbers of calcium ions, as well as sodium and potassium ions, to cross the membrane when its associated ion channel is open, but this opening requires strong depolarization. The receptor is activated by glutamate, but the mouth of the associated channel is normally blocked by magnesium ions present in the extracellular fluid, which prevent the flow of ions through it. The magnesium ions are repulsed from the channel opening when the membrane depolarizes, but if there are only one or two input spikes, or if the input frequency is low, little ion flux will occur because the membrane potential is already moving back to resting levels by the time the Mg^{2+} has moved away from the channel. On the other hand, if there is a strong, simultaneous input from another source, the stronger depolarization of the nerve terminal in the region of the synapse causes the Mg^{2+} to move away from the mouth of the channel of the NMDA receptor, opening it while there is still sufficient depolarization to allow a substantial influx of ions.

The key to the function of the NMDA receptors is their regulation of Ca^{2+} entry through the channel (Figure 8-11). When the stimulus conditions do not release the channel from Mg^{2+} block, there will be no significant flow of Ca^{2+} into the neuron (Figure 8-11A). On the other hand, if conditions bring about a strong depolarization of the postsynaptic terminal just before or at the same time that NMDA receptors are opened by the presence of glutamate, significant amounts of Ca^{2+} will enter the neuron (Figure 8-11B). Calcium ions activate Ca^{2+}/calmodulin-dependent protein kinase II, protein kinase C, and Ca^{2+} dependent proteases. The likely targets of these kinases are the AMPA and K receptors or their associated channels. The effect is to make the receptor-channel complex more responsive to glutamate, and therefore to enhance the response of the neuron to subsequent synaptic input.

Early studies of LTP dealt exclusively with postsynaptic mechanisms. Recent evidence, however, has suggested that presynaptic events also contribute to LTP. It is now known that the influx of Ca^{2+} that accompanies the opening of NMDA channels induces the synthesis of the gas nitric oxide (NO), as well as activating the protein kinases discussed previously. This synthesis is induced by Ca^{2+}/calmodulin-dependent protein kinase II, which activates the enzyme NO synthase and hence stimulates the production of NO. The NO acts as a **retrograde messenger**, a molecule that diffuses from a postsynaptic cell back to the presynaptic one, where it influences subsequent synaptic transmission. In neurons that show LTP, NO stimulates the release of additional neurotransmitter the next time the synapse is active. As a gas, NO has the advantage of being able to diffuse rapidly through cell membranes. Furthermore, it has

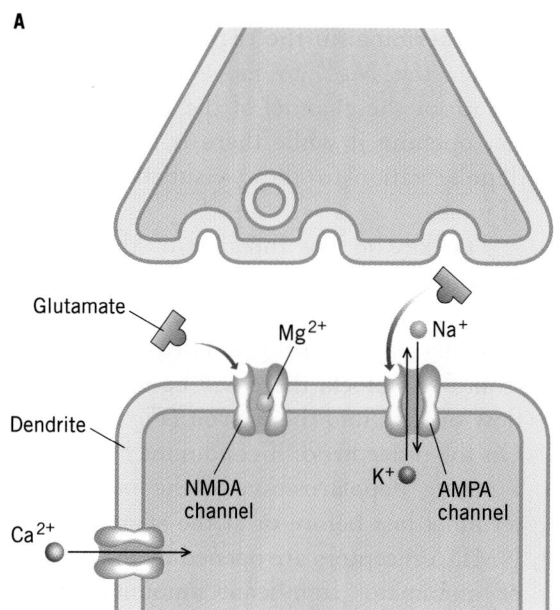

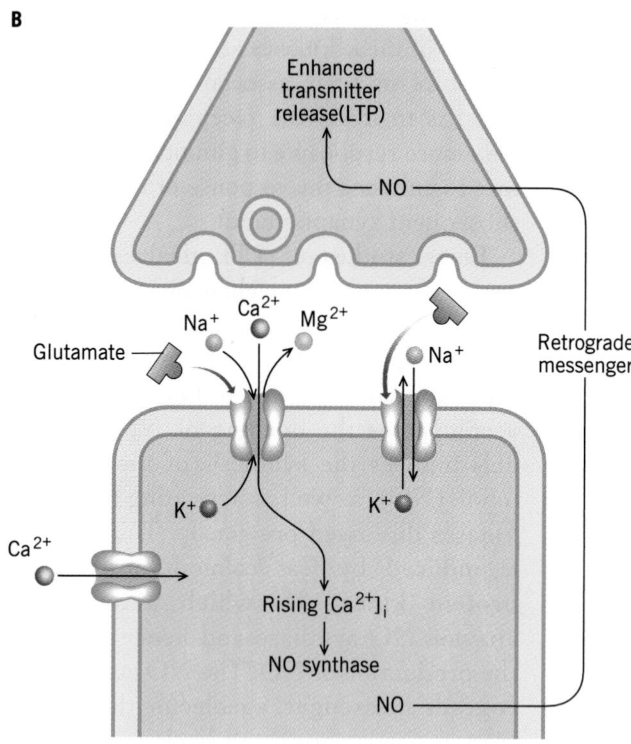

FIGURE 8-11. Schematic diagram showing some of the molecular elements involved in LTP at the cellular level. (A) Ordinary transmission. Magnesium ions block the NMDA channel but not the K channel. (B) Transmission during high-frequency stimulation. The sustained depolarization allows the Mg^{2+} to move out of the NMDA channel, allowing an influx of Ca^{2+}. The Ca^{2+} initiates a cascade of events that enhance subsequent responses of the postsynaptic neuron.

a limited life, spontaneously reacting with other chemicals in the neuron within a few seconds. These features make it an effective messenger for signals that must be restricted in their effects in time and space. Other candidate molecules have been proposed to act as retrograde messengers as well. Which ones perform this function, and exactly how they enhance LTP, remain to be discovered.

Posttetanic potentiation (PTP) is a relatively short-term form of facilitation that may last for some hours. Long-term potentiation (LTP) is facilitation that may last for hours or days. LTP is induced through NMDA receptors, by the entry of calcium ions. The calcium ions initiate a biochemical cascade that causes the neuron to become more responsive to neurotransmitter and, by means of the retrograde messenger NO, to cause more transmitter release from the presynaptic neuron as well.

Integration and Single Cells

It is not hard to understand how cellular integration affects the activity of a single cell. What is less often clear is how integration, by influencing different cells in different ways, really affects the operation of networks of cells and ultimately the behavior of the animal. Some of the complexities of integration and how integration acts at the level of individual neurons can be illustrated by considering a single, identified neuron and the synaptic connections that it makes.

The sea hare *Aplysia*, the marine slug studied by Kandel and Tauc, has been studied extensively by neurobiologists because its nervous system contains many large neurons from which intracellular recordings are relatively easy to make. Kandel and his coworkers have been especially active in studying this animal. In the abdominal ganglion of *Aplysia* are a number of cells that control respiration and the beating of the heart. One neuron, identified as cell L10, forms electrical and at least four different types of chemical synapses with other cells. These four chemical synapses generate different responses in the postsynaptic neurons with which they synapse, responses generated by different ionic mechanisms (Figure 8-12).

One synapse is excitatory, generating its postsynaptic response entirely by an influx of Na^+ instead of the more usual flow of both Na^+ and K^+. The second is a typical inhibitory synapse, in which the ipsp is due to an influx of Cl^-. The third synapse, however, has a dual inhibitory function in which Cl^- influx and K^+ efflux generate two separate ipsps. Kandel and coworkers showed that these are *separate* ionic mechanisms. That is, there are separate channels for each of the ions. Since these channels have different response characteristics, the postsynaptic cell shows two distinct types of postsynaptic responses. The chloride channels open quickly in the presence of the neurotransmitter and yield a long-lasting ipsp of about 500 msec. The chloride channels also inactivate readily. In consequence, these channels are effective mainly at low frequencies of stimulation by cell L10. The potassium channels open relatively slowly and require some facilitation for their effects to become noticeable, at which time they may yield a 10- to 15-second ipsp. (This is not a misprint!) Because it takes several impulses in rapid succession for these channels to open, they tend to be active only during relatively high-frequency stimulation.

The fourth junction is a dual excitatory-inhibitory synapse. The excitatory effect is

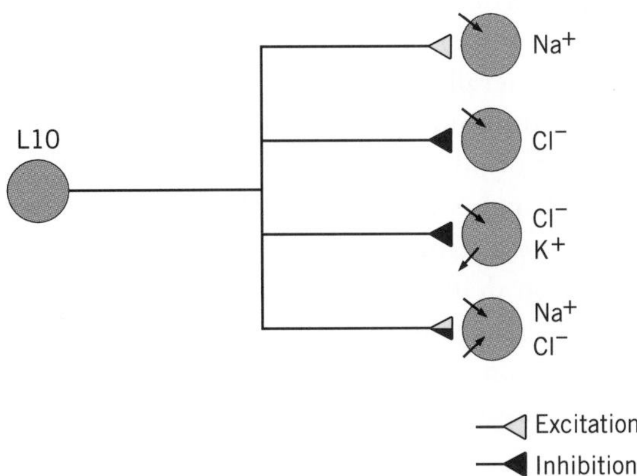

FIGURE 8-12. Four chemical synapses in *Aplysia* between cell L10 and various postsynaptic cells. Two of the postsynaptic neurons have dual ionic mechanisms. These mechanisms result in different postsynaptic effects at low frequencies of stimulation than at high frequencies.

mediated by Na^+ influx, the inhibitory effect by Cl^- influx. When cell L10 fires at a low frequency, the primary postsynaptic response at this synapse is an epsp, due to the influx of Na^+. However, the sodium channels easily desensitize, so that at higher frequencies of stimulation the excitatory response disappears. Meanwhile, the chloride channels give only a weak response at low frequencies of stimulation, a response that is swamped by the strong epsp. At higher frequencies, however, these channels show considerable facilitation, quickly converting the initial epsp into a long-lasting ipsp. Acetylcholine is the neurotransmitter at each of these synapses, illustrating the principle that the postsynaptic effect of a transmitter is due to the properties of the receptor molecule(s) with which it binds, not to the chemical properties of the transmitter itself.

The complexity of synaptic action of cell L10 demands some biological explanation. Kandel and his collaborators have suggested two possible reasons for such a diversity of synaptic actions. First, they suggest that it may allow quantitatively different synaptic action of the same sign. If other things are equal, the effectiveness of a postsynaptic potential in exciting or inhibiting activity in a neuron will depend on the difference between the equilibrium potential of the ion and the membrane potential. Hence, inhibition due to K^+ efflux is usually "stronger" (i.e., more effective) than that due to Cl^- influx because E_K is generally more negative than E_{Cl}. This allows different synapses to have a different amount of influence on the postsynaptic neuron. A potassium-based inhibitory synapse will have a stronger inhibitory effect on spike initiation than will a chloride-based synapse; the receiving neuron may exploit this difference in controlling its own excitatory postsynaptic responses.

A second reason for a diversity of synaptic action might be to permit combinations of conductances in order to produce a variety of single or double actions (as in the dual-acting synapses). This allows neurons to be tuned, so to speak, to the frequency of input they receive so that they will respond in a specific manner to distinct patterns of input. For example, in the specific dual-acting inhibitory synapse discussed earlier, the postsynaptic cell will be inhibited for a relatively short time at low frequencies of stimulation, whereas it will be inhibited for a much longer time at higher frequencies. These differences will be translated into behavioral differences if the postsynaptic cell is involved in the generation of a behavioral response. A similar argument can be made for the effects of the dual inhibitory-excitatory cell.

Whatever the reason for the presence of these complex synapses in *Aplysia*, it seems likely that they play an important role in the biology of the animal. Several other neurons in *Aplysia* have been shown to make similarly complex synapses with postsynaptic neurons, some using serotonin as a neurotransmitter, some using histamine. From these examples it is clear that the possibilities of synaptic integration, and the tuning of synaptic properties to the needs of the organism, are great indeed.

Several neurons in the sea hare *Aplysia* show complex synapses at which more than one ionic mechanism is activated in the presence of the neurotransmitter. These synapses allow for variations in synaptic strength and also for variations in the sensitivity of synaptic transmission to the frequency of input that drives it.

Additional Reading

General

Kandel, E. R. 1976. *Cellular Basis of Behavior: An Introduction to Behavioral Neurobiology*. San Francisco: W. H. Freeman.

Katz, B. 1966. *Nerve, Muscle, and Synapse*. New York: McGraw-Hill.

Levitan, I. B., and L. K. Kaczmarek. 1996. *The Neuron: Cell and Molecular Biology*. 2d ed. New York: Oxford University Press.

Matthews, G. G. 1991. *Cellular Physiology of Nerve and Muscle*. 2d ed. Boston: Blackwell Scientific.

Research Articles and Reviews

Atluri, P. P., and W. G. Regehr. 1996. Determinants of the time course of facilitation at the granule cell to Purkinje cell synapse. *Journal of Neuroscience* 16:5661–71.

Bekkers, J., and C. F. Stevens. 1990. Presynaptic mechanism for long-term potentiation in the hippocampus. *Nature* (*London*) 346:724–29.

Kandel, E. R., and L. Tauc. 1965. Mechanism of heterosynaptic facilitation in the giant cell of the abdominal ganglion of *Aplysia depilans*. *Journal of Physiology* (*London*) 181:28–47.

Katz, B., and R. Miledi. 1968. The role of calcium in neuromuscular facilitation. *Journal of Physiology* (*London*) 195:481–492.

Zucker, R. S. 1989. Short-term synaptic plasticity. *Annual Review of Neuroscience* 12:13–31.

PART 3

Sensory Systems

Sensory systems process stimuli. Some of these stimuli, such as light and sound waves, originate in the external world. Other stimuli, such as the body's internal chemicals or those generated by body movement, originate inside the body. Whether stimuli originate internally or externally, however, they must all be converted into a form that the nervous system can understand. At the same time, information about the nature of the stimulus (whether it is an odor or a touch, for example), its strength, its quality (color of light, pitch of sound), even its source, must be conveyed to the appropriate parts of the central nervous system for interpretation. Detecting and conveying information about sensory stimuli is the function of the sensory system, and is the subject of the chapters in Part 3.

CHAPTER 9

Properties of Sensory Systems

A major function of nervous systems is to detect events both inside and outside the animal. This function is carried out by sensory receptors and sense organs, each specialized to respond best to a particular kind of stimulus. In this chapter and the next, you will learn about these common principles: how sensory systems are organized, how they encode information for transmission to the central nervous system, and how they themselves are influenced by the nervous system.

Perception of the world has been part of life since the evolution of conscious awareness. Only in the last century, however, has a sound scientific basis for the performance of sensory systems enabled researchers to understand how they work. During that time, research on these systems has led to the discovery of common principles that underlie their operation.

Sensory Performance

It is all too easy to take for granted your astonishing senses. After all, you don't have to make a conscious effort to see the words on this page, hear the music from your radio, or smell the coffee in your cup. But take a moment to think about the few minutes before a lecture in a large class. The room is filled with the noise of students coming in and the chat of those already there. Yet despite the babble of conversation all around you, you have no difficulty in following the story your friend is telling you. The casual ease with which you can attend to particular sounds from a background filled with competing noise (Figure 9-1) makes it difficult to appreciate just how enormous the task of extracting the relevant information really is.

Just as astonishing are the sensitivity and range of stimuli that sensory systems can detect. In pitch darkness, a cell in the human retina can respond to a single photon of light, and stimulation of only 10 to 15 cells by single photons in a small area of the retina is sufficient for the light to be detected consciously by an observer. In practical terms, this is like just being able to see the light from a small candle at a distance of nearly a mile, assuming no attenuation by the atmosphere and no other light. Other senses have similar sensitivities. You can smell skunk odor a mile away, and the weakest sound you can hear has so little energy that it displaces your eardrum less than 10 nm, one ten millionth part of a millimeter.

At the same time, many sensory systems have an enormous range. For example, a full moon gives off 2×10^4 times as much light as is required to see anything in pitch darkness. Your ears can also deal with a wide range of sound. The loudness of sound is usually expressed on a logarithmic scale in terms of

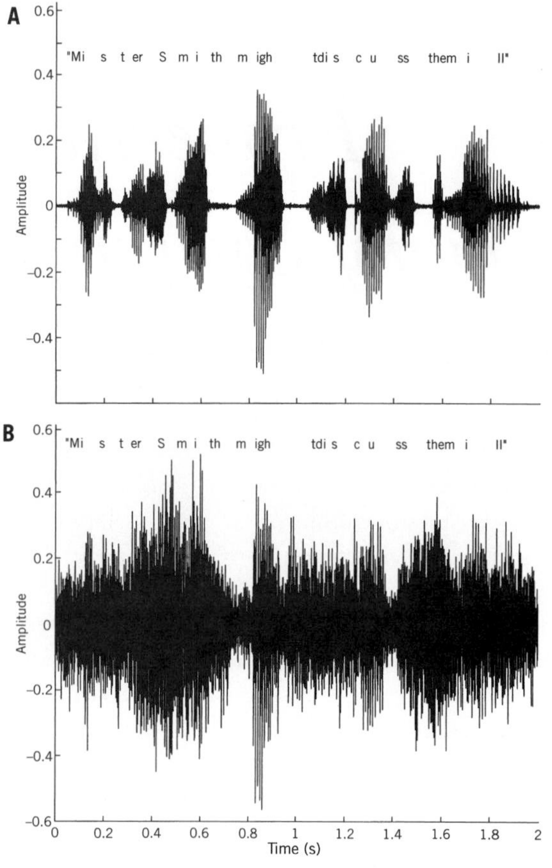

FIGURE 9-1. (A) The electrical signal equivalent of a spoken phrase, "Mr. Smith might discuss the mill," said against a quiet background. (B) The same phrase said against a background of 12 individuals speaking simultaneously. The voices of the 12 individuals are slightly louder than the voice of the single speaker. Although the electrical signals of the single voice and the background voices appear inextricably tangled, a listener has no difficulty in understanding all the words of the phrase. (Courtesy of Chen Liu and Robert C. Bilger.)

the sound's energy. The loudest sound humans can endure without pain is 10^{12} times more energetic than the faintest sound it is possible to detect.

Sensory systems are able to extract relevant information from input that may contain a great deal of competing information. They also have exquisite sensitivity and many can respond to an enormous range of stimulus strengths.

Sensory Specificity

Although modern researchers mostly investigate sensory systems in terms of objective molecular and cellular events, in humans and presumably in many other animals sensory stimuli evoke a subjective experience that is an important component of the way in which stimuli are interpreted. In each of us, a touch, a sound, and a ray of light evoke completely different sensations. These sensations are what allow us to enjoy the feel of a stone smoothed by the surf, the melody of a bird's song, or the beauty of a sunset. What few people stop to consider is how this **sensory specificity**, the separation of distinct subjective sensations arising from different sense organs, is maintained, how it is we never confuse the melody with the sunset, or the feeling of smoothness with the melody. The surprise you probably feel at this idea is a reflection of just how thoroughly we take for granted the separation of our subjective sensations from one another.

Actually, the commonsense rule that a particular stimulus always evokes a particular type of subjective sensation is not completely accurate. Even 150 years ago, systematic investigation of sense organs and sensation revealed that the actual situation is a little different. If a particular sensory nerve is stimulated through its sense organ, the subject always reports a consistent, specific sensation that depends on the type of sense organ that is stimulated and not on the type of stimulus used. For example, light is the normal stimulus that produces visual sensations. However, pressure on the eyeball will also produce a subjective sensation of light. Conversely, if nerves from different sense organs are subjected to the same stimulus, the subjective sensation is always different. For example, although brief pressure on the eyeball gives the sensation of flashes of light, similar pressure on the arm gives the sensation of touch.

From these and similar observations, Johannes Müller, in about 1838, formulated what he called the **law of specific nerve energies**. Müller proposed that the sensation created by different stimuli is independent of the stimulus type, and is dependent only on which particular nerve is carrying the response. In other words, the information carried by any one sensory nerve is normally interpreted in only one way, no matter how activity in the nerve is evoked. Modern physiological methods applied during operations to patients under local anesthetic allow neurologists to verify this law easily today. Electrical stimulation of any sensory nerve produces the subjective sensation corresponding to the kind of stimulus that normally evokes activity in that nerve—light flashes for stimulation of the optic nerve, sound for stimulation of the auditory nerve, and so forth.

In the early part of the nineteenth century, the law of specific nerve energies represented a significant advance in scientists' understanding of sensory system function. At that time, action potentials and the cellular nature of the nervous system were still unknown. Researchers now understand the basis of the observation. Sensory neurons always transmit their signals via specific pathways to specific brain centers. It is the brain center that interprets the incoming signals and determines the nature of your subjective impressions. If it were somehow possible to reroute sensory signals from, say, your auditory to your visual center, you would see sound. Rock bands would not need a light show—you would perceive the music itself as a kaleidoscope of rhythmically shifting colors!

This notion may seem wildly far-fetched, but actually about ten out of a million people have a neurological condition, synesthesia, in which specific stimuli give rise to subjective sensations of more than one type. For example, one man reports feeling various shapes with his hands whenever he tastes different foods. A woman sees diffuse colored figures when she hears certain sounds. The crossover is usually confined to one type of stimulus, sometimes even just to a small portion of one type, such as a certain sound frequency. Synesthesia also appears briefly in some individuals with epilepsy just before the onset of an epileptic seizure. It is known that in everyone there is integration of sensory stimuli from different sense organs in the limbic part of the brain. This mixing is what allows us to fuse the different visual, tactile, or even auditory sensations we have of the same objects into a unified impression of the object. What is different in a synesthete is that he or she becomes consciously aware of the mixing of different sensory modalities.

The law of specific nerve energies states that each sensory nerve carries information about only one particular subjective sensation. The basis of this specificity is that nerves from different sense organs send their messages to different parts of the brain, which determine how the message is interpreted subjectively. In synesthesia, an individual may have more than one subjective sensation arising from a single type of stimulus.

The Organization of Sensory Systems

The distinctness of the sensations mediated by different types of sense organs led the Greek philosopher Aristotle to suggest that sensory function be organized by **modality**, that is, by subjective sensation. He listed five sensory modalities common to all humans—sight, hearing, touch, taste, and smell. This is not a complete list of the sensory capabilities of humans or indeed of other animals. It omits the sense of balance, an important "sense" that allows us to stand and walk without falling over, the sensations of hot and cold, the feeling of pain, and several senses possessed by other animals but not by humans, to say nothing of internal senses that help the body to maintain homeostasis, and others as well.

If the skin is considered an organ for touch, then all of Aristotle's five sense are mediated by **sense organs**, which are complex, multicellular structures specialized to detect one particular type of stimulus. However, not every kind of sensation is mediated by a sense organ. Relatively simple sensory structures such as hairs and their associated sensory neurons are called **sensory receptors**, or just **receptors**. These, too, are specialized to respond primarily to a single type of stimulus, such as heat or a certain type of chemical. The sensory neurons that are actually responsible for responding to a stimulus are usually called **receptor cells**. Sometimes, however, this term is applied especially to nonspiking, usually anaxonal neurons that are specialized to respond to specific kinds of stimulus energy. In sense organs, receptor cells synapse with other neurons, which directly or indirectly send the sensory signal back to the central nervous system.

A useful way to group sense organs or sensory receptors is according to the type of energy to which each is the most sensitive. The types of energy that can be detected by various types of receptors are chemical, mechanical, light, thermal, electrical, and magnetic. The receptors that detect specific types of energy are given corresponding names. Hence, a **chemoreceptor** responds to one or more chemical substances, a **mechanoreceptor** responds to the mechanical energy of physical movement, a **photoreceptor** responds to radiant energy in the visible wavelengths, and a **thermoreceptor** responds to thermal (heat) energy. Other receptors respond to types of energy that humans do not normally associate with specific sensations. Some fish and amphibians have **electroreceptors**, which respond directly to electrical energy. Some animals have been shown to have **magnetoreceptors**, which respond to the energy of a magnetic field.

Grouping receptors according to the energy of the stimulus is useful but does not apply equally well to all receptors. For example, **hygroreceptors**, which are sensitive to the water content of air, are known to exist, but they do not readily fit into any of the categories given previously. Receptors for pain, called **nociceptors**, are also difficult to fit into this scheme, since they may be activated by chemical, mechanical, or thermal stimuli. Nevertheless, organizing most receptors by the type of energy that stimulates them is both practical and useful. Table 9-1 lists the main types of receptors.

Nervous systems can detect many kinds of stimuli, both external and internal. Receptor cells and sensory neurons are cells that respond to sensory stimuli. Sensory receptors and sense organs are one or more cells and associated structures that are specialized for sensory reception. At present, neurobiologists group receptors according to the type of energy to which they are most sensitive-chemical, mechanical, light, thermal, electrical, or magnetic.

Sense Organs as Filters

It might seem obvious that sense organs and sensory receptors allow an animal to form an accurate and objective impression of its external world, but actually they do not. Instead, sense organs actually shape an animal's view of its world.

Any neuron will respond to more than one kind of stimulus if the stimulus is strong enough. An electrical shock, a strong chemical, even strong mechanical stimulation, can

TABLE 9-1. Receptor Cells and Sensory Neurons

Name of receptor	Energy of stimulus	Example(s)
Chemoreceptor	Chemical	Taste receptor on tongue or olfactory receptor in nose
Electroreceptor	Electrical	Electric receptor in skin of electric fish
Hygroreceptor	Water in air	Humidity receptor in insect antenna
Magnetoreceptor	Magnetic field	Magnetic receptor in honey bee brain
Mechanoreceptor	Mechanical	Touch receptor in skin or auditory receptor in ear
Nociceptor	Mechanical/chemical	Pain receptor in skin
Photoreceptor	Radiant (light)	Rod cell or cone cell in eye
Thermoreceptor	Radiant (heat)	Temperature receptor in skin

evoke action potentials in a neuron. The cells of sense organs are specialized to respond to extremely weak stimuli of one particular type. They are filters, informing the animal about the external world only in terms of a single sensory modality such as sound or light. Hence, if an animal lacks a sense organ or cells that can respond to sound energy, it will be entirely oblivious to sounds that to other animals are important signals.

Sense organs act as filters even within a single modality. That is, they filter out all but a relatively small portion of even the type of stimulus to which they are most sensitive. For example, eyes respond only to light of certain wavelengths, and ears only to sounds of particular frequencies. By acting as filters, sense organs go beyond being merely passive windows on the world in which an animal lives; they actually shape the animal's perception of that world. We humans, for example, tend to think that what we see is an accurate representation of what actually exists. However, our eyes are sensitive to light only in the wavelengths between red and violet (Figure 9-2). Ultraviolet (UV) and infrared light are invisible to us. Most insects, on the other hand, are insensitive to red light but are able to detect UV light. Many flowers reflect both red and UV light. These flowers will obviously appear quite different to a human than they will to an insect. For example, to human eyes a flower like a silverweed flower is uniformly colored. To an insect sensitive to UV light, on the other hand, it will appear to have two colors, one near the center and the other in the periphery (Figure 9-3) because the outer parts of the petals reflect UV whereas the inner parts do not.

Sense organs are filters that influence an animal's view of the world. By limiting the stimuli that an animal can detect, they determine and shape the way the animal perceives the world.

Transduction

In order for the nervous system to detect the presence of a stimulus, the stimulus energy must be transformed into usable neural electrical activity, a process called **transduction**. In physics, transduction is the transformation of one form of energy into another, and transducers are devices that carry out this process. Receptor cells behave as transducers

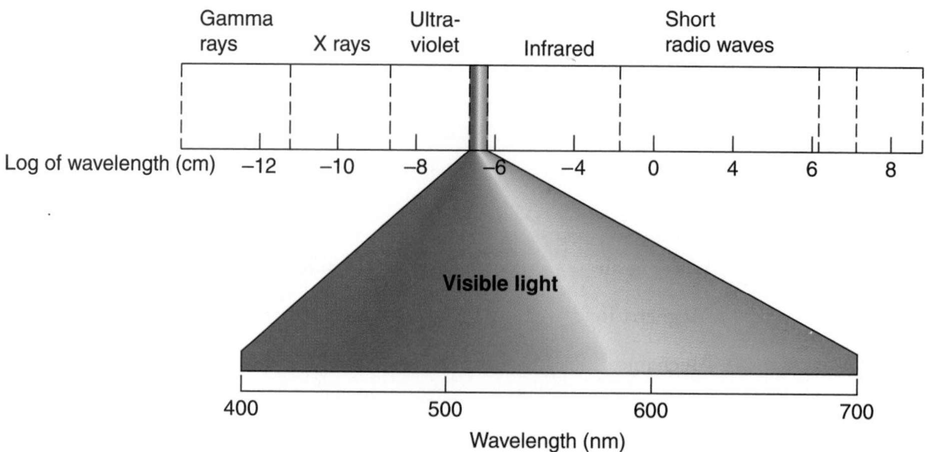

FIGURE 9-2. The electromagnetic spectrum, showing the range of wavelengths of light to which humans are sensitive.

when they transform the energy of a stimulus into electrical energy in the form of a neural response.

Transforming the Stimulus

At one level, all transduction mechanisms are fundamentally the same, irrespective of the type of sensory cell that does the transformation and the type of energy that is being transformed. The energy of a stimulus causes a change in the permeability of the neuronal membrane to one or more ions, and the resulting ion flow changes the neuron's membrane potential. The change in membrane potential, typically called a **receptor potential**, constitutes the cell's response to the stimulus.

The molecular mechanisms that underlie transduction, however, are not the same in all receptor cells. In the case of photoreceptors and most chemoreceptors, a stimulus initiates a biochemical cascade similar to the biochemical

FIGURE 9-3. A silverweed flower photographed in (A) yellow and in (B) ultraviolet light. The dark area in the center of the UV-photographed flower represents an area that is UV-absorbent. An insect with color vision that can see light at UV wavelengths would see the light and dark areas as different colors.

reactions induced by neuromodulatory neurotransmitters. The stimulus increases or decreases the concentration of a second messenger, often one of the cyclic nucleotides, which in turn affects the gates of certain ion channels, thereby increasing or decreasing the flow of ions across the cell membrane. Transduction in other receptor cells requires no second messenger intermediary. Mechanoreceptors respond as a result of a deformation of the cell membrane, which causes ion channels to open or close. The ion channels in the membranes of some chemoreceptors respond directly to the presence of chemical substances, causing ion flow across the membrane.

The first response of a receptor cell to a stimulus is always a change in membrane potential that is graded to the strength of the stimulus. However, some receptor cells generate action potentials if the stimulus is strong enough. For this reason, some researchers have adopted the phrase **generator potential** to refer to the potential that arises in receptor cells that can generate action potentials, and to restrict the phrase receptor potential to the potential that arises in nonspiking receptors. In this book, receptor potential will be used in its broad sense as applied to any receptor cell. The term generator potential will be reserved for the receptor potential that appears in spiking receptors.

Transduction of a stimulus produces a transient change in the membrane potential of the receptor cell due to a flow of ions. This change in membrane potential is called a receptor potential or generator potential. The molecular mechanisms of transduction may vary in different types of receptors.

Encoding Stimulus Strength

The function of transduction is to transform the energy of a stimulus into a train of action potentials that is passed on to the appropriate processing region of the central nervous system. Nonspiking receptor cells do not generate action potentials but instead pass their information on, directly or indirectly, to neurons that do; the neurons then relay the information to the CNS via a train of spikes. In general, the relationship between stimulus and response is that as one increases in strength, the other increases in frequency. In other words, the frequency of firing of a sensory neuron carrying information to the central nervous system is proportional to the strength of the stimulus. Figure 9-4 shows an example.

Qualitative studies are helpful as a first step toward understanding a physiological process, but quantitative studies provide data that can be used to test hypotheses more rigorously. One sensory receptor that has been studied quantitatively is the crustacean stretch receptor. Just under the hard plates along the back of the lobster tail that you may eat for dinner lie the abdominal stretch receptors. Each sense organ consists of two or more fine strands of muscle, called receptor strands, that stretch between each pair of adjacent plates. Embedded in each receptor strand is a complex network of dendrites from a single sensory neuron (Figure 9-5). When the receptor strand is stretched, as it is when the lobster curls its tail underneath itself, the dendrites of the sensory neuron will also be stretched, and a series of action potentials will be generated. Intracellular recordings can be made from the cell body of the sensory neuron during controlled stretch of the receptor strand (Box 9-1). Such recordings show that in this receptor, once the stimulus exceeds some minimum threshold amount, the frequency of firing of the sensory neuron is linearly proportional to stretch (Figure 9-6).

The crayfish stretch receptor is somewhat unusual in that it exhibits a linear relationship between stimulus strength and the amplitude of the cell's response over essentially its entire range. In most sensory neurons, increasing the strength of the stimulus has a smaller incremental effect on the neuron's response when the neuron is already being strongly stimulated. The result is a

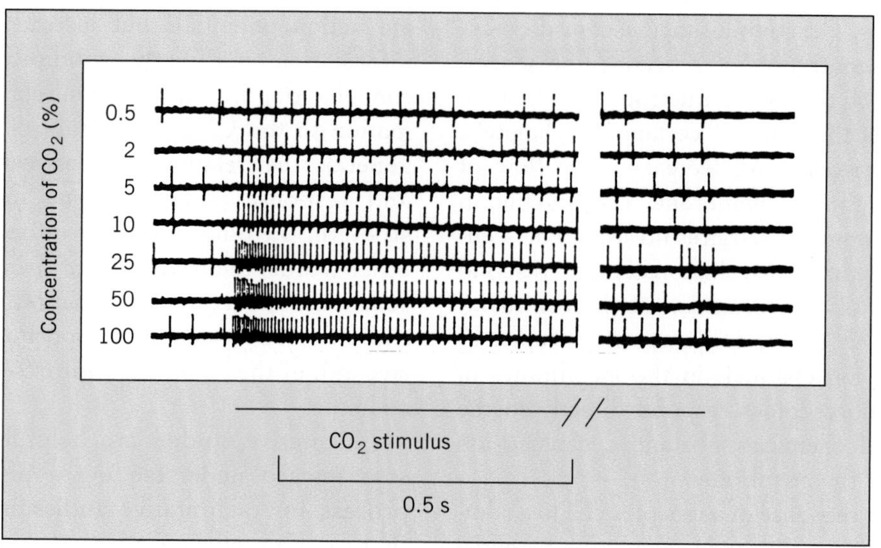

FIGURE 9-4. The response of a chemoreceptor in a honey bee to increasing concentrations of CO_2. The strength of the response, as measured by the frequency of firing of the sensory neuron, increases as the strength of the stimulus, measured by the concentration of CO_2, increases. The break in the stimulus line represents about four seconds in time.

curve that flattens as the stimulus reaches its maximum possible strength. This has been demonstrated clearly in the pacinian corpuscle, a type of mechanoreceptor found in the skin of mammals (discussed further in Chapter 14). When they are stimulated strongly enough, pacinian corpuscles normally generate action potentials. These spikes can be suppressed by application of a pharmacological agent that interferes with the opening of voltage-gated sodium channels and hence reveals the size and shape of the underlying generator potential. Careful application of various levels of pressure to the corpuscle shows that the amplitude of the generator potential increases logarithmically as the strength of the stimulus increases (Figure 9-7).

The ionic mechanism that underlies a receptor potential is different from that which underlies action potentials or postsynaptic potentials. This is evident by the inability of tetrodotoxin, other blockers of voltage-sensitive sodium channels, and drugs that prevent postsynaptic responses to block the production of receptor potentials. Pharmacological and voltage clamp experiments have shown that in most sensory neurons the channels that open in response to a stimulus are permeable to nearly all cations. This includes not only small cations such as K^+ and Na^+ but also larger ions such as Ca^{2+}. Entry of positive ions into the cell leads to depolarization. There is no difference between receptor potentials in nonspiking receptor cells and those that are typically called generator potentials, which can elicit action potentials.

Just as the membranes of dendrites that receive synaptic input usually have no voltage-sensitive ion channels, there are usually no such channels in the membranes of the dendrites of sensory neurons either. When a sensory neuron is stimulated, the generator potential arises as a result of the opening of ion channels that are sensitive to the stimulus or to the second messenger that the stimulus produces. In invertebrate sensory neurons, in which the cell body lies peripherally, the potential spreads passively down the dendrite, over the cell body, and to the region of spike initiation at the place where the axon joins the cell body. Only if the generator potential exceeds threshold at this site will one or more action potentials be generated. In vertebrate sensory neurons, the region of spike initiation is a short distance along the dendrite. Action potentials generated there sweep along the dendrite into the dorsal root ganglion, then along the axon and into the spinal cord.

The strength of a stimulus is encoded in the amplitude of the receptor potential evoked in a receptor cell, or in the frequency of action potentials in a sensory neuron. Receptor potentials are usually caused by the opening of ion channels that allow cations to pass through the membrane. If generator potentials are above threshold, they can initiate action potentials at the neuron's region of spike initiation.

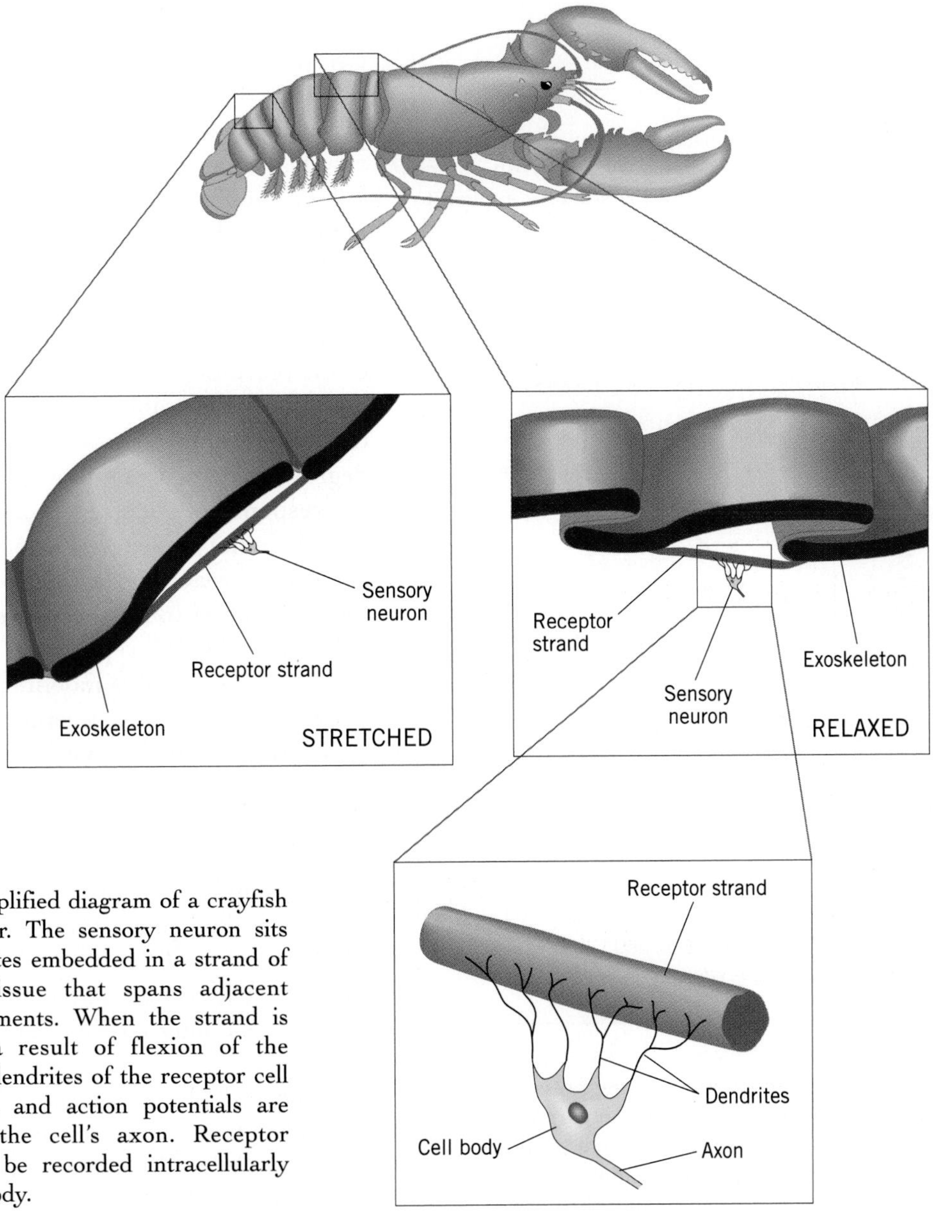

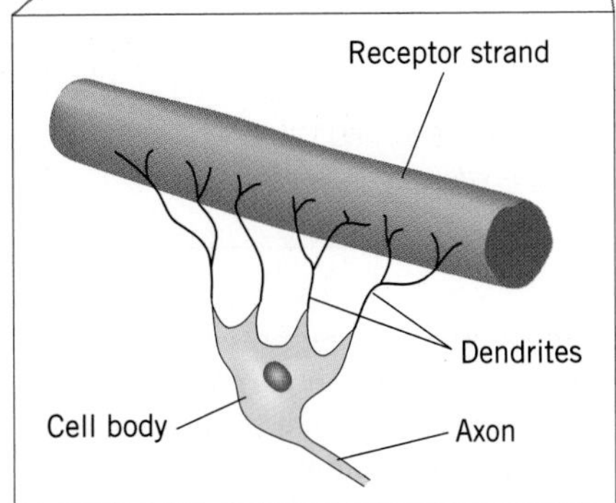

FIGURE 9-5. Simplified diagram of a crayfish stretch receptor. The sensory neuron sits with its dendrites embedded in a strand of fine muscle tissue that spans adjacent abdominal segments. When the strand is stretched as a result of flexion of the abdomen, the dendrites of the receptor cell are stimulated, and action potentials are generated in the cell's axon. Receptor responses can be recorded intracellularly from the cell body.

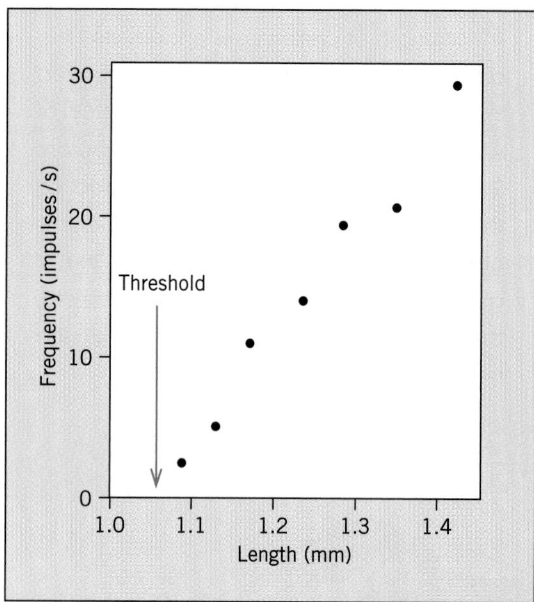

FIGURE 9-6. The relationship between the frequency of action potentials generated by a crayfish stretch receptor and the strength of the stimulus applied to it. The abscissa of the graph shows the stretch applied to the receptor strand.

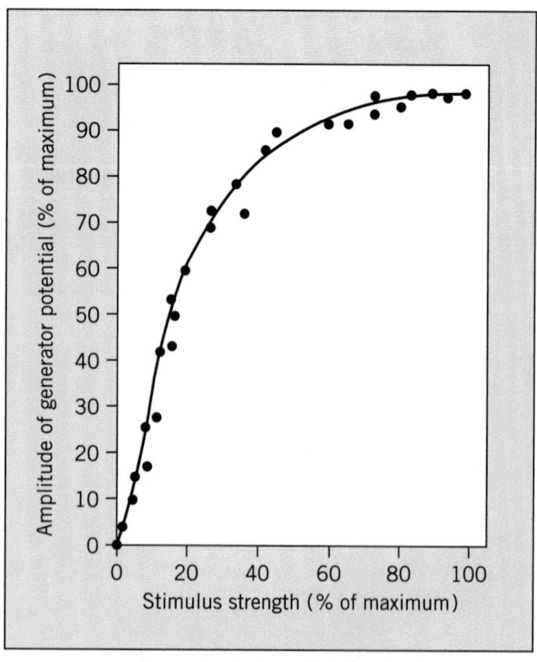

FIGURE 9-7. The effect of increasing pressure (stimulus strength) applied to a pacinian corpuscle on the amplitude of the generator potential. Action potentials were suppressed by application of tetrodotoxin, a blocker of voltage-gated sodium channels.

Temporal Variation of Responses

The electrical responses of most receptors usually declines as time passes, even if the stimulus is maintained at a constant strength. This change, called **adaptation**, is apparent as a gradual decline in the frequency of spikes generated in a sensory neuron (Figure 9-8A) and a gradual fall in the amplitude of a receptor potential (Figure 9-8B). You have probably experienced adaptation when you walked into a room containing a distinctive odor. After some minutes, you found that you could no longer smell the odor, even after someone else who entered the room called your attention to it. This loss of sensitivity to the stimulus happens as chemoreceptors in your nose adapt to the odor and cease responding. When your chemoreceptors no longer respond to the odor, you can no longer smell it. In most cases, adaptation seems to be caused by desensitization of one or more ion channels, causing a reduction in the response of the receptor to the stimulus.

With the possible exception of nociceptors, all receptors show adaptation to some degree. Some neurons show little adaptation; the response of such receptors declines only slightly in the presence of a constant stimulus and never reaches zero. Receptors of this type are called **tonic receptors**, in reference to their ability to report the presence and strength of a stimulus indefinitely (tonically) (Figure 9-9A). Other receptors adapt rapidly. These are called **phasic receptors**, due to their transient responses (Figure 9-9B). Tonic and phasic receptor types are not mutually exclusive but instead represent the two extremes of a continuum. Many sensory neurons, called **phasi-tonic receptors**, have characteristics of both. Such cells show a strong phasic response at stimulus onset, followed by a long-lasting but weaker tonic response.

Tonic and phasic receptors have different functions. Tonic receptors signal the presence or absence of a stimulus over time. They also convey relatively accurate information about

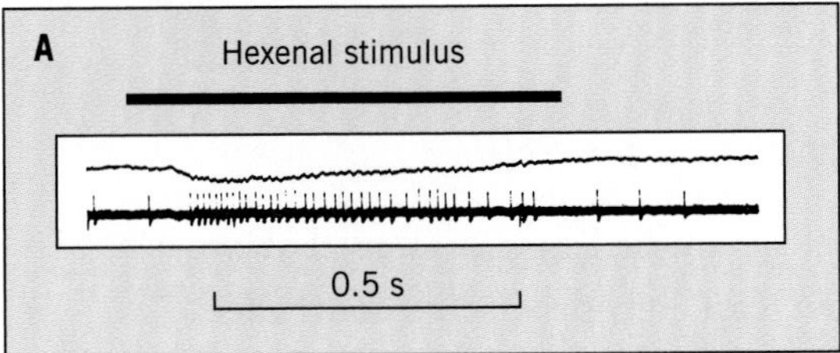

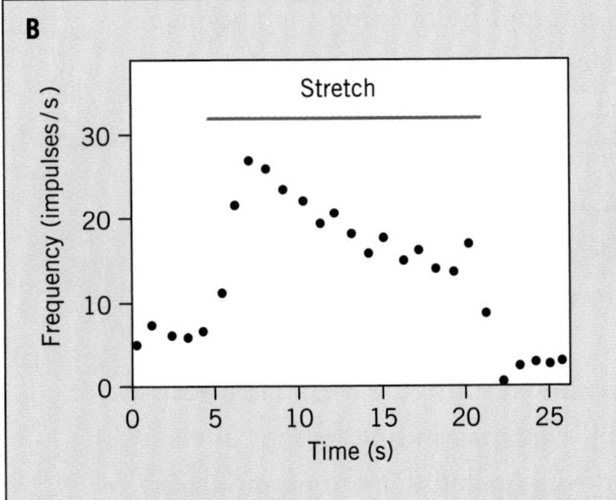

FIGURE 9-8. Adaptation in sensory receptors. (A) Response of a locust chemoreceptor to application of the chemical hexenal. Notice that the frequency of the response declines during presentation of the stimulus. (B) Graph showing the instantaneous firing frequency of a frog muscle spindle organ during a long applied stretch.

the strength of that stimulus. For this reason, they are well suited to provide information about the positions of body parts relative to one another to help the animal maintain a particular posture. Receptors that provide such positional information are called **proprioceptors**. Tonic receptors also provide information about the presence in the body of biologically significant chemicals such as CO_2.

Phasic receptors, on the other hand, provide no information about static position or long-term stimulation. In fact, they often stop firing altogether less than a second after a stimulus is presented, even if the stimulus persists. Instead, phasic receptors provide dynamic information about a change in condition and are often used for moment-to-moment adjustments of muscle force during movements.

Muscle spindle receptors are phasi-tonic and hence illustrate the responses of both phasic and tonic receptors. The generator potential of a phasi-tonic receptor has two components, a transient phasic component followed by a steady tonic component. The tonic component of the generator potential is proportional to the length to which the sense organ is stretched, that is, to the strength of the stimulus (Figure 9-10). The phasic component, on the other hand, is influenced not by the length but by the rate at which the receptor is stretched (Figure 9-11). When the spindle is stretched quickly, the initial depolarization of the generator potential is large (Figure 9-11A); when the spindle is stretched slowly, the initial depolarization of the generator potential is small (Figure 9-11D). Intermediate rates of stretch produce intermediate initial depolarizations (Figure 9-11B, C). Hence, the amplitude of the receptor's phasic response is dependent on the velocity of stretch, and it will vary significantly when different velocities of stretch are applied, even if the receptor is stretched to the same final length.

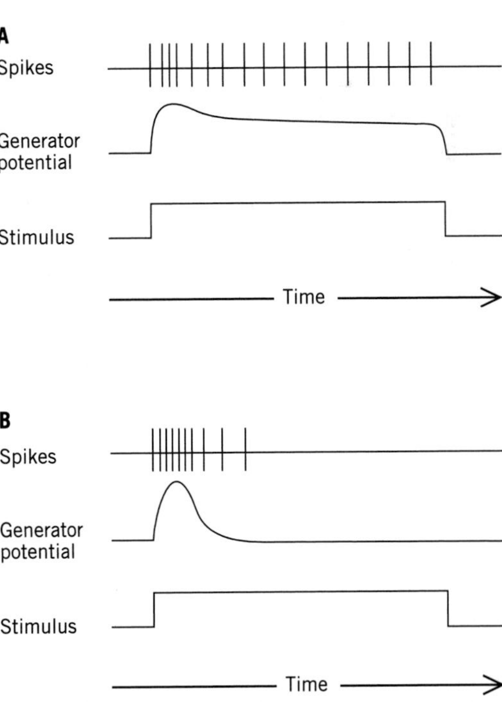

FIGURE 9-9. Idealized responses of tonic (A) and phasic (B) neurons to a stimulus. Top traces: spike response recorded from the axon; middle traces: the underlying generator potential (seen with action potentials suppressed); bottom traces: stimulus.

BOX 9-1

Investigating Sensory Function: Stretching for Science

The responses of sensory cells can best be studied in receptors that are readily accessible, with stimuli that can be precisely controlled. Mechanoreceptors like the abdominal stretch receptor found in crayfish or lobsters are ideal. These receptors can be found by opening the abdomen and removing the large abdominal muscles. Carefully dissecting away the surrounding tissue exposes the stretch receptor organ, consisting of one or two fine strands of muscle and the nerve cell or cells whose dendrites penetrate them. Since stretch of the receptor strand is the stimulus that excites the sensory neuron, a researcher must be able to apply small and carefully controlled stretch to the strand in order to study the response of the sensory neuron to this stimulus.

One way in which this can be done is shown in Figure 9-A. The receptor strand for a stretch receptor is carefully grasped in fine forceps at each end. Each of the forceps is held in a device called a micromanipulator, which allows the forceps to be moved in extremely small increments in any

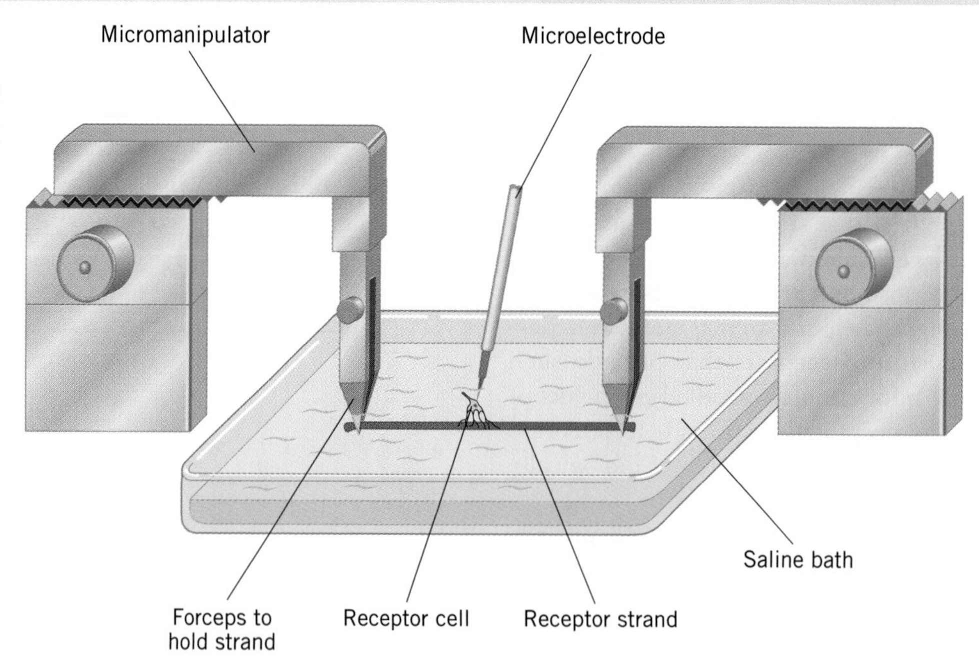

FIGURE 9-A. Recording from a stretch receptor. The receptor strand is clamped at each end and gently pulled by moving apart the forceps holding the ends. The micromanipulators allow controlled movements of fractions of millimeters. At the same time, intracellular recordings can be made from the cell body of the sensory neuron. The receptor strand and the receptor are submersed in oxygenated saline during the experiment.

direction. When the sensory cell is impaled with an intracellular microelectrode (see Box 4-1) and the receptor strand is carefully stretched by moving one of the forceps with its micromanipulator, the researcher can record the response of the sensory neuron to the stretch. Careful stretch in small increments will yield information about how the receptor responds to different amounts of stretch. The generator potential that underlies action potentials can be studied as well. Application of a drug or other pharmacological agent that blocks sodium channels suppresses the generation of action potentials. Uncovering the underlying generator potential allows study of the quantitative relationship between a stimulus and the ion currents that produce the generator potential.

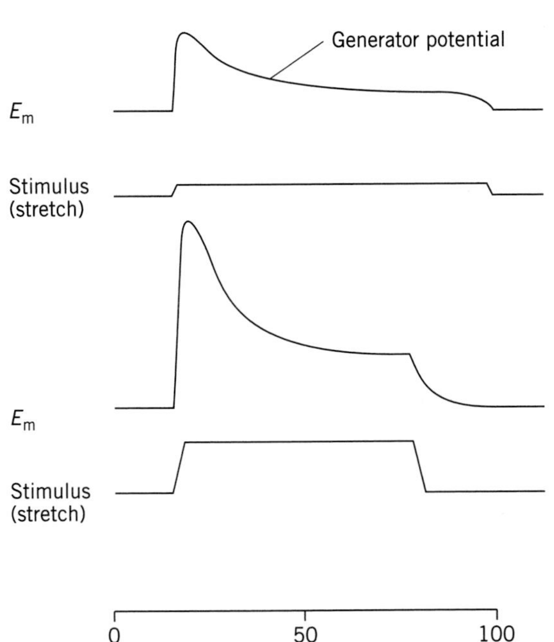

FIGURE 9-10. The generator potential (top traces) of a muscle spindle sensory neuron responding to stretch (bottom traces). The sensory neuron was bathed in lignocaine, the British term for the anesthetic lidocaine, in order to suppress action potentials. Notice that the response of the sensory neuron to stretch is stronger when stretch is first applied. The steady amplitude of the generator potential is linearly related to the degree of stretch.

Adaptation is the decline in response of a receptor over time in spite of the continued presence of a stimulus of constant strength. Tonic receptors adapt slowly and hence provide information about body position or the presence of stimuli. Phasic receptors adapt rapidly and hence give information about the rate of change of a stimulus, such as the velocity or acceleration of moving body parts. Some receptors are phasi-tonic, showing characteristics of both tonic and phasic receptors.

Common Features of Sensory Systems

Sensory systems have features in common at levels above that of individual neurons. These features, which are independent of the type of stimuli the system normally processes, arise as a way for the system to provide the most useful information possible to the central nervous system about the environment.

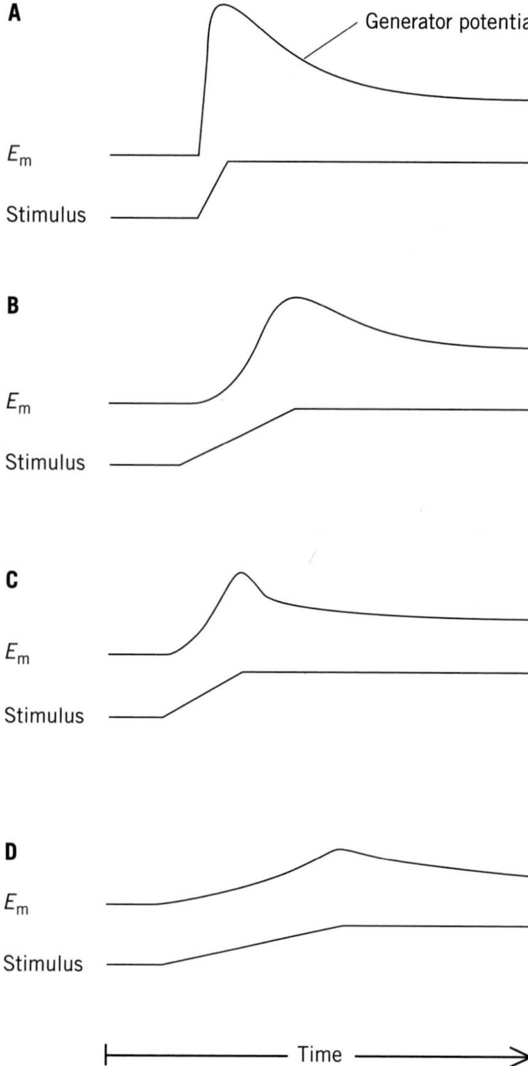

FIGURE 9-11. The phasic component of the response of a muscle spindle receptor to stretch. In each panel, the length of the stretch applied to the spindle is the same, but the velocity varies from rapid (A) to quite slow (D). The response is greatest when the velocity of the stretch is the highest.

Receptive Fields

When you burn your hand on a hot stove, you feel the pain in your hand, not all over your body; you are able to localize the pain to a specific place. Localization is possible because a stimulus must interact with a receptor in order to excite it. Stimuli that are too far away will have no influence on a receptor and hence will not be detected by it.

The region or area of a sensory surface that must be stimulated in order to elicit a change in activity in a neuron is called the neuron's **receptive field**. A stimulus that falls within a sensory neuron's receptive field excites the neuron (as long as the stimulus is above threshold), whereas a stimulus that is outside the field does not (Figure 9-12A). Receptive fields have several important properties. First, they may overlap. If two sensory receptors are close together, a single stimulus may excite both of them (Figure 9-12B). Second, receptive fields are not static, fixed entities. The area of the field depends on the strength of the stimulus used to elicit a response. A strong stimulus has a more widespread effect than does a weak one and hence may influence a neuron that a weak stimulus cannot reach (Figure 9-12C).

Receptive fields are not restricted to the primary receptor of a sensory structure or surface. Every neuron in a sensory pathway has a receptive field, even neurons in the brain. This is because for every neuron along a sensory pathway that responds to sensory input there is some area of the sensory surface that must be stimulated in order to elicit a change in that neuron's activity.

Receptive fields, the regions of a sensory surface that must be stimulated to cause a change in activity in neurons, allow the position of a stimulus to be localized. Receptive fields may overlap, and their areas depend on the strength of a stimulus.

Contrast Enhancement

In principle, a stimulus is easy to localize if it is so weak that it stimulates only a single sensory neuron or receptor cell. However, most stimuli are strong enough to stimulate many receptors, resulting in uncertainty about their exact location. The problem is especially acute in the visual system because being able

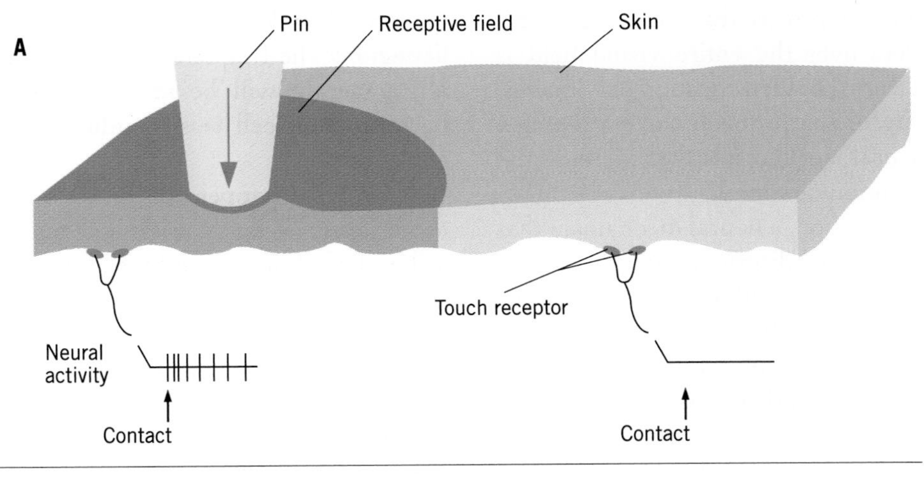

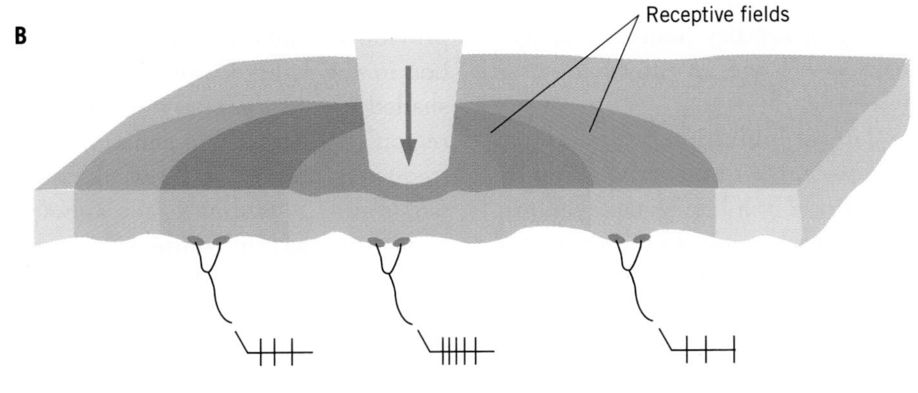

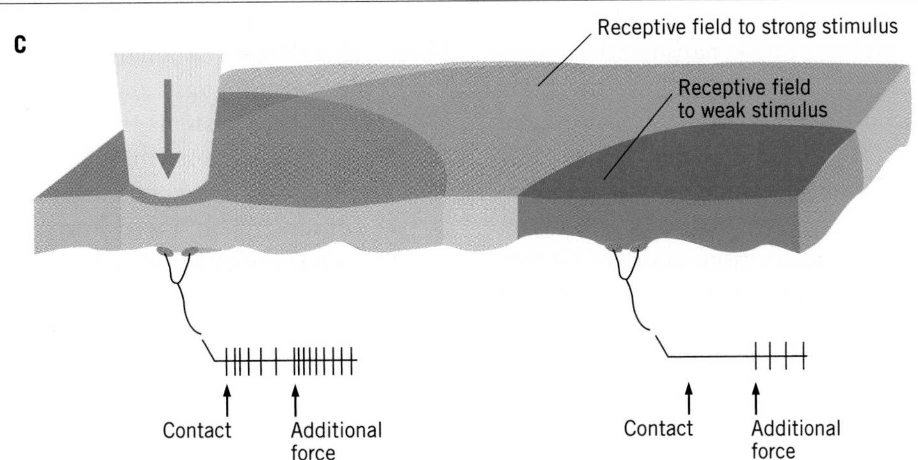

FIGURE 9-12. The properties of receptive fields, using touch receptors in human skin as an example. (A) The receptive field of a neuron is the region of the sensory surface that must be stimulated to activate a neuron. The tip of the pin is touching the skin within the receptive field of the sensory cell on the left but not the receptive field of the cell on the right, as shown by the lack of response by this cell. (B) Receptive fields may overlap. Here, the pin is touching the skin within the receptive fields of three different sensory neurons. (C) The size of a receptive field is influenced by the strength of the stimulus. A light touch activates only the neuron on the left, whereas a stronger one also activates the neuron on the right. Hence, the location of the pin is within the second cell's receptive field as long as the stimulus is strong enough.

to determine the precise distribution of light-dark borders over the entire visual field is important for the identification of images. The severity of the problem can be reduced through neural mechanisms that increase the spatial resolution of stimuli.

Lateral inhibition, a neural mechanism that can increase spatial discrimination, was carefully studied by H. Keffer Hartline and his colleagues in the 1950s. Because of the fundamental importance of Hartline's work on this phenomenon, and because of his contribution to such basic concepts of sensory physiology as the receptive field and the relationship between stimulus strength and frequency of firing of sensory neurons, he was awarded a Nobel Prize in Physiology or Medicine in 1967.

Formally, **lateral inhibition** is the mutual inhibition of adjacent neurons in an array by means of reciprocal inhibitory connections. How this mechanism works to increase spatial contrast can best be explained by considering an example from the simple visual system of *Limulus*, a primitive arthropod that Hartline used for many of his studies. Each eye of *Limulus* consists of a number of facets, each of which contains a single photoreceptor. The photoreceptors synapse with sensory neurons that send their axons into the brain and also send branches to neighboring neurons. The arrangement is shown in Figure 9-13. (For the sake of simplicity, the photoreceptors and the sensory neurons in the figure and in the subsequent discussion are treated as if they were single units, photoreceptive cells with axons.)

Consider how such a system can work when part of the eye is brightly lit and an adjacent part is shaded and hence only dimly lit. Each receptor exposed to the light (A–C) will tend to fire spikes at a certain frequency due to the stimulation it receives from the light. However, since each receptor also receives inhibition from its neighbors, the frequency of spikes each will actually generate will be somewhat reduced. Receptors in the shaded region (F–H) receive much less light and therefore tend to generate a lower frequency of spikes than cells in the brightly lit region. The frequency that each will actually generate will be reduced even further because each cell is also inhibited a bit by adjacent cells.

Now, what happens at the border between the brightly lit and the shaded regions? The cell in the light just next to the border with the shaded region (D) will tend to generate the same frequency of spikes due to stimulation as would cells A–C, but its actual output frequency will be greater than that of these others because it does not receive as much inhibition. This is because one of its neighbors is in the shaded region and hence does not provide as much inhibition as does the neighbor in the light. Conversely, the cell in the shaded area just next to the border with the lit region (E) will tend to generate the same low frequency because of its weak stimulation as would cells F–H, but its actual output will be less than that of these others because it receives more inhibition from its neighbor in the light than these other cells do from their dimly lit neighbors. In consequence, the two border cells, D and E, have a greater difference between their rates of firing than they would have without the effects of lateral inhibition. This difference enhances perception of the border between light and dim.

Other sensory systems that require spatial discrimination, such as the skin tactile system, also show contrast enhancement through lateral inhibition. In these systems, lateral inhibition has the effect of reducing the number of neurons that are active when a touch is applied to a particular location, hence making precise location of the touch easier.

Lateral inhibition is a mechanism for increasing the spatial resolution of stimuli. It works through reciprocal inhibition between adjacent elements in a receptor array and accentuates the difference in firing rate of receptor cells on two sides of a border or edge of a stimulated region.

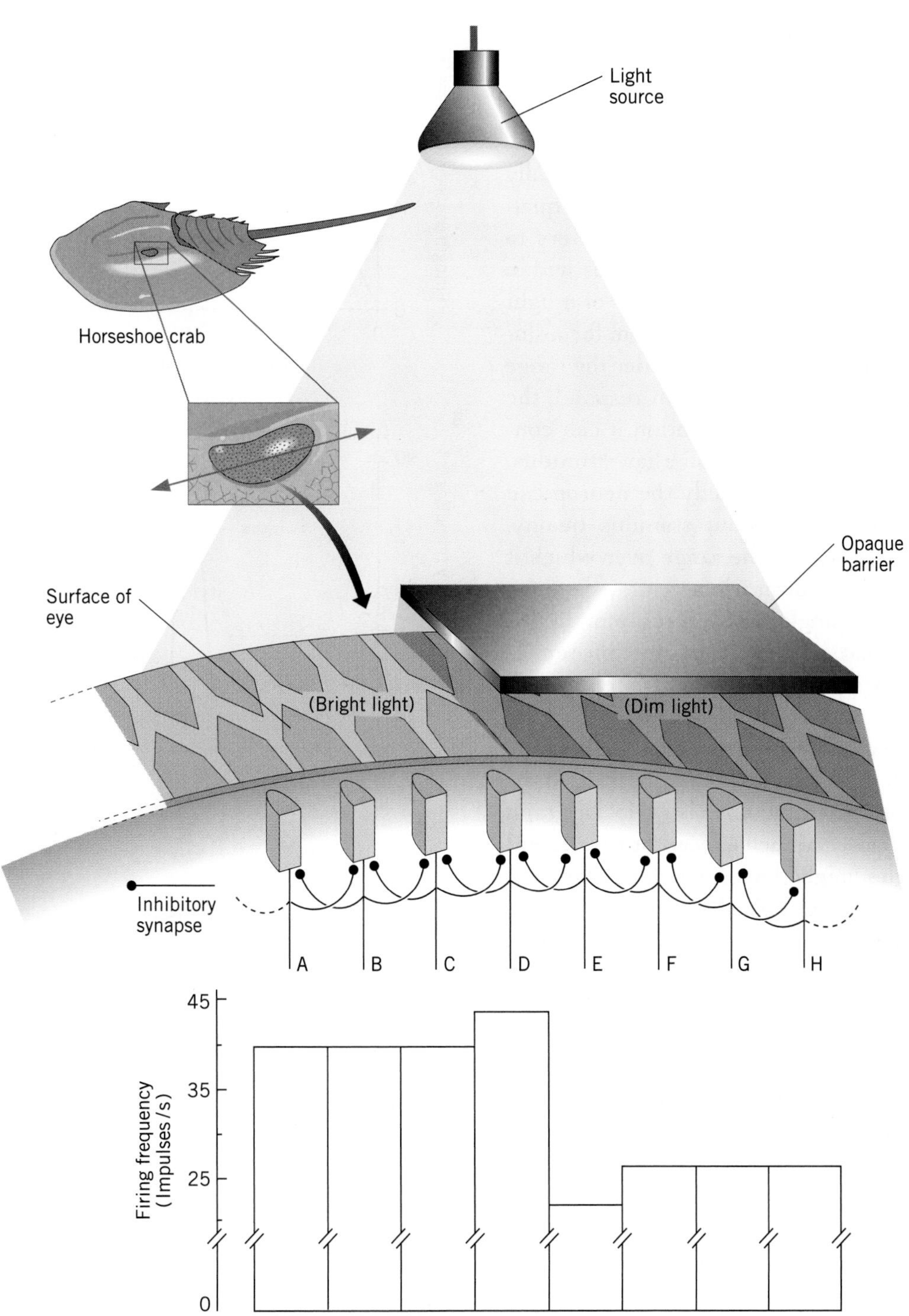

FIGURE 9-13. Lateral inhibitory neural interactions between photoreceptors in the eye of *Limulus* (the horseshoe crab) that allow for contrast enhancement. Each cell inhibits its neighbor in proportion to the strength of stimulus it receives. Hence, cells that are exposed to bright light depress their neighbors' activity more than cells that are exposed only to dim light. The inhibition is unbalanced at the bright/dim border, so that the receptor in the bright light receives less inhibition than its neighbors and the receptor in the dim light receives more inhibition than its neighbors. Hence, the difference in output (bottom line) between these two border cells will be greater than if there were no inhibition.

Range Fractionation

Most receptors in large animals are grouped into sense organs. An advantage of sense organs is that several sensory neurons can work together to overcome the functional limitations of single neurons. This is especially true when it comes to differentiating the quality of a stimulus. Stimulus **quality** refers to the "type" or subdivision of a stimulus within a single modality, such as the color of a light or the tone of a sound. The problem for a single sensory neuron is that the wider the range of stimulus type to which it can respond, the less exact will be the information it can convey about the quality of that stimulus. Conversely, the more exactly the neuron can convey information about stimulus quality, the narrower will be the range over which it can convey this information.

Suppose a single mechanoreceptor in the knee of a cat will respond over the entire range of motion of the knee joint, about 90°. If you assume that the sensory neuron can fire from a low of 1 spike/sec to a high of 90 spikes/sec, and that the neuron fires at its highest rate when the knee is at 45°, a graph of the responses of such a receptor to different knee angles might look something like that shown in Figure 9-14A. If the mechanoreceptor's response were as shown in the figure, then a change in firing frequency of about ten impulses per second would signal a change in the angle of the knee of about 5°. For example, if the knee is at an angle of about 25° and extends to an angle of 30°, a change of 5°, the sensory neuron would increase its firing rate from 65 to 75 spikes/sec, a change of 10 spikes/sec.

But suppose the mechanoreceptor were sensitive over only a 10° range, from 40° to 50°, and yet had the same range of response, from 1 to 90 spikes/sec (Figure 9-14B). That is, the receptor might not start responding until the knee reached an angle of 40°, and would no longer respond once the knee reached an angle of 50°. Because the receptor now has to cover only one-ninth the angular distance that it did in the previous case, a change in the firing frequency of about 10

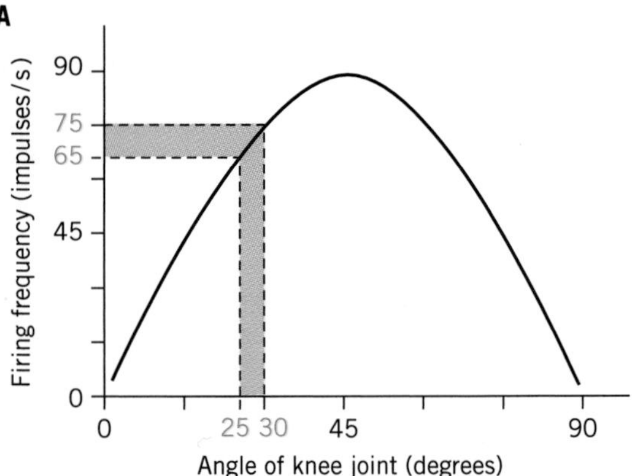

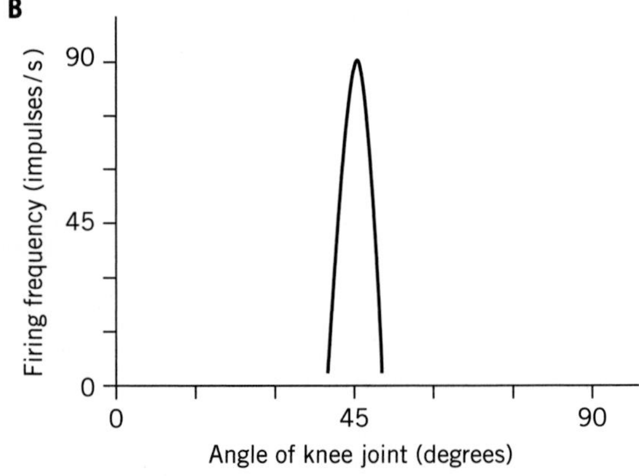

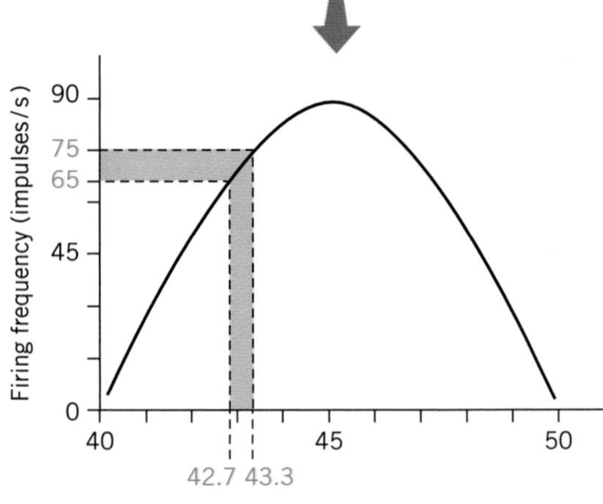

FIGURE 9-14. The hypothetical response of two individual sensory neurons that could signal bending of a knee joint. (A) The neuron fires over a wide range of joint angles. As a consequence, it would be difficult for this neuron to give an unambiguous signal representing a small change in joint angle. (B) A different neuron responds over a considerably more restricted range of joint angles. Hence, even a small angular change of joint angle will be signaled by a measurable change in the rate of firing of the neuron.

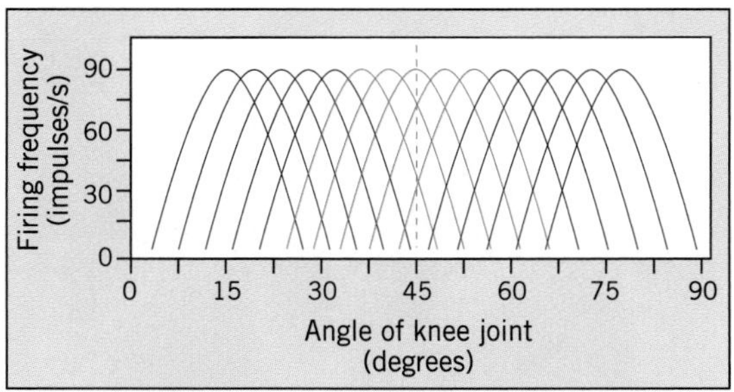

FIGURE 9-15. Idealized representation of range fractionation in a knee-joint sense organ of a cat. Each curve represents the response of a single sensory mechanoreceptor to stretch over a small range of angles. Even though each sensory neuron is sensitive to only a small range, the cat can determine the angle of the knee joint quite accurately by having available a large set of neurons with different but overlapping ranges.

impulses/sec would now signal a change in the angle of the knee of only about 0.55°, and the receptor cell would reliably be able to convey information about much smaller changes in joint angle. This gives it considerably greater resolution of joint angle, but at the cost of a decrease in the range of joint angles that can be covered.

By grouping several sensory neurons together in a sense organ, it is possible to cover a wider range of stimulus quality without sacrificing sensitivity. Each cell can be highly sensitive to a small part of the stimulus range of the whole organ. If different cells cover different parts of the range, no part will be left out. Such a division of labor among neurons in a sense organ is called **range fractionation,** and is shown in Figure 9-15.

In addition to increasing the resolution of sense organs without sacrificing range, range fractionation also reduces the ambiguity that is inherent in the responses of single mechanoreceptors to stimuli such as joint position. Consider the cell whose response is depicted in Figure 9-14A. Considering only the information from this one sensory neuron, it is impossible for the nervous system to determine whether the knee is at an angle of 10° or 80° because the receptor responds with the same frequency to both angles. The cat would have to rely on additional information from other receptor cells in order to resolve the ambiguity. In a range-fractionated system, on the other hand, there is little ambiguity because different neurons are responding at 10° than at 80° (see Figure 9-15), so the cat would have no difficulty in identifying the actual joint angle.

Range fractionation is division of labor among receptor cells in a sense organ such that each cell is sensitive to only a small part of the range of stimuli to which the entire sense organ can respond. This increases the resolution and sensitivity of the organ to stimulus quality without sacrificing its range, as well as reducing the ambiguity inherent in interpreting responses from only a single sensory neuron.

The Organization of Brain Sensory Regions

Just as receptors or sense organs have specific anatomical and functional features that enhance their abilities to extract useful information from sensory stimuli, so too do the parts of the vertebrate central nervous system that are devoted to interpreting sensory information. These features transcend sensory modality and offer clues to the way in which the brain makes sense of the information it receives.

Topographic Organization

A striking feature of nearly every region of a vertebrate brain devoted to the processing of sensory information is its **topographical organization,** the point-to-point representation of a sensory surface in the nervous system, especially the brain. Topographical organization helps the animal handle sensory input from complex sense organs by arranging it in an orderly manner, so that the specific part of the sensory surface that is stimulated will cause a response in a specific part of the brain, and adjacent regions of the sensory surface will be represented in adjacent regions of the brain.

The **somatosensory system**, which deals with sensory information from the surface and interior of the body, provides perhaps the simplest example of topographical organization. Each area of the skin contains receptors for stimuli such as touch and pressure. Receptor cells that are located in adjacent parts of the skin pass their signals on to adjacent regions of the **somatosensory cortex**, the part of the cortex devoted to input from the somatosensory system. The result is a functional mapping of the surface of the skin onto the cortex, as shown diagramatically in Figure 9-16. This maplike arrangement is called **somatotopic organization,** or just **somatotopy**. The brain map does not have to (and indeed does not) keep the same proportions as the skin itself. Skin regions that are heavily invested with touch receptors, such as the fingers and lips, have a relatively greater representation in the brain than do regions such as the middle of the back, where there are fewer receptors.

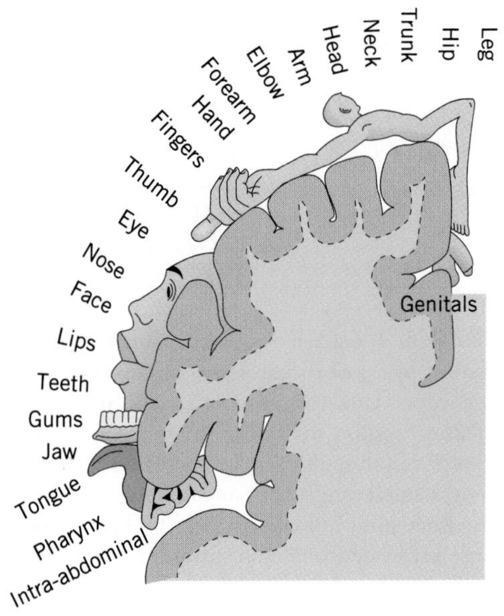

FIGURE 9-16. A section through the left hemisphere of a human brain at the level of the somatosensory cortex. The distorted human figure (called a *homunculus*) drawn over the cortical surface represents the parts of the skin surface that are represented on the cortex. Notice that some areas, like the fingers and lips, are disproportionately large in the homunculus; relatively more cortex is devoted to input from these skin areas than from regions like the back that are physically larger. This is because there are more receptors per unit skin area in the fingers and lips than there are in the back.

Similar maps exist for the eye and the ear, each with its own name. In the visual system, there is a point-to-point representation of the retina onto the visual cortex, a representation called **retinotopic organization** or **retinotopy**. Hence, light that stimulates a particular spot on the retina will cause activation of neurons in a particular region of the visual cortex. In the auditory system, the cochlea is mapped point-to-point onto the auditory cortex, a mapping called **tonotopic organization** or **tonotopy,** and stimulation of different regions of the cochlea will cause responses of different regions of the auditory cortex.

As originally conceived, topographical organization referred only to the mapping of a sensory array onto some cortical brain region in vertebrates, although from the beginning it was recognized that different maps were associated with different cortical layers. However, more detailed investigation has revealed that not only are the sensory cortexes of the brain topographically organized, but so also is any brain nucleus that serves as a relay station in the pathway between a group of sensory neurons and the sensory region of the brain to which their information is being carried. For example, both the lateral geniculate nucleus, a relay center for visual information in mammals, and the spinal columns carrying somatosensory information from the body up to the brain, are topographically organized. Topographical organization is present even in invertebrate ganglia. Sensory neurons from hairs on an insect appendage show a topographical projection into the closest ganglion. Hence, an array of adjacent hairs makes synaptic connections with neurons at an array of adjacent locations within the ganglion.

Furthermore, the entire concept of topographical organization has changed from a map based strictly on anatomy to one that implies a functional organization of whole sensory systems. Consider the vertebrate retina. Since there is a point-to-point representation of the retina on the visual cortex, and since the structure of the eye allows the stimulation of a particular spot in the retina by light originating only from one particular place in the visual field, the topographically organized anatomical map in the brain actually represents a map of visual space as much as it does a map of the physical organization of the retina. A similar situation exists with the vertebrate ear. Because different parts of the cochlea respond to sounds of different frequency, the anatomical map of the cochlea in the auditory cortex is functionally a map of sound frequency.

This idea of functional maps has found new applicability in recent work on sensory systems, especially the auditory system, as you will learn in Chapter 12. In some animals, parts of the auditory cortex are organized so as to provide specific information about the location of sounds relative to the animal. These cortical regions are "topographically" organized, so that sounds that originate close to one another spatially will excite neurons in the cortex that are also close to one another physically. There is even some evidence from chemosensory systems in insects that the olfactory (smell) or gustatory (taste) regions of the brain may shuttle incoming information about stimuli to specific brain regions depending on the nature of the chemical stimulus that caused the input. The basis of the functional organization that may exist in the chemosensory system, however, is still a mystery.

Based on how widespread it is among sensory systems in many types of animals, topographical organization is thought to be quite important for the proper analysis of sensory information. It may be that topographical organization helps a sensory region of the nervous system to interpret the input it receives by organizing that input in a spatial array that bears a precise relationship to some aspect of the stimulus that is being interpreted. Hence, stimuli that have some similarity of location (like adjacent points in visual space) or quality (like similar frequencies of sound) are sent to brain regions that are adjacent to one another, thereby perhaps accentuating the similarity. This kind of organization may help an animal to build a coherent picture of its environment from the point of view of each sensory modality, or may allow it to process sensory information that is functionally or anatomically related in a more efficient way.

Topographical organization is the point-to-point representation of a sensory surface like the skin or the retina onto a particular region of the central nervous system. It is present in both vertebrate and invertebrate animals, and is thought to help the animal to build a sensory view of its environment.

Columnar Organization

Topographical organization is not the only functional feature of a sensory region of the vertebrate brain that helps it to interpret the information it receives. The discussion of the brain in the preceding section may have suggested to you that in vertebrates the sensory regions of the brain are strictly two-dimensional, having area but no depth. In reality, the cortex is a complex three-dimensional structure that consists of up to six distinct layers of cells. In the sensory regions of the cortex the layers often have a **columnar organization** in addition to a topographical one. Columnar organization refers to an arrangement of neurons into functionally distinct columns of cells, each of which processes a particular type of information. Columnar organization is strictly functional; there is usually no discernible anatomical difference between the neurons that make up one "column" and those that make up another,

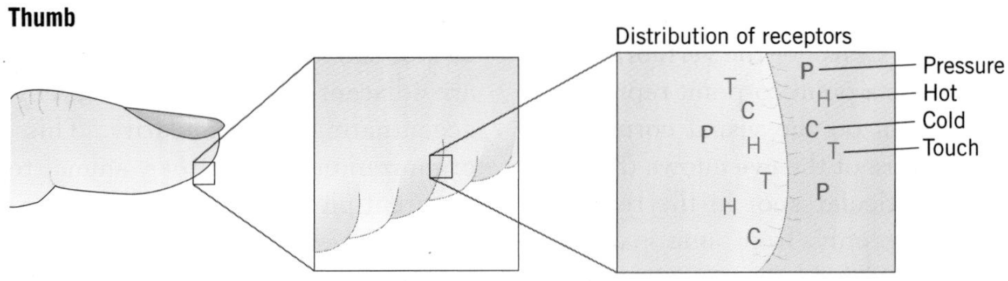

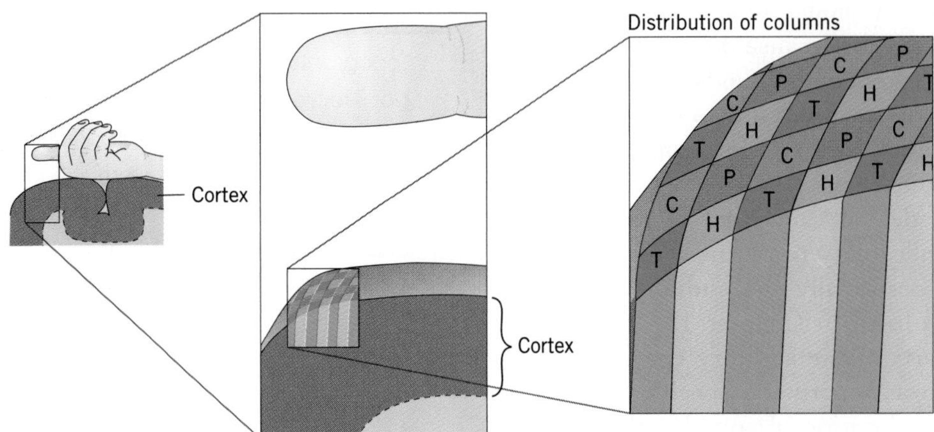

FIGURE 9-17. Columnar organization of the somatosensory cortex. The skin of the thumb (top) is mapped to one region of the somatosensory cortex (bottom). Within one small region of the skin lie receptors for different senses, such as pressure, touch, and temperature. In the corresponding cortical region, separate columns of neurons process input from each type of receptor. The diagram is highly schematic and does not represent the actual density or distribution of somatic sense organs or cortical divisions.

although different columns can and do receive different inputs. However, all the neurons in one column in the cortex receive the same types of inputs, so that they all process the same kind of information. Columns of adjacent neurons receive related but different input, and hence produce different output.

Columnar organization works within the context of the topographical organization of the sensory cortex. For example, consider a region of somatosensory cortex that receives input from the skin at the tip of the thumb. The skin has receptors that are adapted to respond to many types of stimuli—pressure, touch, temperature, and so forth. Each type of receptor will send its signals to one specific column of cells in the part of the cortex that receives input from the tip of the thumb (Figure 9-17). The columns that process these different kinds of input are distinct, yet adjacent to one another. In this way, the somatosensory system can keep separate the different types of sensory modalities that it deals with; at the same time, it can keep intact the topographical organization of the cortex as a whole so that activation of each receptor type can be associated with a particular region of the skin.

In other sensory regions there may be separate columns of neurons devoted to different aspects of a single type of stimulus. In the visual system, for example, separate columns are devoted to different orientations of a line stimulus (see Chapter 11). Neurons in one column will process the input from the eyes if the line is vertical, an adjacent column will process it if the line is slightly off vertical, and so forth. In rats, separate columns are devoted to input from each whisker on the face. In this case, the columns are distinct from one another anatomically, and (because of their shape) are called barrels.

The region of the cortex devoted to a particular type of sensory input is often organized into functional columns, an arrangement in which a vertical cylinder of neurons receives information from a single type of sensory neuron or about a specific aspect of a sensory stimulus.

Additional Reading

General

Aidley, D. J. 1989. *The Physiology of Excitable Cells*. 3d ed. Cambridge: Cambridge University Press.

Barlow, H. B., and J. D. Mollon, eds. 1982. *Cambridge Texts in the Physiological Sciences*. Vol. 3, *The Senses*. Cambridge: Cambridge University Press.

Corey, D. P., and S. D. Roper, eds. 1992. *Sensory Transduction*. Society of General Physiologists, 45th Annual Symposium. New York: Rockefeller University Press.

Cytowic, R. E. 1993. *The Man Who Tasted Shapes*. New York: Putnam.

Schmidt, R. F., ed. 1986. *Fundamentals of Sensory Physiology*. 3d ed. New York: Springer-Verlag.

Somjen, G. 1972. *Sensory Coding in the Mammalian Nervous System*. New York: Plenum.

Research Articles and Reviews

Eyzaguirre, C., and S. W. Kuffler. 1955. Processes of excitation in the dendrites and in the soma of single isolated sensory nerve cells of the lobster and crayfish. *Journal of General Physiology* 39:87–119.

Hartline, H. K., and F. Ratliff. 1957. Inhibitory interaction of receptor units in the eye of *Limulus*. *Journal of General Physiology* 40:357–76.

Ottoson, D., and G. M. Shepherd. 1971. Transducer properties and integrative mechanisms in the frog's muscle spindle. In *Handbook of Sensory Physiology*. Vol. 1, *Principles of Receptor Physiology*, ed. W. R. Loewenstein, 442–99. New York: Springer-Verlag.

Shepherd, G. M. 1991. Sensory transduction: Entering the mainstream of membrane signaling. *Cell* 67:845–51.

Terzuolo, C. A., and Y. Washizu. 1962. Relation between stimulus strength, generator potential and impulse frequency in stretch receptor of *Crustacea*. *Journal of Neurophysiology* 25:56–66.

Torre, V., J. F. Ashmore, T. D. Lamb, and A. Menini. 1995. Transduction and adaptation in sensory receptor cells. *Journal of Neuroscience* 15:7757–68.

CHAPTER 10

The Coding and Control of Sensory Information

It is not enough that a sensory receptor report the presence of a particular type of stimulus. It must also convey to the appropriate sensory processing center other information about the stimulus. In this chapter, you will learn how information about a stimulus is encoded as series of action potentials. You will also learn that sensory receptors are not merely passive sensing devices, but that they can be strongly influenced by the central nervous system to affect the kind of information that they pass on.

As you learned in Chapter 9, different stimuli such as sound and light are distinguished from one another in part by being processed in different parts of the brain. However, there is more to a stimulus than its type. Stimuli of a single type can vary in quality. For example, light comes in different wavelengths, which give rise to different colors; sounds come in different frequencies, which give rise to tones of different pitches; and chemicals come in different molecular configurations, which give rise to different odors or tastes. Stimuli can also vary in strength. Variations in **stimulus strength** give rise to variations in the subjective sense of **intensity**. Hence, you can speak of light as being dim or bright, sound soft or loud, odors weak or strong. The phrase **stimulus intensity** is often used as a synonym for stimulus strength.

How does the nervous system convey information about stimulus quality and stimulus strength to the brain? Sensory information is conveyed to the central nervous system by action potentials. Even nonspiking receptor cells synapse with one or more neurons that send action potentials into the central nervous system. Since there is typically no difference between one action potential and another in a sensory neuron, information about stimulus quality and strength must be contained in the pattern of action potentials. Humans have also devised ways of conveying information via a pattern. In the days of the telegraph, the Morse code, brief patterns of short and long electrical pulses representing letters of the alphabet, was used to transmit messages over long distances. In information theory, the term coding is used to refer to the rules by which information is transmitted by patterns like a Morse code. By analogy, **coding** is also used in sensory physiology to refer to the mechanisms by which information about sensory quality or strength is conveyed to the central nervous system. Hence, sensory physiologists speak about how the loudness of a sound or the identity of an odor is coded by the sensory system.

Stimuli differ not only in type but also in quality and strength. Differences in stimulus strength give rise to differences in subjective intensity. Information about stimulus quality and strength is conveyed to the brain in patterns of action potentials, a process referred to as coding.

Coding of Stimulus Strength

You learned in Chapter 9 that the amplitude of a generator potential increases as the strength of a stimulus increases; strong stimuli cause the opening of more ion channels, and therefore the flow of more current in the receptor cell, than do weak stimuli. You also learned that larger generator potentials in turn generate a higher frequency of action potentials. The net result of these relationships is that in a sensory neuron stimulus strength is coded by the frequency of action potentials.

To understand just how this comes about, consider what happens when a stimulus is applied to a sensory neuron (Figure 10-1A). The interaction of the stimulus with the sensory neuron causes the opening of ion channels and the subsequent depolarization of the neuron. The depolarization (the generator potential) spreads to the spike initiation zone and, if it is greater than the neuron's threshold, generates an action potential by inducing the opening of voltage-sensitive sodium channels. If the stimulus is brief, a single action potential will be generated. If the stimulus is sustained, however, the generator potential will also be sustained (assuming a low rate of adaptation). Following the first action potential, the membrane potential of the neuron will once more rise to the level of the generator potential because the ion channels opened by the stimulus are still open, allowing depolarizing current to flow into the cell. The threshold of spiking of the neuron, which had become infinite when the first action potential was generated (see Chapter 5), falls toward its normal level following the action potential. When the threshold intersects the generator potential, another action potential will be generated.

Stronger stimuli produce higher frequencies of action potentials because stronger stimuli produce larger generator potentials, due to the greater number of ion channels that are opened by these stimuli. A larger generator potential means that the cell is more depolarized. This, in turn, means that as the cell's threshold falls back to its normal level following an action potential, it will intersect the membrane potential sooner than it did when a weaker stimulus was applied. In con-

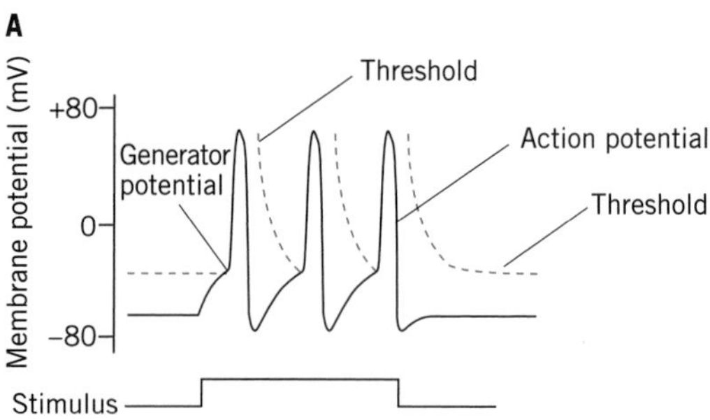

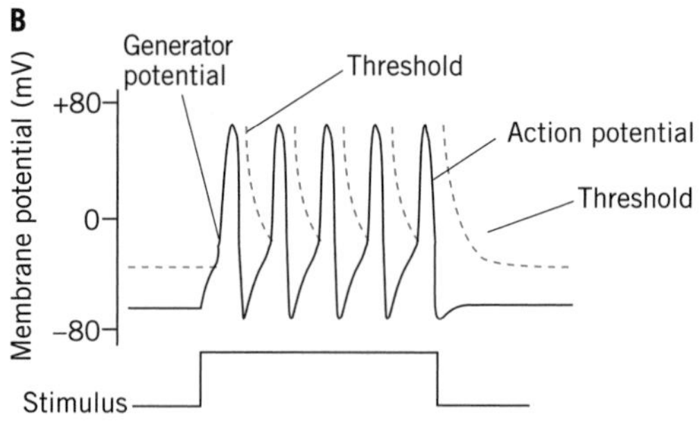

FIGURE 10-1. The relationship between the generator potential and the spikes produced by a sensory neuron. (A) A sustained stimulus will generate a series of action potentials because the generator potential brings the membrane potential above the threshold for spike initiation after each spike. (B) A strong stimulus produces a higher frequency of spikes because the recovering threshold intersects the elevated membrane potential sooner than it does when the stimulus is weaker. Adaptation has been ignored in these examples for the sake of clarity.

sequence, the second action potential appears sooner when a strong stimulus is applied than it does in the presence of a weak stimulus (Figure 10-1B).

Two factors limit the frequency at which the sensory neuron will fire. First, there is a finite number of ion channels available in the membrane. Once they are all open, further increases in stimulus strength can have no effect. Second, the absolute refractory period that follows each action potential as a result of sodium inactivation (see Chapter 5) represents a minimum interval between successive action potentials, and hence sets another upper limit on the frequency of spikes. Different sensory neurons typically have different upper limits on spike frequency, which may range from less than 100 impulses/sec to over 300 impulses/sec.

Stimulus strength is encoded in the rate of firing even when a nonspiking receptor cell is interposed between the stimulus and the sensory neuron that conveys action potentials to the central nervous system. A stimulus impinging on a nonspiking cell produces a receptor potential whose amplitude rises as the strength of the stimulus rises. As you learned in Chapter 7, synaptic transmission from nonspiking neurons is graded, so the receptor cell will release a proportionate amount of neurotransmitter as long as it is stimulated. The postsynaptic sensory neuron will then fire at a rate that is proportional to the amount of neurotransmitter that it receives, thereby encoding stimulus strength.

Although firing frequency is usually a logarithmic function of stimulus strength, the change in the rate at which spikes are produced is not the same for a given change in stimulus strength for all sensory receptors. Furthermore, the immediate past history of the sensory receptor is also critical in determining its response to a particular stimulus. A simple experiment will illustrate this point. Set out three bowls of water, one containing hot water, one warm, and one cold. If you place one hand in the hot water and the other in the cold water for a few minutes, then move both hands into the bowl of warm water, you will find that the water feels warm to the hand that had been in cold water, and cool to the hand that had been in hot water. The temperature receptors in your two hands will fire at different rates because they had adapted to the temperature they had been exposed to just moments before, even though they are now exposed to the same temperature. For most receptors, this kind of shift in the range over which they report stimulus strength is quite common. As you will learn in a later section of this chapter, the nervous system can actively control this aspect of sensory receptor function in many sense organs and sensory receptors.

For tonic receptor cells such as the crustacean stretch receptor, the meaning of a change in stimulus strength is easy to visualize. Stronger stimuli are those that stretch the receptor strand farther, hence inducing firing at a higher frequency in the sensory neuron. But for phasic stretch receptors, the "strength" of a stimulus is not related to the distance the receptor is stretched. As you saw in Figure 9–9, phasic mechanoreceptors respond differently to different rates of stretch, even if the degree of stretch is the same. Hence, for a phasic mechanoreceptor, a weak stimulus is one that is applied slowly, whereas a strong stimulus is one that is applied quickly.

Information about stimulus strength is usually encoded by the firing frequency of the sensory neuron. However, this information may be relative to the neuron's past history. Phasic receptors encode the speed at which a stimulus is applied.

Coding of Stimulus Quality

For the nervous system to distinguish between different aspects of a stimulus, it must be able to encode these differences into specific and recognizable patterns of neural activity. In principle, there are two ways in which the nervous system might be able to do this. On the one hand, the nervous system might use a **labeled line code**. In this type of coding, each individual sensory neuron conveys information about only a single aspect of a stimulus. In a sense, the nervous system bypasses the problem of how to use patterns of action potentials to convey differences between stimuli by dedicating each individual neuron to just one aspect of a stimulus. In this way, the sensory processing center in the brain would "know" which aspect was present merely by recognizing the particular sensory neuron that was bringing in the signal. Suppose, for example, that the receptive fields for all the sensory neurons in your skin were nonoverlapping. If you then stepped on a pebble with your bare foot, you would know exactly what part of your foot was touched because the neuron or neurons that were stimulated by the pebble uniquely identify the location of the stimulated part of the skin; they send their signals to just the one location in the somatosensory cortex devoted to input from that part of your foot.

Labeled line coding usually refers to the coding of the type of stimulus within a particular modality, such as the color of a visual stimulus. In a sense, however, sense modalities use a labeled line code to encode the modality itself. For example, all information from the eye is routed to the visual centers and thereby is "labeled" visual, and all information from the tongue goes to taste centers and hence is labeled taste. This is the reason the law of specific nerve energies is valid.

An alternative method of encoding stimulus quality is called an **across-fiber code** or a **population code**, in which the pattern of response of all the sensory neurons in a sense organ is used to convey information about the stimulus. In this case, many sensory neurons will respond to any one stimulus. However, each neuron will respond in a different way to each stimulus, so that the identification of the stimulus comes from a uniquely coded pattern of response of many neurons rather than the firing of only one or two particular ones. For example, if you smell banana and orange in a fruit salad, many thousands of odor receptors in your nose will become active. Some receptors will respond to both odors, some to only one or the other. The brain will be able to identify each odor by the pattern of activity of the population of receptors that respond to it.

Both methods of encoding information about sensory stimuli are used in the nervous systems of all animals.

Sensory systems can encode information in two ways. In a labeled line code, each sensory neuron unambiguously conveys information about only a single aspect of a stimulus. In an across-fiber code, many different sensory neurons respond to each specific stimulus, and the precise information about the stimulus is carried as a pattern of response of all the neurons together.

Labeled Line Code

In a labeled line code, each individual sensory neuron conveys information about just one particular aspect of a stimulus. How can such a code work? Consider the specialized hairs on the feet of a fly such as the blowfly, *Phormia regina* (Figure 10-2). Each hair is open at the tip and contains the dendrites of four chemoreceptive neurons. Each of the chemoreceptors is extremely sensitive to one particular substance—salt, sugar, water, or amino acids—but is hardly sensitive to anything else. Imagine the fly walking about on a picnic table on which some lemonade had been spilled. When the fly steps in a drop of lemonade, the liquid will come into contact with the chemoreceptors through the opening at the tip, exciting the "sugar" and "water" receptor cells. The fly would know it had encountered sugar and water because only the two neurons that respond to those substances would be sending signals into the nervous system. On the other hand, if the fly had encountered a drop of fluid from a hamburger, the

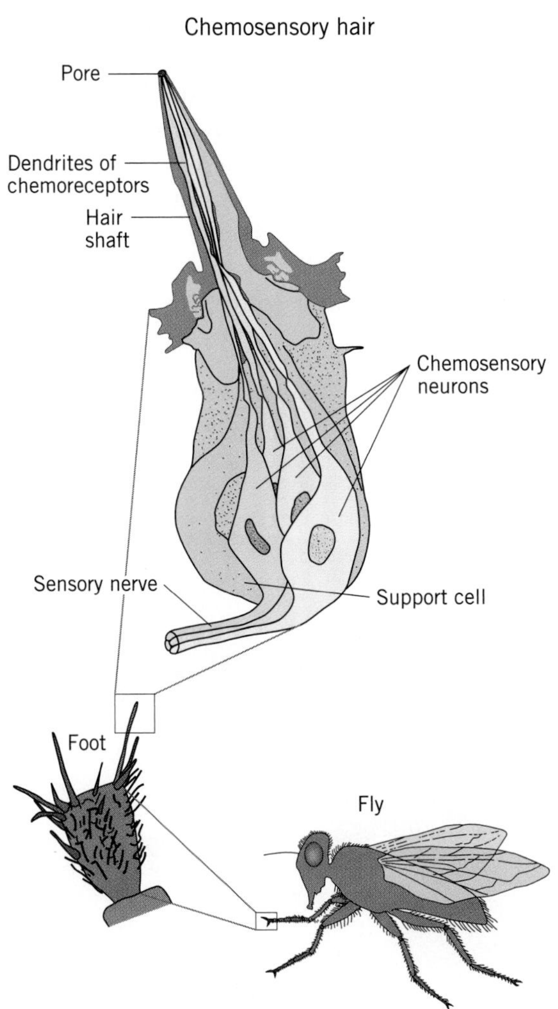

FIGURE 10-2. A chemoreceptive hair from the foot of a blowfly. Each hair contains four chemoreceptive sensory neurons, each primarily sensitive to a different type of chemical substance. The sensory neurons are exposed to the outside through a small pore at the tip of the hair. Insect chemoreceptors like this often contain a mechanoreceptor that is sensitive to movement of the hair as well, but this mechanoreceptor has been omitted from the illustration.

amino acid receptor would be excited, telling the fly that there was protein in the fluid. Substance identification in this case is quite simple. All the fly needs to know is which of the chemoreceptors is active in order to know what substance it has encountered. Substances that do not excite any of the chemoreceptors not only cannot be identified, they cannot even be detected.

The vertebrate somatosensory system also uses a labeled line code to encode information about stimuli. The somatosensory system conveys information about two special aspects of each stimulus: where on the surface of the skin the stimulus is located, and the nature of the stimulus. In each case, the information about these aspects of the stimulus is inherent in the specific neurons that are carrying it; since each neuron sends its information to a specific brain region, the identification of stimulus location is implicit in the activated region. For example, a touch receptor on the finger will send its message to the finger area of the somatosensory cortex, whereas one on the back will send its message to the back area, as described in Chapter 9. Identification of the place on the skin at which the touch occurs is based on where in the cortex the neural message is sent. Identification of the type of stimulus that has been encountered, such as cold, heat, pressure, or touch, is based on the specific column in the appropriate somatosensory region that receives the signal. Input from touch receptors is sent to the column devoted to touch, input from cold receptors is sent to the column devoted to cold, and so forth.

A labeled line code works because each "labeled" sensory neuron sends its information to a specific location in the central nervous system. There is nothing intrinsically different about the neurons themselves. In fact, if a labeled line neuron is induced to fire by some stimulus other than the one by which it is normally stimulated, the animal will respond as if the stimulus with which that neuron is normally associated were present, as indicated in the discussion of the law of specific nerve energies in Chapter 9. For example, when you accidentally strike the nerve that runs across the outside of the elbow, you may feel a tingling sensation in your hand, even though your hand was not struck. This is because the mechanical blow has stimulated the neurites of sensory neurons in your hand, and your brain has interpreted the resulting barrage of action potentials as input from the area from which these receptor cells receive their input. The phenomenon occurs because your nervous system uses a labeled line code for the identification of sensory modality.

The so-called phantom limb often experienced by people who have had an arm or leg amputated after an accident is another manifestation of labeled line coding. Apparent sensations in a missing hand or foot are due to activity of the neurons in the brain that previously had received sensory input from the hand. Even though the hand and its sense organs are no longer present, the brain will interpret activity in the region of the somatosensory cortex that was devoted to the hand as input from the hand.

Labeled line coding works well in many cases, but it is not well suited to encode every type of sensory stimulus. The disadvantage of labeled line coding is that for stimuli such as odors that have many different qualities, a very large number of sensory neurons would be required to ensure that they could all be detected. It also leaves no room for the detection of novel stimuli. Furthermore, loss of individual sensory neurons through injury or natural attrition could impair the animal's ability to detect some aspect of a stimulus by reducing the numbers of neurons that are able to respond to that aspect.

In a labeled line code, each neuron conveys information about a particular aspect of a stimulus, and the information is sent to a specific part of the central nervous system. All neural activity at that location in the central nervous system is interpreted as originating from a type of stimulus that normally excites sensory neurons that send their signals there.

Across-Fiber Code

In contrast to the highly specific responses of sensory receptors in a labeled line system, the receptors in a sensory system that uses an across-fiber (population) code will respond to many different stimuli, and each stimulus will excite many different receptors. Because each sensory neuron responds to more than one stimulus, activity in any individual neuron will not serve to identify it. Instead, the sensory processing area of the brain must look at the pattern of response of all the information it receives in order to identify the stimulus.

As an example of how across-fiber coding works, consider the paired **statocyst organs** in the head of a lobster (Figure 10-3A). A statocyst organ consists of a hollow chamber lined with mechanoreceptive hairs, containing

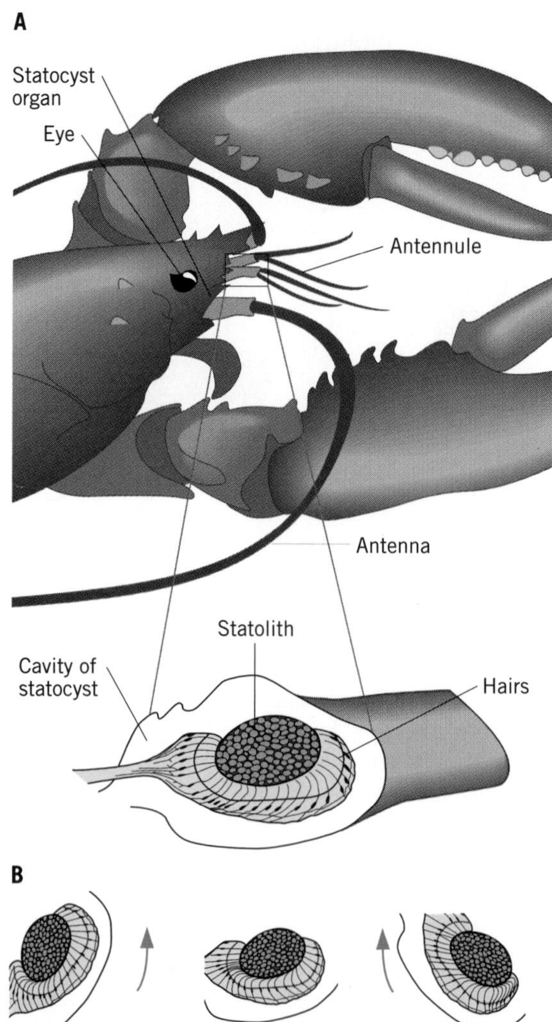

FIGURE 10-3. (A) A statocyst organ in the head of a lobster, with an enlarged view of the statolith inside the hair-lined statocyst cavity. (B) The effect of tilting on the statocyst organ. When the cavity is tilted, the statolith shifts to a different position, causing the bending of a different population of hairs.

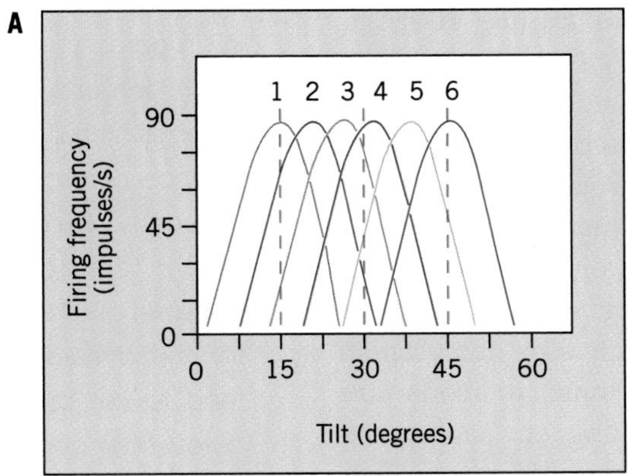

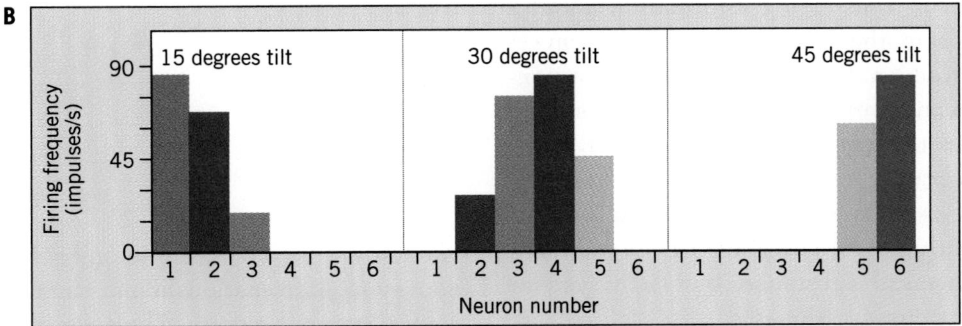

FIGURE 10-4. Across-fiber coding in a mechanoreceptive sense organ. (A) The response profiles of a few of the neurons in a lobster statocyst organ, showing their reactions to different degrees of tilt. Each curve represents the increase and decrease in rate of firing of a single neuron associated with a hair in the statocyst organ. The numbers refer to six individual neurons. (B) The profiles of response for each of the six neurons when the lobster tilts to different degrees. The precise tilt of the animal can be determined by looking at the combined profiles of responses of all the neurons together.

a stonelike concretion called a **statolith**. The statolith lies on the hairs in the chamber, each hair associated with a single sensory neuron. When the animal's body is level, the statolith rests on the hairs on the floor of the chamber, bending them and stimulating the associated sensory neurons.

As the lobster climbs over rocks on the ocean floor, its body will tilt up and down. This tilting causes the statoliths to move inside the statocysts, stimulating some of the hairs in the chamber (Figure 10-3B). When the animal climbs up onto a rock, the statolith will slide to the back of the chamber and bend the hairs there, which stimulates their sensory neurons. As the animal climbs over the rock and starts down the other side, the statolith will slide to the front of the chamber, bending the hairs and stimulating the sensory neurons there. Each sensory neuron will be stimulated maximally when the body is at some particular angle because the hair it is associated with is maximally bent at that body angle. If you examine the pattern of firing of many hairs, you will find that each hair responds to a slightly different range of tilt angles than the other hairs (Figure 10-4A). The brain of the animal can accurately assess the angle of its body by looking at the profile of response of all the neurons at once (Figure 10-4B). Different body angles excite slightly different populations of sensory neurons, and excite each individual neuron to a different extent, so that the total pattern of response of all the neurons is quite different even for tilt angles that do not vary by much.

You may have noticed that the pattern of response of sensory neurons in the lobster statocyst organ shown in Figure 10-4A is similar to the pattern of response in range-fractionated systems, as shown in Figure 9-15. This is because an across-fiber code is always used to relay information about stimulus quality in cases in which range fractionation is used in the sense organ. Range fractionation is the splitting of the range of response of a sense organ into many small parts, in which only some neurons are responsible for responding to each part. Across-fiber coding, on the other hand, is a method of transferring information about a stimulus to the central nervous system. A range-fractionated sense organ cannot convey meaningful information about stimuli without the use of an across-fiber code. However, the converse is not necessarily true. For example, chemosensory systems frequently use an across-fiber code to identify a chemical substance, but such systems are not range-fractionated.

For the purposes of clarifying their differences, this discussion of labeled line and across-fiber coding has treated them as either-or systems, as if in each sensory modality the nervous system uses either a labeled line code or an across-fiber code. In actuality, although sensory systems may use primarily one or the other, all sensory systems use at least a little of both. In the foot of the fly, for example, the four chemoreceptors are not labeled exclusively for amino acids, salt, sugar, and water. Each will respond to other substances if these are present in high enough concentration, thereby requiring the brain to identify the stimulus by taking into account the pattern of activity of all the neurons, as well as the identity of the neuron(s) firing at the highest frequency. In the case of the lobster statocyst organ, although no single hair receptor signals a particular body angle, some receptors clearly indicate body angles in one direction, whereas others indicate body angles in another. Hence, the brain receives a gross determination of body angle through activity of a particular (labeled) set of neurons, but it determines the exact angle by decoding the (across-fiber) pattern of activity of the entire population.

In an across-fiber code or population code, each sensory neuron responds to many types of stimuli. These stimuli cause unique patterns of response in different populations of sensory neurons, and the central nervous system then interprets the pattern of response of all the neurons together to identify the stimulus. Although sensory systems tend to use predominantly labeled line or across-fiber coding, no system uses one or the other exclusively.

Extending the Code

An across-fiber code can be quite efficient in conveying information about stimuli to the central nervous system. However, the ability of a sense organ to discriminate different stimuli from one another is reduced in sense organs that have few sensory neurons. Nervous systems of invertebrates and lower vertebrates, where such sense organs are common, have evolved mechanisms that help to minimize or overcome this drawback.

One characteristic of sensory neurons that enhances the ability of a sense organ to convey information about a sensory stimulus is the maintenance of a low level of background firing even in the absence of any specific stimulus. Sensory neurons with this feature increase their rate of firing when some stimuli are applied but decrease their rate of firing when other stimuli are applied. This pattern of response increases the amount of information that the sense organ can convey to the central nervous system.

An example of a sense organ whose sensory receptors show this type of response is the **neuromast organ** in the lateral line system of teleost fish (Figure 10-5). The receptor cells in this organ, called **hair cells**, possess fine, hairlike projections embedded in a gelatinous matrix called a **cupula**. The hair cells are nonspiking and synapse with sensory neurons that send information to the central nervous

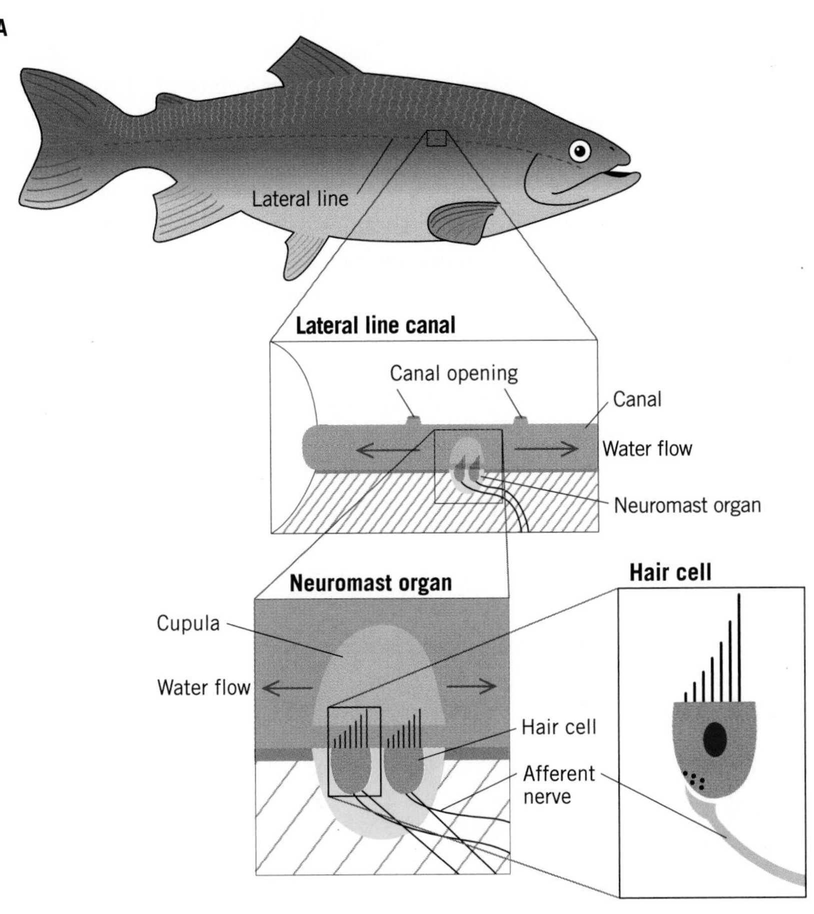

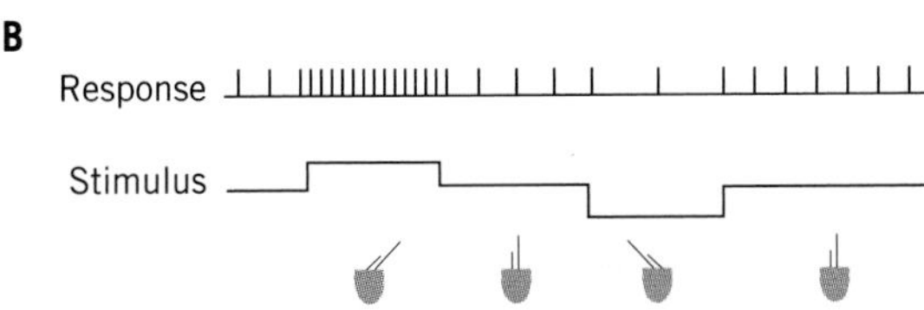

FIGURE 10-5. Neuromast organs from the lateral line of fish. (A) The lateral line system and the neuromast organ, showing a single hair cell from the organ. (B) The response of the afferent nerve from a neuromast cell in the lateral line. Depolarization of the neuromast cell increases the rate of firing of the sensory neuron, whereas hyperpolarization of the neuromast cell decreases the rate of firing.

system. A disturbance near the fish will create pressure waves in the water that will travel into and through the lateral line, where they will bend the cupula of each neuromast organ back and forth as the pressure waxes and wanes. Bending the cupula moves the hairs of the receptor cell. When the cupula is bent in one direction, a hair cell will depolarize, release more neurotransmitter, and hence cause an increase in the rate of firing of the sensory neuron. When the cupula is bent in the other direction, the hair cell will hyperpolarize, release less neurotransmitter, and cause a reduction in the firing rate of the sensory neuron. Even when the cupula is not disturbed, the hair cell continuously releases neurotransmitter, generating continual spontaneous activity in the sensory neuron.

The presence of spontaneous neural activity in the absence of a stimulus enables the sensory neurons to encode more information about the stimuli that excite or inhibit them than if they only increased their rate of firing. In the case of the fish, the ability of the neuromast organ to signal the back-and-forth movement of water increases the ability of the fish to pinpoint the source of any disturbance in the water. In the case of, say, an insect olfactory system, the ability of an olfactory receptor to signal the presence of a chemical by either increasing or decreasing its rate of firing enables the insect to identify more kinds of olfactory stimuli than it otherwise could.

A second mechanism for increasing the ability of a sense organ to convey information about a sensory stimulus is to superimpose a **temporal code** on top of a labeled line code or an across-fiber code. A temporal code conveys information by variations of a response over time (besides the decline of a response due to adaptation). One sense organ that shows temporal coding is the antenna of the tobacco hornworm caterpillar. Hornworms have a number of olfactory receptors distributed on their antennae that allow them to discriminate between odors from many different plants and chemical substances, even though each receptor contains relatively few sensory neurons. Apparently, the caterpillar can make these discriminations by analyzing the firing pattern of the sensory neurons over a period of several seconds (Figure 10-6).

Examination of the pattern of response to different substances shows how a temporal analysis might help the insect identify a chemical substance. Exposure of two of the sensory neurons in one olfactory receptor of the caterpillar to the odor of geranium, tomato, and linalool (an essential oil found in many plants) will produce responses such as those shown in Figure 10-6.

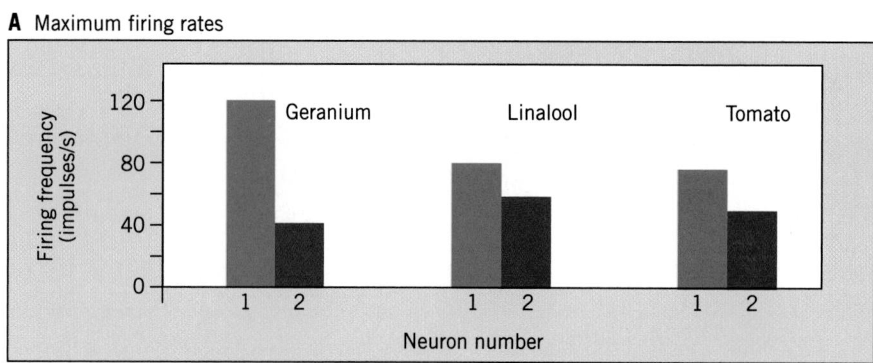

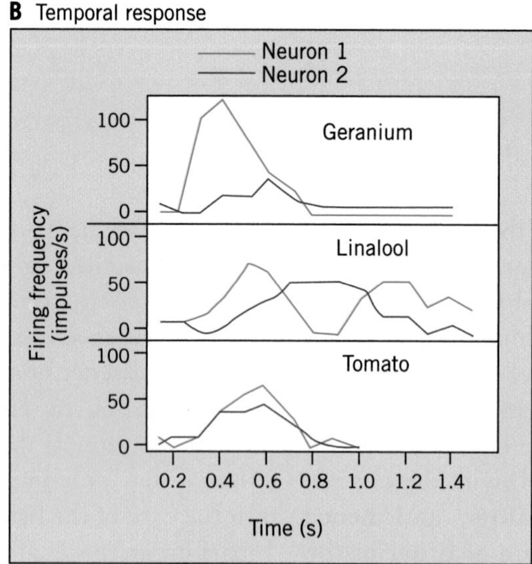

FIGURE 10-6. The responses of two of the sensory neurons in an olfactory (smell) receptor organ of a caterpillar to three different substances. (A) Histograms of the maximum rate of firing of the neuron to each substance. Note that the maximum firing rates induced by linalool and tomato in each of the two receptor cells are nearly the same. (B) The time course of the responses of the neurons, showing the significant temporal variation in their responses. Note the large temporal variation in the neuronal response to linalool.

Even considering only the maximum rates of firing achieved by the two sensory neurons (Figure 10-6A), the sensory response to geranium is easily distinguishable from the responses to the other two substances. On the other hand, the maximum rates of firing achieved by the sensory neurons on exposure to the odor of linalool and tomato are quite similar, making these difficult to distinguish from one another on this basis alone. However, if the brain is able to analyze the firing rates of the two receptors over several seconds, the substances can readily be distinguished. Figure 10-6B shows that although the peak firing rates of the two receptors are similar, the firing rate of receptor 1 at 800 msec is quite low compared with the firing rate of receptor 2 at the same time, hence serving as a basis of identification of the substances.

In some sensory receptors or sense organs, sensory neurons may show a decrease in firing to some stimuli and an increase in firing to other stimuli. In other receptors, sensory neurons show fluctuations over time in their response to a brief exposure to a stimulus. These mechanisms increase the possible types of responses that sensory neurons can use to encode information.

Efferent Control of Sense Organs

The sensory mechanisms discussed so far have been those that sensory systems use to ensure accurate reporting about the stimulus that excites them. However, you have already learned about several features of sensory receptors that make them less than perfect detecting devices. For example, sensory receptors act as filters (you and a bee see flowers quite differently), and sensory receptors can be influenced by past history (dipping your hand in warm water after it has been in cold water gives a different sensation than after it has been in hot water).

You might think that sensory systems would be adapted to reduce or eliminate such "distortions," but this is not always the case. In fact, there are many instances in which the central nervous system actively modulates the information that a sensory neuron or sense organ delivers to a sensory processing area in the brain. This modulation is called **efferent control**. Recall that the term efferent refers to neural signals that travel from the central nervous system to the periphery. In this context, efferent control refers to the neural impulses that are sent from the central nervous system to a sensory receptor or sense organ to modulate its sensitivity to stimuli. Control or modulation of sensory signals can also occur in the central nervous system by the modulation of sensory information in brain stem or thalamic nuclei by input from other brain regions. The term **centrifugal control** is sometimes used to refer to modulation of sensory information within the central nervous system, but some researchers consider this synonymous with efferent control. You can consider them functionally the same.

Efferent or centrifugal control may serve any of four functions in a sensory system. First, it may help to generate smooth muscle action via sensory feedback while the muscles are contracting under voluntary control. Second, it can help compensate for reafference, sensory signals that arise as a consequence of an animal's own movements. Third, it can provide limited protection of a sensory system against damage or adaptation. Fourth, it can be the means by which an animal can selectively suppress unimportant input. These functions are not mutually exclusive.

Sense organs or sensory receptor cells are often under efferent control, meaning that their output is influenced or modulated by the central nervous system, as well as by the stimuli they respond to. This modulation may help smooth reflexes, compensate for the animal's own movements, protect the sensory system, or allow for selective suppression of unimportant stimuli.

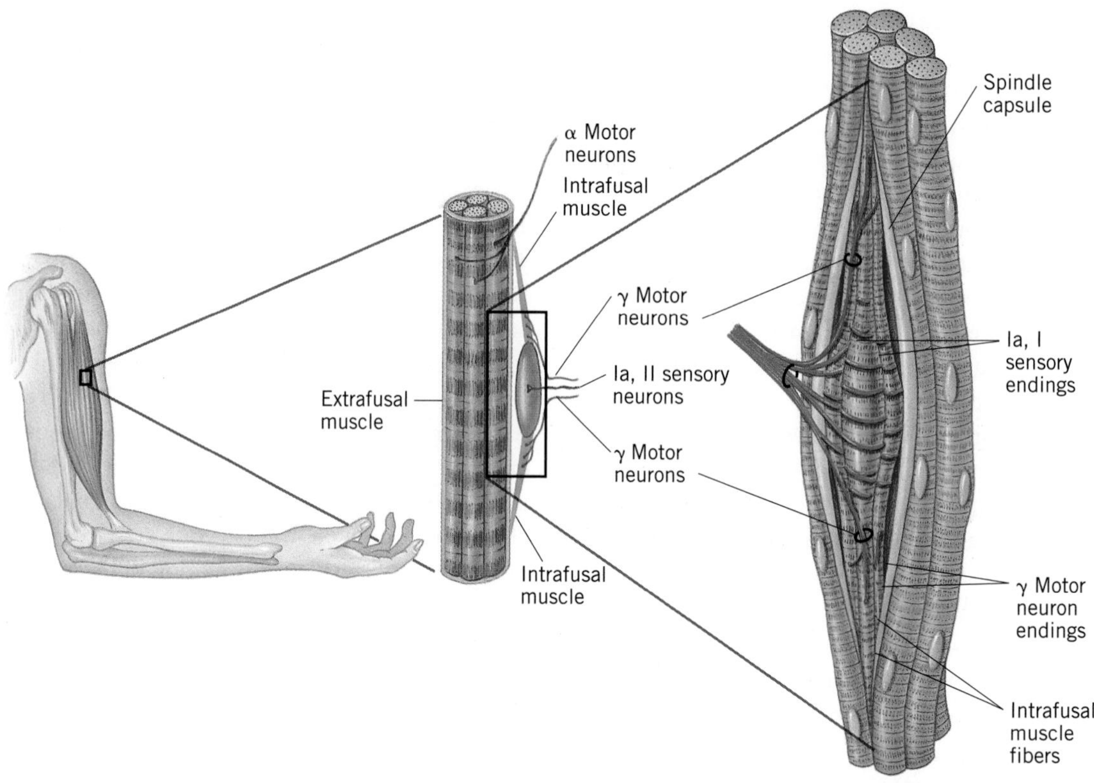

FIGURE 10-7. Diagrammatic and simplified view of the structure of a mammalian muscle spindle and its relation to the muscle in which it is found. Each spindle is attached along the length of one of the fibers of the main muscle, in parallel with them, and consists of fine strands of muscle with a noncontractile central capsule. Stretch of the central regions of the strands excites mechanoreceptors wrapped around the strands in the capsule, which causes firing of group Ia and group II sensory neurons. The main muscle receives signals from α motor neurons, and the spindle muscle strands receive signals from γ motor neurons.

Muscle Spindles and Reflex Smoothing

Muscle spindle organs, located in the skeletal muscles of most vertebrates, provide an excellent example of the smoothing of muscle action by efferent control. In mammals and birds, muscle spindles are attached along the length of individual muscle fibers, as shown diagramatically in Figure 10-7. Functionally, muscle spindle organs behave like stretch receptors, sending a strong barrage of sensory signals into the central nervous system when they are stimulated. These sensory signals excite the motor neurons that control the skeletal muscles, helping the muscle respond smoothly and appropriately to a stretch stimulus.

Muscle spindles consist of several fine strands of muscle, the middle regions of which are surrounded by a central capsule. The strands are attached at both ends to fibers of the skeletal muscle in which the spindle organ is located (Figure 10-7). The central regions of the muscle strands, which are enclosed by the central capsule, are noncontractile, but the rest of the strands can contract. Because they are part of the spindle organ, the muscle strands are referred to as **intrafusal muscle** fibers. To distinguish the main mass of the skeletal muscle from the intrafusal muscle that is attached to it, the skeletal muscle is referred to as **extrafusal muscle**. The intrafusal muscle fibers are a key element in the function of the spindle organ.

Within the central capsule, the terminals of two types of mechanoreceptor neurons wrap around the noncontractile central regions of the intrafusal muscle fibers. When the *extrafusal* muscle cell to which a spindle organ is attached is stretched, the mechanoreceptors are excited. Stretch of the muscle cell pulls on the ends of the intrafusal fibers, and pulling on the ends of these fibers pulls on the central, noncontractile part inside the capsule to which the sensory neurons are attached. The pull stimulates the sensory neurons by stretching the ends of their neurites. The two types of sensory neurons are known as type Ia and type II afferents. (They are called afferents because they carry action potentials toward the central nervous system. The names Ia and II stem from a classification of action potentials recorded extracellularly from the nerve in which they travel, and are not important here.) Both type Ia and type II afferents respond to spindle stretch, but type Ia fibers are phasic and respond more strongly to fast stretch.

Both extrafusal and intrafusal muscle fibers receive input from motor axons from the central nervous system. However, the inputs of the two are completely separate. The extrafusal muscle fibers receive input from large axons called alpha (α) motor neurons, which control contraction during voluntary movements (see Chapter 15). The intrafusal muscle fibers, on the other hand, receive input from much smaller-diameter axons called gamma (γ) motor neurons. These are sometimes called γ efferents, and constitute the pathway for efferent control of the spindle organ, as described in the following section. None of the α motor neurons make synaptic connections with intrafusal muscle fibers, and none of the γ efferents make synaptic connections with extrafusal muscle fibers.

Muscle spindle organs are functional stretch receptors of vertebrate skeletal muscle. They consist of strands of intrafusal muscle fibers, the noncontractile middle regions of which are surrounded by a central capsule that contains the ends of mechanoreceptive sensory neurons. The intrafusal muscle fibers of the spindle and the extrafusal skeletal muscle fibers are controlled by separate sets of motor neurons.

The Action of Muscle Spindles

An understanding of how muscle spindles work requires an understanding of the conditions that will activate them. Since the spindle organ is attached alongside an extrafusal muscle fiber, it is said to have an "in parallel" arrangement (Figure 10-8A). This kind of arrangement can be analyzed as if the extrafusal muscle and the spindle were independent but attached to the same sites, as shown in the figure. When the extrafusal muscle is stretched, as it will be when an opposing muscle contracts, the spindle organ will be stretched as well (Figure 10-8B). Stretching the spindle, which is referred to as "loading" it, pulls on the mechanoreceptors in the center and stimulates an increase in the rate of firing of the group Ia and II afferents. On the other hand, contraction of the muscle in which the spindle lies will bring the ends of the extrafusal muscle closer together. The ends of the spindle will also be brought closer together, reducing the firing of the afferents due to the reduced stretch of the mechanoreceptors (Figure 10-8C). This reduction of stretch is referred to as "unloading" the spindle.

The location and response characteristics of the spindle organ allow it to provide information about stretch of the extrafusal muscle. Even when the extrafusal muscle contracts, which unloads the spindle and

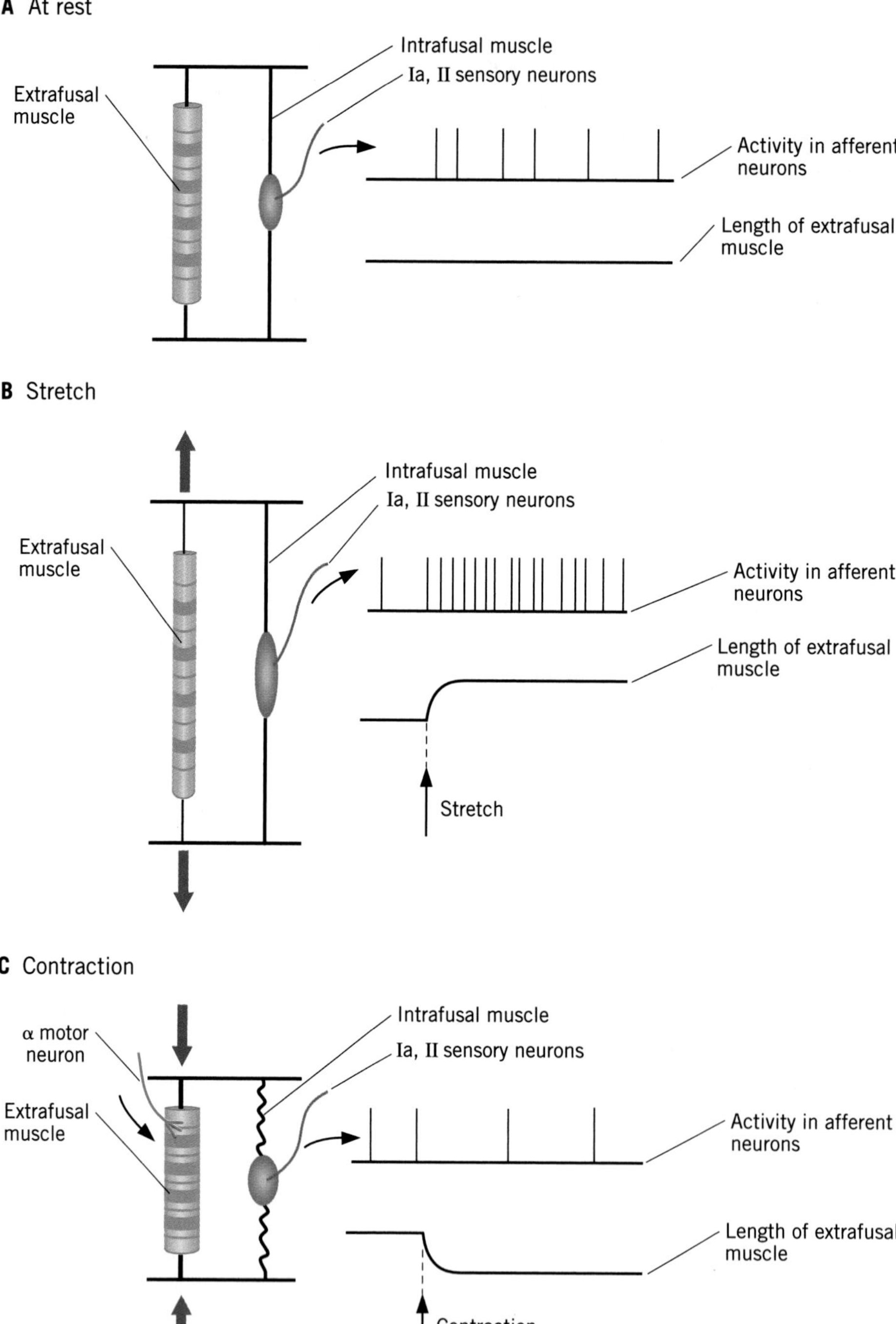

FIGURE 10-8. The response of the muscle spindle afferents to contraction or stretch of the extrafusal muscle when there is no concurrent activity in the γ efferent motor neurons. (A) The muscle at rest. The spindle fires at a low level. (B) The muscle stretched. The spindle is stretched as well, and responds by increasing its rate of firing. (C) The muscle contracts due to stimulation by the α motor neurons. The spindle, being in parallel with the extrafusal muscle, is unloaded and therefore stops firing for a time until its natural elasticity restores it to a condition of mild stretch.

hence reduces the rate of firing of the afferents, the spindle has a natural elasticity that causes it to shorten even without any action by the intrafusal muscle fibers. Hence, a spindle that is "unloaded" by the contraction of its extrafusal muscle will shorten by itself so that the afferents can signal stretch of the contracted extrafusal muscle.

The period between the unloading of the spindle and the time at which it can once again respond to stretch of the extrafusal muscle is a time during which the spindle afferents cannot report any changes in muscle length. To understand why this matters, imagine that you are walking through a forest when your foot strikes the end of a fallen branch protruding into the path. At the start of the step, your brain had sent out signals to your quadriceps (thigh) muscle for it to contract with enough force to lift your leg and swing the foot forward. The branch, however, impedes the forward movement of your foot. If the muscle spindles in your quadriceps muscle were only passive detectors of stretch as described so far, they would have become unloaded as you began your step. At the instant your foot hit the stick, they would be producing no output, not yet having had time to shorten to a length at which they could again signal stretch of the extrafusal muscle. As a consequence, they could provide no information to your nervous system that there was a mismatch between what you had intended to do (swing your leg forward) and what had actually happened (the swing of your leg impeded by the branch). The result would probably be a fall.

However, muscle spindles do not actually become unloaded when the extrafusal muscle to which they are attached contracts. At the same time that the extrafusal muscle contracts, the intrafusal muscles of the spindle organ contract, too, driven along with the α motor neurons by input from the motor cortex that descends in the lateral corticospinal tract of the spinal cord. This contraction is adjusted to the expected rate of shortening of the extrafusal muscle so that the spindle stays under constant tension throughout the entire period of contraction of the extrafusal muscle (Figure 10-9A).

To go back to your forest adventure, if you have properly operating muscle spindles when your foot encounters the branch, the spindle afferents will instantly increase their rate of firing, because when the movement of your foot is impeded, the shortening of your thigh muscle is also impeded. The muscle does not stop contracting; it merely shortens at a lower rate after your foot hits the obstruction than before. However, the rate of shortening of the intrafusal muscle fibers is not altered. The intrafusal muscle fibers therefore pull on the spindle mechanoreceptors, exciting them to fire at a higher rate. The spindle afferents make excitatory synapses on the α motor neurons in the spinal cord and hence cause them to increase their rate of firing (Figure 10-9B). This produces a greater strength of contraction of the quadriceps muscle, and the stronger contraction allows your foot to push the branch out of the way, preventing your fall. There are, of course, many other reflexes that are evoked under these circumstances, but the action of the muscle spindles is an important component.

The reflex activation of extrafusal muscle by excitation of the muscle spindle works to smooth muscle action in day-to-day activities as well. For example, if you hold out a cup and someone pours coffee into it, the weight you must hold will gradually increase as the coffee fills your cup. Without the muscle spindles in your arm and shoulder muscles, you would only be able to keep the cup steady by the use of information from your eyes and from receptors in your joints. If you are not looking at the cup while the coffee is being poured, the chances are that you would spill the drink because information from your joints would not be received and interpreted fast enough for you to keep the cup steady. Muscle spindles, on the other hand, respond as soon as the extrafusal muscle stretches, as it does when the arm is slightly depressed by the additional weight of the coffee in the cup. This stretch activates the spindle afferents, whose reflex excitation of the α motor neurons that control the extrafusal muscle will be added to

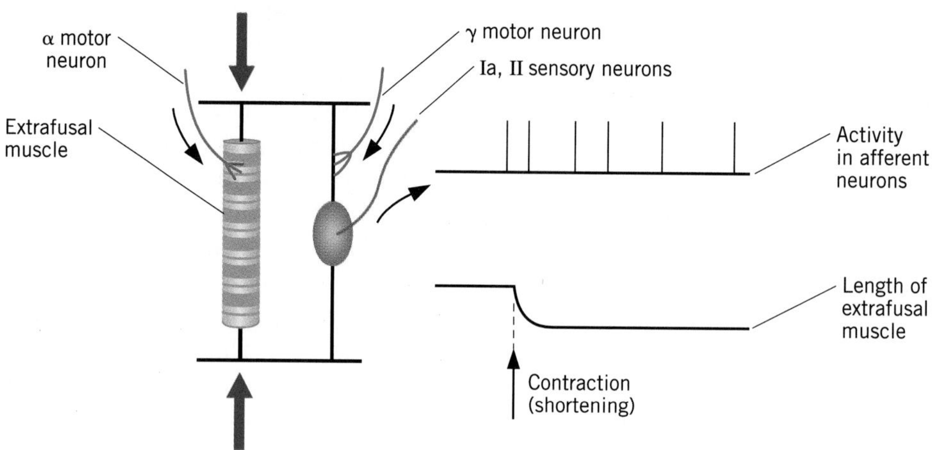

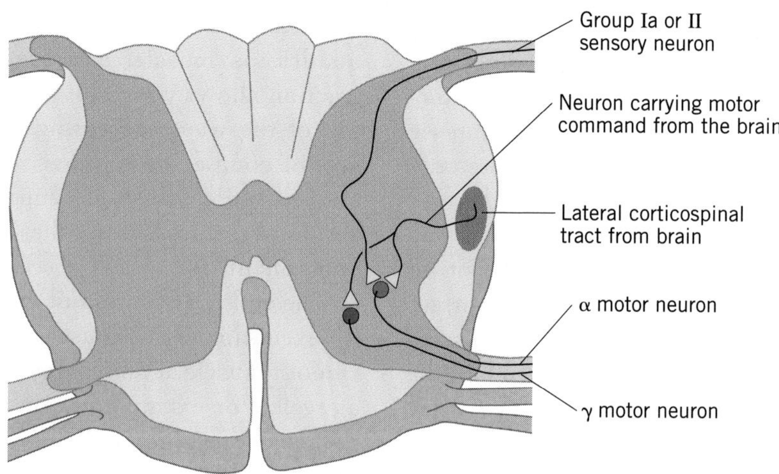

FIGURE 10-9. (A) The effects of extrafusal muscle contraction stimulated by α motor neurons when accompanied by stimulation of the spindle intrafusal muscle by the efferent (γ) motor neurons. If the contraction of the intrafusal muscle is timed right, the tension on the spindle sensory region will stay constant, and the output of the group Ia and group II afferents will also stay constant, ready to signal any deviation from the intended movement. (B) The reflex circuit that regulates the activity of muscles via stimulation of the muscle spindles.

their ongoing firing and immediately compensate for the extra weight, hence allowing you to keep the cup steady.

The key to both these events is the efferent control of the muscle spindle by the stimulation of the *intrafusal* muscle fibers during any contraction of the *extrafusal* muscle. Because the tension in the spindle is adjusted to stay relatively constant as the length of the extrafusal muscle changes, the spindle is able to detect any difference that arises between what is intended to happen and what actually happens in the extrafusal muscle. The spindle organ would be virtually useless if it acted only passively and gave information only about an actual stretch of the extrafusal muscle to which it was attached.

Muscle spindle organs provide information to the central nervous system about the length of a muscle compared with its intended length, so that reflex adjustments can be made to the excitation of the muscle. The spindle is controlled by γ efferents, so that the tension on the spindle is kept relatively constant throughout the contraction of the extrafusal muscle in which it is located.

Compensation for Reafference

A second function of efferent control is to compensate for **reafference**, sensory signals that an animal receives as a consequence of its own movements. The feeling of wind blowing on your face as you run is an example of reafference. In many circumstances, an animal must be able to suppress sensory signals that arise from reafference. For example, many fish show a strong escape response in the presence of predators. The escape response, which consists of vigorous swimming produced by strong back-and-forth bending movements of the body, is triggered by disturbances in the water produced by the movement of the predator toward the fish. These water disturbances are detected by the fish's lateral line system. However, similar water disturbances are generated by the fish's own vigorous movements during escape. If there were no way for the fish to compensate for the sensory input generated by its own movements, the sensory input might easily trigger yet another escape sequence. The fish's normal swimming might also stimulate the lateral line strongly enough to trigger escape swimming.

Fish have several mechanisms for avoiding these problems. Signals from the lateral line go through several nuclei in the medulla. Recordings of neural activity in these nuclei in resting fish exposed to an object vibrating in the water show a certain amount of activity due to the sensory input. However, if the fish is made to swim, the level of activity in these nuclei is significantly reduced when the same stimulus is presented. Furthermore, stimulation of a set of neurons (Mauthner cells, described in Chapter 16) whose activity can trigger escape elicits strong activity in *efferent* neurons that go to the lateral line sense organs. Intracellular recordings from the hair cells in the lateral line show that the effect of this efferent activity is inhibition of activity in the hair cells (Figure 10-10). Hence, fish have efferent control of lateral line signals at two levels, in the brain stem before it reaches the parts of the brain where the signals are interpreted and in the sense organs themselves.

Similar effects can be demonstrated in other sensory systems. In monkeys, for example, some of the sensory input from the eyes is sent to the superior colliculi in the midbrain, nuclei that help control oculomotor reflexes. Moving an image across a screen in front of a monkey sitting with its eyes and head still elicits strong responses in neurons in the colliculi. However, if the monkey moves its eyes while the image is stationary, an action that will cause movement of the image across the retina, neurons in the colliculi show no response whatsoever, even

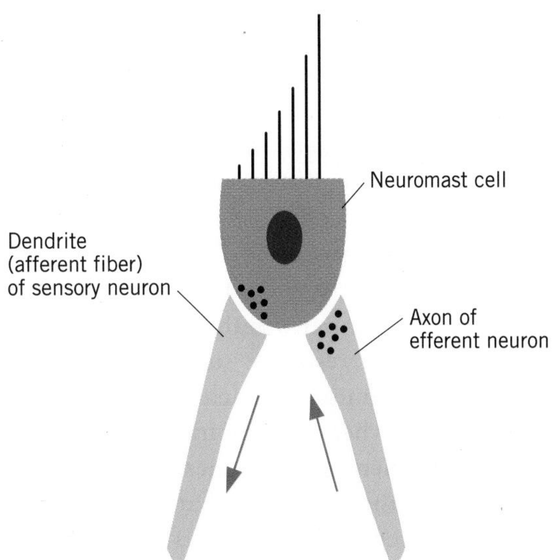

FIGURE 10-10. Diagram of a neuromast cell of a fish, showing the synapses between efferent and afferent fibers and the cell.

though the effect of the image on retinal neurons is the same in the two cases. Centrifugal control signals from elsewhere in the brain suppress the reafferent input from the eyes when the monkey shifts its gaze.

In these and similar cases, the function of the efferent or centrifugal signals is to reduce the magnitude of a neural response to the sensory signal that is generated by the animal's own movements, or even to suppress the response altogether. Not all reafference is reduced or suppressed (you feel wind on your face equally well when you are standing still as when you are running), but in cases in which reafferent input can interfere with the proper interpretation of input, or can generate an inappropriate motor response, it is essential that the sensory system modulate, and indeed alter, the report from the sense organs to take account of the circumstances.

Reafference is the sensory input generated by an animal's own movements. Many animals modulate or suppress reafferent signals by efferent or centrifugal control.

Protection

A third function of efferent control is protection against functional damage. Strong stimuli can physically damage a sense organ. Animals have evolved a number of protective reflexes that reduce the effective strength of the stimulation impinging on some sense organs. For example, in some vertebrates, stapedius and tensor tympani muscles in the middle ear contract reflexly in response to extremely loud sounds. These muscles attach to the bones of the inner ear; when they contract, they dampen the movement of the bones, reducing the strength of the stimulus that reaches the cochlea. The central nervous system regulates the strength of a sensory stimulus by efferent control, even though the control is exerted indirectly on the sense organ rather than directly on the receptor cells.

In these examples, efferent control dampens the sensory response to externally generated signals. However, some mammals also exert protective efferent control over reafferent stimuli. The gray bat, *Myotis grisiscens*, for example, dampens its sensitivity to ultrasonic sound in the inner ear just at the instant that it makes its own ultrasonic cry, which is very loud. In humans as well, sound sensitivity is dampened each time we speak or make a sound. Thus, a shout that you make yourself will seem quieter than a shout made by someone standing very near you, even if objective measures show that the two shouts had the same sound energy at the entrance to your ear.

In addition to possibly causing damage to a sense organ, strong stimuli can also bring about adaptation of the receptor cells, which can result in a reduced responsiveness of the receptor to subsequent stimuli. Efferent control of the receptor, reducing or suppressing its response to a stimulus, will "protect" that receptor from adaptation, leaving it fully able to respond to a stimulus once the efferent inhibition is lifted. Inhibitory efferent input to the cells of the lateral line in sharks and other fish during vigorous escape swimming is thought to play this role.

Efferent control can protect a sense organ against physical damage from strong stimuli. It can also "protect" receptor cells from the loss of responsiveness by adaptation to strong stimuli.

Selective Suppression of Sensory Input

The mechanisms of efferent and centrifugal control discussed in the preceding sections are all entirely automatic reflex actions over which the animal has no conscious control. However, there is evidence that mammals, at least, are able to selectively suppress input. In mammals, all sensory signals pass through several sensory nuclei in the brain before they reach the part of the cortex where they are interpreted. These nuclei serve as sites at which sensory signals can be shaped, modulated, or suppressed. By placing electrodes into one of these brain nuclei, it is possible to record evoked potentials from it, and to demonstrate the modulation of the sensory input. Evoked potentials are generated ("evoked") by an appropriate sensory stimulus and represent the summed action of many neurons in the nucleus, all responding together.

If you insert electrodes into a cat's cochlear nucleus (one of the nuclei in the auditory pathway), you can record evoked potentials in response to click stimuli, just as you can record responses to clicks in the auditory nerve from the ear. Both the evoked potentials and the auditory nerve responses persist for as long as the click stimulus persists. However, if while the click stimulus is still being given, the cat's attention is caught by a different stimulus, such as the sight of a mouse or the odor of fish, then the amplitude of the evoked potential resulting from each click declines dramatically. The amplitude of the auditory nerve response, on the other hand, does not change. It is as if the cat suppresses the boring input in order to pay better attention to the novel one. The cat seems able to exert control over the auditory input when it wishes to concentrate on the new visual or olfactory input. The effect of the suppression is to reduce the number of active neurons in the cochlear nucleus and therefore to reduce the amplitude of the evoked potential. Suppression does not, however, affect the input from the ear, so the recording from the auditory nerve is unaffected.

In some cases, an animal may be able to modulate or suppress sensory signals consciously, thereby allowing it to concentrate on sensory input of greater immediate interest.

Sensory Processing and the Organization of Sensory Systems

This chapter and Chapter 9 have introduced the common features shared by nearly all sensory systems. Whether a sensory system processes light through eyes, sound through ears, or some other stimulus through other specialized sense organs, that system and its component elements will likely exhibit the features you have just read about—filtering, adaptation, contrast enhancement, range fractionation, topographical and columnar organization, labeled line and across-fiber coding, and efferent control. These features play a crucial role in allowing the nervous system to detect and interpret information in a way that serves the needs of the animal.

The descriptions of the individual sensory systems in the next four chapters will provide you with more detail about how each of these systems work. As you read about them, you will learn how these common features play an important role in the effectiveness of sensory systems. They do this in part by providing mechanisms that help each system to shape and process the specific sensory information that it receives. Hence, the complex events that transform sensory information into a useful form may be better understood in the context of these common features.

All sensory systems share common features even though they process different kinds of stimuli. The complex processing of information performed by different sensory systems may be better understood in the context of these common features.

Additional Reading

General

Aidley, D. J. 1989. *The Physiology of Excitable Cells*. 3d ed. Cambridge: Cambridge University Press.

Barlow, H. B., and J. D. Mollon, eds. 1982. *Cambridge Texts in the Physiological Sciences*. Vol. 3, *The Senses*. Cambridge: Cambridge University Press.

Schmidt, R. F., ed. 1986. *Fundamentals of Sensory Physiology*. 3d ed. New York: Springer-Verlag.

Somjen, G. 1972. *Sensory Coding in the Mammalian Nervous System*. New York: Plenum.

Research Articles and Reviews

Adrian, E. D., and Y. Zotterman. 1926. The impulses produced by sensory nerve-endings. 2. The response of a single end organ. *Journal of Physiology (London)* 61:151–71.

Dethier, V. G. 1971. A surfeit of stimuli: A paucity of receptors. *American Scientist* 59:706–15.

Pfaffmann, C. 1955. Gustatory nerve impulses in rat, cat and rabbit. *Journal of Neurophysiology* 18:429–40.

Russell, I. J. 1976. Central inhibition of lateral line input in the medulla of the goldfish by neurons which control active body movements. *Journal of Comparative Physiology* A 111:335–58.

CHAPTER 11

The Visual System

Most photoreceptive organs not only detect and respond to light but also form an image. Information about this image is processed in the visual centers of the brain so that the animal can make a meaningful and useful interpretation of its surroundings. This chapter describes how eyes and the parts of the brain that process visual information are organized to fulfill these functions.

Photoreceptors are specialized to respond to electromagnetic radiation in the wavelengths between ultraviolet and red, and can do so with exquisite sensitivity. But when photoreceptors are organized into eyes, the ensemble does more than simply signal that light is present; it forms images that can convey important information about an animal's surroundings. The function of the visual system is to extract from these images information about the shape, location, and movement of objects in the environment.

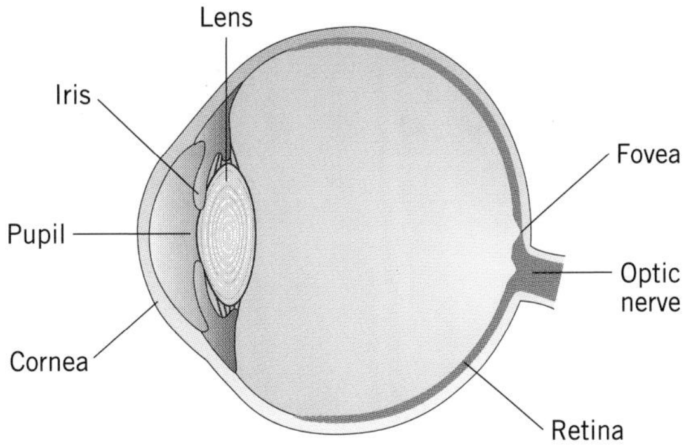

FIGURE 11-1. A human eye showing the path of light to the retina, the photoreceptive layer that lines the inside of the back of the eyeball.

The Performance of Vertebrate Eyes

Photoreceptors in vertebrate eyes are arrayed along the back of the eyeball, in the layer of receptor cells and neurons called the **retina**. Light enters the eye through the transparent cornea, passes through the pupil and the lens, and falls on the retina (Figure 11-1). The pupil, the opening in the middle of the iris, widens in dim light and contracts when light intensity is high, thereby to some extent adjusting the amount of light that falls on the retina. Light is focused on the retina by the lens, whose shape can be changed by intrinsic muscles of the eye. The optic nerve, cranial nerve II, carries visual information to the brain. Primates, as well as some birds and reptiles, have a specialized region of the retina, the **fovea**, which is adapted for acute vision by having an especially dense array of photoreceptors. The fovea is also the region of the retina on which the light falls from objects the animal looks at directly.

The complex and sophisticated structure of vertebrate eyes confers to vertebrate animals the most acute vision in the animal kingdom. Acuity is usually expressed in terms of the angular resolving power of the eye in degrees (°), minutes ('; 60' = 1° of arc), or seconds ("; 60" = 1' of arc). Humans with good eyesight can resolve about 1' of arc, which in practical terms means that you can probably see the period at the end of this sentence at a distance of about 1 meter (just over 3 feet). By comparison, the full moon is about 0.5° (30') of arc in diameter. Some birds of prey, such as eagles, can resolve 20" of arc. Hence, an eagle could see a bird about the width of your hand (10 cm) from a distance of 1000 meters (about 3280 feet)!

Light entering a vertebrate eye passes through the cornea, pupil, and lens before forming an image on the retina. Information is sent from the retina to the brain via the optic nerve. Some vertebrates have a fovea, a region of the retina specialized for acute vision.

The Vertebrate Visual System

Information about light enters the brain via the axons that form the optic nerve (Figure 11-2). In vertebrates such as fish and amphibians, the bulk of these axons terminate in the dorsal roof of the mesencephalon, an area called the **optic tectum**, which is the main visual processing center in these animals. In mammals, most axons in the optic nerve travel to the lateral geniculate nuclei **(LGN)**, a pair of nuclei in the posterior part of the thalamus. From there, information about visual stimuli travels to the **primary visual cortex**, also known as the **visual cortex,** area V1, and **striate cortex**, the first cortical region devoted to processing visual information. Visual signals are also sent to the superior colliculi, a pair of domelike structures in the midbrain that are the mammalian equivalent of the optic tectum, and to the pretectal area that lies just anterior to the col-

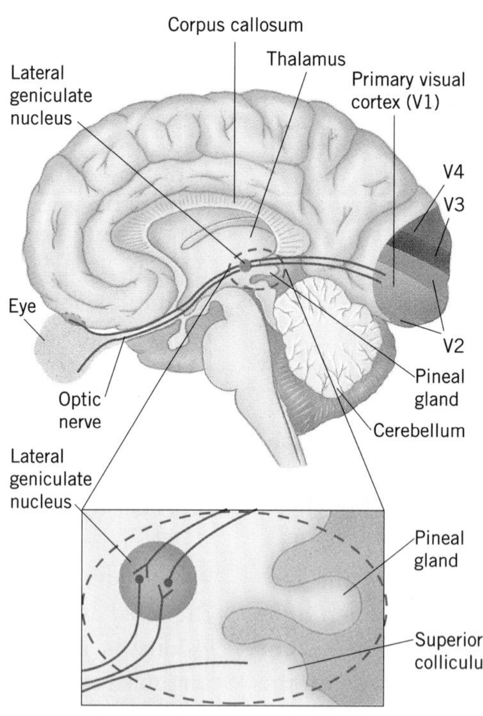

FIGURE 11-2. The main pathways for the flow of visual information in humans, shown in a medial view of the right hemisphere of the brain. In mammals, most of the axons in the optic nerves terminate in the lateral geniculate nuclei of the thalamus. Neurons originating in the lateral geniculate nucleus send information to the primary visual cortex at the back of the head. Some axons in the optic nerves also travel to the pretectum and the dorsal colliculi in the midbrain. Area MT (V5) is not visible in this view.

liculi in the midbrain. Visual input to the colliculi helps to control movements of the eyes to new targets, and input to the pretectum forms part of the circuit for reflex control of the pupils.

Several areas of the cortex near V1 and elsewhere are devoted to the processing of visual information. Four of these areas are numbered, V2 through V5. Area V5 is also known as the **middle temporal region (MT).** As described in a subsequent section, the numbered visual areas are primarily concerned with analysis of different aspects of a visual stimulus, such as form, color, depth, and motion. Other cortical regions also contribute to the extraction of more abstract features of an image, such as the recognition of familiar faces. The locations of the numbered visual areas in the human brain are shown in Figure 11-2.

In lower vertebrates, the optic nerve sends visual information to the optic tectum. In mammals, visual information is sent mainly to the lateral geniculate nuclei in the thalamus, and then to the primary visual cortex and other cortical areas devoted to abstracting information, such as form, color, depth, and motion, from an image. Some retinal fibers also go to the superior colliculi and pretectal areas, which help control eye reflexes and the size of the pupil.

The Photoreceptors

Rod cells and **cone cells** are the photoreceptors of vertebrate eyes. They differ in several respects, especially in their sensitivity to light and to the wavelength of light that excites them, as detailed in the following paragraphs.

The Properties and Distribution of Photoreceptors

Compared with cones, rods are more sensitive to light, are more numerous in the periphery of the eye, and are sensitive to a wider band of wavelengths. Rods are generally more numerous than cones, although some animals (e.g., some fish, many reptiles, ground squirrels) have mostly cones and very few rods. In general, diurnal animals have many cones, while nocturnal ones have relatively few. In primates, the retina of one eye contains about 120 million rods but only about 6.5 million cones. Cones are concentrated in the center of the retina. There are over 150,000 cones/mm^2 in the human fovea but only about 5,000/mm^2 10° away from it. On the other hand, there are no rods in the fovea but as many as 160,000/mm^2 10° to 20° away from it.

Rods are much more sensitive to light than are cones. They are therefore used at night or under other conditions when light levels are too low to stimulate the cones. In moderate light, both rods and cones contribute to vision, but in bright light the rods are saturated (responding maximally) and hence can no longer provide any information necessary for vision except that bright light is present.

Rods and cones contain different photosensitive pigments. As described in detail in the next section, the pigment in rods is rhodopsin, which consists of a light-sensitive

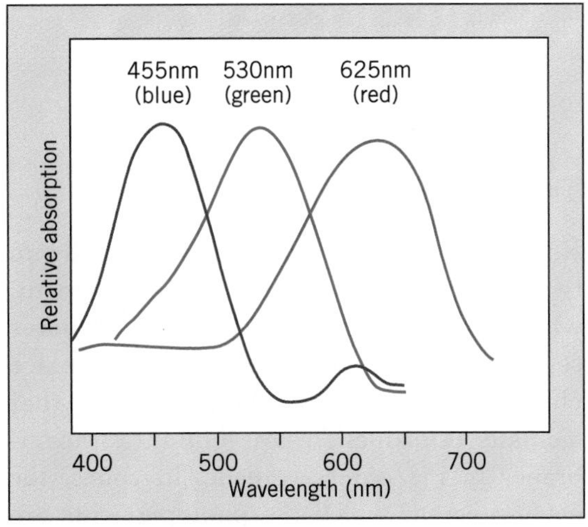

FIGURE 11-3. Spectral sensitivity of three types of cone cells from the eye of a fish. The graphs show the amount of light of different wavelengths absorbed by each cone cell. The more light of a particular color that is absorbed, the better the cone cell responds to light of that color. Cone cells of human eyes have similar absorption curves, except that the red cone cells in humans have a peak absorption near 560 nm, which is closer to yellow light than red.

component, retinal, and a protein called opsin. The photosensitive pigment in cones is also a rhodopsin-like combination of retinal and an opsin. However, the opsins in cones are not quite the same as those in rods, and they also differ in different types of cones.

Opsins can be maximally sensitive to light of different wavelengths; this differential sensitivity forms the basis of color vision (Figure 11-3). Depending on the particular pigment it contains, a cone cell is referred to as a blue, green, or red cone, to indicate the wavelength of light to which it is most sensitive. Cones confer color vision to primates, birds, reptiles, and most fish. The eyes of amphibians and most mammals have relatively few cones. As you may have noticed, you lose your color vision at night. This is the result of the low sensitivity of cones to dim light; there is too little light at night to activate cones, leaving vision entirely dependent on the rods, which are not able to differentiate between light of different wavelengths.

Rods and cones differ in their sensitivity to light and in their distribution in the retina. Cones are less sensitive; different cones respond best to different wavelengths of light.

Transduction

Rods and cones are elongate cells, divided into two main compartments: an **inner segment**, which includes the cell body, and an **outer segment**, where transduction takes place (Figure 11-4A). The photopigments that mediate transduction are bound to membranes in the outer segment. In cones, the membranes to which the photopigments are bound are infoldings of the cell membrane. The cell membrane forms infoldings in rods as well, but here the infoldings detach from the cell membrane to form a stack of free-floating disks to which the photopigments are bound (Figure 11-4B). The photopigment of rods is rhodopsin, a combination of retinal and opsin. It is a 40 kdalton molecule in which the retinal sits in the middle of the opsin, which has seven linked transmembrane domains embedded in the disk membrane of the rod outer segment (Figure 11-5). The photopigments in cones

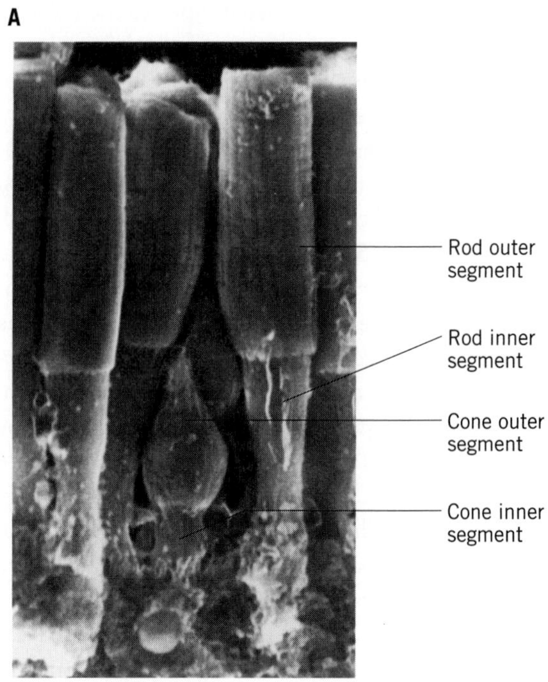

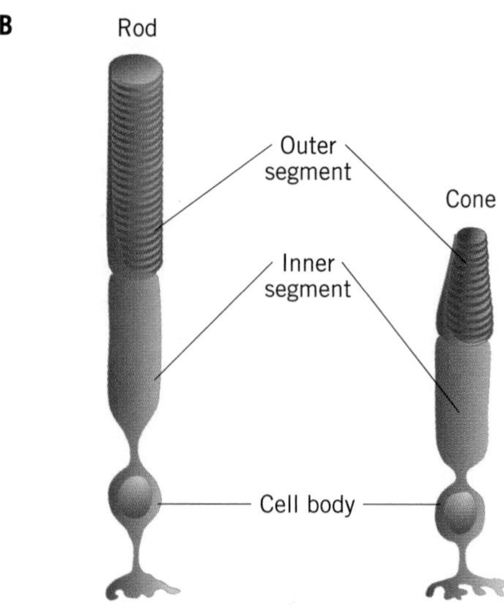

FIGURE 11-4. Structure of typical rod and cone cells. (A) Scanning electron micrograph from the eye of a toad, *Xenopus*, showing rods and a single cone. The epithelial layer that normally attaches the photoreceptors to the back of the eyeball has been removed. (B) Schematic drawing of rod and cone cells. Transduction of light takes place in the outer segments of these cells.

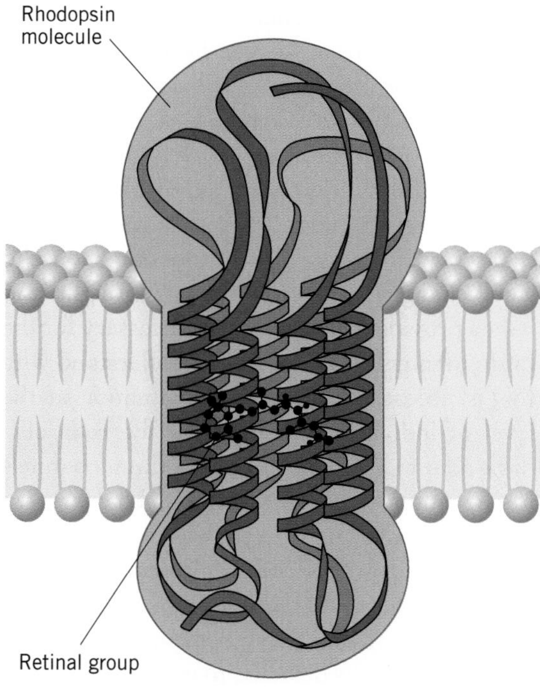

FIGURE 11-5. The molecular structure of rhodopsin from a vertebrate rod. Retinal lies in the interior of the larger membrane-spanning opsin portion.

have a somewhat different structure than opsin and are given names according to the particular form of opsin that they contain.

The critical first step in phototransduction was discovered by George Wald in the late 1950s. He found that interaction of a photon of light with retinal, which is actually a derivative of vitamin A, causes the molecule to isomerize from one structural form (isomer) to another; this isomerization triggers a series of extremely rapid changes in the structure of opsin and further changes in retinol, leading within about a millisecond to a photoexcited state of rhodopsin that in turn initiates a sequence of other biochemical events. These events lead eventually to a change in the membrane potential of the rod. In 1967, Wald was awarded a Nobel Prize in Physiology or Medicine for this discovery.

The response of either a rod or a cone to stimulation by light is *hyperpolarization* of the receptor cell. This electrical response is brought about by a chain of events starting with the interaction of a single photon with a molecule of rhodopsin. The photoexcitation of rhodopsin causes activation of a G protein called **transducin**. Transducin in turn activates an enzyme, phosphodiesterase (PDE), which breaks down free cGMP (3',5'-cyclic guanosine monophosphate) in the cell to 5' GMP (Figure 11-6A).

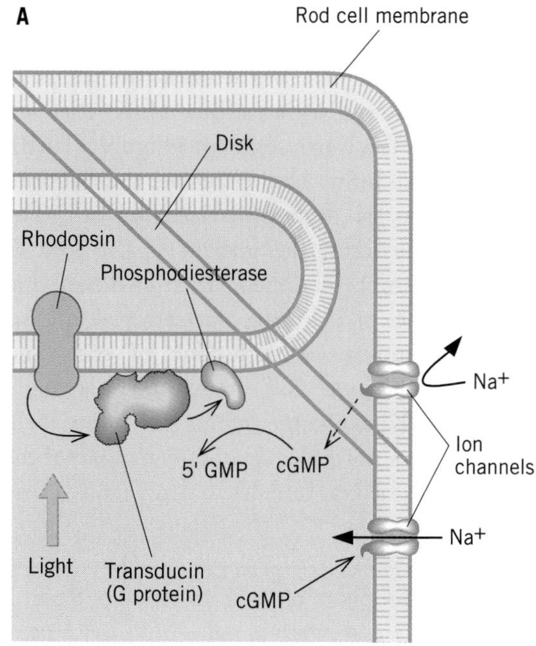

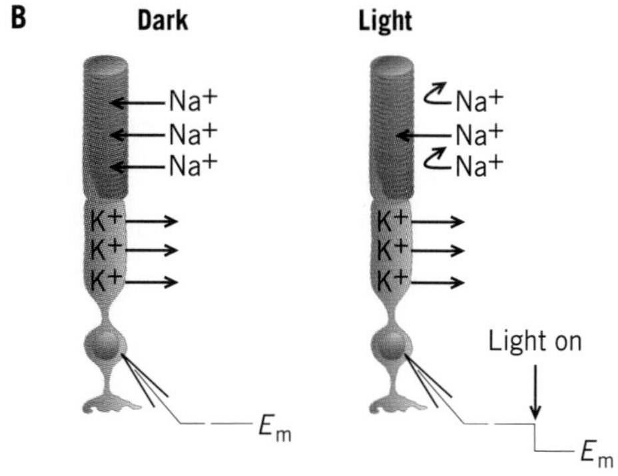

FIGURE 11-6. (A) The pathway for transduction in a typical rod cell. In the presence of light, rhodopsin activates transducin, a G protein, which in turn activates phosphodiesterase (PDE). PDE accelerates the breakdown of cGMP to 5' GMP; the resulting reduction of cGMP levels causes sodium channels to close, hyperpolarizing the rod. (B) The flow of ions into and out of rod cells under dark and light conditions. When the cell is illuminated, some of the sodium channels close. The efflux of K^+ continues, however, hyperpolarizing the cell. Active transport of Na^+ out of the cell and K^+ into the cell maintains the proper ionic balance.

cGMP is a ligand for (cyclic) nucleotide-gated sodium channels in the cell membrane. When cGMP is present, the channels are open, allowing an influx of cations, mainly Na^+ (hence the name) but also a little Ca^{2+}. When light activates rhodopsin, transducin activates PDE and the level of cGMP falls, causing the closure of the sodium channels and a reduction in the flow of Na^+ from the cell. In the dark, the influx of Na^+ is countered by an efflux of K^+ through channels in the inner segment (Figure 11-6B). When the sodium channels close, the continuing efflux of K^+ hyperpolarizes the cell. The brighter the light, the greater is the loss of cGMP, the more channels are shut down, and the more the cell is hyperpolarized, yielding the graded response characteristic of rods and cones. Both rods and cones have strongly active Na^+-K^+ exchange pumps that allow them to maintain a normal level of ions in spite of the large fluxes of Na^+ and K^+ that normally cross their membranes.

The transduction of light to an electrical response takes place in the outer segments of rods and cones. Light activates rhodopsin or a rhodopsin-like photosensitive pigment. This in turn activates a G protein (transducin), which turns on the enzyme PDE. Because of the subsequent reduction in the concentration of the second messenger cGMP, the ion channels for sodium close, producing a hyperpolarizing response in the photoreceptor.

Adaptation

Rods and cones can adapt to an enormous range of light intensity. You have probably experienced the almost painful intensity of your bedside light when you turn it on in the middle of the night, or temporary blindness when you enter a darkened theater from bright daylight. In each case, your rods and cones adapt to the new level of light, and within a few minutes you see normally again. Taking into account adaptation, you can see over a range of light intensity that, from the brightest daylight to the pitch black of a moonless night, covers 12 orders of magnitude.

Adaptation requires an adjustment of the sensitive range of the eye's photoreceptors so that they can change their output in response to small changes in the level of ambient light, and hence provide the information necessary for the brain to interpret an image. To understand why this is necessary, consider the situation at the level of individual channels. In the rods of a dark-adapted eye, most of the sodium channels in the outer segment are open. Under these conditions, a strong light such as that from your bedside lamp will excite so many molecules of rhodopsin that all of these sodium channels will close. Hence, small differences in light intensity, which must be detected to see an image, will have no effect because all the photoreceptors are fully saturated and cannot detect these differences. The converse occurs in a light-adapted eye. Moving into an environment of very low light, such as walking into a darkened theater, causes all the sodium channels to open, making it impossible for the photoreceptors to signal to small differences in light intensity. Hence, adaptation requires a readjustment of the number of open channels so that changes in light intensity can be reflected in changes in the membrane potential of the photoreceptors.

Experiments indicate that the small amount of Ca^{2+} that enters a photoreceptor through open sodium channels plays an important role in adaptation. Free Ca^{2+} in the outer segment inhibits the synthesis of cGMP. In a dark-adapted eye, the Ca^{2+} that enters with Na^+ through open sodium channels puts a brake on the amount of cGMP that is synthesized, keeping the level of cGMP in the cell below what it would have been without the Ca^{2+}. When a dark-adapted eye is exposed to bright light, the transduction pathway quickly saturates; that is, the amount of activated rhodopsin produces such a strong effect on PDE that cGMP levels are reduced enough to shut all the sodium channels. This produces the maximum hyperpolarization possible in the cell. Hence, if the light intensity increases a bit more, the cell will be unresponsive because all the sodium channels are already closed, and if the light intensity is reduced a bit, the cell will be unresponsive because the amount of light is still more than enough to produce a maximal response. The result is an inability to see.

However, the closure of the sodium channels due to the light will also cause a reduction in the amount of free Ca^{2+} that enters the outer segment. A reduction in free Ca^{2+} removes the brake on synthesis of cGMP and hence makes more cGMP available in the cell. Increased levels of cGMP cause some of the sodium channels to reopen, bringing the membrane potential back toward its "normal" level. As this happens, the receptor regains its ability to respond to further increases in light intensity.

The converse happens when you walk from bright light into a darkened theater. At first you cannot see because the light levels in the theater are too low to stimulate the light-adapted photoreceptors—essentially all the ion channels are open, so slight differences in light levels cannot be detected. However, the open ion channels allow the entry of Ca^{2+} as well as Na^+ into the outer segment. The Ca^{2+} reduces the synthesis of cGMP and hence the amount of cGMP available to keep the ion channels open. In consequence, some of the Na^+ channels close, allowing you to see again as the photoreceptor regains the ability to signal changes in light intensity.

The function of adaptation in the eye is to keep the membrane potential of each photoreceptor adjusted to a level so that both increases and decreases in light intensity can be detected and passed on to other neurons. You can think of the mechanism as adjusting the working range of the photoreceptors, the range of light intensity to which they will respond. The mechanism described in this section accounts for some of the adaptation that eyes exhibit, but not all. When you walk into a darkened room, for example, some of the time necessary for your eyes to adjust is due to the time it takes for rhodopsin to be reconstituted from its excited state. Other mechanisms of adaptation, not yet well understood, involve interactions between neurons within the retina, mediated by neuromodulatory transmitters released by retinal neurons.

Adaptation to ambient light conditions requires an adjustment of the number of open sodium channels. At the cellular level, it is affected in part by the level of free Ca^{2+} in the receptor, which dampens the synthesis of cGMP and hence affects the number of open channels.

The Retina

In most sense organs, the receptor cells either send fibers directly to the central nervous system or synapse with sensory neurons that do so. The eye is unusual in containing several layers of cells that perform considerable processing of visual stimuli before passing them on to the brain.

The Structure of the Retina

The vertebrate retina contains five major types of cells: receptor cells (rods and cones), horizontal cells, bipolar cells, amacrine cells, and ganglion cells (Figure 11-7). Receptor cells transduce light into an electrical signal, as detailed in the preceding section. **Horizontal cells** connect the rods and cones in one region of the retina with those of another, conveying information horizontally in the retina. **Bipolar cells** connect one layer of the retina with another; all information from the receptor cells must pass through the bipolar cells. **Amacrine cells** modulate signals from bipolar to ganglion cells. All of these cells give graded responses to synaptic input. Receptor cells, horizontal cells, and bipolar cells are nonspiking, but amacrine cells can generate action potentials. **Ganglion cells** constitute the output from the retina to the brain. They are the only retinal cells with axons that leave the eye, which they do via the optic nerve. They always respond to sufficiently strong excitatory synaptic input by firing action potentials.

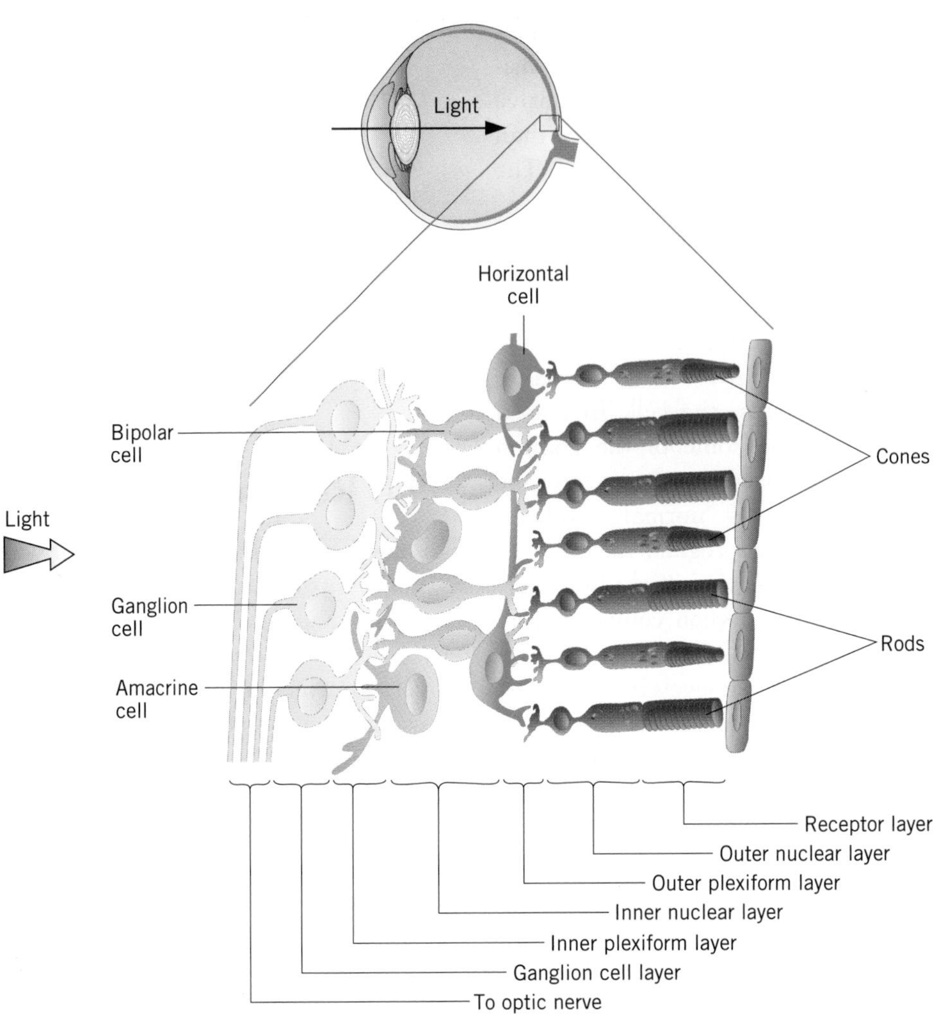

FIGURE 11-7. Semidiagrammatic drawing of retinal structure. Light passes through all the other cell types, which are transparent, before being absorbed in the outer segments of the receptor cells.

The cells in the retina and the connections between them are organized into six morphologically distinct layers, each named according to its distinguishing feature or its location relative to the eyeball (see Figure 11-7). The outermost layer of the retina, the **receptor layer**, lies directly against the inside wall of the eyeball; it contains the photosensitive parts (outer segments) of the photoreceptor cells. The innermost layer of the retina, the **ganglion cell layer**, contains the cells whose axons form the optic nerve. Between these layers, from the outside of the eyeball inward, are the **outer nuclear layer**, which contains the cell bodies of the receptor cells; the **outer plexiform layer**, a layer of neurites and synapses; the **inner nuclear layer**, which contains the cell bodies of the horizontal, bipolar, and amacrine cells; and the **inner plexiform layer**, another layer of neurites and synapses. Table 11-1 lists the properties of mammalian retinal cells. Those of other vertebrates are similar.

As indicated in Figure 11-7, the photoreceptors face the back of the eyeball. This means that light must traverse all of the cells and neurites of the retina before it can reach the photosensitive outer segments of the receptor cells. However, these cells and neurites are essentially transparent and do not block or react to the light.

TABLE 11-1. Characteristics of Mammalian Retinal Cells

Type	Synaptic Inputs From	Synaptic Outputs To	Response Characteristics
Rod	Rods, cones, horizontal cells	Rods, cones, horizontal, bipolar cells	Graded, nonspiking, hyper-polarizing
Cone	Rods, cones, horizontal cells	Rods, cones, horizontal, bipolar cells	Graded, nonspiking, hyper-polarizing
Horizontal cell	Rods, cones	Rods, cones, bipolar cells	Graded, nonspiking, hyper-polarizing
Bipolar cell	Rods, cones, horizontal cells	Amacrine, ganglion cells	Graded, nonspiking, hyper-polarizing or depolarizing
Amacrine cell	Bipolar and amacrine cells	Amacrine, ganglion cells	Graded, depolarizing, non-spiking, and phasic spiking
Ganglion cell	Bipolar and amacrine cells	Brain nuclei	Spiking

There are five main types of cells in the retina: receptor, horizontal, bipolar, amacrine, and ganglion cells. Of these, only amacrine and ganglion cells generate action potentials. The retina is organized in layers of cell bodies alternating with layers of synapses. Light passes through these layers before it interacts with the photoreceptors.

Information Processing in the Outer Retina

Neurons in the retina do more than simply pass on information about visual stimuli passively from one layer of cells to the next. They shape the information that they pass on in two important ways: by the way in which they respond to input, and by the way in which they interact to modulate their responses.

The properties of the receptive fields of bipolar cells provide an excellent example of both ways of shaping information. Recall from Chapter 9 that every neuron in a sensory pathway has a receptive field, a region of a sensory surface the stimulation of which produces a change in the electrical activity or membrane potential of the neuron. Hence, for every neuron in the retina there is some area of the retina that, when properly stimulated, causes a response in the cell. The receptive fields of rods and cones are circular, defined by the area within which either the stimulus itself or scattered light from the stimulus will strike the receptor and produce a response. The receptive fields of other types of retinal cells are determined by the locations of the rods or cones whose activity can influence the cell in question.

Receptive fields of bipolar cells consist of two parts, a circular **center** and a concentric **surround**, the latter so named because it encloses the center like a doughnut surrounding its hole (Figure 11-8A). The response of the neuron to a stimulus on the surround is opposite to the response of the neuron to a stimulus in the center. If the center and the surround are stimulated simultaneously, the cell gives no response if the light energy falling on the center is about equal to that falling on the surround (Figure 11-8B). If one stimulus is stronger than the other, the effect of the stronger stimulus will predominate.

There are several functional types of bipolar cells. Some bipolar cells receive input only from cones and hence carry information about color, whereas others receive input

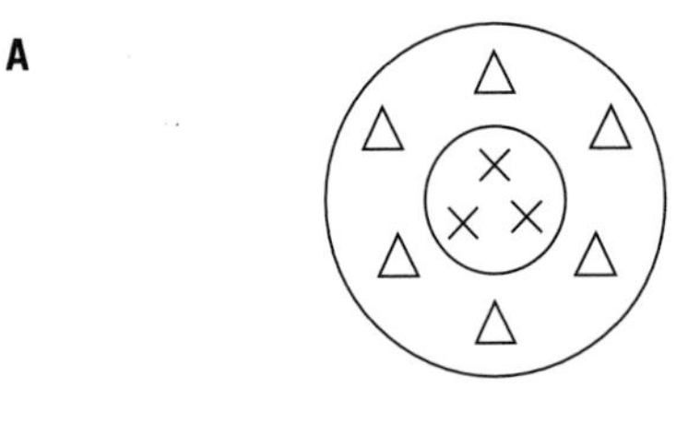

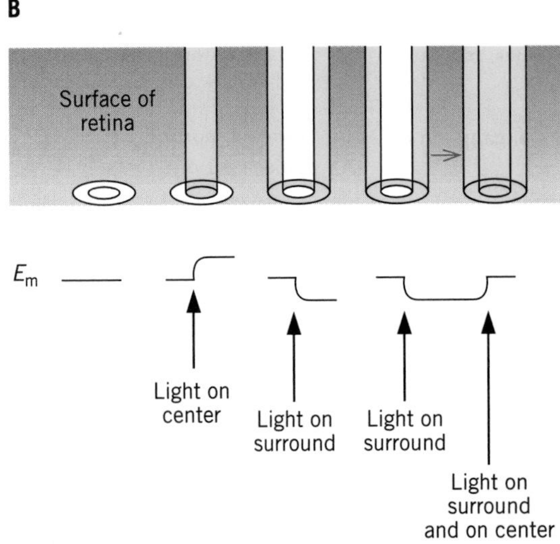

FIGURE 11-8. (A) The receptive field of a bipolar cell in the mudpuppy retina. The excitatory center (about 100 μm in diameter) is marked with *X*s, and the inhibitory surround (about 250 μm in diameter) is marked with triangles. (B) The response characteristics of a bipolar cell when stimulated by a spot of light in the center of its receptive field, a ring of light on its surround, and a spot and a ring together.

only from rods and hence carry no color information. Irrespective of whether a bipolar cell receives input from rods or cones, it may have one of two characteristics. An **off-center bipolar cell** hyperpolarizes when the center of its receptive field, a particular region of the retina, is stimulated with light. An **on-center bipolar cell** depolarizes when the center of its receptive field is stimulated. Responses for both off-center and on-center cells are graded to the strength of the stimulus.

The difference in response is due to a difference in the effect that the neurotransmitter (believed to be glutamate) released by the photoreceptors has on the postsynaptic (bipolar) cell. Glutamate hyperpolarizes on-center bipolar cells but depolarizes off-center bipolar cells. The nonspiking photoreceptors release glutamate more or less continuously. Stimulation of the receptor cells causes them to hyperpolarize and release less neurotransmitter (see Chapter 7). If the neurotransmitter binds with receptor proteins that cause hyperpolarization, then less neurotransmitter impinging on a bipolar cell will cause a *depolarization* of the bipolar cell (Figure 11-9A). Because the

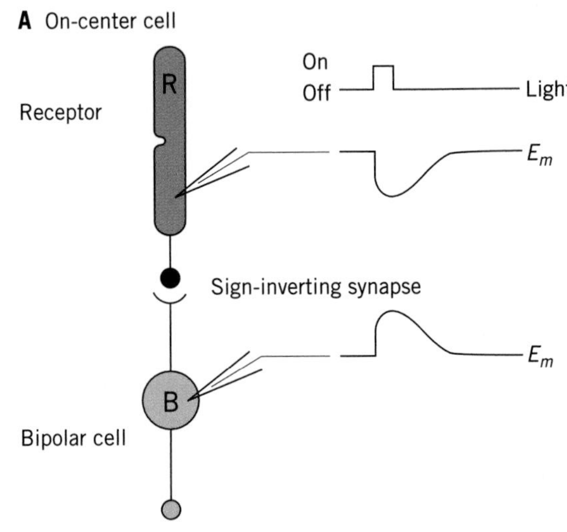

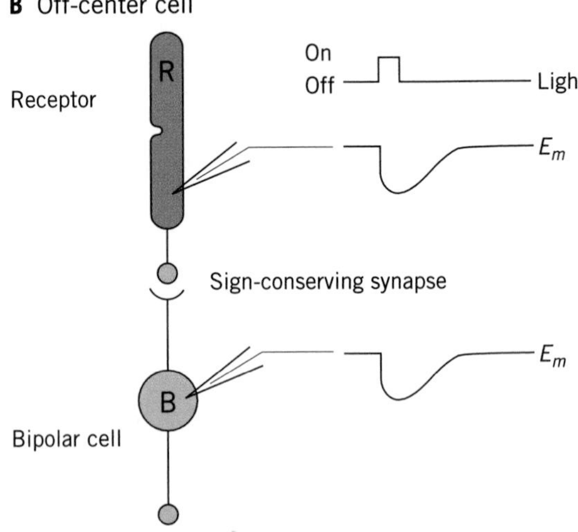

FIGURE 11-9. Synaptic mechanisms of on-center (A) and off-center (B) bipolar cells in the retina. The photoreceptor hyperpolarizes when excited by light. This causes a decrease in the amount of neurotransmitter it releases. If the transmitter is inhibitory (A), the synapse is sign-inverting and the effect is to depolarize the bipolar cell. If the transmitter is excitatory (B), the synapse is sign-conserving and the effect is to hyperpolarize the bipolar cell.

postsynaptic cell depolarizes when the presynaptic cell hyperpolarizes (and vice versa), this is called a **sign-inverting synapse**. On the other hand, if the neurotransmitter causes depolarization, then a reduction in the amount released will cause *hyperpolarization* of the bipolar cell (Figure 11-9B). Because the postsynaptic cell hyperpolarizes when the presynaptic cell hyperpolarizes (and vice versa), this is called a **sign-conserving synapse.**

The characteristic response of on-center or off-center bipolar cells to a spot of light falling on the center of the receptive field is intrinsic to the cell itself. On-center cells produce a different kind of receptor protein for binding with glutamate than do off-center cells. Hence, on-center cells generate a different kind of response to the presence of the neurotransmitter than do off-center cells. In contrast, the effects of light falling on the peripheral part of the receptive field of a bipolar cell arise from interactions between receptor cells and horizontal cells.

Horizontal cells make lateral connections in the retina. They receive input from receptor cells in one part of the retina and send their output to receptor cells in a nearby part of the retina. The synapse between a presynaptic receptor cell and a postsynaptic horizontal cell is sign-conserving, whereas that between a presynaptic horizontal cell and a postsynaptic receptor cell is sign-inverting. These features account for the surround of a bipolar cell, as shown in Figure 11-10. Review first the effects of light shining on the center of the receptive field of an on-center bipolar cell (Figure 11-10A). The bipolar cell depolarizes when light strikes the center of its receptive field because of the reduction of neurotransmitter released by the receptor cells above it when the receptors hyperpolarize. The effect is depolarizing because of the sign-inverting synapse and is uninfluenced by the horizontal cell shown in the figure because that horizontal cell receives its input from a receptor cell that is not stimulated.

Now consider the effect of light on the surround part of the bipolar cell's receptive field. This light does not stimulate the receptor cell that synapses with the bipolar cell. It does, however, stimulate one a short distance away that forms a sign-conserving synapse with a

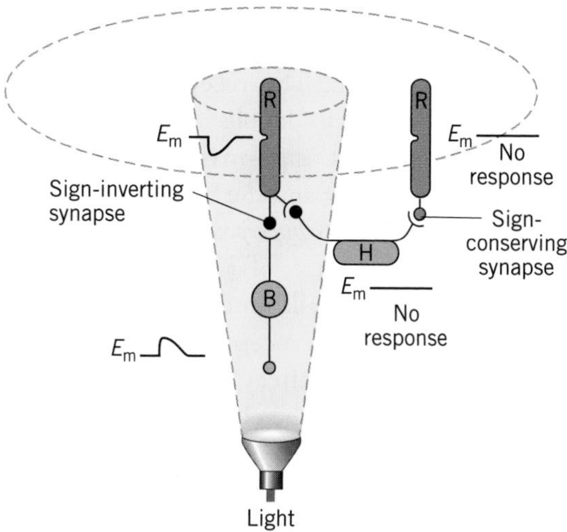

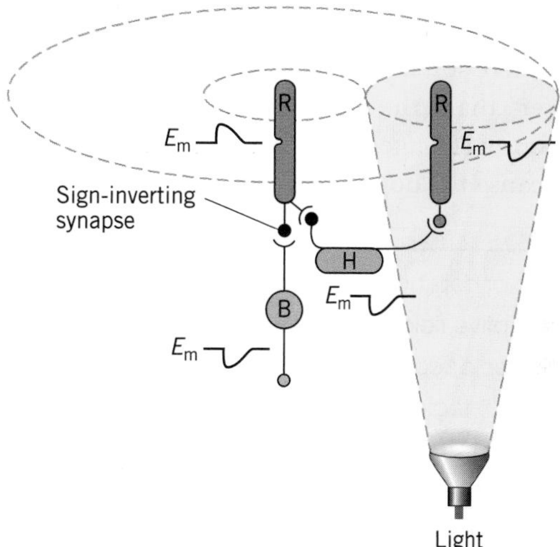

FIGURE 11-10. Synaptic mechanisms that underlie a center-surround receptive field of an on-center bipolar cell. (A) Light in the center of the receptive field of the bipolar cell stimulates receptors that make direct synaptic contact with it. Hyperpolarization of the receptor caused by stimulation by light leads to depolarization of the bipolar cell due to the sign-inverting synapse between the two cells. (B) Light on the surround of the receptive field does not reach the receptors that make direct synaptic contact with the bipolar cell. Instead, it stimulates nearby receptors, which make excitatory synaptic connections with horizontal cells via sign-conserving synapses, hence hyperpolarizing the horizontal cells. Each horizontal cell makes a sign-inverting synaptic connection with nearby receptors, causing them to *de*polarize. Depolarization of the receptor causes hyperpolarization of the bipolar cell via a sign-inverting synapse.

horizontal cell. The horizontal cell therefore hyperpolarizes in response. The horizontal cell, in turn, forms a sign-inverting synapse on the *receptor* cell that synapses with the bipolar cell under consideration. The receptor cell therefore *de*polarizes. This depolarization, working through a second sign-inverting synapse between the receptor and the bipolar cell, causes the bipolar cell to *hyperpolarize* (Figure 11-10B). Hence, because of the sign-inverting effects of horizontal cells on the receptors, light on the surround of the receptive field of a bipolar cell will cause an effect on the bipolar cell that is the opposite of the effect caused by light on the center. Of course, in the retina, more than one receptor cell and horizontal cell will be stimulated by a spot of light; there may also be interactions between direct stimulation of a receptor cell and input from surrounding horizontal cells. Nevertheless, the principles outlined here still hold. Stronger stimulation of the center than the surround will cause one response; stronger stimulation of the surround will cause the opposite.

Bipolar cells have concentric center-surround receptive fields and may be on-center or off-center, according to whether stimulating the central region depolarizes or hyperpolarizes the cell. The center-surround organization is produced by the effects of horizontal cells on receptor cells.

Processing by Ganglion Cells

The axons of ganglion cells form the optic nerve and hence constitute the only pathway from the retina to the brain. Ganglion cells were first investigated in depth in the 1930s in frogs by H. K. Hartline (see Box 11-1) then later in cats by Stephen Kuffler. Kuffler described two categories of ganglion cells: **on-center** and **off-center ganglion cells** (Figure 11-11). Like bipolar cells, these ganglion cells have a concentric center-surround receptive field. An on-center cell gives an "on" response (by increasing the rate at which it fires action potentials) when the center of its receptive field is stimulated, and an off-center cell gives an "off" response (by reducing its background rate of firing). Stimulation of the surround always gives a response opposite to that of the center.

On-center and off-center ganglion cells acquire their properties mainly from the types of bipolar cell that provide input to them. On-center ganglion cells are excited by on-center bipolar cells, whereas off-center ganglion cells are excited by off-center bipolar cells. In addition, amacrine cells in many retinas help to shape the responses of ganglion cells to more specific stimuli.

One of Kuffler's important contributions was the demonstration that the receptive fields of ganglion cells were organized in concentric and antagonistic center-surround areas. (Kuffler's work on cat ganglion cells predated the work on bipolar cells described in the preceding section.) However, study of the functional features of cat ganglion cells revealed that they were also specialized in other ways. Some cells have small receptive fields and hence have high spatial resolution of stimuli; they also respond well to sustained stimuli. These cells, called **X cells,** constitute roughly half of all ganglion cells. They are distributed throughout the retina, although they are somewhat more numerous in the central retina. Other cells have much larger receptive fields and are especially sensitive to moving stimuli. These cells, called **Y cells,** constitute roughly 5% to 15% of ganglion cells and are more numerous in the periphery of the retina. **W cells,** a third category, have mixed properties. Less is known about this group. On-center and off-center ganglion cells are found among all three of these functional classes.

Careful study of ganglion cells in primates has revealed similar functional groups of cells that cut across the on-center and off-center classification. Cells known as P cells are somewhat similar to the cat X cells—they have small receptive fields and high spatial resolution. They are also color-sensitive. In contrast, M cells have large receptive fields and are quite sensitive to

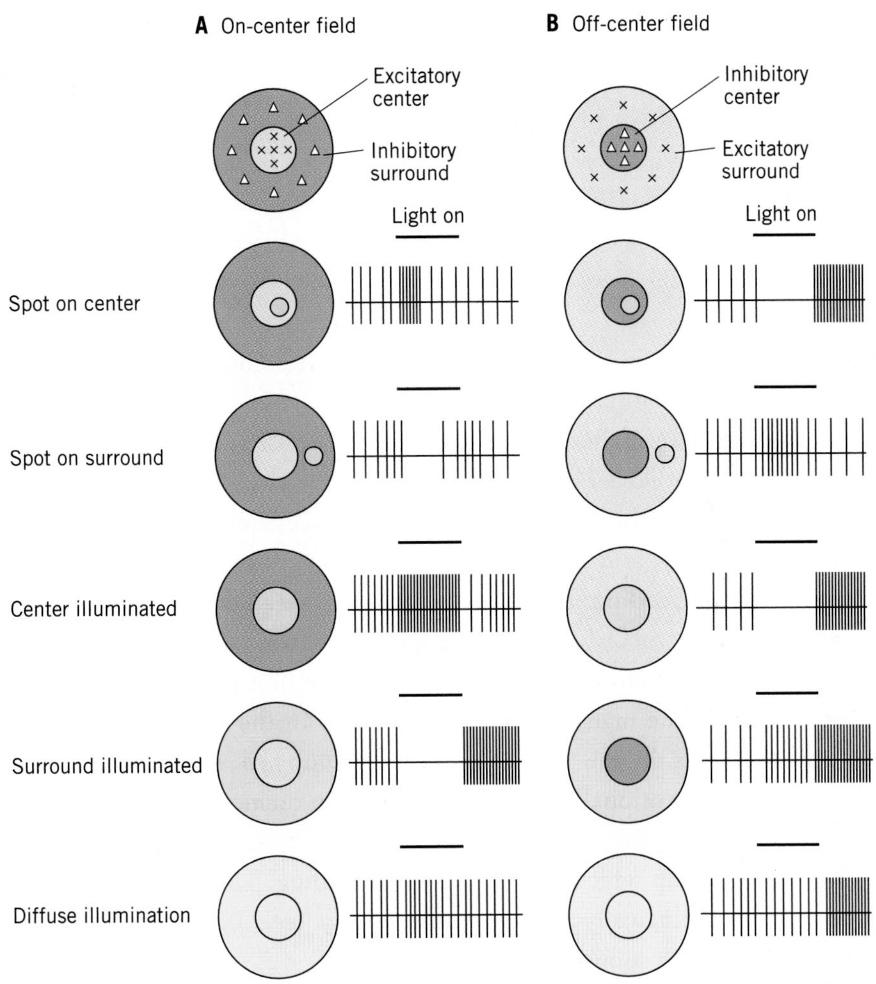

FIGURE 11-11. The responses of on-center and off-center ganglion cells in the cat retina. The bars show the period of stimulation.

movement and to small differences in contrast, hence sharing some of the properties of cat Y cells. They show no color sensitivity.

The details of how ganglion cells in various animals respond to stimuli should not obscure the importance of these response properties. The photoreceptors in vertebrate eyes are exquisitely sensitive to variations in the intensity and, in the case of cones, color of light, but they show no special sensitivity to contrast, patterns, or movement. Ganglion cells, in contrast, respond less to absolute intensity and report instead on such factors as contrast (on-center, off-center cells) or movement (Y cells). Hence, the information sent by the ganglion cells to the brain is already highly processed, and contains an abstraction of some of the features of an image that might be important to an animal.

Ganglion cells have on-center or off-center receptive fields. The type of field is determined by the type of bipolar cell from which the cell receives its input. Other classification schemes, in which cells are grouped into X, Y, and W types, or P and M types, are also used in some mammals.

The Processing of Visual Stimuli

A major issue in research on visual systems is how information about important aspects of an image,

BOX 11-1

Eyes on the Eye: Recording from Retinal Neurons

Recording the electrical activity of individual neurons has been a key to advances in the understanding of many neural systems, and the visual system is no exception. The earliest cellular insights into retinal function came when Hartline and his associates were able to record the activity of single ganglion cells in frogs. The understanding of phototransduction was advanced when it became possible to record from individual rods or cones from small pieces of retina (Figure 11-A).

The other main retinal cell types—horizontal cells, bipolar cells, and amacrine cells—are harder to record from because they are not the primary photoreceptors, they are quite small, they have short axons that are not readily accessible, and their normal synaptic connections must not be disturbed by the recording. For investigations of these cells, the development of intracellular recording and staining technology has been essential. Frank Werblin and John Dowling were among the first to make intracellular records from the main types of retinal cells. In the late 1960s, they recorded from cells in the retina of the mudpuppy, a salamander-like amphibian with exceptionally large (5 to 30 μm in diameter) retinal neurons. After each physiological recording, they injected the dye Niagara blue into the cell from which they had been recording. This allowed them to identify morphologically the cell under study (see Box 3-2) and to match the cell to its response characteristics.

Further refinements in technology have allowed researchers to record from quite small cells in the retinas of many animals, cells whose diameters

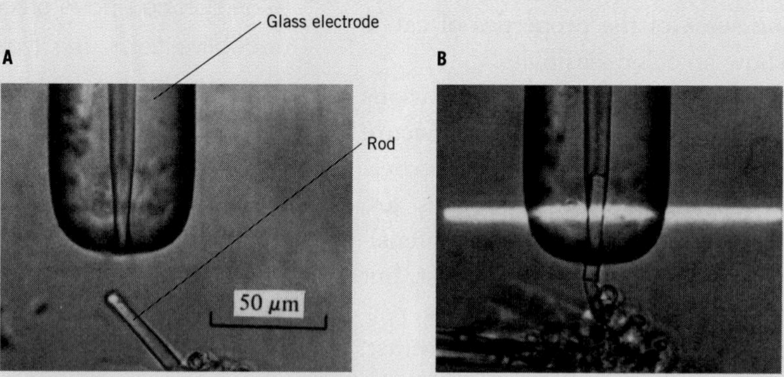

FIGURE 11-A. Recording from a single rod in turtle retina. (A) A glass microelectrode is moved to the vicinity of a piece of detached retina in a recording dish. (B) After the electrode has been manipulated over a single rod, a small, highly focused beam of light is directed through the electrode and the rod. Because the rod is essentially sealed inside the electrode, the electrical currents generated by the light can be recorded by the electrode.

are typically in the 5- to 10-μm range. The development of stains that are easier to use and that can be detected in small quantities, most notably the marker horseradish peroxidase (HRP) and the fluorescent stain Lucifer yellow, has also made many modern studies of retinal cell types possible by assuring positive identification of the cell after each experiment (Figure 11-B). Modern experiments combine these anatomical and physiological techniques with pharmacological and other methods to study such phe-

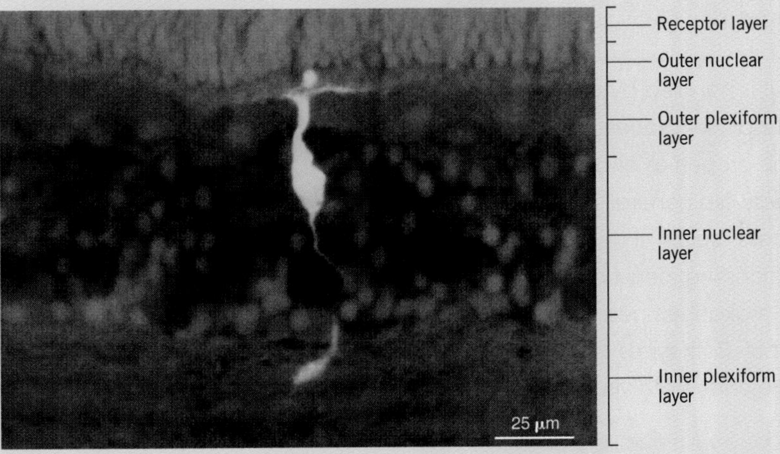

FIGURE 11-B. A bipolar cell from the retina of a carp stained with the fluorescent dye Procion yellow, a forerunner to dyes such as Lucifer yellow that are used today. The same electrode that contains the dye can be used as a recording electrode, making it possible to identify positively the cell from which the recording is being taken. The receptor layer and the ganglion cell layer are not visible in this view.

such as movement or contrast, is extracted by the system. It is clear that by the time visual signals reach the ganglion cells, such information has already been identified and specific cells have been assigned to carry this information to higher centers. In general, the nervous system uses two complementary mechanisms for identifying and reporting specific sensory information, not just in the visual system but in other sensory systems as well. The response characteristics of bipolar cells and ganglion cells in the retina serve as good examples of these two types of processing.

One mechanism is called **serial processing**. In this mode, cells in a system are arranged in a hierarchy, and interactions between cells at one level of the hierarchy shape and determine the characteristics of the cells at the next higher level. For example, the center-surround receptive fields of bipolar cells are produced by the inputs these cells receive. The center region is caused by input from receptor cells, and the antagonistic surround is caused by input from horizontal cells; neither is the result of some intrinsic property of the cells themselves.

The other mechanism is called **parallel processing**. In this mode, at least some of the characteristics of the different types of cells are intrinsic to the cells themselves. For example, on-center bipolar cells and off-center bipolar cells receive the same kinds of inputs from photoreceptors and horizontal cells. The response characteristics of each cell type are the result of the kind of postsynaptic receptor proteins for glutamate the cell has. Cells in which the presence of glutamate opens a channel and causes depolarization of the bipolar cell are off-center cells (because the photoreceptor-to-bipolar cell synapse is sign-conserving), whereas those in which the presence of glutamate closes a channel and causes hyperpolarization of the bipolar cell are on-center cells (because the photoreceptor-to-bipolar cell synapse is sign-inverting). As is apparent from these examples, both serial and parallel processing are essential for processing of visual information and appear to be critical for the complete and proper interpretation of visual images.

Visual input is handled by a combination of serial and parallel processing. In serial processing, the properties of a neuron in a sensory pathway are generated by the interactions of cells below it in a hierarchy. In parallel processing, cells have intrinsic properties that make them respond in particular ways to input. Both types of processing take place in the vertebrate visual system.

Visual Processing in the Thalamus

Information from the retina is sent to the brain via the axons of ganglion cells. In the brain, this information is distributed to several centers for processing. In mammals, most ganglion cell axons terminate in the lateral geniculate nuclei (LGN) of the thalamus. Within each LGN, the axons terminate in an orderly fashion on various layers of cells, preserving the spatial relationship between axons arising from different locations in the retina. In other words, each LGN has a topographical organization representing visual space. In addition, input to the LGN is segregated according to the functional type of the ganglion cell, such as X or Y, or M- or P-type ganglion cells.

The optic nerves also distribute information from each eye to specific halves of the brain. Axons from each eye cross to the opposite side of the brain, so that the left side sees the right side of the world and vice versa. In mammals with laterally placed eyes, this means that the optic nerve from each eye sends fibers to the LGN on the opposite side of the brain. In animals with forward-looking eyes, each eye sees portions of both the left and the right visual space, and each optic nerve splits so as to send fibers to both halves of the brain. Information from the right visual field (which falls on the left half of the retina) is sent to the LGN on the left side of the brain, whereas that from the left visual field (which falls on the right half of the retina) is sent to the LGN on the right side (Figure 11-12).

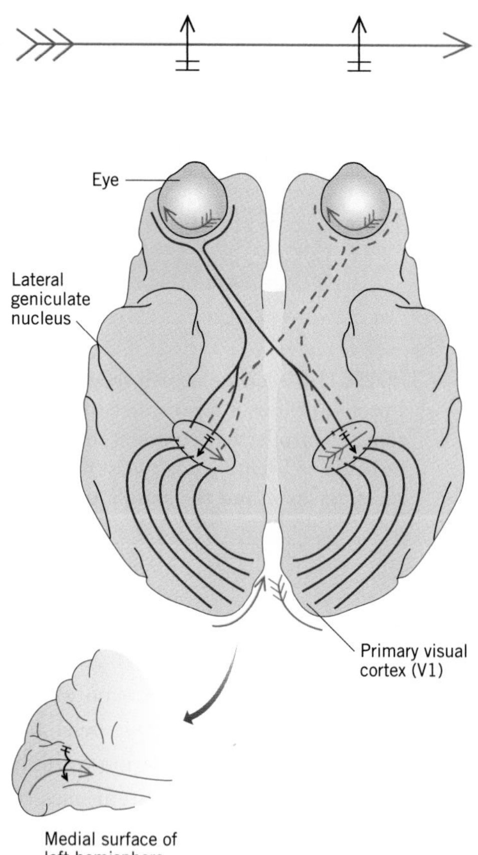

FIGURE 11-12. Pathways of visual information from the retina to the lateral geniculate nucleus (LGN) of the thalamus, and from the LGN to the visual cortex. In primates, information about the visual field is segregated so that the right half of the field is processed by the LGN and cortex on the right side of the brain, and the left half of the field is processed by the LGN and cortex on the left side of the brain. Each half image is projected topographically to the posterior and posterior medial suface of the occipital lobe.

The segregation of ganglion cell input to the LGN has been studied extensively in the LGN of primates (Figure 11-13). In these animals, each nucleus contains six distinct layers; layers 1 and 2 consist of relatively large cells and hence are called the **magnocellular layers** (magno = large). Layers 3 through 6 consist of smaller cells and hence are called the **parvocellular layers** (parvo = small). The M-type and P-type ganglion cells in primate retinas are distributed to the magnocellular and parvocellular layers of the LGN, respectively. In fact, it is from this distribution that the ganglion cells are named. Within these two main regions, inputs from the right and left eyes alternate.

The responses of LGN cells to visual stimuli falling on their receptive fields in the retina are for the most part quite similar to the responses of the ganglion cells from which they receive their input. That is, they exhibit characteristic concentric center-surround receptive fields, with both on-center and off-center cells being present. LGN cells in the parvocellular and magnocellular layers exhibit the features of P-type and M-type cells, respectively. Hence, based on these data, the LGN seems to be mostly a relay station whose specific function is to send signals to higher centers more or less unchanged, except that information about certain aspects of a visual stimulus (detail and color, or movement) and gross location in the visual field (right or left side) is segregated into physically distinct pathways.

However, 75% of the input to the LGN is from the visual cortex and the brain stem; only 25% is from the retina. This input apparently has little direct influence on the electrical activity of LGN neurons, but it seems able to modulate responses to visual stimuli. That is, it affects the magnitude of a cell's response to light but not the type of response it gives. The function of this input is not known, but it has been suggested that it plays a role in directing attention to specific stimuli. Hence, the LGN may play a more important role in the processing of visual input than the physiological characteristics of the LGN neurons themselves might suggest.

In mammals, most axons of the optic nerve terminate in the two lateral geniculate nuclei (LGN) in the thalamus. Visual information is segregated in the LGN according to whether it is about movement (the magnocellular layers) or about detail and color (the parvocellular layers), and then is sent to the visual cortex. Cells of the LGN have center-surround receptive fields like those of ganglion cells.

Cortical Processing of Visual Input

Lateral geniculate neurons project directly to the primary visual cortex in the occipital lobe of the brain. There they make synaptic connections mainly in the region known as layer 4,

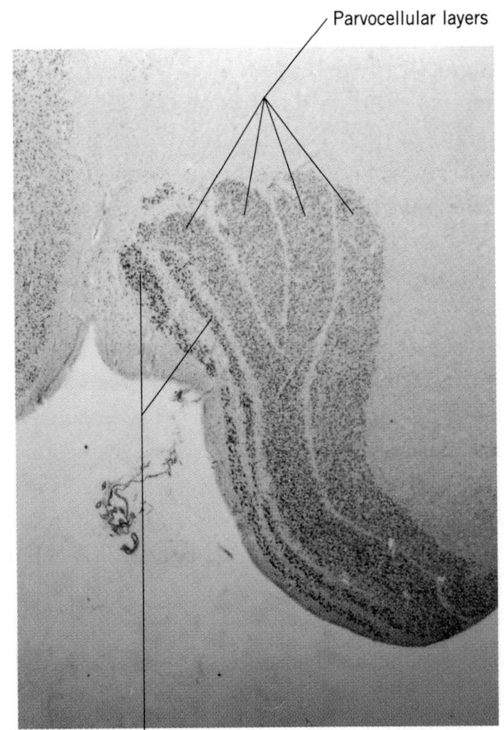

FIGURE 11-13. Cellular organization of a human lateral geniculate nucleus, showing the parvocellular and magnocellular layers. (Courtesy of Jonathan W. Marks and Joseph G. Malpelli.)

where visual input undergoes extensive additional transformation. The properties of cells in the visual cortex remained a mystery for some years after the properties of retinal cells had been described because the steady spots and rings of light that worked well to stimulate retinal cells had minimal effect on cortical cells. Not until the pioneering work by David Hubel and Torsten Wiesel in the early 1960s did a picture of these neurons begin to emerge. For their work, Hubel and Wiesel shared the Nobel Prize in Physiology or Medicine in 1981.

Physiology of Cortical Cells

Hubel and Wiesel described two main types of visual cortical neurons in cats. They called the first type a **simple cell** because they could map its receptive field by the weak responses of the cell to flashing spots of light. Stimuli that elicit strong responses in cells of this type are a straight bar of light or a dark-light border. Two features of a stimulating border or bar are critical: its *position* and its *orientation*. That is, a bar of light has to be positioned correctly in the receptive field and also has to be oriented properly in that field in order for the cortical cell to show a strong response (Figure 11-14).

The receptive fields of simple cells do not all have the same characteristics (Figure 11-15). Like ganglion cells, simple cells have receptive fields with specific regions that excite the cell and others that inhibit its spontaneous activity. In some cells, regions that give an ON response are flanked by regions that give an OFF response, so that a bar of light must run through the center of the receptive field in order to elicit strong firing from the neuron (Figure 11-15A). Illumination of the outer regions of the receptive field elicits no response when the light is turned on but elicits action potentials when the light is turned off. In other simple cells, ON regions flank an OFF region (Figure 11-15B). Asymmetrical receptive fields (Figure 11-15C) and receptive fields divided into ON and OFF halves are also known (Figure 11-15C, D). All of these types of receptive fields may be found with every possible orientation of the center or the border: vertical, horizontal, and all angles in between.

Hubel and Wiesel also found neurons whose receptive fields could not be mapped by flashing light; they called these **complex cells**. Some complex cells respond mainly to an edge (that is, a light-dark border) crossing their receptive fields in the retina. The orientation of this edge stimulus is critical in eliciting a response in these cells, just as it is in simple cells, but the position of the edge within the receptive field is relatively unim-

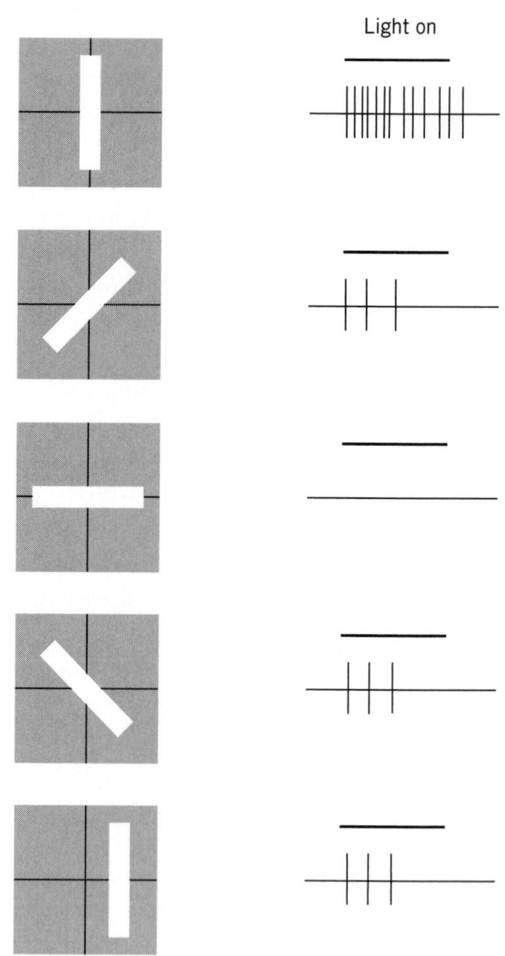

FIGURE 11-14. The responses of a simple cortical cell to a bar of light with different orientations. The receptive field of the neuron is represented only by the crossbars (shown for purposes of reference). The position and orientation of the stimulus are shown superimposed on this reference framework (leftmost of each pair of diagrams). The response of the neuron to the stimulus is shown as a series of action potentials in an oscilloscope trace (right side). Note that the cell shows very little response to the stimulus if it is not properly positioned and properly oriented.

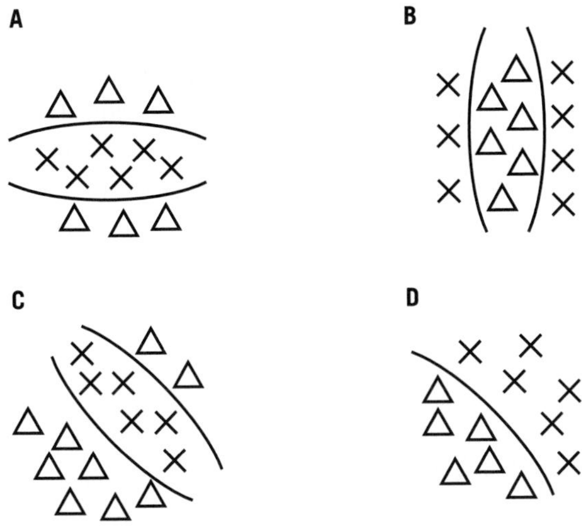

FIGURE 11-15. Receptive fields of simple cortical cells in a cat visual cortex. (A) An on-center cell. (B) An off-center cell. (C) A cell with an asymmetrical receptive field. (D) A cell lacking a true center; the receptive field is divided into on and off halves. Symbols as in Figure 11-8A.

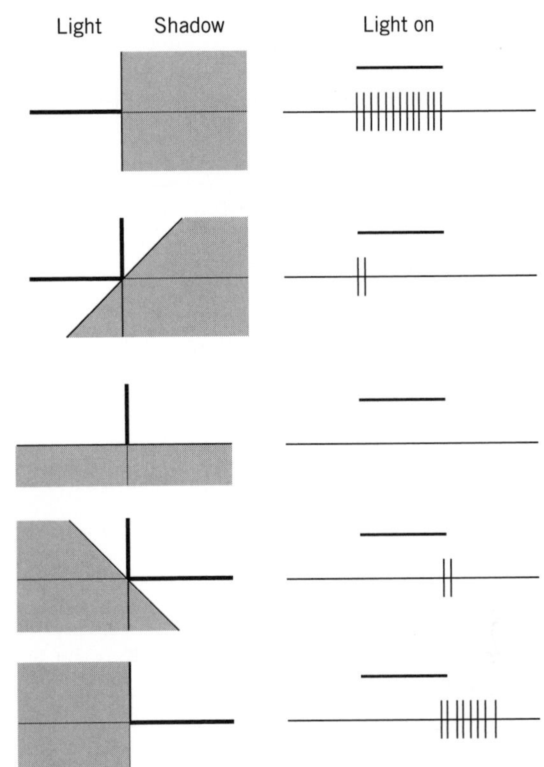

FIGURE 11-16. The responses of a complex cell to an edge with different orientations. Conventions as in Figure 11-14.

portant. These features are illustrated in Figures 11-16 and 11-17. Orientation selectivity is shown by the more vigorous response of the neuron to a stimulus with one particular orientation. Position independence is shown by the strong response of the neuron to a properly oriented stimulus positioned almost anywhere in the receptive field. Given appropriately placed stimuli, cells can show either ON or OFF responses and can be selective to the direction of movement of the stimulus as well. In the case of a directionally selective cell, strong excitation requires that a stimulus be moving in a particular direction.

Cells in the visual cortex require more complex visual stimuli to be strongly excited than do cells in the retina. A simple cortical cell requires a properly positioned and properly oriented bar of light in its receptive field. A complex cortical cell requires a properly oriented edge that can be positioned almost anywhere in the receptive field.

Emergent Properties of the Visual System

It should be apparent from this description of the visual system that neurons at different levels in the system require rather different stimuli to excite them, ranging from simple stationary spots to the edges of shadows. An important question is how neurons can come to have such different response properties. According to the idea of serial processing, the output of a cell is determined by the types of input connections it receives. To put it another way, the type of stimulus parameters necessary for excitation of a neuron depends in large part on the stimuli that excite the cells that make synaptic contact with it. The key to this idea is that, because of the types and pattern of inputs it receives, a neuron may show response characteristics not found in any of the cells that synapse with it. This appearance of a neural feature in one neuron that is not present in any of the neurons that synapse with

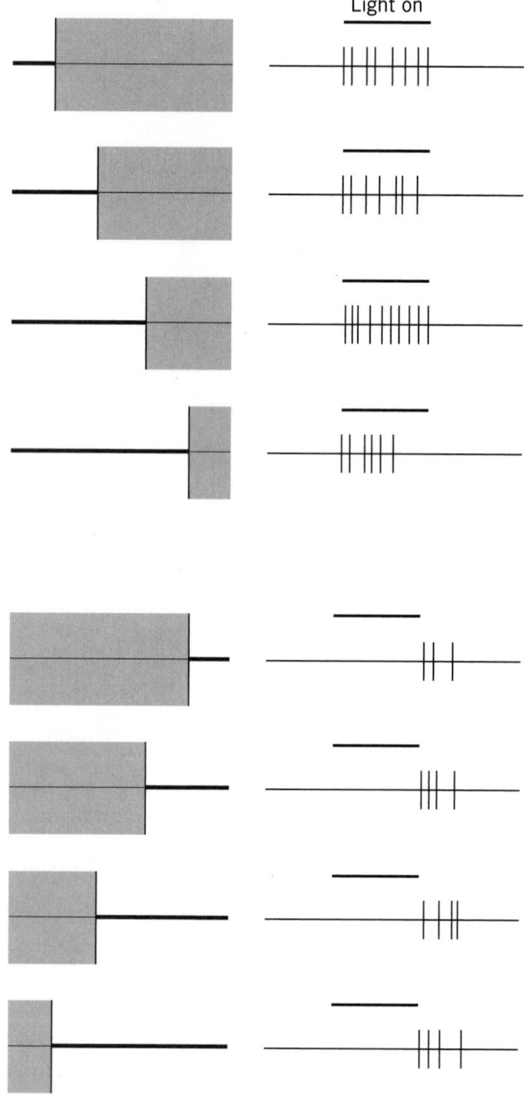

FIGURE 11-17. The responses of a complex cell to an edge in the "best" orientation but in different positions. Responses of a single cell to light on the left and light on the right are shown.

it is called an **emergent property** of the network of neurons responsible for it. That is, a new property emerges out of the interactions of neurons that individually do not have that property.

The response of a simple cortical cell to a properly oriented, properly positioned bar of light in its receptive field is a good example of an emergent property. The cells from the lateral geniculate nucleus that drive the cortical cells have receptive fields similar to those of ganglion cells: circular, with an on- or off-center region and an antagonistic, concentric surround. Hence, the responsiveness of cortical cells to a completely different type of stimulus must derive from the specific interactions of neurons from the LGN with the cortical cells.

Hubel and Wiesel suggested a mechanism by which this unique responsiveness might arise (Figure 11-18). If LGN cells whose receptive fields are aligned in a straight row converge onto each simple cortical cell, then the only adequate stimulus for that cortical cell (assuming that input from several cells simultaneously is necessary to excite it) will be a bar of light that is aligned through the centers of the receptive fields of the relevant LGN cells. A small spot of light will not do because it will be unable to excite many LGN cells. A larger spot will not do because it will excite too much of the inhibitory surround and therefore reduce the firing of the LGN cells. An improperly oriented or positioned bar of light also will not do for either or both of these reasons. Only a bar that goes through the centers of the receptive fields of most of the LGN cells will excite the simple cortical cell to give its maximum output.

Other patterns of interaction are possible, too. The convergence of LGN cells with a different alignment or shape onto specific cortical cells makes it possible to generate simple cortical cells that have asymmetrical receptive fields or a different preferred orientation. Furthermore, the properties of complex cells can be developed in a similar fashion from selective input from particular simple cells. You will encounter further examples of emergent properties of groups of neurons when you study motor systems (see Chapter 16).

Many of the complex analytical functions of the visual system result from the emergent properties of groups of neurons, in which the neuron shows characteristics not exhibited by any individual cell that synapses with it. For example, the response of a simple cortical cell to a properly oriented bar or line of light in its receptive field arises as an emergent property from the connections between the cortical cell and a particular group of cells in the LGN whose receptive fields are arrayed in a line.

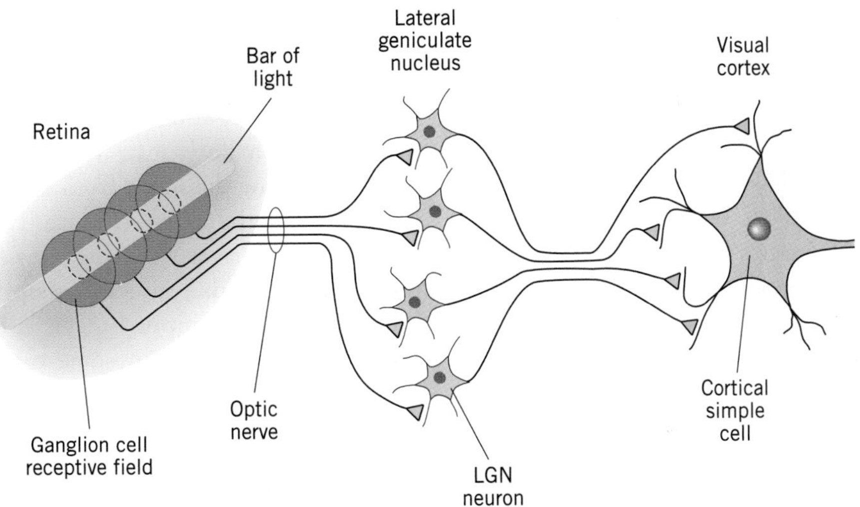

FIGURE 11-18. Diagram of hypothetical interactions between a series of neurons in the visual pathway that will produce the emergent property of sensitivity to a properly positioned and properly oriented bar of light in a simple cortical cell. Retinal ganglion cells with adjacent, aligned receptive fields send their axons via the optic nerve to neurons in the LGN. If the corresponding LGN cells send their axons to a single cortical neuron, that cortical cell will be excited only when a bar of light excites each of the ganglion cells and hence the corresponding LGN cells.

Organization of Input to the Visual Cortex

The cortex of any mammal is far from being a homogeneous sheet of tissue; histological examination generally reveals six readily distinguishable layers of cells. The primary visual cortex, visual area V1, actually has about nine easily recognized layers in histological preparations, but to conform to the convention of six used everywhere else in the cortex, four of these are considered subdivisions of layer 4 rather than bearing numbers of their own.

Nearly all the input to V1 from the LGN terminates in one of two sublayers of layer 4, segregated according to the part of the lateral geniculate nucleus from which it originates. Neurons that originate in the magnocellular regions of the LGN terminate in one specific subregion of layer 4. Because of their origin, these neurons are called the **magnocellular pathway,** sometimes shortened to the **M channel.** The magnocellular regions of the LGN mainly process information concerning motion, and cortical neurons that receive input via the M channel are frequently quite sensitive to the direction of motion of a stimulus.

Neurons that originate in the parvocellular regions of the LGN terminate in a different subregion of layer 4. Because of their origin, these neurons are called the **parvocellular pathway,** sometimes shortened to the **P channel.** The parvocellular regions of the LGN mainly process detailed information about objects and their colors, and these features are also conserved in the cortex. Neurons that receive input via the P channel are often sensitive to the fine detail of an image as well as its color, and hence convey information about the shape of an object in the visual field.

From layer 4, visual information is distributed directly or indirectly to the other layers of area V1. This distribution of information is mainly vertical. That is, most local interneurons project from an area in one layer to an area that is above or below it in another layer. There is relatively little horizontal connection between areas within a layer, or indeed between areas in different layers. Consequently, the retinotopic organization of layer 4 that arises from the orderly input it receives from LGN neurons is preserved in the other five layers of V1.

Input to the visual cortex enters at layer 4. It is segregated according to the part of the lateral geniculate nucleus from which it originates. Input from the magnocellular layers of the LGN constitutes the M channel and conveys information about motion. Input from the parvocellular layers of the LGN constitutes the P channel and conveys information about color and shape. From layer 4 information is distributed to other layers.

Structural and Functional Organization of the Visual Cortex

Area V1 is functionally organized in three important ways: topographically, by eye, and by simple visual features. First, as you learned in Chapter 9, the visual cortex is organized topographically. Input from the retina via the LGN is mapped onto layer 4 of the visual cortex so that stimulation of adjacent regions of the retina evokes responses in neurons in adjacent regions of layer 4. Because most connections within V1 are vertical, the topographical map formed in layer 4 is preserved throughout the other five layers. The map is not linear, being distorted to allow greater representation in the brain of those regions of the retina with the greatest concentration of receptor cells. Hence, the fovea or other central regions of the retina that subsume the greatest visual acuity are overrepresented in area V1 at the expense of more peripheral areas.

Area V1 is also organized by eye. Imagine a vertical section taken through the cortex. If the cells in such a section could be labeled according to whether the main input they receive comes from the right or left eye, you would see alternating "columns" of cells spanning the six cortical layers. Each column, called an **ocular dominance column**, directly or indirectly receives input mainly from one of the two eyes (Figure 11-19A). A vertical section taken at right angles to the one that shows the alternating ocular dominance columns would show the columns aligned in rows, with each row receiving input from the same eye. In layer 4, which receives the input from the LGN, the segregation by eye is particularly distinct, and a horizontal view of this layer, treated to show the input from one eye, strikingly reveals the rows (Figure 11-19B). In animals with forward-looking eyes (i.e., those that have binocular, or stereoscopic, vision), the segregation of input by eye is blurred in other layers, resulting in some overlap. This overlap is the basis of binocular vision. The appearance of ocular dominance columns and binocular vision during development has been the subject of considerable research. You will learn more about it in Chapter 23.

A third feature of area V1 is its organization into columns based on simple visual features. In their investigations of simple and complex cells, Hubel and Wiesel found that all cells in a vertical column through the cortex tended to respond to lines or edges oriented at the same angle. For this reason, they called such a column an **orientation column**. They found that adjacent orientation columns contain cells that respond to lines at a slightly different angle, so that the preferred angle varies systematically until a full a circle has been covered (Figure 11-20A). Subsequent work has shown that orientation columns are commonly arranged in a pinwheel-like array around central points scattered throughout the cortex; linear arrays of orientation columns have been reported in some animals as well (Figure 11-20B).

Physiological studies have provided significant insights into the organization of the visual cortex, and indeed of other regions of the brain, but they present a biased sample of neurons because they direct the attention of researchers to cells that show significant activity in response to some stimulus. Cells that do not respond to the stimulus will be missed. It can therefore be useful to apply other techniques to the study of brain regions, even if these regions seem fully mapped already. In 1978, Margaret Wong-Riley stained area V1 for the presence of the enzyme cytochrome oxidase, which is widely used in cells for general oxidative metabolism. To her surprise, she discovered columns of cells extending

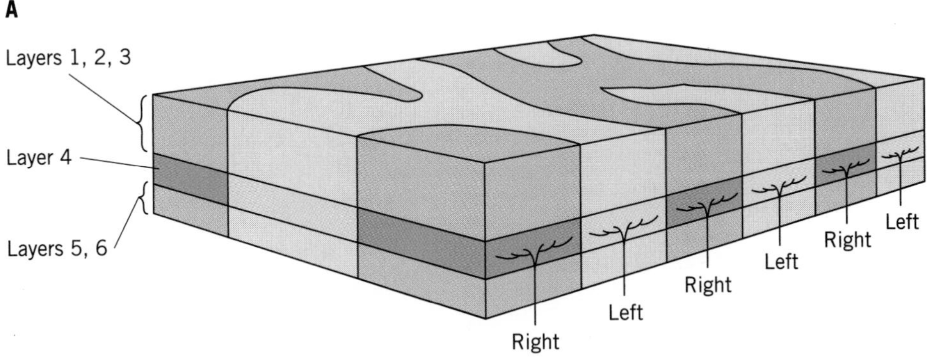

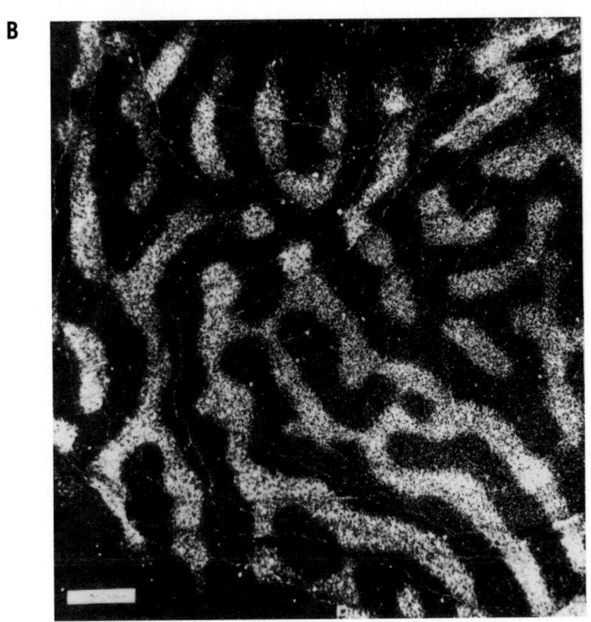

FIGURE 11-19. Ocular dominance columns in the primary visual cortex. (A) Diagrammatic view of a section of the cortex, showing alternating columns of cells that receive information exclusively (layer 4) or primarily (other layers) from one eye. The columns are aligned to form stripes when viewed from another angle. (B) Ocular dominance stripes in layer 4 of the visual cortex of a monkey, demonstrated by injecting into one eye a marker that is transferred via the LGN to the visual cortex. Scale bar: 1 mm.

through layers 2 and 3 that stained heavily for this enzyme. When cut in cross section, these columns appear as spots or **blobs** (Figure 11-21A), and for want of a more elegant name, they are now known as blobs. Blobs are situated in the middle of orientation columns (Figure 11-21B) but the cells in them show no sensitivity to spots, lines, or edges. Instead, they respond well only to differences in color on different parts of their receptive fields.

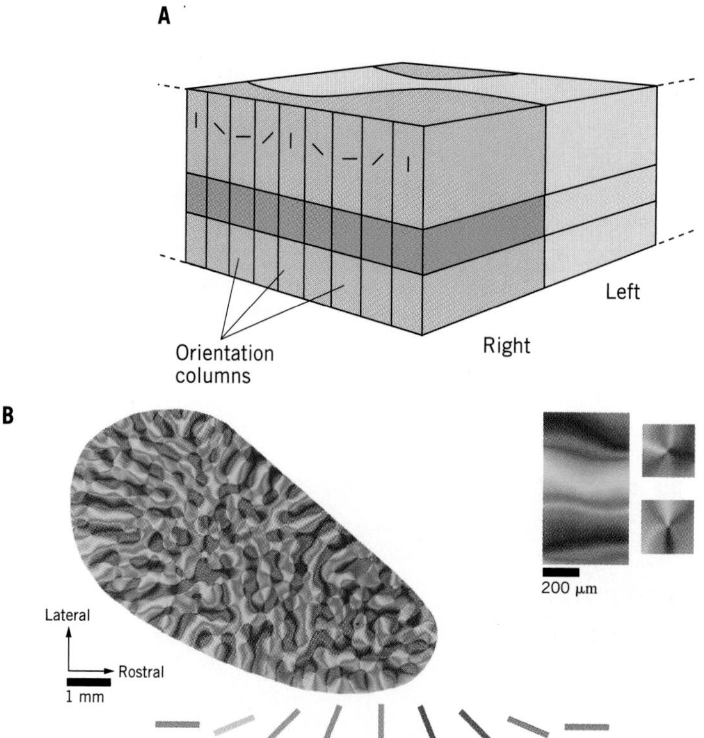

FIGURE 11-20. Orientation columns in the primary visual cortex. (A) Columns of cells showing preferences for bars of light at different orientations, shown arrayed within each ocular dominance column. (B) Color-coded image of the visual cortex of a tree shrew showing the parts of the cortex that become active during presentation to the retina of a grating with a particular orientation. The image was obtained by optical measuring of small increases in oxygenated blood flow to the active regions of cortex. Orientation columns are arrayed linearly or in a pinwheel array, as shown in the inset on the right.

The primary visual cortex is organized topographically, each layer containing a map of the retina. Input from the eyes is segregated and distributed in ocular dominance columns throughout the cortex such that each column of cells responds mainly to input from one eye. The cortex also contains orientation columns (columns of cells, all of which respond to a line or edge at a particular angle) and, within these columns, blobs (regions containing cells with high cytochrome oxidase activity that are sensitive to color contrasts).

Other Visual Areas of the Cortex

Area V1 is only one of several regions of the brain that are intimately involved in processing visual information. Area V2 lies adjacent to V1, and areas V3, V4, and the middle temporal area (MT, the same as V5) lie adjacent to or near V2. At a gross level these five areas can be considered a hierarchy, since neurons from one area project to the next higher-numbered area (V1 to V2, V2 to V3, etc.). Furthermore, to a certain extent each area extracts more complex aspects of a scene from the image that falls on the retina than does the area "below" it. On the other hand, many axons do not follow a serial path. For example, V1 projects directly to MT as well as to V2, and there are pathways from "higher" levels to "lower" ones. Hence, neurobiologists today consider each area a part of the visual system, each having a specific role to play in interpreting scenes but not arranged functionally in a hierarchy of "higher" and "lower" functions.

The question of what V2, V3, V4, MT, and other visual areas actually do is a difficult one. It is not possible to assign one single function to any particular area, in part because the separation of information in the LGN into P and M channels is maintained not only in V1 but in these other areas as well. Hence, the processing of color, motion, and form is carried out in parallel in most of these areas rather than being confined to one or the other. Furthermore, there is considerable exchange of information between the two channels within an area, making it all the more difficult to assign a specific function to one part of the brain.

Nevertheless, the rather astonishing finding has emerged that features of an image are handled separately in different brain areas. This view is vividly supported by medical case histories of individuals who have suffered brain lesions restricted to one or more of the visual areas. For example, it is thought that area MT is critical for a proper perception of motion. This idea is supported not only by animal experiments but also by the case of an individual who suffered damage in this area and subsequently found she could not distinguish between objects that were stationary and those that were in motion. She reported seeing the world essentially as a series of still pictures, causing her great trouble with a simple action like pouring coffee into a cup because she could not see that the cup was filling. Similar losses for other specific parts of the normal visual process have been reported, such as loss of color vision, loss of depth perception, even loss of the ability to recognize faces, suggesting that these elements are processed by one or more parts of the brain quite distinct from parts that process other aspects of vision. You will learn more about the brain's modular mode of processing information in Chapter 21.

Although there is a considerable body of experimental data to support the view that the brain interprets pieces of an image in a modular fashion, it is also true that the brain reassembles the pieces to give us a coherent and seamless visual impression of the world. An idea that appeals to many neurobiologists is that rather than consisting of separate areas that serve exclusive functions, the brain is organized to extract specific kinds of information from an image. Hence, some researchers speak of processing information to discover "what" an image represents separately from processing information to discover "where" it is. Features such as color, shape, and even motion can obviously contribute to both tasks.

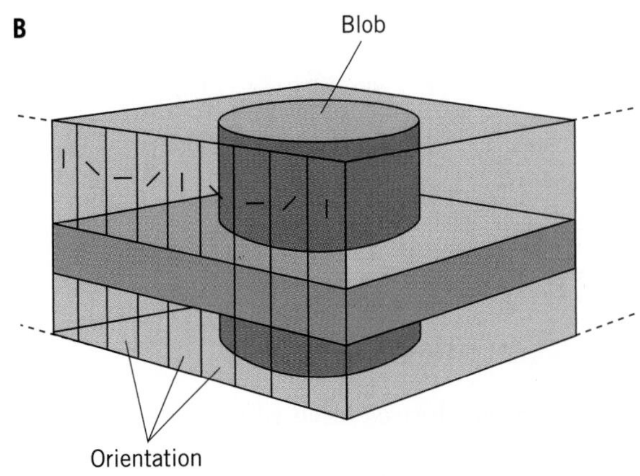

FIGURE 11-21. Blobs in the primary visual cortex. (A) The arrangement of blobs, columns of cells that respond to color contrast but not to the orientation of a bar of white light, within ocular dominance and orientation columns. (B) A horizontal section of monkey visual cortex stained for cytochrome oxidase. The dark spots are the cytochrome oxidase-rich blobs.

The areas of the brain that process visual information show a rough serial order of connection, but enough connections that violate a strict hierarchical order are present that neurobiologists do not consider them arranged in a true hierarchy. Parallel channels devoted to information about color and shape and about motion are present in all areas. Visual areas beyond V1 seem to handle specific aspects of vision, such as motion, depth, or recognition of faces.

Molluscan and Other Invertebrate Eyes

Most invertebrates have photoreceptors that allow them to detect light. However, image-forming eyes are for the most part confined to the annelids (segmented worms), mollusks (snails, scallops, squid, and the like), and arthropods (spiders, crabs, insects, and other invertebrates with jointed legs). Some image-forming eyes are similar to the eyes of vertebrates, each having a single lens that focuses an image on the photoreceptor cells. Many mollusks, a few annelids, and some spiders and insect larvae have eyes of this type. Other image-forming eyes have multiple lenses, each lens directing light to one or a few photoreceptive elements. This kind of eye, called a **compound eye**, is found in nearly all adult arthropods, as well as in a few species of annelids and mollusks; it is discussed in the next section. A few invertebrates have lensless eyes. These may form images in a manner analogous to that of a pinhole camera.

The single-lens eyes of squid and octopuses are remarkable, conferring to these animals an acuity of viszion apparently close to that of many vertebrates. Their eyes are also an excellent example of convergent evolution, showing several structural and functional similarities to the eyes of vertebrates (Figure 11-22). The eyeball is roughly spherical, and light is focused onto the retina by a lens, which in most of these animals can be moved to focus on objects at various distances. In squid and octopuses, the eyeballs can be moved in the head by three pairs of ocular muscles. There is evidence that squid and octopuses have reflexes for stabilizing a moving image on the retina. All of these are features of vertebrate eyes as well. The retina itself also has similarities to a typical vertebrate retina. In some squid there is a specialized area in the retina in which the photoreceptors are packed at a higher density, just like in the fovea of some vertebrate eyes. Absolute numbers of photoreceptors are also similar (e.g., about 20,000 to 50,000/mm^2 in shallow-water squid). Color vision is absent, however; all the photoreceptors are equivalent to rods.

In spite of these similarities, there is no doubt that these eyes evolved quite independently, as indicated by several fundamental differences between them. First, the retinas of squid and octopuses are arranged inside-out compared with those of vertebrates. That is, the photoreceptors face the pupil of the eye rather than the back of the eyeball. There are also fewer cell types in molluscan retinas; more processing of visual information therefore takes place in the brain. Perhaps more telling, the photosensitive parts of the photoreceptors are arranged in rhabdoms like those of an insect (see next section), not like the rods typical of a vertebrate. These differences point to a separate evolutionary origin for vertebrate and molluscan eyes.

Some annelids, mollusks, and arthropods have true image-forming eyes. These eyes may have single lenses as in vertebrate eyes, or they may have multiple lenses as in compound eyes. The eyes of squid and octopuses are remarkably similar in structure and function to the eyes of vertebrates but are thought to have evolved independently.

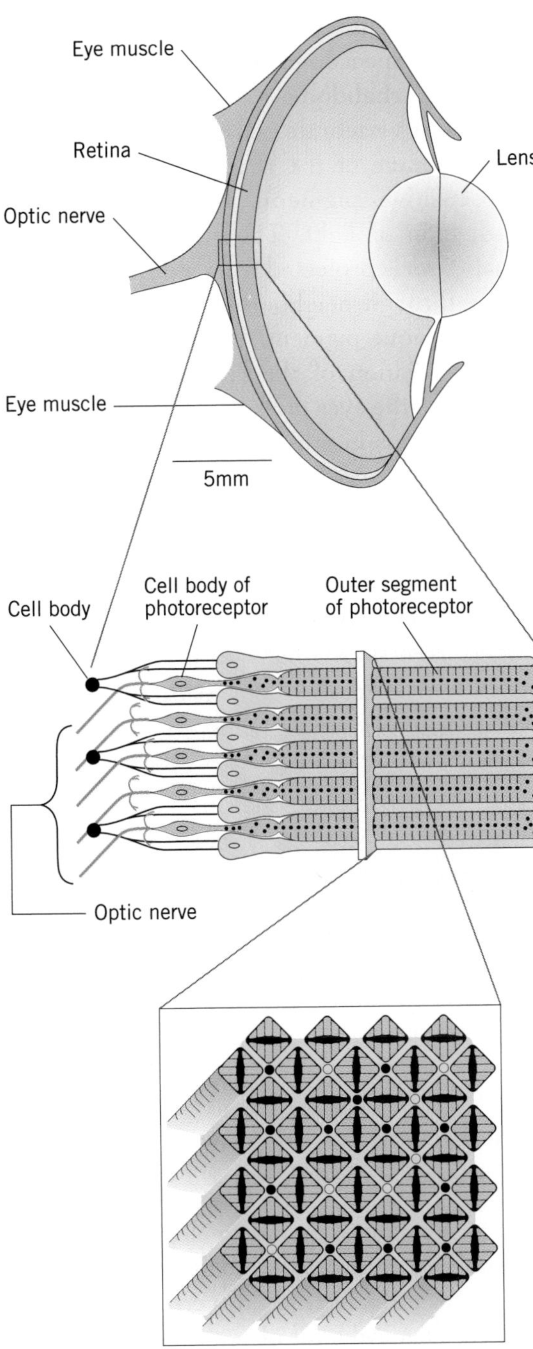

FIGURE 11-22. The eye of a deepwater squid. The lens can be moved relative to the retina, allowing for some adjustment of focus. Photoreceptors are oriented facing the lens, unlike photoreceptors in vertebrate eyes. The photoreceptive outer segments have an appearance in cross section similar to that of insect photoreceptors.

The Eyes of Insects

An insect may have as many as three kinds of eyes over the span of its life. In its immature, larval form, an insect may have one or more single-lens eyes, called **stemmata**, that in some species are capable of providing information about the size, shape, movement, and color of objects in its visual field. In addition, both larvae and adults may have simple, single-lens eyes, called **ocelli**, that are not capable of forming usable images. These detect the presence of light for the regulation of physiological functions that depend on cycles of day and night and changes of the seasons. They also provide information about the position of the horizon, since ocelli are able to detect the border that the horizon forms between the bright sky and the darker ground. Finally, as adults, most insects have a pair of compound eyes. The immature stages of these insects that undergo incomplete metamorphosis (insects like grasshoppers that look similar as young and adults), lack stemmata and have compound eyes from the time they emerge from the egg.

In this and the remaining sections, you will learn about the main features of the insect compound eye as a representative of a complex invertebrate eye. You will see that although the structure and physiology of a compound eye are quite different from those of a vertebrate eye, the vertebrate and insect visual systems nevertheless share many features.

Insects may have three types of eyes: stemmata, ocelli, and compound eyes. Stemmata are found only in larvae; they can form images. Ocelli detect light and dark, and also can detect the horizon. Compound eyes are the main image-forming organs of some immature insects and nearly all adult insects.

The Compound Eye

If you look closely at a compound eye, you can see that the surface is broken into a number of small, hexagonally shaped facets. Each facet is the cornea of a single photoreceptive structure called an **ommatidium** (Figure 11-23A). The number of ommatidia, their spacing, and the sizes of the facets in each eye determine the acuity of the eye, and vary considerably from species to species. Many insects have small, inconspicuous eyes; the eyes of some beetles contain only about 300 ommatidia each. At the other extreme, a fast-moving insect like a dragonfly can have enormous eyes, each containing well over 25,000 ommatidia. About 3000 to 9000 ommatidia per eye is a more typical number for most insects.

The eyes of many insects, especially good flyers like flies or honey bees, or prey-catching insects like praying mantises, have specialized regions of acute vision that serve the same function as the fovea of primates and some birds. In these acute zones, the surface of the eyeball tends to be flattened, and the facets are larger than facets elsewhere in the eye (Figure 11-23B). Flattening of the eye allows more ommatidia to receive light from a particular spot, which improves the resolution of the image. Larger facets improve resolution by ensuring that each ommatidium receives an adequate signal. In spite of these specializations, the best insect eye has a resolving power of only about 1° of arc, compared with the 1' of arc that humans can resolve and the 20" of arc that some birds of prey can resolve. This means that from a distance of a meter (just over 3 feet), a fly could resolve the individual fingers of your hand if you held it up with the fingers slightly spread.

Each ommatidium consists of several elements (Figure 11-24). Just under the cornea lies a transparent wedge of material called the **crystalline cone,** which helps focus light onto the underlying sensory neurons. The sensory neurons form the **retinula,** the light-sensing element of the ommatidium. The retinula typically contains eight sensory neurons, often referred to as **retinal cells** or more simply **visual cells,** arranged around the central photoreceptive region, the **rhabdom.** The rhabdom is formed by a specialized portion of each retinal cell, called a **rhabdomere,** that is equivalent to the disks of vertebrate rods (Figure 11-24B). The membrane of the rhabdom contains the photosensitive pigments that mediate the transduction of light. The rhabdom of each ommatidium is protected from light that enters the eye through neighboring ommatidia by a layer of opaque pigment. In some insects, the size and position of the rhabdoms and pigments make the eyes most suitable for daytime vision, whereas in other insects the rhabdoms and pigments make the eye better suited for night vision. The eye depicted in Figure 11-24 is specialized for daytime vision.

A compound eye may be composed of anywhere from several hundred to more than 25,000 ommatidia. Each ommatidium consists of two main parts, the light-focusing elements at the surface and the photoreceptive retinula below. Phototransduction takes place in the rhabdom, which is typically formed by specialized parts of eight sensory neurons that together form the retinula.

Transduction and Spectral Sensitivity

The transduction of light into electrical activity in the eyes of insects has many striking similarities to transduction in vertebrate eyes. First, the chemical basis of transduction is similar in the two. Insect retinal cells contain high concentrations of rhodopsin or rhodopsin-like pigments. These pigments consist of an opsin that anchors them to the cell membrane, and retinal or some similar photosensitive component. Activation by light of the photosensitive component initiates a biochemical cascade that results in a change in the membrane potential of the sensory neuron.

The second similarity is that the biochemical mechanisms of transduction in insect eyes

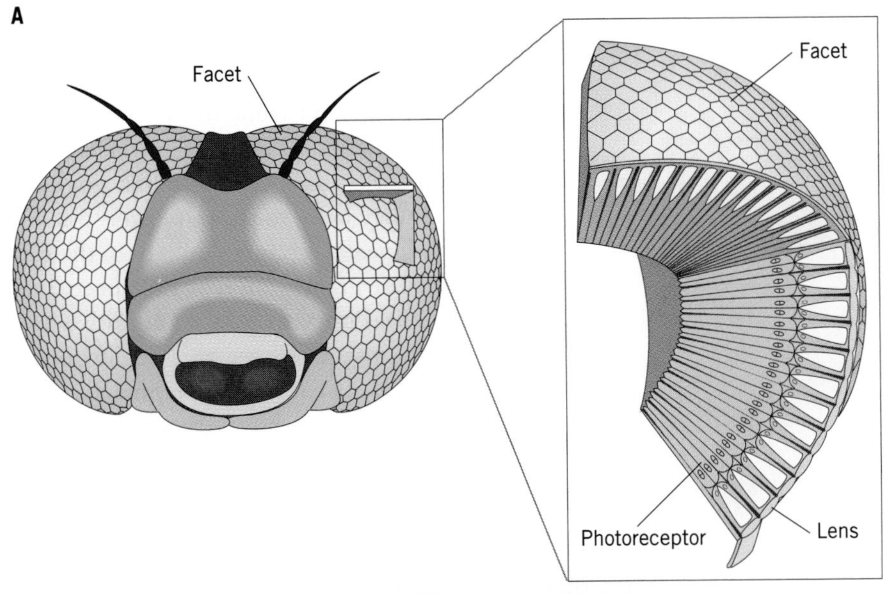

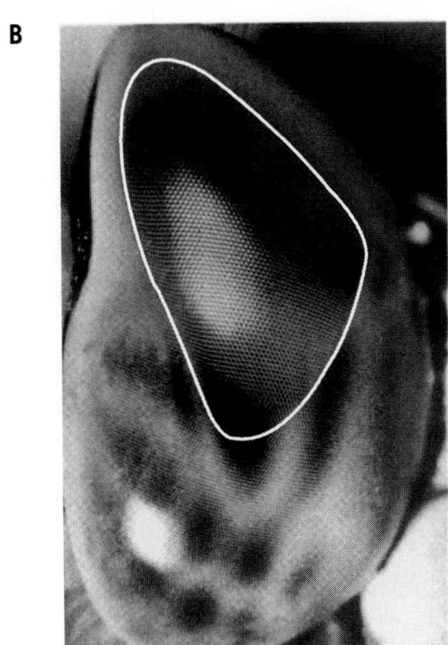

FIGURE 11-23. An insect compound eye. (A) The eyes are placed on the sides of the head, as shown in this head-on view of a dragonfly. Inside, the eye consists of groups of photoreceptive cells, each group associated with a single lens and a single exterior facet. (B) A micrograph of a compound eye of a different species of dragonfly, showing a specialized region of larger facet size that serves as an acute zone analogous to a vertebrate fovea.

are similar to those of vertebrate eyes. Activation of the retinal part of rhodopsin activates a G protein. It is thought by some researchers that this activation is followed by a change in cGMP levels in the photoreceptor, as occurs in vertebrate rods and cones. However, it is also clear that inositol 1,4,5-triphosphate (IP_3), a second messenger that will induce the release of Ca^{2+} from intracellular stores (see Chapter 6), is also activated when light strikes the photoreceptor. For this reason, it has been suggested that insect (and other invertebrate) photoreceptors contain two separate and parallel pathways for the regulation of ion channels.

Researchers agree that light on an insect retinula or on any other invertebrate photoreceptor changes the state of one or more ion

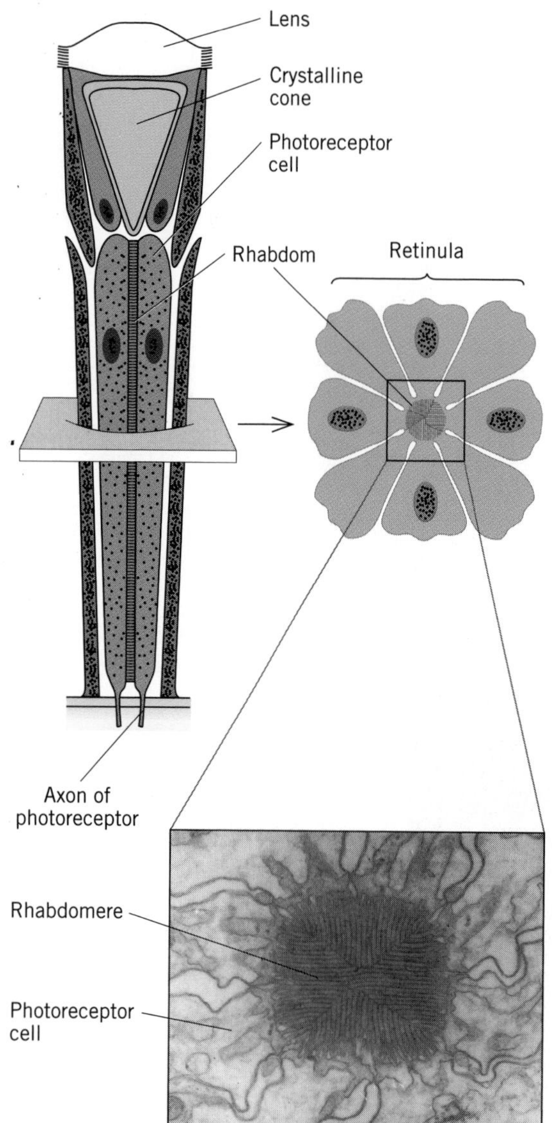

FIGURE 11-24. An ommatidium of a compound eye. The lens and the crystalline cone focus incoming light onto the rhabdom. The rhabdom is formed by rhabdomeres from individual photoreceptor cells, the cell membranes of which are outlined in the electron micrograph.

channels. A major difference between vertebrate and invertebrate photoreceptor cells, however, is that whereas vertebrate rods and cones *hyperpolarize* when they are exposed to light, insect or mollusk photoreceptors *depolarize*. Hence, whereas channels for the passage of Na^+ *close* in response to light in a rod or cone cell, they *open* in response to light in an invertebrate photoreceptor. The mechanism by which this opening occurs is not yet understood.

Most insects are able to perceive color. This ability results from the presence of different kinds of photopigments in the individual retinula cells that contribute to a single rhabdom. In honey bees, for example, a retinula cell may show peak responses to light in the ultraviolet, blue, or green range of the spectrum (Figure 11-25). Pigments that are sensitive to UV light allow insects to see this region of the spectrum, as discussed in Chapter 9.

Insects also have the ability to detect the plane of polarization of light, which allows them to orient to and navigate by the sun even on cloudy days. The ability does not reflect the presence of any specific photopigment, however. Instead, it is due to the orientation of the visual pigment within each individual rhabdomere and to the makeup of each rhabdom by rhabdomeres with different orientations.

Transduction in photoreceptors of insects and other invertebrates occurs by activation of rhodopsin or a rhodopsin-like pigment by light. This initiates a biochemical cascade, which causes opening of ion channels and depolarization of the photoreceptors. Insects can see color and can detect the plane of polarization of light.

The Processing of Visual Information

Information from the compound eyes is sent to the extensions of the brain known as the **optic lobes** and is processed mainly in four regions there. These regions are retinotopically organized and extract important features of the visual stimulus. In order inward from the eye, they are the **lamina**, the **medulla**, and the two-part **lobula complex** (the **lobula plate** and the **lobula**) (Figure 11-26). The lamina, which lies just under the layer of receptor cells of the eye, receives direct input from the photoreceptors.

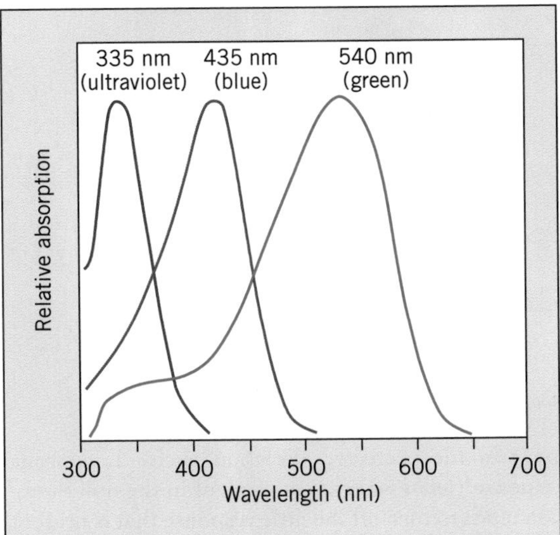

FIGURE 11-25. The spectral sensitivity of three retinal cells of a honey bee.

Axons from the lamina form the thick optic nerve; they invert the image from front to back as they project to the medulla. That is, axons from the anterior part of the lamina project to the posterior part of the medulla and vice versa. The medulla, in turn, sends information to the lobula plate and lobula, from where it projects to higher brain centers and to descending neurons that carry information to motor centers in the thoracic ganglia.

Intracellular recordings from neurons in the medulla and the lobula complex have revealed several features of cells in insect visual systems. The most thoroughly studied feature is motion detection, since the detection of motion is critical to most daytime flying insects. A large number of cells in both the medulla and the lobula complex show sensitivity to the motion of an image across the surface of the eye. Many neurons in these brain regions also show other features that are typical of neurons in the vertebrate visual system, such as restricted receptive fields, tonic or phasic responses, and on-off responses to light stimuli.

Insect eyes can extract more complex aspects of visual stimuli from their environment than was first thought. For example, specialized neurons that respond only to visual stimuli with specific features have been described from the lobula complex of several types of insects, ranging from dragonflies to house flies. These units show four features. First, like many other lobular neurons, these specialized neurons are sensitive to the motion of a visual stimulus. Second, they are motion-selective, responding strongly to motion in one particular direction, the **preferred direction**, and only weakly to motion in the opposite direction, the **null direction** (Figure 11-27). Third, specific neurons respond to stimuli of a particular size only, some to small stimuli, some to larger stimuli. Finally, some of these neurons will respond best to a specific, simple image like a

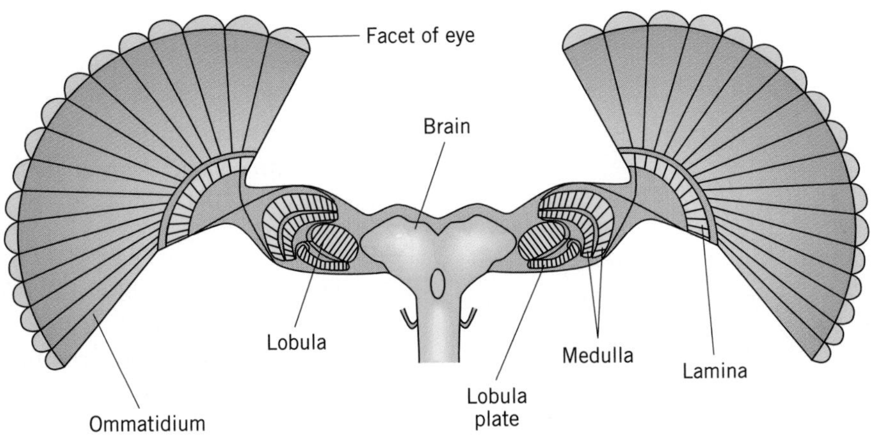

FIGURE 11-26. Schematic view of the organization of the visual system of a fly (*Calliphora erythrocephala*), showing the main centers for the processing of visual information. Neurons in the lamina, medulla, lobula plate, and lobula all receive input organized topographically relative to the surface of the compound eye.

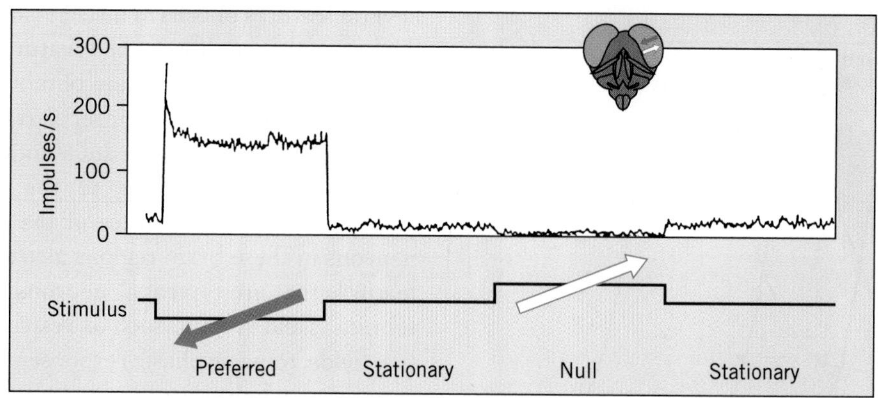

FIGURE 11-27. Selectivity to direction of motion of a neuron in the lobula plate of a house fly. The stimulus, a short bar of light, is held stationary or moved over a restricted region of the retina (inset), as shown in the lower part of the figure. Slow movement in the preferred direction elicits a vigorous response (left), whereas movement in the null direction tends to turn off the little response that is present when the bar is held stationary. Each movement lasts about 10 seconds.

bar of light, just like simple visual cortical cells in cats or monkeys. These features reveal the functional similarities between visual systems that at first glance seem to be completely different.

Visual information in insects is processed in the lamina, the medulla, and the two parts of the lobula complex. Individual neurons in the medulla show sensitivity to motion in any direction. Neurons in the lobula complex may be sensitive to directed motion, to stimuli of a particular size, or to stimuli of a particular shape.

Additional Reading

General

Buser, P., and M. Imbert. 1992. *Vision*. Cambridge, Mass.: MIT Press.

Dowling, J. E. 1987. *The Retina: An Approachable Part of the Brain*. Cambridge, Mass.: Harvard University Press.

Hubel, D. H. 1988. *Eye, Brain, and Vision*. Scientific American Library. New York: W. H. Freeman.

Zeki, S. 1993. *A Vision of the Brain*. Oxford: Blackwell.

Research Articles and Reviews

Hubel, D. H., and T. N. Wiesel. 1959. Receptive fields of single neurons in the cat's striate cortex. *Journal of Physiology (London)* 148:574–91.

Kaas, J. H. 1987. The structural basis for information processing in the primate visual system. In *Visual Neuroscience*, ed. J. D. Pettigrew, K. J. Sanderson, and W. R. Levick, 315–40. Cambridge: Cambridge University Press.

Kuffler, S. W. 1953. Discharge patterns and functional organization of the mammalian retina. *Journal of Neurophysiology* 16:37–68.

Land, M. F. 1984. Molluscs. In *Photoreception and Vision in Invertebrates*, ed. M. A. Ali, 699–725. New York: Plenum Press.

Livingston, M. S., and D. H. Hubel. 1987. Psychophysical evidence for separate channels for the perception of form, color, movement, and depth. *Journal of Neuroscience* 7:3416–68.

Martin, K. A. C. 1988. From enzymes to visual perception: A bridge too far? *Trends in Neurosciences* 11:380–87.

O'Carroll, D. 1993. Feature-detecting neurons in dragonflies. *Nature (London)* 362:541–43.

Strausfeld, N. J. 1989. Beneath the compound eye: Neuroanatomical analysis and physiological correlates in the study of insect vision. In *Facets of Vision*, ed. D. G. Stavenga and R. C. Hardie, 317–59. Berlin: Springer-Verlag.

Tootel, R. B. H., A. M. Dale, M. I. Sereno, and R. Malach. 1996. New images from human visual cortex. *Trends in Neurosciences* 19:481–89.

CHAPTER 12

Hearing

Because there are so many sounds that are important for critical behaviors like mating, prey capture, and escape, hearing can be at least as important to an animal as sight. The auditory system must transduce the mechanical energy of a sound wave into electrical activity in receptor cells; it must encode information about the sound, such as its frequency or the direction from which it comes, into neural signals, and it must interpret complex patterns of natural sounds that are meaningful to the animal. This chapter describes how auditory sense organs and the parts of the brain that process auditory information are organized to carry out these functions.

In many animals, the ability to hear is even more important than the ability to see because so much communication takes place via sound. From an owl trying to catch a mouse to a moth trying to avoid a bat, innumerable animals detect one another as prey or predator by sound. Many animals also use sound to communicate with potential mates. Hence, an ability to detect and discriminate sound waves is critical to the survival of many animals.

Sound and Its Reception

Sound is the subjective sensation produced when our ears are stimulated by sound waves, the physical vibration of air or water. The term also refers to the energy of the sound waves themselves. Solids can transmit sound energy, but we usually detect such energy only when it is transferred to air or water. It is possible, however, to hear sound transmitted directly through the bones of the head. Sound waves in air or water are produced when a solid object vibrates, producing alternating phases of compression and rarefication of the medium. The frequency of a sound, what we think of as its pitch, is determined by the number of times the object vibrates per second. How vigorously the object vibrates determines the amplitude of the sound wave, which in turn determines its energy content and loudness.

The loudness of a sound is subjective, but the energy that a sound contains, called its intensity, can be measured objectively. Because of the enormous range of sound energies humans and other animals can detect, sound intensity is expressed in decibels (dB) on a logarithmic scale relative to some standard, ordinarily one that roughly defines the faintest sound that a human can hear. Hence, a sound that is 10 dB more intense than another is 10 times as intense, one that is 20 dB more intense is 100 times as intense, and so forth. To give you some idea of sound intensity measured in this way, a whisper is about 20 dB, ordinary conversation is about 65 dB, and the sound of a nearby jet engine is more than 100 dB. Sounds over 120 dB are painful.

Any auditory system, whether simple or complex, can respond to the mechanical energy of sound waves only when the waves cause physical movement of a structure attached to a mechanoreceptor. Because these structures may be quite diverse, not all auditory systems capture and encode sound in the same way. Auditory systems may also differ in functional characteristics, such as the frequency range over which the system operates, whether the system is capable of frequency discrimination, and the complexity of sound signals that the system can interpret.

Natural sounds consist of many different frequencies generated more or less simultaneously. Furthermore, the frequency content can vary rapidly. Your voice, for example, has a fundamental frequency that is set by the frequency of vibration of your vocal cords. It also has a series of harmonics whose amplitudes are determined by your oral and nasal cavities. When you talk, both the fundamental frequency and the harmonics may change as often as every few hundred milliseconds, yet the person you are talking to can decode these rapidly changing and complex sounds with ease. Understanding how this decoding is accomplished is a serious challenge.

Sound refers to the sensation produced by sound waves, as well as to the mechanical energy of the waves. Its intensity is measured in decibels. Auditory systems respond to sound waves by physical movement of structures attached to mechanoreceptors; they may differ in structural and functional characteristics yet share many similarities. Natural sounds may consist of many frequencies.

The Vertebrate Auditory System

In the mammalian auditory system (Figure 12-1), sound energy is funneled by the pinna into the auditory canal, where it causes vibration of the tympanic membrane (the eardrum). Movement of this membrane is transferred via the three bones of the middle ear (the malleus, incus, and stapes) to the oval window of the **cochlea**, the spiral-shaped inner ear within which the transduction of sound energy into neural activity takes place. As described in the following section, movements of the membrane at the oval window cause movement of fluid inside the cochlea and subsequently stimulate **hair cells**, the primary receptors for sound in the cochlea.

When the hair cells are stimulated they excite neurons of the **spiral ganglion**, which lies inside the cochlea, and initiate the conduction of auditory information into the brain and ultimately to the auditory cortex (Figure 12-2). Axons of the spiral ganglion neurons enter the brain via the **cochlear nerve** component of the auditory nerve (cranial nerve VIII) and travel to the **dorsal** and **ventral cochlear nuclei** in the medulla. (The rest of nerve VIII contains fibers from the vestibular system of the inner ear, which is discussed in Chapter 14.) From the cochlear nuclei, auditory signals are sent

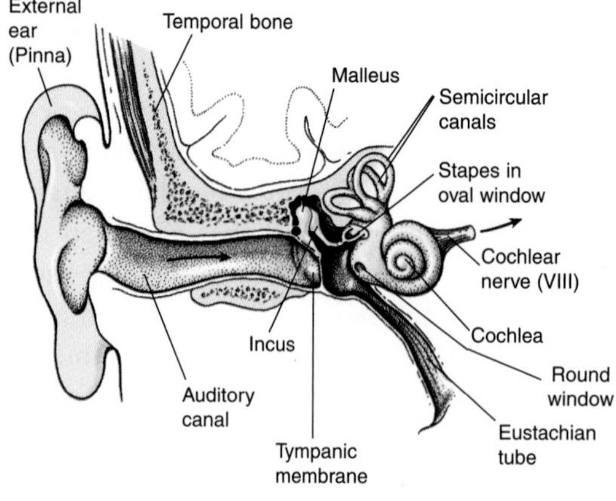

FIGURE 12-1. Structure of the human auditory apparatus. Sounds cause vibration of the tympanic membrane; this vibration is transmitted via the bones of the middle ear to the cochlea, where transduction takes place.

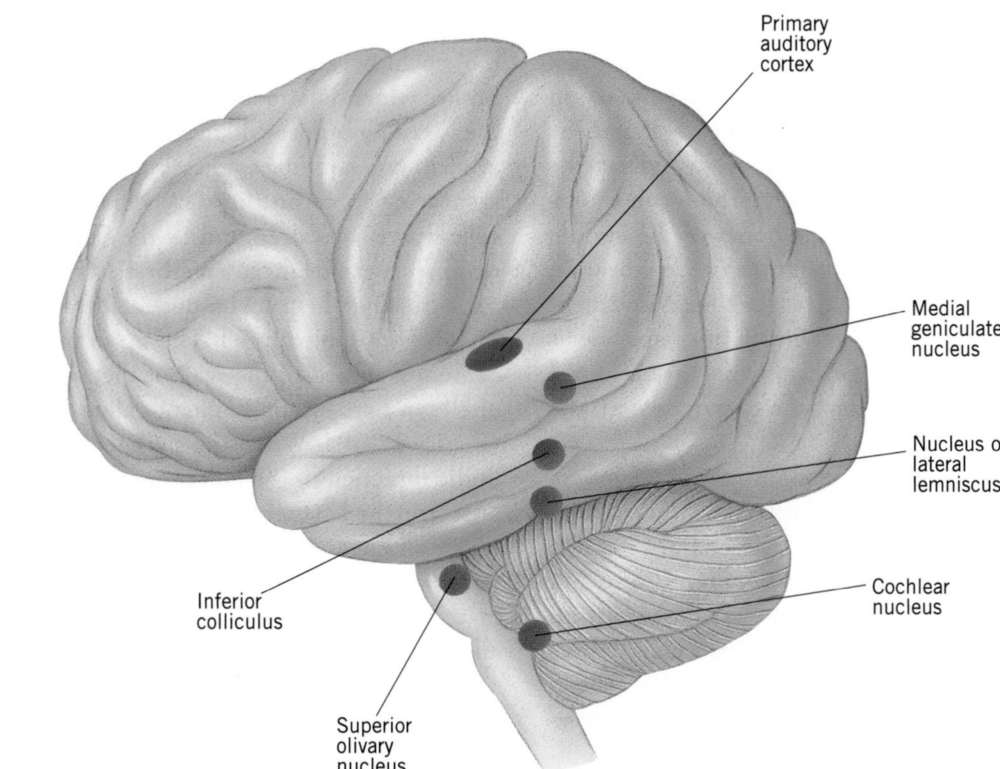

FIGURE 12-2. Diagrammatic and simplified view of the path of auditory information in the human brain. (A) Lateral view, showing the arrangement of auditory nuclei in the brain stem.

via the **superior olivary nuclei** in the medulla and the **nuclei of the lateral lemniscus** in the pons to the inferior colliculi in the midbrain, on to the medial geniculate nuclei of the thalamus, and finally to the auditory cortex.

Sound transduction in mammals is carried out by hair cells in the cochlea of the inner ear. Auditory signals travel from there into the brain, passing through nuclei in the medulla, pons, midbrain, and thalamus, and terminate in the auditory cortex.

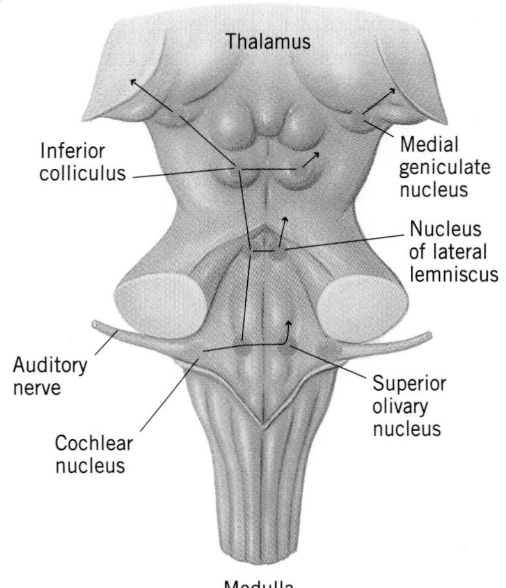

FIGURE 12-2. (B) Dorsal view of the brain stem. The cerebellum and all of the telencephalon except for part of the thalamus have been removed in this view. Auditory information crosses to the opposite side of the brain at the level of the superior olivary nucleus, the nucleus of the lateral lemniscus, and the inferior colliculus.

Transduction

Transduction of sound stimuli takes place in the cochlea, a closed, fluid-filled structure. The process begins when sound waves strike the eardrum, causing it to vibrate. This vibration is transmitted to the bones of the middle ear, whose movements in turn cause vibration of the oval window of the cochlea. Movement of the oval window generates pressure waves in the fluid in the cochlea, which are relieved by movement of the round window at the other end.

The cochlea is divided into three chambers by two membranes. The **basilar membrane** and the lower wall form the chamber called the scala tympani. **Reissner's membrane** and the upper wall form the scala vestibuli (Figure 12-3). Between these two membranes lies the third chamber, the scala media, as well as a flattened, stiff, gelatinous mass called the **tectorial membrane**. The hair cells, the auditory receptors in which transduction occurs, span the small gap between the basilar and tectorial membranes. The ensemble of hair cells, basilar membrane, and tectorial membrane is called the **organ of Corti**.

There are two types of hair cells, **inner hair cells** and **outer hair cells**, named for their location relative to the outer wall of the cochlea. Each type has at its apex small hair-like projections called **stereocilia**, which are arranged in rows of different heights (Figure 12-4). The rows on the outer hair cells are generally arranged in a sharp arc or W shape, whereas those on the inner hair cell form a shallow bow. The number of stereocilia in each hair cell varies from 20 to more than 100 in different species of animals. Both the inner hair cells and the outer hair cells are attached to the basilar membrane. The stereocilia of the outer hair cells are embedded in the tectorial membrane, whereas those of the inner hair cells reach but do not quite touch it.

Transduction of sound energy occurs in the inner hair cells of the organ of Corti. The pressure waves in the fluid of the cochlea cause the basilar membrane to vibrate up and down. Because of the mechanical relationship between the flexible basilar membrane, which is attached on both sides along the length of the cochlea, and the stiffer tectorial membrane, which is attached along one side only, the two membranes will move very slightly relative to one another as they vibrate. This puts a shearing force on the stereocilia of the hair cells by viscous drag and as they are pushed up against the tectorial membrane (Figure 12-5A), causing them to be deflected relative to the body of the hair cell. The tip of one stereocilium is linked to the wall of the next larger one by a fine, elastic filament called a tip link (Figure 12-5B). When the stereocilia are bent in the direction of the tallest ones, the tip links are stretched (Figure 12-5C), which pulls open ion channels located in the walls or tips of the stereocilia. When the stereocilia are bent in the opposite direction, the tip links relax, allowing the ion channels to close.

In the absence of stimulation, about 15% of the ion channels in each hair cell are open, generating a small depolarizing current. When the stereocilia are bent toward the tallest ones, more channels open and the hair cell depolarizes further. When they are bent in the opposite direction, the hair cell hyperpolarizes because of the closure of some of the few channels that are open at rest. The ion channels in the stereocilia are selective to small cations (positive ions), but since the extracellular fluid bathing them always has a high concentration of K^+, inflowing K^+ provide most of the current during depolarization. The hair cells are nonspiking, so the magnitude of the depolarization, and hence the amount of neurotransmitter that is released by the hair cell to the neurons of the spiral ganglion, is graded with the intensity of the sound stimulus.

Diagrams like Figure 12-5A that illustrate the vibration of the basilar and tectorial membranes and the deflection of the stereocilia exaggerate these movements enormously to make them easier to see. The actual movement of the stereocilia at the threshold of hearing is incredibly minute, amounting to only about 0.3 nm at the tip. Considering the height of the hairs, this is like moving the top of the Eiffel Tower back and forth the width of your thumb.

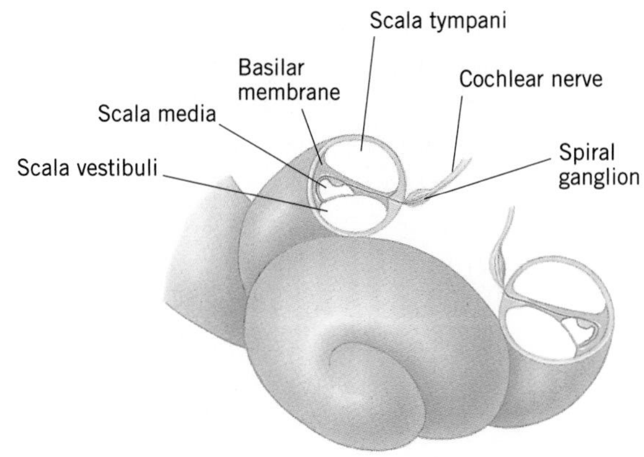

FIGURE 12-3. Inner structure of the cochlea. The basilar membrane and Reissner's membrane divide the cochlea into three chambers. The organ of Corti consists of the basilar membrane, the tectorial membrane, and the inner and outer hair cells. The neurons forming the spiral ganglion constitute the auditory nerve. Their axons terminate in the ipsilateral (same side) cochlear nucleus in the medulla.

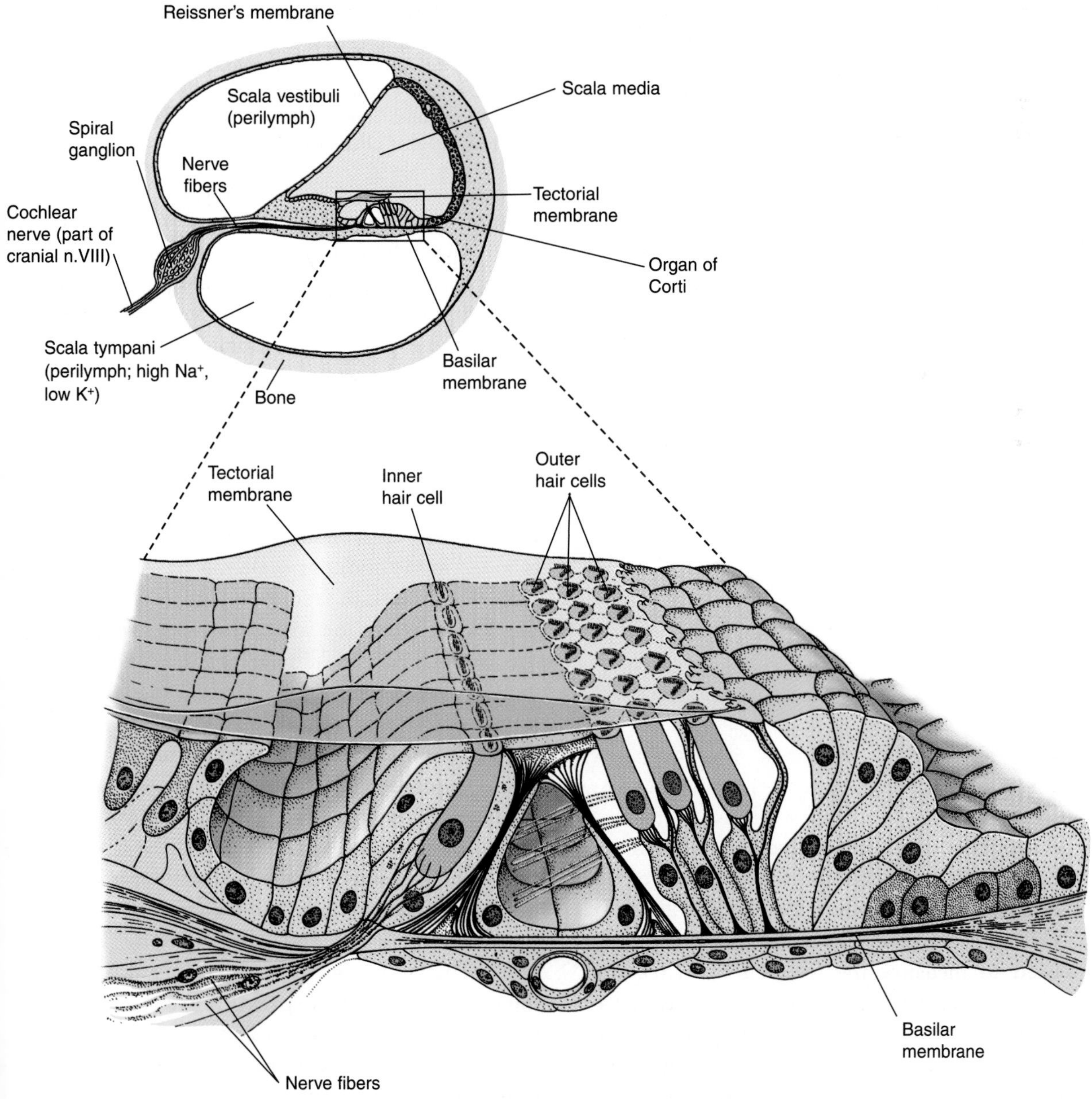

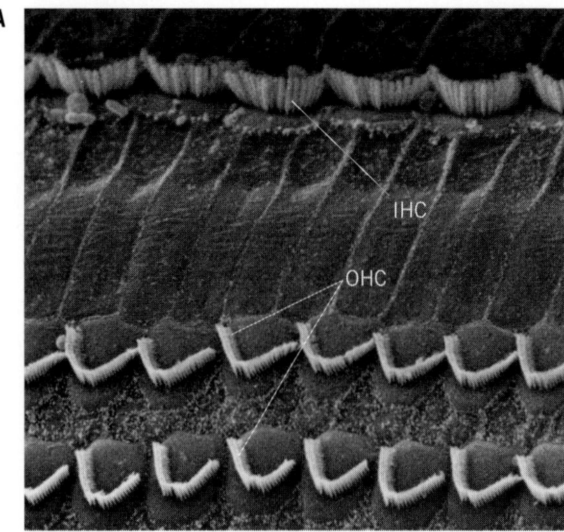

FIGURE 12-4. Scanning electron micrographs of hair cells of the inner ear of a chinchilla. (A) View of the hair cell surface of the organ of Corti. The tectorial membrane has been removed. The single row of inner hair cells (IHC) and the two rows of outer hair cells (OHC) are clearly visible. (B) The stereocilia of outer (top) and inner (bottom) hair cells. Note the different heights of the stereocilia.

Transduction of sound energy to nerve activity takes place in the inner hair cells in the cochlea. Sound energy causes vibration of the basilar membrane, which causes mechanical stimulation of the hair cells anchored there by deflecting their stereocilia. This deflection causes the opening of ion channels in the stereocilia, producing a receptor potential that is graded to the strength of the sound stimulus.

Frequency Discrimination

At its best, the human auditory system can hear and discriminate between sounds of frequencies in the range 20 Hz to 20 kHz. (Hertz [Hz], or cycles per second, is the unit used to measure the frequency of a sound wave.) Auditory systems of other animals can hear even higher frequencies, greater than 100 kHz in some cases. Not only can a wide range of frequencies be heard, but many vertebrates can distinguish between tones that differ by only a few Hertz. Humans, for example, can distinguish between pure tones of 1000 Hz and 1003 Hz. Several mechanisms, some passive and some active, allow animals to make such fine distinctions.

Frequency Discrimination and the Coding of Sound

Our current understanding of how frequency is coded in the auditory system is based on the pioneering work of Georg von Békésy during the 1950s, work for which he was awarded the Nobel Prize for Physiology or Medicine in 1961. Working with dead animals and human cadavers, von Békésy observed that all parts of the basilar membrane do not vibrate equally well to sounds of different frequencies. Movement of the oval window sets up a pressure wave in the cochlea that causes a wave of movement of the basilar membrane. Think of holding one end of a rope that is fastened at the other end, then flicking your wrist. This action sets up a traveling wave in the rope. Similarly, the pressure wave sets up a traveling wave in the basilar membrane.

However, the basilar membrane is not as uniform as a rope. Instead, it is narrow and

A

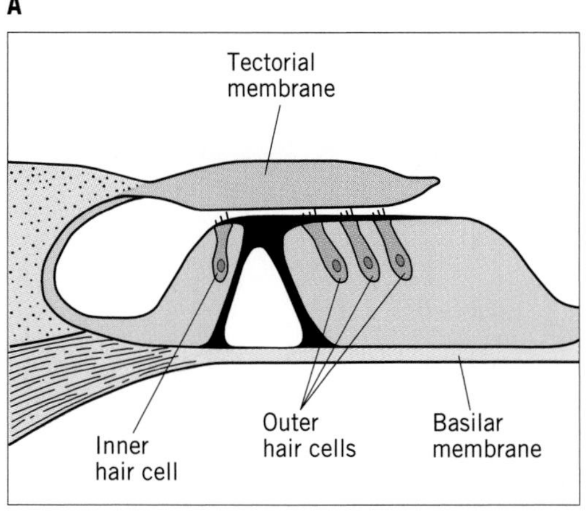

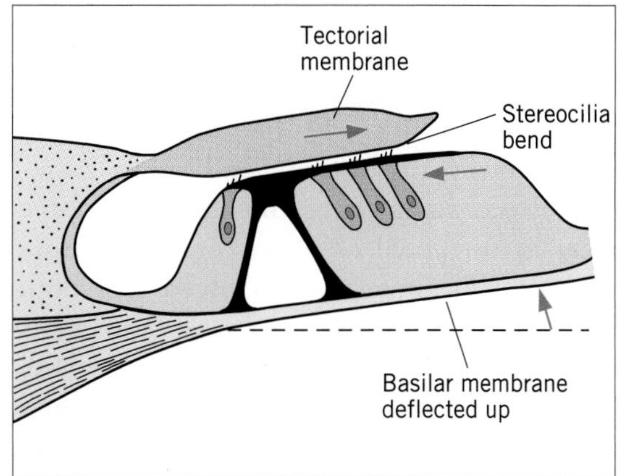

B

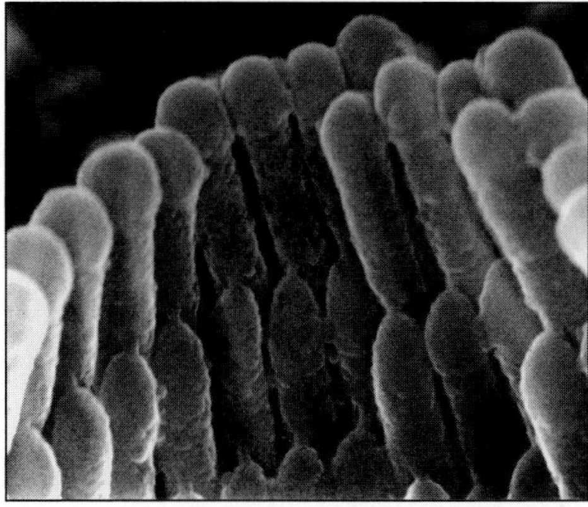

C

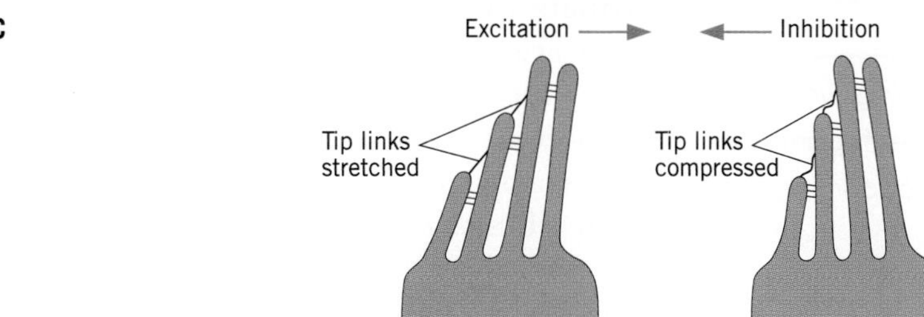

FIGURE 12-5. Stereocilia and their movement. (A) The stereocilia of the hair cells are bent by movements of the basilar and tectorial membranes due to the progress of the sound pressure wave through the channels of the cochlea. (B) Scanning electron micrograph of the stereocilia of an outer hair cell, showing the fine tip links (arrows) that connect the tip of one stereocilium to the wall of the next larger one. (C) The effect of stereocilia deflection on the tip links. When the filaments are stretched, ion channels in the wall of the longer of the stereocilia or in the tip of the shorter stereocilia to which each tip link is attached will open.

stiff at the base of the cochlea near the oval window and broad and more flexible at the apex. When it is generated by sound, the traveling wave has different effects on different parts of the membrane. Exposure to low frequencies of sound causes the broad (far) end of the basilar membrane to vibrate at greater amplitudes than other regions; at high frequencies, the narrow (near) end vibrates the most. This differential response gives the basilar membrane a rough tonotopic organization, in which each region vibrates most strongly to sound stimuli of specific frequencies. Hair cells located on the parts of the basilar membrane that are most strongly vibrating are most strongly stimulated.

Von Békésy's observations provided a basis for frequency discrimination. However, they were not sufficient to account fully for the extraordinarily fine discrimination demonstrated by humans and other mammals. Recordings from the auditory nerve fibers in the cochlear nerve, and later from the hair cells themselves, showed that both the auditory neurons and the hair cells have the property of **tuning.** That is, these cells respond only to a relatively narrow band of sound frequency, as shown in Figure 12-6. The **tuning curves** shown in the figure represent the frequency and intensity range to which each hair cell will respond. The **best excitatory frequency**, sometimes called the **characteristic frequency**, of each cell is the sound frequency to which the cell will respond at the lowest stimulus intensity. Comparison of the tuning curves of auditory neurons and hair cells with the sharpness of the traveling wave (i.e., how much of the basilar membrane moves in response to sound of a particular frequency) that von Békésy had demonstrated showed that the neural tuning was much sharper than his data would have allowed.

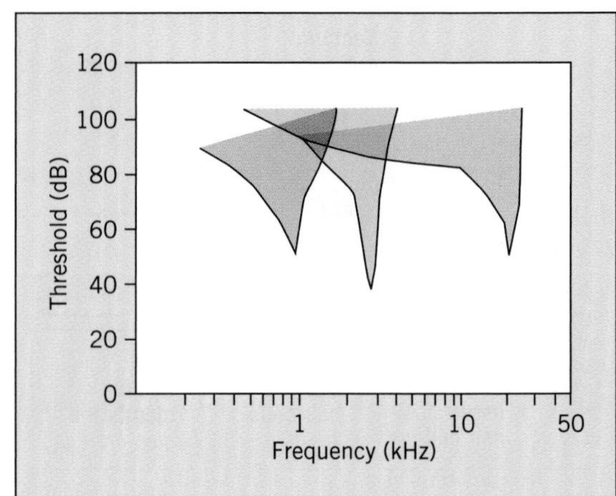

FIGURE 12-6. Tuning curves for three different auditory neurons. Each shaded area represents the range of intensity and frequency of a sound that will cause the neuron to fire. There is no clear upper limit to the curves because very intense sounds damage the ear. Note that the three tuning curves are not shaped exactly the same. The best excitatory frequency of each cell, roughly 1 kHz, 3 kHz, and 20 kHz for the cells shown here, is the frequency at which the threshold is the lowest.

> Sound stimuli produce a traveling wave in the basilar membrane, different parts of which vibrate more strongly according to the frequency of the stimulus. Both the hair cells and the auditory neurons with which they synapse are sharply tuned. This means they respond to a rather narrow band of sound frequency. The tuning is sharper than predicted from consideration only of the sharpness of the traveling wave.

Passive and Active Factors in Frequency Discrimination

It is not completely clear how sharp tuning of hair cells is achieved, but it is clear that several passive factors and active processes contribute to tuning. Passive factors are those that arise from the mechanical properties of the inner ear; two have been shown to be important. First, work with live animals has shown that the length of basilar membrane that vibrates in response to a sound stimulus of any given frequency is much narrower than von Békésy had thought. This is because the mechanical properties of the living membrane are different from those of the dead membranes with which von Békésy worked. Hence, fewer hair cells are stimulated by a given sound than was thought to be the case.

Second, study of the stereocilia of the hair cells has shown that not all have the same mechanical properties. The stereocilia of hair cells at the narrow base of the cochlea are generally shorter and stiffer than those at the broad apex. Short, stiff stereocilia resonate at high frequency, whereas long, more flexible stereocilia resonate at low frequency. To understand what this means, consider the rope example again. If you flick your wrist repeatedly, you can generate standing waves in the rope. The frequency at which you must flick your wrist in order to get the largest and most stable waves depends in part on how long and stiff the rope is. The frequency that generates the largest-amplitude standing wave for a given amount of energy is its natural resonant frequency.

With regard to the hair cells, the length and stiffness of the stereocilia also impart to these stereocilia a certain natural resonant frequency. If the frequency of a sound is at or close to the resonant frequency of the stereocilia, they will vibrate with a higher amplitude than if the sound has some other frequency. Hence, the type of stereocilia borne by a hair cell will help tune that cell to a particular frequency of sound and will limit its response to other frequencies; when the stimulus frequency is the same as the resonance frequency of the cell, the cell's response will be augmented compared with that of nearby cells that are also being stimulated but whose resonant frequency is slightly different. This effect tends to sharpen the tuning curve for the hair cells.

Passive processes alone cannot entirely account for the sharp tuning of hair cells, as is apparent when the metabolic activity of the hair cells is temporarily suppressed by application of a metabolic inhibitor. Under these conditions, the tuning of the hair cells broadens, indicating that some active process that requires metabolic energy normally contributes to tuning as well. Two active processes that help sharpen hair cell tuning have been identified, one electrical and one biomechanical.

Electrical tuning of hair cells was recognized in the early 1980s in lower vertebrates; whether it is present in mammalian ears is still unresolved. All hair cells contain voltage-sensitive calcium channels and potassium channels that are influenced by the level of free Ca^{2+} in the cell. When a hair cell is depolarized, the currents mediated by these channels interact to produce oscillations of the membrane potential. The frequency at which the membrane potential oscillates when a hair cell is depolarized, which is not the same for all cells, is called the **electrical tuning frequency** of the cell.

Hair cells in different locations along the basilar membrane differ in the amplitude and kinetics of the potassium currents that are produced when the cell is stimulated. Consequently, the frequency of the oscillations of the membrane potential varies systematically along the cochlea. As the membrane potential of a hair cell rhythmically depolarizes and hyperpolarizes under the influence of stimulation from the movements of the tectorial and basilar membranes, the amplitude of the membrane oscillations will be augmented if the stimulation frequency is the same as the electrical tuning frequency. Conversely, the farther the stimulating frequency is from the electrical tuning frequency, the more the oscillations of the membrane potential due to bending of the stereocilia will be dampened. Ion movements through the voltage-sensitive calcium and potassium channels will be out of synchrony with the oscillations produced by ion flow from the mechanical stimulation, and will therefore oppose them, making them somewhat smaller than they otherwise would be.

Biomechanical tuning of the hair cells also occurs by means of a feedback mechanism that involves just some of the hair cells. As you saw in Figure 12-3, there are two sets of hair cells in the organ of Corti, inner hair cells and outer hair cells. Transduction of sound stimuli occurs in the inner hair cells. The outer hair cells do respond to sound, but they have a different role to play. These cells physically shorten when they depolarize and lengthen when they hyperpolarize in response to a sound stimulus. It is thought that this effect, which increases or decreases tension on the tectorial membrane, may influence the transfer of sound energy to the inner hair cells. If the mechanical response of the outer hair cells is tuned in some way to the frequency of stimulation, the change in length may indirectly help sharpen the responses of the inner hair cells. The outer hair cells also receive considerable efferent innervation, suggesting that the brain can influence the process.

The tuning of inner hair cells is aided by passive factors such as the stiffness of the basilar membrane and the length and stiffness of stereocilia of hair cells at different locations in the cochlea, and by active processes such as electrical tuning and the biomechanical action of the outer hair cells.

Auditory Neurons

The approximately 15,000 hair cells in each human ear synapse with approximately 30,000 sensory neurons of each spiral ganglion. More than 90% of the synapses are between inner hair cells and spiral ganglion neurons; the rest are between outer hair cells and the sensory neurons. There is also considerable divergence and convergence of synaptic contacts. The inner hair cells diverge, each cell making synaptic contact with as many as 20 neurons in the spiral ganglion. The outer hair cells converge, with several outer hair cells forming synapses with a single spiral ganglion neuron. The axons that leave the spiral ganglia form the auditory nerve (part of cranial nerve VIII; the other part carries information from the vestibular apparatus of the inner ear, as described in Chapter 14) and carry all information about auditory stimuli to the brain. Hence, the neurons of the spiral ganglia are often referred to as auditory neurons.

Auditory neurons convey information about sound to the brain in several ways. First, auditory neurons represent a rough labeled line system, in which individual neurons convey information about sounds of specific frequencies. Because of the overlap of hair cell receptor synapses on auditory neurons, it is obvious that there is more than one auditory neuron that will respond to sound at any particular frequency. However, the number that respond will not be large at low intensities of sound because auditory neurons are narrowly tuned, just as are the hair cells that drive them. This narrow tuning is illustrated in Figure 12-6. As shown there, a neuron that has a best excitatory frequency of 3 kHz and a threshold of just under 40 dB will not respond at all to a 40-dB stimulus at 2 kHz but will respond to a sound of 2 kHz if its intensity is 80 dB.

An auditory neuron may also convey information about sound frequency through synchronization of its firing with the beat of the sound. Sound waves push repetitively against the eardrum at a rate that is equal to the frequency of the sound. Each push stimulates hair cells by causing the generation of a pressure wave that moves the basilar and tectorial membranes relative to one another. At low to medium sound frequencies, up to about 4 kHz, the responses of auditory neurons are synchronized with the frequency of the stimulating sound (Figure 12-7), thereby encoding the frequency of the sound in the frequency of the repetitive firing of the neurons. At low sound frequencies, each neuron fires a short burst of spikes (Figure 12-7A). As sound frequency increases, the number of spikes per burst decreases, until just a single spike fires in time with each wave of sound (Figure 12-7B). At still higher sound frequencies, the neurons cannot generate action potentials fast enough to keep up with the frequency of the stimulus, so they will occasionally skip a cycle or two. When they do fire, however, their spikes are properly timed to the sound wave (Figure 12-7C).

As you learned in Chapter 9, most sensory systems encode intensity (strength of stimulus) in the frequency of firing. The stronger the stimulus, the higher the frequency of firing. In the auditory system, however, the repetitive, synchronized firing of the auditory neurons means that single auditory neurons cannot convey reliable information about stimulus intensity. Instead, intensity is encoded in the number of neurons that are activated.

The response properties of auditory neurons are usually studied by using pure tones, that is, tones consisting of a single frequency. Such pure tones are rare in nature, where auditory signals are nearly always contaminated by many other sounds, and sounds may contain mixtures of several frequencies. Many animals are aided in interpreting sounds with several frequencies through the mechanism of **two-tone suppression**, the reduction of the response of a hair cell or auditory neuron to one tone by the presence of a second tone of a different frequency. Not just any tone will do; the second tone must have a frequency near the best frequency of the neuron being investigated in order for the phenomenon to manifest itself (see Box 12-1). Two-tone suppression has been found in various places in the auditory pathway, even in the cochlea itself. Study of the phenomenon in the cochlea has shown that it arises from mechanical interference between the two tones in the basilar membrane. Two-tone suppression is functionally a way for the auditory system to increase the discrimination of sounds of one frequency in the presence of other tones with similar frequencies by sharpening the tuning curves of the neurons that respond to the sound of interest.

Information about sound frequency is conveyed to the brain through labeled lines and by the timing of action potentials. Intensity is encoded in the total number of auditory neurons that are active. Two-tone suppression leads to a reduction of the response to one tone in the presence of another. It arises in part from mechanical interference between the two tones in the basilar membrane.

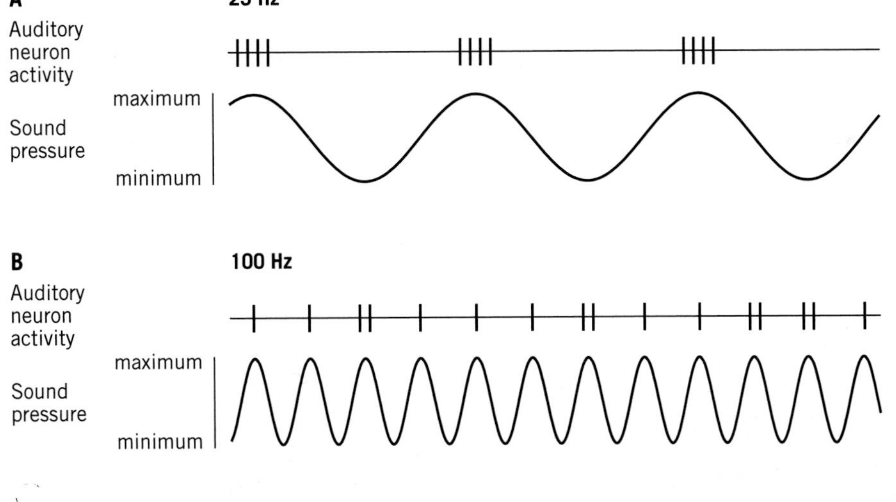

FIGURE 12-7. Synchronization of responses of a single auditory nerve fiber with a sound wave. (A) At low sound frequencies, 25 Hz in this example, the fiber fires short bursts of action potentials in synchrony with the peak sound pressure of the sound wave. (B) At 100 Hz, the fiber fires just one or two spikes per wave, still synchronized with it. (C) At 200 Hz, the fiber begins to miss peaks of the sound wave, but the spikes that are generated are still synchronized to the wave.

BOX 12-1

Exciting Sounds: Mapping an Auditory Tuning Curve

Tuning curves are important for understanding the function of auditory neurons because they represent the combined frequency and intensity of a sound stimulus that will excite each hair cell or auditory neuron. Tuning curves are mapped by sounding a tone having a particular frequency and a particular intensity, then noting whether the cell from which a recording is being made gives a response. In the case of simple tuning curves such as those shown in Figure 12-6, weak sounds evoke a response over only a very narrow range of frequencies, whereas louder sounds can evoke responses at frequencies farther away from the neuron's best excitatory frequency.

The effects of one tone on a neuron's response to another, as in two-tone suppression, can also be mapped (Figure 12-A). To map the area of two-tone suppression, for example, a researcher must first plot the tuning curve for the neuron of interest. Then, a tone in the response range of the neuron, referred to as the probe tone, is sounded. A second, louder tone at another frequency, but one that is outside the tuning curve of the neuron, is then turned on briefly. If presented by itself, this tone would have no effect on the neuron, but it might cause a reduction in the response of the neuron to the first tone. By testing many combinations of frequencies and intensities of second tones, the researcher can identify the combinations that cause a reduction in the response of the neuron to the probe tone. Plotting the border of the suppression area shows the areas of two-tone suppression of the neuron. Plots of tuning curves and two-tone suppression can be obtained for any neuron in the auditory pathway that will respond to a pure tone.

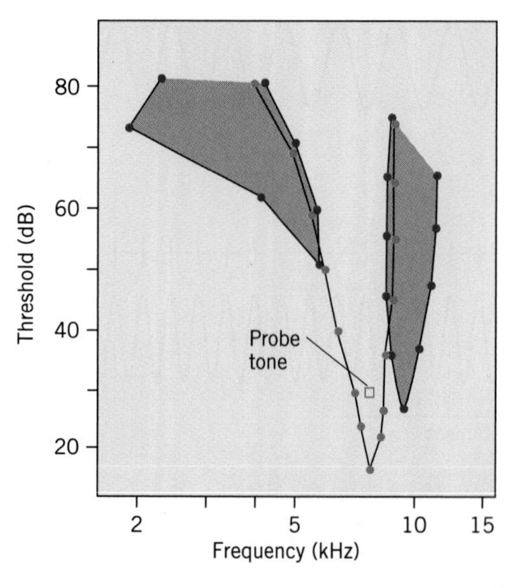

FIGURE 12-A. Two-tone suppression in an auditory neuron. The purple circles define the tuning curve of the neuron. Any tone with a combination of frequency and intensity that falls within the area of this tuning curve evokes a response in the neuron. Tones outside the tuning curve ordinarily elicit no response. When a continuous probe tone whose frequency and intensity are indicated by the square is sounded, presentation of another tone at a frequency and intensity within one of the shaded areas results in a reduction in the response of the neuron to the probe tone.

Interpreting Sound Stimuli

Pure tone stimuli can be useful in determining the fundamental response properties of the hair cells and auditory neurons of the vertebrate auditory system, but they are far removed from the complex sounds that animals actually hear. Animals employ sounds for many purposes, such as communication and detection of potential predators and prey. To accomplish such tasks, animals must be able to detect a sound, identify it, locate its source, and pick out its relevant features. You have already learned how sound is detected. In the following sections, you will learn how these other functions are carried out by the auditory system and the brain.

Identifying a Sound

If you have ever visited a forest pond on a spring night, you can appreciate some of the difficulties an animal like a frog can have in identifying a particular sound. Not only is the cacophony of frog calls overwhelmingly loud, but calls from several different species of frog will be sounding at the same time. How can an individual frog recognize the call of a potential mate?

Researchers have found two mechanisms of call identification. Some species of frogs rely primarily on the frequency components of a song to identify it. Every natural sound contains sound energy concentrated at one or more frequency bands. For songs generated by different species of frogs, the sound energy may be concentrated at different frequencies. For example, the sound energy of the mating call of a green tree frog has peaks at 900 Hz to 1100 Hz and 2.8 kHz to 3.2 kHz, as well as a small amount around 5 kHz (Figure 12-8A). On the other hand, the sound energy of the mating call of a large bullfrog peaks at about 250 Hz and 1.3 kHz to 1.7 kHz (Figure 12-8B). Tests with artificial calls of male bullfrogs that are presented to males of the same species have shown that both peaks must be present in order for the male to recognize the song. (Males are used to test song recognition because they usually give a very clear behavioral response to the call of another male.)

Frogs may use three physiological processes to help them identify the appropriate calls. First, the structures that in frogs correspond to the mammalian cochlea are tuned to the sound frequencies that the animal must be able to hear. Hence the ear of the tree frog will respond well to sounds in the 2- to 3-kHz range, whereas bullfrogs cannot hear such high sounds very well. Second, in some

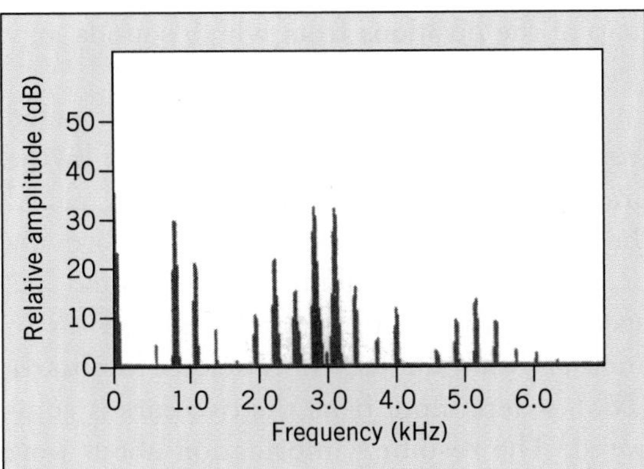

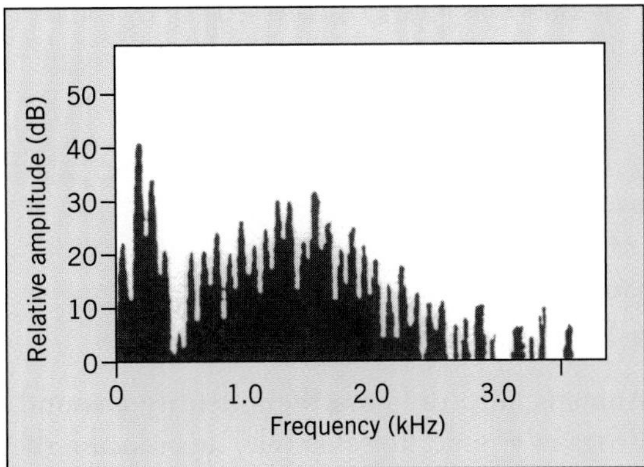

FIGURE 12-8. Amplitude spectra of mating calls of two different species of frogs. (A) Green tree frog. (B) Bullfrog. Each graph shows the sound energy of the call at various frequencies. Calls that have different distributions of sound energy sound different to our ears, as well as to the ears of the frogs. The call of the green tree frog will sound higher-pitched than the sound of the bullfrog, which has few components above 2.5 kHz.

species a call must have a particular temporal structure to be recognized. Many frog songs consist of sequences of short "pulses" of sound, much like the ratcheting sound made by a piece of cardboard stuck between the spokes of a child's bicycle as the wheel turns. Calls of various species differ in their temporal structure, that is, the pulse repetition rate and the duration of each pulse in the sequence. The auditory nucleus in the thalamus of ranid frogs contains neurons that respond only to sound pulses that contain the proper temporal pattern.

The third physiological aid to call identification is the presence of neurons that respond only to calls that contain one or a combination of particular frequencies of sound. Such neurons can also be found in the thalamus. Some of these neurons will not respond to single tones; they require two or three tones at the right frequencies to sound together before they will respond. Hence, in ranid frogs, there is parallel processing of sound stimuli. One pathway decodes the frequency components of the stimulus, and the other decodes its temporal components.

Frogs can distinguish the calls of members of their own species even if these calls are intermingled with the calls of many other species. They do so by being able to hear only certain frequencies, by recognizing the temporal structure of the pulses that make up the call, by recognizing the presence of specific frequencies together in the call, or by some combination of these means.

Locating a Sound

Animals must do more than identify a sound. Frogs in a pond, for example, must locate the call to which they wish to respond. Animals use two mechanisms to locate sound: differences in the sound's time of arrival at the two ears, and differences in the sound's intensity. The ear farthest from the source of the sound will receive the sound wave later than will the closer ear because the sound has to travel farther to reach it. Similarly, the sound will be weaker when it reaches the farther ear because sound loses energy as it travels through air or water. Frogs have good localization abilities, being able to place a sound to within 8° to 12° of its source, but other animals can localize sounds even better. Barn owls probably have the most acute ability for sound localization of any animal. These night-flying birds, which prey on mice and other small rodents, are able to hunt in total darkness. They can locate the rustling sound of a mouse to within just 1° to 2°, both horizontally and vertically. The special adaptations that allow them to pinpoint sound sources so accurately have been studied by Mark Konishi, Eric Knudsen, and their collaborators.

The owl's ability to localize sound is based on neural features of the medulla and the midbrain, especially the auditory centers in the **mesencephalicus lateralis dorsalis (MLD)**, the bird homologue of the mammalian inferior colliculus. Investigation of neurons in the MLD has revealed an auditory "space" map of the bird's surroundings (Figure 12-9). That is, sounds that originate from a particular point in three-dimensional space relative to the bird's head cause neural activity in a particular region of the MLD. Sounds that originate from a neighboring point cause activity in an adjacent region of the MLD. Hence the brain of the owl contains an accurate neural map of the positions from which sounds may originate.

The space map in the MLD is constructed from input received from two nuclei in the brain stem. Neurons in one nucleus, called the **nucleus magnocellularis (NM)**, provide precise information about when a sound has occurred. This information is passed on to another brain center, the **nucleus laminaris (NL)**, where input from the two ears is compared. The resulting information about time differences is sent to the MLD. Differences in the time at which a sound strikes the owl's ears represent the horizontal location of the sound, called its azimuth. If the sound is to

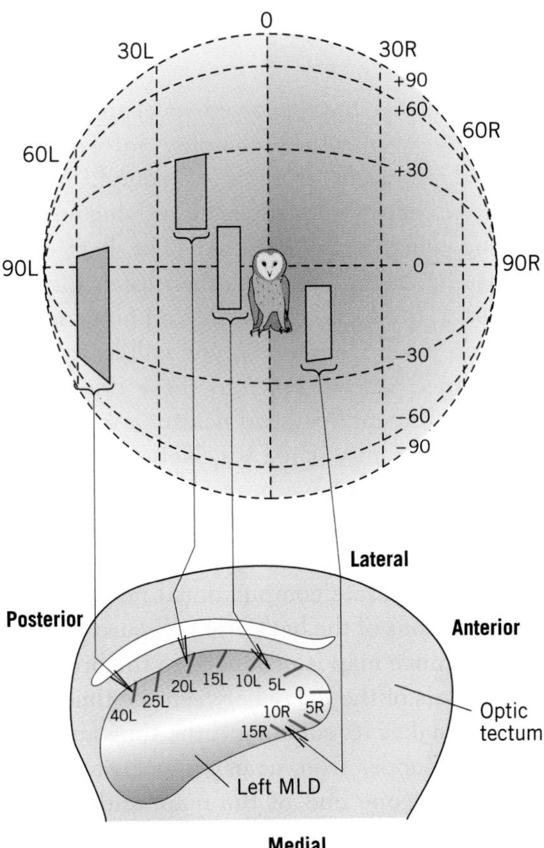

FIGURE 12-9. A space map for sound in a barn owl. The sphere represents the auditory space around the owl in the center. Recordings from cells in different locations in the MLD of the owl show that they respond to sounds located in different places in front of the owl. The rectangles represent the locations of sounds that evoke responses in cells at the sites of the electrodes in the MLD cortex. Sounds to the left (L) or right (R) evoke sounds in different positions along the MLD, as shown by the lines drawn from the rectangles to the pictures of recording electrodes. The numbers and letters shown in the MLD indicate the lateral angle and direction of the sites mapped at that location. The designation 10L, for example, indicates the spots that are 10° to the left of center. The entire hemisphere in front of the owl is mapped onto the MLD.

the owl's left, the stimulus will reach the left ear before it reaches the right ear, and the resulting time difference, which may be less than a millisecond, will be interpreted to mean "sound to the left." If the sound is directly in front of the owl, there will be no time difference, which will be interpreted to mean "sound straight ahead." If the sound is to the right, sound will arrive at the right ear first, and the time difference will be interpreted to mean "sound to the right."

The owl can locate sound in the vertical plane, called elevation, as well. To do this, it uses an anatomical anomaly: one of its ears is higher on the face than the other. Hence, sounds coming from above or below the owl's head will be louder in one ear than the other. Input from the auditory nerve, in addition to traveling to the NM, also travels to another brain stem nucleus, the **nucleus angularis (NA)**. There the intensity of input from the two ears is encoded and sent to a brain region where the signals are compared. The time and intensity signals, representing the azimuth and elevation of the sound source, are brought together in the MLD, where they converge on neurons to generate the space map. This separate handling of time and intensity information is another example of parallel processing.

Topographical maps, as you have already learned in Chapter 9, are present in many sensory systems. Maps of visual space, skin surface, and sound frequency are all well known. In these familiar cases, the map is formed by projection of a sensory surface (the retina, the skin, or the cochlea in the examples) onto a specific brain region. The owl's space map is different, however, being what is called a **computational map**. This kind of map is topographical, since it represents a systematic representation of a sensory parameter on the surface of the brain. However, rather than being a map of a sensory surface like the retina or cochlea, it is formed from the precise and systematic convergence of two inputs. These inputs represent separate parameters and are projected onto an array of neurons that forms the map (Figure 12-10). Hence, the new parameter is "computed," as it were, from the convergence of the two inputs. As you will learn in the next section, other types of computational maps exist as well, suggesting that a maplike organization is useful for the brain to process spatial information.

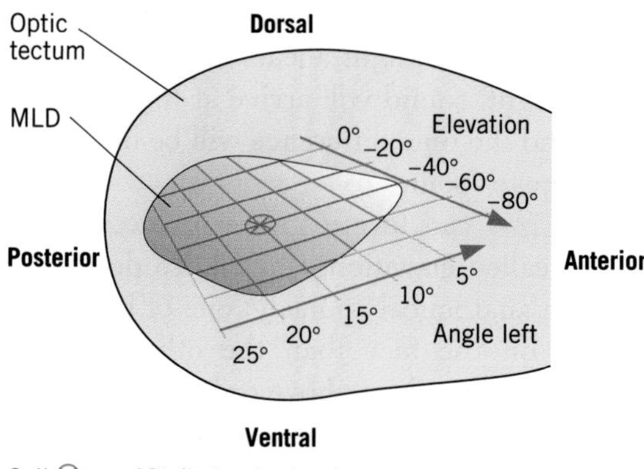

FIGURE 12-10. Mechanism of formation of a computational map of auditory space in a barn owl. The diagram represents a section through the MLD, showing how part of the space map is formed. This is a vertical section; the section shown in Figure 12-9 is a horizontal section through the MLD. The elevation of a sound is mapped on a slant from dorsal to ventral in this section. Neurons that respond to sounds below the owl are located ventrally (at the bottom region) in this part of the MLD; those that respond to sounds directly in front of it are located dorsally (at the top). Left-right location is mapped from posterior to anterior. Neurons that respond to sounds to the left in the diagram are located posteriorly (left in the illustration) in the MLD; those that respond to sounds directly in front of the owl are located anteriorly (on the right). Since in the MLD an individual neuron will respond only if elevation and right-left orientation coincide for that neuron, the owl has effectively computed the location of any sound, as shown for the neuron marked **x**. The rest of the owl's auditory space is mapped in parts of the MLD not shown in this particular section.

Barn owls are able to localize sound in space. To do this, they separately process information about differences in the timing and intensity of the sound as it reaches the two ears. The information is united in the MLD to form a map of auditory space; this map allows the bird to pinpoint the azimuth and elevation of the sound source.

Analyzing a Sound

Some animals have evolved mechanisms for handling even more complex information derived from an analysis of sound. Many species of bats, for example, hunt flying insects on the wing, using biosonar. The bats emit ultrasonic cries as they fly, then listen for the echoes of these cries to locate and home in on their prey. (In Chapter 19, you will learn how the hunted insects react to these cries.) By studying the auditory and nearby areas of the cortex in the mustached bat, Nobuo Suga and his collaborators have been able to show that information about distance, direction, and velocity of the prey relative to the bat is represented in separate computational maps in different regions of the bat's brain (Figure 12-11).

A distance map is produced in the brain by an analysis of the delay between the time that the bat makes its sound and the time an echo returns. Echoes from nearby objects are represented at one end of the map, and echoes from distant objects are represented at the other end. A map of direction is produced from an analysis of differences in the times at which an echo strikes the two ears. Directions to the right are represented at one end of the map, distances to the left at the other end.

A velocity map is produced by an analysis of the Doppler shift in the frequency of the cry's echo. A Doppler shift is the change in frequency of a sound as the source of the sound approaches or recedes from the listener. For example, the pitch of a train whistle sounds higher when the train is coming toward you than when it is moving away. In the case of a bat, the frequency of an echo reflected back from an insect flying toward the bat will be higher than that of an echo reflected from a stationary object, whereas one reflected from an insect flying away from the bat will be lower (Figure 12-12). In the brain, neurons at one end of the map are excited when an object is approaching the bat rapidly, and neurons at the other end of the map are excited when an object is receding rapidly. As in the barn owl, both parallel and serial processing are used to generate these computational maps.

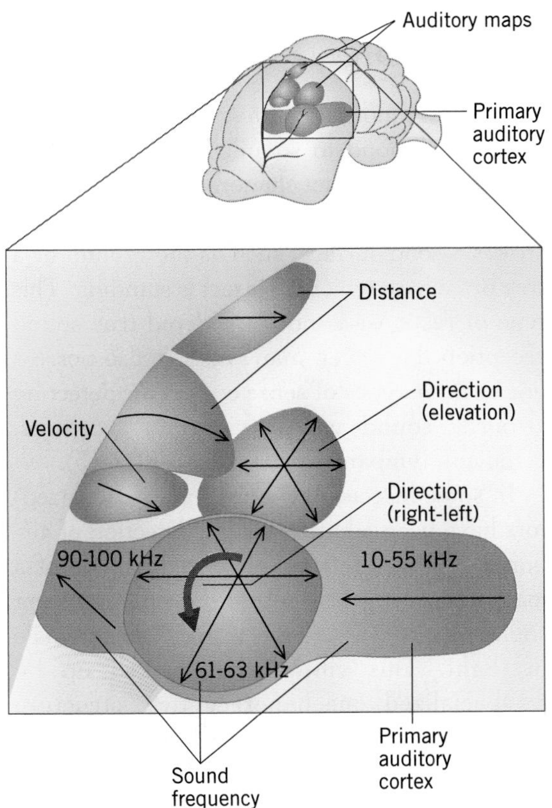

FIGURE 12-11. Simplified schematic of maps for sound frequency and other aspects of sound analysis in the brain of the mustached bat. Several areas of the brain are used to process sound echoes, as shown. The detail shows the approximate regions devoted to the main types of information that the bat can extract from time delays, frequency shifts, and intensity differences of the returning echoes. Most of the regions are linearly arranged, but some have a circular or polar orientation. Sound frequency as such is discriminated in the primary auditory cortex. Note the overrepresentation of space dedicated to frequencies of sounds that the bat itself makes.

Obviously, the ability of the mustached bat and other insect-hunting bats to extract complex information from sound echoes is specific to this group of mammals. However, one aspect of Suga's findings in this animal might have broader applicability. None of the complex processing described here takes place in the primary auditory cortex, the region of cortex that shows a tonotopic representation of sound frequency. All the computational maps of sound features that are biologically important to the bat are formed in surrounding auditory areas. This organization, in which separate areas of the brain are set aside for the processing of sounds especially relevant for such behaviors as

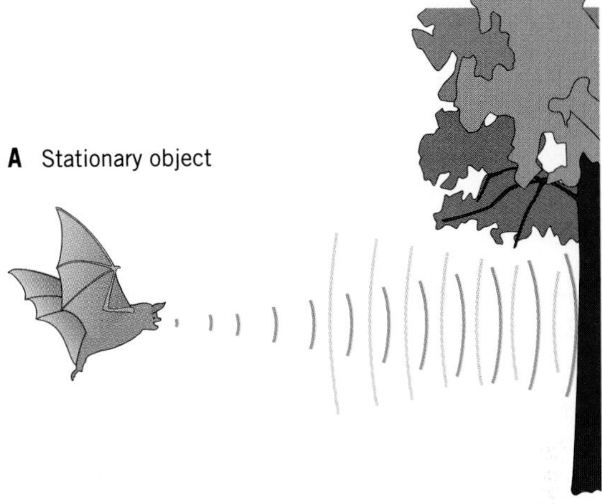

FIGURE 12-12. The Doppler shift in sound frequency. In these examples, the shift in frequency produced by the bat's own motion is ignored for the sake of simplicity. (A) A flying bat detects a stationary object. The sound waves of the bat's cry will reflect off the surface with the same spacing as they had when they arrived, so there is no shift in the sound frequency of the cry and its echo. (B) The bat detects an insect flying away from itself. Here the sound waves of the echo will have slightly wider spacing than the sound waves of the cry itself because the insect moves slightly farther away from the bat between waves. (C) The bat detects an insect flying toward itself. Here the sound waves of the echo will be slightly compressed compared with the sound waves of the cry itself because the insect moves slightly toward the bat between waves.

prey capture or communication, seems to have parallels in the way the sounds of speech are processed in humans. As discussed in more detail in Chapter 21, speech is processed not in the tonotopically organized primary auditory cortex but in other auditory areas. Recognition of this division of labor in the processing of sound between biologically relevant sound and everything else may speed neuroscientists' understanding of how complex sounds are analyzed in many other mammals.

Some animals can analyze sound in sophisticated ways to obtain information about their environment. For example, insectivorous bats analyze the echoes of the ultrasonic cries that they make for information about the insect prey they hunt. Information about distance, direction, and velocity of the insect relative to the bat is represented in computational maps in different regions of the bat's brain. These regions are separate from the primary auditory cortex.

Hearing in Insects

Most invertebrates can detect the vibrations that are produced by low-frequency sound in solids or fluids, but only insects have true ears that can detect high-frequency sound. In many insects, the sense of hearing is essential for survival. For example, noctuids, a family of night-flying moths, depend on rather simple ears to detect the approach of their bat predators. Crickets and some of their relatives use more sophisticated ears in an elaborate system of communication between individuals that allows males and females to find one another to mate. Hence, ears and communication by sound can be essential for survival and successful mating.

Insect Auditory Organs

Insects have more than one type of sense organ that can respond to sound. Nearly all insects have generalized mechanoreceptors that can respond to the vibration produced when sound strikes a solid surface, such as the ground or a tree branch on which an insect is standing. This type of response is not considered true sound reception. However, many insects also possess one of three types of sense organs for detecting airborne sound waves: vibration receptors, cerci, and tympanic organs (Figure 12-13).

In some insects, specialized vibration receptors have been adapted for the detection of airborne sound. For example, the antennae of a male mosquito will vibrate to the buzzing sound produced by the wings of a female in flight. This vibration is picked up by a specialized mechanoreceptive structure (Johnston's organ) at the base of each antenna, allowing the male to hear the female. A more bizarre example of a vibration receptor used to detect sound is found among members of a family of night-flying hawk moths. In these insects, a mouthpart has become a detector for ultrasonic sound, presumably as a defense against predation by bats (see Chapter 19).

Cerci are hair-covered, peglike structures located at the end of the abdomen in insects like crickets and cockroaches. In some of these insects, the hairs are so extremely fine that even the displacement of air molecules by sound is enough to move them, causing activity in the mechanoreceptor at the base of each hair. Functionally, the ability of these organs to detect sound seems to be merely a by-product of their extreme sensitivity to air disturbances, a sensitivity that is necessary for them to escape from predators (see Chapter 19). There is no evidence of any ability to discriminate frequency.

Tympanic organs, the true ears of insects, always consist of a membrane stretched over an air sac. The sensory neurons are attached directly to the membrane in such a way that they are repetitively excited when the membrane vibrates. In some insects, all the auditory neurons innervating the tympanic organ respond equally well to sounds over a wide

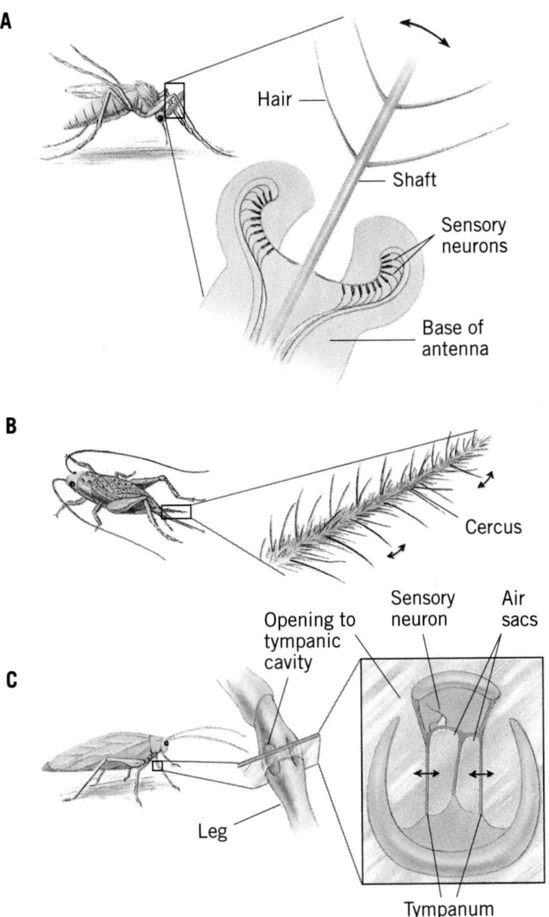

FIGURE 12-13. Insect sound receptors. The double-headed arrows show the movement produced by sound. (A) Johnston's organ, at the base of the antenna of a mosquito. Vibration of the antennal shaft causes excitation of the sensory neurons in the base. (B) A cercus of a house cricket, with its fine filiform hairs. The hairs are sensitive enough to vibrate to sounds of certain frequencies, which excites mechanoreceptor cells at the base of each hair. (C) A true insect ear from the front leg of a bush cricket. Sound causes vibration of the membrane (the tympanum) stretched over an air sac and excites sensory neurons attached at one end.

range of frequencies, leaving the insect literally and completely tone-deaf. In other insects, the auditory neurons are tuned to specific sound frequencies, allowing the insect to discriminate sounds of different frequencies. The most common location for tympanic organs is at the junction between the thorax and the abdomen, but in crickets and some species of grasshoppers they are on the front legs. Tympanic organs are present in insects such as grasshoppers, crickets, moths, and others, but they are not ubiquitous; many insects lack them.

Three different kinds of sound receptor organs are found among insects: specialized vibration receptors, cerci, and tympanic organs, the true ears of insects. Tympanic organs are common in some groups of insects but absent in others. They may be located on the body or on the front legs.

Auditory Processing in Insects

In spite of the many differences between the ears of insects and those of mammals, auditory systems in the two groups of animals show

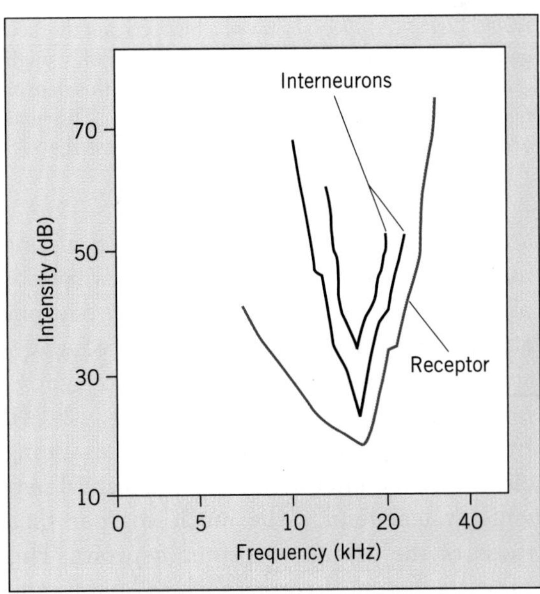

FIGURE 12-14. Tuning curves for sensory neurons and interneurons in the auditory system of a cricket. Note that the interneurons are much more sharply tuned than are the sensory neurons.

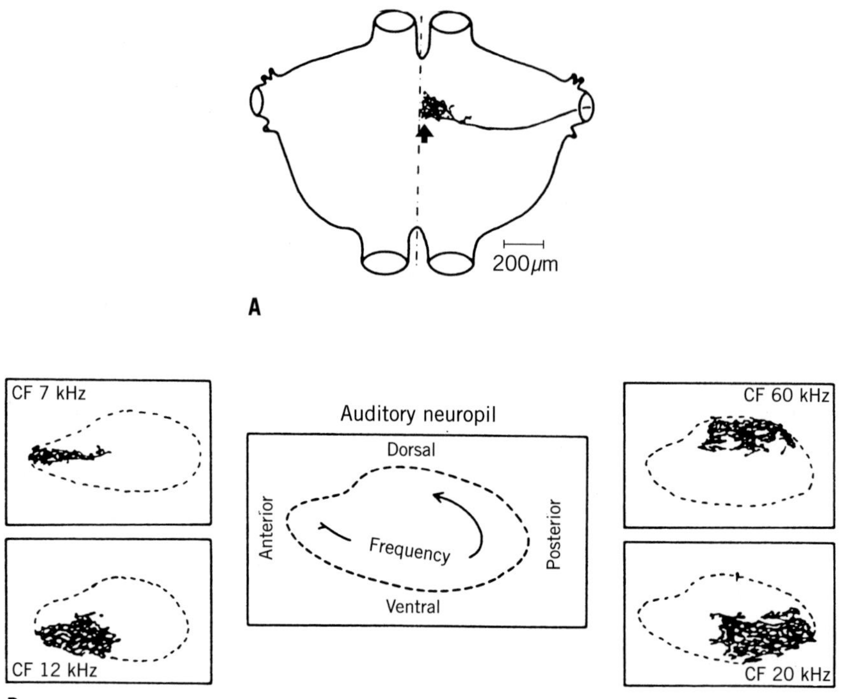

FIGURE 12-15. (A) Dorsal view of a bush cricket thoracic ganglion, showing the highly branched end of a single afferent auditory neuron. The region where this neuron terminates is known as the auditory neuropil. The small arrowhead near the midline indicates where the sections shown in (B) were taken. (B) The tonotopic organization of the auditory neuropil, showing its circular arrangement. The blackened areas are the sites of axonal endings of auditory afferents whose best excitatory frequencies (BEF) are indicated in the diagrams.

a number of similarities. For example, among insects that can discriminate between sounds of different frequencies, both sensory neurons and auditory interneurons show frequency-selective tuning similar to that of vertebrate auditory neurons (Figure 12-14). Furthermore, as in vertebrates, the tuning curves for the interneurons in the central nervous system tend to be much sharper than those of the primary receptor neurons. This suggests that mechanisms such as lateral inhibition are at work to help sharpen discrimination between sounds of different frequencies, just as there are mechanisms for sharpening tuning in vertebrate systems. Mechanisms similar to two-tone suppression have also been described. For example, the response of one neuron may be completely suppressed in the presence of a louder tone at another frequency. This suppression allows an insect to distinguish between songs that have different frequency components, and that would normally generate different behavioral responses.

There are also similarities in the organization of regions of the central nervous system devoted to analyzing auditory information. Tonotopic organization of auditory nuclei or auditory cortex is a consistent feature of vertebrate systems. Similarly, insects also show tonotopic organization in the auditory areas of central ganglia. For example, in the bush cricket *Mygalopsis marki,* the ganglion that receives input from the ears contains a region called the auditory neuropil, where the primary auditory neurons terminate. The auditory neurons, which are tuned to specific sound frequencies, synapse in specific regions of the neuropil in the order of their best excitatory frequency, counterclockwise around the auditory neuropil, producing a circular tonotopic arrangement (Figure 12-15).

Tonotopic organization of auditory input may help a male insect determine its distance from another singing male. The calls of these and other bush crickets typically consist of a wide spectrum of sound energy, from 10 kHz up to over 40 kHz. Low-frequency sounds propagate farther than do high-frequency sounds. If one insect is close to another, most of the auditory neuropil will be stimulated by input from the ears, since all the frequency components of the song are audible. However, if the two insects are some distance apart, only the low-frequency components of the song will be audible, and only the anterior and ventral parts of the auditory neuropil will be stimulated. Hence, in theory, a male can estimate the distance of a rival by gauging the area of the auditory neuropil that is excited by the rival's song.

It is also important for insects that communicate acoustically to be able to determine the direction from which a sound originates. Specific neurons have been found in the central ganglia that help mediate direction finding. In crickets, for example, neurons have been found that are extremely sensitive to differences in sound pressure (intensity) at the two ears. A single sound will produce different sound pressures at the two ears if it originates anywhere but in a plane through the center of the insect (i.e., if it is off to the right or left of the animal). The greater the difference in sound pressure, the greater the activity of these neurons. The neurons exist in symmetrical pairs, so that one of the pair is active if the louder sound is on the right side, and the other is active if the louder sound is on the left. Similar neurons have been reported in other insects that communicate by sound.

Insect auditory systems show many similarities to those of vertebrates. The tuning of receptor neurons and interneurons, the recognition of sounds produced by members of the same species, tonotopic organization, distance estimation, and sound localization are all features of many insect auditory systems.

Additional Reading

General

Bailey, W. J. 1991. *Acoustic Behaviour of Insects: An Evolutionary Perspective*. London: Chapman and Hall.

Feng, A. S., and J. C. Hall. 1991. Mechanoreception and phonoreception. In *Neural and Integrative Animal Physiology*, ed. C. L. Prosser, 247–316. New York: Wiley-Liss.

Jahn, A. F., and J. Santos-Sacchi, eds. 1988. *Physiology of the Ear*. New York: Raven Press.

Pickles, J. O. 1988. *An Introduction to the Physiology of Hearing*. 2d ed. New York: Academic Press.

Research Articles and Reviews

Fettiplace, R. 1987. Electrical tuning of hair cells in the inner ear. *Trends in Neurosciences* 10:421–25.

Hudspeth, A. J. 1989. How the ear's works work. *Nature* 341:397–404.

Konishi, M. 1993. Listening with two ears. *Scientific American* 268(4):66–73.

Schildberger, K., F. Huber, and D. W. Wohlers. 1989. Central auditory pathway: Neuronal correlates of phonotactic behavior. In *Cricket Behavior and Neurobiology*, ed. F. Huber, T. E. Moore, and E. Loher, 423–58. Ithaca, N.Y.: Cornell University Press.

Suga, N. 1990. Biosonar and neural computation in bats. *Scientific American* 262(6):60–68.

Yates, G. K., X. M. Johnstone, R. B. Patuzzi, and D. Robertson. 1992. Mechanical preprocessing in the mammalian cochlea. *Trends in Neurosciences* 15:57–61.

CHAPTER 13

The Chemical Senses

Subjectively, taste and smell are quite different from one another and could be treated separately. However, both are based on the detection of chemical stimuli by chemoreceptors. In addition, the mechanisms employed by the nervous system to identify chemosensory stimuli have many similarities not only among different animals but also for different chemical stimuli. In this chapter you will learn how chemical stimuli are processed and about the similarities among and differences between chemical sensory systems.

Scientists' understanding of **olfaction**, the sense of smell, and **gustation**, the sense of taste, has lagged behind their understanding of sight and hearing. This is in part because smell and taste play a more subtle role in human behavior than do sight and hearing. It is also due to the experimental difficulty of delivering precisely timed and quantitatively accurate amounts of chemical stimuli to the receptors. However, researchers now recognize that in many animals, the ability to locate and identify a variety of chemical substances can be far more important for survival than the ability to see or hear. Along with scientists' increased appreciation of the importance of chemoreception has come a better ability to deliver controlled chemical stimuli to the receptors. The payoff has been results that are increasing neuroscientists' understanding of olfaction and gustation across the animal kingdom.

Do Animals Have More Than One Chemosensory System?

In land animals, it seems clear enough that olfaction and gustation are handled by two rather different systems. Olfactory receptors are extremely sensitive and detect volatile chemical stimuli that may come to the animal from some distance away. Gustatory receptors, on the other hand, are often up to six orders of magnitude less sensitive and detect mainly dissolved chemical stimuli that are in direct contact with an animal's oral cavity.

In aquatic animals, however, the distinction is less clear. Gustatory receptors in fish, for example, can equal olfactory receptors in sensitivity and may also act as long-distance receptors. Nevertheless, most sensory physiologists who deal with chemoreception make a distinction between olfaction and gustation, for the most part because the two senses are used differently. Gustation is used mainly to detect and identify food-related chemicals in the near environment of the animal. Olfaction, on the other hand, is used not just for the identification of things to eat or not to eat but for the detection of enemies or mates, or cues for migration and homing. Furthermore, distinctly different regions of the nervous system and brain are used to process olfactory signals than are used for gustatory ones.

However, the chemosensory systems that we normally think of as mediating taste and smell are not the only chemosensory systems that animals possess. For example, some animals have two gustatory systems, the familiar one consisting of chemoreceptors in and around the mouth, and an external one distributed on the body surface. Catfish have dense arrays of chemoreceptors on the entire surface of the skin that help the fish to detect the presence of foodstuffs in the water. Neuroanatomical studies suggest that these body chemoreceptors form a distinct sensory system, since taste information from the mouth and that from the body are processed in different areas of the brain. Many insects also have external taste receptors. Some insects have chemoreceptors on their feet (see Figure 10-2), where they help to detect the presence of food; some female insects carry them on the ovipositor, where they help the female to identify a suitable site to lay her eggs. It is not known where input from these receptors is processed, but it does not appear to be the same brain locations where input from the antennae and mouth is handled.

Other chemosensory capabilities are even more remote from what we typically think of as taste or smell. Most animals, including humans, possess what is called a common chemical sense. In vertebrates, receptors for this sense are free nerve endings that respond to a variety of noxious chemicals. These nerve endings are concentrated in areas that are normally moist, like oral and nasal cavities. If you have ever had soap in your eyes, you have experienced the powerful response that a noxious chemical can evoke in these receptors. Receptors for the common chemical sense are also present all over the body surface, as the sensitivity of the skin in a broken blister will attest. Nearly any fluid will cause a strong, painful reaction due to stimulation of the exposed neurons. Furthermore, in addition to detecting substances through the common chemical sense, most animals can measure the levels of internal body chemicals, such as O_2 and CO_2, that are important for maintaining life and health. Information about these chemicals is not processed in the same regions of the nervous system as is information from gustatory or olfactory receptors; indeed, the signals from internal chemoreceptors are not even perceived consciously in humans or, presumably, in other vertebrates.

There is no doubt that based on the criterion of where input from chemoreceptors is processed in the central nervous system, animals actually have several chemical senses besides the classical ones of taste and smell. However, since our understanding of these other chemosensory systems is still sketchy, the emphasis in this chapter will be on the traditional modalities of taste and smell.

Animals have several chemoreceptor systems. All animals have the well-known senses of smell and taste, mediated by receptors in the nose and mouth, respectively. They may also have separate systems for taste on the surface of the body, for internal chemoreception, and for a common chemical sense. These are usually considered separate chemosensory systems since input from them is processed in different places in the central nervous system.

Vertebrate Chemosensory Systems

The common features of vertebrate chemosensory systems are much more apparent in their functional organization and mechanisms of transduction than in their anatomical organization. Anatomically, both the pathways by which taste and smell information enters the brain and the structure of the primary receptor cells of the two systems are different.

The Gustatory System

Gustatory systems in vertebrates show an anatomical organization broadly similar to that of other vertebrate sensory systems. Gustatory receptor cells on the tongue and in the mouth and throat synapse with afferent fibers that travel to the brain stem via three different cranial nerves. In humans, afferent fibers from the tip of the tongue enter the brain via the facial nerve (cranial nerve VII), those from the back of the tongue via the glossopharyngeal nerve (IX), and those from the pharynx via the vagus nerve (X) (Figure 13-1A). After entering the brain stem, all afferent fibers synapse in the **solitary nuclear complex** of the medulla. In most vertebrates, gustatory information is sent from the medulla to the **parabrachial nucleus** in the pons and then to the appropriate sensory area in the cortex. In primates, however, gustatory information is instead processed in the **ventral posteriomedial nuclei** in the thalamus, then passed on to the cortex. Two areas of the cortex appear to be involved in taste discrimination in primates. These are the facial part of the somatosensory cortex and the **insula**, the infolded part of the brain that lies just under the lobe of the temporal cortex (Figure 13-1A, B).

Gustatory receptor cells in vertebrates are typically found in clusters called taste buds that lie in the sensory epithelium, often depressed from the surface so that the opening into the taste bud is seen as a small pit. The receptor cells are anaxonal but are nevertheless capable of generating action potentials. They synapse with afferent sensory neurons at the base of each taste bud (Figure 13-2).

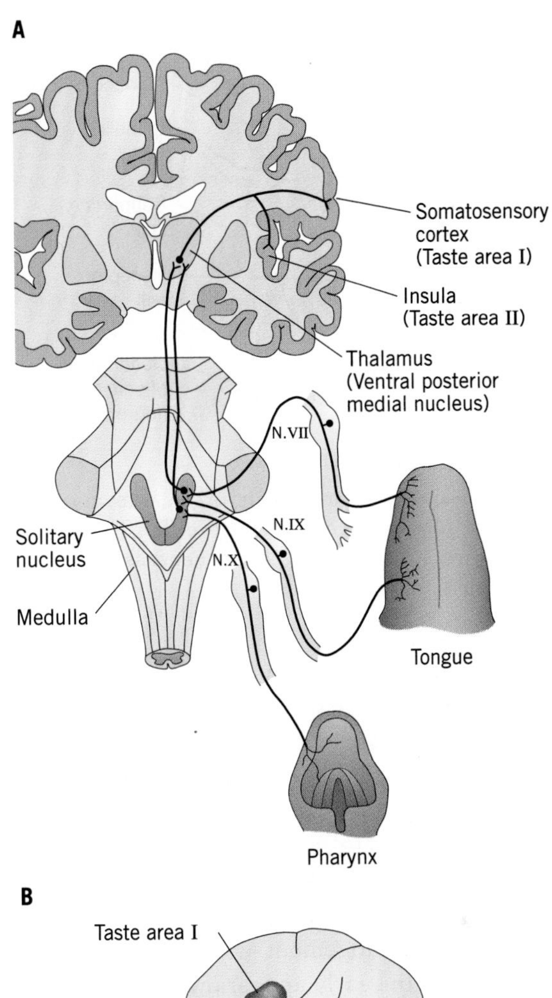

FIGURE 13-1. (A) Pathways for gustatory information in a primate. Gustatory input from the pharynx and tongue travels through nuclei in the medulla and the thalamus, then to the two cortical areas devoted to processing gustatory information. (B) Lateral view of a monkey brain, showing the two regions of gustatory cortex.

In vertebrates, afferent fibers from gustatory receptors enter the brain stem via cranial nerves VII, IX, and X, and synapse in the medulla. Tracts carry gustatory information from there to the pons or, in primates, the thalamus, and then to two areas of the cortex devoted to taste discrimination. The receptor cells are clustered in taste buds, each consisting of several anaxonal neurons and afferent sensory neurons.

The Olfactory System

The olfactory system is organized a bit differently than most other vertebrate sensory systems. Olfactory receptor cells are located in the tissue lining the nasal cavity. These receptors send their axons into the **olfactory bulb**, a projection of the brain that in most primates lies over the nasal cavity and in other vertebrates lies posterior to the nose, in the anterior region of the brain. Axons from the olfactory bulb enter the brain via the **olfactory tract** (cranial nerve I) and project

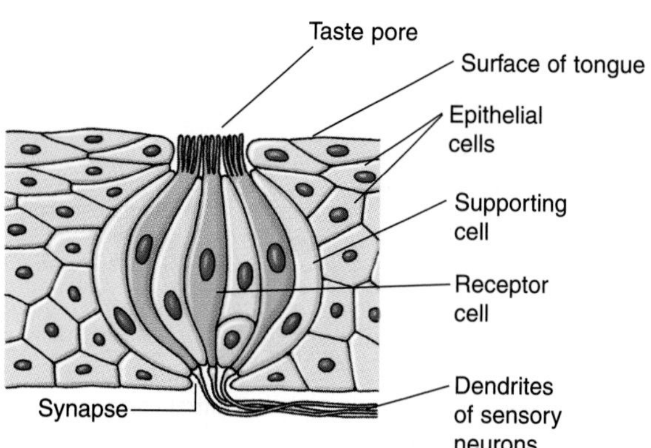

FIGURE 13-2. A vertebrate taste organ, showing the anaxonal gustatory receptors (taste cells). The receptor cells synapse with the sensory neurons that send information to the medulla.

to the primary **olfactory cortex**, also called the **pyriform cortex**, which is located on the anterioventral surface of the telencephalon (Figure 13-3). Because of this pattern of projection of olfactory tracts, in contrast to that in other vertebrate sensory systems, olfactory information travels to the primary olfactory cortex without first passing through the thalamus. This arrangement is thought to be a reflection of the early appearance of the olfactory system in the evolutionary history of vertebrates.

This unusual anatomy does not preclude involvement of the thalamus or other brain regions in processing olfactory input, however, for olfactory information is widely distributed in the brain. Together with axons from the olfactory tract that bypass the pyriform cortex, axons from the pyriform cortex project to the **entorhinal cortex**, which lies posterior to the pyriform cortex. Both pyriform and entorhinal axons also project to the limbic system, where they help to link the perception of odors to emotional reactions. Other axons form a tract to the mediodorsal nucleus of the thalamus. From the thalamus, axons travel to the **orbitofrontal cortex**, on the ventral surface of the frontal lobe, lateral to the olfactory bulb. It is here that primary discrimination of odors (and, in primates, conscious recognition) takes place.

Elongate olfactory receptor cells line the nasal epithelium. At one end, these cells have short axons that extend through pores in the bones of the skull that form the floor of the nose; at the other, they have a robust dendrite. The dendrites have a number of fine branches, called olfactory cilia (Figure 13-4). The cilia contain the receptor proteins for odorants. These proteins must be in contact with air or water in order for odor molecules to be able to interact with them, but they are protected from desiccation or a salt imbalance by a thin layer of mucus or other fluid.

In vertebrate olfactory systems, information travels from the olfactory receptor cells to the olfactory bulb and then directly on to the cortical area devoted to smell without passing through the thalamus. Information is also passed to several other brain areas for conscious recognition of odors. The receptor cells in olfactory systems are clustered in nasal epithelium, and their receptor regions are protected by liquid or mucus.

Transduction of Chemical Stimuli

The sensitivity of chemoreceptors varies a great deal. In general, gustatory receptors are not nearly as sensitive as olfactory receptors. The lowest threshold values for gustatory receptors are in the 10^{-6} M to 10^{-7} M range for substances such as saccharin (minnow) or quinine (human). The lowest threshold values for olfactory receptors can be more than a billion times lower: α-ionone (cedar wood odor) can be detected by dogs at a concentration of 5.5×10^{-17} M. At a practical level, such sensitivity means that a fox, for example, can smell meat at a distance of nearly a mile. Even humans, notoriously poor at detecting odors, can smell skunk odor at a similar distance. At the molecular level, this kind of sensitivity translates into responses of olfactory

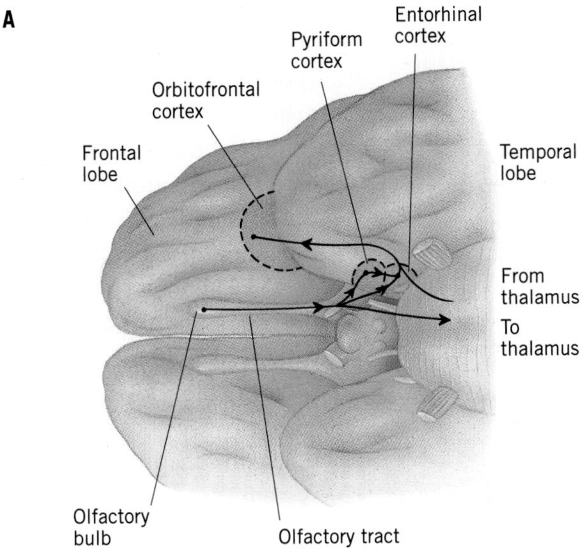

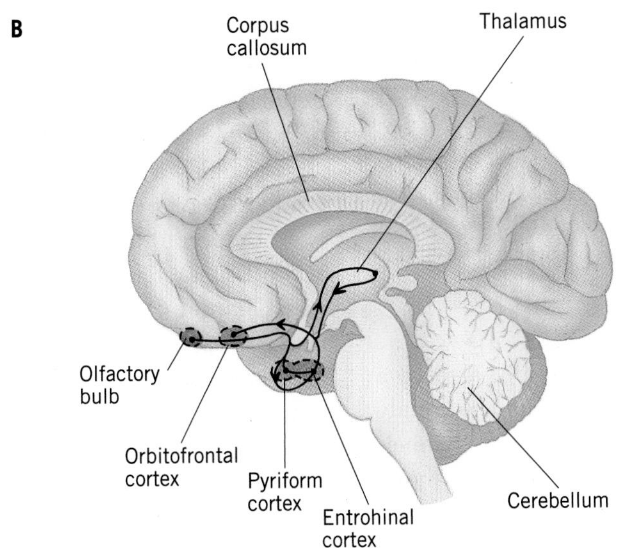

FIGURE 13-3. Processing of olfactory information in primates. (A) Ventral view of a human brain, showing the olfactory bulb, the olfactory cortical areas, and the main pathways connecting them. (B) Section through the center of the human brain, showing the same structures and pathways. Note that information can travel from the olfactory bulb to the three olfactory cortical regions without passing through the thalamus. The pathways drawn represent functional connections and do not show the actual path the tracts take. Pathways to the limbic system are also present but are not shown here.

receptor cells to single molecules of odorant. In insects, it has been calculated that the response of a male moth to the sex pheromone of the female requires only one to two molecules per receptor cell in order for a behavioral response to be elicited (Box 13-1). Activation of several receptors appears to be necessary for conscious recognition of an odor, at least in humans.

Gustatory Transduction

There are two main types of transduction mechanisms in gustatory receptor cells. Transduction of a substance that evokes a sour or salty taste in humans is mediated by a direct effect of the chemical on one or more ion channels in the receptor. Transduction of substances that evoke

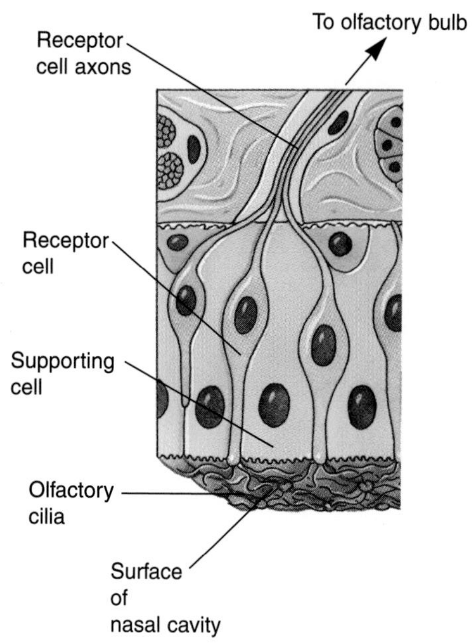

FIGURE 13-4. A vertebrate olfactory receptor. The mucus film covers the epithelium in the nose, where the receptor cells are located. Transduction takes place in the olfactory cilia.

a sweet or bitter taste is usually mediated by a second messenger pathway that indirectly changes the flow of ions through ion channels (see Chapter 6). Some bitter substances, however, interact directly with ion channels and do not require a second messenger. Responses to amino acids, some of which do not readily fall into any of the four traditional taste categories, may be transduced with or without a second messenger. Although no single taste receptor can respond to all types of chemical substances, it is not unusual for cells to be able to respond to more than one type.

All substances that evoke a sour or salty taste exert their effects by directly influencing ion channels in the membrane of the receptor cell. Weak acids like vinegar evoke a sour taste because they ionize in water to produce protons (H^+) and anions (negative ions). In the mud puppy, the hydrogen ions block a specific type of potassium channel in the apex of the taste cell, causing depolarization by shutting down the efflux of K^+ (Figure 13-5A). For salty substances like table salt, Na^+ or other cations (positive ions) act as stimuli. Sodium ions enter taste cells directly through specialized sodium channels, depolarizing the cell (Figure 13-5B). The transduction of salty stimuli, however, is not as simple as that of sour stimuli. Although the anion of a salt does not have a direct effect on ion channels, it seems to be able to modify the response of a cell to the accompanying cation, as evidenced by the different tastes of sodium chloride and sodium bicarbonate.

Many sweet substances produce a response in receptor cells via a second messenger system, but there are many varieties of receptors for sweet substances, even in a single species. In hamsters, for example, at least eight different receptor molecules that bind sweet substances have been found. Only a few of these have been studied in enough detail to determine the transduction mechanism. One mechanism that has been demonstrated in a variety of vertebrates involves the familiar G protein and adenylyl cyclase pathway that you have already encountered in second messenger neuromodulatory systems (see Chapter 6). When sucrose binds with the receptor protein in the cell membrane, the receptor activates a specific G protein, which in turn activates the enzyme adenylyl cyclase. The enzyme catalyzes the conversion of adenosine triphosphate (ATP) to cyclic adenosine monophosphate (cAMP), which then turns on cAMP-dependent protein kinase A. The protein kinase phosphorylates a potassium channel in the cell membrane, thereby closing it. The resulting reduction in the efflux of K^+ from the cell results in depolarization of the receptor cell membrane and can evoke an action potential in the cell if the stimulus is strong enough (see Figure 6-14).

Because of the variety of receptor types, responses of receptor cells to sweet substances vary. You have probably noticed that not all sweet substances taste exactly the same. Sugar and many artificial sweeteners, for example, can easily be distinguished. Furthermore, the rate of adaptation to different sweet stimuli is not the same. For example, the perception of the sweet taste of saccharin will decline more than twice as

BOX 13-1

An Insect's Nose: Recording an Electroantennogram

It is not always easy to determine the absolute sensitivity of an olfactory system. For humans, experimenters can simply ask if a subject smells an odor. For an animal like a dog or a fish, on the other hand, it is necessary to train the animal to carry out some action when the odorant is present, so that the animal shows by its behavior whether it can detect the stimulus. A much more direct way is by electrophysiological recording from the sensory receptors, although the presence of electrical activity in a cell does not guarantee that the animal can behaviorally respond to the stimulus. Intracellular recording is the most precise method, but this is a demanding technique. However, for insects, a technique developed by Dietrich Schneider in the mid-1950s has offered two irresistible assets for the evaluation of olfactory capability: it is simple to carry out, and it is extraordinarily sensitive.

The antennae of insects are richly invested in chemoreceptors (Figure 13-A). In moths that communicate with sex pheromones, there may be as many as 75,000 receptor cells per antenna. Schneider reasoned that when

a

b

FIGURE 13-A.
(A) Front view of the head of a male silkworm moth, showing the large, feathered antennae. (B) Each filament of the antenna is covered with fine sensory hairs, within which lie chemoreceptive cells with their olfactory receptors. Most of these cells are specialized to detect the pheromone emitted by the female to attract the male.

a substantial number of these receptor cells respond to the presence of an appropriate stimulus, he ought to be able to record a summed potential that represents their massed action. This turned out to be the case. When he placed glass recording electrodes into and over opposite ends of an isolated antenna of a male silkworm moth and exposed the antenna to the sex pheromone of the female, he was able to record an electrical potential that varied in amplitude in proportion to the concentration of pheromone (Figure 13-B). Schneider named this potential the **electroantennogram**,

(continued)

An Insect's Nose: Recording an Electroantennogram *(continued)*

and the method of recording an electroantennogram has become standard for evaluating the olfactory capability of insects.

The electroantennogram method had a strong impact on the field because it demonstrated the extraordinary sensitivity of the receptor cells to substances such as sex pheromones. In fact, later studies with radioactively labeled synthetic pheromone showed that in the silkworm moth, only one or two molecules per receptor cell over a 2-second period were sufficient to evoke a behavioral response in the insect.

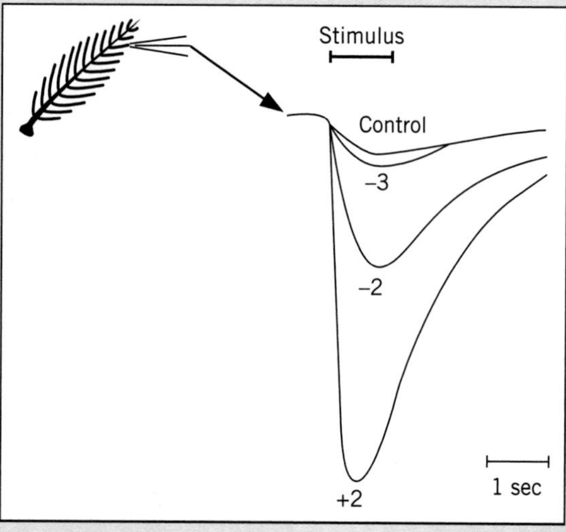

FIGURE 13-B. Diagram of the arrangement for recording an electroantennogram, the summed, simultaneous activity of many olfactory receptor cells, in response to a stimulus. The recorded response is shown on the right. By convention, a response is depicted as a downward deflection. The larger the deflection, the stronger the response. The small numbers represent the powers of ten of the concentration of female sex pheromone used as a stimulus, in μg/ml.

fast as will the sweet taste of sugar when the taster is given successive samples of the two substances. This has the practical effect of making it more difficult to cook desserts in which you must "sweeten to taste," because repeated tasting will have different effects on your ability to taste sweetness if you use saccharin rather than sugar. The result may be that you use too much saccharin in the dessert. The mechanism by which this variation in adaptation occurs is not known, but it could be due to different binding affinities of some substances, to different numbers of receptors available to bind with the substances, or to other means.

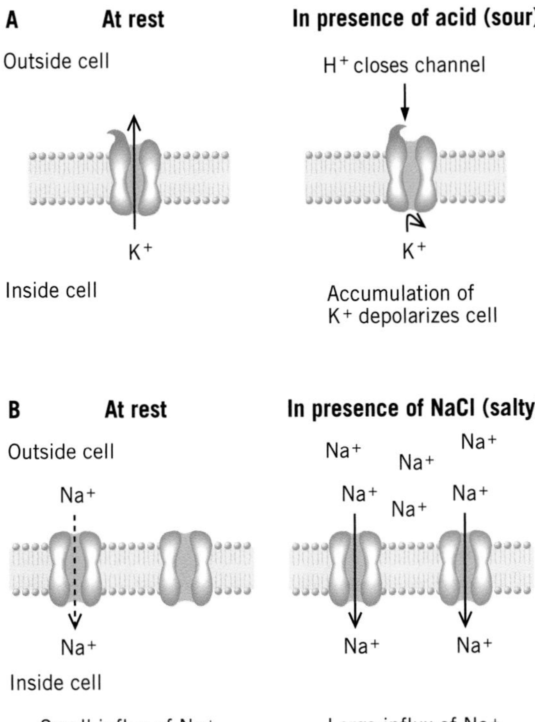

FIGURE 13-5. Mechanisms of transduction in sour and salt receptor cells. (A) Receptors that respond to acids (sour taste) contain ligand-sensitive potassium channels that close in the presence of free protons (H^+). (B) Receptors that respond to salts generally contain channels that allow entry of the cation of the salt. Entry of the cation, Na^+ in the example, depolarizes the cell.

Substances that evoke a bitter sensation have extremely diverse chemical structures. They range from relatively complex molecules like the alkaloid quinine, the local anesthetic novocaine, and the potassium channel blocker tetraethylammonium (TEA), to simple calcium or magnesium salts like $CaSO_4$ and $MgSO_4$. As befitting such diversity of structure, receptor cells that respond to bitter substances exhibit a diversity of transduction mechanisms, summarized in Table 13-1. These range from direct interactions with ion channel or ion pump proteins to second messenger-mediated responses. For example, a number of bitter substances block potassium channels just as sour substances do. Others exhibit a novel mechanism. In frogs, transduction takes place by enhancing the activity of a chloride pump in the membrane of a taste cell, which depolarizes the cell.

Other transduction mechanisms for bitter substances involve second messenger systems. One that has been studied in detail includes two complementary pathways. A stimulating molecule at the receptor activates a G protein, which in turn activates the enzyme phospholipase C (PLC). PLC catalyzes the synthesis of two products, inositol 1,4,5-triphosphate (IP_3) and diacylglycerol (DAG), from phosphatidylinositol 4,5-biphosphate (PIP_2). In one pathway, IP_3 promotes the release of Ca^{2+} from intracellular stores, and the increased levels of free Ca^{2+} directly stimulate transmitter release at synapses without membrane depolarization. In the other pathway, DAG activates protein kinase C, which is thought to close a potassium channel by phosphorylating it. Shutting a potassium channel depolarizes the receptor cell and hence causes the release of neurotransmitter. The pathways are summarized in Figure 13-6.

TABLE 13-1. Mechanisms of Bitter Taste Transduction

Type of Effect	Mechanism	Example Substance
Direct effect on ion channel	Block of potassium channels	Barium ions
Receptor-controlled ion pump	Chloride pump activation	Quinine
Receptor-activated G protein	Activation of PLC, leading to elevated Ca^{2+} and block of potassium channels	Strychnine
Receptor-activated G protein (e.g., gustducin)	Inhibition of PDE and reduction of cAMP levels	Denatonium

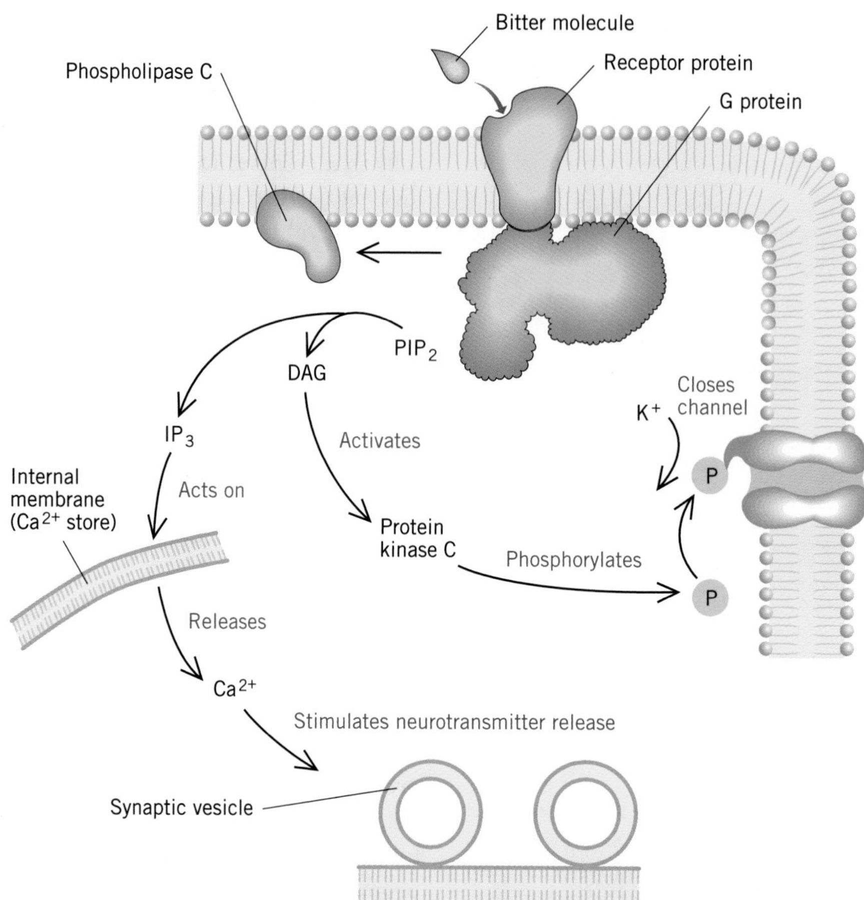

FIGURE 13-6. One mechanism of transduction in bitter receptor cells involves two complementary pathways, one involving release of neurotransmitter without depolarization, and the other employing more common depolarization-dependent release mechanisms. The stimulating bitter compound binds with a receptor, which activates a G protein. The G protein activates the phospholipase C (PLC) system and eventually, through inositol 1,4,5-triphosphate (IP_3), causes the release of Ca^{2+} from Ca^{2+} stores in the cell, promoting the release of neurotransmitter without membrane depolarization. Through diacylglycerol (DAG), the G protein also stimulates an increase in protein kinase C. This brings about a block of specific potassium channels, depolarization, and the release of more neurotransmitter.

The application of molecular techniques to the study of G proteins in taste cells has revealed a remarkable and unexpected transduction pathway for some bitter and sweet substances that is closely related to that found in rods and cones. A G protein with about 90% similarity to rod and cone transducin has been cloned from taste cells. By analogy to transducin, it has been named **gustducin** (for gustatory transducin). Like transducin, gustducin activates a phosphodiesterase enzyme (PDE), in this case one that catalyzes the breakdown of cAMP. Rod transducin itself has also been found in taste cells. Because high levels of PDE activity are characteristic of many taste cells that respond to bitter substances, gustducin and rod transducin were originally proposed to function in the transduction of bitter stimuli. However, mice in which the gene that directs synthesis of gustducin was inactivated (see Box 24-2) showed considerably reduced sensitivity to sucrose, as well as to certain bitter substances, indicating that gustducin mediates at least some sweet taste as well.

Some animals have gustatory receptor cells that will respond to the presence of amino acids. To humans, most amino acids taste bitter. However, alanine and serine taste sweet, and a few, such as the salts of glutamic acid, have a unique taste, called umami, that does not resemble sour, salty, sweet, or bitter tastes. A variety of transduction mechanisms have been identified for amino acids. In catfish, for example, arginine activates a receptor that is directly coupled to a nonselective cation channel. Activation of the channel allows the entry of positive ions, which depolarizes the cell. Alanine, on the other hand, binds with a G protein-coupled receptor that activates both adenylyl cyclase and IP_3. Alanine hence activates transduction pathways that are similar to those of some sweet and some bitter substances.

Transduction mechanisms in gustatory receptor cells vary for different types of chemical substances. Acids exert their effects by direct action of hydrogen ions, and salts exert their effects by the action of cations such as Na^+, on ion channels in the receptor cell. Sugars exert their effects via a second messenger system involving cAMP and protein kinase A. Responses to bitter substances are transduced by a variety of mechanisms; a common one is a G-protein-to-IP_3 mechanism. Responses to amino acids are transduced via direct coupling to an ion channel, or via a second messenger.

Olfactory Transduction

Transduction in olfactory receptors is similar in many respects to transduction in gustatory ones. In land animals, however, a preliminary step is necessary to ensure contact between the odorant and the receptor protein on the surface of the olfactory cell. Because the sensory receptor is protected by a layer of fluid consisting of mucus (vertebrates) or liquid (invertebrates), odorants must first dissolve in this fluid. In both vertebrate and invertebrate animals, this protective fluid contains a class of proteins called **odorant binding proteins**. These are specialized molecules that bind with the odorant and move it to the receptor cell membrane that contains the receptor molecules. They are thought to help concentrate the odorant at the receptors.

There are two main transduction pathways in many vertebrate olfactory neurons, each involving its own specific G protein second messenger cascade and ligand-gated ion channel (Figure 13-7). Both pathways may be present in any single olfactory neuron. One type of G protein is linked to adenylyl cyclase and hence promotes the synthesis of cAMP when activated (Figure 13-7A). The cAMP, in turn, opens a type of cyclic nucleotide-gated ion channel that is nonspecific for cations, and the ensuing influx of positive ions depolarizes the cell. The ion channel is structurally similar to the nucleotide-gated channels of vertebrate photoreceptor cells that were discussed in Chapter 11. Sufficiently strong stimulation will cause the generation of action potentials in the olfactory receptor cell.

The other type of G protein activates PLC and hence stimulates the production of IP_3 and DAG (Figure 13-7B). The precise mechanism by which a change in membrane potential is produced by this pathway is still not fully understood. The second messenger IP_3 may act on IP_3-gated calcium channels or nonspecific cation channels. Opening these channels will depolarize the receptor cell and produce an action potential if the depolarization is strong enough. There is also evidence, however, that in some animals the influx of Ca^{2+} may open calcium-dependent potassium channels, so that the cell hyperpolarizes instead. Although all the possible responses of an olfactory neuron to a single odor have not yet been fully described, it is nevertheless already clear that receptor cells may respond to an odorant in a variety of ways. Multiple responses and dual transduction pathways presumably play important roles in the identification of odors by themselves or as one component of a complex mixture.

You may recall that cAMP and the PLC-IP_3 second messenger pathway can activate protein kinase A (PKA) and protein kinase C (PKC), respectively. In gustatory receptor cells, PKA and PKC promote phosphorylation of ion channels and hence directly affect the membrane potential of the cell, as described in the previous

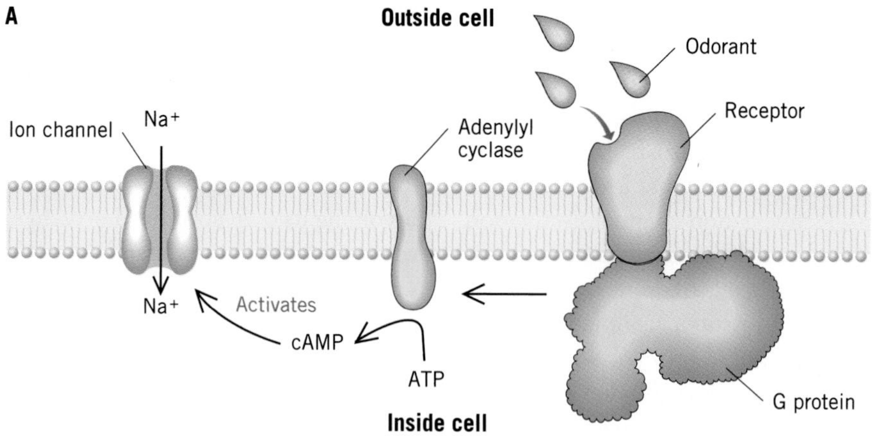

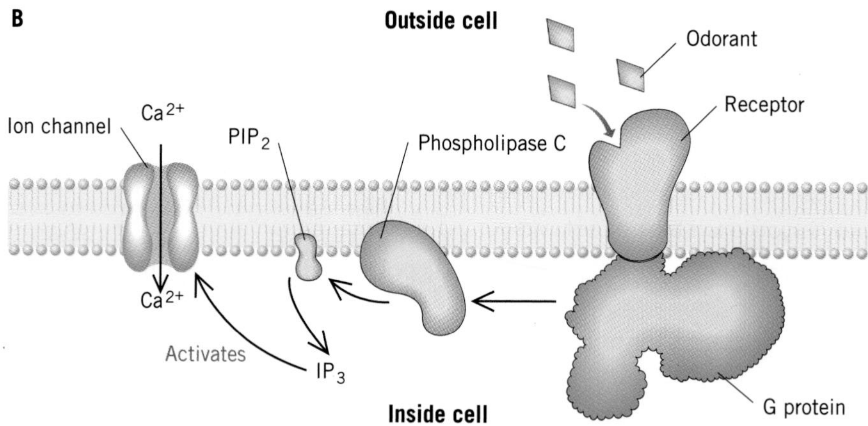

FIGURE 13-7. The two main pathways of olfactory transduction in vertebrates. (A) One type of receptor activates a G protein specific for activation of adenylyl cyclase, promotion of cAMP synthesis, and the opening of a cAMP-sensitive ion channel. Opening the channel depolarizes the cell. (B) The other type of receptor activates a different G protein, which activates phospholipase C (PLC) and thereby promotes synthesis of IP_3. The IP_3 binds with an IP_3-sensitive ion channel, opening it and depolarizing the cell. In some cells, the PLC pathway may have different (inhibitory) effects.

section and shown in part in Figure 13-6. In olfactory receptor cells, PKA and PKC have a different role—they limit the response of the olfactory cell to the presence of a sustained stimulus by desensitizing the receptor to the presence of a stimulating molecule. As shown for PKA in Figure 13-8, the presence of the kinase promotes the phosphorylation of the receptor protein. This phosphorylation desensitizes the receptor, making it less responsive to the presence of odorant. Hence, this is a mechanism for adaptation to odors—the loss of sensitivity to an odor upon sustained exposure to it.

In land animals, odorant molecules are brought to the membrane of the receptor cell by odorant binding proteins. At the cell membrane, transduction is mediated by a second messenger system involving adenylyl cyclase and cAMP, or the IP_3 system, or both. The second messengers cAMP and IP_3 open specific ligand-gated ion channels, causing cell depolarization and possibly an action potential, or hyperpolarization.

Transduction and Cell Biochemistry

The preceding two sections have outlined the main transduction pathways in gustatory and olfactory receptors. These range from the direct effect of a stimulating molecule on an ion channel to complex, second messenger-mediated biochemical pathways. Second messenger-mediated pathways involve activation of a specific type of G protein followed by activation of an enzyme such as adenylyl cyclase, phospholipase C, or phosphodiesterase. These enzymes initiate a biochemical chain reaction that results in an increase or decrease in the concentration of a ligand that interacts with a protein ion channel, activation of a protein kinase that phosphorylates a channel, or an increase in internal Ca^{2+} concentration.

It is necessary to know the specific pathway activated by any gustatory or olfactory stimulant in order to understand how the stimulant evokes a particular neural response. From a broader perspective, however, the importance of these pathways lies not so much in their role in sensory transduction as it does in revealing the rich possibilities that neurons have of regulating their responses to any outside event, be it the arrival of a photon or the arrival of a neurotransmitter from a neighboring cell. The key players in transduction—G proteins, adenylyl cyclase, PLC, PDE, the cyclic nucleotides, and the protein kinases—are by no means restricted in their functions to transduction. They are involved in every process that can influence the electrical or metabolic activity of a neuron, which is to say virtually every time communication between neurons occurs, and therefore are central to the function of the entire nervous system.

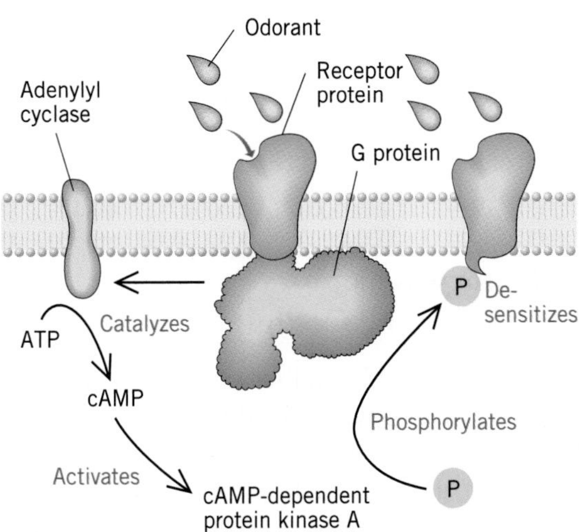

FIGURE 13-8. A mechanism of adaptation in olfactory neurons. The synthesis of cAMP, in addition to its other effects, also activates cAMP-dependent protein kinase A, which promotes phosphorylation of the receptor, hence reducing its sensitivity to the stimulant.

There are many kinds of chemosensory transduction. Transduction that employs a G protein and a second messenger may also involve adenylyl cyclase, phospholipase C, phosphodiesterase, a cyclic nucleotide, and protein kinases. These diverse biochemical pathways are important not just in transduction but in communication between neurons.

The Identification of Chemical Substances

How do chemosensory systems encode the identity of a chemical compound? Research has shown that gustatory and olfactory systems do not generally use the same mechanism. Gustatory systems, which identify relatively few substances, commonly use a partly labeled line code. Olfactory systems, on the other hand, which in many animals discriminate thousands of odors, tend to use an across-fiber or population code, although elements of both codes are used in each system.

Gustatory Coding

Gustatory systems in many animals use a partly labeled line coding system. As described in Chapter 10, such a system is used in some insects to identify the four main classes of compounds to which the animal is sensitive: sweet and salty substances, water, and amino acids. Chemoreceptive hairs on the feet of blow flies contain sensory neurons that are sensitive to these substances. When the fly steps in a drop of fluid, stimulation of the water receptors will allow it to recognize the fluid as water, since none of the other receptors will be activated at the same time. Simultaneous activation of the receptor for salt will tell the fly that it has stepped into a fluid, such as sweat from your arm, that contains salt.

Identification of substances on the tongue of a mammal is also made by a partly labeled line code. As in the case of the blow fly, different receptor cells respond strongly to one of the four main types of substances—salty, sour, sweet, or bitter—and not so strongly or not at all to the others. The basis of this selectivity is the type of receptor molecule present in the cell membrane and the biochemical pathways the activated receptor molecules can trigger in each receptor cell. Receptors that are sensitive mainly to one type of chemical synapse with specific sensory neurons that send axons to the brain stem. Hence, different substances will activate specific sets of these sensory neurons, each set more or less devoted to a particular class of compound. In chimpanzees, for example, some nerve fibers in cranial nerve VII from the tongue respond strongly to bitter substances, but weakly, if at all, to sweet or salty substances, and vice versa.

This separation of taste by type of substance is maintained in the brain stem. Recordings from the parabrachial nucleus in the rat have revealed neurons similar to those in chimpanzee cranial nerve VII that also respond strongly to specific types of substances and weakly or not at all to others. Anatomical techniques that reveal neural activity have shown that neurons activated by different types of substances tend to be clustered in certain parts of the parabrachial nucleus (Figure 13-9).

Although the concept that there are separate lines for the coding of different types of compounds is well supported experimentally, it is also clear that the separation is not absolute and that the pattern of response across different sets of neurons is also important in taste. Most gustatory receptors in vertebrates will respond to more than one type of substance, even if weakly. Furthermore, experiments with mixtures of compounds indicate that the response to one type of substance can be modulated by the presence of another, even at the receptor level. Although it is possible to pick out the separate components of a mixture of sugar and a bitter substance, the subjective sweetness of the sugar will be less in the presence of the bitter substance than it will be alone. Some of this modulation is due to interactions of the two stimuli at individual receptors, but some is also due to interactions in the parabrachial nuclei between fibers carrying information about sweet and bitter substances.

Coding in both insect and vertebrate gustatory systems is based on a partly labeled line system using chemoreceptors specialized to respond mainly to one type of chemical substance. The resulting separation of lines carrying information about different substances is, in vertebrates, maintained in the brain stem. However, separation of different taste qualities is not absolute, and the pattern of response of many different fibers is also important for substance identification.

Olfactory Receptors and Olfactory Coding

Understanding the coding of olfactory stimuli presents an especially difficult challenge because animals can discriminate among thousands of different odors. At present, it is believed that olfactory systems use a combination of labeled line and across-fiber coding to convey information about odor. This is shown by study of the response characteristics of individual olfactory neurons. Single neurons respond to several, even many, odorants, yet

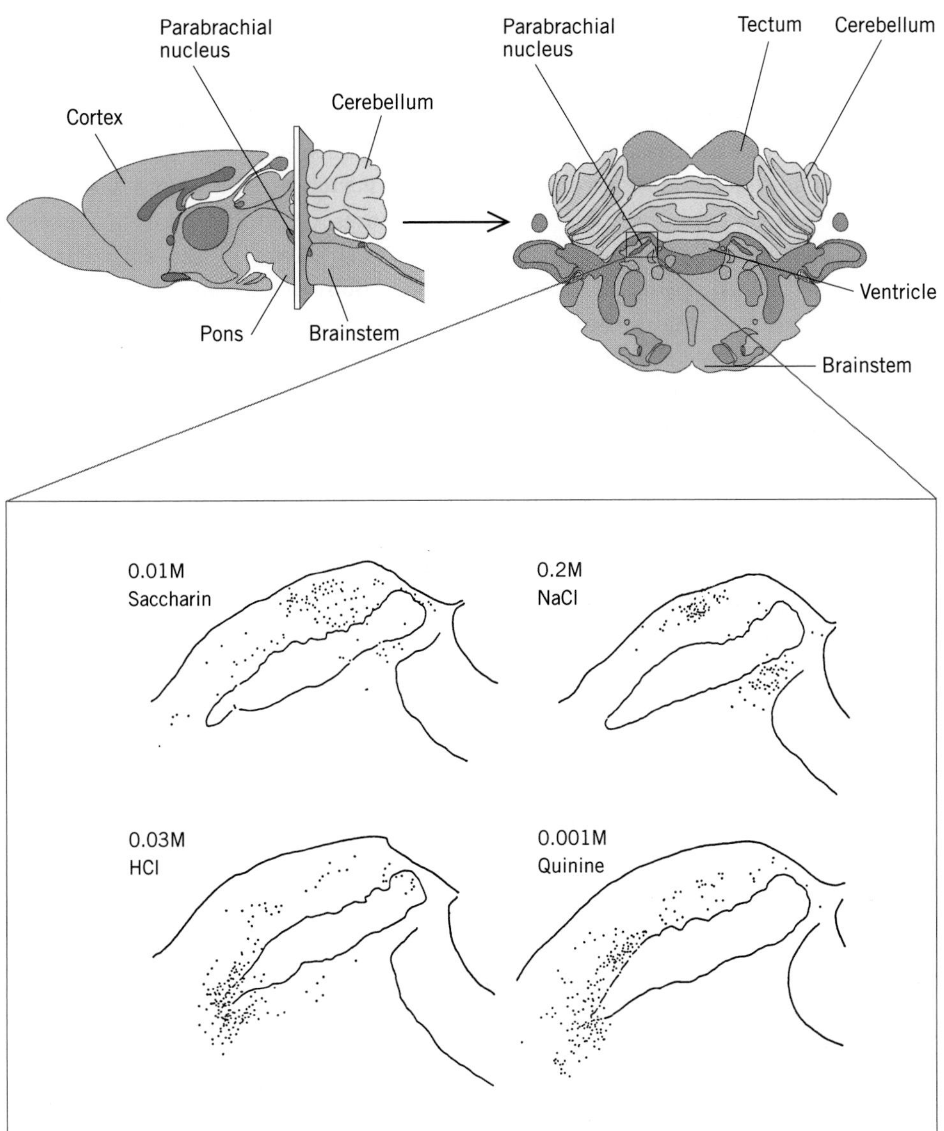

FIGURE 13-9. The specificity of response sites in the parabrachial nucleus of the rat to different substances. The sagittal and cross sections (see Appendix) of the brain show the location of the nucleus. The enlargements of the parabrachial nucleus below show the regions of neural activity that accompany exposure to solutions of different tastes—sweet (saccharin), salty (NaCl), sour (HCl), and bitter (quinine).

they usually respond to only a small subset of all the odorants to which the animal is sensitive. Hence, the identity of a class of odorants may be made by the specific receptor cells that are stimulated by an odorant, a labeled line system, whereas the exact identification of the odor may be made by taking account of the pattern of activity of all the olfactory neurons that respond to it, some being more responsive than others.

The molecular basis of the responsiveness of individual olfactory neurons to particular odorants has begun to emerge. Olfactory receptor proteins belong to a large, multigene family of receptors, consisting of perhaps as many as 1000 genes in humans and rodents. All of the receptors coded for by these genes have the seven transmembrane domains characteristics of receptors that interact with G proteins. Each gene is expressed in only a small number of olfactory neurons, ranging from roughly 0.01% to 1.00% of all the neurons in the olfactory epithelium, depending on the particular gene. Comparing the number of genes and the number of neurons in different species of animals has led some researchers to suggest that each neuron expresses only a single gene, which means that each neuron will have only a single type of receptor protein. The binding properties of that protein then determine to which odorants the neuron will respond. Each specific odorant will activate all the neurons containing a receptor protein that will bind with that odorant. In theory, the particular subpopulation of cells that respond to the odorant can provide a preliminary identification of the odorant by labeled line coding, whereas the final determination would be made by the pattern of activity in the activated neurons by across-fiber coding.

The olfactory system uses a combination of labeled line and across-fiber coding to identify odorants. A large number of genes code for different receptor proteins with distinctive response properties, but each neuron may contain just a single type of receptor protein. Thus each receptor cell can respond to a limited number of odors, and each odor will activate a limited number of receptor cells.

Brain Structure and Olfactory Coding

The organization of the brain regions devoted to processing olfactory information is intriguing because the olfactory systems of many animals, including vertebrates, arthropods, mollusks, and annelids, have a remarkably similar anatomical organization. In all of these animals, the olfactory bulb, or the equivalent part of the brain that processes olfactory information, contains as many as several thousand **glomeruli**, discrete globular tangles of densely packed dendrites and axon terminals. The structural organization and the function of these glomeruli and related parts of the brain in vertebrates are discussed in this section. Insect systems are considered in the next section.

A vertebrate olfactory bulb contains the terminals, cell bodies, and synapses of several types of neurons (Figure 13-10). These neurons, plus the axons of several other types that enter the bulb, constitute the neural substrate for the processing of olfactory information. Input to the glomeruli comes from axons of the olfactory neurons, each of which terminates in a single glomerulus. Since there may be tens of millions of olfactory neurons and only about 2000 glomeruli in a mammal such as a rabbit or rat, 5000 to 10,000 olfactory neurons may converge on each glomerulus. Within each glomerulus, olfactory neurons synapse with three other types of neurons. The axons of two of these, **mitral cells** and **tufted cells,** provide the main output of the olfactory bulb. Since it has been estimated that there are about 70,000 mitral cells and 160,000 tufted cells in the rat olfactory bulb, on average there will be 35 mitral cells and 80 tufted cells in a single glomerulus. Periglomerular cells are the third type of glomerular neuron.

Processing of sensory information in the olfactory bulb is aided by two kinds of cells, both of which provide inhibitory signals. One of these, **periglomerular cells,** provide a communications link between glomeruli. Each periglomerular cell receives input from olfactory neurons and makes reciprocal dendrite-to-dendrite synaptic contact with mitral and tufted cells; it also sends its axons to another glomerulus, where it inhibits mitral cells. The other type of inhibitory neuron is called a **granule cell,** which does not enter the glomeruli but instead receives excitatory input from and sends inhibitory output to mitral and tufted cells outside the glomeruli. Granule and periglomerular cells also receive input from axons originating in other brain regions, mainly the contralateral olfactory bulb and the anterior olfactory nucleus. One of the main functions of these inhibitory cells is to sharpen the responses to chemical stimuli of neurons

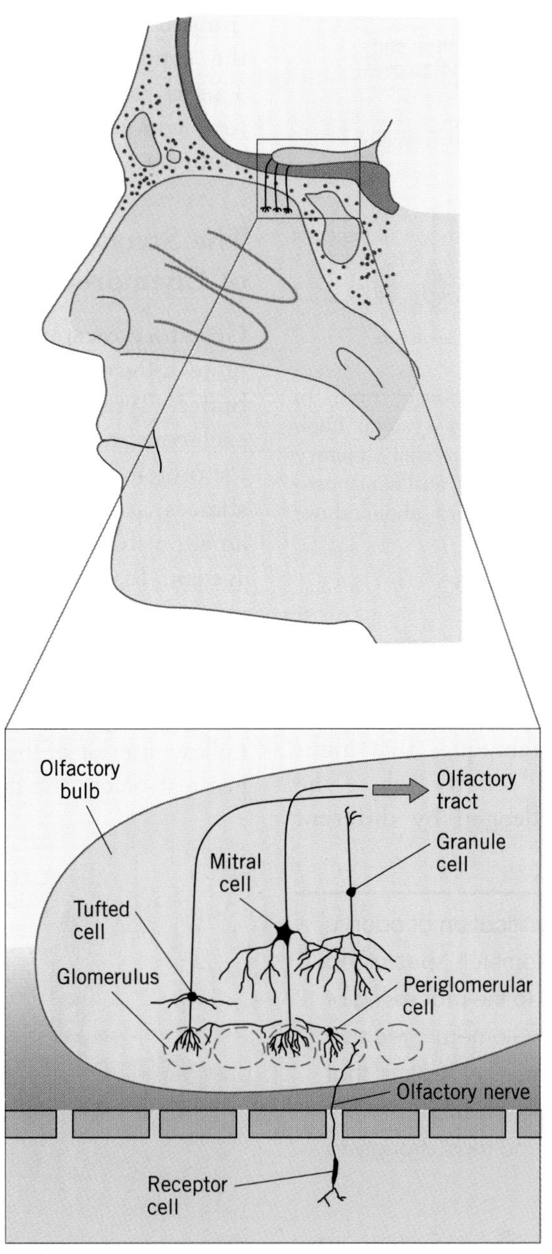

FIGURE 13-10. The human olfactory bulb, showing some of the main neural pathways. The insets show the relationship between the different cell types and the glomeruli.

leaving the olfactory bulb. Hence, stimuli that may excite a large number of primary olfactory neurons will excite far fewer mitral and tufted cells in the glomeruli.

The rules by which primary olfactory receptor cells converge on glomeruli in the olfactory bulb are not completely understood, but there is clearly some order to the arrangement. It is thought that olfactory neurons that respond similarly to odorants send their axons to the same glomerulus. Hence, specific glomeruli will be activated preferentially by specific types of odorants (Figure 13-11). In other words, one particular stimulus will tend to activate one set of glomeruli, whereas another stimulus will activate a different set, although there may be some overlap between the subsets. A nonrandom distribution of

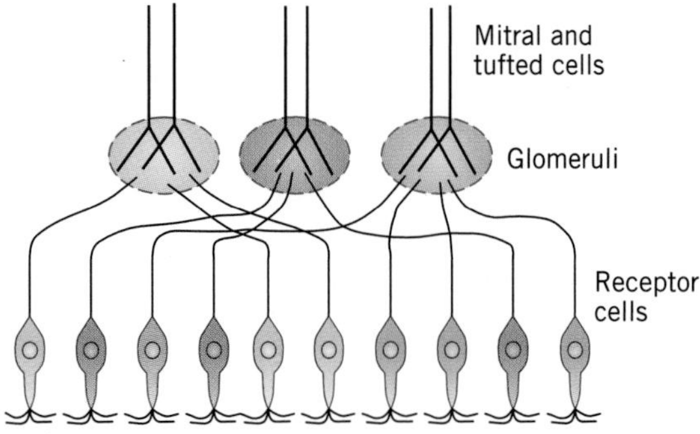

FIGURE 13-11. Hypothetical organization of inputs to individual glomeruli in the olfactory system. Each glomerulus is thought to receive input from olfactory neurons that respond to similar odors and hence provides information to higher centers about those odors.

glomerular activation to the presence of different odors, as is required for this kind of system, has been demonstrated experimentally through the use of techniques that highlight centers of neural activity when the olfactory system is challenged by different odorants (Box 13-2).

In olfactory systems, identification of odors is carried out in part by glomeruli, specialized brain regions dedicated to the processing of olfactory information. Glomeruli receive input from sets of olfactory receptors that respond similarly to odors, so that different groups of glomeruli respond most strongly to different types of stimuli.

Chemoreception in Invertebrates

Although many chemoreceptive organs and receptor cells of invertebrate animals look quite different than their vertebrate counterparts, there are also many remarkable similarities between the systems of the two groups of animals. When you consider the physiology of chemosensory systems, from mechanisms of transduction to their functional organization, the similarities are even more astonishing. This is especially true for the olfactory system, which is similarly organized in animals as diverse as horses and earthworms.

The Structure of Chemoreceptors

Gustatory receptor cells in invertebrates look quite different than the taste buds of vertebrates. Typically, they are bipolar neurons, each consisting of dendrites (containing receptor proteins for the detection of chemical substances), a cell body near the body surface, and an axon that projects into the central nervous system. In insects, the receptive ends of the gustatory receptors are usually located in a raised peglike or hairlike structure that has a single opening at the tip (Figure 13-12). The dendrites of the receptor cells extend into the hollow interior of the peg or hair; the cell body lies just below the cuticle. The chemosensory

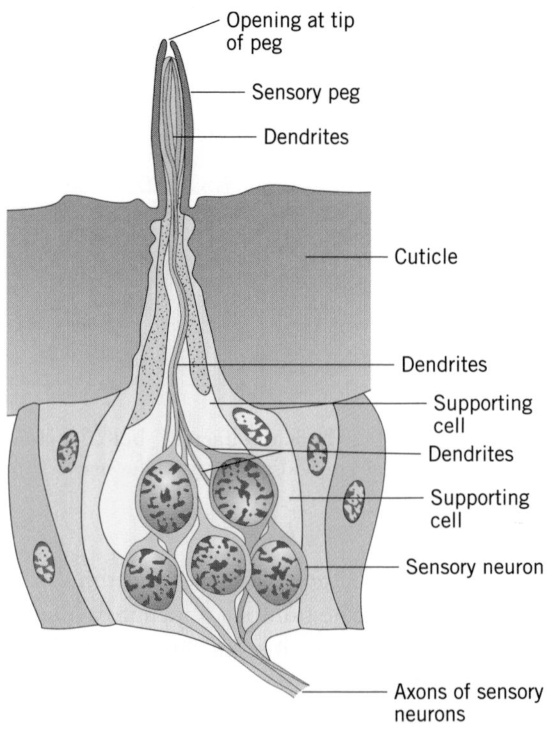

FIGURE 13-12. An insect taste organ. The cell body of each receptor cell lies at the base of the organ, sends its dendrites into the sensory peg, and sends its axon into the central nervous system.

hairs on the feet of blow flies (see Figure 10-2) contain gustatory receptor cells.

Olfactory receptor cells, in contrast, show a remarkable structural similarity across phyla. In the invertebrates as in the vertebrates, they consist of a peripherally located cell body, a dendrite with fine, cilia-like projections reaching to the receptor surface or opening, and an axon that projects to the central nervous system (Figure 13-13). In insects, olfactory receptor cells are located in peglike or threadlike structures on the antennae or body (Figure 13-14). Each structure contains many individual receptor cells. Unlike gustatory receptors, olfactory receptors generally have many minute openings by which odorant molecules can reach the receptor cells.

Gustatory receptor cells are quite different in vertebrates and invertebrates, exhibiting in the latter the bipolar structure typical of invertebrate animals. In contrast, olfactory receptor cells are remarkably similar in vertebrates and invertebrates. In insects, gustatory receptor organs have a single opening to the environment, whereas olfactory organs have many.

Transduction

Gustatory transduction has been studied only minimally in invertebrates, but in olfactory neurons the process has been investigated in depth in some animals. The most important similarity between olfactory transduction in vertebrate and invertebrate animals is that transduction in invertebrates appears to be based on G protein-linked second messenger systems similar to those that have been described in vertebrate olfactory receptor cells.

However, there are differences between vertebrate and invertebrate systems as well. First, no evidence of cAMP action has been found in olfactory receptors from *insect* antennal organs. It may be that insects do not have a dual transduction pathway, or that a second pathway is not present in the receptor cells that have been studied so far. Second, the two second messenger pathways found in some invertebrate olfactory receptors act on different channels and have different effects than do the same pathways in vertebrate cells. For example, receptors on the antennae of lobsters show both cAMP and IP_3 second messenger systems. In contrast to the situation in vertebrates, however, cAMP opens a cyclic nucleotide-gated potassium channel rather than a nonselective cation channel. The effect of activation of the cAMP second messenger

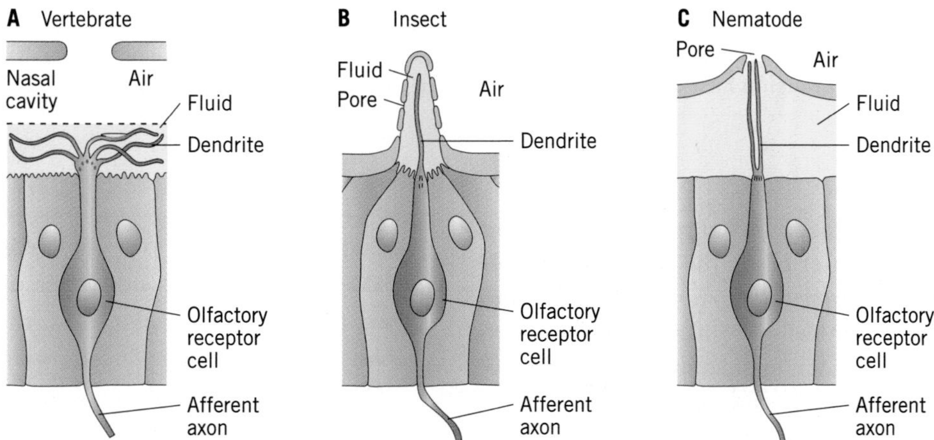

FIGURE 13-13. Olfactory receptor cells from: (A) a typical vertebrate, (B) an arthropod, and (C) a simple roundworm (nematode). In each case, the cell consists of a cell body, a rather thick dendrite with thinner, filament-like dendrites projecting from it, and, at the other end, an axon that projects into the central nervous system.

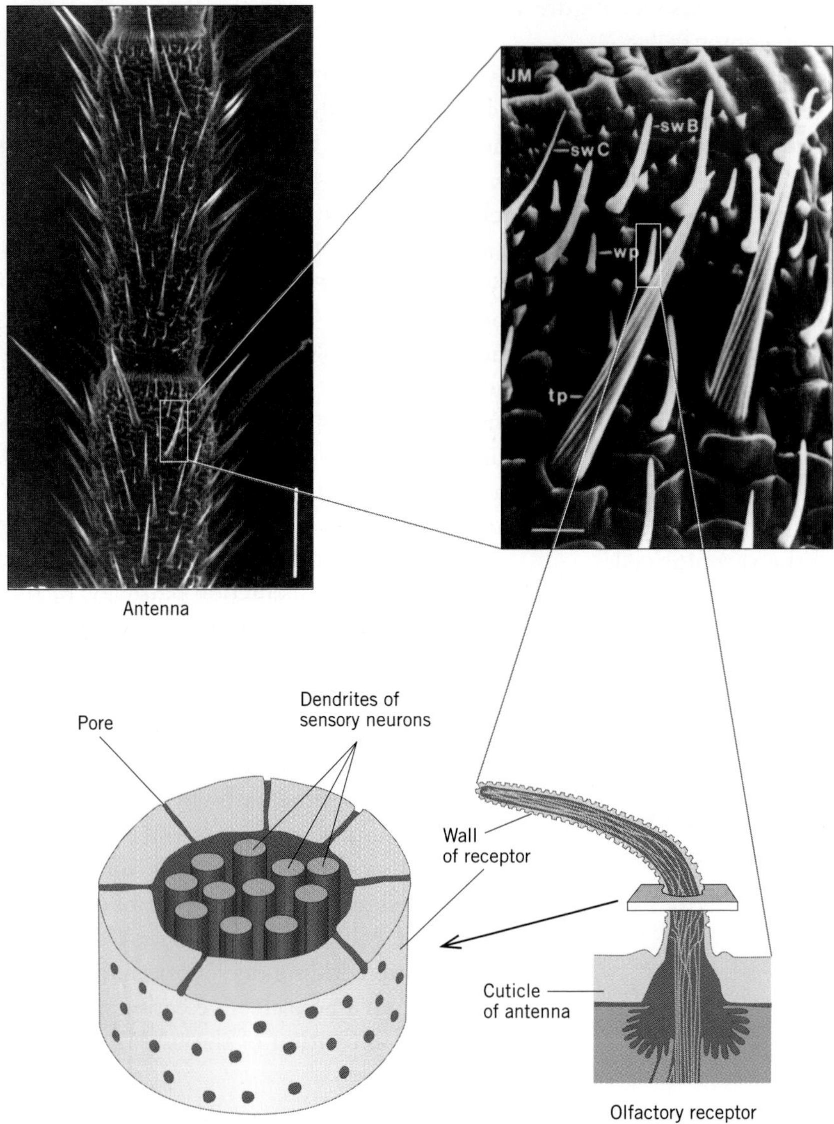

FIGURE 13-14. An olfactory receptor from the antenna of a male cockroach. The peglike structure has numerous pores through which chemical substances can enter. Inside the structure, the sensory neurons show considerable dendritic branching. The scale bar in the uppermost left photomicrograph is 100 μm.

system is therefore a loss of positively charged K^+ from the interior of the cell, and consequently *hyperpolarization*, not depolarization, of the receptor cell. IP_3 has been shown to open either a calcium-selective or a nonselective cation channel, leading to depolarization. The depolarizing and hyperpolarizing effects have been shown to coexist in single receptor cells, where they are activated by different odorants.

Olfactory transduction involving IP_3 and cAMP occurs in most invertebrates, although insects' olfactory receptors show no cAMP mechanism. In some invertebrates, the cAMP transduction pathway leads to hyperpolarization of the receptor cell, whereas the IP_3 mechanism leads to depolarization.

Insect Olfaction

In a most remarkable example of convergent evolution, the part of the brain that processes olfactory information in animals as diverse as annelids, mollusks, and arthropods is organized into glomeruli that are structurally similar from species to species and similar to the glomeruli found in the olfactory bulb in vertebrates. Among the invertebrates, the structure and function of this system has been especially well studied in insects.

Like vertebrate glomeruli, the glomeruli in insect antennal lobes receive input from afferent olfactory neurons that provide information about odorants from each antenna. Output from the glomeruli is conveyed by neurons termed projection neurons that send their axons to other brain regions. These projection neurons are thus analogous to mitral and tufted cells. In addition, local interneurons confined to the antennal lobe connect glomeruli with one another, hence apparently serving a function similar to that of the periglomerular cells of vertebrates. The arrangement of these cells in an insect brain is shown in Figure 13-15.

Nevertheless, the insect and vertebrate systems are not identical. Some of the differences result from differences in the size of the brain in insects and vertebrates. The smaller insect brain has fewer receptor cells and fewer glomeruli; only about 1000 to 2000 receptors converge on each glomerulus. Furthermore, on average only one to two projection neurons synapse in each glomerulus. Other differences represent functional specializations in the insect that are not so common in vertebrates. One such specialization is the presence of paired **macroglomeruli**, giant glomeruli that are found in the antennal lobes in the males of insect species in which females release a volatile sex pheromone into the air to attract the male. These macroglomeruli receive input exclusively from receptor cells in the antenna that respond selectively to the sex pheromone of the female of the species.

Study of the response characteristics of receptor cells in insects has shown that elements of both a labeled line and an across-fiber code are used to convey information about odorants. For example, several types of receptor cells have been identified. Some, which respond to a wide variety of odorants, are called **broad generalists**. At the other end of the range of specificity, some olfactory neurons respond only to a single chemical substance, such as the sex pheromone of the species; such neurons are called **narrow specialists**. Other neurons fall between these extremes, in that they will respond to several chemically related substances, such as the aroma of fruits, certain alcohols, and the like.

In a few specialized cases, the relationships between specific chemosensory receptors and particular glomeruli are quite precise. For example, the pair of macroglomeruli in the antennal lobes receives input from receptor cells that convey information about the sex pheromone. Some of these receptors respond just to certain components of the pheromone

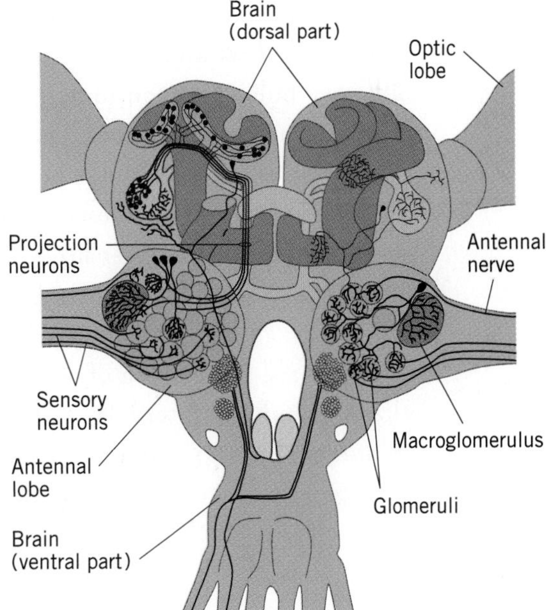

FIGURE 13-15. Frontal view of the brain of a male cockroach, showing the main olfactory pathways. The olfactory receptor neurons send their axons in from the antenna through the antennal nerve and terminate in the glomeruli or macroglomerulus in the antennal lobe of the brain. From there, projection neurons carry odor information to other regions of the brain for further processing.

BOX 13-2

Seeing Neural Activity: Recording with Voltage-Sensitive Dyes

Recording neural activity from single cells has revolutionized researchers' understanding of many aspects of neural function, including that of sensory systems. But because it allows an investigator to deal with only a few cells at one time, the technique limits what can be learned about the action of several neurons simultaneously. One approach to the study of massed neural activity has been to use fluorescent voltage-sensitive dyes to track the neural activity of many neurons at once. The technique was pioneered by I. Tasaki and Larry Cohen and their collaborators, who demonstrated that it was possible to record the electrical activity of neurons by such optical methods. The technique has produced some interesting results, but until recently its widespread application was hampered by the lack of devices that had sufficient spatial and temporal resolution to study electrical events that take place very rapidly in a small region.

In 1987, however, John Kauer improved the technique by taking advantage of newer, more sensitive voltage-sensitive dyes and better equipment, and applied the method to study the processing of sensory information in the olfactory bulb of a salamander. To make his recordings, Kauer first applied the voltage-sensitive dye to the neurons of the olfactory bulb. He then illuminated the bulb with ultraviolet light and, through a microscope, took video images of the exposed olfactory bulb when pure air was wafted over the receptor cells and again when a particular odorant was introduced into the air (Figure 13-Ca). The UV light stimulates the voltage-sensitive dye to give off light when the cell in which it is located depolarizes. By using a computer to subtract the background image from the image obtained when the odor is present, Kauer is able to see the pattern of activity of many hundreds of neurons in the olfactory bulb, at intervals of about 33 msec (Figure 13-Cb). Using this technique, he has been able to show that different odorants have different latencies and evoke clearly distinct patterns of activation of neurons in the olfactory bulb.

FIGURE 13-C (opposite). (A) Arrangement for the optical recording of neural activity in the olfactory bulb of a salamander. An image of the olfactory bulb is magnified and recorded by a video camera through a filter. The filter keeps out all light except that at the wavelength of the voltage-sensitive dye, which fluoresces when it is in a depolarized cell and exposed to UV light. (B) The pattern of response of neurons in the bulb to three different substances. The outline of part of the olfactory bulb has been drawn in the first frame (left) to help orient you. Note the changed pattern of response by frame 11 (arrows) of each of frames.

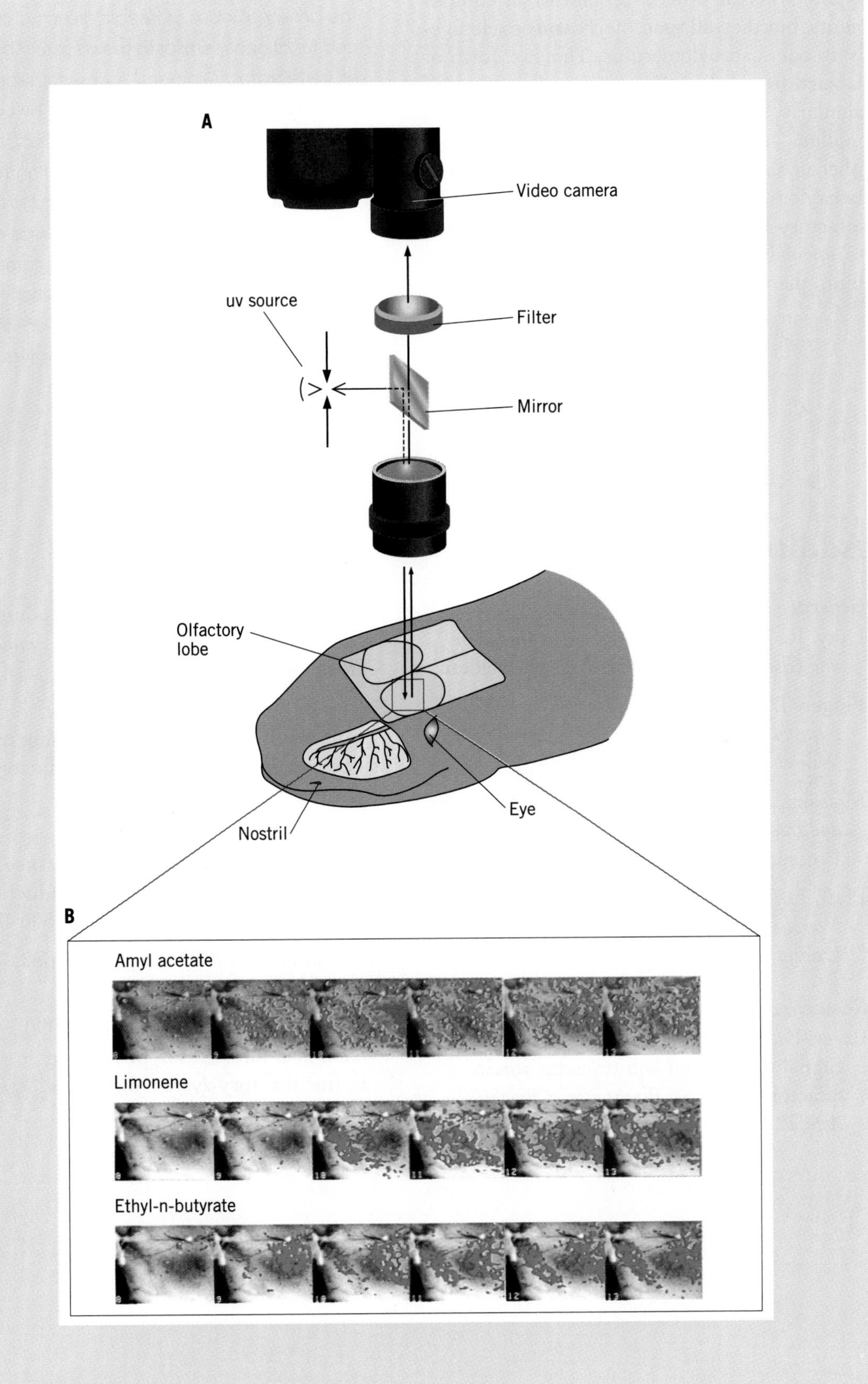
A
Video camera
uv source
Filter
Mirror
Olfactory lobe
Eye
Nostril
B
Amyl acetate
Limonene
Ethyl-n-butyrate

blend, whereas others respond to all components, but they all send their axons exclusively to one macroglomerulus. This is a form of labeled line coding. There are a few cases of similarly specialized regions of the mammalian olfactory bulb. For example, pheromones that evoke special odor-related behaviors, such as suckling in rat pups, excite olfactory receptor cells that send their axons to specialized glomeruli in the rat olfactory bulb, the modified glomerular complex.

The antennal lobes of insects are structurally and functionally similar to the olfactory bulbs of vertebrates. Glomeruli in insects receive input from receptor cells, process it by comparing inputs with other glomeruli, and relay it to other brain regions via projection neurons. Some olfactory neurons respond to only one or two types of odorants, whereas others respond to many types. Macroglomeruli are large, specialized glomeruli that are devoted to processing information from pheromone-sensitive olfactory receptors.

Additional Reading

General

Doty, R. L., ed. 1995. *Handbook of Olfaction and Gustation*. New York: Marcel Dekker.

Finger, T. E., and W. L. Silver, eds. 1987. *Neurobiology of Taste and Smell*. New York: Wiley.

Meisami, E. 1991. Chemoreception. In *Neural and Integrative Animal Physiology*, ed. C. L. Prosser, 335–434. New York: Wiley-Liss.

Pfaff, D. 1985. *Taste, Olfaction, and the Central Nervous System*. New York: Rockefeller University Press.

Research Articles and Reviews

Breer, H., K. Raming, and J. Krieger. 1994. Signal recognition and transduction in olfactory neurons. *Biochemica et Biophysica Acta* 1224:277–87.

Buck, L. B. 1996. Information coding in the vertebrate olfactory system. *Annual Review of Neuroscience* 19:517–44

Kauer, J. S. 1991. Contributions of topography and parallel processing to odor coding in the vertebrate olfactory pathway. *Trends in Neurosciences* 14:79–85.

Kinnamon, S. C., and R. F. Margolskee. 1996. Mechanisms of taste transduction. *Current Opinion in Neurobiology* 6:506–13.

Lindemann, B. 1996. Taste reception. *Physiological Reviews* 76:719–66.

Masson, C., and A. Mustaparta. 1990. Chemical information processing in the olfactory system of insects. *Physiological Reviews* 70:199–245.

CHAPTER 14

Somatic and Other Senses

Sensory structures that allow animals to detect touch, heat, cold, and even pain are distributed all over the skin. Together with specialized internal receptors, they constitute an animal's somatosensory system. Animals also receive sensory input that does not involve conscious perception, like the information used to maintain balance. Some animals have senses that humans lack completely, such as electrical or magnetic senses. These sensory systems are the focus of this chapter.

Aristotle stopped counting at five senses, of which only one, the sense of touch, was concerned with the surface of the body. However, all of us are quite familiar with other sensations that are associated with the skin, like tickle, vibration, hot, cold, and pain. Furthermore, in addition to this information from the body surface, the brain also receives information about the internal state of the body, such as the positions and movements of its parts, and the state of the blood and internal organs. The sense organs and sensory receptors that provide this information are all part of what is called the somatosensory system.

The Somatosensory System

The somatosensory system differs from other sensory systems in two principal ways. First, it provides information about more than one modality. For example, your somatosensory system mediates sensations like touch, cold, and pain. In contrast, although your eyes can respond to different colors, they mediate only the single modality of vision. Second, receptors for the somatosensory system are found all over your body. Every inch of your skin and many of your internal organs bear receptors that are part of the somatosensory system. In other sensory systems all the receptors are generally located in a pair of discrete sense organs like the eyes or confined to one part of the body like the nose or tongue.

In spite of the variety of modalities that are represented by the somatic senses, and the scattered distribution of receptors that mediate these modalities, the somatic senses collectively constitute a single sensory system. Anatomically, the somatosensory system employs just two pathways to convey information to the brain, and one main region, the somatosensory cortex, for processing that information. The somatic senses also have two common functions. First, they convey information about the nature of a stimulus: whether it is soft or hard, dull or sharp, hot or cold, or has other features. Second, they provide information about the location of a stimulus on or within the body. These two types of information allow the identification and localization of a stimulus so an animal might take appropriate action.

The somatosensory system is different from other sensory systems in mediating several modalities and by having its receptors distributed inside and on the surface of the body. Nevertheless, anatomically and functionally it is a single sensory system that provides information about the nature and location of stimuli.

Vertebrate Somatosensory Pathways

In vertebrates, the cell bodies of most somatosensory afferent neurons are located in the dorsal root ganglia, which lie just outside the spinal column. Their axons enter the spinal cord or brain via the spinal or cranial nerves. Where they terminate in the cord depends on the type of information they carry and where the receptor end of the sensory neuron is located. Neurons that convey information from proprioceptors and from touch and vibration receptors branch immediately after entering the spinal cord. One set of branches terminates on interneurons in the cord and forms the neural circuits necessary for reflexes, as described in Chapter 17. The other set of branches continues to the medulla via dorsally located bundles of axons on the ipsilateral side of the spinal cord. These axons terminate in the **cuneate** and **gracile nuclei** in the medulla, which are collectively known as the **dorsal column nuclei**, where they form synapses with interneurons that cross over to the contralateral (opposite) side of the brain and ascend via the **medial lemniscus tract** to the thalamus. Other neurons then carry the signals to the somatosensory cortex. This route is referred to as the **lemniscal pathway** (Figure 14-1A).

Input from temperature, pain, and other receptors takes a different route into the brain. Axons from these receptors also enter the spinal cord via the nearest spinal nerve. However, all of these axons terminate in the cord, forming synapses in the gray matter with neurons whose axons cross the cord and ascend in several lateral tracts. Most of these ascending neurons synapse in the thalamus with neurons that project to the somatosensory cortex. This route is called the **spinothalamic pathway** (Figure 14-1B).

The two pathways are similar in that in both, sensory information originating from one side of the body crosses over to the other side. Information that ascends via the lemniscal pathway crosses over in the medulla. Information that ascends via the spinothalamic pathway crosses over just after it enters the spinal cord. The result is that sensory

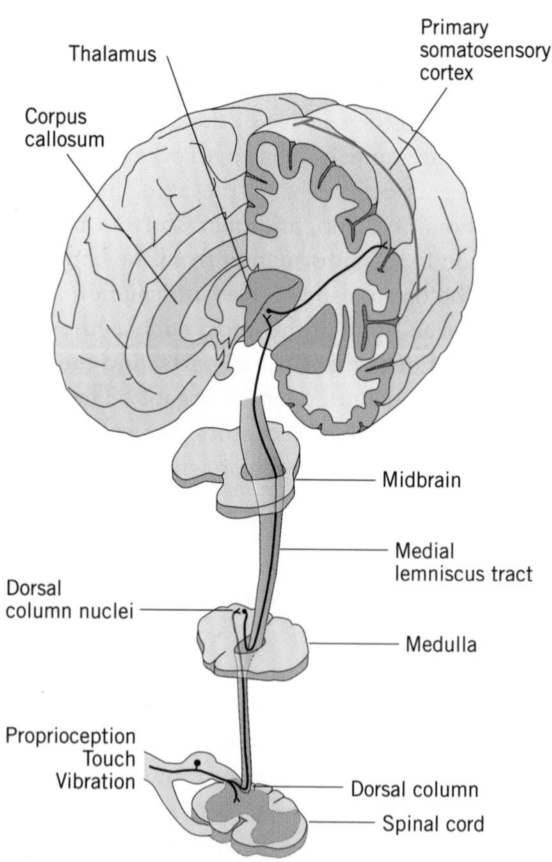

FIGURE 14-1. The two pathways by which somatosensory signals ascend to the somatosensory cortex in a human. (A) The lemniscal pathway, in which crossover of sensory signals to the other side of the nervous system occurs in the medulla.

information is processed on the side of the brain contralateral to the side from which the stimulus originates. Hence, individuals who suffer a stroke in part of the left somatosensory cortex will lose feeling in the right part of the body that is mapped to the damaged region.

The pathways differ in several respects. First, fibers in the lemniscal pathway conduct action potentials rapidly, in the range of 30 meters/second (m/sec) to 110 m/sec, whereas those in the spinothalamic pathway conduct signals more slowly, from a few meters per second to about 40 m/sec. Second, the two pathways have corresponding functional differences. The fast-conducting lemniscal pathway transmits information that may require rapid action, such as information from skin mechanoreceptors and from muscle spindle organs and other proprioceptive receptors. The slower spinothalamic pathway transmits information that can be interpreted less urgently, such as information about heat, cold, and some kinds of pain.

Somatosensory information enters the spinal cord or brain via the spinal and cranial nerves. From the cord, somatosensory signals travel to the brain via the lemniscal or spinothalamic pathway. Information about touch and related parameters travels via the lemniscal pathway, which crosses over in the medulla, then goes to the thalamus. Information about pain, temperature, and other information crosses over in the cord before it goes to the thalamus.

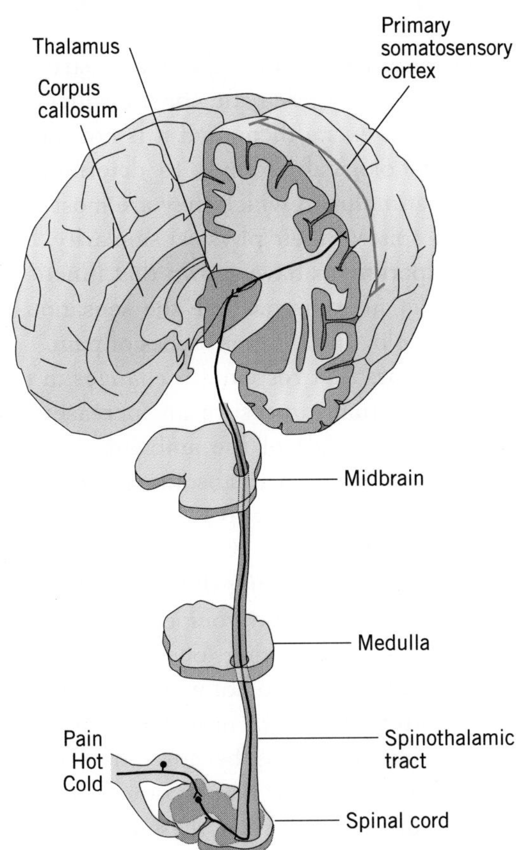

FIGURE 14-1 (continued). (B) The spinothalamic pathway, in which the cross-over occurs in the spinal cord.

Vertebrate Somatosensory Receptors

The different modalities mediated by the somatosensory system are sensed by a variety of receptors. The skin of a mammal, for example, contains receptors for temperature (hot and cold receptors), pain (nociceptors), and several kinds of contact (touch, vibration, pressure, and tickle receptors). The receptor cells that mediate different sensations differ not only in the route by which they send their signals to the brain, as described in the preceding section, but also in the characteristics of the neurites that carry information into the central nervous system. Peripheral neurons have been classified according to the velocity of the action potentials they conduct, designated by Greek and Roman letters. Fibers designated $A\alpha$ (A alpha) are fastest; $A\beta$ (A beta), $A\gamma$ (A gamma), and $A\delta$ (A delta) are progressively slower; C fibers are slowest. The A fibers are myelinated, which contributes to their greater conduction velocity, whereas the C fibers are not. Table 14-1 lists the main fiber types, their physical characteristics, and the modality or function of each.

TABLE 14-1. **Fiber Characteristics of Skin Receptors**

Modality	Fiber Type	Myelinated?	Conduction Velocity	Fiber Diameter
Skin contact	Aβ	Yes	30–70 m/s	6–12 μm
Cold	Aδ	Yes	5–30 m/s	1–5 μm
Sharp pain	Aδ	Yes	5–30 m/s	1–5 μm
Touch, tickle	C	No	0.5–2 m/s	0.2–2 μm
Hot	C	No	0.5–2 m/s	0.2–2 μm
Dull, burning pain	C	No	0.5–2 m/s	0.2–2 μm

Less familiar than the somatic receptors of the skin are receptors that provide the somatosensory system with information from muscles and other internal organs. Many of these receptors, such as those associated with joints, form distinct sense organs, but others are merely specialized sensory neurons like skin touch receptors. Some somatic information from internal organs, especially from the gut, reaches consciousness. The sensations of a full or upset stomach are familiar examples. Other information, however, rarely reaches consciousness in a direct way, although you can use it consciously. This information contributes to **kinesthesis**, the sense of body position in space. This sense, which allows you to close your eyes and touch the tip of your nose, is based on somatosensory information your brain receives from muscles and joints.

The skin and internal organs of vertebrates contain many kinds of receptors that provide information for the somatosensory system. Skin receptors may differ in the type of information they carry, the route they follow to the brain, or the velocity at which they conduct action potentials. Internal receptors contribute to the kinesthetic sense, the sense of body position in space.

Temperature Receptors and Nociceptors

Sensory neurons that respond to hot, cold, or damaging stimuli have one common feature: the receptive ends of the neuron always terminate in the skin or internal organs as **free nerve endings**. None exhibit any structural specializations for detecting or transducing the stimuli that excite them, and it is not possible on morphological grounds to determine the type of stimulus to which they are most sensitive. In spite of their physical similarity, however, experiments have shown that functionally distinct neurons mediate the sensations of hot and cold, and different types of pain.

Receptors that respond to changes in temperature actually measure an increase (hot) or decrease (cold) of the ambient skin temperature rather than an absolute temperature. You have probably experienced the consequences of this after returning inside on a winter's day and finding that even lukewarm water feels too hot on your cold hands. The experiment of adapting one hand to cold water and the other to hot water, then putting them both into lukewarm water as described in Chapter 10, works because the temperature receptors in your hands provide information about relative rather than absolute temperature. Although the receptive endings of hot and cold receptors cannot be distinguished morphologically, the fibers that carry information about hot and cold stimuli can be.

Relatively slowly conducting C fibers (Table 14-1) convey information about warm stimuli, and faster conducting Aδ fibers convey information about cold.

Because information about temperature is carried by fibers that conduct action potentials much more slowly than the Aβ fibers that carry information about ordinary touch, you become aware of the temperature of an object only after you have become aware of its contact with your skin. For example, a drop of hot water on the back of your hand will evoke two distinct sensations, first contact from the drop, then, perceptibly later, the heat. For a person about 5 feet 8 inches (173 cm) tall, the distance from the back of the hand to the spinal cord is about 74 cm. Assuming an average rate of conduction of 30 m/sec for action potentials from touch receptors, touch information will reach the spinal cord in just under 25 msec. If impulses travel along fibers from heat receptors at approximately 1.4 m/sec, information about heat will arrive at the spinal cord in 532 m/sec, more than half a second later. This time difference will be preserved or even increased as information travels to the brain.

Like thermoreceptors, nociceptors also fall into two categories according to the characteristics of the fibers that convey information to the spinal cord and the subjective sensations they produce. Activity in slow C fibers evokes a dull, burning feeling, whereas activity in the somewhat faster Aδ fibers produces a sharp, stinging sensation. Nociceptors with Aδ fibers are generally activated by stimuli that produce strong shear forces in the skin, like a cut, a strong blow (such as from a hammer on your thumb), or a tug on a hair.

Nociceptors that convey information via C fibers cannot easily be grouped based on the types of stimuli that excite them because many of these receptors respond to more than one kind of stimulus. For example, a single receptor might respond to a sharp blow, damaging heat, or even to chemicals that are released locally from mechanically damaged tissues. Because they respond to more than one modality of stimulus, these neurons are referred to as **multimodal receptors**.

Thermal receptors and nociceptors terminate in free nerve endings rather than having specialized receptive endings. Receptors for heat and for noxious stimuli that evoke dull, burning sensations send information along slowly conducting, unmyelinated C fibers. Receptors for cold and for noxious stimuli that evoke a sharp pain send information along slightly faster conducting, myelinated Aδ fibers. Some nociceptors are multimodal, capable of responding to several types of stimuli.

Skin Mechanoreceptors

Of all skin mechanoreceptors, those in mammalian skin have been studied most thoroughly. They consist of six main types: free nerve endings, **hair follicle receptors**, **Meissner's corpuscles**, **Merkel's nerve complex** (Merkel cells), **pacinian corpuscles**, and **Ruffini's end organs** (Ruffini's endings). These receptors can be differentiated by their locations, response properties, and types of stimuli to which they respond. Their locations in skin are illustrated in Figure 14-2, and their characteristics are summarized in Table 14-2. They are discussed in more detail in the following paragraphs.

Different types of receptors respond to different kinds of stimuli. Free nerve endings are thought to respond to very light touch. Researchers have also suggested that stimulation of some free nerve endings is responsible for the sensations of tickle or itch. Receptors associated with hair follicles will respond to any touch that moves the hair with which they are associated. The remaining four mechanoreceptors all respond to contact between the skin and a solid object, but not in the same way. Meissner's corpuscles will respond best to light touch, but neurons associated with Merkel's nerve complex and Ruffini's end organ will respond best to pressure rather than light touch, and the pacinian corpuscle will respond best to vibration.

Some skin mechanoreceptors adapt extremely rapidly; these are referred to as fast-adapting (FA) receptors (see Table 14-2). These phasic (rapidly adapting) receptors provide accurate, detailed information about changes in skin contact. Other receptors adapt quite slowly; these are referred to as slow-adapting (SA) receptors. These tonic receptors provide information about long-term contact with the skin. Some receptors (type I in Table 14-2) are located superficially in the skin and have small, highly localized receptive fields. Others (type II) have their sensitive endings positioned more deeply and because of the mechanical properties of the skin therefore have relatively wide and diffuse receptive fields.

Although knowledge of the characteristics of individual receptor types is necessary for a full understanding of the skin somatosensory system, concentrating only on these individual characteristics can be misleading. Skin receptors rarely act alone. Contact of the skin with a solid object almost always produces a myriad of responses in many kinds of receptors. It is the total pattern of response of these receptors that provides an animal with a rich and detailed picture of the nature of the object. For example, your ability to identify the texture of different surfaces, or to identify shapes by feel alone, depends on the interpretation of all the information from the different mechanoreceptors in your hands. What is important is how these receptors respond as your hands move over a surface or structure, not whether they respond mainly to light touch, pressure, or vibration.

Vertebrate skin contains many kinds of mechanoreceptors. Most provide detailed information about the characteristics of physical objects that contact the skin, such as its roughness and whether it is moving relative to the skin.

Internal Receptors

Skin receptors provide only part of the input to the somatosensory system. Both the internal organs of the gut and the muscle system of the body contain receptors that provide information about the internal state of the body. Many

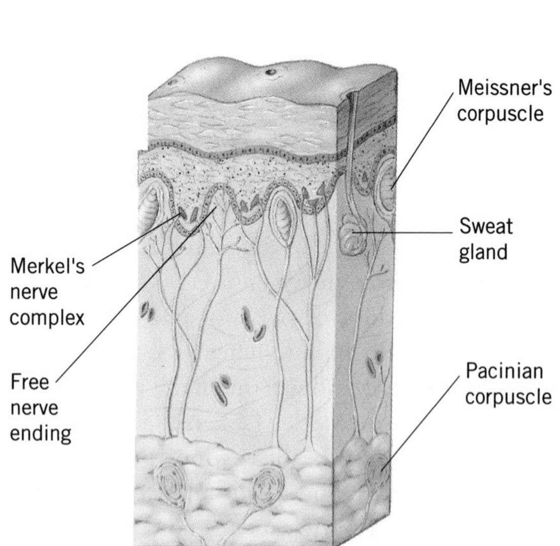

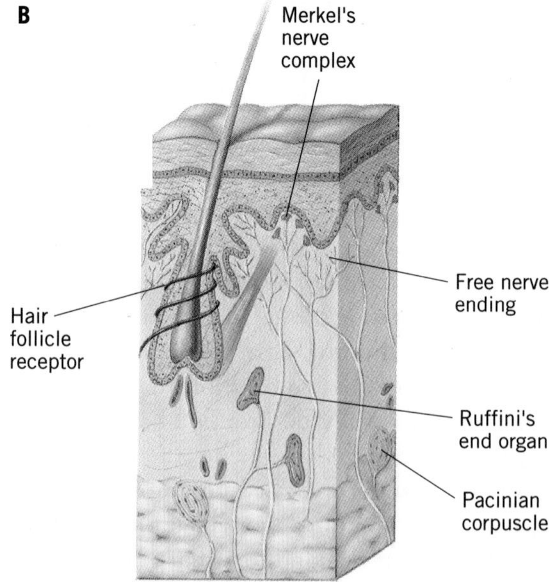

FIGURE 14-2. Sensory neurons of mammalian skin. (A) Hairless skin contains three types of mechanoreceptors (Meissner's corpuscle, Merkel's nerve complex, and the pacinian corpuscle) and free nerve endings for sensing heat, cold, and pain. (B) Hairy skin lacks Meissner's corpuscles but contains two additional mechanoreceptors, hair follicle receptors and Ruffini's end organs.

TABLE 14-2. Mechanoreceptive Structures in Mammalian Skin

Receptor	Appearance	Skin Type	Where Found	Adaptation (Type)	Modality
Free nerve ending		Hairy Hairless	Superficial	Slow	Tickle (light touch)
Hair follicle receptor		Hairy	Superficial	Fast	Light touch
Meissner's corpuscle		Hairless	Superficial	Fast (FA I)	Light touch (vibration)
Merkel's nerve complex		Hairy Hairless	Superficial	Slow (SA I)	Pressure (touch)
Pacinian corpuscle		Hairy Hairless	Deep	Fast (FA II)	Vibration
Ruffini's end organ		Hairy	Deep	Slow (SAII)	Pressure (touch)

of these receptors are components of the autonomic nervous system, and their activity does not produce conscious sensations. Others are part of the musculoskeletal system and provide proprioceptive information, reporting on the positions and actions of the parts of the body relative to one another and to the external world. Some proprioceptors contribute to kinesthesia, the sense of body position in space, but little is known about them.

The gastrointestinal tract contains at least two types of somatic receptors. Pacinian corpuscles are found in the walls of the stomach and intestines of both cats and humans, where they presumably provide information about touch and vibration. Their axons enter the spinal cord, where they ascend to the cuneate and gracile nuclei in the brain stem via columns in the dorsal part of the spinal cord. They synapse in the medulla with neurons that continue to the thalamus via the medial lemniscus tract. Mechanoreceptive free nerve endings are also present in the gut. Some authors have suggested on physiological grounds that these mechanoreceptors serve both ordinary mechanoreceptive and nociceptive functions, since they seem to respond to both normal and noxious stimuli. These fibers enter the spinal cord and are believed to cross over to the other side and ascend to the brain in the spinothalamic somatosensory tract. An interesting feature of information from the gut is that it is not organized in detailed somatotopic maps like information from the skin. You may have noticed that a stomachache is a fairly diffuse feeling, not localized to any one part of the viscera like a pinprick is localized on the skin.

Two major types of receptors can be found in joints and muscles: nociceptors and mechanoreceptors. Nociceptors consist of free nerve endings that respond to noxious stimuli, such as tears, excessive stress, or other extreme mechanical disturbance, and produce the sensation of pain. They are distributed throughout joints and muscles, and convey their information to the brain stem, the thalamus, and the somatosensory cortex via the spinothalamic pathway.

Joints and muscles contain four main types of mechanoreceptors, two of them similar to skin receptors: Ruffini's endings, pacinian corpuscles, Golgi tendon organs, and muscle spindle organs (Figure 14-3). Free nerve endings that mediate pain are also present. A thorough and systematic study of the distributions of these receptors in the body and in different vertebrates has never been done, but it is known that these receptors are widely distributed.

They are generally fast-conducting type Aα or Aβ neurons (see Table 14-1). In humans, Ruffini's endings are equally distributed in tendon and muscle tissue at the junction between a muscle and its tendon. These receptors adapt only slowly, and hence will provide information about steady state force conditions at a joint. Pacinian corpuscles are relatively rare in muscle but more common in tendons. They are rapidly adapting and hence provide information about dynamic changes in force or movement. **Golgi tendon organs** are located at the junction of a muscle and its tendon. They are activated by stretch or contraction of the muscle, and hence provide information about muscle force. Muscle spindle organs are located exclusively in muscle, where they signal any discrepancy between the rate of which a muscle actually contracts and the rate at which it is intended to contract, as you learned in Chapter 10.

Information from the muscle and joint mechanoreceptors enters the brain through various lateral and dorsal spinal columns, and is distributed to the cerebellum, the thalamus, and the cortex. The role of input from these sense organs to the cerebellum and the role of the cerebellum itself in the control and coordination of movement are described in Chapter 18. For now, it is enough to understand that these four types of mechanoreceptors provide an animal with detailed information about limb position, limb movement, muscle force, and actual muscle length compared with intended length. This information serves as the basis of the kinesthetic sense. Some of it is directly available as a conscious sensation (you can readily feel tension in your muscles), but much of it is not. It is difficult to describe how you know where your hand is when your eyes are closed. Yet you can easily use this information to move your hand to a specific place.

In addition to sense organs that are part of the autonomic nervous system, the viscera contain mechanoreceptors and nociceptors that provide information about the state of the gut and other internal organs. Receptors in joints and muscles are proprioceptors, and provide the information used by the cortex and cerebellum in the sense of kinesthesia to coordinate and guide movements of parts of the body.

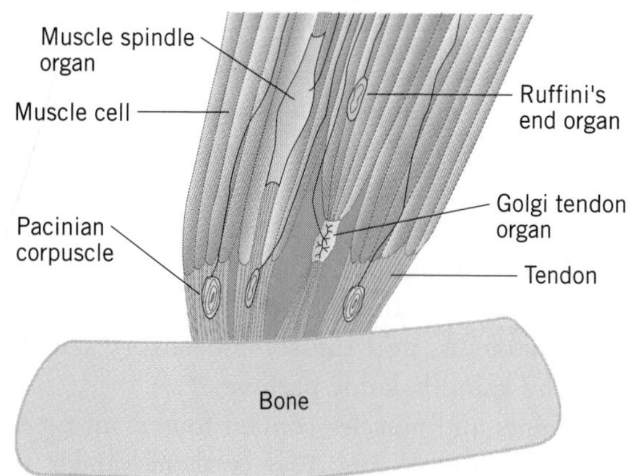

FIGURE 14-3. Mechanoreceptors in muscle and tendons in mammals. Four types are typically present: Golgi tendon organs and muscle spindle organs, which are mostly (Golgi) or exclusively (spindle) found in the muscle; pacinian corpuscles, which are mostly found in the tendon; and Ruffini's end organs, which are found with about equal frequency in tendon and muscle. Free nerve endings, which mediate pain, are also present but are not shown.

Transduction

Transduction of mechanical stimuli in somatosensory mechanoreceptors is mediated by two types of mechanically sensitive ion channels. **Stretch-activated channels** open when mechanical stimuli are applied to the cell, whereas **stretch-inhibited channels** close when the cell is stimulated. There appear to be several types of each kind of channel that differ in their conduction properties and ion selectivity. Both kinds of channels are fixed to the internal cytoskeleton anchored to the membrane and, in some cells, to an additional external skeleton consisting of carbohydrate chains. Molecular genetic studies of the mechanism of transduction in the roundworm *Caenorhabditis elegans* have suggested that some channels contain gates that are linked with springlike proteins to an external structure. When the external structure moves, the movement pulls the links, opening the channel gate. Such a mechanical arrangement may allow many channels to be affected by even a weak stimulus, giving the sensory neuron great sensitivity. The mechanism is similar to the proposed opening of channel gates on the stereocilia of auditory hair cells by pulling of tip links due to movement of the stereocilia in relation to one another (see Chapter 12). However, simple mechanoreceptive channels that are not associated with an external structure have been discovered in bacteria, so such structures are not always necessary for transduction.

All skin and deep mechanoreceptors except those consisting of free nerve endings are associated with or encased in some type of specialized structure that shapes the cell's response to mechanical stimuli. For example, the tips of the nerve endings of vertebrate pacinian corpuscles are encased in an accessory structure at the tip. In cross section, the structure appears layered, like the leaves of an onion (Figure 14-4A). Intact pacinian corpuscles are strongly phasic and extremely sensitive to repetitive stimuli (vibration). If the corpuscle around the tip of the neurite is removed, the bare tip will produce a tonic rather than a phasic generator potential when it is stimulated (Figure 14-4B), and will elicit a series of action potentials as long as the stimulus is applied. Hence, accessory structures can have a strong influence on the response characteristics of the sensory neuron.

The mechanisms of transduction in most types of free nerve endings are not well understood. These sensory neurons may respond to thermal, mechanical, or noxious stimuli. How hot or cold induces a response in the unspecialized terminals of free nerve endings is not known. Mechanical stimuli presumably act as described in the preceding paragraphs, but there are no substantive experimental data. Noxious stimuli are thought to work through chemical intermediaries. Intense mechanical stimuli damage tissue, which releases molecules that activate nociceptive neurons. It is known that adenosine triphosphate (ATP), a putative neurotransmitter in the peripheral nervous system, is released by cells in tumors, by skin cells, and by Merkel's cells, under the appropriate conditions. At high concentrations, ATP excites nociceptors through purinergic receptors, and the nociceptors convey the message of damage to the spinal cord.

In mechanoreceptors, stimuli are transduced by the opening or closing of stretch-sensitive channels in the cell membrane. The response of a mechanoreceptor is often shaped by an accessory structure. Nothing is known about the transduction of thermal or noxious stimuli in free nerve endings.

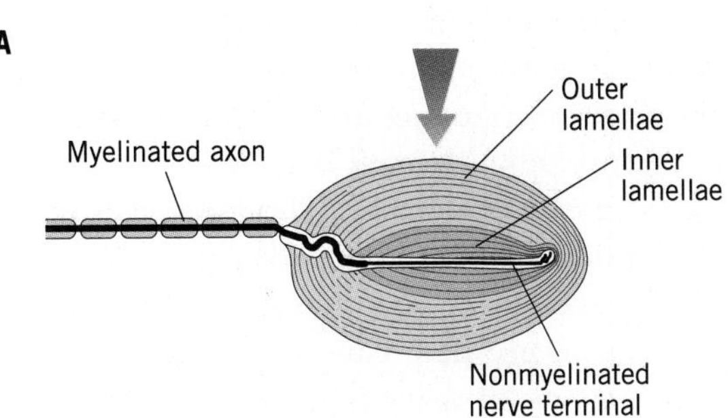

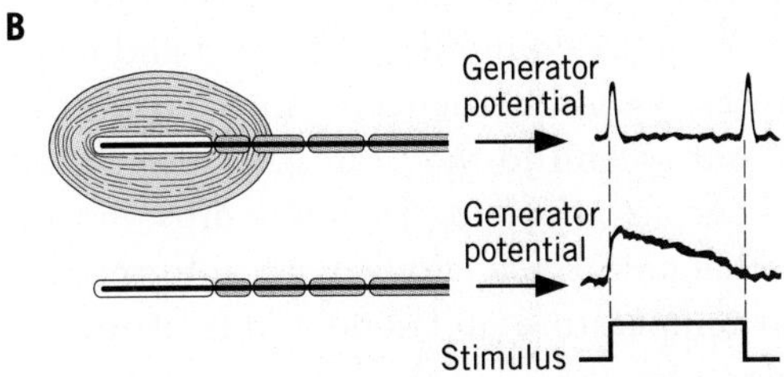

FIGURE 14-4. (A) A vertebrate pacinian corpuscle, showing the specialized capsule that covers the tip of the sensory neuron. The arrow shows the direction of force applied to the sensory receptor to produce deformation and hence depolarization of the dendrite. (B) Generator potentials evoked by application of brief pressure (marked stimulus) to an intact pacinian corpuscle (top), and to one from which the capsule has been removed (bottom). Note how the presence of the capsule dampens the neuron's response.

Central Processing of Somatosensory Information

As described previously, somatosensory information enters the spinal cord or the brain stem and progresses to higher brain centers. Some efferent or centrifugal modulation of the signals is possible before they reach the somatosensory cortex, however (see Chapter 9). This modulation helps shape the signals for interpretation by the brain. In addition, the somatosensory area of the cortex has two regions devoted to somatosensory information, which will be discussed later in this section.

Pathways for Pain

Every animal must be able to detect noxious or potentially damaging stimuli. Complex animals can have complex pathways to conduct information about noxious stimuli to the brain and more sophisticated ways to interpret it. These can include the ability to sense pain, which on behavioral grounds simple animals like insects do not have. Furthermore, because severe pain can incapacitate an animal in a time of danger, animals that are able to feel pain also must be able to modulate or ameliorate it so they can act in spite of it.

Nociceptors enter the spinal cord and form synapses with interneurons that cross the cord and ascend to the brain. Some of these interneurons travel to the thalamus in the spinothalamic tract, along with information about temperature and some kinds of touch. Others terminate in the reticular formation of the pons and midbrain, where they may synapse with other fibers that project to the thalamus. From the thalamus, other neurons carry information about noxious stimuli to the somatosensory cortex.

Modulation of information about noxious stimuli can occur at synapses along these pathways. For example, it has been found that stimulation of large-fiber skin mechanoreceptors such as Merkel's complex and Ruffini's end organs (type Aβ fibers) can reduce the subjective intensity of pain. Young children take advantage of this phenomenon by rubbing the skin around a scraped knee to alleviate pain. In the **gate control theory of pain**, Ronald Melzack and Patrick Wall suggested that this reduction of pain occurs in the spinal cord. They proposed that interneurons that receive input from nociceptors also receive inhibitory input from Aβ fibers. When both the nociceptors and the nearby Aβ fibers are active simultaneously, the activity of the interneurons that carry information about the noxious stimulus is reduced.

There is also strong evidence for descending control by the brain of the pathways that carry information about noxious stimuli. For example, pain input can be blocked at the level of synapses in the spinal cord by endorphin neuromodulators. In the spinal cord, interneurons that descend from the brain stem release endorphins at the synapses between the incoming afferent nociceptors and the neurons that send the signals to higher brain centers. The endorphins suppress transmission across these synapses both by presynaptic depression of transmitter release from synaptic terminals and by postsynaptic inhibition of the spinal interneurons (Figure 14-5). Endorphins also work at synapses in the pathway carrying information about noxious stimuli in the brain stem. It is at these endorphin-sensitive synapses that morphine acts to prevent information about pain from reaching consciousness.

In some instances, neural mechanisms cause an increase rather than a decrease in pain sensitivity. Inflamed skin, for example, becomes extremely tender in part because of increased firing of nociceptors and in part because of amplification of these signals in the central nervous system. In some circumstances, even input from Aβ mechanoreceptors that normally evokes a sensation of touch can be perceived as painful. The biological usefulness of such increased sensitivity is presumed to be reduced use of an injured part of the body during a period when it is more vulnerable to further damage. Since increased pain is a condition often associated with various medical conditions, considerable research is currently

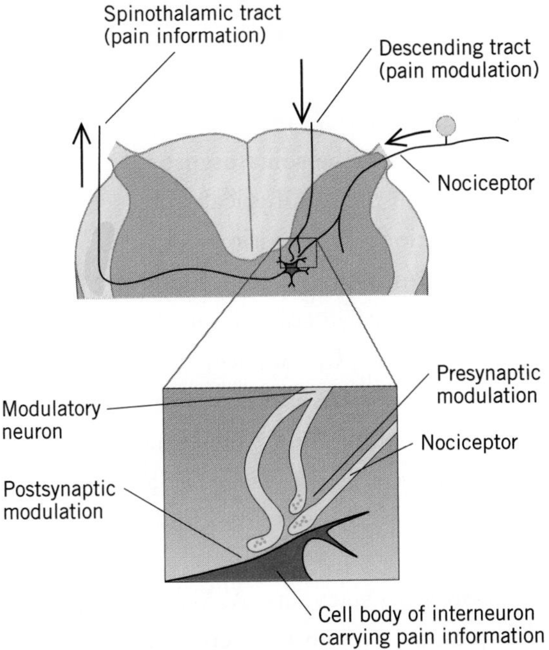

FIGURE 14-5. Spinal mechanisms of pain modulation. Neuromodulatory neurons from the brain synapse on the axon terminals of nociceptors and on the cell bodies of interneurons that carry pain information. These neuromodulatory neurons release endorphins and presynaptically or postsynaptically inhibit the transfer of input from the pain receptors.

directed toward understanding and controlling the mechanisms that underlie it.

Nociceptors send information about noxious stimuli into the spinal cord, from where it ascends to the brain stem, thalamus, and somatosensory cortex. Pain information can be modulated along these pathways by suppression via large mechanosensitive fibers that also enter the spinal cord, or by descending control from higher brain centers via endorphins.

Touch Discrimination

Presenting complex visual and auditory stimuli to the eyes and ears is more informative than presenting simple spots of light or pure tones. Similarly, presenting "complex" stimuli to the skin is more informative than simply using a fine pin. In studies of the somatosensory system, complexity can be introduced by moving a small brush or a complex surface across the skin. Using the method of microneurography (Box 14-1), it is possible to determine the effects of these stimuli in humans as well as in other primates.

Movement of a brush across the skin produces a consistent pattern of neural response in individual receptor cells that signals rate and direction of movement. That is, the responses of individual receptors vary according to how fast and in which direction the stimulus is applied. Fast-adapting receptors (FA type, Table 14-2) show greater increases in firing rate when the brush is moved faster than do slow-adapting receptors (SA type), which indicates that these phasic neurons encode information about the velocity of the stimulus better than the tonic SA receptors do. In addition, some receptors respond better to movement in one direction than to movement in the opposite direction, presumably because of differences in the way they are stimulated when the skin is gently pulled in different directions by the brush. Hence, the somatosensory system is able to determine the speed and direction of a stimulus from the responses of individual receptors in the skin.

Bringing a complex surface in contact with the skin can reveal the astonishing resolving power of the somatosensory system. Recordings from individual mechanosensitive afferents in human subjects while a pattern of embossed dots is rubbed across the skin of a finger reveal activity in both FA and SA receptor types. Type I receptors (FA I and SA I, Table 14-2), which lie close to the surface and have small, discrete receptive fields, are able to resolve dots only 1.5 mm apart. That is, individual dots spaced that close together will evoke discrete bursts of action potentials in type I receptors as they are moved over the skin; dots spaced closer than that will evoke more or less continuous activity, indicating that individual dots cannot be distinguished. Even the deeper type II receptors with larger and more diffuse receptive fields can resolve dots that are 3.5 mm apart.

BOX 14-1

Needles in Your Arm: Recording Human Neural Activity

Fifty years ago, investigators interested in receptors of the somatosensory system had two choices. First, they could carry out psychophysical studies in which human subjects reported their sensations when various stimuli were applied to the skin. These studies could provide valuable information about the relationship between various controlled stimuli and the sensations they evoked. Second, they could carry out physiological studies in animals, recording from peripheral nerves that were exposed surgically. These studies could provide valuable information about the actual responses of individual neurons to somatic stimulation. What was missing, however, was a way to relate subjective sensations to the activity of individual types of receptors.

The situation changed in 1968 when two Swedish scientists, A. Vallbo and K.-E. Hagbarth, introduced the technique now known as **microneurography**, a method for recording activity in single axons from nerves in awake human subjects. The method involves the insertion into the skin of a sharpened tungsten electrode (5–15 μm at the tip). Except for the tip, the electrode is coated with several layers of lacquer to serve as an electrical insulator, so that electrical activity will be picked up only at the end. The electrode is gently manipulated through the skin and near a nerve from which recordings are to be taken, then adjusted until a good recording of a single unit is obtained.

The method is not without its risks. The electrode must penetrate the sheath of the nerve bundle so that recordings from single axons can be obtained without damaging the axons in the nerve. But inserting even a fine metal shaft into a nerve bundle is bound to cause some damage, even if only temporary. In fact, the symptom of the approach of the electrode to a good recording site is brief paresthesia, a burning or prickling sensation in a localized area of the skin, caused by mechanical stimulation of the nerve. In spite of the risks, a small number of dedicated researchers interested in the somatosensory system have used this method on themselves to obtain fundamental information about the relationship between mechanical stimuli and the responses of individual cutaneous mechanoreceptors in humans, as described in the section entitled "Touch Discrimination."

This ability to discriminate detail has been demonstrated in monkeys as well, both at the level of individual receptors in the skin of the hand and in the somatosensory cortex. For example, rubbing a cylinder containing embossed letters of the alphabet over a fingertip generates a pattern of response of skin mechanoreceptors that effectively conveys the shape of the letter (Figure 14-6A). Receptors with different properties (such as SA vs. FA types) respond with different levels of precision (Figure 14-6B, C). Furthermore, neurons in the finger area of the somatosensory cortex also respond precisely enough to stimulation of the finger to allow the letters to be identified (Figure 14-6D).

Receptors in the skin of the hand convey precise information not only about the strength and duration of an applied stimulus but also about the speed and direction of its movement. Furthermore, the pattern of activation of individual receptors by a complex moving stimulus like an embossed letter is accurately reflected in the activity of somatosensory cortical neurons, as well as in the activity of primary skin receptors.

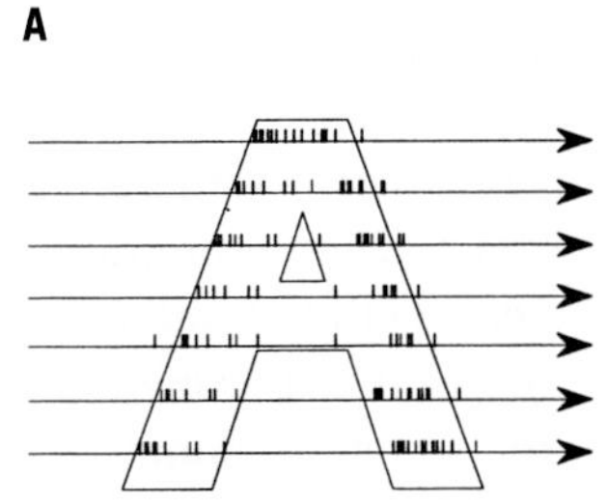

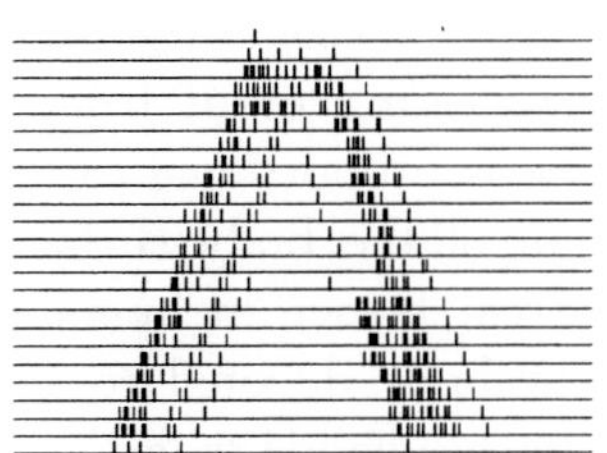

FIGURE 14-6. Discrimination of detail by skin mechanoreceptors in a monkey. (A) Repeatedly rubbing an embossed letter over a finger produces action potentials in a single tactile receptor in the finger. The letter is much larger than the receptive field of the stimulated receptor. Here (left), each horizontal line represents a single swipe of the embossed letter across the finger, and the vertical tick marks on each line represent the action potentials generated in the tactile receptor as the letter moves across its receptive field. The vertical position of each horizontal line relative to the outline of the letter shows the part of the letter that is moved across the receptive field. Notice that the action potentials occur primarily as the embossed surface of the letter crosses the field. Rubbing the letter across the finger about 40 times rather than just 4, moving it down just a little after each rub, and then marking action potentials with tick marks as different parts of the letter are rubbed over one skin receptor yields a picture of the letter (right). (B) Response of a single slowly adapting peripheral tactile receptor to a series of embossed letters. (C) Response of fast-adapting peripheral tactile receptor. (D) Response of the finger region of the somatosensory cortex.

Cortical Representation of Somatosensory Information

In vertebrates, somatosensory information is sent via the thalamus to the primary somatosensory cortex, designated SI (S one). Since the discovery of the somatotopic organization of this cortex, it has served as the quintessential example of topographical mapping in the vertebrate brain (see Chapter 9). The entire surface of the body and many of the internal organs are mapped onto this cortical region so that adjacent regions of skin are represented in adjacent regions of the brain (see Figure 9-13). The size of the brain region devoted to processing input from a particular area of skin is proportional to the density of innervation of the body surface, not the size of the area. Hence in humans, more space in the brain is devoted to the heavily innervated lips and hands than to the sparsely innervated back. The somatosensory cortex is also a good example of the columnar organization of some cortical areas (see Chapter 9). The various types of receptors (mechanoreceptors, thermal receptors, nociceptors) in one part of the skin, say the thumb, all send their information to the same region of the cortex, the thumb region in this case. Fibers carrying information from each receptor type, however, project to specific columns devoted to that type of receptor within the general cortical region (see Figure 9-14).

Just as the visual, auditory, and chemosensory systems have more than one region of cortex in which the appropriate sensory signals are processed, so too does the somatosensory system. The secondary somatosensory cortex, designated SII (S two), is located laterally and somewhat posteriorly in the brain relative to SI, partly inside the lateral sulcus (Figure 14-7). In the visual and auditory systems, most output from the thalamus is directed to the appropriate primary cortex, and input to the secondary and other sensory cortexes is mainly from the primary cortex. This somewhat simplified description should nevertheless call to mind both serial and parallel processing of sensory information in these systems. The situation in the somatosensory system is more complex. In cats and other mammals, both SI and SII receive substantial input directly from the thalamus. This arrangement is also present in primates, but in these animals SII receives considerable input from SI as well. Hence, the somatosensory system also shows both parallel and serial processing of sensory input.

The role of SII in processing somatosensory information and its relationship to SI are not fully understood. For example, it is not clear to what extent SII is topographically organized. Some researchers find little indication of mapped responses to stimulation of skin receptors. Others, however, report that at least a crude topographical map is present. The conflicting results may be due to the use of different experimental animals or stimuli. As for function, some researchers have suggested that SII extracts more abstract features from a stimulus than does SI, just as in the visual and auditory systems the primary sensory cortex does mainly preliminary processing, whereas the secondary and other sensory cortical regions perform functions like recognizing faces and putting a meaning to words. In the somatosensory cortex, SII might abstract features of a stimulus such as

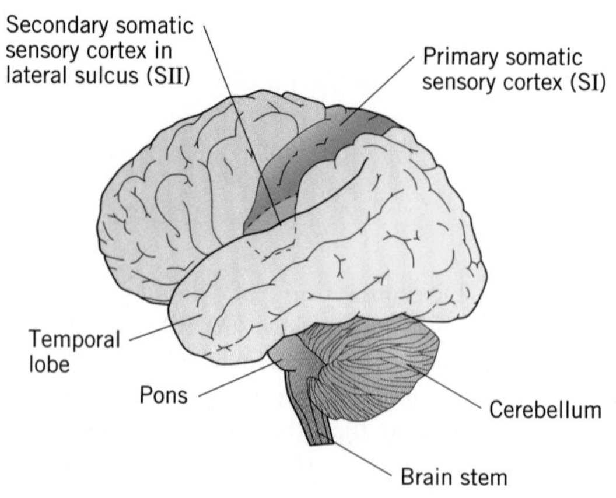

FIGURE 14-7. The human brain, showing the locations of the primary (SI) and secondary (SII) regions of the somatosensory cortex.

texture, and hence not be concerned with precise spatial localization. In these circumstances, a precise map of skin surface would not be expected for some types of receptors.

Vertebrate brains contain two regions of cortex devoted to processing sensory information from the body, designated the primary (SI) and the secondary (SII) somatosensory cortex. Region SI is organized topographically, and input is segregated into columns. Region SII is only weakly topographically organized; it receives input from the thalamus and, in primates, from SI. Region SII may abstract surface features such as texture from complex stimuli.

The Somatic Senses of Invertebrates

Invertebrate animals have at least as rich a diversity of somatic receptor types as do vertebrates. Annelids, mollusks, and arthropods all have a variety of receptors that respond to mechanical stimuli from the body surface, from joints, and from around joints, as well as receptors that provide information about modalities such as temperature. Invertebrates seem to have proprioceptive receptors for a kinesthetic sense, as well as somatic receptors for information about the internal and external environment. Furthermore, many receptors in or near the skin of invertebrates, especially in annelids and mollusks, are free nerve endings. Little is known about them. In leeches, however, it has been shown that some free nerve endings are nociceptors, meaning that they respond to mechanical stimuli like pinching that are strong enough to damage tissue. Stimulation of these nociceptors evokes swimming. Other invertebrates also show aversive behavior to a variety of noxious or potentially damaging stimuli. Locusts, for example, can be trained to maintain a particular leg position to avoid strong heat focused on the head. For these reasons, it is believed that nociceptors are widespread throughout the animal kingdom. Nevertheless, invertebrates do not have a brain region devoted to noxious or other somatic input, nor is there any evidence of self-awareness among them, so it seems unlikely that invertebrates feel pain in the same way we do.

In invertebrates, input from the body surface, the muscles, and the joints is not funneled to a single associative region of the brain for interpretation. In insects, for example, input from each appendage or body segment is processed in the ganglion that controls that appendage or segment. Somatic receptors, such as mechanoreceptive hairs on a leg, project axons into the ipsilateral ventral neuropil (see Chapter 2), where they form a topographical map of the leg. Distal parts of the leg are represented in the posterior part of the ganglion, and proximal parts of the leg are represented in the anterior part. Proprioceptors, such as those that provide information about joint angle or leg movements, terminate in intermediate levels of neuropil. The sensory afferent fibers make extensive synaptic connections with local interneurons, which in turn connect with motor neurons to complete reflex circuits. These reflexes are organized so that stimulation of a particular sensory receptor will evoke an appropriate neuromuscular response, as described in greater detail in Chapter 17.

Invertebrates have a distributed somatosensory system that includes somatic and proprioceptive receptors, including nociceptors. Somatosensory information is processed mainly at the level of the local ganglion that receives it. In insects, somatic information projects in a topographical pattern from a leg into the nearest ganglion.

The Sense of Balance

The role that human perceptions play in advancing or retarding our understanding of biological systems is most clearly manifested in studies of sensory systems. Even prehistoric peoples must occasionally have speculated about the nature or basis of the sense of touch or smell. Appreciating a sense, to say nothing of understanding it, becomes much more difficult when the "sense" involves no external object that can be perceived. When you see, hear, smell, taste, or touch something, you can usually identify the source of the stimulus external to yourself. In the case of the "sense" of balance, however, this is not so. This sense allows you to walk upright by providing information about the position of the body with respect to gravity, and about body movements. However, you are usually entirely unconscious of its actions; even when you think about it, there is nothing you can point to as it acts.

In most animals, information about body orientation is provided by statocyst organs, as described in Chapter 10. You will recall that these organs consist of one or more stonelike concretions called statoliths, attached to or resting on sensory hairs in a fluid-filled hollow chamber. Tilt or other movement of the animal results in movement of the statolith and therefore a change in the pattern of impulses from the sensory hairs. Nearly all animals have statocysts, even simple jellyfish, which use them to keep floating upright in the ocean.

The sense of balance is present in all animals, even simple ones. It is usually mediated by statocyst organs containing mechanoreceptors that respond to changes in the position of the statolith.

The Vertebrate Vestibular System

Vertebrates also have statocyst organs, but in this group of animals the statocyst is a more complex structure than that found in most invertebrates. Because they are located in a hollow opening in the skull known as the vestibule, the statocyst organs and the other structures involved with balance are known as **vestibular organs** or, collectively, the **vestibular system** (Figure 14-8). The vestibular system is part of the inner ear (see Figure 12-1) and is contiguous with the cochlea. Like the cochlea, it is filled with the fluid called endolymph, which is important in the process

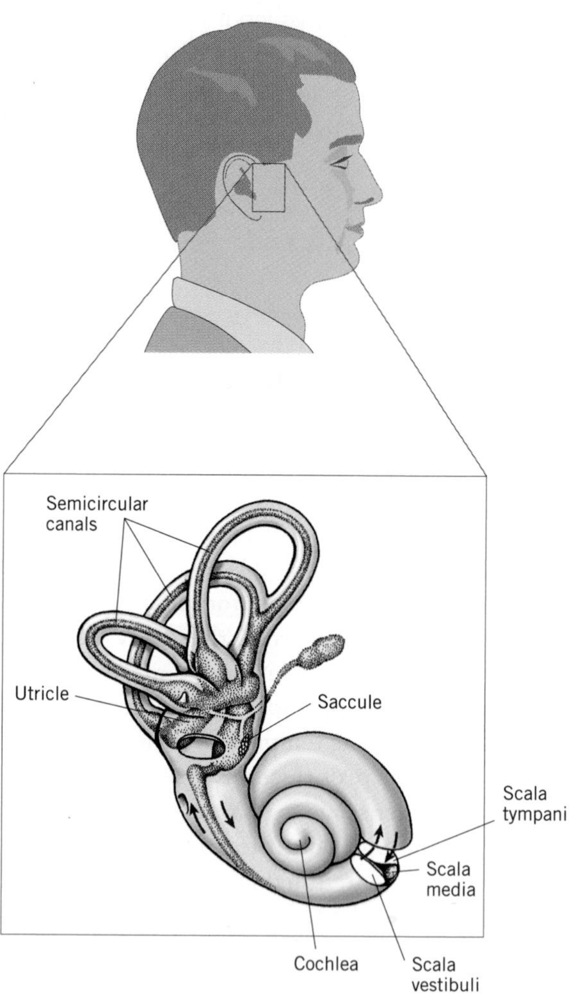

FIGURE 14-8. The human inner ear, showing the cochlea and the vestibular organ. The vestibular organ consists of two parts, the semicircular canals and the otolith organs. Each otolith organ itself consists of the utricle and saccule and responds to linear acceleration. The semicircular canals (ducts) are arranged orthogonally to one another; they respond to angular (rotational) acceleration.

of transduction explained in a following section. The mechanoreceptors of the vestibular system are hair cells like those in the cochlea. These cells are nonspiking, anaxonal neurons that release neurotransmitter continually to afferent neurons of the **vestibular ganglion.** The afferent neurons form the **vestibular nerve,** then join the fibers from the spiral ganglion to form cranial nerve VIII, the auditory nerve, through which they enter the brain.

The vestibular system of a vertebrate animal consists of several parts and performs several functions. Its two main parts are the statocyst organs, known in vertebrates as **otolith organs,** and the **semicircular canals.** In most vertebrates, each otolith organ itself has two parts, the **utricle** and **saccule.** In each part, a region called the **macula** (pl. maculae) is densely covered with mechanoreceptive hair cells like those found in the cochlea. As described in the next section, the stereocilia of the receptor cells are attached to a gelatinous mass in which small, stonelike particles are embedded. These particles, called **otoliths,** serve the same function as the statoliths in the statocyst organs of invertebrates. In mammals, the utricular macula is oriented horizontally, and the saccular macula is oriented vertically. The orientation of the semicircular canals in three planes is described in a following section.

Vertebrate organs of balance consist of two vestibular organs, collectively called the vestibular system. The two organs are the paired otolith organs, each composed of a utricle and a saccule, and the semicircular canals.

The Otolith Organs

The otolith organs have two functions: to monitor the position of the body relative to the force of gravity, and to provide information about linear acceleration of the body. Strictly speaking, since the otolith organs are in the head, it is actually the position or motion of the head that is detected. In most cases, however, forces applied to the body are transmitted to the head as well, so the otolith organs in effect serve the whole body.

The utricle and saccule both contribute to these two functions, but each is specialized for one. Monitoring the direction of gravity is done mainly by the utricular macula, which is oriented horizontally (Figure 14-9). Imagine a mass of gelatin glued to the palm of your hand. Within the mass are a number of small marbles. If you hold your hand horizontally, palm up, you will feel the weight of the marbles directly on your palm. If you now rotate

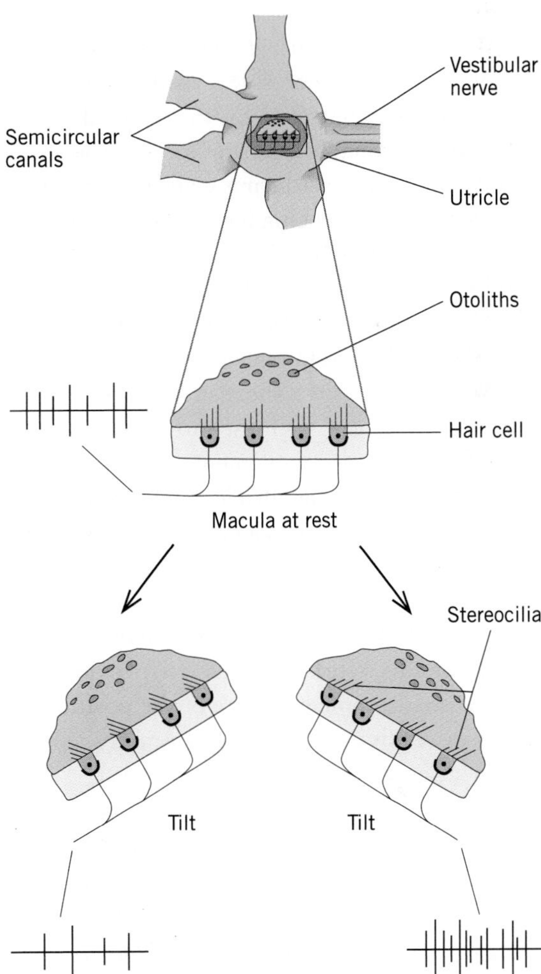

FIGURE 14-9. The effects on the stereocilia of tilting the utricle. Tilt in different directions will cause deflection of the stereocilia of the hair cells in different directions and changes in the frequency of firing of the vestibular afferents, as shown.

your hand until your palm is vertical, the weight of the marbles will be felt as a shearing force rather than weight. If you rotate your hand again, until your palm is facing down, you will feel the pull of the marbles away from your palm. The utricle operates on a similar principle. The horizontal orientation of the utricular macula allows it to be especially sensitive to changes in the direction of gravity due to changes in head angle. Any tilt of your head will cause a change in the direction that gravity is pulling on the otoliths, and hence a change in the direction in which force is being applied to the stereocilia. Consequently, each possible position of the head will generate a unique pattern of sensory impulses that can be interpreted to determine the exact position.

Linear accelerations of the whole body, on the other hand, either side to side or up and down, are best detected by the vertically oriented saccular macula. Linear accelerations can be self-generated, as during walking, or externally generated, by being jostled in a crowd, for example. Even small movements of the body and head will generate strong signals from the saccule, because the otoliths essentially hang as weights attached to the macular surface. Think of a ball attached by a short string to a vertical support. Even the slightest movement of the support will cause a movement of the ball.

The saccule and the utricle will both respond to all movements. The saccule will respond to tilting as well as to linear acceleration, and the utricle will respond to linear acceleration as well as to tilt. The two components of the otolith organs work together to provide the brain with a complete picture of the position of the head relative to the force of gravity and of the linear acceleration of the body.

In vertebrates, the vestibular system is responsible for detecting body acceleration and for maintaining balance. The utricle primarily monitors the direction of the force of gravity (head tilt), and the saccule detects linear acceleration of the body.

The Semicircular Canals

The semicircular canals, which are specialized to respond to angular (rotational) acceleration, constitute the other main part of the vertebrate vestibular system. As the name implies, these canals are hollow, roughly circular tubes. Like the otolith organs and the cochlea, they are filled with endolymph. All the canals have a common opening that leads to the utricle and saccule. In most vertebrates, there are three canals in each vestibular organ. One is horizontal, and the other two are vertical, at right angles to each other. In humans, the two vertical canals are oriented at about 45° in front of and 45° behind a transverse plane through the head (Figure 14-10). Each semicircular canal contains an enlargement, called the **ampulla**, near its base. Inside the ampulla is a field of hair cells, with their stereocilia projecting into a small, gelatinous **cupula** that is similar to the cupula found in fish lateral line organs (see Figure 10-5) and that partly occludes the lumen of the canal.

The semicircular canals work on the principle of conservation of inertia. If you shake your head, the fluid in your horizontal semicircular canals will tend to stay in place because of inertia while the walls of the canals move past it (Figure 14-11). You may have seen this principle at work if you have twisted a pail of water by its handle. When you rotated the pail, the water in the bucket stayed in place while the wall of the pail moved because the friction between the pail and the water was not sufficiently strong to overcome the inertia of the water as you rotated the pail. The same situation prevails in the semicircular canals. The endolymph in the canals will tend to stay in place as the walls of the canals move. This relative motion between the endolymph and the wall of a canal will cause the cupula in the ampulla to be deflected, thereby stimulating the hair cells in the cupula by bending the stereocilia.

The semicircular canals are phasic sense organs. The strongest deflection of the cupula will be due to acceleration at the start of

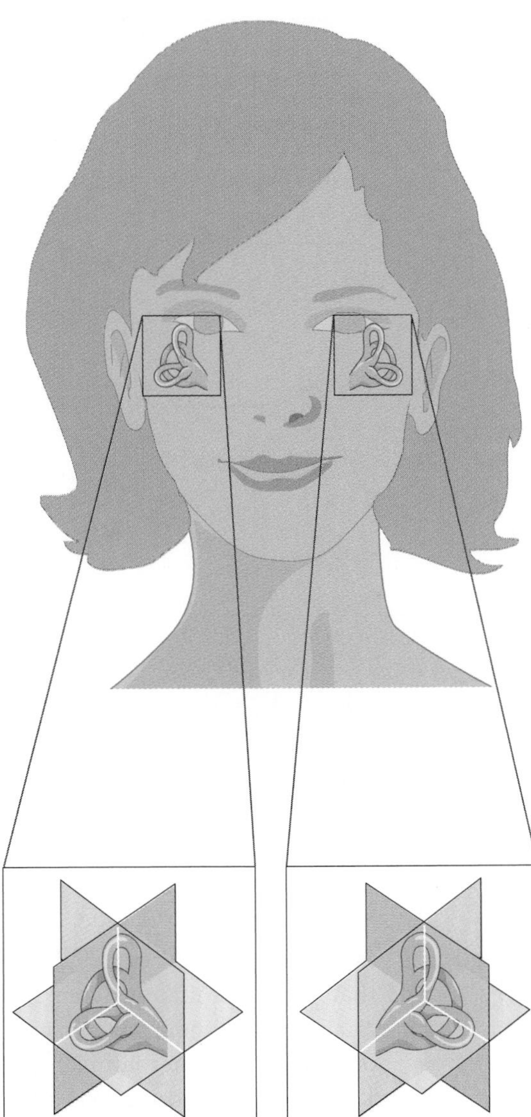

FIGURE 14-10. The orientation of semicircular canals in a human head. One canal is horizontal, and the other two are vertical. Each canal is oriented at right angles to the other two.

the rotation. If rotation is maintained, as for example when a skater twirls on ice, viscous drag will soon bring the endolymph into motion along with the canal. When this happens, there is no longer any relative motion between the endolymph and the wall of the canal, the hair cells are no longer stimulated, and they no longer depolarize (Figure 14-12). When the skater stops twirling, the canal walls stop rotating, but the endolymph keeps moving, once more providing a strong push to the cupula. This time the cupula is pushed in the opposite direction, and the hair cells are inhibited, reducing their release of neurotransmitter below background. Because there are three canals on each side of the head oriented orthogonally to one another, every possible rotational movement of the head will cause a unique pattern of impulses from the afferent neurons of the semicircular canals.

The semicircular canals contribute to postural reflexes, but their main function is to stabilize gaze when the head moves. An animal needs to operate in a stable visual environment. It will not do for the visual field to move every time the animal moves its head just because the eyes are in the head. The semicircular canals play a critical role in stabilizing gaze by providing information about head movements to several nuclei in the brain stem. Reflexes involving neurons in these nuclei evoke movements of the eyes that compensate precisely for head movements, as you will learn in Chapter 20.

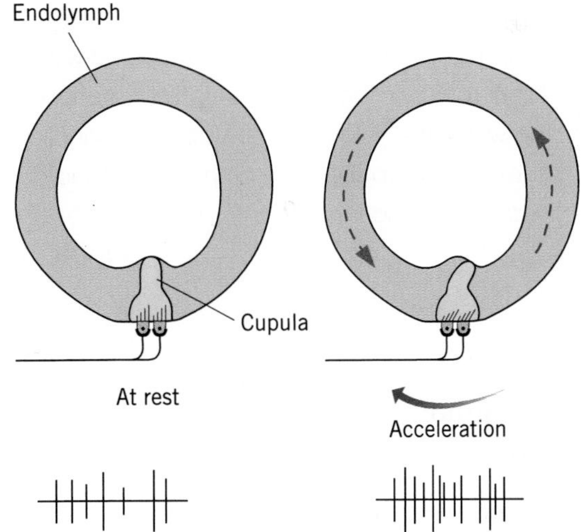

FIGURE 14-11. The effects of rotation on fluid in a semicircular canal. The wall of the canal rotates (large solid arrow), causing relative motion between the wall and the stationary fluid (endolymph) in the canal (small dotted arrows). The relative motion deflects the cupula, thereby depolarizing the hair cells attached to it, and exciting the afferent neurons with which the hair cells synapse.

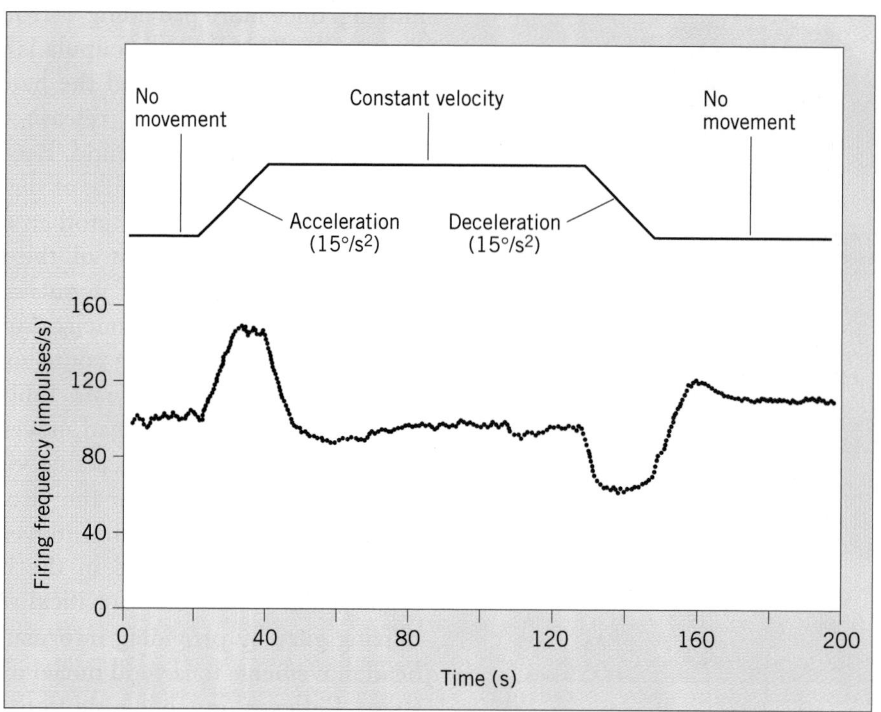

FIGURE 14-12. The response of a single vestibular afferent from a semicircular canal of a squirrel monkey in response to rotation of the whole body in the plane of the canal. Note that the average rate of firing of the sensory afferent transiently increases when the body is rotated in one direction and held there, and transiently decreases when the body is rotated back again.

The semicircular canals are fluid-filled tubes that help detect angular acceleration (rotation) of the head by relative motion between the fluid and the wall of a canal. They help keep the gaze of the eyes steady when the head is moved.

Vestibular Pathways

The axons in the vestibular nerve enter the medulla and synapse in one of four pairs of **vestibular nuclei** there. From these nuclei, fibers descend via several **vestibular tracts** to motor centers in the spinal cord, and ascend to other nuclei in the pons and the cerebellum, and to the cortex as well.

The two main descending vestibular tracts carry axons that serve different functions. The **medial vestibulospinal tracts (MVSTs)** carry signals that originate mainly from the semicircular canals. MVST neurons synapse directly on motor neurons innervating muscles of the neck and trunk, whose action they affect to ensure that the body remains upright (Figure 14-13). It is activity in these neurons that keeps your back straight against the pull of gravity when you stand. The **lateral vestibulospinal tracts (LVSTs)** carry signals from the receptors in the maculae of the otolith organs. LVST neurons synapse indirectly with motor neurons controlling limb muscles. It is activity in these neurons that controls movements of your arms as you swing them to keep your balance when you walk along a narrow beam.

Tracts also ascend from the vestibular nuclei. Most ascending vestibular information goes to the cerebellum or to the brain stem. Ascending fibers synapse in the abducens nuclei in the pons and in the trochlear nuclei and oculomotor nuclei in the midbrain. In these nuclei, vestibular information participates in reflexes that stabilize the eye or, via the cerebellum, control posture (see Chapter 20). In addition, there is also a small vestibular tract that ascends to a nucleus in the thalamus (not shown in Figure 14-13). Neurons in this tract synapse with neurons that project to two cortical regions. One is a small area in the somatosensory cortex devoted to input from the face, and the other is a region near the auditory area of the temporal lobe. It is thought that in humans these areas produce subjective sensations relating to balance, since electrical stimulation of the areas during brain surgery produces subjective sensations of body movement and vertigo.

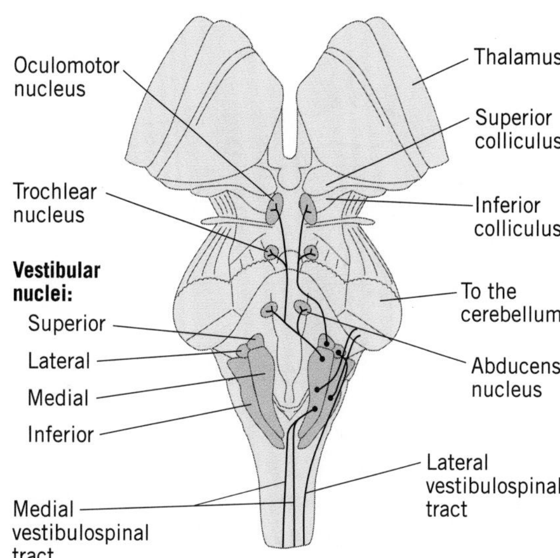

FIGURE 14-13. The main vestibular pathways in the human brain. Input from the vestibular system and synapses in the vestibular nuclei in the brain stem. From there, pathways distribute vestibular information down to spinal motor centers and up to various other brain nuclei. Minor tracts to the thalamus and cortex are not shown.

Afferents from the vestibular apparatus enter the brain via the vestibular nerve and synapse in vestibular nuclei in the brain stem. From there, vestibular signals descend via two pathways to spinal motor centers. Signals from the semicircular canals control neck and trunk muscles, and signals from the utricle and saccule control limb muscles. Vestibular signals ascend to the pons, midbrain, and cerebellum to control eye movement and influence balance. They also ascend to the cerebral cortex, where they may mediate sensations associated with movement.

Transduction in Vestibular Cells

Vertebrate vestibular receptors are like the receptor cells of the lateral line neuromast organs (see Chapter 10) and the hair cells of the mammalian cochlea (see Chapter 12): nonspiking, anaxonal, and bearing a tuft of stereocilia of different sizes. As in hair cells in the ear, bending the stereocilia in a direction toward the tallest depolarizes the receptor cell, whereas bending them in the opposite direction hyperpolarizes it. Transduction is also similar, taking place via mechanical opening or closing of nonspecific cationic channels on each stereocilium as a consequence of the movements of the stereocilia relative to one another (see Figure 12-5C).

Because the hair cells in vestibular systems are morphologically polarized, they must be properly oriented in order to respond to movements of the head or body. In the ampullae of the semicircular canals, the receptor cells are aligned so that the rows of stereocilia from tallest to shortest all face the same direction along the length of the canal. With this arrangement, all the stereocilia will be bent toward the tallest one when the endolymph moves in one direction, and toward the shortest when the endolymph moves in the opposite direction (see Figure 14-11). These movements will result in increases and decreases, respectively, of the frequency of firing of the vestibular afferents with which the hair cells synapse.

In contrast, hair cells in the maculae of the utricle and the saccule are not all oriented in the same direction. Each macula is divided

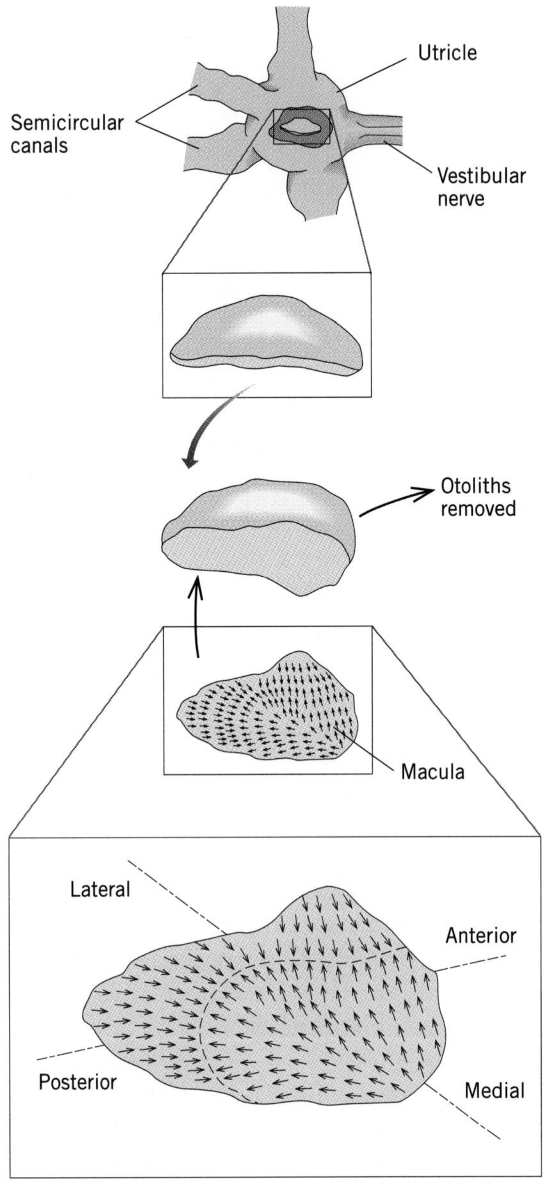

FIGURE 14-14. Top view of the macula of the utricle, showing the orientation of the hair cell stereocilia, which are visible after removal of the otoliths. The arrows point in the direction of deflection of the stereocilia that will excite each hair cell. The hair cells are oriented in different directions, so that any movement of the overlying otoliths will cause excitation of some hair cells.

into two zones (Figure 14-14). As shown by the small arrows in the detail in Figure 14-14, the receptor cells in one zone are aligned in rows, with all the stereocilia in one row facing the same direction. Adjacent rows of hair cells are oriented so their stereocilia face slightly more to the right or left, so that within one zone, the hair cells are arranged with their stereocilia pointing to different directions covering about 90°. Hair cells in the other zone are aligned in rows so that their tallest stereocilia face in the opposite direction from the stereocilia rows in the first zone. In the utricle, the tallest stereocilia in the two zones face toward one another, whereas in the saccule the tallest stereocilia in the two zones point away from one another. In either case, this arrangement ensures that there are stereocilia facing in almost every direction. Consequently, the slightest tilt or acceleration of the head will stimulate at least some of the hair cells, and that stimulation will lead to excitation of the vestibular afferents with which they synapse.

Anaxonal and nonspiking, vestibular hair cells are depolarized by deflection of the tuft of stereocilia in one direction, which opens mechanosensitive channels on the stereocilia, and hyperpolarized by deflection in the opposite direction. In the semicircular canals, hair cells are oriented with their stereocilia all arranged in the same way. In the two maculae, the stereocilia are oriented in various directions, which increases the sensitivity of the vestibular organ to accelerations in every direction.

The Electric Sense

Some animals have sensory capabilities that have no human counterpart. **Electroreception**, the process by which the presence of an electrical field is sensed, falls into this category. Electroreception, and hence an electric sense, occurs in a variety of animals. Among fish, eels, catfish, electric fish, sharks, rays, and lampreys have an electric sense. Amphibians like salamanders and newts, as well as a primitive mammal, the Australian platypus, also have this sense. Animals that can sense an electrical field but cannot generate and control one of their own are said to possess a **passive electric sense**. Sharks and rays, the platypus, and amphibians that are capable of electroreception all have a passive electric sense.

Some fish can generate an electrical field of their own; such fish are called **electrogenic** and are known as electric fish (Figure 14-15). Electrogenic fish are said to possess an **active electric sense**. Some electrogenic fish, like the electric eel, generate a very strong field that is used to stun prey or warn off predators. Others, like the aptly named weakly electric fish, can generate only a weak electrical field. Weakly electric fish generate their electrical fields continuously. They use these fields for **electrolocation** (navigation or orientation through the use of an electric sense) and for **electrocommunication** (communication with other weakly electric fish through the use of an electrical field). You will read more about electrocommunication in Chapter 20.

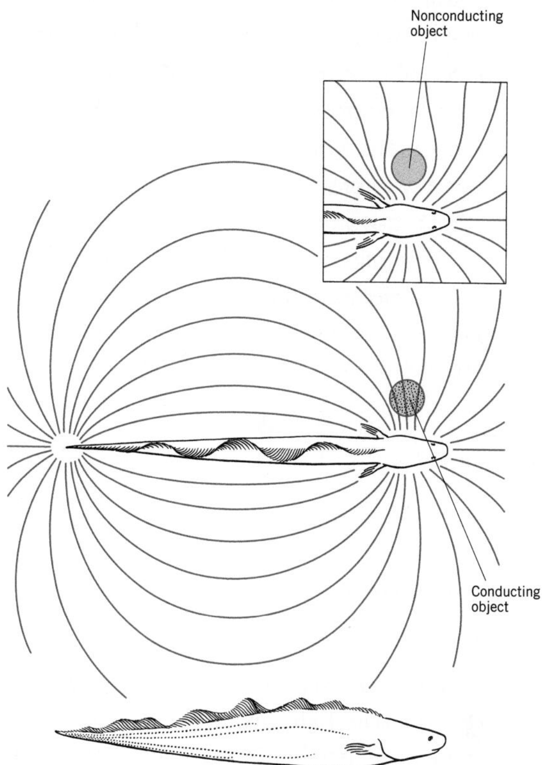

FIGURE 14-15. The electrical field generated by a weakly electric fish. The field can be used for locating and identifying objects in the murky water in which these fish live. An object that does not conduct electricity distorts the field in a characteristic way (inset) compared with the effect of a conducting object. The side view of the fish at the bottom shows the location of the organs that generate the electrical field (solid bars).

Some fish, some amphibians, and a primitive mammal have an electric sense that allows them to detect the presence of weak electrical fields. Animals that can only detect an external field have a passive electric sense; fish that can also generate a field of their own have an active electric sense. Electric fish use the electric fields they generate for electrolocation or for electrocommunication with other electric sensing animals.

The Anatomy of Electroreception

Electroreceptive sense organs are evolutionarily derived from the lateral line organ, and in fish are of two main types: ampullary organs and tuberous organs. All electric fish have **ampullary organs** (Figure 14-16A), not to be confused with the ampulla of vertebrate semicircular canals. Ampullary organs are located just under the skin and consist of a small cavity also called an ampulla, which is lined with sensory receptors and connected to the surface of the skin by a canal. Ampullary receptors are specialized to detect weak, low-frequency variations in electrical fields, and hence are commonly used to detect the minute electrical activity produced by prey or predators in the water. **Tuberous organs**, in contrast, are present only in fish with an active electric sense. They are located in a sac formed by epidermal cells, and typically do not have any physical connection with the skin surface (Figure 14-16B). Tuberous organs respond best to high-frequency fluctuations in voltage such as those caused by

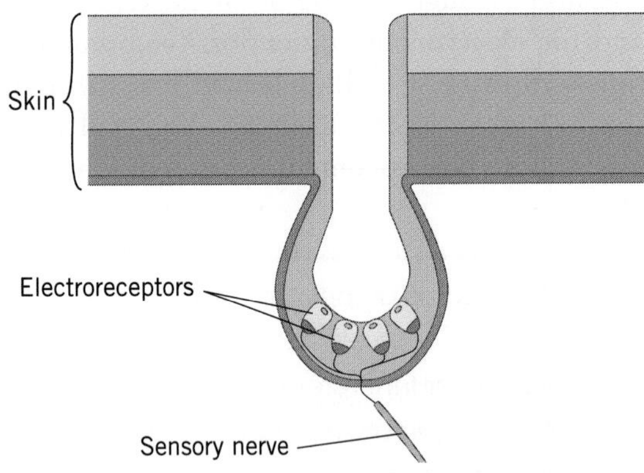

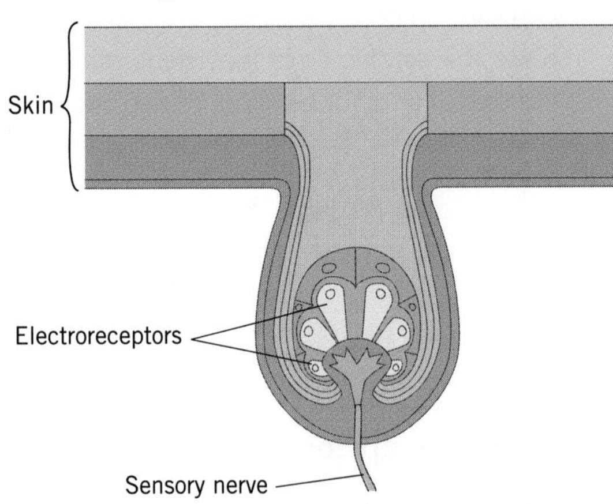

FIGURE 14-16. The two types of electroreceptive organs in fish. (A) Ampullary organs, which have an opening to the exterior of the animal via a canal to the surface of the skin. (B) Tuberous organs. These are entirely beneath the skin and have no opening to the skin's surface.

discharge of the fish's own electric organ, and hence are used in active electrolocation and in communication. In both ampullary and tuberous organs, many receptor cells synapse with a single afferent sensory neuron.

The sensory afferents from electroreceptors typically terminate in a specialized region of the medulla, the **electrosensory lateral line lobe (ELL)**. (The terminology of the brain region devoted to electroreception is not the same in all types of animals that have the capability.) From the ELL, information flows to the **torus semicircularus**, and from there to the optic tectum in the mesencephalon, and to the cerebellum. The optic tectum lies draped over the torus like a cloth over a fresh loaf of bread. In some of the weakly electric fish, the projection into the cerebellum is enormous, and part of the cerebellum has expanded to cover much of the rest of the brain. Several areas of the brain devoted to processing electrosensory information are organized topographically according to the location of the electrosensory organs on the surface of the body. Typically, the brain contains separate maps for the ampullary and the tuberous receptors, each specialized to represent different aspects of the electric field.

All electric fish have ampullary receptors, which detect mainly weak, low-frequency electrical fields. Tuberous organs, present in fish with an active electric sense, detect high-frequency changes in voltage, and hence are used in electrolocation and electrocommunication. Input from the electric sense organs is typically sent to the electrosensory lateral line lobe. Several areas of the brain devoted to processing electrosensory information are organized topographically according to the location of the electrosensory organs on the surface of the body.

Transduction in Electroreceptors

How sensitive are electroreceptors? The weakly electric fish produce electric fields with electrical gradients of several millivolts per centimeter. Since these fields are detected by other fish, the animals can clearly detect fields of this strength. The sharks and rays, however, have truly astonishing sensitivity, being able to detect electrical gradients of only 0.005 to 0.02 μV/cm. This is like being able to detect the electrical current generated by a 1.5-volt battery if one lead were placed in the ocean 3000 km (about 1800 miles) from the other. The ability to detect such a weak gradient has several practical consequences. First, it allows these fish to navigate in the ocean because electrical gradients of this magnitude are produced by the movement of the fish through the magnetic field of the earth. Swimming at right angles across the lines of force of the magnetic field will produce an electrical gradient stronger than that produced by swimming along the lines of force, providing the basis of an orientational ability.

The second consequence of this ability is that it allows the fish to find prey. Muscle in all animals produces a minute electrical signal when it contracts. These muscle potentials are large enough to be detected by a shark or ray. For example, researchers have been able to condition a ray to respond to a single muscle potential in a fish more than 10 cm away. Such sensitivity to bioelectric potentials allows the predator to detect a flounder buried just under the sand on the ocean floor from the minute electrical potential generated between its gills and seawater.

Transduction mechanisms have been studied in electroreceptors. In ampullary organs, an electrical field induces a depolarization that is mediated by inward flow of Ca^{2+} and is followed by an outward potassium current that repolarizes the receptor cells. The receptors are tonic and, being nonspiking, continuously leak transmitter. Depolarization increases the amount of transmitter released, and hence the rate of firing of the afferent sensory neuron. Ampullary organs respond in proportion to the strength of the electric field that is stimulating them.

Tuberous organs also depolarize with Ca^{2+} currents. However, the receptor cells in most tuberous organs generate action potentials. It is quite unusual for anaxonal receptor cells to be able to generate spikes (vertebrate taste receptors are the only other known example), and the functional significance of this property is not known. Tuberous organs respond phasically to an electrical field, and hence do not encode its strength. Their function is to provide dynamic information about changes in field strength rather than static information about absolute strength. Most tuberous receptors transmit signals to their afferent sensory neurons by chemical transmission, but a few (nonspiking) types use electrical transmission.

Some electroreceptors are amazingly sensitive, being able to detect electrical gradients of only 0.005 to 0.02 $\mu V/cm$. Ampullary organs are nonspiking, tonic receptors that depolarize by an inward flow of Ca^{2+}, followed by an outward flow of K^+ that repolarizes the receptor cells. Tuberous organs are spiking, phasic receptors that also depolarize via calcium currents.

Magnetic and Other Senses

A **magnetic sense**, the ability to detect the presence and orientation of a magnetic field, has been postulated for animals for many years, but a full understanding of it has been elusive. Some bacteria orient their swimming in a magnetic field. These bacteria contain **magnetite**, minute crystals of ferrous ferrite, an oxide of iron. The magnetite is arranged in chains, which presumably gives the cell greater sensitivity to any change in its orientation in the earth's magnetic field (Figure 14-17).

The discovery of magnetically sensitive bacteria and an elucidation of the cellular basis of this sense have encouraged researchers to search for similar abilities and mechanisms in animals, especially in those such as honey bees and migrating and homing birds that have excellent direction-finding ability. Behavioral experiments showing that bees have a magnetic sense are convincing, and the evidence for some birds, especially pigeons, is also good. The basis of the magnetic sense, however, is not clear. Magnetite has been found in many animals capable of some sort of direction finding, including humans, but the extent of its contribution to the ability is not known.

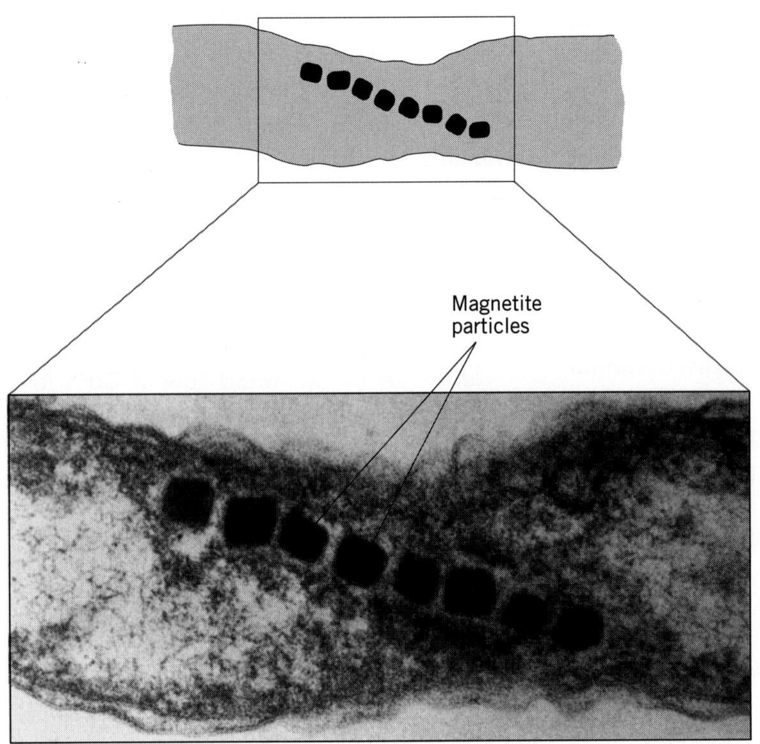

FIGURE 14-17. Magnetite particles in a magnetically sensitive bacterium. Each particle is surrounded by a thin membrane, presumably to anchor it in place in the cell.

Magnetite in honey bees can be demagnetized without affecting the insects' ability to use a magnetic sense. On the other hand, small magnets placed on either side of the head of a homing pigeon, where magnetite is located, will disorient the bird during homing, presumably because the magnets interfere with the operation of the magnetic sense organs in the head.

Another possible source of a magnetic sense is the visual system. According to one hypothesis, the magnetic field of the earth could influence certain specialized photoreceptors that utilize electron spin resonance. The idea is that the response of the photoreceptors to light will be modulated by the orientation of the receptor in the magnetic field of the earth. In consequence, the animal would see a slightly different hue of certain colors when facing one direction than when facing another. Behavioral experiments support the hypothesis, but physiological evidence has yet to be obtained.

A true magnetic sense has been shown to exist in sharks and rays. It is a consequence of the exquisitely sensitive electric sense of these fish, which allows them to detect the weak electrical fields induced by their own movement through the magnetic field of the earth.

The discovery of additional senses over the past few decades suggests that we have not necessarily identified the full sensory repertoire of animals. Some insects, such as caterpillars, bees, and cockroaches, can directly measure the humidity of air, via an antennal sense organ known as a hygroreceptor. There is evidence that separate hygroreceptors signal increases and decreases in humidity; some receptors increase firing as the air becomes more moist, and others increase firing as it becomes dryer. Some snakes are able to locate and capture prey through the use of infrared (heat) detecting organs located on the face. Functionally, these organs act much like eyes, but at the much longer wavelength of infrared light. Some animals may well have yet other senses that are mysterious from a human perspective.

Some animals have a magnetic sense that allows them to determine their orientation within the earth's magnetic field. The presence of a magnetic sense has been correlated with the presence of magnetite, an oxide of iron, but it is not known whether this material is always necessary for the sense. Sharks and rays have a magnetic sense based on their ability to detect weak electrical fields. Other unusual senses detect the humidity of air and infrared (heat) radiation.

Additional Reading

General

Schmidt, R. F., ed. 1986. *Fundamentals of Sensory Physiology*. 3d ed. New York: Springer-Verlag.

Research Articles and Reviews

Brown, F. A., Jr. 1981. Unusual senses. In *Sense Organs*, ed. M. S. Laverack and D. J. Cosens, 348–65. London: Blackie.

Devor, M. 1996. Pain mechanisms. *The Neuroscientist* 2:233–44.

Edin, B. B., G. K. Essick, M. Trulsson, and K. A. Olsson. 1995. Receptor encoding of moving tactile stimuli in humans. I. Temporal pattern of discharge of individual low-threshold mechanoreceptors. *Journal of Neuroscience* 15:830–47.

Feng, A. S. 1991. Electric organs and electroreceptors. In *Neural and Integrative Animal Physiology: Comparative Animal Physiology*, 4th ed., ed. C. L. Prosser, 317–34. New York: Wiley-Liss.

Sachs, F. 1992. Stretch-sensitive ion channels: An update. In *Sensory Transduction*, ed. D. P. Corey and S. D. Roper, 241–60. New York: Rockefeller University Press.

Vallbo, A. B., H. Olausson, J. Wessberg, and N. Kakuda. 1995. Receptive field characteristics of tactile units with myelinated afferents in hairy skin of human subjects. *Journal of Physiology (London)* 483:783–95.

PART 4

Motor Systems

An obvious feature of animals is their ability to move. It is the coordinated contraction of muscles that enables effective action. The nervous system must also take the environment into account when it issues motor commands by modulating those commands according to incoming sensory signals. And actions must be adjusted to take account of the physical structure of the animal itself.

This part describes the organization of motor systems, including the neural control of muscle, the organization of simple movements, and the planning of complex motor tasks. It is difficult to make generalizations about motor systems because they are often highly specialized, and may vary quite significantly depending on the type of animal in which they are found.

CHAPTER 15

Muscle and Its Control

The function of the motor system is to send to muscles a pattern of nerve impulses that will produce a desired movement. Just as sensory systems are shaped by the physical properties of the stimuli to which they are sensitive, so motor systems are shaped by the properties of the muscles whose contractions they control. In this chapter, you will learn about these properties, as well as the ways in which motor neurons control muscles.

Muscles are effector organs of the body, the means by which the nervous system is able to direct action. The action may be any type of movement imaginable, from the casual wave of a tentacle to the graceful leap of a ballerina. All movement generated by an animal is brought about by the contraction of muscle. Although there are circumstances in which muscle contracts without receiving any neural input (as when your leg cramps, for example), contraction is generally initiated by the nervous system. However, different muscles are not alike in their contractile characteristics, and even a single muscle may have different properties when it is well rested compared with when it is fatigued. Hence, in order to assure a particular result, the nervous system must take into account the properties of a muscle when it commands it to contract.

Muscle Tissue

All muscle shares the ability to exert a pulling force by contracting. This feature, which arises from the presence of arrays of actin and myosin filaments (see next section), is the basis for considering muscle as a single type of tissue. Nevertheless, because of significant differences between muscles, muscle tissue is normally subdivided into three types: cardiac muscle, smooth (visceral) muscle, and skeletal muscle (Figure 15-1). Cardiac muscle is the muscle of the heart, smooth muscle is the muscle of the gut and other internal organs (including blood vessels), and skeletal muscle is the main muscle mass of the body, attached to the internal or external skeleton if the animal has one. Each type of muscle has its own distinctive properties that determine how the muscle will respond to stimulation. Each type also differs from the others in how it is controlled by the nervous system.

Cardiac muscle and smooth muscle are similar in being controlled by the autonomic nervous system, and hence are not ordinarily under voluntary control. Animals cannot deliberately speed up their hearts or make their stomachs stop churning. However, cardiac and smooth muscle cells are quite different in appearance and physiological properties. Cardiac muscle cells form a branching network, with individual cells in any one part of the heart (around one chamber) interconnected by electrical gap junctions (Figure 15-1A). These junctions ensure that each cell contracts with all the rest as a functional whole. The gap junctions are formed by channel proteins in the cell membrane, as described for neuron-to-neuron electrical junctions in Chapter 2.

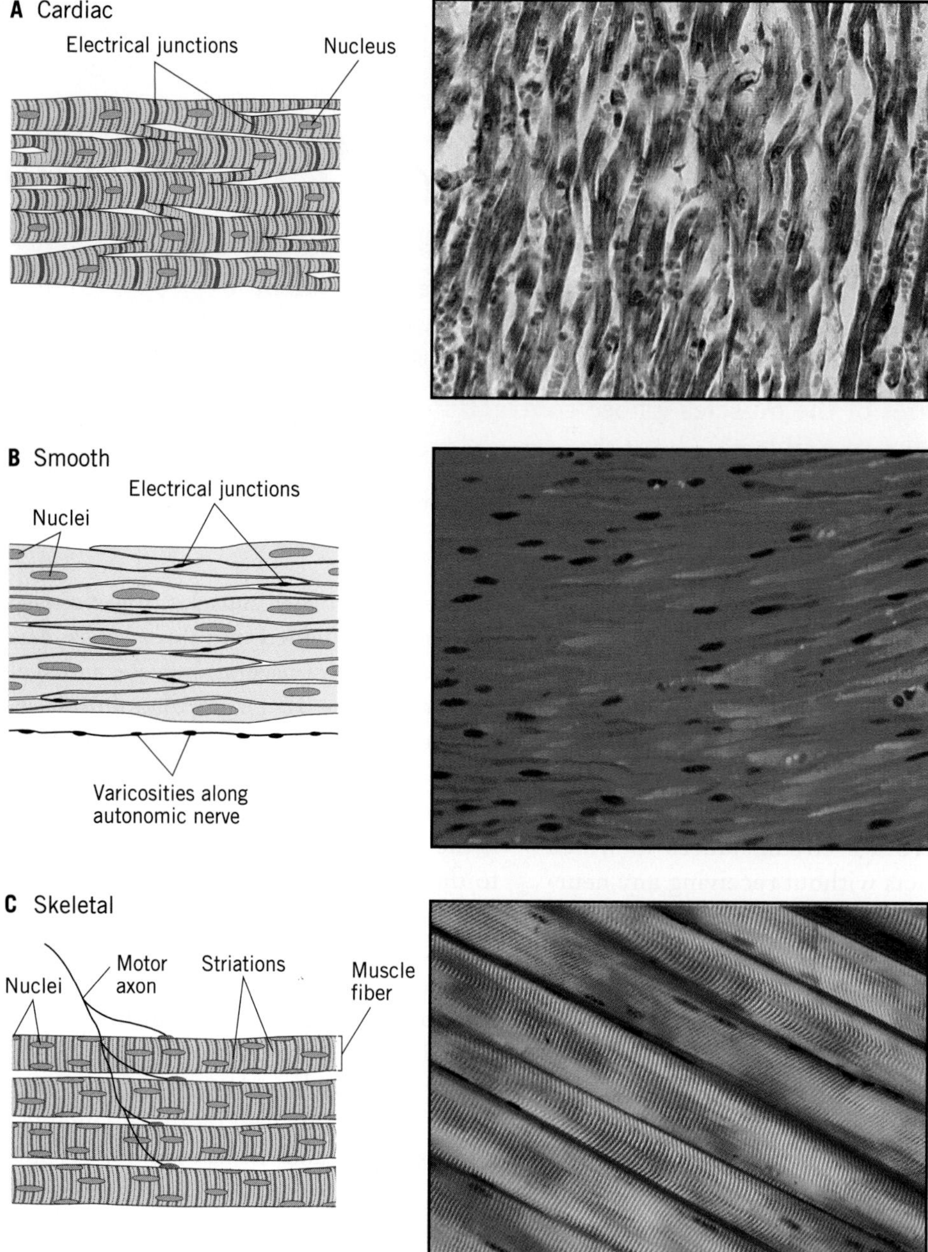

FIGURE 15-1. The typical appearance of individual cells of (A) cardiac muscle, (B) smooth (visceral) muscle and (C) skeletal (striated) muscle. The photomicrographs on the right show the actual appearance of the different types of muscle cells.

Cardiac muscle cells are adapted for continual, rhythmic contraction and have specific properties, such as a long action potential and a long postcontraction refractory period, that help them to maintain their rhythmicity.

Smooth muscle is formed by sheets of interdigitating, spindle-shaped cells that are also frequently joined by electrical junctions (Figure 15-1B). Generally it is slow to contract and frequently it can sustain tension for a considerable time after neural stimulation has ceased. It can also shorten more than skeletal muscle but generates less force.

Skeletal muscle, also called striated muscle because of its striped appearance under a microscope, is controlled by the somatic nervous system. It is used by animals to make desired movements, and hence has often been called voluntary muscle. Skeletal muscle differs morphologically from cardiac muscle and smooth muscle (Figure 15-1C). Instead of being composed of individual cells, skeletal muscle consists of long, cylindrical fibers (muscle fibers) that often run the entire length of the muscle and contain many nuclei from the embryonic cells that fused to form them. Each muscle fiber itself consists of smaller contractile filaments called myofibrils. Figure 15-2 shows the relationship between individual muscle fibers and a whole muscle. Skeletal muscle fibers are controlled by motor neurons from the central nervous system. It is the precise pattern of impulses in motor neurons that regulates and controls skeletal muscle. Some of the most important questions in motor systems research concern how these patterns of impulses are generated within the central nervous system.

Muscle tissue occurs in three forms: cardiac, smooth, and skeletal. Cardiac muscle and smooth muscle are controlled by the autonomic nervous system. Cells are linked to one another via electrical junctions. Skeletal muscle is controlled by the somatic nervous system and generates voluntary movements. It is organized into multinucleated muscle fibers.

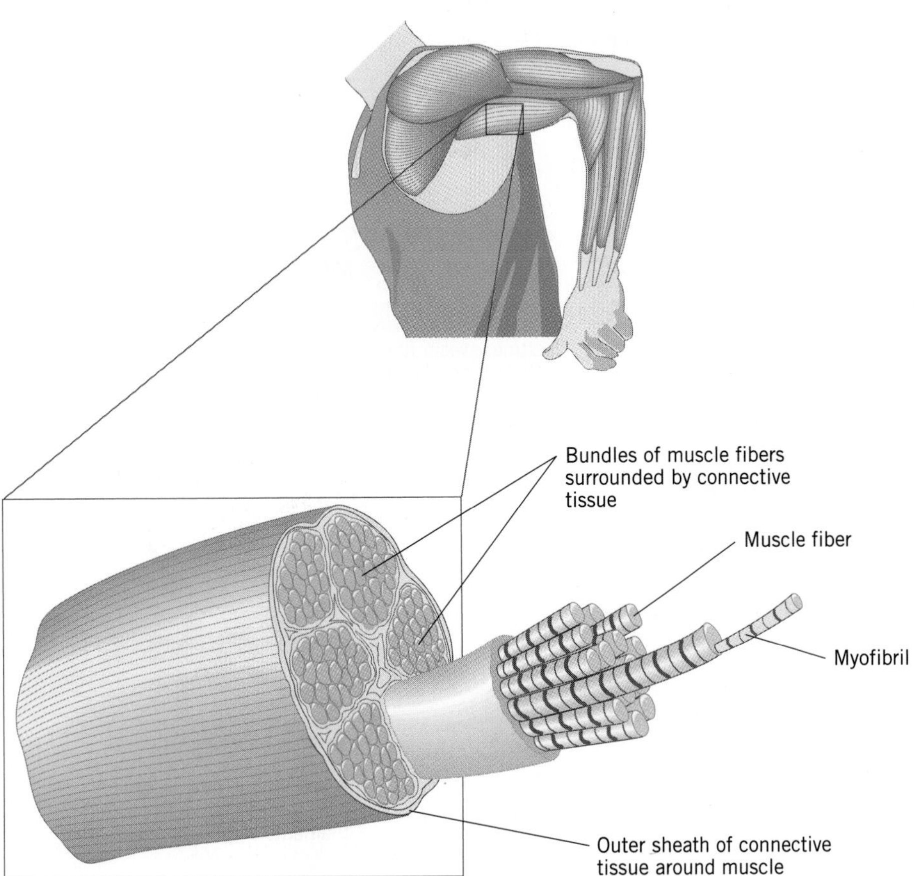

FIGURE 15-2. The relationship between muscle fibers and a whole muscle. Muscle fibers are usually bundled into groups surrounded by a connective tissue sheath.

The Contraction of Muscle

The events that bring about the contraction of a muscle are reasonably well understood. The contractile elements of muscle are the sarcomeres, which are arranged end to end to form myofibrils. Sarcomeres contain two types of filamentous proteins, called thick filaments (consisting of myosin) and thin filaments (actin), arranged in interdigitating arrays that allow them to slide past one another. Movement of the filaments past one another causes the sarcomeres, and hence the myofibril, to shorten (Figure 15-3A).

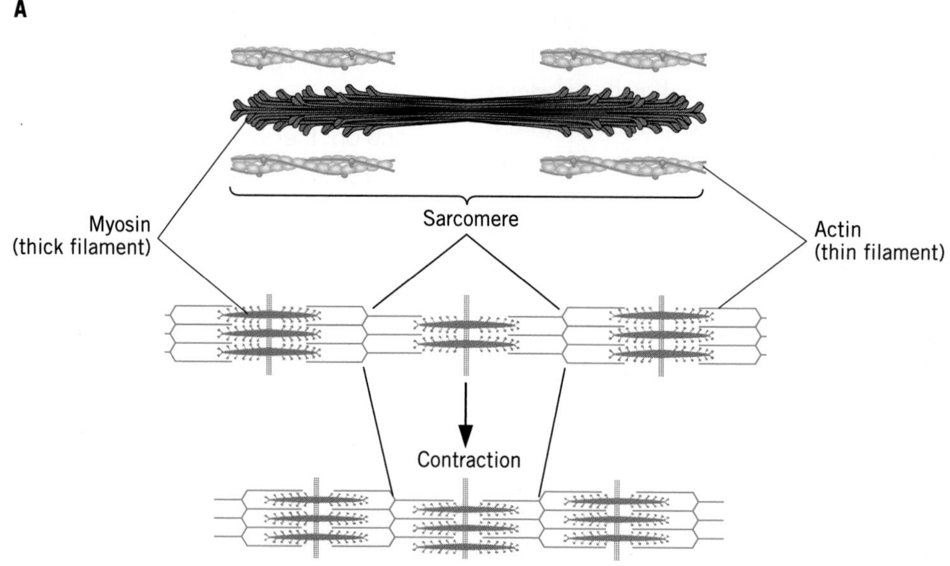

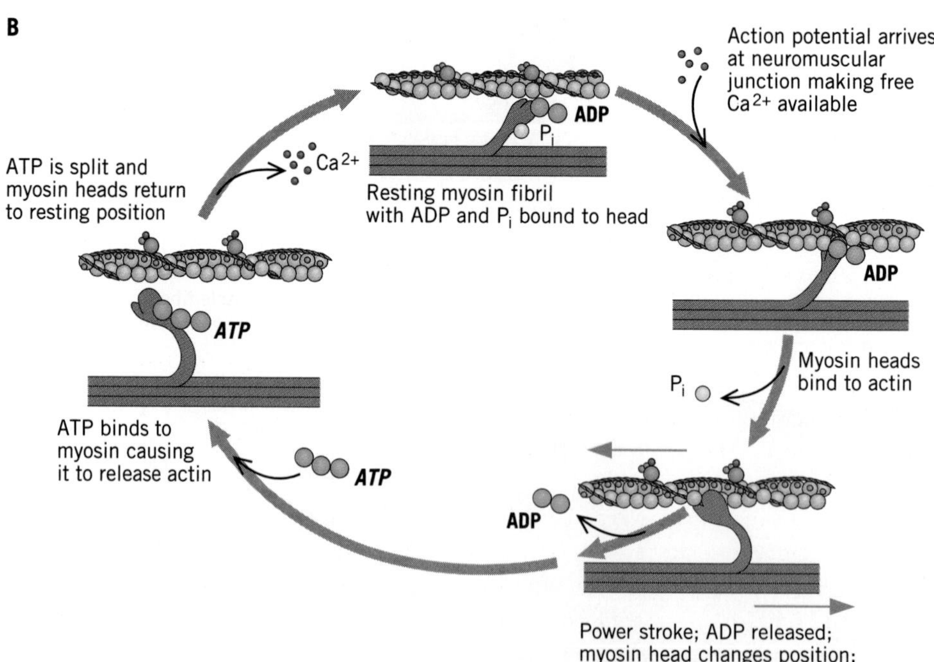

FIGURE 15-3. Mechanism of contraction of muscle. (A) Thick and thin filaments move past one another during contraction. (B) This movement is caused by the pulling of cross-bridges from the thick myosin filaments on the thin actin filaments.

Movement takes place when small, fingerlike extensions of the thick filaments, called cross-bridges, bind to specialized sites on the actin thin filaments. The cross-bridges change their shape when they bind to the actin, pulling the actin and myosin filaments past one another about 12 nm. In the presence of adenosine triphosphate (ATP), the cross-bridges release the actin, regain their original shape, and then bind to new sites. This cycle of binding, pulling, and releasing is repeated in the muscle fibers to produce shortening and hence movement (Figure 15-3B). This sequence of events is common to cardiac, smooth, and skeletal muscle.

Muscles are not in a perpetual state of contraction. Hence, there must be some method for regulating the cycle that produces shortening. This occurs through regulation of free Ca^{2+} in the muscle fiber. In relaxed skeletal muscle, the specialized sites on actin to which the cross-bridges bind are blocked by the presence of another molecule, tropomyosin. Since under these conditions the cross-bridges cannot bind with actin, there can be no sliding of the filaments past one another, and hence no shortening. However, free calcium ions in the muscle fiber interact with a complex of other molecules in such a way as to pull the tropomyosin molecules away from the specialized sites they cover, exposing these sites for binding with myosin cross-bridges. Contraction is regulated, through regulation of free Ca^{2+} in the cytoplasm of the muscle. The level of free Ca^{2+} regulates force production in smooth muscle as well, but via a different mechanism.

How is Ca^{2+} regulated? When a muscle is relaxed, nearly all Ca^{2+} is sequestered in the sarcoplasmic reticulum (SR), a meshwork of internal membranes in the muscle cell or fiber. That is, they are bound in the SR so that they cannot move around and cannot reach the actin molecules. Depolarization of the external membrane of the muscle cell, however, causes the affinity of the SR for Ca^{2+} to be reduced for a short while. This allows Ca^{2+} to diffuse out to the thin filaments, where they cause exposure of the active sites for binding with the myosin cross-bridges. When the membrane of the muscle cell repolarizes once more, the SR regains its high affinity for Ca^{2+}, calcium ions are removed from the cytoplasm, the binding sites on the actin filaments are covered again, and the muscle relaxes. Muscles differ dramatically in the kinetics of Ca^{2+} release from and sequestration in the SR. These differences in turn strongly affect the speed of contraction of the muscle.

Muscle contraction occurs as a result of the sliding of thick and thin filaments past one another in the presence of ATP and free Ca^{2+}. Contraction is regulated by withholding or making available free Ca^{2+}. At rest, Ca^{2+} is sequestered in the sarcoplasmic reticulum. When the cell depolarizes, Ca^{2+} is released and allows contraction to take place. Differences in the rate at which Ca^{2+} is released or sequestered influence the contractile properties of the muscle.

Cardiac Muscle

Although vertebrate cardiac muscle superficially resembles skeletal muscle, it has three features that set it apart functionally. First, heart muscle of all vertebrates, in common with heart muscle of most invertebrates, is myogenic. It will contract on its own without any input from motor neurons. Second, the membrane of vertebrate heart muscle develops a very long-lasting action potential (250 to 300 msec), accompanied and followed by a relatively long refractory period. Consequently, heart muscle can sustain a contraction for less than half a second, after which it is forced to relax. Hence, individual contractions cannot summate. Third, the extensive electrical coupling among the heart muscle cells that constitute one chamber of the heart makes each chamber contract essentially as a single unit. Contraction in one chamber spreads to other chambers with a slight delay.

The contraction of cardiac muscle is influenced by the autonomic nervous system. Neurons that make synaptic contact with fibers in a muscle are said to **innervate** that muscle, a term that applies to all three types of muscle. The heart is innervated by neurons of the parasympathetic system via cranial nerve X (the vagus nerve) and by neurons of the sympathetic system via the thoracic sympathetic ganglia (see Figure 3-27). Acetylcholine (ACh), the neuromodulatory neurotransmitter released by the vagus nerve, binds with two kinds of muscarinic ACh receptors. One type of receptor initiates a second messenger cascade that reduces the magnitude of a voltage-sensitive calcium current. The other type, via a G protein, directly opens K_{ACh} channels (see Chapter 6). These effects result in slower depolarization and prolonged hyperpolarization, which combine to slow the heartbeat. Norepinephrine, the neuromodulatory neurotransmitter released by sympathetic neurons, acts on ß-adrenergic receptors via a different G protein to increase the rate of depolarization of the cardiac muscle cells, hence speeding up the heartbeat.

Cardiac muscle is myogenic, with long action potentials and long refractory periods that prevent sustained contractions. Electrical coupling among cardiac muscle cells ensures simultaneous contraction of all the muscle surrounding one chamber. Parasympathetic neurons slow the heartbeat by the action of ACh on muscarinic receptors. Sympathetic neurons speed the heartbeat by action of norepinephrine on ß-adrenergic receptors.

Smooth Muscle

Smooth muscle differs from cardiac muscle and skeletal muscle in lacking obvious cross striations. Although tension in smooth muscle is generated as thick and thin filaments move past one another, the filaments are not aligned with one another as they are in cardiac muscle and skeletal muscle. For this reason, striations cannot be seen in light microscopic examinations of smooth muscle.

Vertebrate smooth muscle generally falls into one of two categories, multiunit or unitary. **Multiunit smooth muscle** is found in places such as the iris of the eye, arteries, and the skin of mammals. Each muscle cell is typically innervated by several autonomic motor neurons. There is little electrical coupling between cells, so the nervous system can readily regulate the activity of this muscle. **Unitary smooth muscle**, in contrast, is located principally along the gut, and hence is sometimes called visceral muscle. There is extensive electrical coupling via gap junctions (see Chapter 2) among the cells, so large areas of visceral muscle tend to contract together.

Like cardiac muscle, smooth muscle is innervated by the autonomic system. However, the relationship between motor neuron activity and muscle contraction in unitary smooth muscle is indirect. There are no typical neuromuscular junctions; instead, each autonomic motor neuron contains varicosities along its length (see Figure 15-1B). These are small swellings of the axon, typically 2 µm wide by 4 µm long, that contain synaptic vesicles. Neurotransmitter is released from these varicosities and diffuses to the nearby muscle cells. Corresponding to this somewhat loose functional relationship, neural control of smooth muscle is also somewhat loose, meaning that autonomic motor neuronal activity controls layers of smooth muscle rather than individual, discrete muscles as occurs in the skeletal muscle system. In fact, there is considerable local control of smooth muscle contraction via neurons of the enteric nervous system (see Chapter 3) and the many different neuromodulatory neurotransmitters released by the neurons of this system (see also Chapter 7).

Smooth muscle lacks cross striations, even though it does generate contraction by movement of thick and thin filaments past one another. Vertebrate smooth muscle may be multiunit, in which relatively few muscle fibers are controlled by each motor neuron, or unitary, in which extensive electrical coupling between muscle fibers produces widespread contraction of the muscle when it is excited by any motor neuron innervating it. In unitary smooth muscle, neurotransmitter is released from varicosities in the motor neurons.

Skeletal Muscle

Skeletal muscle can readily be distinguished morphologically from the other two types of muscle. It is different from smooth muscle in having an obvious striped appearance. The cross striations, from which it comes by its other name, striated muscle, are formed by the alignment of thick and thin filaments in adjacent myofibrils within each muscle fiber (muscle cell). Skeletal muscle differs from cardiac muscle in lacking electrical junctions and in consisting of long, spindle-shaped cells that are formed from many embryonic muscle cells that fuse together. Individual fibers do not exhibit the branching structure characteristic of muscle in a vertebrate heart.

As mentioned in a previous section, skeletal muscle is under the direct control of the somatic nervous system, and hence is the type of muscle used by animals to make voluntary movements. However, in spite of this uniformity of use, skeletal muscle exhibits enormous structural and functional diversity, which reflects the great differences that may exist between muscle fibers, even within a single muscle.

Muscle Fiber Diversity

Functionally, a muscle fiber may be classed as tonic or twitch. A **tonic muscle fiber** contracts slowly and relatively weakly in response to neural stimulation, typically taking many hundreds of milliseconds to contract and then relax again when it is excited by a motor neuron. However, tonic fibers are rich in mitochondria and the enzymes of oxidative metabolism, and can therefore contract for long periods without significant fatigue. Muscles consisting of tonic fibers are most common among the lower vertebrates and invertebrates; they are relatively rare among mammals. They serve mainly as postural muscles that must be active continuously for long periods to maintain a particular body position.

Twitch muscle fibers, in contrast, generally contract more strongly and quickly in response to neural stimulation, in some cases taking only 5 to 10 msec to contract and relax. Muscles containing twitch fibers, which constitute the bulk of skeletal muscle in all animals, are used in a variety of actions, from quick and powerful movements to slower, weaker ones. Twitch muscle fibers themselves show considerable diversity, as has been demonstrated by histochemical and other techniques that reveal the biochemical makeup of fibers. Each muscle typically contains a mixture of cell types.

Mammalian muscle has been studied in the greatest depth. One well-established classification scheme for mammalian twitch muscle fibers is based on the main metabolic pathway for the production of ATP that is present in the fiber. On this basis, three types of muscle fiber are recognized: **slow-twitch oxidative** (SO), **fast-twitch oxidative glycolytic** (FOG), and **fast-twitch glycolytic** (FG) fibers. The slow and fast designations have been added to call attention to the contractile properties of the different fiber types. Muscle fibers that contain a plentiful supply of enzymes for oxidative metabolism will be able to contract repeatedly for long periods without significant fatigue, as long as an adequate supply of oxygen is available. These muscle fibers are generally modest in size and therefore are not especially strong. Those that contain enzymes for both oxidative and glycolytic (anaerobic) metabolism will fatigue somewhat more readily; however, they tend to be somewhat stronger. Fibers containing predominantly enzymes for glycolytic metabolism tend to be large; hence, they are able to generate considerable power but can do so only for short periods of time. Some important features of mammalian muscle fibers are given in Table 15-1.

Mammalian muscle fibers have been classified in other ways as well. One classification scheme is based on their ATPase activity—that is, the rapidity with which they split ATP during a contraction cycle. Some muscles split ATP extremely rapidly, and hence can contract and relax rapidly. Others take a little longer, and hence are slower to develop tension and to relax. Recently, muscles have also been classified according to the type of thick filaments they contain. With the advent of molecular biological techniques, investigators can now describe significant differences in the structure of the myosin that makes up thick filaments within each muscle fiber. This work has revealed the presence of several myosin isoforms (structurally similar but not identical forms of myosin) in different types of muscle fibers; these isoforms have been used as the basis of classification of muscle fibers.

Significant differences exist between twitch muscle fibers in non-mammalian vertebrates and in invertebrates as well. For example, researchers have described three fast and two slow types of twitch fibers in amphibian muscle. Several types also have been described in reptiles and fish. Even invertebrate muscle shows clear distinctions between fast- and slow-twitch fibers, and there are indications that these two main groups may have further subdivisions as well.

The biochemical and molecular variability among muscle fibers confer specific properties that adapt the fibers to specific functions. Differences in resistance to fatigue, for example, reflect differences in the abundance of the enzymes of oxidative metabolism. Differences in speed of contraction reflect differences in the rate at which the cross-bridges step through their cycles of attachment to and release from actin filaments. This, in turn, is determined by the ATPase activity of the fiber and by the particular myosin isoform that it contains.

Some decades ago, it was common to speak of fast-twitch or slow-twitch muscle as if all the fibers in the muscle had the same characteristics. However, it is now recognized that all but a small number of muscles have a mixture of fiber types, which ensures that each muscle can be used efficiently for a variety of purposes (Figure 15-4).

TABLE 15-1. Characteristics of Mammalian Twitch Muscle Fibers

Feature	Slow-twitch Oxidative	Fast-twitch Oxidative Glycolytic	Fast-twitch Glycolytic
Speed of contraction	Slow	Fast	Fast
Twitch duration	Long	Short	Short
Fatigability	Low	Low	High
Glycogen content	Low	Intermediate	High
Myoglobin content	High	Intermediate	Low
Mitochondrial content	High	High	Low
Glycolytic capacity	Low	Intermediate	High
Oxidative capacity	High	High	Low

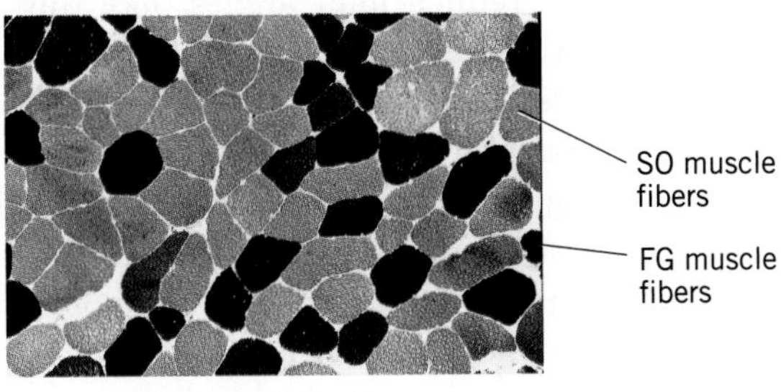

FIGURE 15-4. Diversity of fiber types within a human muscle. A cross section through a muscle, showing individual muscle fibers in profile. The lightly stained cells are slow-twitch fibers; the darkly stained ones are fast-twitch.

Skeletal muscles consist of different types of muscle fibers. Tonic fibers are weak, slow, and relatively rare. Twitch fibers are stronger, faster, and common. Twitch fibers in mammals are divided into several subtypes, according to biochemical differences between them. These differences give the fibers somewhat different contraction characteristics, and adapt them to specific uses. Whole muscles contain a mixture of fiber types.

The Neural Control of Skeletal Muscle

Muscles are usually controlled by the motor neurons that synapse on the fibers of each muscle. Except in very small muscles, innervation is not uniform—that is, all axons that go to a particular muscle do not innervate all the fibers in that muscle. Instead, each motor neuron controls the contraction of only some of the fibers in the muscle.

The muscle fibers controlled by one axon (i.e., all those fibers with which the axon makes synaptic contact), plus the axon and the rest of the motor neuron, are collectively referred to as a **motor unit**. Since a motor unit is defined in terms of the motor axon that innervates the muscle fibers in it, the number of motor units in a muscle is simply the number of excitatory motor neurons that innervate the muscle. A large muscle in a mammal may contain thousands of motor units, each consisting of a single motor neuron and the muscle fibers it innervates. The muscle fibers of a single motor unit are usually scattered randomly throughout the entire muscle (Figure 15-5). In invertebrates, some muscles are so small that they consist of only one or two muscle fibers. In such cases, a single axon will innervate all the fibers, and therefore the entire muscle is a single motor unit. This arrangement, however, is unusual.

In vertebrates, each muscle fiber usually receives synaptic contact from only one motor neuron. Hence, in these animals, motor units generally do not overlap. In invertebrate skeletal muscle, however, such overlap is common. This means that in invertebrate animals, a particular muscle fiber may receive innervation from more than one motor neuron, and hence may be a member of more than one motor unit.

Muscle fibers may also differ in the number of synaptic contacts made by a single motor axon. Most vertebrate muscle fibers are capable of conducting true action potentials. Hence, a single synaptic contact from a motor neuron can excite the entire fiber as the action potential generated by it sweeps along the fiber membrane, causing the release of Ca^{2+} from the SR along the whole muscle fiber. In vertebrate tonic muscle and all skeletal muscle of invertebrates, however, the muscle membrane cannot sustain an action potential; excitation generated at one site decays as it spreads away from the site of excitation. Thus it is necessary for each motor neuron to make multiple synaptic contacts with the muscle fiber to ensure that it depolarizes, and hence contracts, uniformly when it is excited.

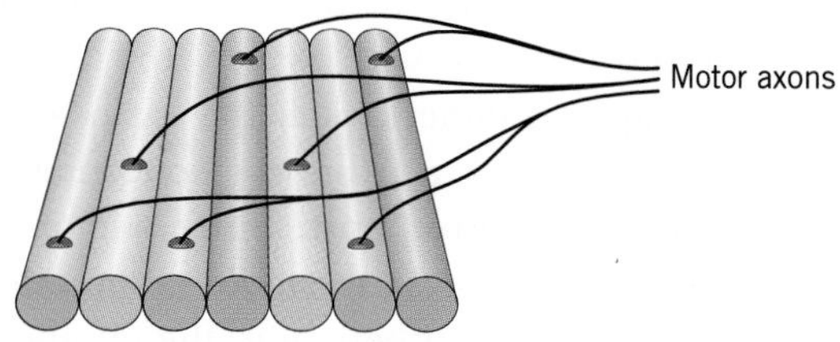

FIGURE 15-5. Motor unit organization in a hypothetical vertebrate muscle innervated by three axons. Each axon innervates several muscle fibers, scattered randomly in the muscle. In this diagram, each motor unit is designated with unique shading. Large vertebrate muscles may have thousands of motor units.

Motor units are the functional elements of muscle; a motor unit consists of a motor neuron and all the muscle fibers it innervates. Muscles usually contain many motor units. Motor units usually overlap in invertebrate muscle, so that each muscle fiber belongs to more than one unit. Motor neurons make more than one synaptic contact along muscle fibers in invertebrates.

Functional Organization

Muscle function in animals without skeletons can be difficult to study, as you can imagine by thinking about the hundreds of muscles in each arm of an octopus. However, the study of muscle function is helped by the organization of muscles into functional groups. In elongate animals such as leeches and earthworms, body shape and even locomotion are controlled by the opposing action of circular muscles, which wrap around the body, and longitudinal muscles, which are arranged lengthwise. Contraction of circular muscles decreases body diameter and lengthens the body. Contraction of longitudinal muscles has the opposite effect (Figure 15-6A). Muscles that act against one another like this are called **antagonistic muscles**.

In animals with skeletons, many muscles are arranged to provide motion around joints. Muscles that reduce joint angles, like those that bend an insect leg, are called **flexor muscles**. Muscles that increase joint angles, like those that straighten a leg, are called **extensor muscles** (Figure 15-6B). Since the biceps and triceps muscles act against one another around the elbow joint, they are antagonists. At many joints several muscles can act together to move a limb in a particular direction. In your lower leg, for example, the gastrocnemius, soleus, and posterior tibialis muscles act together to extend your ankle. Muscles that act together like this are called **synergistic muscles** (Figure 15-6C).

You will learn in Chapter 16 that in many repetitive behaviors the nervous system must arrange for an alternating action of antagonistic muscles and a simultaneous action of synergistic muscles.

Functionally, muscles are often grouped into antagonistic sets that increase either body diameter or length in soft-bodied, wormlike animals or flex or extend a limb in animals with skeletons. Synergistic muscles act together to effect a single type of movement.

Functional Control of Skeletal Muscle

Muscle is controlled by the central nervous system through the actions of the motor neurons that innervate it. In order for muscle to be used effectively, the nervous system must regulate its force production, so that it can generate force appropriate to the intended use. Furthermore, the diversity of muscle fiber types would be of little use if the nervous system of an animal were not able to make full use of the functional characteristics of the muscle fibers as they are activated during different tasks; in this section you will learn about some of the ways in which the

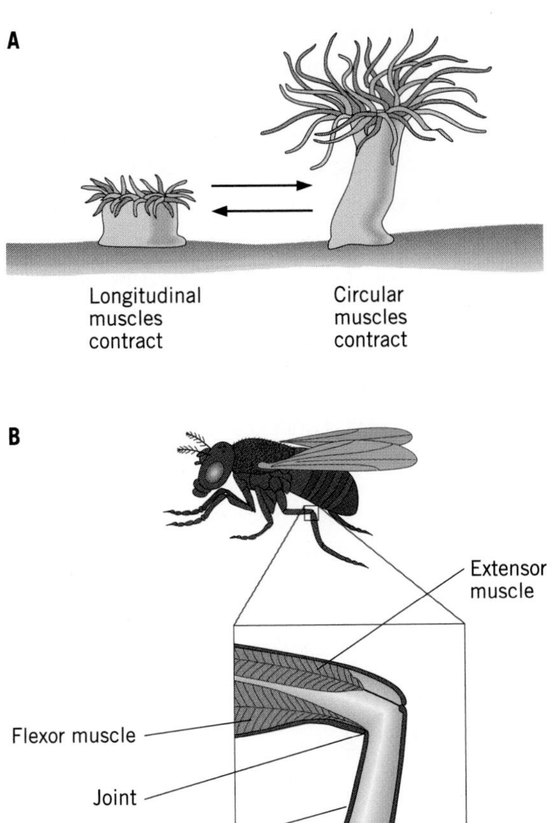

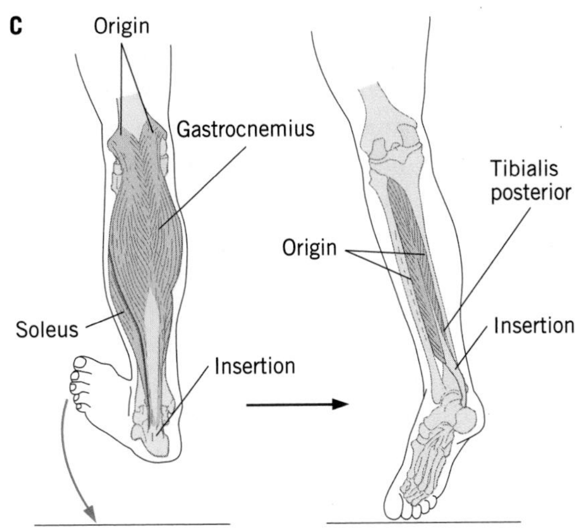

FIGURE 15-6. The main functional arrangements of muscles. (A) In a soft-bodied animal, muscles may act either to increase the body diameter or to lengthen the body. (B) In a jointed limb, muscles may act to flex or extend the limb. (C) In some cases, several muscles may act synergistically to bring about a particular movement. The gastrocnemius, soleus, and tibialis posterior muscles shown here work together to extend the foot.

nervous system matches the use of the muscle it controls with the means by which this control is exerted.

Neural Control of Tension

An ability to regulate the force (tension) produced by a muscle is clearly a necessity for any animal. Sometimes a muscle must produce maximum force rapidly, as does a leg muscle of a deer running to escape a cougar. Sometimes it must produce minimum force over a long time, as does that same muscle when the deer is merely grazing in a meadow. Since the same muscle is used in two quite different behaviors that require quite different levels of force, there must be some way for the deer to regulate the force the muscle generates.

Muscle force can be regulated in two ways: by recruitment and by frequency control. Regulation by recruitment controls the total force by regulating the number of motor units that are active at any one time. **Recruitment** refers to the addition of one or more motor neurons to the pool of those that are already active in a muscle. Consider the deer standing in the meadow. Only a few motor units will be active in its leg muscles, since not much force is required merely to support the animal. When the deer starts to run, however, it can generate more force by activating additional motor neurons. The force generated by the whole muscle can be regulated over a wide range by controlling the total number of motor units that are active at any one time. Low levels of force are produced by activation of just a few motor units, and higher levels are produced by recruiting additional motor units. Force regulation by recruitment works best in muscles with relatively large numbers of motor units.

The second method of regulating force is by **frequency control**. With this method, differences in the frequency of firing of an axon generate different forces in the muscle fibers innervated by the axon. At low frequencies of firing of a motor neuron, a muscle fiber relaxes just a bit between the contractions caused by each neuronal action potential. Hence, the muscle generates a lower total force than if the motor neuron were firing at a higher rate, during which there would be little or no relaxation between action potentials.

Force is regulated differently in different muscles. In small muscles consisting entirely of slow-twitch fibers, frequency control is the exclusive method of force regulation. For this reason, it is common in invertebrate animals. In large muscles, force is usually regulated by frequency control when only small forces are required and by recruitment when large forces are needed. In humans and other mammals, even large forces are produced by frequency control in those muscles that have relatively few motor units; recruitment plays only a minor role in increasing force in these muscles. The method by which muscle force is regulated appears to be intrinsic to the muscle, and is related to its fiber composition and size. Nevertheless, human studies have revealed significant differences between individuals in the relationship between the rate of firing and the force produced.

The control of force has other complexities. At one time it was thought that in mammals there was significant load sharing among motor units. At low levels of force, motor units were thought to share the load; as one unit lost power because of fatigue, another would be recruited to take up the slack. The concept was based on the assumptions that motor units were equal in strength and that they were used in a more or less random fashion. However, work with human muscle has shown that this idea is entirely false. Instead, there is a highly regulated use of motor units, some always being activated first and others always being activated later, in a particular order, independent of the final amount of force required. Using the prin-ciples of biofeedback it is possible for human subjects to learn to regulate the activity of individual motor units (Box 15-1). However, it is virtually impossible for subjects to change the order in which motor units are turned on or off.

Control of muscle force can be achieved through recruitment, in which more motor units are activated when more force is required; by frequency control, in which the rate of firing of motor neurons in creases when more force is required; or by a mixture of the two. In humans, it has been shown that motor units are recruited in a specific order.

Size Principle and Matching

As pointed out previously, most muscles consist of a combination of FG, FOG, and SO fibers. The particular combination is a reflection of the main function of a muscle: a muscle that consists mostly of FG fibers is generally fast and powerful, whereas a muscle that consists mostly of SO fibers is usually postural and needs to act for long periods without significant fatigue. Muscles consisting mostly of FOG fibers are intermediate in function. Investigation of the neural control of mixed muscles has revealed that each type of muscle is controlled to maximize the efficiency of muscle use while at the same time allowing it to be used for different functions.

Muscle can be used for different functions in part because of the fixed order of recruitment referred to in the previous section. This fixed order, which has a morphological basis, allows a muscle to be used to produce slow or weak movements as well as quick, powerful ones. Slow-twitch muscle fibers are always innervated by motor neurons that are physically smaller, in both soma size and diameter of axon, than those that innervate fast-twitch muscle fibers. Furthermore, there is some size variation even within axons innervating only

FG or SO muscle fibers. This is important because small neurons are always excited before larger ones.

This size-related order of recruitment is called the **size principle**; it is a consequence of the higher input resistance and greater density of current flow from excitatory postsynaptic potential (epsps) across the membrane of cells with smaller volumes. These features lead to a larger epsp at the spike initiation zone and hence to easier firing of small neurons (Figure 15-7). Hence, for a given level of excitatory input to the motor neuron pool that innervates a particular muscle, the most excitable neurons (i.e., the smallest ones) will fire first. As the level of excitatory input goes up, the frequency of firing of the neurons that are already firing increases, and new ones become excited and add their muscle fibers to the active pool. When a reduction of force is required, the order in which neurons drop out of the active group is the reverse of that in which they were recruited. The size principle has been described in both vertebrate and invertebrate animals, and so seems to be a general feature of motor systems.

The size principle allows even large, powerful muscles that contain mostly FG fibers to be used for relatively slow and weak movements, and for small muscles consisting mainly of SO muscle fibers to have some capability for fast, powerful movements. The smaller motor neurons innervating the SO fibers will always be activated first, generating relatively slow, weak movements no matter what the predominant type of muscle fiber. The large motor neurons innervating FG fibers will always be activated last, ensuring that these fibers will be used only when there is demand for considerable force, even if the muscle is generally used for slow, sustained force production.

Innervation patterns are also organized to maximize the efficiency of muscle use. Just as there may be differences in size between motor neurons, so may there be differences in function as well. Investigation has shown that in mixed-function muscles there is often a matching of the properties of each motor neuron (e.g., size, tendency to fire phasically) and those of the muscle fibers each innervates (e.g., speed of contraction). For example, a motor neuron that tends to fire easily and for

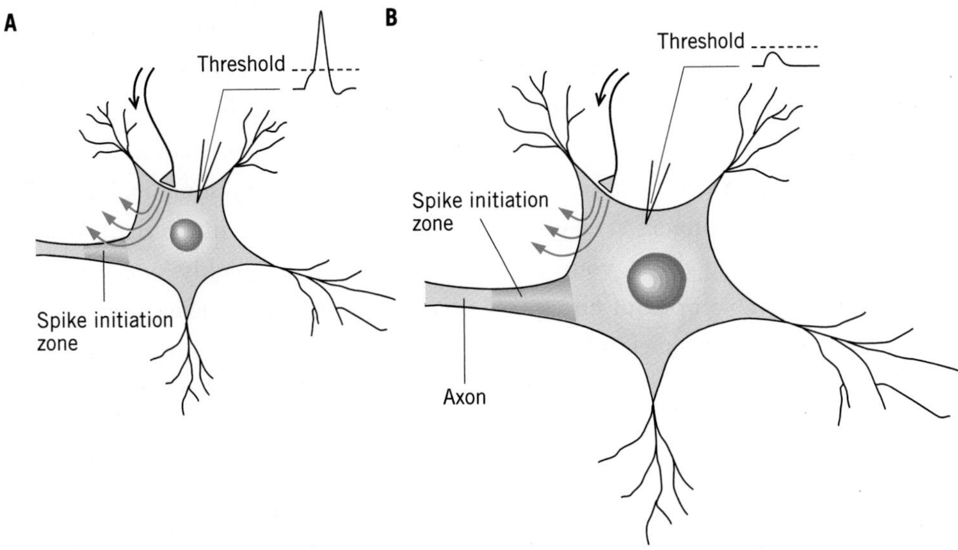

FIGURE 15-7. Operation of the size principle that for a given level of excitatory synaptic input, small neurons are excited first. (A) In a small neuron, activation of the excitatory synapse leads to generation of an action potential because the epsp exceeds threshold at the spike initiation zone. (B) In a larger neuron, the same level of excitatory input generates only a subthreshold epsp because the synaptic current spreads out over a larger area.

BOX 15-1

Twitching on Command: Demonstrating Motor Control in Humans

It is easy to think that studying the actions of motor neurons or muscles always involves complex operations to expose the motor nerves or for the implantation of recording electrodes. Complicated procedures are not always necessary, however. Contraction of a muscle is always accompanied by an electrical signal, because the membrane of the muscle cell depolarizes when the cell is excited by the motor neuron. Hence, it is possible to detect the contraction simply by placing electrodes in or even near the muscle and picking up the associated electrical activity. The technique of recording muscle action by recording its electrical activity is called **electromyography**, and the resulting record of muscle action is referred to as an **electromyographic recording (EMG).**

EMG recordings can be made from human subjects. The cleanest recordings are made via fine wire electrodes that are placed in or near the muscle through small, hollow needles inserted under the skin. For large surface muscles, EMG recordings can be made through electrodes placed on the surface of the skin.

Remarkably, whatever recording technique is used, it is relatively easy for nearly every subject who tries this to learn to control the action of individual motor units. If the subject is given a cue as to performance, either a picture of an action potential on an oscilloscope or the sound of an action potential through a loudspeaker (Figure 15-A), he or she can learn to turn on a single motor neuron. This is the principle of **biofeedback**, learning to control a bodily process by observing some external effect that the process has.

Learning to turn on specific motor units may sound quite easy since you do so every time you make a movement. However, the brain usually organizes movement in terms of motions to be carried out rather than in terms of individual muscles to activate. To see the difference, try some manipulations with your fingers. You can easily close your fingers into a fist. You can probably also drum your fingers, striking them quickly one after another on a tabletop. But now place your palm on a flat surface and try to lift each finger by itself while you keep the rest flat on the surface. You can probably do it but not as easily as the other manipulations. On the other hand, if you happen to play a musical instrument like a piano, you probably found manipulating your fingers as easy as making a fist, showing that it is possible to learn to control individual muscles that are not usually used by themselves. So, too, can you learn how to turn on individual motor units.

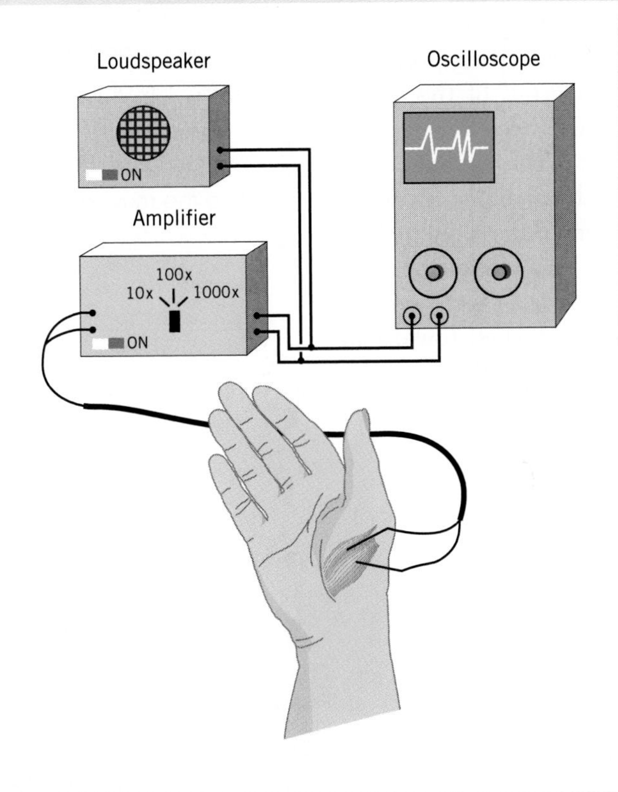

FIGURE 15-A. The arrangement for biofeedback training of a human subject for control of individual motor units. Recording electrodes are placed in or on a muscle, and the electrical signals from the muscle are amplified and fed to an oscilloscope and/or a loudspeaker. When the subject activates the muscle, the sight or sound of the action potentials gives the subject an indication of performance.

How much control over motor units can be learned? An amazing amount. The shape and size of extracellularly recorded EMG signals will vary from one motor unit to another because of the location of the different muscle fibers that respond. Hence, individual motor units can be distinguished from one another by differences in the appearance and sound of their recorded electrical signals. Not only can subjects learn to activate individual motor units; they can control the activity of those units to such an extent that they can even produce rhythmic sounds in the loudspeaker! What cannot be learned, however, is how to change the order in which the motor units are recruited, since this is fixed within the central nervous system.

long periods tends to innervate muscle fibers that act posturally and are easy to excite and relatively fatigue-resistant (Figure 15-8).

This matching of neuronal firing properties with the contractile properties of the muscle fibers they innervate is controlled developmentally. If the axons of motor nerves that control predominantly slow and fast types of muscle are cut and crossed over, so the slow nerve is allowed to innervate the fast muscle and vice versa, the fiber characteristics of the reinnervated muscle tend to be adjusted toward the type expected based on the neural innervation, showing that the nerve can exert an effect on the features of the muscle fibers.

Most muscles contain mixtures of fiber types. These muscles are controlled for maximal efficiency and multiple use by several features. Two of these are recruitment according to size (the size principle) and the matching of properties of motor neurons and the muscle fibers they innervate.

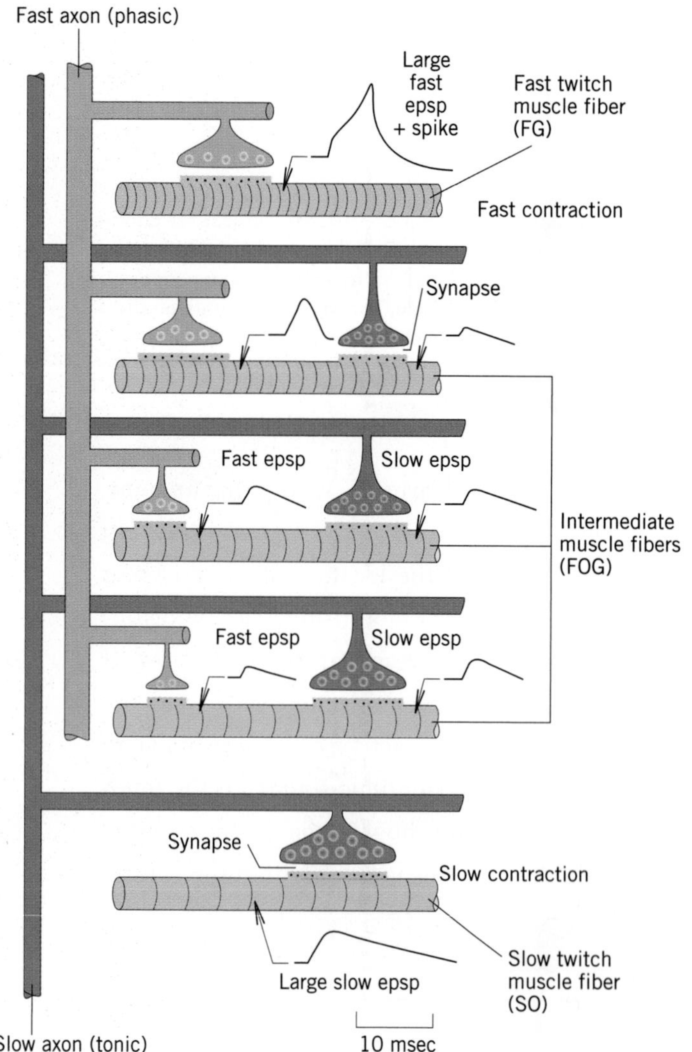

FIGURE 15-8. The matching of innervation with the properties of the muscle fibers that are innervated in an invertebrate. Fast axons are used to evoke quick, powerful movements; they innervate FG muscle fibers (top) that will give fast, strong contractions. Slow axons are used to evoke relatively weak, sustained movements; they innervate SO muscle fibers (bottom) that contract slowly and less strongly but can contract for long periods without significant fatigue. Intermediate types of muscle fibers (FOG) receive innervation from both types of axons but with different synaptic effectiveness, giving them an intermediate range of contractile properties.

Motor Neurons

Just as sensory neurons are the only route by which sensory information can reach the central nervous system, motor neurons are the only route by which patterns of action can be converted to behavior. This is because the movements that constitute behavior can be brought about only by properly timed muscle contractions that are caused by activity in motor neurons. For this reason, motor neurons are sometimes referred to as the final common path for behavior. One of the central questions in research on motor systems is how motor neurons can be made to produce the pattern necessary for the execution of any specific behavior.

One major difference between vertebrate and invertebrate nervous systems is that in the former the sensory input to and the motor output from the spinal cord are mostly separated. Nearly all sensory input enters the spinal cord via the dorsal root, and all motor output leaves via the ventral root. You will learn more about this arrangement in Chapter 16. In invertebrates, there is no separation of motor and sensory roots; all peripheral nerves are mixed.

Motor Neuron Morphology

The morphology of typical invertebrate and vertebrate motor neurons is shown in Figure 15-9. Motor neurons in invertebrates are typically monopolar. The single neurite branches to form a dendritic area in the neuropil and a long axon that leaves the ganglion in which the cell body is located. Vertebrate motor neurons, called α motor neurons, are multipolar, with several dendrites and one long axon arising from the cell body. The axon in these neurons frequently gives off a short branch before it leaves the spinal cord, a branch that curves back into the gray matter to form other connections.

As you learned in Chapter 3, invertebrate neurons are arranged in a ganglion, with the cell body of each neuron at the perimeter of the ganglion and the synaptic area (the neuropil) in the center. Motor neurons are usually located in the ganglion of the body segment in which the muscle they innervate is located. In the vertebrate spinal cord, the cell soma, dendrites, and synaptic areas of motor neurons are located in the gray matter, usually in the part of the cord closest to the innervated muscle. Gray matter, as you may recall from Chapter 3, owes its color to the typical dull gray of unmyelinated nervous tissue.

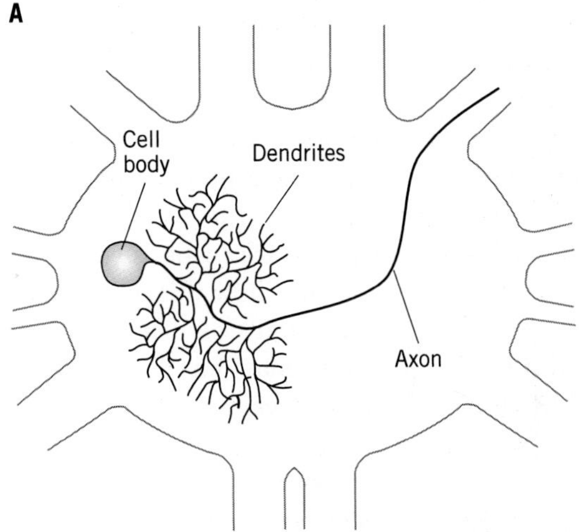

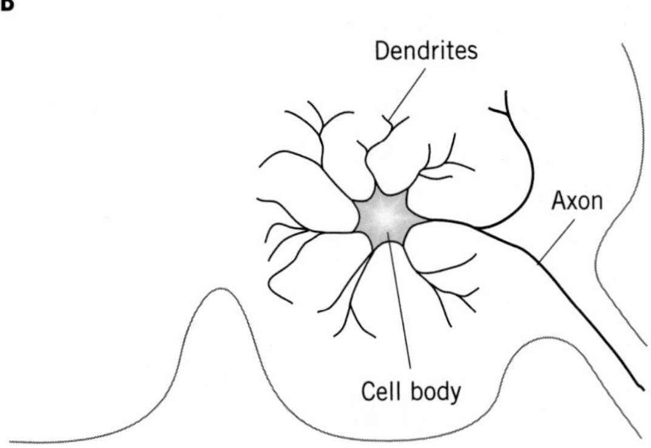

FIGURE 15-9. The morphology of a typical (A) invertebrate and (B) vertebrate motor neuron. The axon branch of the vertebrate neuron allows the motor neuron to synapse with other neurons in the nerve cord. A light outline shows the ganglion in which each motor neuron is located.

Motor neurons in invertebrates are monopolar and are located in the ganglion of the segment in which the innervated muscle lies. Motor neurons in vertebrates are multipolar and are usually located in the spinal cord close to the muscle they innervate.

Control of Motor Neurons

How is the pattern of motor neuron activity produced? You will learn about various mechanisms in subsequent chapters, but it will be useful first to consider the overall problem here. In theory, the brain could be directly responsible for every movement, via pathways between brain centers responsible for selecting the muscles to be activated and the motor neurons that are to be turned on. However, if you consider how difficult it is to play the piano when you are just learning, you can see that this may not be the best method of control. The very time when you are consciously thinking about each movement of the fingers is the time when your playing is at its slowest and most hesitant. The movement of individual fingers is not especially difficult, but somehow having to deal with them directly and consciously seems to be.

If you have ever tried to contract specific muscles in your body, you may have discovered that this can be difficult, unless you think about executing a movement in which that muscle is used rather than trying to activate the muscle directly. This subjective experience is in concordance with current thinking on motor control. Researchers believe that action is organized on a movement-by-movement basis. It is thought that the brain organizes movement by selecting a set of interneurons that will activate the proper motor neurons at the local level. For complex acts, several levels of interneurons can be involved. Learning a new motor skill involves establishing new pathways so that specific interneurons can activate a specific subset of motor neurons for the task. The implication of this idea of motor organization is that the motor system is organized hierarchically, comprising a serial arrangement of neurons from the level of the motor neuron up to the brain. Each level of the hierarchy has more far-reaching control over muscles, from individual motor units controlled by the motor neurons, up through the coordinated action of several muscles around a joint, to activation of an entire sequence of movements. You will learn more about this concept in Chapter 16.

Although motor neurons can to a certain extent be controlled individually by the brain, complex behavior is controlled not by the direct regulation of motor neurons by the brain but by a hierarchy of interneurons that control various parts of the required movement. Further, action is organized on a movement-by-movement basis rather than muscle by muscle.

Additional Reading

General

Basmajian, J. V., and C. J. DeLuca. 1985. *Muscles Alive: Their Functions Revealed by Electromyography*. 5th ed. Baltimore: Williams and Wilkins.

Hoyle, G. 1983. *Muscles and Their Neural Control*. New York: Wiley.

McMahon, T. A. 1984. *Muscles, Reflexes, and Locomotion*. Princeton, N.J.: Princeton University Press.

Research Articles and Reviews

Atwood, H. L. 1973. An attempt to account for the diversity of crustacean muscles. *American Zoologist* 13:357–78.

Henneman, E., G. Somjen, and D. O. Carpenter. 1965. Excitability and inhibitability of motoneurons of different sizes. *Journal of Neurophysiology* 28:599–620.

Pette, D., and R. S. Staron. 1990. Cellular and molecular diversities of mammalian skeletal muscle fibers. *Reviews of Physiology Biochemistry and Pharmacology* 116:1–76.

Seki, K., and M. Narusawa. 1996. Firing rate modulation of human motor units in different muscles during isometric contraction with various forces. *Brain Research* 719:1–7.

Smits, E., P. K. Rose, T. Gordon, and F. J. R. Richmond. 1994. Organization of single motor units in feline sartorius. *Journal of Neurophysiology* 72:1885–96.

CHAPTER 16

Reflexes and Pattern Generation

An important question in neurobiology is how the central nervous system generates the patterns of motor neuron activity that underlie behavior. One way to approach this question has been through the study of the relatively simple movements that constitute reflexes and repetitive behavior like walking or swimming. In this chapter, you will learn about two generalizations that have come from these studies. First, a great deal of the control of simple movement lies at the local level, in the region of the nervous system close to the muscles that are active during the movement. Second, although sensory information plays an important role in motor patterning, the nervous system can generate the coordinated movements that constitute repetitive behavior in the absence of such information.

All behavior consists of a properly coordinated series of movements. Since any movement is the result of a particular sequence and timing of muscle contraction, the question of how behavior is generated becomes one of how the proper sequence of muscle activation is generated by the central nervous system. One way to approach this question is to study simple movements such as reflexes, or simple repetitive behavior such as swimming or walking, in the expectation that understanding how these are generated will aid in understanding how complex movements and complex behavior are generated.

Reflexes

Gently touch the leg of an insect, and the insect will move its leg. Step on a tack, and you will quickly shift your balance and withdraw your foot. Simple, relatively stereotyped actions like these, each caused by a specific stimulus, are called **reflexes**. Reflexes were studied extensively in the first few decades of the twentieth century because they were thought to be the fundamental building blocks out of which more complex behavior could be constructed. The idea was that any action of an animal would generate sensory input, which, in turn, could trigger additional reflexes. In this way, even the most complex behavior could be produced as a chain of reflexes. In the last few decades, however, reflexes have come to be viewed more as modulators and shapers of behavior than as the prime causes, as described in the following sections.

A Simple Reflex

The physiological basis of a simple reflex is quite straightforward and may already be familiar to you (Figure 16-1). The neural circuit that underlies a reflex, called a **reflex arc**, usually consists of at least three elements: a sensory neuron, an interneuron, and a motor neuron. The sensory neuron typically makes synaptic contact with an interneuron, called an **association neuron** because it "associates" the sensory input with the motor response. The interneuron, in turn, makes synaptic contact with a motor neuron. Sufficiently strong stimulation of the sensory neuron will excite first the interneuron and then the motor neuron, hence producing a simple behavioral response when the muscle innervated by the motor neuron contracts. In a few cases, a sensory neuron may bypass the interneuron and make direct synaptic contact with the motor neuron. In this case, the reflex is termed **monosynaptic**, referring to the presence of the single synapse in the arc. Monosynaptic reflexes are usually involved in actions that require great speed. More commonly, one or more interneurons are interposed between the sensory and motor neurons, and the reflex is termed **polysynaptic**. Polysynaptic reflexes may be simple or complex. (The terms monosynaptic and polysynaptic can also be applied to any single or multiple synaptic pathway in the nervous system, not just a reflex arc.)

The knee-jerk reflex, a common component of many physical examinations, is a familiar example of a simple reflex. It is induced by tapping the tendon of the quadriceps muscle just below the knee. The quadriceps is the main muscle of the upper leg, and its tendon runs over the patella (the kneecap) before it attaches to the proximal end of the tibia. The response to the tap is a brief twitch

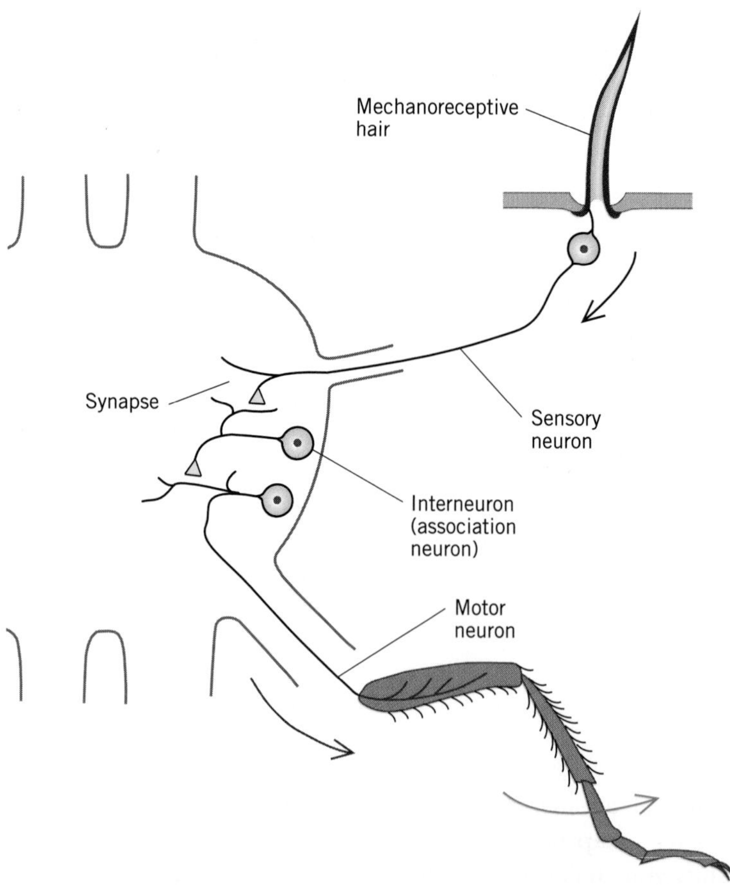

FIGURE 16-1. The neural basis of a simple reflex. Deflection of the mechanoreceptive hair will lead to contraction of the leg muscle and kicking action of the leg. The interneuron is not present in all reflexes.

of the muscle, which makes the lower leg kick forward. The reflex works because briefly tapping the tendon stretches it, hence briefly stretching the quadriceps muscle itself. (This is the reason this and similar reflexes are called stretch reflexes.) Stretch of the muscle excites several sensory receptors, especially the group Ia mechanoreceptors in the muscle spindles (see Figure 10-7, Chapter 10). The afferent fibers of the group Ia sensory neurons run into the lumbar spinal cord, where they synapse directly on the α motor neurons that innervate the quadriceps muscle. The brief volley of action potentials in the group Ia afferents that is generated by quick stretch of the muscle spindles causes the motor neurons to fire, and hence causes the muscle to contract. As a result, the leg kicks.

Stretch reflexes are not just medical curiosities used in a physician's office to evaluate the state of your nerves and synapses; they play important physiological roles. In everyday life they act undramatically and almost unconsciously to assist you in standing erect. Think of what must happen as you stand. If your quadriceps muscle relaxes a bit, your knee will bend from the weight of your body. This will stretch the quadriceps muscle slightly. This stretch will cause excitation and hence the generation of action potentials in the group Ia spindle afferents, which in turn will produce further excitation of the already active motor neurons of the quadriceps muscle, as described in Chapter 10. Additional excitation of the motor neurons will increase the force of the muscle, straightening your knee so your leg does not collapse under you. This kind of minor adjustment of muscle action is continuous not only in the quadriceps muscle but also in all the muscles of your legs and trunk while you stand. Stretch reflexes also are important in other kinds of continuous adjustments of muscle action, such as holding out a cup while someone pours coffee into it (see Chapter 10).

Adjustments of motor output based on small variations of sensory input are an important means of regulating behavior. You will learn more about the role played by these kinds of compensatory neural systems in Chapter 20.

The neural circuit that underlies a simple reflex usually consists at minimum of a sensory neuron, an interneuron, and a motor neuron. Simple reflexes assist the body in making necessary moment-to-moment postural adjustments and in making other adjustments of muscle action.

Complex Reflexes

The stretch reflex described in the preceding section is relatively simple, involving only the sensory fibers and motor neurons of a single muscle. Other reflexes are more complex and may involve the activation of synergistic muscles, the use of reciprocal inhibition between antagonists, and crossed extensor activation. These features and others were described by Sir Charles Sherrington and his collaborators in England during the early part of the twentieth century. Their studies, which provided important insights into the organization of neural circuits in the mammalian spinal cord, earned Sherrington a share of the Nobel Prize for Physiology or Medicine in 1932. Some of the main features of reflexes that he and his colleagues described can best be discussed in the context of specific examples.

One type of more complex reflex, a withdrawal reflex, is probably already familiar to you. If you have ever accidentally jabbed your finger with a pin, you have experienced this type of reflex. The painful stimulus evokes an extremely rapid and largely automatic jerk of your hand and arm away from the pin, a reaction that seems simple enough but actually consists of several events.

First, withdrawing your arm involves activation of synergistic muscles, those in the upper arm that flex the elbow and those around the shoulder that bring the arm back toward the body. As you learned in Chapter 15, synergistic muscles are those that act together around a single joint to effect a particular movement. The term synergism may also be used broadly, referring to a single functional action such as limb flexion, in which muscles at different joints are involved. During a withdrawal reflex, many synergistic muscles around both the elbow and shoulder joints are activated simultaneously to bring about the movement.

Second, motor neurons innervating the antagonist extensor muscles in the stimulated limb are strongly inhibited. Inhibition of the motor neurons of antagonistic muscles is common during many movements that are fast and strong. It is brought about in part by reciprocal inhibition among the pools of antagonistic motor neurons. **Reciprocal inhibition** refers to an arrangement of two neurons so that excitation of one causes inhibition of the other, and vice versa. In the context of reflexes, it means that excitation of one motor neuron causes excitation of an inhibitory interneuron that synapses on another motor neuron, thereby inhibiting it. In motor systems, the inhibited motor neuron always innervates a muscle in the same limb that is antagonistic to the activated muscle, so the inhibition serves the function of preventing force generation in the antagonistic muscle that would oppose the intended movement.

Some reflexes involve muscles distant from the part of the body to which a stimulus is applied. Consider a wolf walking along a forest trail and stepping on a sharp stick. The wolf's response will be to lift its foot from the stick, just as your response to a pinprick is to withdraw your hand from the pin. This movement involves the action of synergists and is also mediated by reciprocal inhibition, as just described. However, since the wolf is walking, it also must make adjustments in its other legs to avoid falling over. Hence, at the same time that flexor muscles are excited and extensor motor neurons are inhibited in the leg to be lifted, *extensor* muscles in the contralateral (opposite side) leg are excited and *flexor* motor neurons are inhibited so that this leg will be able to take the weight of the body when the stimulated foot is lifted. The reflex activation of muscles in the contralateral leg is referred to as a **crossed extension reflex** because it involves the excitation of extensor muscles in the limb on the other side of the body from the one undergoing a reflex flexion. Other reflexes will produce adjustments in muscles in the other legs and elsewhere in the body.

Reflexes like the withdrawal or crossed extension reflexes are possible because many interneurons are usually involved in the neural circuit that is activated by a particular stimulus. The features of complex reflexes described earlier arise due to the actions of interneurons. These actions are illustrated in Figure 16-2, which shows just a small part of spinal circuitry. Interneurons spread the input from sensory afferents to different motor neurons, either on the same side of the spinal cord as the side at which the stimulus enters or across to the other side. They are also responsible for whether the sensory input will have an excitatory or inhibitory effect on

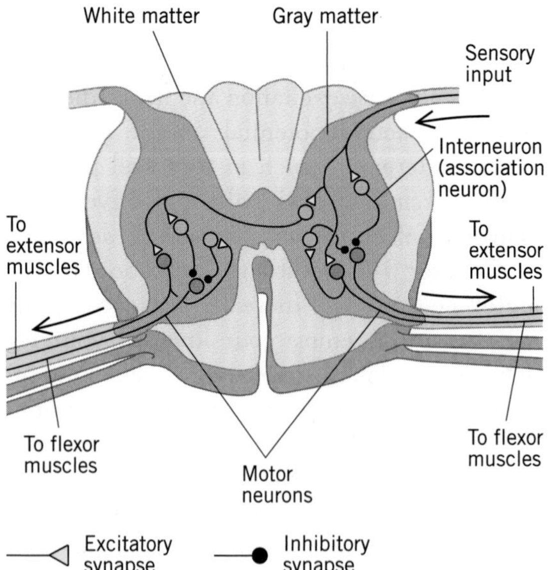

FIGURE 16-2. A reflex pathway in the spinal cord of a vertebrate such as a wolf. Illustrated are the use of interneurons to change the sign of sensory input signals from positive (excitatory) to negative (inhibitory), reciprocal inhibitory interactions between antagonistic muscles (extensor and flexor groups), and a crossed extension and inhibition response to the other side of the nerve cord.

the motor neurons. For example, extensor motor neurons to one leg and flexor motor neurons to the contralateral leg are inhibited by interneurons that are activated by sensory input. Reciprocal inhibition in the spinal cord also is usually mediated by the action of inhibitory interneurons on motor neurons, as shown in Figure 16-2. In general, interneurons are important in reflexes because they allow the coordinated action of several sets of muscles in one limb, or even muscles in different limbs, in response to a single stimulus.

Complex reflexes may involve the activation of synergistic muscles, the use of reciprocal inhibition between antagonists, and crossed extensor activation. These features are mediated by the actions of interneurons in the neural circuits that are responsible for the reflexes. Interneurons allow the coordinated action of several sets of muscles in one limb, or even muscles in different limbs, in response to a single stimulus.

Reflexes and Behavior

Research conducted since Sherrington's time has revealed previously unsuspected subtlety and flexibility in reflexes. Work on scratch reflexes provides a good example. Many vertebrates will scratch a specific spot on the skin when it is stimulated. Moving the leg into position and then scratching requires the activation of most of the muscles of the leg. In order for the foot to be directed to a specific site on the skin, however, these muscles must be precisely activated. For example, more hip flexion is required to bring a rear foot to a dorsally located spot than is required to bring it to a ventrally located spot.

The principle has been demonstrated especially well in lower vertebrates. In the common red-eared turtle, a stimulus applied to the skin of the flank excites many interneurons in the spinal cord. However, not all interneurons are excited to the same extent, and stimulation at a different site causes a slightly different pattern of excitation of interneurons. The interneurons synapse with an array of leg muscle motor neurons in such a way that stimulation of a particular spot on the skin will activate and inhibit leg motor neurons to the degree needed to bring the foot to the spot at which the stimulus is applied. Hence, rather than a stimulus causing a fixed reaction in a limited subset of interneurons, and therefore in a limited number of motor neurons, each stimulus causes widespread activation of interneurons. The interneurons in turn activate and inhibit many different motor neurons. The precise pattern of motor neuronal activation depends on the integration of the interneuronal input on the motor neurons; it will change if the stimulus is applied at even a slightly different location.

Some invertebrates, in spite of their rather different central nervous system organization, show a similar specificity in their reflex actions. For example, if you gently touch a locust on the tibia of its rear leg, the segment just above the foot, the insect will move its leg away from the contact. Touch the tibia on the front, and the leg is moved back. Touch it on the back, and the leg is moved forward. A touch along the side will bring the leg closer to the body. These precise avoidance reflexes are mediated by interneurons within the thoracic ganglion in which the leg motor neurons lie. Different sensory afferents make different patterns of synaptic connection with these interneurons, and the interneurons in turn will excite motor neurons to varying degrees, so that a stimulus in a specific location on the leg will elicit a motor response that is specific to the location of the stimulus.

In vertebrates, the organization of interneurons in the spinal cord into groups that can control and coordinate the movement of a leg in a particular way has been described and studied quite apart from the possible role of these interneurons in reflexes. In frogs, focal microstimulation (stimulation strong enough to excite neurons only within 5 to 15 μm of the stimulating electrode) of interneurons in the spinal cord will elicit quite specific and directed leg movements. Stimulating slightly different regions of the spinal cord elicits slightly different movements. Hence, a flexible and yet highly specific array of spinal interneurons appear to control the motor neurons of each leg of a vertebrate. These interneurons are presumably activated by signals from the brain when the animal wishes to make a particular movement (see Chapter 18), but they can also be activated by sensory input from the skin and used to coordinate an appropriate reflex without involving the brain.

Reflexes may be subtle and flexible. A sensory stimulus can evoke a remarkably precise motor response by activating sets of interneurons that in turn activate the proper sets of motor neurons for appropriate responses. Animals may use these sets of interneurons to organize nonreflexive motor responses as well.

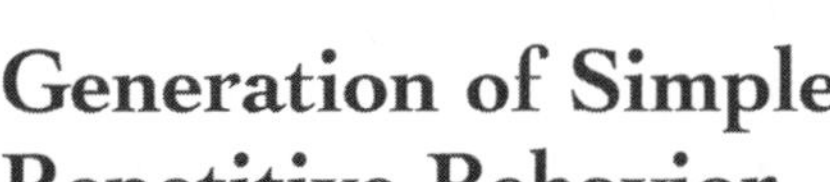

Generation of Simple Repetitive Behavior

The studies of Sherrington and others suggested that sequences of reflexes might be the basis of some behavior. However, not all behavior is explainable solely as a chain of reflexes, so other explanations for behavior have been explored as well.

Feedback and Repetitive Behavior

The role of sensory information in behavior has been studied most intensively in simple, repetitive locomotor behavior such as walking, swimming, and flying. Repetitive behavior is widespread in the animal kingdom. Not only is locomotion itself a common activity, but many other activities are performed repetitively. Scratching, breathing, and even many kinds of sound production involve repetitive movements. If there are general principles in the production of repetitive behavior, they must have considerable importance.

Coordinated repetitive behavior can be generated only if particular motor neurons, and hence the muscles they control, are activated at specific times and in a precise sequence. Strongly influenced by the work of Sherrington, many researchers during the first half of the twentieth century thought that sequential reflex actions were particularly important in establishing the timing of muscle action. Each movement of an animal stimulates a particular set of sense organs, causing these organs to send signals back to the central nervous system. These signals would then serve as the timing cues for initiation of action in specific muscles for the next movement, and so on. In modern terminology, we say that sensory feedback from ongoing movements establishes the selection and timing of subsequent movements. The term **feedback** refers to the functional role of the sensory signals, since in electrical engineering terms they come, or "feed," back into the central nervous system to help select and generate the next phase of the behavior.

Based on early study of reflexes, some researchers in the first half of the twentieth century thought that simple, repetitive behaviors such as walking and flying are coordinated by a chain of reflexes, in which sensory feedback from ongoing movements provides timing cues for subsequent movements in the cycle.

Evidence for Centrally Generated Patterns

In theory, at least, repetitive behavior can be generated without using sensory feedback to provide the timing signals for a rhythmic pattern of motor impulses. Although sensory feedback plays an important role in the generation of repetitive behavior (see Chapter 17), neurons in the central nervous system can be connected to one another in such a way as to produce a repetitive pattern in the absence of any external timing cues from sensory feedback. Behavioral observations also suggest that an animal does not rely exclusively on sensory feedback to coordinate behavior. For example, animals frequently start all the components of a coordinated behavior at once. That is, rather than starting to walk by moving just one leg, then letting the movement of that leg trigger the next by reflex action and so on until all the legs are moving, the animal starts moving all the legs at the same time, and in the proper pattern, an occurrence that is incompatible with control dependent only on chained sensory feedback.

It was not until after the middle of the twentieth century, however, that anyone produced convincing evidence to support the possibility of centrally generated patterns. In 1961, Donald Wilson showed that even after removing all the sensory connections from the wings of a locust, he could still record the repetitive action of motor neurons that controlled the flight muscles (Box 16-1). He did find a significant reduction in the frequency at which the wings beat as the locust flew after this operation, but even after complete deafferentation of the wings, the correct muscles were still being activated at the proper time relative to one another. These experiments had a dramatic impact on researchers because they showed in an unambiguous way that the central nervous system was able to generate a properly sequenced pattern of impulses in a set of motor neurons without the necessity of any sensory input to provide timing information.

Wilson's interpretation of his results was that there exists a group or network of neurons in the nervous system that is responsible for generating the repetitive (rhythmic) motor output. Wilson called his network an **oscillator**. At present, this type of network is more frequently called a **central pattern generator (CPG)**, or just a **pattern generator**. The idea is that this network can produce a repetitive and properly sequenced pattern of motor signals entirely by virtue of the connections that exist among the neurons in it, and without the necessity for external timing cues.

After considerable research, investigators now believe that CPGs are important in the generation of most, if not all, rhythmic behaviors that animals engage in, from the pulsating swimming movements of a simple jellyfish to the running of a child chasing a ball. That is, the nervous system contains networks of neurons that are capable of generating repetitive and patterned sequences of motor impulses to muscles that are normally engaged in a particular repetitive behavior, without having to rely entirely on sensory feedback to provide timing. However, this is not to say that in ordinary circumstances sensory information from moving parts of the body plays no role in the production of the normal behavior. On the contrary, in most cases sensory feedback is an integral part of the mechanism that produces the behavior. It is important to recognize that although the nervous system is capable of producing a rhythmic pattern without any sensory feedback, it normally does not do so; sensory feedback plays an important role in maintaining or otherwise regulating the rhythmic behavior. You will learn more about this role in Chapter 17.

Many rhythmic types of behavior are driven by an oscillator, or central pattern generator, a network of neurons capable of producing a properly timed pattern of motor impulses in the absence of any sensory feedback to provide timing. Sensory information is nevertheless important in producing a fully normal motor pattern.

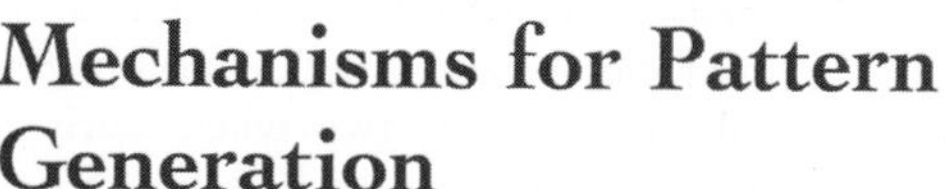

Mechanisms for Pattern Generation

How can a network of neurons produce rhythmic output without any external timing cues? One approach to this question is to develop a **model**, a hypothetical arrangement of neurons that will generate a particular type

BOX 16-1

Pattern without Feedback: Deafferentation as a Tool for the Study of Behavior

A technique that has been the mainstay of research into the neural basis of repetitive behavior such as flying has been **deafferentation**, the elimination of sensory input into the central nervous system. The investigator who undoubtedly gained the most significant results from applying this technique was Donald Wilson. Most researchers will agree that Wilson's experiments in the early 1960s were the most influential in convincing researchers that nervous systems could generate repetitive output in the absence of rhythmic timing cues. At the beginning of his career, Wilson set out to investigate the neural basis of flying in locusts. Locusts have two pairs of wings. The wings of each pair beat in synchrony, but the rear wings lead the front wings in the beat cycle by about 10%. By recording the electrical activity in the wing muscles during flight, Wilson demonstrated a regular rhythm associated with the wing movements, with the hind wing elevator and depressor muscles leading the activity in the corresponding forewing muscles by about 10% (Figure 16-A).

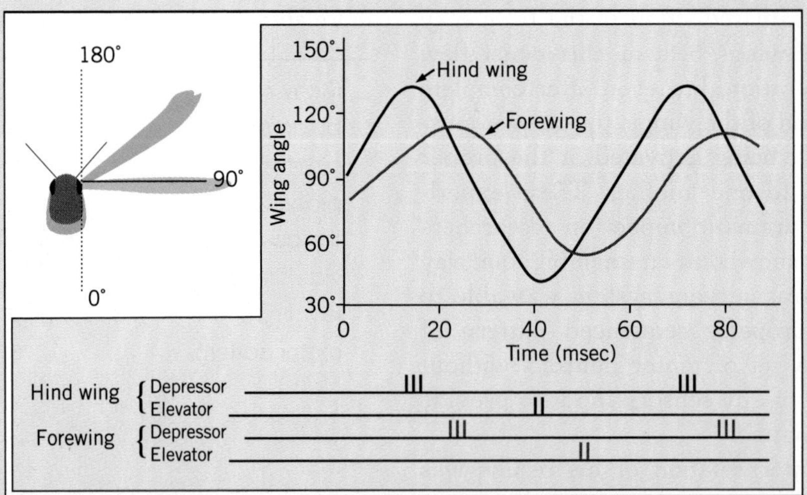

FIGURE 16-A. The flight activity of a locust. The sketch on the left shows the outline of a locust head-on, with a left wing held horizontal. The graph shows the angular movements of the two wings during flight, using the conventions shown in the sketch. The lines at the bottom represent electrical recordings from the muscles that move the wings up (elevators) and down (depressors) in each wing beat. The movements of the hind wings and the motor activity in the hind wing muscles lead the forewings and forewing muscles by about 10% of the cycle. The beat frequency in an intact locust is about 17 beats per second, giving a beat period of about 60 msec.

At the time, most researchers thought sequential reflexes were the basis of repetitive behavior. So Wilson, having seen the neural pattern underlying flight behavior in an intact locust, set out to identify the reflexes that he thought would account for the flight pattern. He did this by interfering with normal flight behavior in two ways. First, he carried out a series of deafferentation experiments: removing sense organs at the bases of the wings, cutting off the wings, or removing other parts of the locust's body that contained sense organs. The results were entirely unexpected. Whatever he did, he found no effect on the timing of the activity of the wing muscles relative to one another, even though he thought that sense organs in the wings ought to be providing timing cues for the motor pattern. He did find a reduction of up to 50% in the overall frequency of beating of the wings if certain sense organs were removed, but the motor signals to the flight muscles still came at the proper time to keep the wing beat correctly synchronized.

In his most extreme experiments, Wilson reduced the animal to a head, the floor of the thorax (the part of the body to which the wings are attached), and the thoracic nerve cord. He could still record the electrical activity of motor nerves by placing electrodes on the stumps of the nerves that had innervated the removed flight muscles. A motor pattern recorded in the absence of any movement of part of the animal is called a **fictive pattern.** Hence, Wilson is said to have recorded fictive flight. By blowing wind on the head of this "animal," he found that the pattern of motor impulses he recorded in the stumps of the motor nerves retained the proper alternation of activity in nerves that went to elevator and depressor muscles, and the proper delay between activity in nerves innervating the muscles for the front and rear wings.

Wilson also carried out a second series of experiments, in which he stimulated the sense organs at times inappropriate to their normal pattern of stimulation. For example, he mechanically fastened together the front and hind wings on each side of the body. The hind wings, being driven by larger muscles, drag the front wings along with them as the insect flies, causing sensory signals from the front wing sense organs to become synchronized with the movements of the hind wings, and not to have the normal delay between their activity and hind wing movement. If reflex effects from these sense organs were important in establishing the proper delay between the contractions of the front and rear wing muscles, this experiment should have disrupted the flight pattern. It did not. All the experiments pointed to the same conclusion—the locust flight system did not *require* sensory feedback to provide timing cues for rhythm generation. However, as discussed in Chapter 17, sensory information nevertheless plays an important role in flight.

of output given certain characteristics and input conditions. By exploring the specific connections and characteristics of simulated neurons, investigators can quickly and easily eliminate many possible arrangements of the network without having to carry out difficult and time-consuming animal experiments. However, just because a particular model produces an output that is similar to the motor neuronal pattern that the animal generates does not mean that the modeled arrangement of neurons is the same arrangement actually found in the animal. Only experimental verification of the presence of particular neurons with certain connections and characteristics will substantiate the existence of a particular neural pattern generator. Models may take some effort to understand, but the reward for those who persevere is a much better understanding of how neurons working together in groups can generate specific kinds of movements.

Models of a Pattern Generator

Consider first a model proposed by Wilson (Figure 16-3). Wilson wanted to determine the simplest configuration of the fewest neurons that might produce part of the flight pattern, the alternating activation of the flight muscles powering the up-and-down movements of a single wing (see Figure 16-A in Box 16-1).

Wilson's model network operates to produce alternating activation of the antagonistic wing muscles in the following way. First, it is assumed that the network receives input from elsewhere in the nervous system, presumably the brain, to signal that flight is to take place. This "fly" signal simultaneously excites two neurons (A and B in Figure 16-3). Cells A and B, which have slightly different thresholds for excitation, reciprocally inhibit each other. (Wilson assumed they did so without any intervening interneurons such as those present in crossed extension reflexes, described previously in this chapter.) Hence, when the cells are excited, one, say A, for example, will begin to fire shortly before the other, thereby shutting off B via A's inhibitory branch. The output from A will drive C, but because of C's electrical properties, only a few spikes will be generated. As cell A continues to fire, it becomes fatigued, allowing B to escape from inhibition. Cell B then shuts off any remaining tendency that A still has to fire and excites its own follower cell, D, which also fires only a few spikes. In a short while, B fatigues and A turns on. The output thus switches back and forth between neurons C and D.

Wilson conceived that A and B represented interneurons, and that C and D represented motor neurons. He experimented with computer models to determine whether he could construct a model consisting of two neurons only, but he found that four were required. Cells C and D in the model were necessary to reproduce the locust flight pattern because, without them, the bursts in A and B were always too long and too close together in time

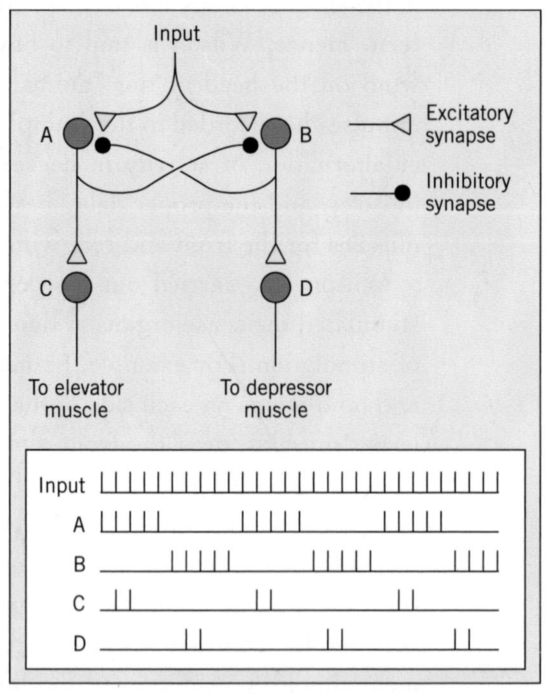

FIGURE 16-3. Wilson's model of the network of neurons (the CPG) controlling the activity of elevator and depressor muscles of one wing during a locust's flight. The lower part of the figure represents hypothetical recordings from such a neural network while it is in operation. See further explanation in text.

to simulate the actual pattern of motor output to flight muscles.

The model shown in Figure 16-3 is not very stable. The electrical properties of the model neurons must lie within a narrow range of possible values in order for the model to work properly. One feature that can be added to the neurons to stabilize their activity is **postinhibitory rebound**, the tendency of a neuron to fire a brief series of action potentials upon being released from inhibition, in the absence of any excitatory input at all. Not all neurons show this property, but some, notably in mollusks, are known to.

Postinhibitory rebound can best be illustrated in another model. In this hypothetical arrangement of neurons, a transient *inhibitory* input to one neuron in the network (the CPG) is all that is required to initiate reciprocal activity in the network cells. Figure 16-4 illustrates the model, showing the arrangement of the component cells. When the input to the CPG is terminated, postinhibitory rebound produces a burst of activity in cell A, which inhibits B at the same time that it excites other cells to which it delivers signals. When A ceases to fire, B is released from inhibition and starts its activity. This in turn inhibits A, and so on indefinitely. The system will continue to generate reciprocating activity until another burst of input to A turns it off, as shown in Figure 16-4.

The cellular basis of postinhibitory rebound is straightforward. In spiking neurons, any change in membrane potential causes a concomitant change in the threshold of the neuron. The threshold increases slightly if the neuron is depolarized and decreases slightly if it is hyperpolarized (Figure 16-5A). This change in threshold is due to changes in the number of voltage-sensitive sodium or other channels that are in an inactive state. Hyperpolarization reduces the number of inactivated channels, and hence lowers the neuron's threshold, making it easier to produce a spike in it. In some neurons, the change in threshold is great enough that if the cell is hyperpolarized by inhibitory postsynaptic potentials (ipsps) for a brief period, the membrane potential will cross the new, lowered threshold when the cell is released from inhibition, causing the generation of one or more action potentials (Figure 16-5B). The network depicted in Figure 16-4 can be shut off by a long burst of activity that trails off in frequency at the end, because the longer intervals between successive ipsps allow the membrane potential to return to resting level slowly, allowing the threshold to recover as well without being crossed (Figure 16-5C).

Models are useful as formal expressions of ideas about how neurons might interact. They can provide an important theoretical basis for ideas about how a particular neural function can be carried out, as Wilson's model did for the idea of a CPG. In this context, however, it is important to understand that a model may not be based on any actual known interactions between neurons in a living animal. In particular, Wilson's model was an attempt to find the simplest arrangement of neurons that would produce rhythmic output. Its importance lay entirely in its theoretical implications. For models to aid in the understanding of nervous systems, they need to be tested against nervous systems in living animals.

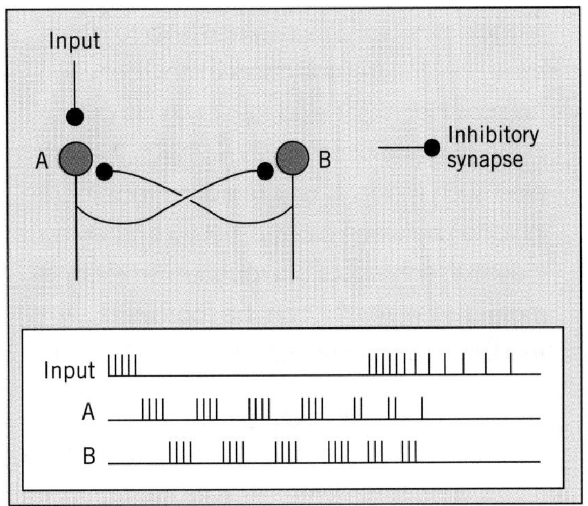

FIGURE 16-4. Model of a pattern generator consisting of only two cells, driven by postinhibitory rebound. The lower part of the figure represents hypothetical recordings from the neurons in such a neural network while it is in operation.

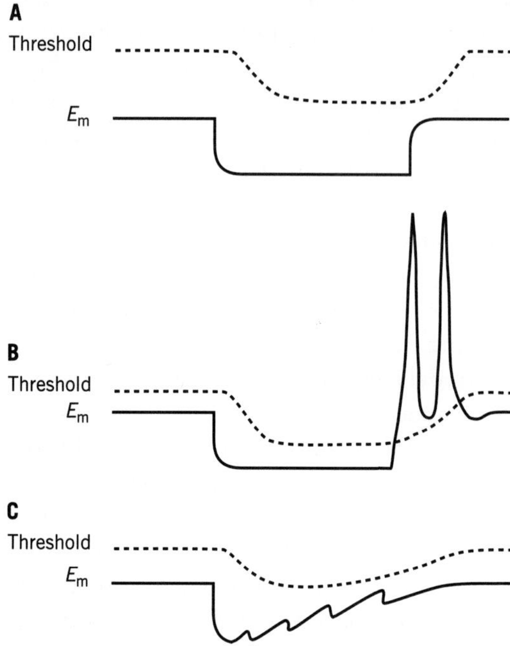

FIGURE 16-5. The effect of a hyperpolarizing stimulus on the threshold of a stimulated cell. (A) In most cells, the threshold is reduced slightly during the stimulus, but no other consequences follow. (B) In cells that show postinhibitory rebound, the threshold is lowered enough by the stimulus so that, on release from hyperpolarization, the membrane potential exceeds the threshold and generates action potentials. (C) Action potentials can be prevented from firing if an inhibitory input is removed gradually, allowing the threshold to recover gradually.

Models of neural networks can help to determine the theoretical connections between neurons that might lead to a rhythmic output in the absence of any rhythmic input. The simplest such model is one featuring reciprocal inhibition between a pair of neurons receiving identical, continuous neural input. Similar and more stable results can be obtained with models incorporating postinhibitory rebound.

Properties of Neural Pattern Generators

The specific neurons that form a CPG have been identified in species from four different phyla: annelids, mollusks, arthropods, and chordates. Comparison of these CPGs with the models proposed by theorists has revealed both differences and similarities at several levels.

The most obvious difference between actual neuronal networks and theoretical model networks is that networks in animals are much more complex than the simple models that have been described here. One of the best-understood CPGs is the network that controls movements of the stomachs of decapod crustaceans, like lobsters and crabs. The stomach of a lobster is a complex organ, consisting of three main parts: the cardiac sac, the gastric mill, and the pyloric region (Figure 16-6). Food is brought from the esophagus into the cardiac sac, moved on to the gastric mill, where it is ground into small pieces by the teeth there, then churned in the pyloric region to aid digestion. Esophageal movements, as well as the chewing and churning movements of the gastric mill and the pyloric regions of the stomach, are regular and repetitive, controlled by the several dozen neurons in the esophageal and stomatogastric ganglia that lie on the wall of the esophagus and stomach. The neurons of the stomatogastric ganglion have been studied most intensively; it is known that the CPG for the gastric rhythm consists of 11 neurons and the CPG for the pyloric rhythm consists of 14 neurons.

A simplified diagram of the two CPGs showing the interconnections between the neurons in them is given in Figure 16-7. The left parts of the figure show the repetitive patterns of activity that are typically exhibited by the neurons of the CPGs. The right parts show schematics of the connections between the neurons. The letters identify the different types of neurons. Some cells, such as GM in Figure 16-7B, occur in multiples. The lines connecting the cell types represent synaptic interactions. Lines with a sawtooth pattern in the middle represent electrical connections. The others represent chemical synapses. The complexity of this pattern generator is typical of CPGs in other animals as well.

A similarity between animal and model CPGs is that CPGs in animals operate in a motor hierarchy. That is, the CPG is a network of interneurons that is usually positioned between the motor neurons to which it sends its commands and higher neural centers such as the brain from which it receives its input. The lack of involvement of the brain in the production of the repetitive output of the pattern generator has been demonstrated in many animals by severing the connections between the brain and the rest of the nervous system. With proper stimulation, an appropriate fictive behavior (see Box 16-1) can be generated even after the brain has been removed. On the other hand, except in a few cases, motor neurons are also not directly involved in generating the pattern produced by the CPG. Hence, the pattern generator is interposed between higher neural centers that control it and the motor neurons that produce the appropriate muscular actions. You will learn more about the role of the brain in behavior in Chapter 18.

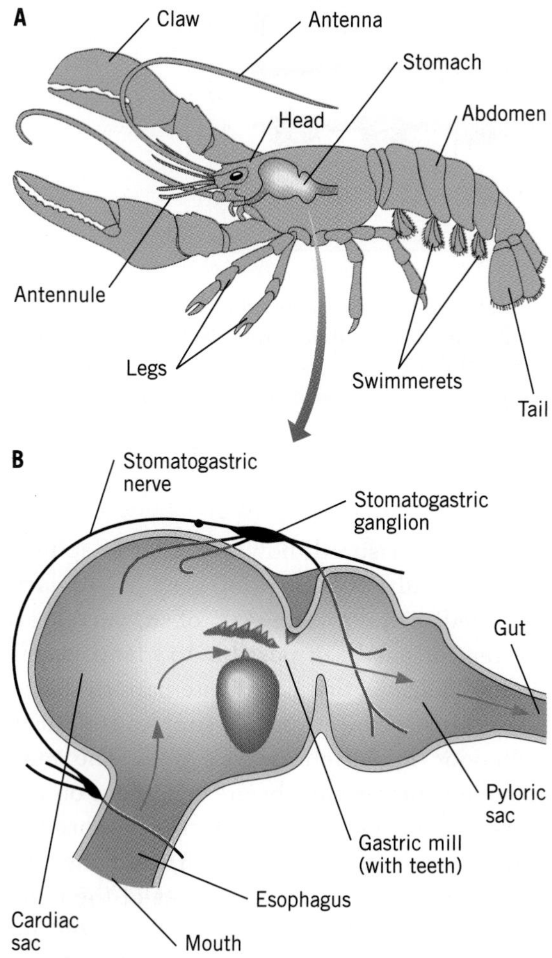

FIGURE 16-6. The stomach of a decapod crustacean. (A) A crayfish, showing the location of the stomach. (B) The stomach, showing some of the main nerves and, in outline, the locations of the teeth of the gastric mill. Arrows show the path of food.

There are similarities and differences between animal CPGs and the models at the cellular level as well. In virtually every case, researchers can identify cells that interact via reciprocal inhibition, which, as you may recall, is the mutual inhibition of each of a pair of cells by the other; this mechanism is at the heart of the model in Figure 16-3 and seems to be at the heart of pattern generation in all animals. An important difference, however, is that in actual CPGs many pairs of cells, not just one, are used to produce a pattern. The CPGs controlling the pyloric and gastric patterns in lobsters are excellent examples. In the pyloric CPG, cells PD and LP, VD and LP, VD and IC, and LP and PY reciprocally inhibit one another (Figure 16-7A), and in the gastric CPG, cells DG and LG, LG and Int 1, Int 1 and MG, LG and MG, and MG and LPG all reciprocally inhibit one another as well (Figure 16-7B).

The wealth of data currently available about different pattern generators and the cellular basis of pattern production may give you the impression that pattern generation is completely understood. This is not the case. First, the individual cells that constitute the pattern-generating network have been identified for only a few types of behavior in a handful of animals. Although there are similarities between CPGs in different animals, CPGs are by no means identical. It is to be expected, therefore, that the CPGs that underlie pattern generation in other animals or for other types of behavior will reveal new details or even entirely new principles as to how such systems work.

Second, even for the pattern generators that have been described in detail, there is still much to be learned. For example, study of the locust flight pattern generator had suggested that there was a single pattern-generating

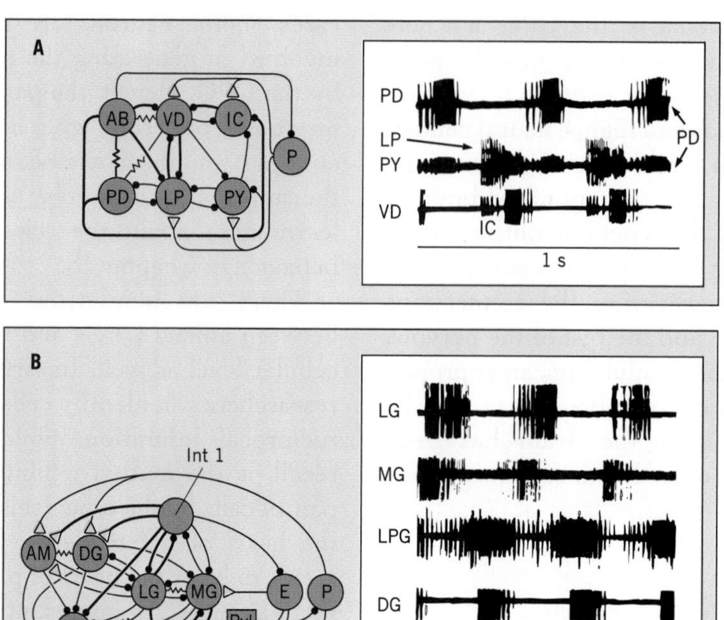

FIGURE 16-7. The main groups of neurons involved in generation of the pyloric (A) and gastric (B) rhythms of a lobster stomach and examples of their rhythmic output. Each circle represents several individual neurons of a particular type. The initials refer to the names that have been given to different classes of neurons. Solid circles represent inhibitory synapses; triangles represent excitatory ones. Lines with resistor symbols represent electrical coupling.

network controlling all four wings together, distributed over the two ganglia containing the motor neurons for the flight muscles. Yet later experiments showed that under the proper circumstances, it is possible to cut apart these two ganglia without eliminating the ability of the ganglia to produce rhythmic alternation of motor output to the flight muscles that they still control. Hence, there are still aspects of the pattern generator for locust flight that are not fully understood.

The emphasis in this section has been on common features of pattern generators. The prime example, the CPG that controls movements of two parts of the stomach of a lobster, may seem esoteric to you. It has been chosen because it is the pattern generator about which most is known at the cellular level. Rest assured, however, that in spite of the difficulty of carrying out a similarly detailed cellular analysis of a vertebrate system, it has been possible to identify individual neurons that constitute pattern generators for generation of simple repetitive behavior in vertebrates as well. The most detailed studies have been carried out on two types of lower vertebrates, the lamprey (a type of primitive jawless fish) and the tadpoles of certain toads, in studies of the neural basis of their swimming.

Although swimming is somewhat different than flying, involving undulating movements of the whole body rather than up-and-down movements of wings, the features of the CPGs that underlie it are similar to those already described. In lampreys and tadpoles, two sets of interneurons in the spinal cord control the motor neurons for muscles in one section of the body. Some of these interneurons reciprocally inhibit one another, leading to alternation of muscle activity on the two sides of the body. When the body bends in one direction by the action of one set of muscles, the muscles on the opposite side relax because their motor neurons are inhibited. As the swimming wave sweeps down the body, the first set of muscles is turned off and the contralateral muscles are activated. A diagram of some of the underlying neural circuitry is shown in Figure 16-8. Note the reciprocal inhibition between the right and left cell I, and that in this animal a single pattern generator controls muscles on both sides of the body. Excitatory input from the brain is not shown.

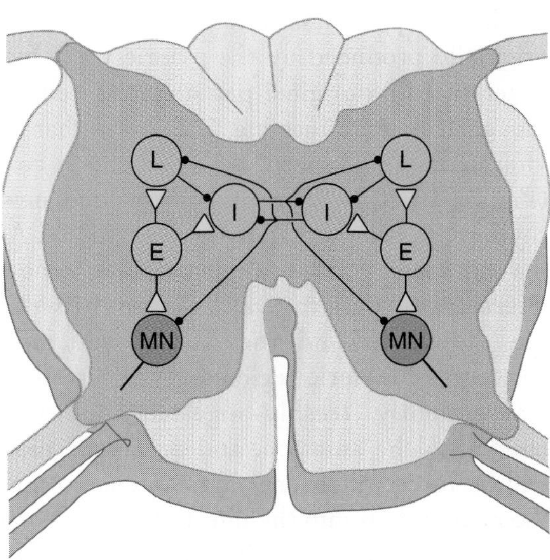

FIGURE 16-8. Model of the CPG for control of muscles in one body segment of the lamprey during swimming, shown against an outline of the spinal cord in cross section. Each circle represents a group of neurons with similar properties, named according to the group's functional features or its location: L, lateral interneurons; I, inhibitory interneurons; E, excitatory interneurons; MN, motor neurons. The CPG controls activation of antagonistic pairs of body muscles on opposite sides of the body. Note the reciprocal inhibition between the inhibitory interneurons.

Pattern generators in living animals are more complex and diverse than the models that were first constructed to explain repetitive motor output. Yet, as suggested by the models, the CPGs of both vertebrate and invertebrate animals are part of a motor hierarchy and involve reciprocal inhibition between interneurons.

Mechanisms of Cross-Segmental Coordination

So far you have learned mainly about the generation of the alternating activity of antagonist muscles associated with a single appendage or segment of the body. Yet a single set of antagonistic muscles rarely acts alone in producing a repetitive behavior. Usually, not only several sets of muscles but also several appendages or other body parts are involved. These also require proper coordination. How is such coordination achieved?

In most systems, it is thought that each segment of the body that contains a pair of appendages also contains a pair of pattern generators, and that each pattern generator is coupled with its neighbors to produce a well-coordinated motor pattern. Take crayfish, for example. Crayfish have pairs of paddlelike abdominal appendages called swimmerets (Figure 16-9), which beat repetitively to help stabilize the animal as it moves about in the water. Each pair moves back and forth about once a second and is slightly out of synchrony with adjacent pairs. The rearmost swimmerets beat first, and more anterior ones beat later. Each swimmeret is known to be controlled by a separate pattern generator. Each CPG is activated by an input signal from the brain and drives the muscles that move the swimmeret back and forth. The individual pattern generators are coupled together by neurons that convey information about the state of one CPG to the next posterior one. If you cut these neurons,

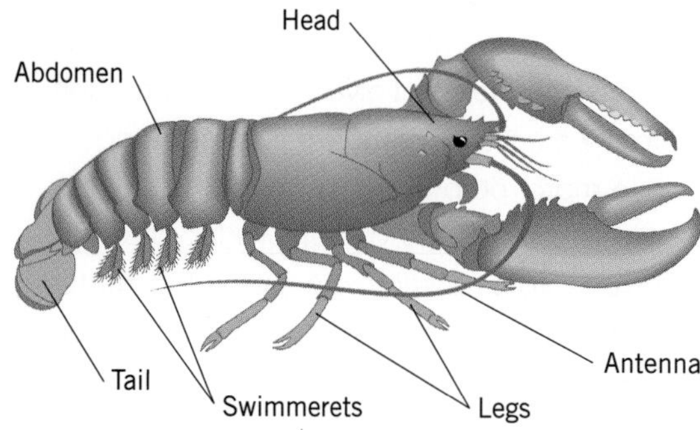

FIGURE 16-9. Lobster showing its swimmerets. All lobsters and shrimplike crustacea have swimmerets, which are used in small animals for swimming and to stabilize the body in the water.

the swimmerets in front of and behind the lesion will beat independently, slowly drifting into and out of coordination with one another.

Similar neurons are thought to coordinate other rhythmic behavior. In lampreys, the CPGs in each body segment receive input from and give output to the CPGs in adjacent segments. These connections keep the CPGs coordinated. In the lamprey, the most anterior CPG has the fastest cycle of activity. Hence, the coordinating interneurons from the first CPG will entrain the second CPG (see Chapter 17), those from the second will entrain the third, and so on, ensuring that the motions of the entire body are synchronized.

Each appendage or part of the body that moves repetitively during a rhythmic behavior is controlled by its own CPG. Individual CPGs are synchronized via coordinating interneurons that couple one pattern-generating network to another, ensuring that the CPGs are properly synchronized.

Neuromodulator Effects on Pattern Generators

Central pattern generators have been described as networks of neurons that control specific rhythmic behaviors by virtue of fixed connections between their member neurons. Since the late 1980s, however, it has been recognized that far from being fixed, immutable units of control, CPGs can actually be shaped and molded in a number of ways by the actions of neurons that make synaptic connections with CPG neurons.

Take the stomatogastric system of a lobster as an example. Earlier in this chapter, you learned that in lobsters and other crustacea there are separate esophageal, gastric mill, and pyloric pattern-generating networks. Sometimes these three CPGs work alone. The esophageal CPG moves food along the esophagus into the stomach, the gastric mill CPG grinds the food into smaller particles, and the pyloric CPG churns the food to aid digestion. However, stimulation of a particular interneuron, called PS, that makes widespread synaptic contact in the ganglia containing these CPGs, causes the three patterns to fuse into a single, quite different pattern, as shown in Figure 16-10A.

The new pattern is slower than the pattern originally produced by the pyloric CPG but faster than the original pattern produced by the gastric mill, forming a pattern that is intermediate in speed between these two (Figure 16-10B). The function of the new rhythm has been shown by several studies. At the same time that the gut engages in the new intermediate pattern, valves open between the esophagus and the cardiac sac, and between the pyloric region and the intestine. Consequently, freshly ingested food will move into the stomach, and processed food will be forced from one stomach chamber to the next and on into the intestine. After some seconds, during which the combined rhythm is expressed, the pattern-generating networks decouple, and each goes back to operating on its own.

The neural basis for this switch in the operation of the pattern-generating networks is still not fully understood, but it appears that neuron PS, whose activity induces the switch, has strong effects on specific neurons of the pattern generating networks. These effects include suppressing intrinsic rhythmic prop-

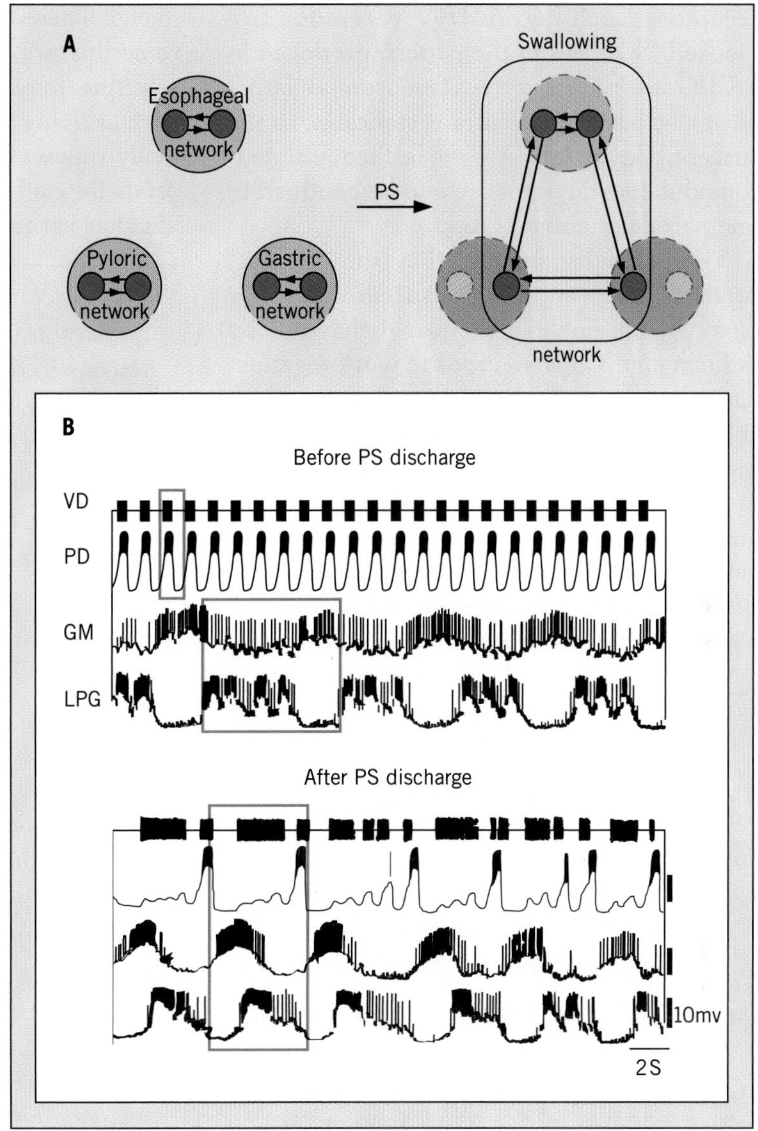

FIGURE 16-10. (A) Reorganization of the esophageal, gastric mill, and pyloric networks of a lobster into a single "swallowing" pattern-generating network following activity of a single neuron, PS. (B) The patterns of activity in the individual pyloric and gastric mill CPGs (top) are quite different from one another. The pyloric pattern (shown by cells VD and PD) is fast, and the gastric mill pattern (cells GM and LPG) is slow. However, after these two CPGs have been incorporated into a single "swallowing" network, the constituent neurons express entirely different patterns (bottom). The repetition rate of VD and PD bursts is much lower, and that of GM and LPG neurons is higher. There are also differences in the durations of the bursts.

erties of some neurons and overriding some of the synaptic interactions that keep the networks working on their own patterns. The net effect is to disrupt the individual patterns at the same time that the neurons of the CPGs are coupled in the new pattern. It is not yet known by what means these effects are brought about, but in crabs, similar behavioral switches of the pyloric network can be mediated by the neuromodulatory neurotransmitter serotonin, which along with dopamine and proctolin can have dramatic effects on the membrane or synaptic properties of individual neurons. These effects change the properties of the entire network, yielding a different pattern of movement of the walls of the pyloric region of the stomach.

Neuromodulators have been shown to affect CPGs in vertebrates as well. For example, application of glutamate to neonatal rats induces stepping. In tadpoles, it induces swimming. Swimming is also induced by glutamate in lampreys. In lampreys, different receptors seem to be involved in different expressions of the behavior. Fast swimming is induced by glutamate when NMDA receptors (see Chapter 8) are blocked, and slow swimming is induced when non-NMDA receptors such as AMPA receptors are blocked. Switches in the pattern produced in a CPG after application of neuromodulators have also been reported in vertebrates, so the susceptibility of CPGs to the influence of neuromodulators does not seem to be confined to one particular animal group.

These results suggest that the "higher" brain centers can select a specific behavior from an ensemble of similar behaviors by inducing individual neurons to work together. The neurons are "activated" and formed into an operational CPG by the action of one or more neuromodulators. Hence, CPGs are not at all fixed, with unchangeable connections among the neurons in them. Instead, neurons can be functionally added to or taken from a pattern-generating network to shape the nature of the motor output that will be produced. There is even some evidence that neuromodulators may be directly involved in selecting between quite different behaviors, such as flying and walking, by turning on normally quiescent networks of neurons appropriate for each specific act. The validity of this idea has yet to be proved, however.

The neural networks that constitute pattern generators are not immutably fixed. Neuromodulatory neurotransmitters can act on a group of neurons, changing synaptic strengths to select a subset of neurons from the entire network that will work together as a CPG for a particular behavior under a particular set of circumstances.

Additional Reading

General

Cohen, A. V., S. Rossignol, and S. Grillner, eds. 1988. *Neural Control of Rhythmic Movements in Vertebrates*. New York: Wiley.

Jacklet, J. W., ed. 1989. *Neuronal and Cellular Oscillators*. New York: Marcel Dekker.

Patton, H. D., A. F. Fuchs, B. Hille, A. M. Scher, and R. Steiner, eds. 1989. *Textbook of Physiology*. Vol. 1, *Excitable Cells and Neurophysiology*. 21st ed. New York: Saunders. See Section V, Control of Movement.

Research Articles and Reviews

Bizzi, E., S. F. Giszter, E. Loeb, A. F. Mussa-Ivaldi, and P. Saltiel. 1995. Modular organization of motor behavior in the frog's spinal cord. *Trends in Neurosciences* 18:442–46.

Delcomyn, F. 1980. Neural basis of rhythmic behavior in animals. *Science* 210:492–98.

Friesen, W. O. 1994. Reciprocal inhibition: A mechanism underlying oscillatory animal movements. *Neuroscience and Biobehavioral Reviews* 18:547–53.

Grillner, S., P. Wallén, and G. Viana Di Prisco. 1990. Cellular network underlying locomotion as revealed in a lower vertebrate model: Transmitters, membrane properties, circuitry, and simulation. *Cold Spring Harbor Symposia on Quantitative Biology* 55:779–89.

Harris-Warrick, R. M. 1992. Neuromodulation of stomatogastric networks by identified neurons and transmitters. In *Dynamic Biological Networks: The Stomatogastric Nervous System*, ed. R. M. Harris-Warrick, E. Marder, A. I. Selverston, and M. Moulins, 87–138. Cambridge, Mass.: MIT Press.

Marder, E., and R. L. Calabrese. 1996. Principles of rhythmic motor pattern generation. *Physiological Reviews* 76:687–717.

Wilson, D. M. 1961. The central nervous control of flight in a locust. *Journal of Experimental Biology* 38:471–90.

CHAPTER 17

Sensory Influence on Motor Output

Sensory information about any ongoing motor activity is available to the central nervous system in abundance. In this chapter, you will learn how this sensory information is used by the nervous system to help produce and maintain coordinated motor patterns, even centrally generated repetitive ones like those discussed in the previous chapter.

You may have formed the impression from Chapter 16 that sensory feedback is of little importance in generating a simple behavior like flying or swimming, and perhaps even in more complex behavior as well. This is far from the case. Sensory input actually plays a critical role in the generation and proper execution of all kinds of movements, even simple repetitive actions. Specifically, it can compensate for disturbances of movement and stabilize behavior, it can provide timing cues for coordination, and it can modulate behavior in other ways. These effects will be discussed as if they were independent, but sensory input may have effects that are not mutually exclusive and not easy to separate from one another.

Sensory Effects on Behavior

By adjusting motor output as necessary for external circumstances, animals continually use sensory input to stabilize a motor act. You do this yourself to keep your balance when you merely walk down the sidewalk. Sometimes an animal may encounter a circumstance that requires more than minor adjustments, as when you stumble over a curb while crossing a street. In these cases, sensory input helps correct the motor output to compensate for the disturbance. Although related, these stabilizing and compensatory actions of sensory input can conveniently be discussed separately.

Stabilization of Performance

An excellent example of the stabilization of motor performance by sensory input is the contraction of a muscle regulated by the vertebrate muscle spindle. You learned in Chapter 10 (see Figure 10-7) that the intrafusal muscle fibers of a muscle spindle organ receive efferent innervation from γ motor neurons. The activity of these motor neurons during contraction of the extrafusal muscle allows the spindle afferents to provide an accurate measure of the actual length of a muscle relative to the expected length during a contraction.

The system works like this. When you lift a weight, the γ and α motor neurons of your biceps muscle are excited together. The γ

motor neurons innervate the intrafusal muscle fibers of the muscle spindle organs in the muscle, and the α motor neurons innervate the extrafusal muscle. The contraction of the intrafusal fibers resulting from activation of the γ motor neurons maintains a relatively constant tension on the spindles even as the extrafusal muscle contracts to cause your elbow to bend. Any slight reduction in the rate of contraction of the biceps muscle compared with the expected rate, due, for example, to underestimation of the power required to lift the weight or to fatigue, will cause the intrafusal fibers, which are not carrying any weight, to shorten faster than the extrafusal fibers. The intrafusal fibers will consequently pull on the spindle, causing increased excitation of the spindle afferents, increased excitatory drive onto the α motor neurons, and hence increased force output by the muscle. If the rate of contraction is too high, the output of the spindle afferents will drop, resulting in a reduction of excitatory input to the α motor neurons, and hence a reduction in the force being generated by the extrafusal muscle. These conditions are depicted in Figure 17-1.

Ice-skating is another behavior that is stabilized by sensory input. As you skate, the otolith organs in your inner ears are continually providing information about the movements of your head. If your head starts to tilt even slightly forward because you are starting to fall in that direction, sensory signals from the inner ears to motor centers in the brain stem will activate neural circuits that control muscles in your arms and trunk. By activating appropriate motor neurons, these neural circuits will adjust your trunk posture and cause your arms to move so that you can maintain your balance. You will learn more about these circuits in Chapters 18 and 20. The neural circuits involved are not innate but must be forged as you learn to skate. However, once they are in place, they are exquisitely sensitive. The corrective movements of an experienced skater are so slight that only quantitative analysis of his or her movements on video or film will reveal them.

Sensory input helps to stabilize an animal's motor performance by providing information about what movements are actually taking place compared with those the animal intends to make. Muscle spindles and the vestibular system are two sources of input used by the vertebrate motor system to stabilize its performance.

Compensation for Disturbance

Stabilizing reactions allow an animal to make smooth movements time after time in spite of variations in muscle power or the weight of a moving limb. Sometimes, external events can produce perturbations that, if uncorrected, would seriously disturb a behavior. Continual small adjustments of muscle forces in your leg muscles as you walk help you to keep your balance. However, if you trip on a curb as you cross a street, stronger and more widespread reflexes come into play. Not only does your leg react with a quick step forward but your trunk will bend and your arms may flail, depending on how severe the disturbance to your walk is. These effects are caused by strong sensory input from your vestibular organs that is distributed widely to the motor neurons innervating many muscle groups throughout the body.

All animals that walk or fly, and many that swim as well, have neural circuits that help them compensate for unexpected deviations from their intended movements. Consider a locust in flight (Figure 17-2). As the insect flies, it must be able to maintain its position and orientation in three-dimensional space, just as a flying aircraft must. There are specific terms for movements around each of the axes of three-dimensional space during flight, whether you are speaking of an airplane or a flying animal. If air turbulence catches and lifts the wings on one side of the body, the

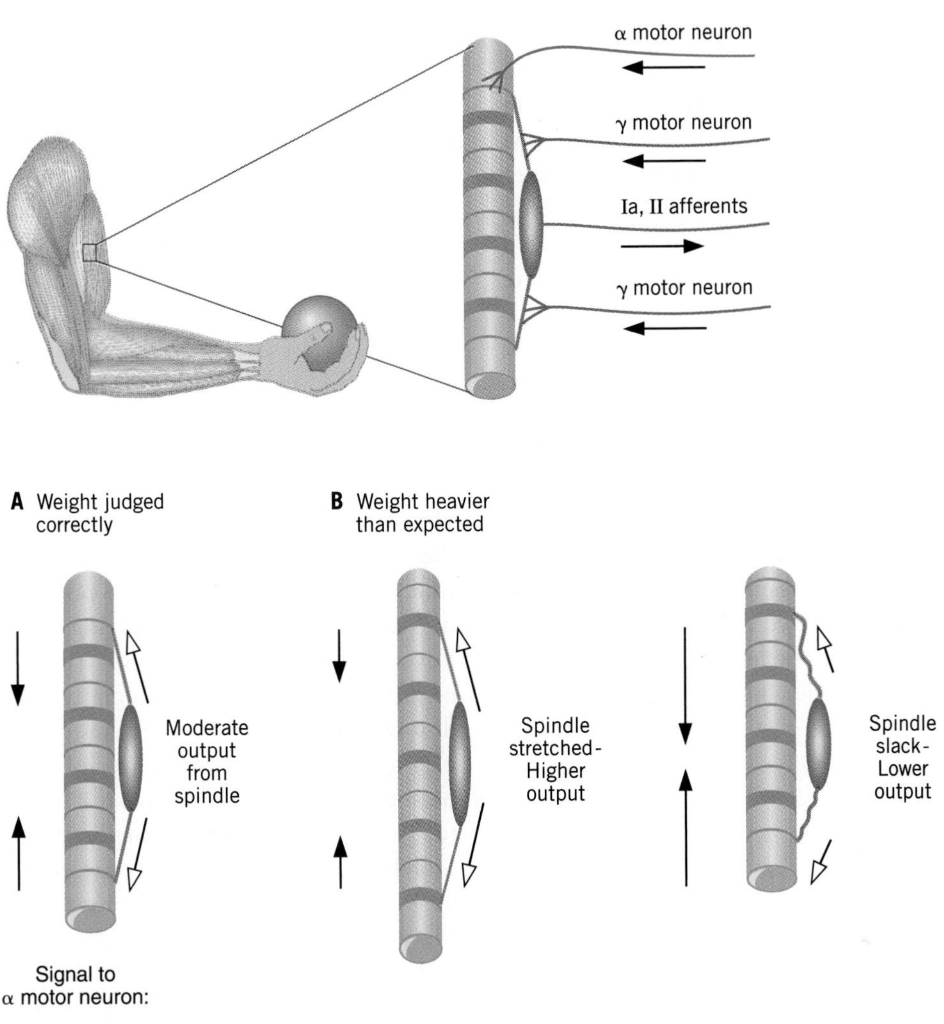

FIGURE 17-1. The relationship between estimation of a weight and the output of muscle spindle organs to the motor neurons that control lifting the weight. (A) If the weight is judged correctly, the contraction of intrafusal muscle fibers under the influence of γ motor neurons will match the contraction of the extrafusal muscle under the influence of α motor neurons (solid arrows), and the pull on the spindle afferents (open arrows) will cause them to fire at a moderate rate. (B) If the weight is heavier than thought, the extrafusal muscle will not be able to contract (solid arrows) as fast as the intrafusal fibers, so there will be a stronger pull on the spindle (open arrows), the spindle will be stretched, and a strong feedback from the spindle afferents will stimulate the α motor neurons to fire more frequently, making the extrafusal muscle contract more strongly. (C) If the weight is lighter than thought, the extrafusal muscle will contract (solid arrows) faster than the intrafusal fibers, there will be a weaker pull on the spindle (open arrows), and the output of the spindle afferents will be reduced, reducing the drive on the α motor neurons and weakening the contraction of the extrafusal muscle.

insect is said to **roll**. If air turbulence causes the front of the insect's body to tilt up relative to its normal position, the insect's **pitch** is said to increase. And if the locust is turned horizontally in the air so that it no longer points in the original direction of its flight, the movement is called **yaw**.

Several cues inform the locust that a change of its orientation in the air has occurred. One important cue is the direction from which air strikes the head during flight. On its surface, the head contains a number of short, wind-sensitive hairs. The sockets of these hairs allow them to bend easily in one plane only. Further, thesockets are oriented on the head so that different hairs can move in different directions. This allows some hairs to respond to air striking the head from one angle, other hairs to respond to air from other angles. When hairs are bent by wind striking the head at an angle other than the preferred angle, the locust responds with appropriate compensatory movements.

The insect has several mechanisms that allow it to make compensatory movements. Some of these involve the legs and abdomen rather than the wings. If the insect is swung around the z-axis by a gust of wind, producing a yaw (see Figure 17-2), there will be compensatory movements of the abdomen and legs in the direction opposite to the direction of the yaw. A yaw to the right induces the legs on the left to be extended out from the body, and the abdomen to bend toward the left. This will swing the body back toward the left in much the same way that sticking an oar into the water around a moving rowboat will cause the boat to swing toward the side on which the oar is inserted. Similar abdominal movements, this time in the xz plane, occur in response to changes in pitch. If the locust is tilted upward, an increased pitch, the abdomen will bend downward, bringing the insect back toward its preferred orientation.

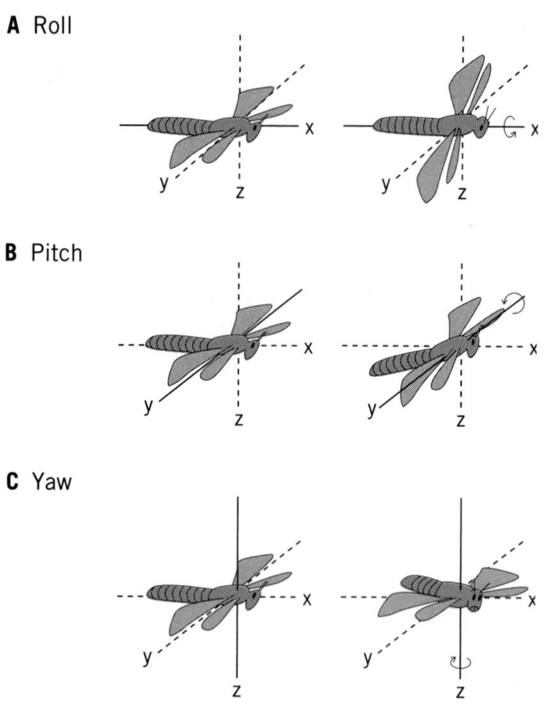

FIGURE 17-2. A diagram of a flying locust and the three axes around which displacement can take place. Rotation around each axis is referred to by a specific name: a roll is rotation about the long axis of the body (x), pitching is rotation about the axis through the wings (y), and a yaw is rotation about the up-down axis (z).

Other compensatory reactions involve the muscles that power movements of the wings. During flight, an insect's wings do not simply move up and down like inflexible paddles. Instead, the wings twist in such a way that the leading (front) edge points up during elevation and down during depression. Several muscles help depress the wing and regulate this twisting, especially during the downstroke of the wing. Some of these muscles are attached to the anterior part of the base of each wing and cause the leading edge to twist down more sharply. Others are attached to the posterior part of the wing base and decrease the downward angle of the wing.

Locusts use wing twisting to help compensate for yawing, rolling, and pitching flight disturbances. If the insect finds itself flying with a pitch that is too great, it must reduce lift, which it does by increasing the downward twist of left and right wings together. On the other hand, yawing and rolling involve changes in course, so both of these must be corrected by a turn back to the correct direction. This is accomplished by an increase in the downward twist of the front wing on the side toward which the insect will turn, reducing thrust and lift on that side and inducing a banking turn in that direction.

All these compensatory movements are induced by input from wind-sensitive hairs on the locust's head. Each hair sends its information into the central nervous system, where it is distributed to motor neurons that innervate muscles that control wing twisting, as well as to those that control abdominal and leg movements. Hairs that respond to wind from the left excite sensory neurons that send their information to appropriate wing, leg, and abdominal motor neurons on the left side of the body. Stimulation of these hairs causes more twisting of the left wing and causes the abdomen and the left side legs to swing toward the left. Similarly, hairs that signal wind from the right send their information to motor neurons that cause twisting of the right wing and bring about rightward extension of the legs and abdomen. Hairs that signal wind from above or below the insect send their information to motor neurons that bring about upward or downward movements of the abdomen, respectively. They also send information to the motor neurons of flight muscles of both right and left wings, inducing an appropriate twist of the wings to change lift.

One role of sensory feedback is to produce actions that help regulate motor performance by compensating for unexpected or unwanted perturbations. A flying locust can compensate for disturbances in pitch, roll, or yaw, in part by moving its abdomen or legs into the airstream, and in part by twisting its wings to a greater or lesser extent.

The Coordinating Effects of Sensory Feedback

Sensory feedback does more than help stabilize a behavior or compensate for disturbances by interacting with motor neurons. Although it is possible for a repetitive pattern of motor signals to be produced by the central nervous system without needing sensory feedback to provide timing cues, in nearly all animals sensory signals actually do feed back to the pattern-generating network and influence its output. The sensory information hence plays an integral role in fine-tuning the motor pattern. Repetitive behaviors in different animals vary widely in their ability to be influenced by sensory feedback in this way.

Pattern Generation and Sensory Feedback

One of the most striking examples of the influence of sensory input on a pattern generator comes from studies of swimming in the small sharks known as dogfish. In an early set of experiments, Sten Grillner and his colleagues deafferented dogfish by severing the dorsal roots along the entire spinal cord, hence preventing sensory signals from the body from entering the spinal cord. When properly stimulated, animals treated in such a way nevertheless swim steadily in a holding tank. This demonstrated that no sensory feedback is required for the production of the properly timed motor output of swimming and showed that the fundamental rhythm of the motor pattern for swimming in these sharks could be generated and coordinated by a series of pattern generators in the spinal cord.

In later experiments, Grillner and his coworkers studied the possible role of sensory feedback in swimming. After all, it was clear that all the muscles of the body wall contained a rich array of mechanoreceptors that could provide feedback to the central nervous system. If feedback from these receptors was not necessary for generation of rhythmic muscle activity, what was its purpose? To answer this question, the researchers did not merely deprive the nervous system of sensory feedback, since they already knew that a rhythmic pattern could be generated without feedback. Instead, they reasoned that if sensory feedback were capable of influencing the swimming pattern, they ought to be able to reveal such an influence by showing that sensory signals were able to change the timing of the motor pattern of swimming. To do this, they arranged for a dogfish shark to swim continuously in a tank, then physically bent the body back and forth at a frequency other than that of the ongoing swimming to see if this manipulation had any effect on the motor pattern for swimming.

The researchers prepared the fish for the experiment by ensuring that only the manipulation they themselves applied was responsible for any effects they noted. This involved four steps. First, they severed the brain stem at the midbrain. In sharks, the cerebral cortex serves to inhibit swimming; severing the connection to the cortex hence releases continuous swimming. Second, they injected the fish with curare, a neuromuscular paralytic agent that blocks the acetylcholine receptors on muscle, thereby preventing excitation of the muscle fibers, and hence preventing contraction. Injecting curare into the shark renders it motionless but does not alter the production of the motor pattern of swimming by the motor neurons. The animal is thus producing fictive swimming, generating the motor pattern of swimming without the actual movements. Third, they severed the spinal cord near the tail. This was done to make certain that sensory signals arising from the mechanical grasping device they used to move the tail would not have any influence on the swimming rhythm being recorded from the rest of the animal (Figure 17-3).

The final step was to move the tail back and forth at various rates. Curare blocks neuromuscular transmission but not the ability of sense organs in the body to respond to movement. Therefore, the imposed movement generates a rhythmic sensory feedback that is independent of the rhythm of the motor pattern for swimming that is already being produced. The effect of this sensory feedback is to entrain the fictive swimming rhythm, that is, to make it follow the rhythm of the moving tail rather than retaining its own rhythm. Without any intervention, one full cycle of the fictive swim rhythm usually takes about 3.5 to 5 seconds. If the tail is moved more slowly than this, so that a complete swim cycle takes as long as 9 seconds, the fictive swim pattern slows down and follows the imposed rhythm. Similarly, if the tail is moved more quickly, as fast as 1.5 seconds per full cycle, the fictive swim pattern speeds up to follow the imposed rhythm. At imposed rates faster than 1.5 or slower than 9 seconds per cycle, the motor output retained its own inherent rhythm and did not follow the imposed cycle.

Grillner's early experiments had shown that a coordinated swimming motor pattern could be generated by the central nervous system in the absence of any sensory input. His later experiments showed that sensory input can completely dominate the intrinsic rhythm-setting mechanisms of the pattern generators. Because strong sensory signals will be generated as a freely moving shark swims, the swimming rhythm of a free and intact shark must be a consequence of the interaction of the intrinsic pattern and the influence of the sensory feedback that is produced as it swims, the sensory signals perhaps helping to stabilize the swim rhythm.

Even locust flight, the classic example of a centrally generated motor pattern, has been shown to be susceptible to modulation by sensory feedback. However, this susceptibility is much less than it is for swimming in the shark. For example, if a researcher clamps one wing of a locust in a mechanical

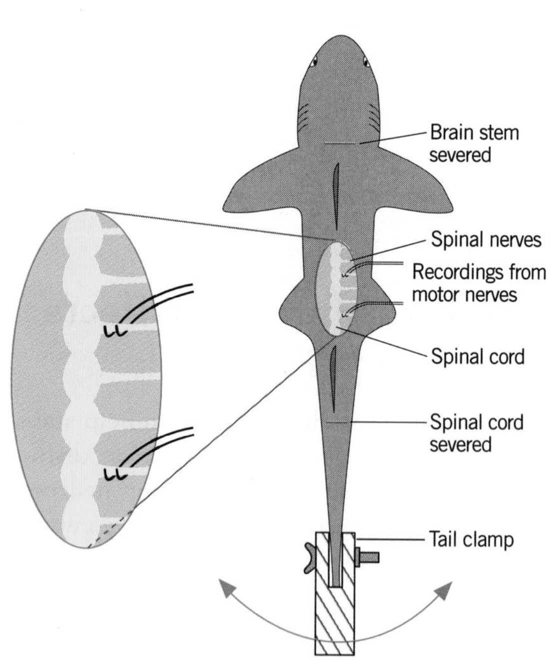

FIGURE 17-3. The experimental setup used by Grillner and coworkers. A small shark with lesions of the spinal cord and brain is held in a tank. The central region of the spinal cord is exposed for extracellular recordings from motor nerves to the body musculature as the tail is moved back and forth.

device so that it can be manipulated like the shark's tail was manipulated and then moves that wing while the locust is flying on a tether, the movements of the free wings will be entrained to the imposed movement of the clamped wing, but only if the frequency of the imposed movement is within 10% to 15% of the natural wing beat frequency. Other repetitive stimuli, such as a flashing light, are also capable of entraining flight.

Why should there be such a difference between a locust and a shark in the sensitivity to sensory feedback of its central pattern generator (CPG) for locomotion? In the locust, the CPG is clearly not very sensitive to sensory input. Only when the CPG is in a susceptible stage of its cycle can sensory input influence the wing beat cycle. At other stages, the CPG seems to ignore sensory signals. In the shark, the CPG can be influenced throughout its cycle. Perhaps the reason lies in differences in the nature of the control of the various behaviors. During flight, the wings beat at a relatively constant frequency, and flight stability will be achieved by making adjustments in body posture and wing twisting, as discussed previously. During swimming in the shark, on the other hand, changes in speed or direction can be carried out only by changing the rhythm that also generates propulsion. It may therefore be essential that the pattern generators for swimming be quite sensitive to sensory feedback to ensure proper coordination between muscles in different parts of the body during changes in swimming speed.

In some animals, sensory feedback can have a strong influence on rhythmic behavior. In these cases, the sensory feedback can drive the rhythm of the behavior over a wide range. In other animals, sensory feedback is able to modify an ongoing rhythm only slightly.

Sensory Feedback and Timing

Why a shark has CPGs for swimming as well as great sensitivity to sensory feedback is still a matter of speculation. For other repetitive behavior the usefulness of such a dual system is clearer. Walking in insects is such a case. Freely walking insects generally adopt a characteristic gait (the timing of leg movements relative to one another) in which the front and rear legs on one side of the body and the middle leg on the opposite side move toge-ther as a unit. In this so-called alternating tripod gait, all the legs move at precise times relative to one another, as they must if the gait is to be maintained. The stepping movements of each leg are thought to be produced by a pattern generator that controls that leg, and the sequence of movement of the six legs together is believed to be generated by interactions among the individual CPGs (see Chapter 16). Sensory feedback from the moving legs is not only capable of influencing the timing of the pattern generators for each leg, it is also essential for the proper timing of the movements of one leg relative to the others.

For example, if insects are made to carry or pull weights, most respond by walking more slowly. This result suggests that sensory feedback has an important role. Experiments with the type of insect known as a walking stick have further supported this view. If the insect is fixed in space over a pair of treadwheels that it can grasp with its feet and rotate under itself, it will walk quite normally. Increasing the friction of turning of one of the wheels, thereby effectively increasing the load on the legs on that side, causes slowing of the movements of all the legs, even the unloaded ones (Figure 17-4). Hence, sensory feedback from the loaded legs must be able to influence the neurons in the CPG network itself to produce a slower rhythm.

Other experiments demonstrate the critical role of sensory feedback in timing the leg movements in insects. If the two middle legs of an insect are removed, the insect would obviously have a difficult time walking if it maintained its normal gait, since it would be without support

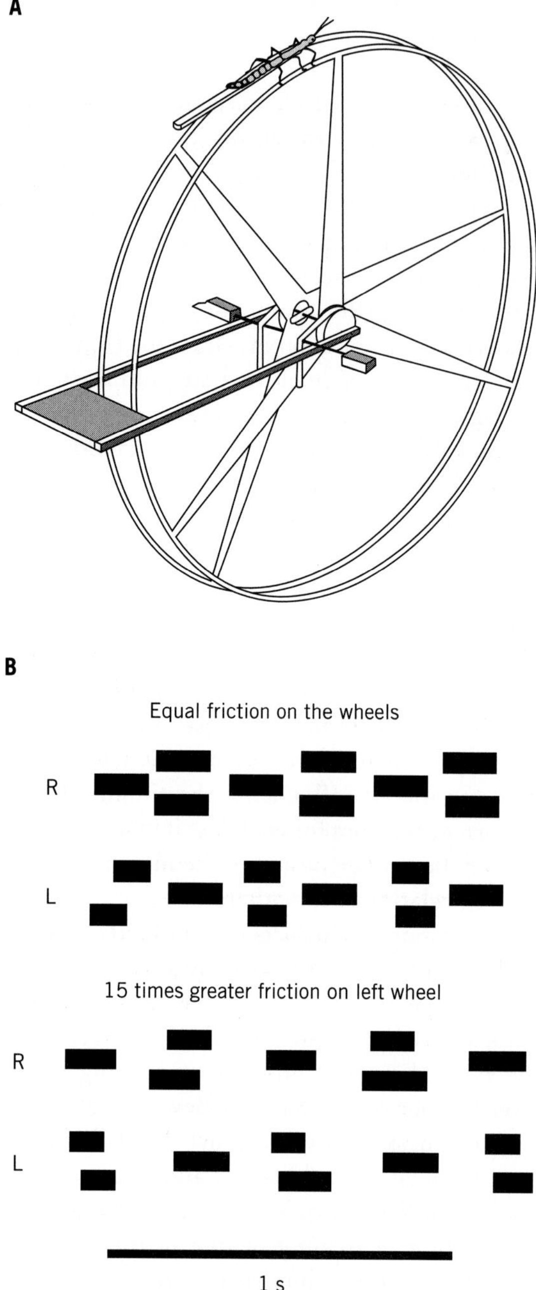

FIGURE 17-4. (A) A walking stick suspended over a double treadwheel and walking by rotating the wheels underneath itself. The friction of rotation of each of the two wheels can be changed independently of one another, thereby changing the force the insect is required to produce in order to move them. (B) The patterns of leg movements of the insect's legs when friction is minimal and equal on the two wheels (top), and when the friction is greater on the left wheel (bottom). The solid bars represent the times when each leg swings forward during a step. When friction is applied asymmetrically, the insect will slow down, but the right and left legs remain coordinated.

first on one and then on the other side of the body. Instead of literally stumbling along, a double amputee insect switches to a gait in which the front and rear legs on one side of the body step alternately rather than together. Hence, the loss of sen-sory input from the missing legs, or the changed sensory input from the remaining intact legs, has a profound effect on the gait the insect uses. Because the gait after loss of the legs is different from the gait used by the intact insect, it is inferred that, in the intact insect, sensory feedback is essential to setting the timing between legs that leads to the alternating tripod gait.

Even Wilson's observation that removal of the stretch receptors in the wing hinges of locusts results in a reduction of the frequency of beating of the wings (see Box 16-1) can be explained in terms of interaction of sensory feedback with neurons of the pattern-generating network. Wilson had thought that the stretch receptors merely provided a general excitatory input to the flight "motor," the network of neurons responsible for generating the motor pattern of flight. Work in the 1980s, however, showed that input from the stretch receptors actually resets the flight rhythm on a cycle-by-cycle basis. That is, excitation of any receptor at the base of a wing feeds back onto the flight pattern generator and causes it to begin the next cycle a little earlier than it normally would, thereby increasing the frequency at which the wings beat. The cumulative effect of all four stretch receptors is to bring the wing beat frequency up from about half of normal to normal.

The effect can be shown in a locust that has had all the sensory feedback from the wings eliminated by cutting the sensory nerves. When the proximal stumps of the stretch receptor nerves are stimulated in such a locust, the wing beat frequency increases cycle by cycle until a stable frequency is attained (Figure 17-5). Locusts in which the stretch receptors have been destroyed beat their wings more slowly because the flight pattern generator goes through its cycle more slowly, due to the lack of input from the missing stretch receptors.

The current view of the role of sensory feedback is that although feedback certainly

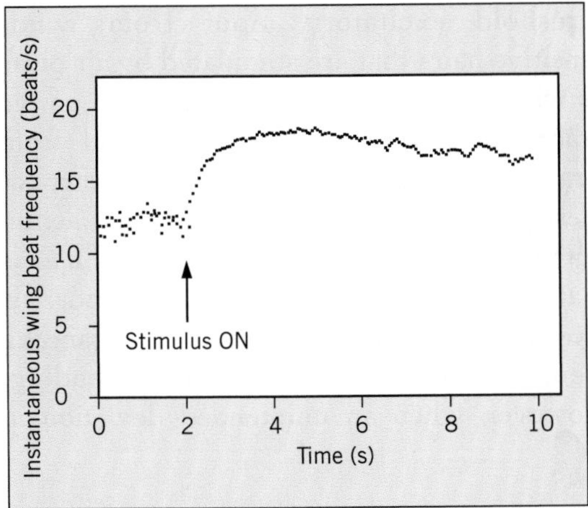

FIGURE 17-5. The effect of stimulating the stumps of stretch receptor nerves on the frequency of the flight pattern in a locust with sensory feedback from the wings removed. The flight frequency increases cycle by cycle as stimulation is applied until a new, higher frequency is maintained.

is not *necessary* for the generation of a pattern of rhythmic motor output in any animal, it is nevertheless possible for it to influence the pattern, and in many cases it is essential for expression of the full, properly timed pattern to appear. There are now several well-documented examples in which sensory input influences the frequency, vigor, and strength of the performance as well as the timing of its various components relative to one another. Furthermore, sensory input is absolutely necessary for the animal to be able to make corrections in its behavior on the basis of events in the external world. In the 1950s and 1960s, the main issue was whether sensory or central mechanisms were at the foundation of repetitive motor patterns. Researchers now recognize that the question itself is wrong. Both CPGs and sensory feedback play critical roles in coordinating repetitive motor activity.

In many animals, sensory feedback is used not only as a means of correcting for disturbances, or for coordinating changes in the rhythm of an activity, but also to provide important timing signals for the proper execution of the behavior itself.

Other Sensory-Motor Interactions

You learned in Chapter 10 that the sensory system is far from a passive recipient of information, and that efferent signals play an important role in shaping the type of sensory information that will be sent to the sensory centers for processing. Similarly, the effects of sensory signals on the behavior of an animal are not always passive either. Instead, the effects of sensory input can themselves be influenced by the ongoing activities of the animal.

Reflex Gating

One way in which an animal's behavior can influence the effect of sensory input is known as reflex gating. **Reflex gating** is the process by which a reflex is produced only when both the stimulus that evokes it and some other behavior are present at the same time. Put another way, the behavior itself acts like a gate, allowing a specific sensory input to exert an effect. In behavioral terms, this means that a specific sensory input has one effect if the animal is engaged in a particular behavior but no effect (or a different effect) if the animal is not engaged in that behavior.

Here again the locust flight system provides a good example. You have already learned about the abdominal movements made by a flying locust in response to changes in the angle of incident wind on its head. The response is mediated by particular wind-sensitive hairs on the front of the locust's head, which because of their orientations are able to detect wind hitting the head from the side. However, if these same wind-sensitive hairs are stimulated in a standing locust by wind of the same force and from the same direction as that which evokes abdominal movements in a flying locust, the animal shows no response whatsoever. In fact, there is no wind at any speed or direction that will elicit any abdominal movements in a locust that is not flying. Hence, flying "gates" the reflex bending of the abdomen observed when the direction of wind on the head changes during flight.

Study with intracellular microelectrodes of the motor neurons controlling the abdominal muscles has revealed that all of these neurons receive a low (subthreshold) level of excitatory input from the flight pattern generator during flight but no such excitatory input when the insect is not flying. Similarly, abdominal motor neurons also receive subthreshold excitatory input from wind-sensitive hairs that are stimulated by air puffs or wind striking one side of the head or the other but not by air puffs or wind from straight ahead. By themselves, neither of these inputs is strong enough to elicit action potentials in the motor neurons. This means that neither flying straight ahead nor air blown on the side of the head of a standing locust evokes any abdominal bending. However, when an unintended deviation in

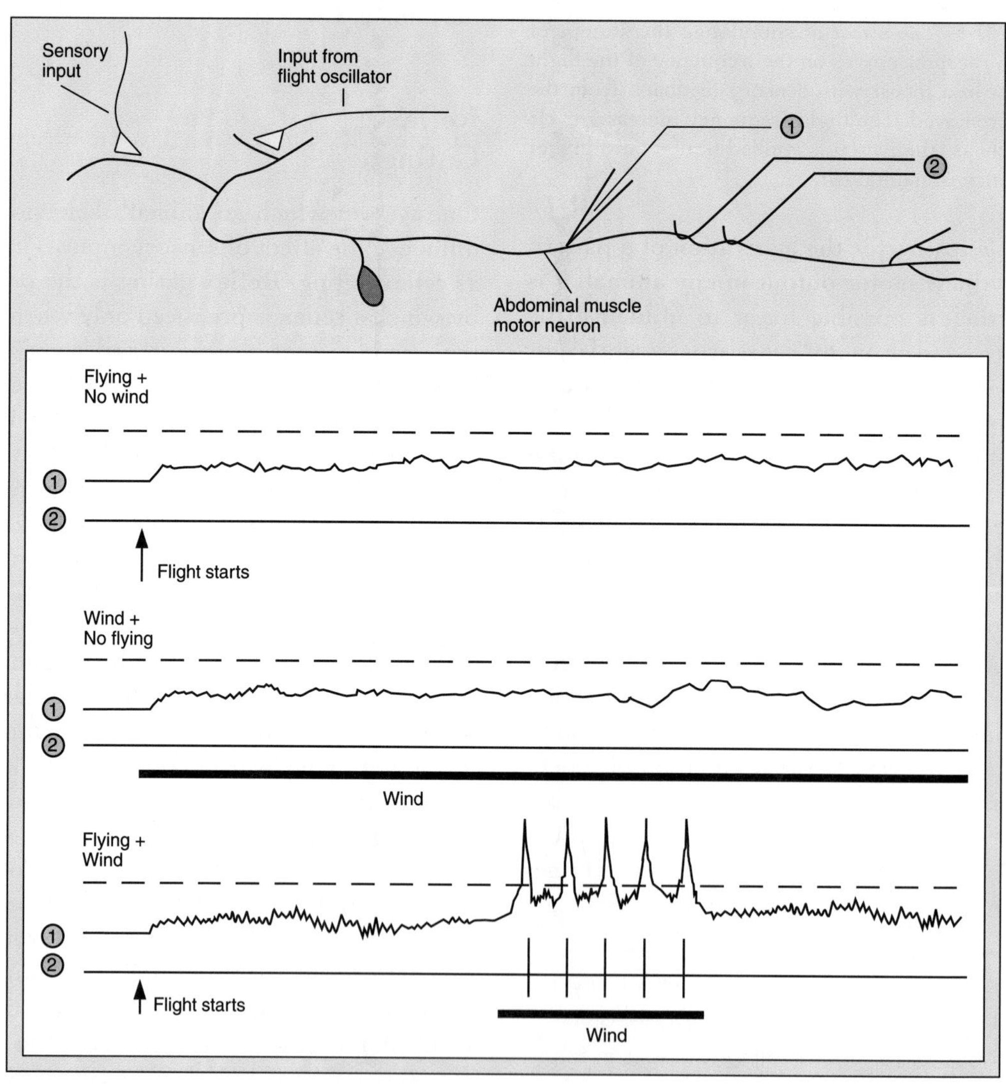

FIGURE 17-6. Reflex gating. An abdominal motor neuron of a locust receives input from a sensory neuron and from a neuron from the flight CPG. The hypothetical recordings show what you might expect to see in an intracellular and extracellular recording from the motor neuron under different conditions. Wind refers to wind on the side of the head, not straight on. Only when flight and wind occur together is the motor neuron activated above threshold (dotted line).

the flight path activates the hairs on the side of the head of a flying locust, the input from these hairs and from the pattern generator for flight will combine to evoke action potentials in the motor neurons for the abdominal muscles, hence causing an appropriate abdominal bending to bring about a course correction (Figure 17-6).

Flight itself, therefore, gates the effects of the sensory input from the head in an all-or-none fashion. If the locust is flying, the reflexive bending of the abdomen resulting from certain types of head stimulation can take place; if it is standing, it cannot. As in the case of an action potential, there is no intermediate condition; you either have abdominal bending or you do not.

Reflex gating refers to the influence of an animal's ongoing behavior on the effect of sensory input. The behavior acts like a gate, allowing a particular sensory stimulus to trigger a reflex when the animal is engaged in the behavior. The sensory stimulus has no effect when the animal is not engaged in the behavior.

Reflex Reversal

A second type of influence of ongoing behavior on the effect of sensory information is called reflex reversal. **Reflex reversal** refers to a reversal of the effect of a sensory stimulus when the animal is engaged in a particular behavior. That is, a stimulus will have one effect when the animal is at rest and a different (often opposite) effect when the animal is engaged in some activity. Reflex reversal is a feature usually found in association with **resistance reflexes**, in which muscles are activated that oppose a movement being imposed on the body. For example, suppose you held a locust and flexed one of its hind legs. You would feel the leg pushing against your fingers as you did this. If you then tried extending the leg, you would again feel some resistance. The muscle actions that oppose your movement of the leg are the result of resistance reflexes. They are present in vertebrates as well, although they are usually suppressed or masked by other effects in these animals. You would probably not feel anything if you tried to evoke resistance reflexes in your dog or in a friend.

Resistance reflexes help an animal maintain a particular posture, since they help hold a joint at a particular angle. However, you can well imagine that they would interfere with movements the animal itself makes. In fact, resistance reflexes typically disappear when an animal starts to move, often to be replaced by reflexes that actually assist the movement instead of resisting it. During walking, extension of a limb activates extensor muscles that extend it farther, rather than activating flexor muscles that oppose extension.

Although reflex reversal is widespread in the animal kingdom, it has been demonstrated especially clearly among the arthropods. One of the first demonstrations was in crabs, where it was shown that resistance reflex activation of a particular leg muscle by imposed movement of the leg, which occurred when the crab stood still, was completely suppressed when the crab walked. In some insects it has been shown that movement itself is not actually necessary. If the animal merely readies itself to move, by putting the central nervous system into what is called an "active state," the effects of various stimuli become completely different than they are when the animal is inactive.

Reflex reversal is a change in the effect of a sensory stimulus when the animal is engaged in a particular behavior. For example, an imposed movement of a limb during walking may cause an acceleration of the movement rather than a resistance to the movement. Resistance to imposed movement is typical when an animal is at rest.

Reflex Modulation

Based on the examples discussed so far, you may think that the influence of an animal's ongoing behavior on the effect of sensory input is always straightforward. Reflex gating is an all-or-none process, and even reflex reversal is usually easy to recognize. But the effect of an animal's behavioral state can be more complicated and subtle than that illustrated by these two examples. For instance, rather than turning a reflex on or off, or reversing it, a behavior may continuously modulate a response. This is called **reflex modulation**, defined as a continuous change in the strength and sometimes the sign of a response as an animal moves through one cycle of an ongoing repetitive behavior. Reflex modulation has been demonstrated in many species.

Consider the response of a cat to a tap on the top of the paw at different stages in the cycle of stepping of a leg (Figure 17-7). The movements of a cat's leg, like those of the limbs of other walking animals, are divided into the swing phase, when the limb is off the ground moving forward, and the stance phase, when the limb is on the ground bearing the weight of the animal. When the leg is in its stance phase, a tap on the dorsum of the foot results in increased activity in extensor muscles in the leg as well as a quickening of the step. When the leg is in its swing phase, a similar tap results in increased flexor activity in the stimulated leg and increased activity in the extensor muscles of the leg on the other side of the body (which is in its stance phase).

This is not simply an example of an all-or-none reversal of a reflex. Instead, the effects of the stimulus are modulated according to the stage of the stepping cycle the leg is in when the stimulus is applied. If the stimulus is applied when the leg is in the middle of its stance phase, extensor activity is strongly increased and there is a strong effect on the duration of the step cycle. On the other hand, if the leg is near the end of its stance cycle, nearly ready to swing forward, there is only a small effect on extensor muscles and little effect on step cycle duration. As the leg moves into its stance phase, the reflex reverses completely and the stimulus causes activity in extensor muscles. Here again, however, the strength of the reflex is modulated according to the stage of the stance phase the foot is in. It is relatively weak at the very beginning and very end of the stance phase, and much stronger near the middle.

Similar effects have been described in many animals. In walking sticks, for example, a leg will move differently if it strikes an obstruction early in the swing phase than if it strikes it late. Even in humans, mild electrical stimulation of the skin over the ankle at different times during a step while the subject is

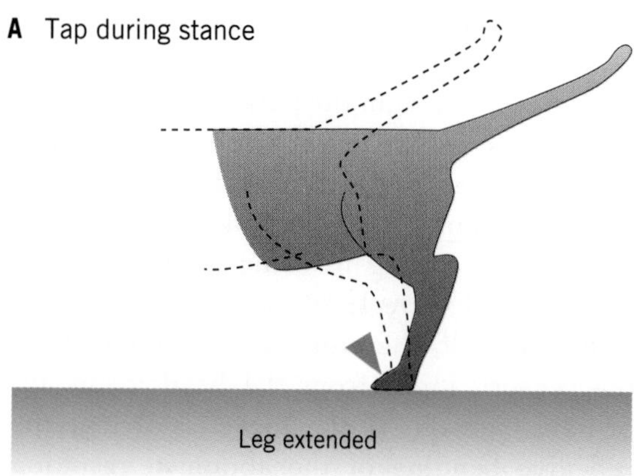

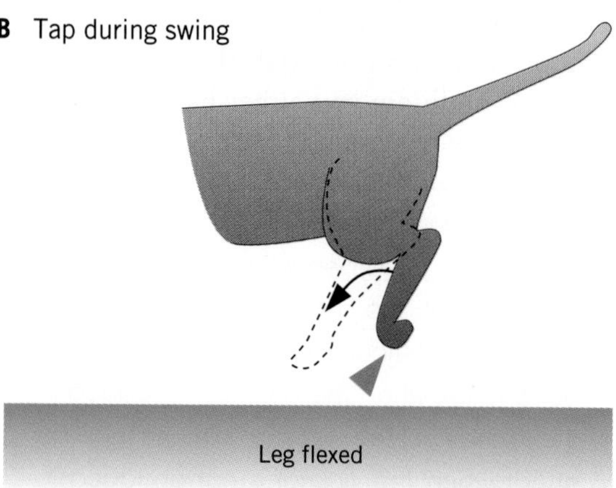

FIGURE 17-7. The reactions of a cat whose brain stem has been severed to a tap on the dorsum of a rear foot while it is walking. (A) A tap applied when the leg is in stance phase on the ground results in extension of the foot. (B) A tap applied when the leg is in swing phase and moving forward results in flexion of the foot.

running causes different reflex effects in various muscles depending on the phase of the step cycle at which the stimulation is delivered, and independently of the intensity of the stimulus (as long as it is not painful).

What is the use of reflex modulation for the animal? One way to look at the phenomenon is as a kind of **gain control** of a response. Gain is a term from electronics that refers to the level of amplification of a signal in a circuit, such as in an amplifier. In the context of a neural circuit, it refers to the magnitude of the response relative to the strength of the stimulus. Hence, if a particular stimulus produces a strong response at one time but only a weak response at another time, it can be said that the gain of the reflex has been reduced.

Gain adjustments can serve a useful function. In the case of the cat, the functional effect of the reflex response to touch on the top of the foot is to move the paw away from a stimulus. However, when the cat is walking it must also maintain balance. When the leg is swinging forward, the best response to the stimulus will be to increase the activity of flexor muscles and lift the foot over the obstacle. On the other hand, when the leg is on the ground bearing the weight of the body, the best response will be increased extension, so the stance phase of the step can be finished quickly and the foot lifted. Reflex modulation, then, is a means for the animal to adapt the response to a particular stimulus to the behavior in which it is engaged at a particular time.

How can the processes of reflex reversal or reflex modulation be carried out by the nervous system? One common mechanism is presynaptic inhibition of the axon terminals of the sensory afferents. For example, experiments with cats in which the brain stem has been severed (to eliminate voluntary actions) have shown that the axonal ends of sensory afferents from cutaneous receptors in the foot receive strong cycles of depolarization during fictive locomotion, in time with the locomotor pattern. As you recall from the discussion of presynaptic inhibition in Chapter 6, depolarization of an axonal terminal will cause a reduction of the amount of neurotransmitter released at the synapse, and hence will effectively produce inhibition. The stronger the depolarization, the more effective the inhibition. Inhibitory input to the sensory afferents will dampen the effects of a sensory signal at certain times during the cycle of leg movement, and will allow it to pass through during other times, hence modulating the effects of the input and modulating the response that is elicited.

Reflex modulation differs from reflex gating and reflex reversal in that there is a continuous modulation of the strength and sometimes the sign of a reflex as the animal goes through a cycle of behavior. The effect of this modulation of reflex gain is to adapt the reflex to the exact behavioral needs of the animal at the moment when the stimulus that evokes the response is encountered.

Additional Reading

Research Articles and Reviews

Grillner, S., and P. Wallén. 1977. Is there a peripheral control of the central pattern generators for swimming in dogfish? *Brain Research* 127:291–95.

Pearson, K. G. 1993. Common principles of motor control in vertebrates and invertebrates. *Annual Review of Neuroscience* 16:265–97.

Prochazka, A. 1989. Sensorimotor gain control: A basic strategy of motor systems? *Progress in Neurobiology* 33:281–307.

Tax, A. A. M., B. M. H. van Wezel, and V. Dietz. 1995. Bipedal reflex coordination to tactile stimulation of the sural nerve during human running. *Journal of Neurophysiology* 73:1947–64.

CHAPTER 18

The Brain and Motor Output

The study of simple reflexes, repetitive behavior, and sensory effects on behavior gives important insights into how simple motor actions are coordinated. It does not, however, provide much help in understanding how the nervous system organizes complex behavior like playing a piano, nor how it selects and initiates such behavior. These functions, which are thought to be carried out by the motor centers in the brain, are the subject of this chapter.

Motor systems comprise more than a collection of reflexes and central pattern generators (CPGs). They also include many neural circuits and centers whose function is to select the behavior to be expressed, control the vigor with which it will be expressed, and decide how long it will last. Specific parts of the brain are necessary for these functions. Because motor systems are considered to be arranged hierarchically, with monosynaptic reflexes at the "bottom" of the hierarchy, the brain is often said to represent a "higher" level of control. It is this higher level that you will learn about here.

The Role of the Brain

The selection of muscles to be used in a behavior and the timing of their use represent the lowest level of the brain's control over motor output, what we might call the **execution level**, of a motor performance. We can also identify the **executive level** of motor control, the level at which the behavior itself is selected and where its vigor, speed, frequency (if repetitive), and duration are determined. Logically, none of these executive functions depends in any way on what particular muscles are to be used or when they will be active, just as whether a computer's output goes to a video screen or a printer has nothing to do with the main functions of the program that is running.

Animals with only simple nervous systems, such as the cnidarians and flatworms, have little executive control over their behavior. Nearly all of their actions are generated and controlled at the local level by reflex responses to specific stimuli. Although these responses can be quite complex, as you saw in the case of the sea anemone described in Chapter 1, no one part of the nervous system seems to control them. A free-living flatworm has a brain that acts as a decision-making center. It may help the animal choose to engage in some behavior. However, many of the actions that in higher animals require active participation of the brain can proceed normally in brainless flatworms. For example, a flatworm can continue to feed after its brain has been removed, something more complex animals cannot do.

Complex animals like mollusks, arthropods, and vertebrates show a much more fully developed executive function than do simpler animals. In higher animals, loss of the brain renders the animal helpless to make any decision. As implied by the phrase "running around like a chicken without its head," even some vertebrates can continue to function for a short time using only reflexes. However, the chicken moves around only for a short while, and its "running" is not especially well coordinated, being the result of disinhibition of the gross body musculature rather than of purposeful action. Chickens well illustrate the fact that in all higher animals, removal of the brain leaves the animal unable to initiate any action or to control a reflex action started by any stimulus.

How is behavior organized in the brain? This is a complex question that is far from having a full answer, but some suggestions have emerged from recent experiments. For example, evidence suggests that in both vertebrate and invertebrate animals the brain is organized around discrete behaviors rather than around specific movements. A fascinating demonstration of such organization is the elicitation of a particular behavior by electrical stimulation of specific regions of the brain. In many animals, it is possible to insert fine wires into the brain. Passing mild current through wires implanted into the brain of a chicken, for example, will cause the chicken to engage in a particular behavior. Depending on where the wires are placed, the chicken can be made to cluck, sit or stand up, or stand still or walk (Figure 18-1). Even a cricket can be made to walk or to sing its courtship song as a result of electrical current sent to specific regions of the brain. An interesting feature of these experiments is that the vigor of the behavior usually depends on the strength of the applied current, which must exceed some minimum threshold before any response is seen. Experiments of this type strongly suggest that the brain plays an important role in regulating behavior.

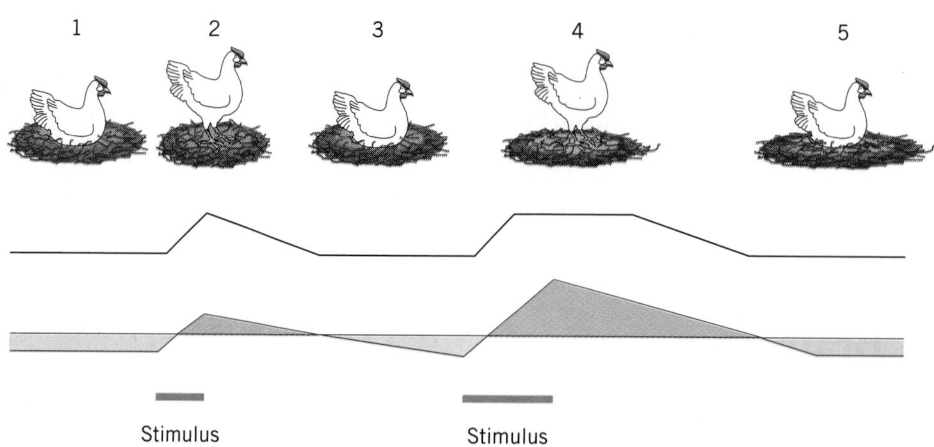

FIGURE 18-1. The effect of electrical stimulation of the brain of a chicken. The line under the sketches of the chicken represents the chicken's posture: up is standing, down is sitting. The shaded areas formed by the second line represent a hypothetical tendency for the chicken to sit (below the line) or to stand (above the line) generated by the electrical stimulus. In the experiment, the chicken is sitting quietly before the stimulus is applied and would remain sitting without intervention. A brief, mild stimulus causes the chicken to stand for a short while, then sit again. The same stimulus applied for a longer time causes the chicken to stand for a longer time before it sits down again.

The executive level of behavioral control, that which regulates the selection, vigor, and duration of a behavior, resides in the brain. Simple animals have relatively little executive function, most behavior being organized at the local level. In higher animals, executive control of different types of behavior seems to be localized in different parts of the brain.

Invertebrates have been the subjects of much research on motor control at the executive level for several reasons: they use relatively few neurons for this function; they often have larger neurons; and it is often possible to study the activity of individual, identifiable neurons during the execution of a particular behavior.

Motor Control in Invertebrates

For many neurobiologists interested in how the nervous system controls behavior, invertebrate animals have been particularly attractive subjects for research. There are several reasons for this. First, most invertebrates have fewer neurons in their nervous systems than do vertebrates. If there are fewer neurons overall, then fewer are available for controlling behavior, suggesting that it might be easier to understand the neural basis of the behavior than if more neurons were involved. Second, the larger size of neurons in many invertebrates makes intracellular recording easier (see Chapter 5). Third, in many invertebrates it is possible to record the activities of single neurons while an animal is executing fictive or real behavior. This makes it possible to study the contribution of individual neurons to the control of a behavior.

The usefulness of invertebrates as subjects for studies of motor control has been furthered by the recognition that many neurons in invertebrates, and even a few in the lower vertebrates, are identifiable (Box 18-1). This means that they can be recognized as individuals on the basis of their unique combination of physiological and morphological characteristics, and researchers can therefore thoroughly study the possible role of these identifiable neurons in the control of a particular behavior. These features, coupled with a simpler organization of motor control, have allowed researchers to gain important insights into the means by which behavior is controlled at the executive level in invertebrates.

Single Neurons and the Command Concept

With the advent of intracellular recording and the recognition that individual neurons are identifiable came a concept that had a powerful influence on researchers' views of how behavior was controlled and regulated. In 1964, C. A. G. Wiersma and K. Ikeda reported interesting results from studies of the swimmeret system of crayfish. As described in Chapter 16, swimmerets are paddlelike appendages found on the ventral side of the abdomen of crustacea like crayfish and lobsters (see Figure 16-9). They often beat in a coordinated sequence, the caudal swimmerets beating first, followed one after the other by the more anterior pairs.

Wiersma and Ikeda were interested in finding neurons that controlled this simple behavior. To do this, they gently teased apart the connectives of the ventral nerve cord to isolate just a few axons, then stimulated these axons electrically while at the same time looking for movements of the swimmerets. In the course of their studies they found several neurons in the cord that, when stimulated, caused the swimmerets to beat in an apparently normal fashion. Perhaps their most compelling finding was that the frequency of beating of the swimmerets could be determined by the frequency at which the axons were stimulated, presumably by regulating the frequency at which the output from the CPGs to antagonistic muscle groups alternated (see Chapter 16).

Because their results suggested that the individual neurons they stimulated could control

BOX 18-1

Cells as Individuals: Identifying Individual Neurons

The introduction of new techniques often brings about a revolution in researchers' views of neural function. The advent of routine intracellular recording in the 1960s and the introduction of intracellular staining techniques shortly thereafter certainly brought about one such revolution. At the time it was believed that all neurons in any one region of the central nervous system were structurally and functionally quite similar. For all types of animals, it was believed that nervous systems consisted of tangled masses of similar neurons whose function could be understood only in terms of the interactions of more or less interchangeable parts.

Intracellular recordings combined with the staining of individual neurons revealed quite a different picture. Recording intracellularly from individual neurons allowed researchers to characterize the physiology of these neurons in considerable detail, often relating their response characteristics to some specific neural function. When the recording electrode is filled with a dye that can be injected into the neuron and then later made visible (see Box 3-2), it becomes possible to study the relationship between how individual neurons look and what they do. Studies of this kind in arthropods and other invertebrates showed that many neurons could be differentiated on the basis of their unique combination of structure and function. This meant that a particular neuron could be located repeatedly in different individual animals, and hence studied more thoroughly than was possible in a single experiment. A neuron with such a unique combination of characters is called an **identifiable neuron**, which can be recognized by its similar morphological and physiological characteristics in all individuals of a species.

Neurons in many different animals are identifiable. This makes it possible to study the role of individual neurons in particular behaviors. For example, a set of neurons in cockroaches is intimately involved in the escape of these insects from predators, as described in Chapter 19. Although these neurons contribute to a single behavior, they differ morphologically (Figure 18-A). Recognition of their unique structure helped researchers discover subtle differences in their functional roles in escape as well. Identifiable neurons in other invertebrates, such as the mollusk *Aplysia,* have made possible a number of important advances in our understanding of the relationship between neurons and behavior. Identifiable neurons are present even in vertebrate animals; the **Mauthner cells**, a pair of extremely large neurons in the hindbrains of fish (Figure 18-B), are clearly distinguishable from all other neurons of the fish brain. They play a unique role in escape swimming in fish, similar to the role played by the giant interneurons in cockroaches.

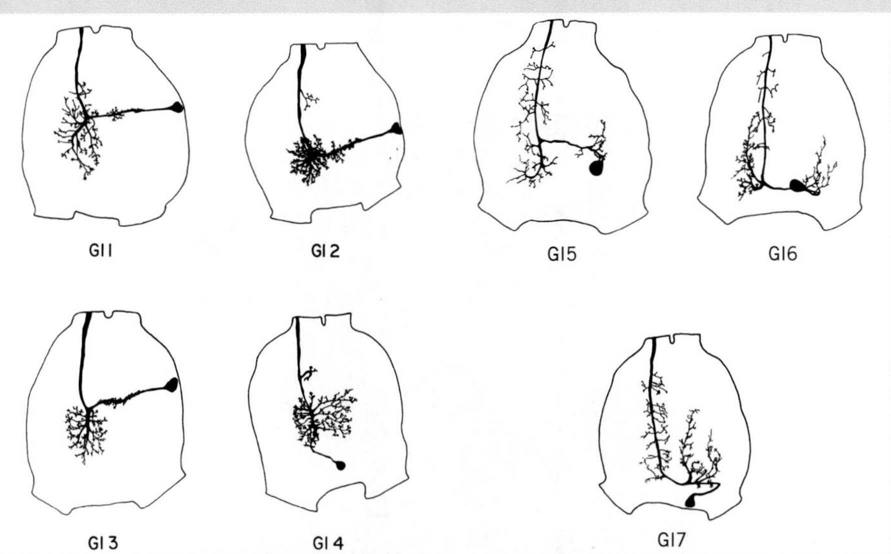

FIGURE 18-A. Neurons giving rise to the seven large-diameter axons found in the last abdominal ganglion of the American cockroach, shown in an outline of the ganglion. These neurons, called giant interneurons (GIs) because of the sizes of their axons, are distinct in shape and in their positions in the ganglion. They receive input from wind-sensitive hairs on the cerci and are involved in the insect's escape reaction, as described in Chapter 19.

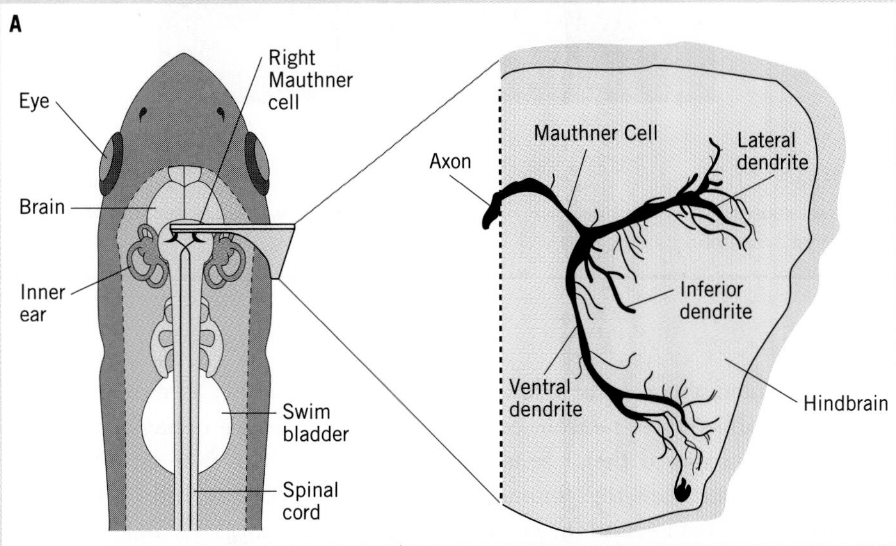

FIGURE 18-B. (A) Mauthner neurons from fish. The dorsal view of a zebra fish at the left shows the location of the paired Mauthner neurons in the brain stem. The cross section through the brain stem at the right shows a drawing of one of the Mauthner cells filled with horseradish peroxidase. No other pair of neurons in the brain of fish looks like these cells.

(continued)

Cell as Individuals: Identifying Individual Neurons (continued)

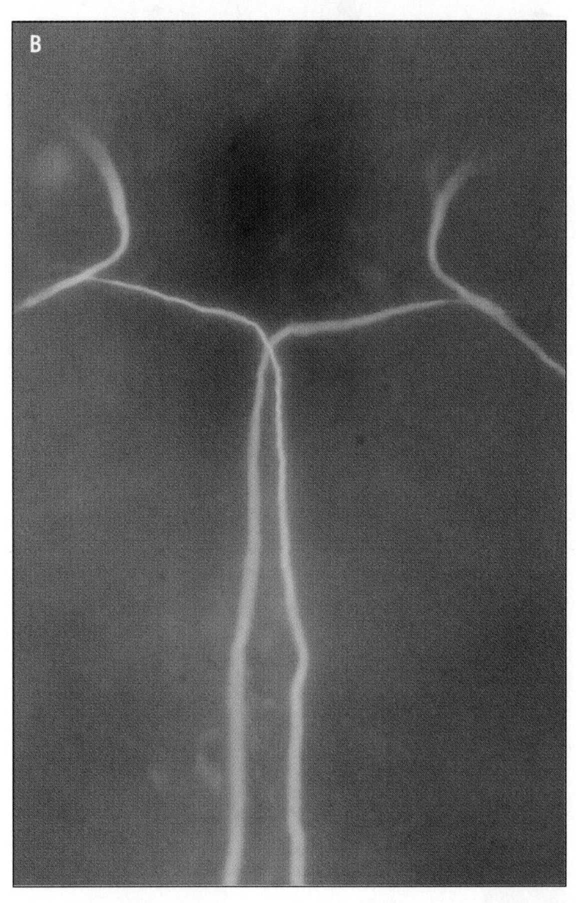

FIGURE 18-B. (B) Mauthner neurons from fish. A whole mount of brain and spinal cord of a goldfish, showing the Mauthner cells filled with the fluorescent dye Lucifer yellow. (Courtesy of Steven I. Zottoli.)

the onset, duration, and vigor of a behavior, Wiersma and Ikeda dubbed these neurons **command neurons**. They proposed that a sensory stimulus would merely activate the command neuron for a particular behavior, and the behavior would be executed automatically. It was an elegant scheme, not the least because it implied that for researchers to have a complete explanation of a behavior, they needed only to find the command neurons that controlled it, characterize the stimuli that excited these neurons, and find the output connections they made.

Unfortunately, as more neurons that were able to influence behavior were investigated, it became clear that most did not by themselves control the full expression of a complete behavior. Because of the confusion resulting from the loose application of the term command neuron to cells that did not actually have a command function, the term has been redefined more rigorously to refer only to neurons that are both *necessary* and *sufficient* for the complete and full expression of a behavioral act.

Using this more rigorous definition, few neurons can truly be designated as command neurons. Even the neurons described by Wiersma and Ikeda do not qualify as command neurons by the necessary and sufficient criterion because in these experiments the researchers were not able to activate the full range of the complete behavior by stimulating only single neurons. For example, in some cases only some of the swimmerets would beat, and in others they could not be made to beat over their entire natural frequency range. Therefore, none of the neurons stimulated in these experiments was sufficient for expression of the complete behavior.

Of all the neurons that have at one time or another been proposed to be command neurons, probably those that come closest to meeting the necessary and sufficient criterion are the **giant interneurons (GIs)** of the common crayfish. GIs have large-diameter axons, specialized for rapid conduction of action potentials. Crayfish have two pairs of such interneurons, each of which makes specific synaptic connections with other interneurons and with motor neurons that control the abdominal musculature (Figure 18-2). If you touch a crayfish, you will elicit a rapid escape reaction: backward swimming if you touch it around the head, upward swimming if you touch it on its tail. Each escape swim is produced by one of the pairs of GIs. Backward swimming is produced by the medial giant and upward swimming by the lateral giant. The GIs may be considered command neurons because each one organizes the complete escape sequence for which it is responsible, inhibiting some muscles and activating others to elicit the appropriate behavior, and making sure there are no competing instructions from other neurons.

Researchers now recognize, however, that no behavior is exclusively initiated by comand neurons. Most behavior involves more than one type of neuron for its control and regulation, and can be elicited through more than one pathway. In crayfish, escape swimming can be triggered in the absence of the giant interneurons, although it then takes longer to elicit. Hence, the giant interneurons are not *necessary* for the expression of escape swimming. The

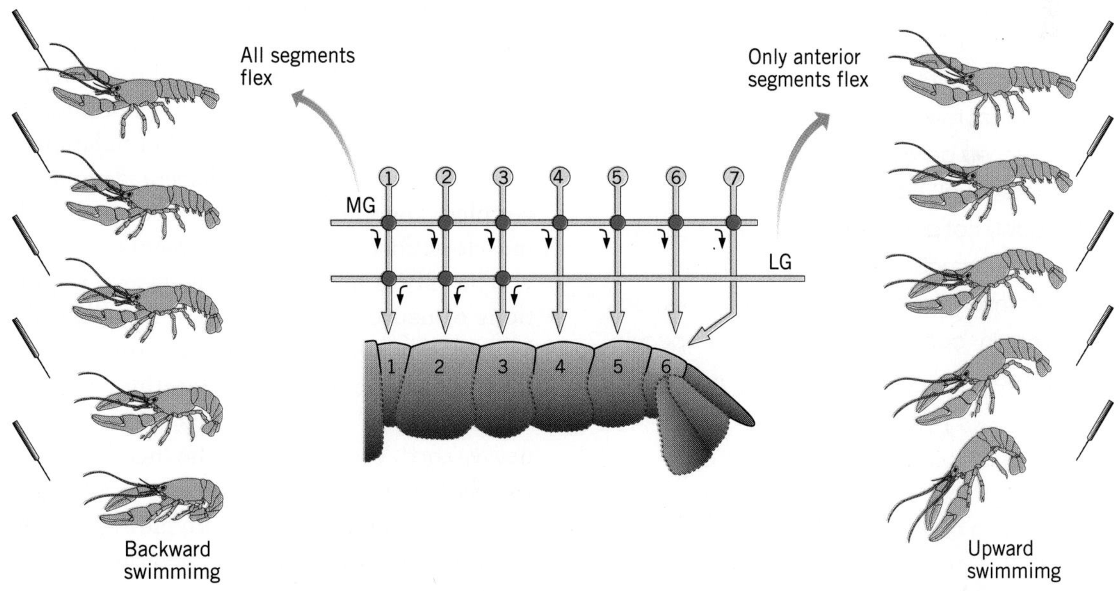

FIGURE 18-2. Touch on different parts of the body of a crayfish evokes escape behavior appropriate to the place that is touched. Touch on the head activates the medial giant (MG) interneurons, which excite motor neurons for all the abdominal flexor muscles, causing a backward tail flip. This results in backward swimming. Touch on the tail activates the lateral giant (LG) interneurons, which excite only the anterior abdominal flexor muscles, causing an upward tail flip and upward swimming. Numbered circles represent motor neuron cell bodies.

same multiplicity of pathways for the initiation of a behavior is found in other animals. In leeches, for example, neurons that help trigger swimming, and that had been considered candidates for command neurons, originate in the head. Nevertheless, because headless leeches can be induced to swim, there must be other ways to initiate swimming as well.

In other animals, behavior is initiated by activation of members of a population of neurons rather than by a small number of "trigger" neurons. Some researchers refer to these neurons collectively as a command system. Other terms have been used as well. In locusts, for example, behavioral decisions are made not so much by single "command" neurons that absolutely determine what the animal is going to do but by a kind of committee vote. Consider walking. A population of neurons, termed "recommendation" neurons, can influence the animal's decision to walk. If many neurons that promote walking are active, the animal will walk. If only a few recommend walking, the locust will stay where it is. According to this concept, behavior is the outcome of a kind of contest between different neurons that tend to generate different types of behavior.

At one time, it was thought that many simple types of behavior were controlled by just one or a few command neurons. Today, researchers believe that absolute command by individual neurons over a behavior is probably not present, and that among invertebrate animals control of behavior in some cases involves small populations of neurons acting in concert to activate a specific behavior.

Turning Motor Patterns On and Off

Although the excitement generated by the idea that one or two "command" neurons might be able to turn on an entire behavioral act has been tempered by the lack of evidence to support the notion, the idea that relatively small numbers of neurons might be responsible for controlling a behavior continues to drive much research in the field of motor control.

Study of simple behavior in invertebrates has yielded new insight into how an animal controls some behavioral acts. Take swimming in the medicinal leech as an example. A leech swims by flattening its body, then undulating through the water like a snake, except that a leech undulates up and down rather than from side to side. The undulations are caused by alternating contractions of dorsal and ventral flexor muscles. The muscles in one particular body segment contract slightly before the muscles in the next posterior segment, causing the wave to move backward along the body.

Recordings from neurons in the ganglia along the ventral nerve cord during fictive swimming have shown that each ganglion (one per body segment) contains a network of neurons, a CPG (see Chapter 16), that is responsible for controlling the motor neurons for muscles in that segment. Just as the neural network for flight controls the alternating activity of elevator and depressor muscles that drive the wings in locusts (see Chapter 16), the CPGs in leech ganglia control the alternating activity of dorsal and ventral flexor muscles that drive the dorsal and ventral undulations in those animals. CPGs in adjacent ganglia are coupled together so as to coordinate the muscle actions in adjacent segments.

The CPGs are excited by several populations of neurons. A population of cells called swim gating neurons make excitatory connections with the pattern-generating neurons. Stimulation of these neurons in an isolated nerve cord (one without the head and tail ganglia) elicits fictive swimming for as long as they are stimulated. The swim gating neurons in turn are influenced by several kinds of excitatory and inhibitory neurons (Figure 18-3). Trigger neurons receive excitatory input from mechanoreceptors in the body wall and are themselves strongly excitatory on the swim gating neurons. Hence, they are responsible for initiating swimming in response to a strong mechanical stimulus. Other neurons, called swim excitor neurons, activate the neurons of the CPGs, as well as exciting the swim gating neurons, but they receive only weak input from sensory neurons. At least some of the inhibitory neurons are part of a swim inactivating system; they can stop ongoing swimming or suppress the initiation of swimming.

Figure 18-3 provides an overview of the leech swimming system and places the details within the context of the animal's swim behavior. Starting with the behavior itself, the lower part of the diagram shows the actions of muscles that produce the swimming undulations of the animal. Muscles are alternately excited and inhibited by the motor neurons that innervate them, due to the pattern of excitation and inhibition that these motor neurons receive from the CPG in the segment and elsewhere. The CPGs receive excitatory input from several sources. If the swimming is initiated by a mechanical stimulus, then the behavior is activated via the trigger and gating neurons. However, the leech can also initiate swimming spontaneously. The neural basis of this mechanism is less well known, but it appears that when the leech decides to swim, swim excitor neurons are activated, stimulating both swim gating neurons and neurons of the CPG for the swim rhythm, while at the same time inhibiting the neurons that stop swimming. Stimuli that stop swimming, or perhaps a decision by the leech to stop, turn on the swim inhibiting neurons. These neurons suppress activity in both the gating neurons and the CPG neurons themselves.

The emerging picture of the control of swimming in leeches is of a system that is both hierarchical (serial) and parallel in organization, analogous to the serial and parallel organization of many sensory systems (e.g., vision; see Chapter 11). The pathway from trigger neurons to gating neurons, and from gating neurons to CPG neurons is hierarchical, whereas the pathway from swim excitor neurons and gating neurons to CPG neurons is parallel. There are also parallel systems for initiating and terminating the behavior.

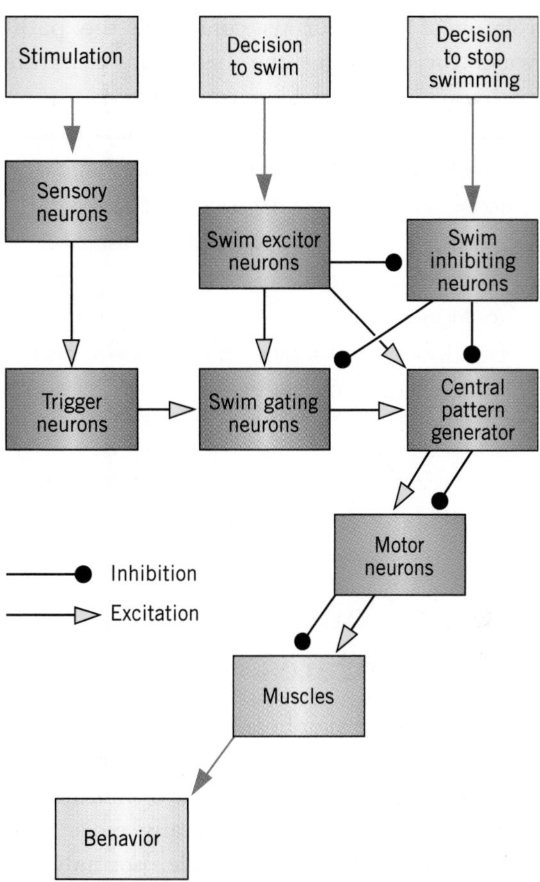

FIGURE 18-3. Schematic diagram of the neural pathway for the control of swimming in a leech. For simplicity, only the main connections between types of neurons are shown here. Swimming may be initiated by stimulation of the skin or by a spontaneous "decision" in the brain. In either case, swim gating neurons activate the CPG for swimming, which controls the coordinated action of body muscles. Swimming stops when there is a "decision" to stop. Swim inhibiting neurons then inhibit both the gating neurons and the neurons of the CPG.

A similar neural organization has been found in other invertebrates. Flying and walking in locusts, escape swimming in various mollusks and crayfish, and even grinding movements in the crustacean stomach have all been shown to be initiated via specific neural pathways. Usually one of several sensory stimuli will elicit activity in a specialized set of trigger neurons, which in turn will organize and activate the appropriate pattern generators for the movement. In each case, relatively few neurons constitute the pathways for initiating the behavior. Separate pathways also exist for terminating it.

Relatively small numbers of neurons are responsible for both initiating and terminating many simple behaviors of invertebrates. Specific sets of neurons may be responsible for triggering a behavior, for maintaining the behavior after the stimulus has ended, and for terminating the behavior.

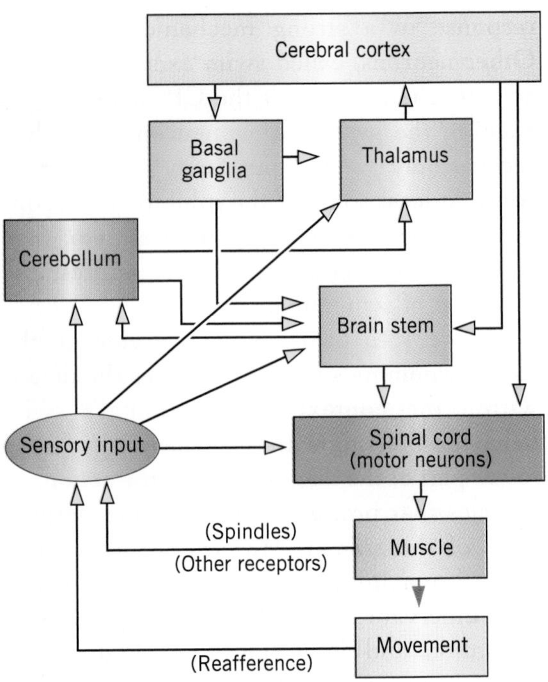

FIGURE 18-4. Simplified diagram showing the main elements and control circuits in the mammalian motor system. Note that in addition to the expected hierarchical organization (e.g., cerebral cortex to brain stem to motor neurons to muscle), there are also parallel pathways (e.g., from the cortex directly to the spinal cord) and feedback loops (e.g., cortex to cortex via the basal ganglia and the thalamus) in the system.

Motor Control in Vertebrates

An obvious and significant difference between the motor control systems of invertebrates and vertebrates is that the latter are enormously more complex and involve many more neurons. In mammals, for example, not only are the motor centers in the spinal cord and the reflex centers in the brain stem involved in regulating muscle activity, but parts of the cerebrum, the cerebellum, the basal ganglia, and the thalamus also play roles in controlling movement (Figure 18-4). A mammalian motor system, like most sensory systems, can be described as hierarchical, but it contains elements of parallel as well as serial organization. However, whereas the function of a sensory system is to extract a pattern from the detail of sensory input impinging on it, the function of a motor system is to organize a temporal and spatial pattern of movement, and then to generate the detail of specific muscle actions that is required to bring it about. In other words, the sensory system makes a pattern out of detail, whereas the motor system makes detail out of a pattern.

Unfortunately, whereas this concept provides a useful way to look at the motor system, it does little to help sort out the specific functions of its individual components. Part of the difficulty is that neurobiologists do not fully understand just what is involved in organizing a pattern of movement. Researchers understand well how reflexes work to maintain balance in a person standing still, but far less is known about how the nervous system organizes a movement like a pirouette. Reflexes are obviously used during such a movement, but they cannot account for it completely.

The following discussion will concentrate on mammalian motor systems because they are the most complex and they have been well studied. Different parts of these systems play different roles in producing behavior. For this reason, they will be discussed individually. You can refresh your memory about the basic structures of the brain motor systems by referring back to Chapter 3. The discussion here will provide sufficient additional anatomical detail to help you understand the broad functional relationships between the parts of the brain involved in motor control.

The motor control systems of vertebrates involve many parts of the central nervous system. They are both hierarchical (serial) and parallel in organization.

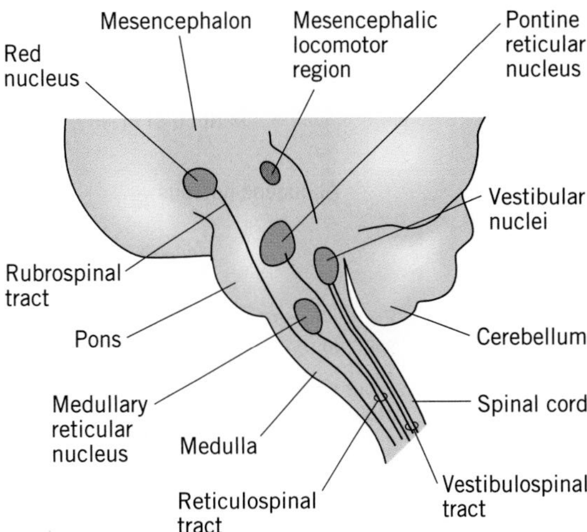

FIGURE 18-5. The brain stem, showing the main motor nuclei. This sketch shows the nuclei as if projected onto an outline of the brain stem. Not all the nuclei are present in the same plane. The nuclei receive input from the motor cortex, the basal ganglia, and the cerebellum, and they send axons to motor centers in the spinal cord. Pathways between the mesencephalic locomotor region and the other motor areas are not shown.

The Brain Stem

When the motor system is considered as a hierarchy, the main motor nuclei in the brain stem are placed just above the motor centers in the spinal cord, and below the cerebellum and the motor centers in the forebrain. Functionally, this means that the brain stem nuclei can influence spinal motor neurons, and that they themselves are influenced by other brain regions. Four of these nuclei provide most of the control of spinal motor centers by the brain stem. They are the two **reticular nuclei** (the pontine reticular nucleus and the medullary reticular nucleus), the vestibular nuclei in the pons and medulla, and the red nucleus in the mesencephalon. The locations of these nuclei in the brain stem are shown in Figure 18-5. The tracts that leave these nuclei and descend to the spinal cord are typically named according to the nucleus of origin, such as the rubrospinal (= red spinal) tract and the reticulospinal tracts. Input enters these nuclei from the cortex, the cerebellum, and the basal ganglia. The reticular nuclei are part of the reticular formation (see Chapter 3). The vestibular nucleus, as you learned in Chapter 14, consists of four parts and receives input from the vestibular apparatus and motor centers (Figure 18-6). Its output is sent to the spinal cord via the medial and lateral vestibulospinal tracts.

In general, these nuclei help to control posture, whether the animal is still or moving, as during standing, sitting, righting, walking, and so on. The nuclei have different roles. The vestibular and reticular nuclei control predominantly the large proximal limb extensor muscles (i.e., the ones mainly involved in postural adjustment). The lateral vestibular nucleus controls extensor motor neurons in the limbs, and the medial vestibular nucleus controls extensor motor neurons that innervate neck and trunk muscles. The red nucleus is less important in posture; it sends nerve fibers mainly to motor neurons that control flexor muscles and distal limb muscles. The function of this nucleus in postural control is not entirely clear, especially in humans.

The roles of the brain stem nuclei can be assessed by experiments in which the brain stem is severed at different levels. If the spinal cord is severed at its junction with the medulla, the motor neurons in the spinal cord lose all of their input from the brain stem nuclei. The animal will no longer be able to stand or maintain any posture, and only reflexes mediated entirely by neurons in the spinal cord will be functional.

If the brain stem is severed between the midbrain and the diencephalon, just anterior to the superior colliculi, what the animal can do varies from species to species. Because

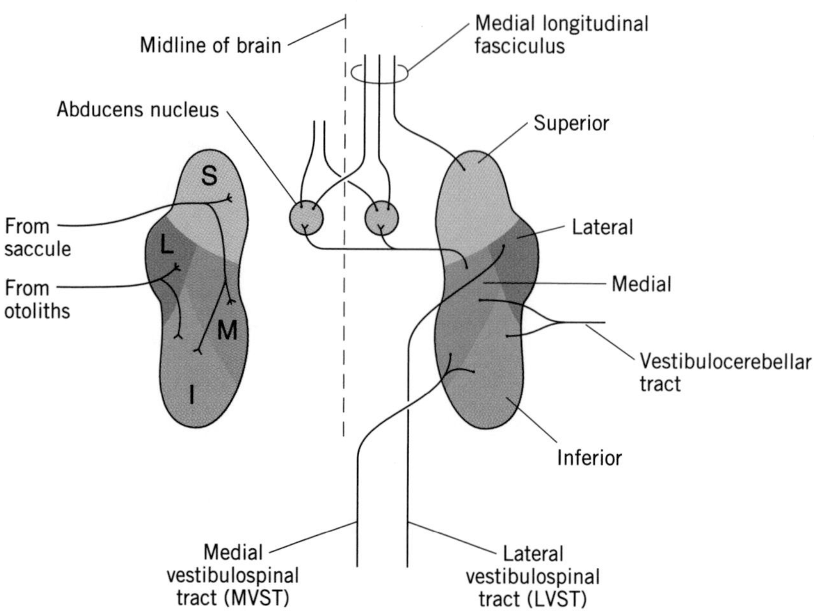

FIGURE 18-6. The vestibular nuclei, showing the four subdivisions and the main input (left) and output (right) tracts. For the sake of clarity, small tracts are shown by dotted lines, and minor tracts are omitted altogether. Ascending output of the vestibular nuclei is used mainly for control of eye movement, descending outputmainly for postural control.

the influence of the cortex is eliminated, all animals lose the ability to make voluntary movements. Cats and some other mammals, however, retain the ability to stand unassisted and can maintain balance even if they are gently pushed. If the cat is supported and placed on a moving treadmill, the legs will step by reflex action. Hence, the brain stem nuclei alone are certainly capable of maintaining posture in some mammals. Monkeys and other primates, on the other hand, cannot easily maintain posture after such a lesion because the cortex exerts greater influence over even simple reflex actions in these animals.

If the brain stem of any mammal is severed below the red nucleus but above the other motor nuclei, the animal exhibits what is known as **decerebrate rigidity**. The term refers to a characteristic posture in which all four limbs are rigid and stiff as a result of overexcitation of the main extensor muscles in the proximal parts of the legs caused in part by exaggeration of the stretch reflex in these muscles. Because the legs are stiff, they can readily support the body. However, because of the loss of control of flexor muscles and the loss of a proper balance between activity in extensor and flexor muscles, the animal cannot adjust to perturbations. If the surface it is standing on is tilted or if the animal is gently pushed, it will topple over, still stiff-legged.

In addition to nuclei that control postural reflexes, the brain stem contains several regions that control walking or other forms of locomotion. These regions are located in the mesencephalic tegmentum. One area, the **mesencephalic locomotor region**, lies in the tegmental area of the pons, ventral to the inferior colliculus, and influences motor performance through the other motor nuclei. In cats, electrical stimulation of the mesencephalic locomotor region will induce walking. The stronger the stimulation, the faster the cat walks; at high levels of stimulation, the cat will run. Locomotion has also been produced by stimulating the corresponding brain regions of lower vertebrates such as fish, in this case resulting in swimming. These observations indicate that specific regions of the midbrain have the capacity to organize the spinal pattern generators to produce coordinated locomotion, irrespective of whether that locomotion is walking, running, or swimming. However, these regions are not capable of initiating locomotion on their own without electrical stimulation; animals without a cortex (decorticate animals) never initiate voluntary movements.

The brain stem contains four nuclei that receive input from other motor regions of the brain and contains other functionally important areas as well. Motor nuclei in the brain stem help to control posture by delivering excitatory and inhibitory signals to spinal motor neurons, and can coordinate standing, righting, and stepping even in the absence of the motor cortex. The brain stem also contains regions that can produce coordinated locomotion.

The Motor Cortex

Three regions of the cerebral cortex, collectively referred to as the motor cortex, are devoted to motor control (Figure 18-7). The **primary motor cortex** lies just anterior to the central sulcus, opposite the somatosensory cortex. The **premotor cortex** lies anterior to the primary motor cortex, extending from the dorsal surface of the brain and down laterally in the parietal lobe. The **supplementary motor area (SMA)** also lies anterior to the primary motor cortex but on the medial surface of the brain. The supplementary motor area is also known as area MII (M two), in contrast to MI (M one), the designation for the primary motor cortex.

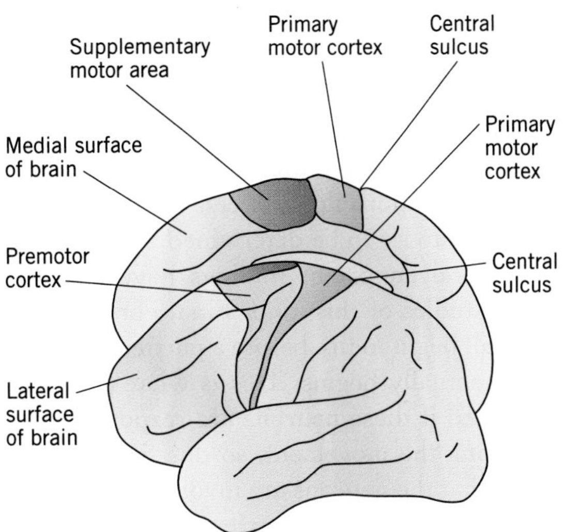

FIGURE 18-7. The three main motor cortical areas in mammals, shown in the human brain. The brain is shown as if sliced in half and separated, so that part of the medial surface of the right hemisphere can be seen. The premotor cortex and the supplementary motor area play important roles in planning complex movements.

Primary Motor Cortex

The primary motor cortex illustrates the principle that the motor system has a parallel as well as a (hierarchical) serial organization. In addition to exerting control via the motor nuclei in the brain stem, it also bypasses these nuclei and controls some spinal motor neurons directly. It exerts this control via axons that descend through the **pyramidal system,** so called because the fibers cross over to the other side of the brain at the level of the medulla–spinal cord junction, in the region of the medulla known as the **pyramids.** However, not all motor neurons are equally influenced by pyramidal input; pyramidal control is confined to those motor neurons that control the most distal muscles of the limbs, that is, in humans the forearms, hands and fingers, and the feet and toes. Shoulder, hip, or trunk muscles are not directly influenced by the pyramidal system.

Studies designed to answer the question of how cortical neurons control individual movements often employ monkeys trained to make specific movements on command. The electrical responses of cortical neurons in these monkeys can be recorded via electrodes implanted painlessly into the brain through the skull under anesthesia (Box 18-2), allowing the relationship between movement and neural activity to be determined.

Several important findings have emerged from studies of this kind. First, cortical neurons all begin to fire before a particular movement actually begins. This is what would be expected if these neurons are responsible for initiating the movement, so this finding indicates that the neurons are involved in producing the movement rather than merely responding to it once it has begun. Second, different neurons respond preferentially to different parameters of a movement. Four main types of neurons have been identified in monkeys, based on the movement parameter they encode. One group, dynamic neurons, codes mainly for the rate of force development. That is, the neurons fire only when the animal changes the level of force it applies with a set of muscles. A second group, static neurons, codes for the steady-state level of force. That is, the neurons fire continually in proportion to the level of force exerted by the animal. A third group shows intermediate properties.

A fourth group of neurons has been shown to encode direction of movement. For this group, it has further been shown that neurons fractionate the range. For example, in a monkey trained to move a lever in one of eight directions, different cortical neurons that respond during a movement are active at different frequencies, depending on the direction in which the monkey moves the lever. Figure 18-8A shows the response profile of a single cortical neuron as a function of different directions of movement. For each neuron there is a single "preferred direction" to which it responds maximally. By analogy with sensory systems, this response profile is called the **directional tuning curve** of the neuron. Note that there is no single neuron that encodes a particular direction of movement. Instead, just as some sensory systems use range fractionation to encode precise information about a stimulus, so here the motor system uses range fractionation to encode the precise direction of an intended movement, which will be determined by the summed action of many directionally tuned neurons acting together. Figure 18-8B shows the responses of groups of cortical cells when the monkey moves its arm in each of three different directions. The summed group vector of the combined activity of all the neurons is shown by the solid arrows. This is quite close to the actual direction of movement, shown by the dotted lines and open arrowheads.

The primary motor cortex projects its axons via the pyramidal system to the brain stem, as well as down to the motor neurons that control muscles of the distal appendages. Different motor cortical neurons encode for different aspects of a motor act, such as rate-of-force development or steady-state level of force. The direction of movement of a limb is determined by an ensemble of cortical neurons that fire together, different combinations of neurons firing when different directions are selected.

Premotor and Supplementary Motor Areas

The primary motor cortex functions mainly in guiding intended movements to their targets. The premotor and supplementary motor areas, in contrast, play important roles in planning the movements in the first place. Experiments with monkeys that have been trained to retrieve a morsel of food or perform some other task, as well as experiments with human subjects, have shown that the premotor and supplementary motor areas serve at least three separate functions related to movement. Each brain area may have a specific role to play in these functions, but for the sake of simplicity the functions will be discussed together here. These three functions are to control visually guided

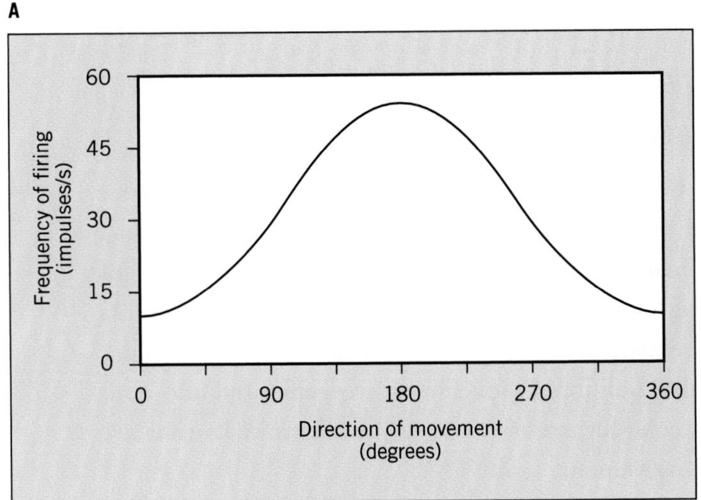

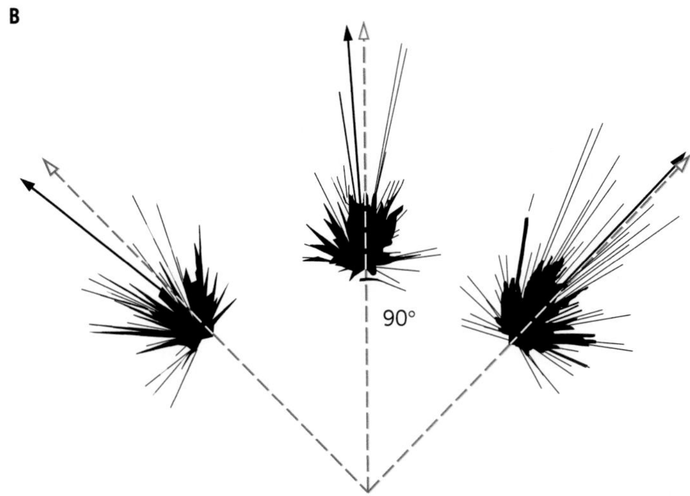

FIGURE 18-8. The responses of motor cortical cells in a monkey during voluntary movement of one arm in different directions. (A) Example of the response profile of one cortical cell, showing its preferred direction of response (peak of curve). Different cortical cells have different preferred directions. The profile of response of a cell is the cell's directional tuning curve. (B) Vectors of peak directional response for groups of cells active during arm movement in three different directions. Each line in a group represents the frequency of response of one cell to arm movement in the direction indicated by the dotted line and open arrowhead. Combining all the lines in one group yields a group vector, shown by the solid line, which points in the direction indicated by the solid arrowhead. The group vector is always close to the direction of arm movement.

movement, to coordinate postural adjustments that must accompany a movement, and to plan a complex physical movement.

Visually guided movements can be disrupted by lesions in the motor cortex outside of the primary motor area. For example, if an intact monkey is presented with a morsel of food placed behind a transparent barrier, it will quickly move its hand around the barrier once it encounters it. However, a monkey with lesions in the premotor area tries repeatedly to grasp the food directly, bumping against the barrier each time. Lesions in the supplementary motor area produce animals that cannot easily retrieve a small morsel of food stuck in a hole drilled through a tabletop. A normal monkey simply pushes the morsel through with a finger of one hand and catches it with the other. The animal with a lesion tries to push the morsel up through the hole with one hand while pushing it down with the other. Both these results indicate a failure of the animal to readjust or plan movements on the basis of specific sensory feedback about a current situation.

The supplementary motor area may help to adjust motor output to the requirements demanded by anticipated external loads, as suggested by experiments on human subjects. Most movements involve some adjustment of musculature elsewhere in the body in order to maintain balance. For example, if

BOX 18-2

Wires in the Brain: The Relationship Between Neural Activity and Movement

From a methodological point of view, studying a sensory system is easier than studying the motor system. This is because an animal can usually be anesthetized and restrained without any serious effect on the processes of transduction or encoding of sensory stimuli. The motor system presents a considerably greater experimental challenge, because if a researcher wishes to understand how volitional movements are generated and controlled, he or she must conduct experiments on awake animals that are capable of at least some movement.

Significant advances in understanding volitional movement were made with the introduction of methods for recording neural activity in unrestrained animals in the 1960s. For example, Edward Evarts was able to record the activity of cortical neurons from the primary motor cortex in monkeys trained to execute a particular limb movement. The technique involved fitting a small sleeve to a hole bored in the skull over the region of the motor cortex

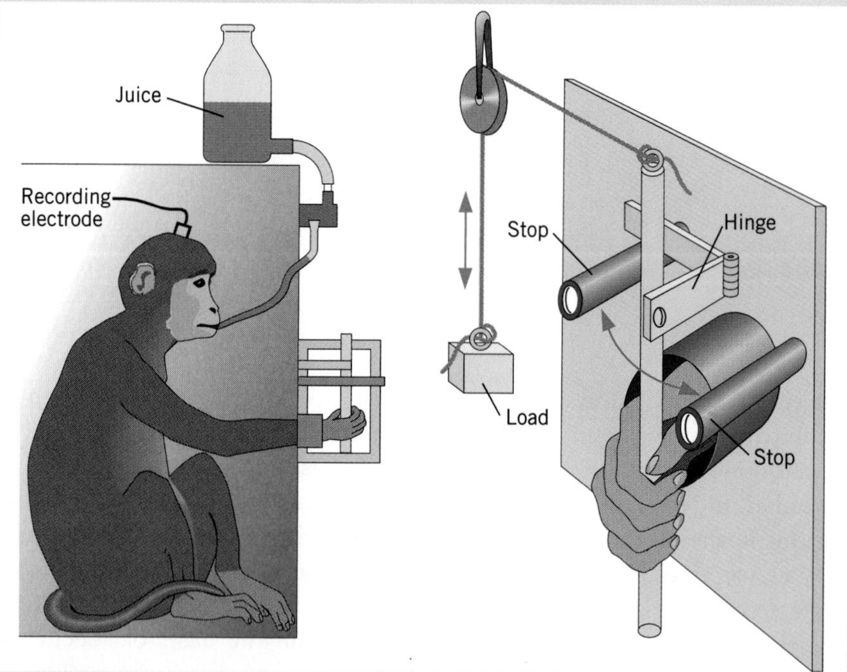

FIGURE 18-C. The arrangement for monitoring the activity of brain neurons during learned movements in a monkey. Left: During the experiment, the monkey sits in a cage and receives a reward of fruit juice when it makes the appropriate movements. Electrodes inserted into the brain record the electrical activity of cortical neurons. Right: The monkey's arm is inserted through a restraining sleeve and its hand allowed to grasp a lever, shown with a load attached. Because of the sleeve, the monkey can move the lever only by flexing or extending its wrist.

devoted to wrist movements. After the monkey recovered from the anesthesia and the operation, a slender strand of wire could be inserted through the sleeve into the brain. (This procedure is painless because there are no pain receptors in the brain.) When connected to an amplifier and other standard neurophysiological equipment, the wire allows researchers to record from one or more individual cortical neurons while the animal moves about.

Evarts and his coworkers conducted experiments on monkeys that had been trained to move a lever by flexing or extending their wrists. If the monkey moved the lever at the right speed, first in one direction and then in another, it obtained a reward of fruit juice (Figure 18-C). The lever was attached by a pulley to a weight, which could be arranged to resist either wrist extension or wrist flexion. Evarts's experiments revealed a population of neurons that were active during movements in one particular direction only and that were sensitive to the load against which the wrist had to work (Figure 18-D). These experiments were among the first to suggest that cortical neurons could encode direction and force.

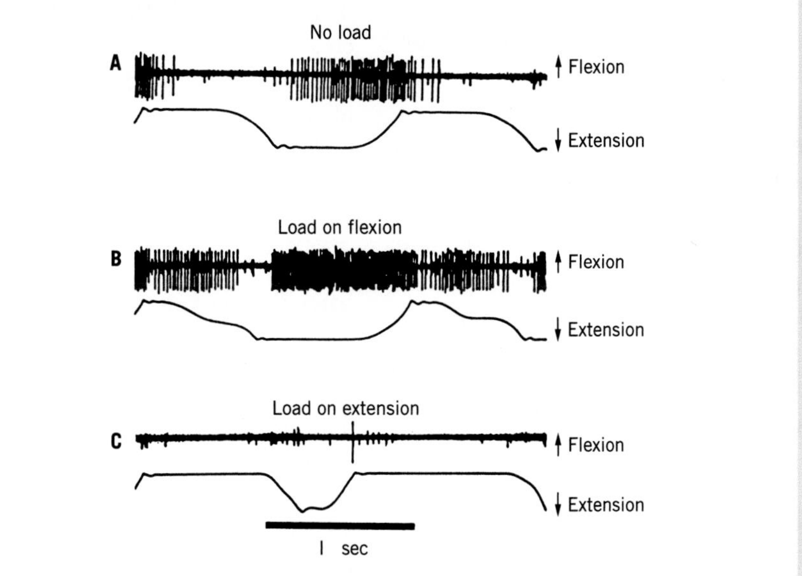

FIGURE 18-D. Recording from a motor cortical neuron during wrist movements. A monkey, arranged as shown in Figure 18-C, is trained to flex and extend its wrist. The top trace of each pair of recordings shows activity of the neuron, and the bottom trace shows the movement of the lever the monkey is holding. (A) The monkey moves the lever alone, without any external weight. The neuron responds before and as the monkey flexes its wrist. (B) A weight is added. This weight must be resisted by increased force in flexor muscles. The neuron now responds much more vigorously before and during flexion, as well as responding during all but the end of extension, presumably because force is required to hold the weight even during extension. (C) The weight is shifted so that now extensor muscles must generate more force to move the lever. The neuron hardly fires at all, since no force is required by flexor muscles to flex the wrist. The neuron hence fires in relation to the force exerted by flexor muscles.

you raise one arm while you are standing erect, you must shift your body to keep the center of gravity over your feet. Hence, the brain must be able to analyze or predict the effects of a particular action on a desired sequence of movements. Imagine a subject seated in a chair, arm held flexed at the elbow so the hand is parallel to the floor. If someone places a small weight on the arm under these conditions, recordings from arm muscles show that there is a short delay before the subject adjusts the motor output to the new load. However, if the subject places the weight, there is no delay. Patients who have damage in one side of the supplementary motor area cannot make this predictive adjustment on the affected side, although they can on the intact side, indicating an important role of the supplementary motor area in planning motor output on the basis of anticipated events.

Planning a complex physical movement is the third function of the nonprimary motor cortex. For example, planning complex movements seems to require action of the supplementary motor area. This can be shown in human subjects by making images of neurally active brain regions (see Chapter 21). If a subject is instructed to make a simple flexion movement of one finger, the primary motor cortex and the corresponding somatosensory cortex become active, as measured by increased blood flow to these areas. (The somatosensory area is activated by the sensory stimulation provided by the movement of the finger.) If the subject is then given a more complex sequence of finger movements to make, not only do the primary motor and somatosensory areas become active, but so also does the supplementary motor area. Finally, if the subject is asked to imagine performing the sequence of finger movements but not actually to make any movement, the supplementary motor area becomes active, but without any corresponding activity in the primary motor or somatosensory cortexes. Electrical recording via surface electrodes also reveals neurons that are active before an animal carries out a specific task or, in the case of humans, "imagines" carrying out some task.

The premotor and supplementary motor areas of the cortex play a critical role in the generation and execution of complex movements. They especially seem to control visually guided movement, to coordinate postual adjustments that accompany a movement, and to plan complex movements.

Controlling Complex Movement

The basal ganglia, the thalamus, and the cerebellum constitute the remaining components of the motor system. Unfortunately, although an important role for these regions in controlling movement is well established, the mechanisms by which they fulfill this role are not clear. Hence, the descriptions of their functions here will be quite general. An overview of these functions is shown diagrammatically in Figure 18-9; they are discussed in the following two sections.

Thalamus and Basal Ganglia

The thalamus and the nuclei that form the basal ganglia lie mostly in the diencephalon. The thalamus consists of several distinct nuclei that receive input from the globus pallidus and part of the substantia nigra, as well as sensory input and input from the cerebellum. The basal ganglia consist of the caudate nucleus, the putamen, the globus pallidus, the substantia nigra, and the subthalamic nucleus (see Figures 3-20 and 3-21). The basal ganglia and the thalamus form several complex neural circuits (Figure 18-10). Information from the somatosensory cortex, and from all three motor cortex areas (MI, MII, and the premotor cortex), enters the basal ganglia via the putamen. From the putamen, neural signals converge onto neurons in the globus pallidus that project to the ventrolateral nucleus of the thalamus. Thalamic neurons, in turn, project back to the supplementary motor cortex. A parallel circuit from the putamen through the substantia nigra and back to the thalamus is also present, as are subcircuits between the globus pallidus and the subthalamic nucleus and between the putamen and the substantia nigra.

The neural circuits involving the basal ganglia and thalamus are critical for normal motor function. This is obvious from the devastating effects of malfunctions of these circuits as evidenced by Parkinson's disease and Huntington's disease (see next section). However, what they actually do can so far be described only in rather vague, general terms, and the neural mechanisms by which their functions are carried out are certainly not clear. Nevertheless, two general findings stand out. First, basal ganglia generally exert an inhibitory influence on the thalamus. When this is disrupted, as in Huntington's disease, movements become exaggerated and uncoordinated. Second, there are important subcircuits involving the subthalamic nucleus and the substantia nigra. Disruption of at least some of these causes an increase in inhibition and a corresponding difficulty in making any movements at all, as in Parkinson's disease.

Functionally, there is evidence that the basal ganglia and thalamus are required for the planning of movements—anything from playing a complex piano arpeggio to a simple reaching movement involving several joints. It is clear that few or none of the functions of the premotor and supplementary motor cortexes that are described in previous sections can be carried out without the basal ganglia and thalamus, as indicated by the motor symptoms exhibited by individuals with Parkinson's disease or Huntington's disease.

In addition to their functions in specific motor tasks, neural circuits involving the basal ganglia and thalamus also seem to be necessary in order for motor acts to be placed in a broader behavioral context. For example, they are thought to be necessary for the automatic performance of learned repetitive tasks. Furthermore, the basal ganglia and inputs to them from elsewhere in the cortex are thought to be important in tying motor action to motivation and emotional reactions. Animal studies, for example, show that neurons in the striatum or the substantia nigra will frequently become active upon the presentation of a stimulus that has become associated in the animal's memory with a particular behavioral response, whereas they will not respond if the

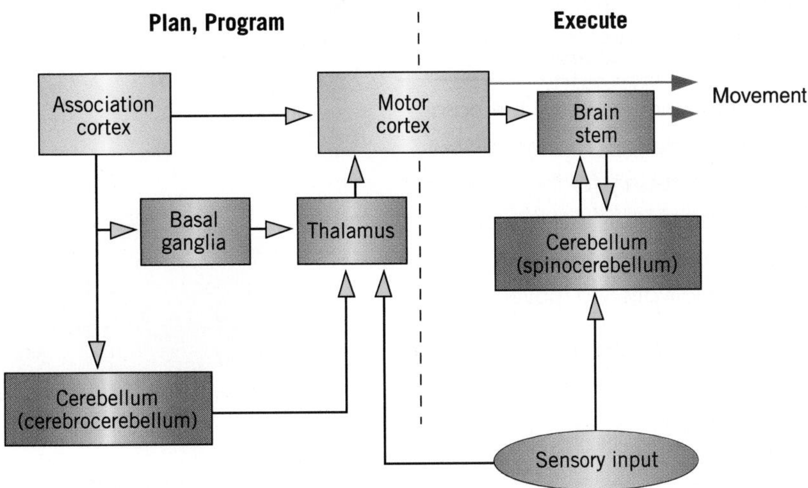

FIGURE 18-9. An overview of the functional roles of the components of a mammalian motor system. Note that the basal ganglia, the thalamus, and the cerebrocerebellar division of the cerebellum are involved in motor planning and programming but not in the actual execution of movements.

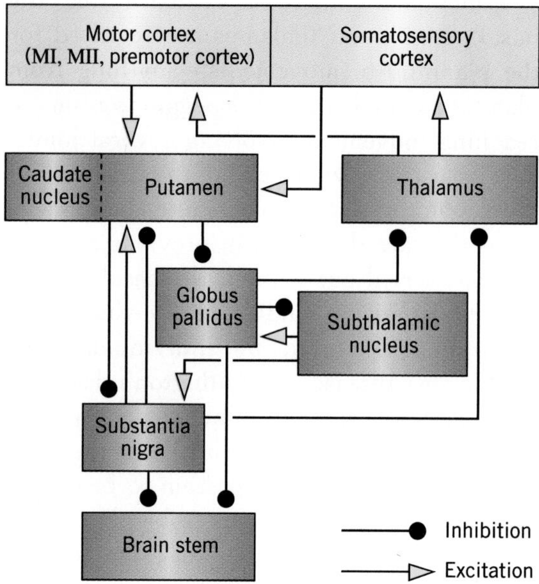

FIGURE 18-10. Schematic diagram showing the main connections formed by the basal ganglia and the thalamus that relate to motor control. The caudate nucleus receives considerable input from nonmotor regions of the cortex, and the thalamus receives input from sense organs, neither of which is shown here. Arrowheads and filled circles indicate known excitatory and inhibitory influences, respectively. See Figure 3-20 for a view of the basal ganglia in the human brain in relation to other structures.

stimulus has no meaning to the animal. Furthermore, some cells will become active when the animal must respond to the stimulus by moving a lever to a remembered target, but not when the animal can see the target.

The functions of the thalamus and the basal ganglia are not understood in detail, but they seem to be necessary for the planning of movements before they are executed. The basal ganglia and thalamus also seem to link emotion and motivation to movements.

The Cerebellum

Structurally, several prominent fissures divide the cerebellum into three lobes. Looking down from above, the primary fissure divides the cerebellum into an **anterior lobe** and a **posterior lobe.** A second deep fissure separates the posterior lobe from the **flocculonodular lobe,** which is tucked underneath the rest of the cerebellum at its posterior margin. These structural divisions are shown in Figure 18-11.

The cerebellum can also be divided into three functional parts. The **vestibulocerebellum** is roughly coincident with the flocculonodular lobe. The other two functional divisions, however, lie medially and laterally, and hence include parts of both the anterior and the posterior lobes. The **spinocerebellum** consists of the medial part of the cerebellum except for a region in the center. The **cerebrocerebellum** consists of the lateral parts of the anterior and posterior lobes, as well as the

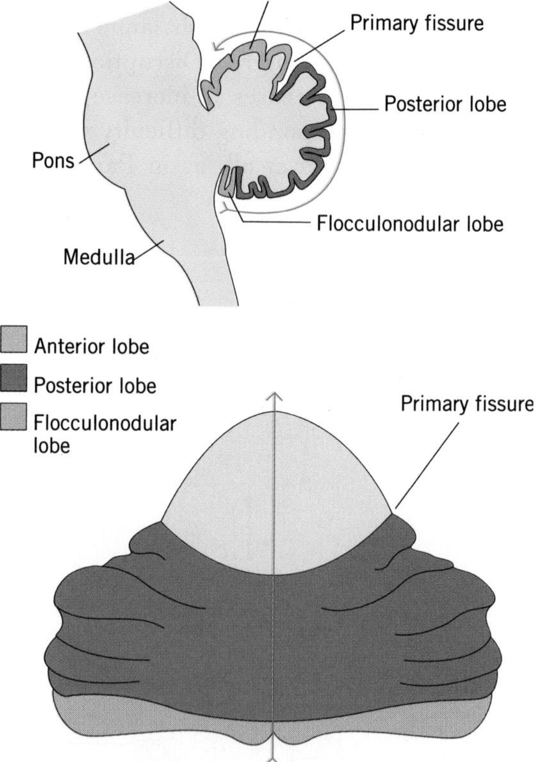

FIGURE 18-11. The structure of the human cerebellum. The bottom diagram shows the entire structure unfolded from posterior to anterior, as indicated by the arrow and as shown in the smaller cross-sectional sketch at the top.

central region of the medial area (Figure 18-12). Each of these functional divisions receives input from specific parts of the nervous system via mossy fibers (see Chapter 3). In addition, all parts of the cerebellum receive input from the **inferior olivary nucleus** in the medulla, the source of all climbing fiber input to the cerebellum.

The specific inputs received by the three functional divisions of the cerebellum, as well as the output connections they make, summarized in Figure 18-13, reflect the role played by each division in the motor system. The vestibulocerebellum receives input from the vestibular and reticular nuclei, via the reticulocerebellar and vestibulocerebellar tracts, respectively. It sends extensive output to the vestibular nuclei. Its main functions are to help maintain balance during ongoing movements and, via oculomotor reflexes, to help stabilize movements of the eyes when the head moves (discussed in more detail in Chapter 20).

The spinocerebellum receives input from the spinal cord via the spinocerebellar tracts, as well as some vestibular input from the vestibulocerebellar tract. It sends output fibers into the fastigial and interpositus

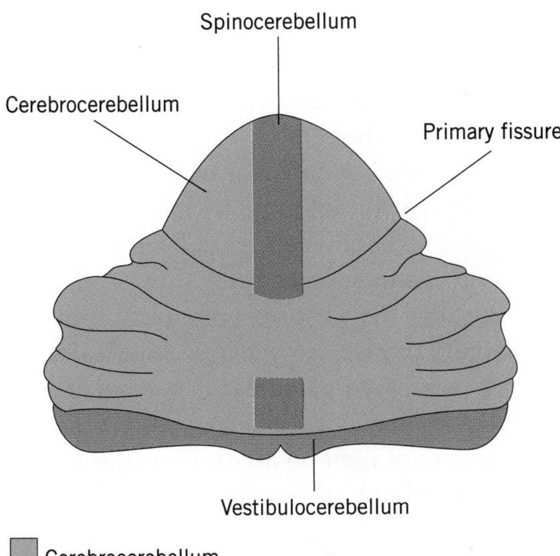

FIGURE 18-12. The functional organization of the human cerebellum. Each of the main functional divisions is shown in a different color. Note that only the vestibulocerebellum corresponds to an anatomical region, the flocculonodular lobe.

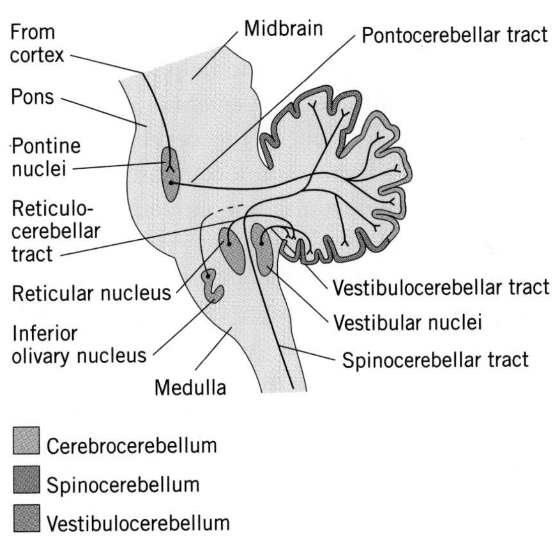

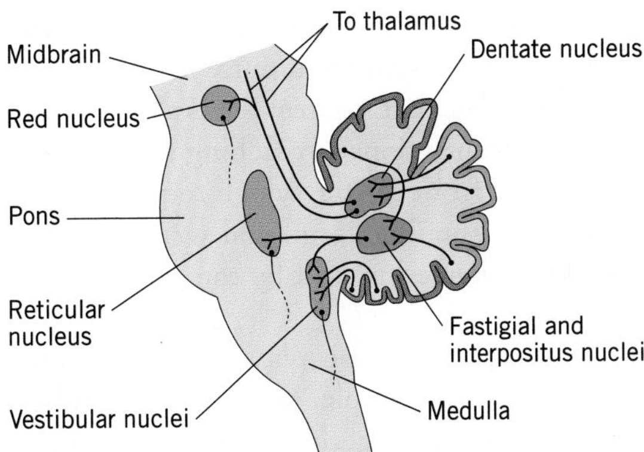

FIGURE 18-13. Connections formed by the cerebellum with the rest of the nervous system. (A) Input to the cerebellum. Most input is via mossy fibers; input from the inferior olivary nucleus is via climbing fibers, which are distributed throughout the entire cerebellum. (The distribution of climbing fibers in the cerebellum is omitted from the diagram for clarity.) (B) Output from the cerebellum. Most output is routed through the three deep cerebellar nuclei. The fastigial and interpositus nuclei have been combined in this diagram for clarity.

cerebellar nuclei, which in turn send fibers to the reticular and vestibular nuclei. Experiments with monkeys have helped uncover the role of the spinocerebellum in motor control. For ex-ample, lesions in specific cerebellar nuclei interfere with the ability of a monkey to stand or walk. Lesions in other nuclei cause severe tremor during reaching. On the other hand, simple movements of any kind that involve only a single joint are not significantly affected. Hence, it is hypothesized that the specific function of the spinocerebellar system is to help an animal to coordinate multijoint movements.

The third division of the cerebellum, the cerebrocerebellum, is important in the planning of complex multijoint movements. The cerebrocerebellum receives input from the cortex via relay nuclei in the pons through the pontocerebellar tract. Output from this division is sent through the dentate nucleus to the red nucleus and the thalamus, thereby making direct connections to the control system of the basal ganglia. In monkeys, lesions of the dentate nucleus have little effect on locomotion or other simple movements. However, they do interfere with the ability of the monkey to reach a specific target or to pick up a small morsel of food, indications of failure in the circuits that plan and direct the coordination of sensory input with the activity of muscles that control several joints. In humans, such actions as reaching for a cup on the table in front of you, which requires coordination of shoulder, arm, and wrist muscles, are likely planned in part by the cerebrocerebellum.

The cerebellum also plays a role in other functions. For example, parts of the cerebellum are crucial for the learning of motor tasks, including conditioned reflexes. This function is considered in more detail in Chapter 24. Perhaps more surprising is recent evidence that the cerebellum may have a cognitive function, at least in humans. For example, studies of schizophrenics have shown that deficits in neural circuits that interconnect the cerebellum and the forebrain and thalamus are correlated with the presence of the disease.

The cerebellum consists of three structural and functional parts. The vestibulocerebellum helps coordinate balance and eye movements, the spinocerebellum helps to coordinate multijoint movements, and the cerebrocerebellum is important in the planning of complex multijoint movements. The cerebellum is also critical for motor learning and may play a role in cognitive functions.

Malfunctions of the Motor System

One of the most compelling lines of evidence that the basal ganglia are intimately associated with movement arises from studies of individuals afflicted with Parkinson's disease. You may recall from the discussion of neurotransmitter systems (see Chapter 7) that the cause of this ailment is the death of dopaminergic neurons in the substantia nigra. These neurons normally provide feedback to the striatum in a neural circuit within the basal ganglia. For many decades, Parkinson's disease had been thought to be environmentally induced (even though researchers have not yet identified the toxin or toxins responsible for it), but recent evidence has been found for a genetic component as well.

Clinically, the problem manifests itself as an inability to plan movements. Take a simple tracing task as an example. A person who is asked to follow with a finger an erratically moving spot of light on an oscilloscope screen can do so only by using visual cues. The subject's finger always lags a bit behind the spot, since the person must see where the spot is before moving his or her finger to that position, by which time the spot has moved on. However, if the movement of the spot is sinusoidal or has some other regular and predictable path, normal subjects anticipate where the spot is going to be and are able to keep a finger directly over the spot at all times. Individuals with Parkinson's disease cannot do this. They behave as if they must rely entirely on visual cues and cannot plan the movement ahead of time.

Other tests and clinical observations lead to the same conclusion: where sensory feedback is required for a movement, persons with Parkinson's disease can execute reasonably well, but under open loop conditions (without any feedback to guide the performance, as in preplanned movements) they have great difficulty. In advanced stages of the disease, it even becomes difficult for afflicted individuals to make any movements at all, and they often become trapped in the "frozen" posture so characteristic of the disease, unable to move. Insight into the complexity of the motor system is provided by a trick that often helps those with Parkinson's disease to break out of a frozen posture when they are trying to walk. This is to have another person place a stick on the ground in front of the "frozen" individual. When the latter attempts to step over the stick instead of just trying to walk, he or she is often able to do so, albeit still slowly and with difficulty. Apparently the brain goes through a different process to move a leg up over something, as in walking up a flight of stairs, than it does simply to move it forward during straight walking. In an individual with Parkinson's disease, this can be the practical difference between being able to move and being immobilized.

The frozen posture of advanced Parkinson's disease is consistent with current ideas about the connections between the basal ganglia and the motor structures influenced by them. As shown in Figure 18-10, it is thought that the basal ganglia exert mainly an inhibitory influence on the thalamus and the spinal motor centers. Loss of the normal influence on the striatum by the substantia nigra is thought to intensify this inhibitory influence, with the result that the afflicted individual exhibits great difficulty in moving.

Huntington's disease, sometimes known as **Huntington's chorea,** is another disease of the basal ganglia. Huntington's disease is genetic and ordinarily does not manifest itself until middle age. At that time the affected individual begins to show chorea (involuntary dance-like movements) that becomes progressively worse. The individual also develops dementia, eventually losing all cognitive function.

Huntington's disease is caused by the death of inhibitory GABAergic neurons in the striatum. Because of this loss, Huntington's functionally can be considered a manifestation of dopamine overload. This is caused by the weakening of the inhibitory influence on the thalamic and spinal motor centers, so that the excitatory neurons, which release dopamine, become overactive. Indeed, antidopamine drugs relieve the chorea to some extent, although there is no long-term cure for the person's loss of mental function due to neuronal degeneration.

Malfunctions of the motor system can give insights into the way the brain organizes movement. Individuals with Parkinson's disease, which is caused by the death of dopamine-containing neurons in the substantia nigra, have great difficulty making purposeful movements. Individuals with Huntington's disease have the opposite problem, making uncontrollable and unwanted movements, as well as eventually suffering a complete loss of cognitive function.

Additional Reading

General

Brooks, V. B. 1986. *The Neural Basis of Motor Control.* New York: Oxford University Press.

Humphrey, D. R., and H.-J. Freund, eds. 1991. *Motor Control: Concepts and Issues.* Dahlem Workshop Reports, Life Sciences Research Report LS 50. New York: Wiley.

Rothwell, J. 1994. *Control of Human Voluntary Movement*. 2d ed. London: Chapman and Hall.

Research Articles and Reviews

Alexander, G. E., M. R. DeLong, and P. L. Strick. 1986. Parallel organization of functionally segregated circuits linking basal ganglia and cortex. *Annual Review of Neuroscience* 9:357–81.

Chesselet, M.-F., and J. M. Delfs. 1996. Basal ganglia and movement disorders: An update. *Trends in Neurosciences* 19:417–22.

Evarts, E. V. 1968. Relation of pyramidal tract activity to force exerted during voluntary movement. *Journal of Neurophysiology* 31:14–27.

Georgopoulos, A. P. 1997. Neural networks and motor control. *Neuroscientist* 3:52–60.

Kien, J. 1983. The initiation and maintenance of walking in the locust: An alternative to the command concept. *Proceedings of the Royal Society of London Series B* 219:137–74.

Kupfermann, I., and K. R. Weiss. 1978. The command neuron concept. *The Behavioral and Brain Sciences* 1:3–39.

PART 5

Integrating Systems: The Neural Basis of Behavior

Interpreting sensory stimuli and organizing responses to them are but two functions of the nervous system. If every stimulus always produced the same response, animal life would not likely have survived long. The nervous system must also be able to integrate and interpret the input it receives, and make decisions about it. Integration and interpretation are the central elements of neural function; at the same time, they are the most difficult to understand.

The chapters in this part build on the concepts of sensory processing and motor control developed in the preceding two parts, and describe emerging principles in the decision-making function of the nervous system, ranging from the simple to the complex.

Part 5 Integrating Systems: The Neural Basis of Behavior *440*

CHAPTER 19

Mechanisms of Escape Behavior

Behavior can be viewed as an overt expression of the integrating action of the nervous system. This is true because behavior is more than just sensory input and motor output; it is shaped and suited to the circumstances in which it is expressed. Even escape behavior, which often seems quite stereotyped, involves many neural decisions. For this reason, and because it is relatively simple and well understood, escape behavior is an excellent model with which to begin examining mechanisms of neural integration.

For many animals, escape is perhaps the most crucial of all behavior, because without an effective way of avoiding predators, an animal will not long have to be concerned about anything else. Escape can take many forms, each appropriate to specific circumstances. Aquatic animals like crayfish have a strong swimming escape. Earthworms and other animals that live in burrows often use a quick withdrawal into their holes. Flying animals like moths also have effective escape mechanisms, frequently using evasive tactics in place of speed. All of these behaviors require an animal to integrate sensory input to determine the most effective response.

Bats and Insects

Many night-flying moths and other insects are preyed upon by bats. For these insects, flying fast is of little use in escape because bats can fly so much faster than the insects. An alternative strategy is for the insect to evade the bats rather than outfly them. You have already read in Chapter 12 about some of the mechanisms that allow a bat to determine the direction, distance, and speed of its prey. Now you will learn about the mechanisms insects use to avoid capture.

How Insects Hear Bats

For a flying insect to evade a bat, the insect must be able to hear the bat and also determine where and about how far away the bat is. Even the rather simple ears of insects are able to provide this information.

In noctuids, night-flying moths of the type preyed upon by several species of bats, each ear typically consists of a tympanic membrane stretched over an enclosed air sac located on the insect's abdomen just where it joins the thorax. The ear contains two sensory neurons, called "acoustic" neurons, attached to the tympanic membrane (Figure 19-1). The neurons, designated A1 and A2, have different thresholds to sound, cell A1 being about ten times more sensitive than A2. This difference in sensitivity means that A1 will respond

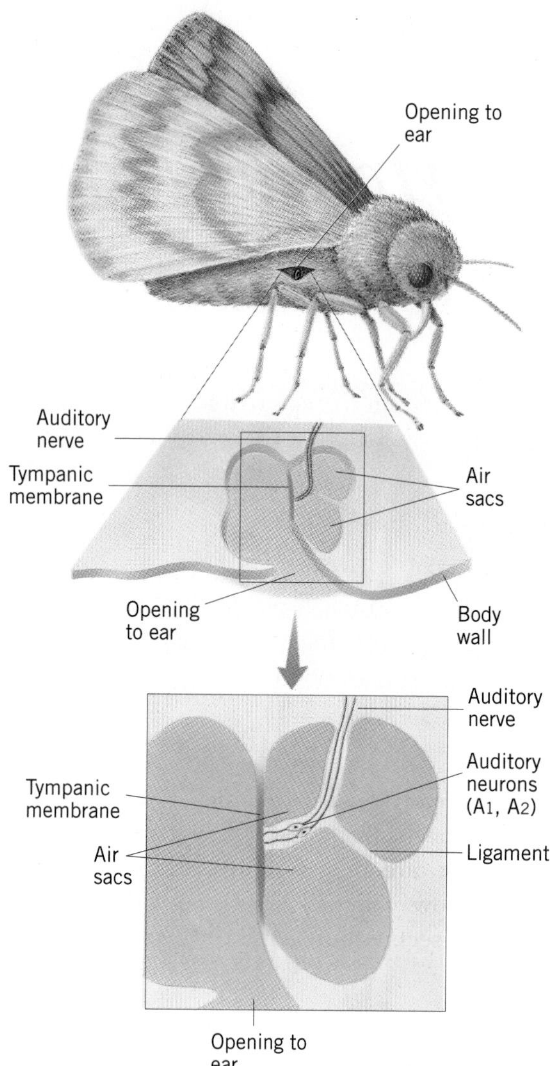

FIGURE 19-1. The ear of a noctuid moth. The external openings of the ears are located under the wings. A longitudinal section through the right body wall shows the internal structure of the ear. The two auditory neurons, A1 and A2, send axons to the central nervous system via the auditory nerve.

to sounds as little as one-tenth the intensity of the weakest sounds to which A2 will respond. The two acoustic neurons of noctuids cannot discriminate sounds of different frequencies; they are both sensitive to sounds in the range 3 to 150 kHz. The exact range of their sensitivity differs from species to species. The best frequency of the acoustic cells in noctuids is about 40 to 70 kHz, well into the ultrasonic range and beyond the range of human hearing. Other insects hunted by bats, such as lacewings, also have ears capable of responding to ultrasonic sound in this same range, although the anatomy of the ears may be quite different from the anatomy of a moth's ear.

Insectivorous (insect-eating) bats use ultrasonic cries to locate their prey, listening for and homing in on the echoes produced when the sound waves of the cry bounce off the insects they hunt (see Chapter 12). Different species of bats emit different kinds of cries. Some bats produce frequency modulated (FM) cries that start at a high frequency and rapidly tail off to a much lower one. The frequencies of these cries may range from a high of as much as 110 kHz to a low of around 30 kHz, depending on the species. Other bats generate a constant frequency (CF) cry, either alone or combined with an FM component. CF cries are also in the ultrasonic range; the horseshoe bat, for example, emits a CF cry of about 83 kHz. The frequency of the sound to which the ear of a typical noctuid moth or lacewing has maximum sensitivity is therefore within or close to the range of frequencies of sound emitted by a hunting bat.

Experiments in which the electrical activity of moth acoustic neurons was recorded during the free flight of a bat have been conducted by Kenneth Roeder and his colleagues. These experiments showed that A1, the most sensitive acoustic neuron in the moth, can just detect the cry of the bat when the bat is about 100 to 120 ft (31–37 m) from the moth. As the bat flies closer, the response of cell A1 becomes stronger, as measured by an increase in the rate of firing of the acoustic neuron. When the intensity of the cry increases further, the second acoustic neuron, A2, begins to respond. When the bat is within about 15 to 20 ft (4.5–6 m) of the moth, both acoustic neurons are entirely saturated, meaning that they are responding at the maximum rate possible.

Laboratory studies have also made it clear that moths have good directional sensitivity to the cries of bats. This is possible even with such a simple auditory system because the body of the moth is large enough that the intensity of the bat cry is lower on the side of the moth's body away from the bat, due to the loss of strength of the sound waves as they diffract around the body of the moth. Hence, as shown by experiments, the responses generated in the left and right ears are different unless the bat is exactly in the plane that vertically bisects the body of the moth. The moth uses the difference in sound intensity to determine the direction from which the bat is approaching.

Night-flying moths and other insects hunted by insectivorous bats have simple ears that can respond to sounds in the ultrasonic range. Noctuid moths have ears with only two sensory neurons. These neurons respond when the moth is exposed to the cry of a bat that may be as far as 120 ft (37 m) away. The auditory neurons also allow the moth to determine the direction from which the sound comes.

How Insects Avoid Capture

Roeder and his colleagues also investigated how moths actually behave when they hear the cry of a hunting bat. They already knew from laboratory studies that the ears of moths could detect artificial bat cries broadcast through loudspeakers. By setting up speakers outdoors and broadcasting bat cries through them, they were able to determine how free-flying moths responded to the cries (see Box 19-1). First, they found that with cries as loud as those of living bats, the moths' reactions begin when they approach to within 100 to 120 ft of the source of the sound. At greater distances, the moth showed no response to the sound. Second, they found that moths show two types of behavioral responses to the bat cries. When the moth is in the range of 120 ft to about 20 ft from the sound source when the cries are turned on, the moth turns and flies directly away from the sound. However, when moths are within about 15 to 20 ft of the sound when it comes on, they perform quite variably and unpredictably. Moths usually execute a wild series of seemingly random loops and turns. Sometimes, however, they show no response, or else they fly down toward the ground, fold their wings and drop to the ground, or sometimes (but rarely) turn and fly directly toward or away from the sound.

These behavioral responses are best understood in the context of a bat's hunting behavior. A bat cannot detect the presence of a moth beyond about 15 to 20 ft because at this distance the echo from such a small object is too weak for the bat to hear. Furthermore, bats rarely fly in one direction for more than a few feet when they are hunting. Instead, they swoop and turn as they fly. For this reason, the best strategy for the moth when it hears a bat beyond about 20 ft is to turn and fly directly away from it. The bat is unlikely to keep flying on that course long enough to detect the moth. On the other hand, closer than 20 ft or so the moth has no hope of escape on a straight path. The bat can fly much faster than the moth and will inevitably catch it in a straight race. Hence, the best strategy for a moth that hears a bat close by is to use its superior maneuverability to escape by engaging in a series of sharp, unpredictable loops, turns, dives, or similar maneuvers. Such maneuvers increase the chance that a moth will escape (Figure 19-2A), but they are no guarantee (Figure 19-2B). Furthermore, sometimes moths that do not change the direction of flight also escape. However, after analyzing over 400 encounters between bats and moths, Roeder found that for every 100 moths that escaped after evasive maneuvers, only 60 escaped without such maneuvers, indicating the high survival value of evasive behavior by noctuid moths.

Bats eat other insects besides moths. Crickets and lacewings are classified in different orders than are moths. Some species of both these types of insects fly at night and may be captured by bats. They also show behavioral responses to ultrasonic batlike

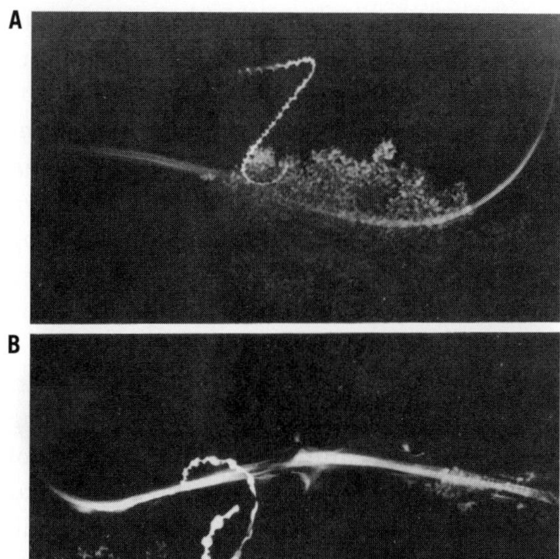

FIGURE 19-2. The interactions of moths and bats. Captured moths were tossed into the air in the vicinity of cruising bats. In these photographs, the track of the moth appears as a fluttery path and that of the bat as a blurry streak. (A) The moth dives as it hears the bat entering from the left and at the last instant makes a sharp turn to escape capture. (B) The moth attempts an evasive maneuver to avoid the bat flying in from the right. Since the track of the moth's flight terminates on the track of the bat, this moth did not escape.

sounds. For example, female Australian field crickets tethered and allowed to fly in a windstream respond to sounds of 30 to 70 kHz by turning away from the sound source. This is not merely an avoidance of any sound; they turn toward the sound source if the sound frequency is 3 to 9 kHz, the frequency of the calling song of males ready to mate.

Lacewings also turn away from ultrasonic sounds when sound intensities are low. However, when a bat is close to a lacewing (within 6 ft, or just under 2 m), and sound levels are presumably quite high, the insect employs a different strategy—it folds its wings and drops to the ground. If the bat detects the lacewing and pursues it as it falls, the lacewing often employs a second strategy, flicking out its wings for just a moment as the bat closes in, in what has been termed a "last-chance" response. This has the effect of suddenly slowing the insect's rate of fall, so that the bat may miss the target. Studies have shown that the behavior is triggered by the characteristic high-frequency "buzz" of ultrasound that bats emit when they are homing in on prey.

The emphasis in this section has been on evasion as an effective strategy for an insect to avoid capture by a bat. However, some insects use different strategies. For example, the dogbane tiger moth is known to emit ultrasonic sound pulses of its own. Bats have been observed to break off pursuit of this species of moth when the moth emits its own sounds, although the reason for this is not entirely clear. The sounds may warn off the bat by advertising that the moth is unpalatable. An alternative suggestion is that the sounds are actually used to interfere with the ability of the bat to locate the moth during the final stages of its attack. Laboratory experiments with restrained moths have shown that the moth times its sound production to the terminal rather than the early phase of the bat's attack. A warning seems likely to be more effective if it is given early rather than late, but no direct experimental evidence to show that bats are confused by the sound pulses has yet been obtained. The issue is likely to remain unresolved until researchers conduct difficult experiments in which free-flying bats can interact with free-flying moths.

Flying noctuid moths behave differently when they hear distant bat cries than when they hear nearby ones. A moth responds to a distant bat by turning and flying away from it, but it responds to a nearby bat by executing a series of random loops and turns or by diving to the ground. Some night-flying insects show other behavioral responses to bat cries; some of these are variants of evasion strategies, and some do not involve evasion.

BOX 19-1

Watching Physiology in Action: Investigating Physiology in a Natural Setting

Physiological investigations that involve studies of behavior or of behaviorally relevant sensory input are common in neurobiology. The hallmark of neuroethological studies, however, is their investigation of a behavior in a setting in which the operation of the system under study can be evaluated in the context of the environment in which the animal normally finds itself. This allows as much expression of the natural behavior as possible. Studies that take physiology out of the laboratory and into the realm of natural history, however, are not easy to carry out. It isn't enough just to observe an animal in a natural setting. A neuroethologist will also investigate the physiology of a behavior while minimizing intrusion on the behavior.

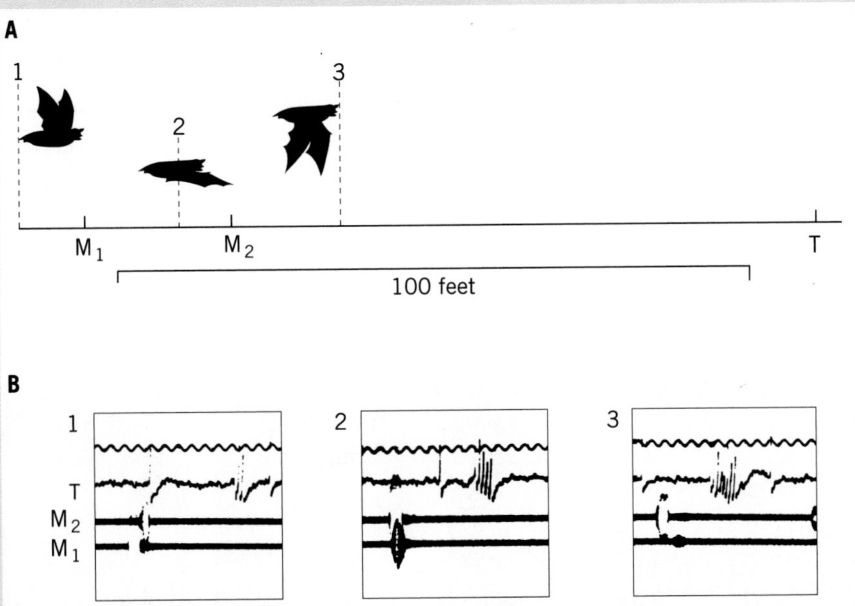

FIGURE 19-A. Estimating the maximum range at which a moth can hear a bat. (A) Diagram of the recording arrangement. A moth from which tympanic (auditory) nerve activity was recorded was placed in an open field (site T, right). Two microphones (M1, M2, at left) were placed some distance away to record bat cries. As a bat (silhouettes) flew toward the moth, the responses of the moth's auditory nerve to the bat's cries were recorded, along with the cries themselves. (B) Neural activity in the auditory nerve of the moth (second trace, T) and the cries of an approaching bat as recorded at the two microphones (third and fourth traces, M2 and M1). The top trace is a time marker. The panels show that the acoustic nerve responds with more spikes as the bat draws nearer (position 3). The position of the bat can be estimated by the times at which the bat's cries are recorded at the two microphones. At position 1, microphone M1 picks up the cry first; at position 3, microphone M2 picks it up first.

(continued)

Watching Physiology in Action: Investigating Physiology in a Natural Setting *(continued)*

Most neuroethological experiments are actually carried out in a laboratory, but experimenters try to allow the animal enough freedom to behave naturally. Kenneth Roeder, in his pioneering studies on the mechanisms by which night-flying moths evade predation by bats, achieved the objective of studying in a natural setting the physiology underlying an animal's behavior.

Roeder and his coworkers knew from laboratory studies that certain moths could hear the echo-locating cries of bats. There were, however, two important questions to which they did not at first have answers: How sensitive are the moths' ears, and what do moths do when they hear a bat? To answer the first question, Roeder and his students (students are always called upon to do the heavy work) hauled several hundred pounds of electrophysiological equipment, plus electric generators, out into the Massachusetts countryside. This was no small feat in the days before integrated circuits and portable equipment. There they captured a moth and set it up to record the electrical activity of the auditory neurons. At the same time, they set out two microphones with which they could capture the ultrasonic cries of cruising bats. By comparing the time at which a bat cry registered in the microphones and the time at which neural activity of the nerve was recorded, they were able to estimate the maximum distance at which the moth could hear the bat (Figure 19-A).

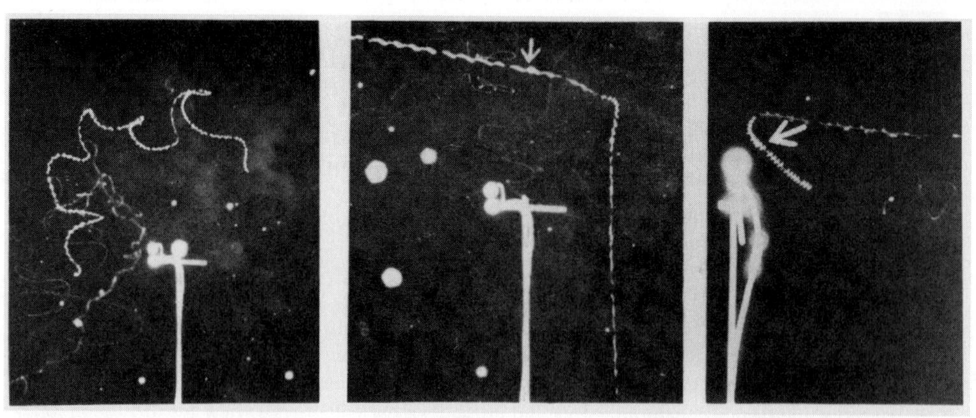

FIGURE 19-B. The paths taken by night-flying moths in the presence of batlike ultrasonic pulses emitted by the speaker seen in outline in each panel. The shutter of the camera was opened when a moth flew nearby, after which an ultrasonic sound pulse was switched on (arrows). Left panel: no sound; middle panel: high-intensity sound; right panel: low-intensity sound.

To answer the second question, Roeder used an open field to set up a loudspeaker through which he could play recordings of the cries that bats make when they hunt. Then he turned on a spotlight and waited until a moth flew close to the loudspeaker. When he turned on the recorded bat cries, he watched what the moth did. By opening the shutter of a camera, he could obtain a record of the moth's response, since the moth would show up as a fluttering streak in the photograph (Figure 19-B). These experiments gave a dramatic visual demonstration of the possible effect of the ultrasonic cries of a bat on the behavior of flying moths.

Although Roeder's experiments in the field are among the most striking examples of the neuroethological approach, they are by no means unique. Whether the subject is a moth escaping from a flying bat or a barn owl hunting a mouse (see Chapter 12), the best neuroethological work always includes a healthy component of behavioral observation that gives context and meaning to the physiological results.

The Neural Basis of Avoidance in Moths

The sensory system of the moth is responsible for detecting the cry of the bat, and the motor system is responsible for producing a particular response. The main question is how the integrating system, that is, the central nervous system of the moth, processes the sensory information and generates the appropriate motor response. The answer to this question is still not entirely clear, but studies by Roeder and others on the actions of neurons in the central nervous system of moths that exhibit evasive responses have provided some clues.

Axons of both A1 and A2 auditory neurons enter the last thoracic ganglion of a moth (Figure 19-3A). A1 neurons branch extensively in both of the thoracic ganglia and even send axons into the brain. A2 neurons branch in the ganglion they enter but do not project anteriorly into the first thoracic ganglion and do not project to the brain. Interneurons that receive input from the auditory neurons, and hence will respond to ultrasound, are called **acoustic interneurons**. Overlap between the incoming auditory fibers and various acoustic interneurons occurs in two local regions of the thoracic ganglia (Figure 19-3B). This is one place where integration of information about bat cries takes place.

Several types of acoustic interneurons have been described in thoracic ganglia of noctuid moths. Relay neurons spread information about the stimulus to regions of the central nervous system not reached by the primary acoustic fibers. These neurons are generally tonic and respond to sustained sound with sustained activity, just like the A cells do (Figure 19-4A) but with more rapid adaptation and for a somewhat shorter duration. Pulse marker neurons are phasic and respond to a brief pulse of ultrasound with only a single spike or two, independent of the intensity or duration of the stimulus (Figure 19-4B). The latency of the response in these pulse marker neurons varies with the intensity of the stimulus, from about 20 msec for weak stimuli down to 6 msec for loud sound pulses.

Different interneurons probably have quite different functions. Relay neurons, for example, because they pass on a signal quite similar to that of the primary auditory neurons, likely function mainly to spread information about a sound stimulus quickly and widely throughout the thoracic ganglia and up to the brain. Pulse marker neurons, on the

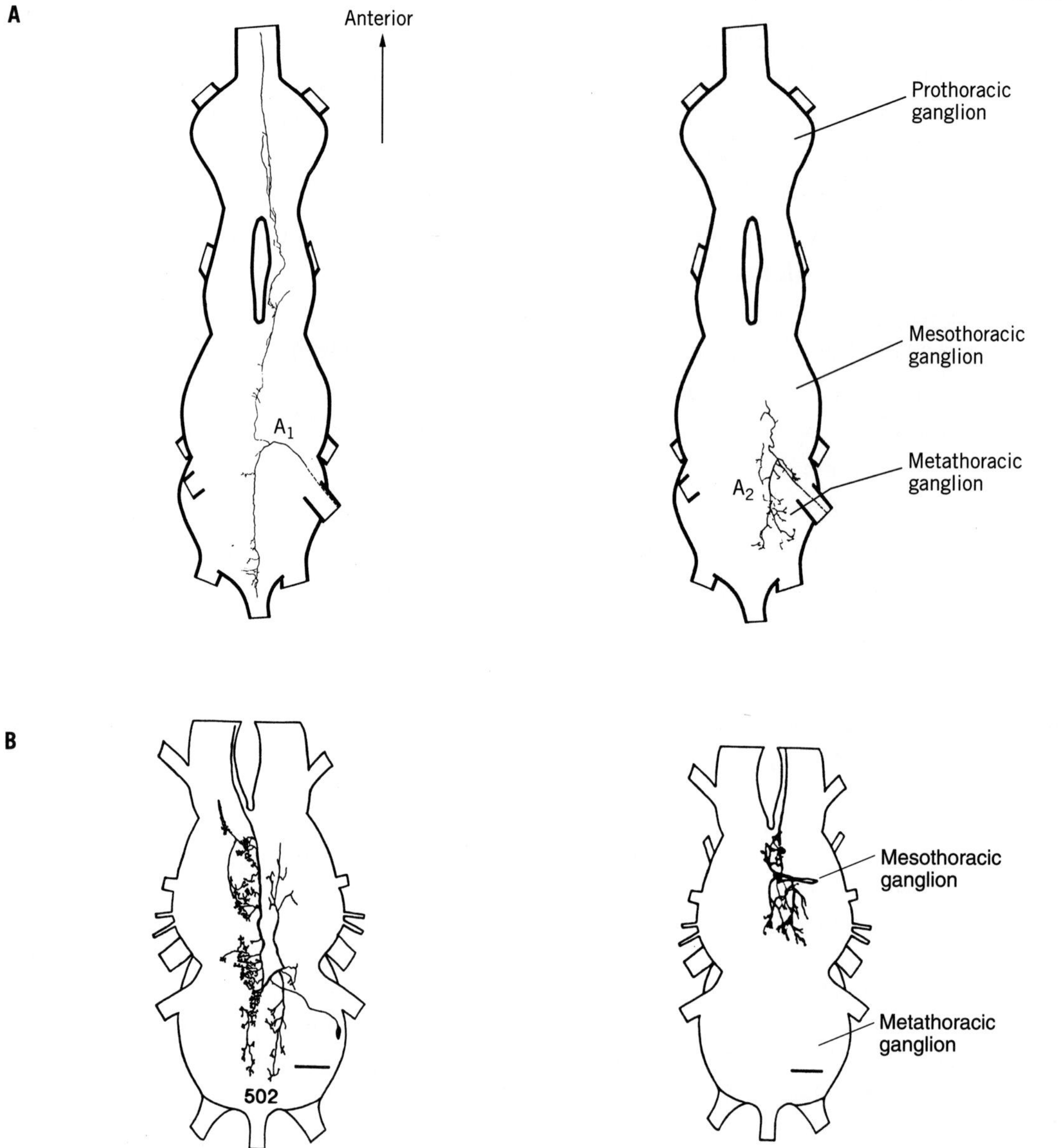

FIGURE 19-3. Part of the neural substrate in the thoracic ganglia for evasive flight in night-flying moths. (A) The location in the thoracic ganglia of the two acoustic neurons from the ear of a noctuid moth, showing the acoustic association areas where they overlap with the acoustic interneurons. For clarity, only cell A1 is shown on the right, and only cell A2 on the left, although both cells are present on both sides of the animal. (B) Two typical acoustic interneurons (from a different species of moth) that receive input from the acoustic neurons, showing their distribution in the nervous system. Only one of the pair of these neurons is shown. The horizontal bars represent 100 μm.

other hand, probably help determine the direction from which a sound stimulus originates because the time they fire relative to the onset of the stimulus is strongly influenced by the intensity of the stimulus. Together with other interneurons, these cells deliver excitatory and inhibitory signals to motor neurons of the flight muscles, exciting those muscles

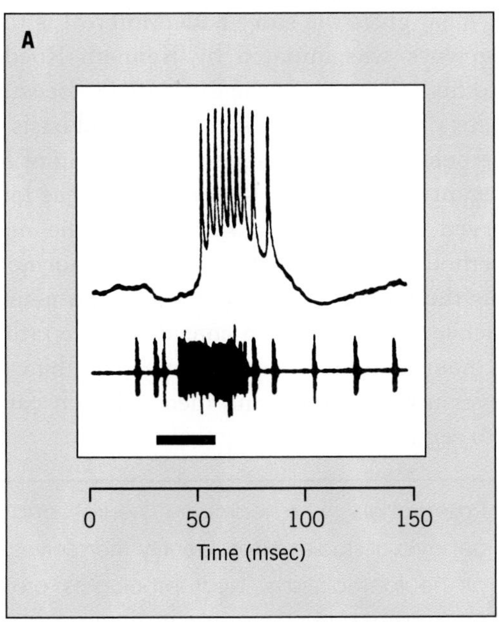

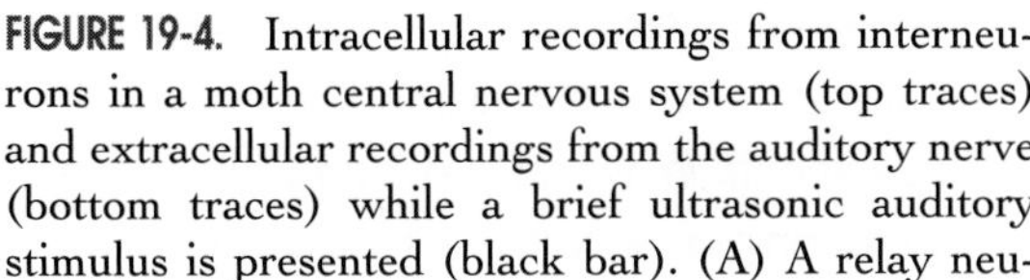

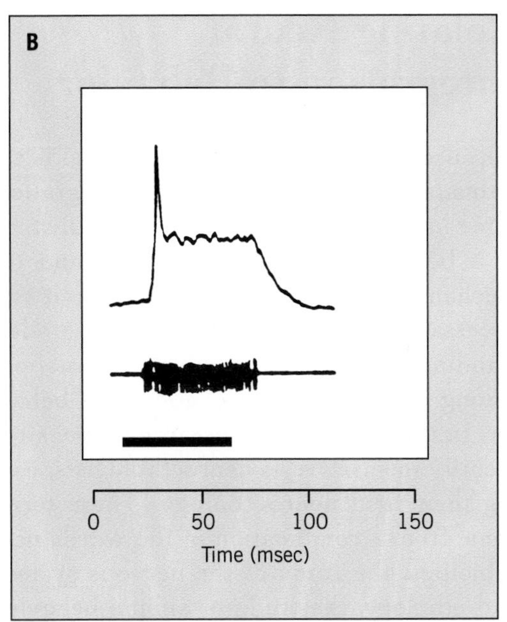

FIGURE 19-4. Intracellular recordings from interneurons in a moth central nervous system (top traces) and extracellular recordings from the auditory nerve (bottom traces) while a brief ultrasonic auditory stimulus is presented (black bar). (A) A relay neuron, which gives a tonic response to an ultrasonic stimulus. (B) A pulse marker neuron, which gives a phasic response, a single spike, to an ultrasonic stimulus. Note that the neuron remains depolarized for the duration of the stimulus.

whose action will help turn the moth away from the stimulus and inhibiting those that would oppose such action.

Knowing the properties of acoustic interneurons is only one step toward fully understanding evasive behavior. It is still necessary to determine how a directed turn away from the cry of a bat or unpredictable evasive behavior is evoked by these and other interneurons. Unfortunately, little direct experimental evidence is yet available to explain how evasive turns are generated. Roeder found two types of brain interneurons that respond to ultrasonic stimuli. He thought these interneurons might help a noctuid moth determine whether a turn away from a stimulus or a series of unpredictable loops and turns might be the most effective response, but the mechanism he proposed was entirely speculative. Working with crickets, other researchers have identified brain interneurons that seem to integrate sound input with flight behavior to influence steering. These interneurons respond to ultrasonic acoustic input, but more strongly when the insect is flying than when it is on the ground, and only to sound in the frequency range of bat cries. They send axons out of the brain to the thoracic ganglia, which contain the flight motor circuitry (see Chapter 16); severing the part of the nerve connectives in which these axons lie considerably weakens the turning response of a flying cricket.

Several types of neurons in the central nervous system of insects respond to the ultrasonic cries of bats. In noctuid moths, these include relay neurons, which seem to relay information about bat cries to different locations within the nervous system, and pulse marker neurons, which may help determine the direction from which a sound originates. Some brain neurons in moths and other insects also respond to bat cries. However, the precise neural mechanisms that bring about the evasive behavior of an insect are not known.

Relating Neural Integration to Behavior

Because the overt action of an animal is the ultimate expression of neural integration, some investigators believe that careful study of a behavior, its natural context, and the mechanisms that are responsible for it is a necessary first step leading to an understanding of the way in which interactions among neurons can bring about the behavior. In the 1970s, these researchers began to identify themselves as neuroethologists, calling their field **neuroethology**. These terms come from a combination of the words neurobiology, the study of the nervous system, and ethology, the study of animal behavior. The hallmark of research in the field is a concentration on the neural basis of biologically important behavior such as communication and reproduction, in addition to prey capture and escape. Neuroethologists generally conduct studies in which the behavior of the whole animal is explicitly considered in their experiments, in which the animal retains at least limited freedom of movement, and, if sensory stimuli are used, in which these stimuli are chosen so as to be similar to the stimuli encountered by the animal in nature.

How insects evade predation by bats is a perfect subject for a neuroethological study because it includes all the elements that exemplify work in this field. Predator avoidance is obviously a biologically significant behavior. The mechanisms that allow an insect to detect a bat and take appropriate evasive action are amenable to physiological investigation. Furthermore, there is great scope for study of the interactions of prey and predator, both in the laboratory and outdoors in the field, so the physiological work can be interpreted in an appropriate behavioral context (Box 19-1). Because of the dramatic nature of interactions between bats and night-flying insects in nature, and the elegant series of experiments carried out on these interactions, the bat-insect story has long been considered a classic of neuroethological research.

What gives the story added interest is that the work was initiated by Kenneth Roeder and his colleagues in the early 1960s. Because of his particular interest in the neural basis of the behavior of animals, to say nothing of his elegant experiments, Roeder's work has long served as the highest standard of the neuroethological approach. In fact, so enormous was the impact of this man's ideas on neuroscience that he is often considered the father of neuroethology, even though he himself never used the term and died before it came into regular use.

Researchers who combine neural and behavioral studies often identify themselves as neuroethologists. Neuroethologists are often especially interested in the neural basis of a behavior like escape, which is important to the survival of the animal, and conduct experiments in which the behavior of the whole animal is taken into account. The escape behavior of night-flying moths has been studied extensively from a neuroethological point of view.

The Flexibility of Escape Behavior

Even though some forms of predator avoidance can be complex, as in the case of insects avoiding bats, most escape behavior seems not only simple but so stereotyped that it could be considered a reflex action. A closer analysis of escape in various animals, however, reveals that the behavior is often remarkably flexible and well adapted to the stimuli that elicit it. This flexibility is the result of specific processes of neural integration in the central nervous system. It is manifested in the directedness of escape movements and in the integration of escape behavior with whatever else the animal may be doing at the time it is elicited.

Directed Escape

The need to direct escape *away* from a triggering stimulus is obvious. You have already learned that directed escape occurs in crayfish, in which touch to the anterior part of the body elicits backward swimming and touch to the tail elicits upward swimming (see Chapter 18). Earthworms and other annelids exhibit similarly directed escape. Touch around the head causes withdrawal toward the tail, and touch around the tail causes withdrawal toward the head. In both crayfish and earthworms, the neural basis of directed escape is differential activation by appropriate stimuli of one of the large-diameter axons known as giant interneurons (GIs), which synapse with sets of motor neurons controlling muscles that produce the escape movement. In both animals, anterior stimuli excite one GI pathway, which excites a set of muscles whose contraction moves the animal in one direction, whereas posterior stimuli excite another GI pathway, exciting muscles whose contraction moves the animal in another direction.

Escape may also be directed in a more precise fashion. Many animals make a rapid but highly controlled turn away from a threatening stimulus. Frogs, for example, will turn from a large, rapidly approaching object, then jump. The angle of the turn depends on the angle of the approaching object relative to the frog. Stimuli approaching the frog from one side and behind elicit small turns, whereas stimuli approaching from one side and the front elicit large turns (Figure 19-5).

Fish show a similarly directed escape behavior, often called a startle response. An acoustic or vibrational disturbance in water elicits a powerful, rapid contraction of the body muscles of a fish; the latency between stimulus and response may be less than 10 msec for some fish (e.g., trout, goldfish). The response is mediated by the paired Mauthner cells (see Box 18-1), the large, identifiable neurons present in the brains of fish and amphibians. These neurons receive considerable synaptic input from sense organs. When a strong, sudden stimulus is detected on one side of a fish, the Mauthner cell on the stimulated side is activated. Action potentials in the Mauthner cell have two effects. First, they inhibit the contralateral Mauthner cell by an electrical process that does not involve chemical inhibitory synapses.

A

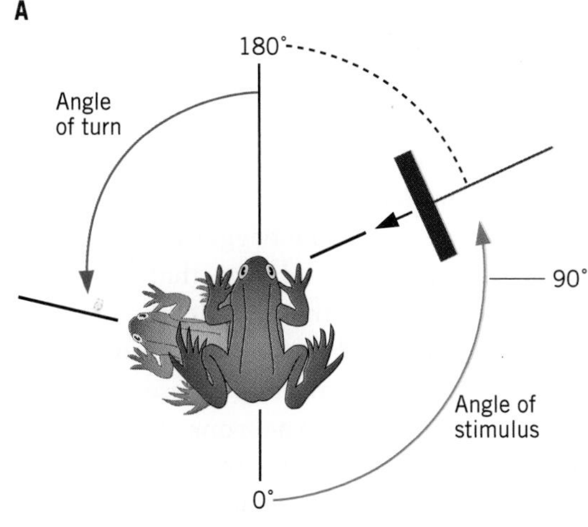

B

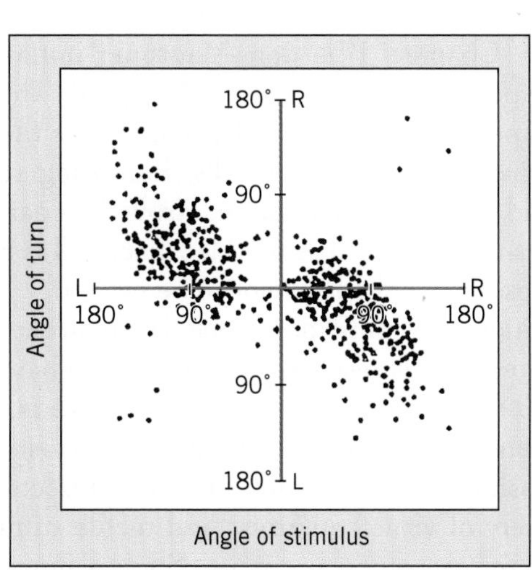

FIGURE 19-5. Frogs turn at an angle away from a threatening stimulus according to the angle at which the stimulus approaches. (A) Diagram showing how the stimulus and turn angles are measured. (B) Plot of the angle of turn as a function of the angle at which a stimulus is presented. The direction of the turn is strongly negatively correlated to the direction from which the stimulus is presented, meaning that the frog turns away from the stimulus.

Second, as they sweep down the axon of the Mauthner cell, they activate motor neurons that innervate body musculature. Since each Mauthner cell sends its axon down the spinal cord on the side opposite the side in which the cell body is situated, the result is a strong contraction of body muscles along the side of the body opposite the stimulated side. This produces a turn away from the stimulus. Although the contractions of body muscles are extremely rapid and powerful, they are nevertheless highly coordinated and will direct the fish away from the stimulus.

In some cases, parallel neural circuits mediate directed escape. For example, if the tactile stimulus to a crayfish is not too sudden or strong, escape can be evoked without activating the GIs. In this case, the escape swim is not just up or straight back but is more precisely directed away from the stimulus. If a crayfish is touched lightly on the right side of its head, it will swim back and to the left rather than straight back. The nongiant escape system operates in parallel to the giant-mediated escape system, activating the appropriate muscles through motor neurons that allow finer control over the response than the all-or-none fast response allows. Even in the acoustic startle behavior of fish, which like crayfish escape has been designated by some to be a behavior driven by command neurons (see Chapter 17), non–Mauthner-initiated escape responses have been described. These escape reactions can be about as fast as those mediated by Mauthner cells, indicating that rapid escape in fish is also controlled by parallel pathways and that the Mauthner cell is not necessary for the behavior to be evoked.

Such flexibility of escape is apparent in many animals, especially when the behavior used in escape is also used when there is no urgency. Consider the jump of a locust or grasshopper, for example. In these insects, a variety of visual, auditory, and tactile stimuli can trigger an escape jump. Such a jump is typically quite strong and will carry the insect some distance away from the stimulus. However, locusts and grasshoppers also use more controlled jumps simply to move from one place to another and to initiate flight.

Jumping involves several stages (Figure 19-6). First, the shank of each hind leg, the tibia, is flexed against the femur. Then the muscles that extend and those that flex the tibia are contracted simultaneously. Because of the geometry of the muscles and the leg, the relatively weak tibial flexor muscles have a significant mechanical advantage over the strong extensors, and hence can keep the leg flexed when both contract. The insect jumps by suddenly inhibiting the flexor motor neurons, leaving the strong extensors unopposed and forcefully extending the tibiae.

Early studies of jumping seemed to suggest that two identified interneurons are especially important in controlling the process (Figure 19-7). One, designated the C neuron, helps to ensure the cocontraction of extensor and flexor motor neurons. The other, the M neuron, which responds to strong auditory or tactile stimuli, seemed to be responsible for triggering the jump by strongly inhibiting the flexor motor neurons. The C neuron was believed to be activated first, essentially cocking the rear legs; then, if the stimulus were strong enough, the M neuron would fire and trigger the jump.

Subsequent study has shown that jump initiation is not so simple. It is possible to suppress activity in the M neuron by passing negative current through an intracellular microelectrode inserted into it, hence hyperpolarizing the cell. In spite of suppression of M neuron activity, a jump reaction can still be elicited. This should not be possible if the M neuron were the sole trigger for jumping.

Further study has shown that there are additional pathways that parallel the inhibitory pathway from M to the flexor motor neurons shown in Figure 19-7. These pathways involve additional interneurons that can also trigger a jump, apparently by turning off the neurons that excite the flexor motor neurons in the first place. Removing the excitation of the flexor motor neurons after the leg is flexed, and while the flexor and extensor muscles are cocontracting, would have the same effect as inhibiting the flexors and hence would initiate a jump. As in the case of the nongiant pathways for swimming in crayfish,

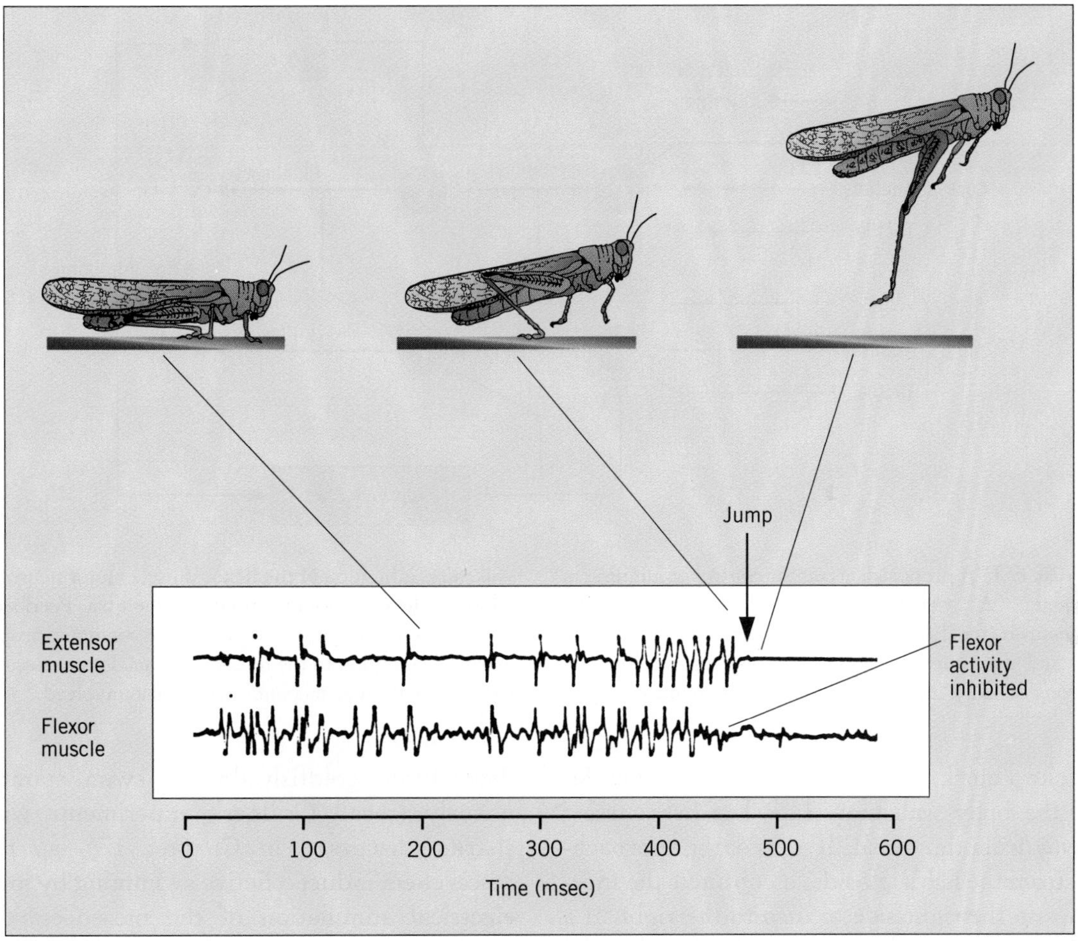

FIGURE 19-6. Jumping in a locust. The locust prepares to jump by flexing the tibiae of the hind legs against the femurs. The extensor and flexor tibiae muscles are then cocontracted, as shown in the electrical records below the drawings, which show electrical activity in the extensor and flexor muscles. Because of the mechanical advantage of the flexor muscles, this activity is sufficient to keep the leg flexed, even though there are several motor impulses to the much stronger extensor muscles. The jump is initiated when the flexor muscles stop contracting, allowing the extensor muscles to straighten the legs.

jumping elicited without evoking the M cell seems to be more variable than that in which the M cell is used.

Escape behavior is often viewed as rigidly stereotyped, but it can be quite variable. Many animals that exhibit escape also have mechanisms to ensure that escape is directed away from the source of the triggering stimulus.

Integrating Escape with Ongoing Behavior

It is one thing for an animal to coordinate escape movements when it is threatened while resting because its only objective is to move away from the threat as rapidly as possible. It is another matter for an animal to coordinate an escape when it is already doing something else. The need for integration of escape behavior with an ongoing activity is especially acute when the animal is already engaged in a form of the behavior used for escape. For example, in a swimming fish, a threatening stimulus will

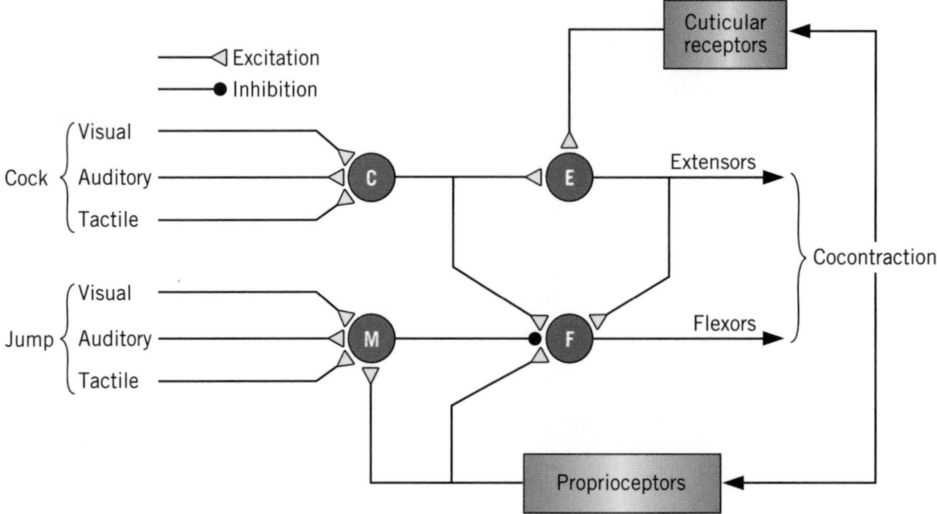

FIGURE 19-7. A proposed circuit for control of jumping in locusts. A variety of sensory input to the C neuron (top, left) will cause cocontraction of the flexor and extensor muscles of the rear legs. A further, strong sensory input will excite the M neuron, which will cause inhibition of the flexor muscle motor neurons, allowing the extensors to straighten the tibia. Feedback from sense organs in the legs helps to maintain proper levels of excitation to the motor neurons. Later research suggests that other mechanisms are also involved.

evoke a quick, powerful contraction of muscles in the body wall that sends the fish darting away from the threat. If a predator approaches from the fish's left side, it contracts the muscles on the right side to turn to the right. If a predator approaches from the fish's right side, it contracts the muscles on the left to turn to the left. Suppose, however, that the stimulus arrives from the right side just at the time during the swimming cycle that the fish is about to contract the muscles on that same side. Without some mechanism for coordination of the ongoing swimming rhythm with the motor pattern that evokes escape, the result could be a brief moment of paralysis of the fish as it contracted the muscles on the right and left sides of the body simultaneously.

The question of how an animal is able to integrate a vigorous escape movement with ongoing locomotion has been investigated in several species. A particularly informative series of experiments on escape swimming has been conducted by Kurt Svoboda and Joseph Fetcho on goldfish. To conduct an experiment (Figure 19-8), the researchers prepared a fish by removing the telencephalon, paralyzing the fish with curare, and inducing fictive swimming. Unlike sharks, decerebrate goldfish do not swim spontaneously (recall Grillner's experiments with sharks, discussed in Chapter 17), so the researchers induced fictive swimming by mild electrical stimulation of the mesencephalic locomotor center (see Chapter 18). They monitored the swim motor pattern by recording from spinal motor nerves. Finally, they inserted a microelectrode intracellularly into one of the Mauthner cell axons so they could stimulate the cell at specific times relative to the ongoing swim rhythm.

The results of these experiments show that the Mauthner neuron has two separate but related effects in the fish and reveal two functions of the neurons. First, a single action potential elicited in the axon of a Mauthner neuron causes strong excitation of the ipsilateral (same-side) body muscles. (At the site of stimulation the axon has already crossed over to the other side of the body relative to the position of the cell body, and hence is ipsilateral to the stimulated muscles.) This result, expected based on the description of action of the Mauthner neuron given previously, shows that the Mauthner neurons, or neurons that they excite, make synaptic contact with the motor neurons innervating the ipsilateral body musculature. In addition, an action potential in a Mauthner neuron also causes *inhibition* of the contralateral body muscles, indicating that the Mauthner neurons, or neurons that they excite, suppress activity in the motor neurons innervating contralateral body muscles. This pattern of excitation and inhibition will override any ongoing swimming activity of the body muscles.

The second function of the Mauthner cells is to reset the timing of the swim rhythm. This function is demonstrated by recordings of neural activity in motor neurons during swimming in which the pattern of the swim cycle is reset by stimulation of a single Mauthner neuron (Figure 19-9). The result shows that the Mauthner cells or neurons that they excite in the spinal cord must synapse with neurons that constitute the central pattern generators for swimming in the cord.

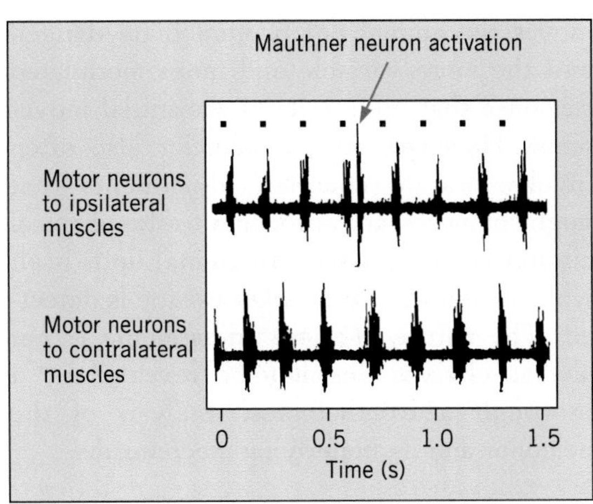

FIGURE 19-9. Excitation of body muscles and resetting of the swim rhythm in a swimming goldfish in which a single action potential is evoked in a Mauthner cell. The figure shows bursts of action potentials in motor neurons recorded from motor roots of the spinal cord. The arrowhead indicates the time at which the Mauthner neuron was activated. The small dots above the bursts show the timing of the motor pattern before the Mauthner neuron was stimulated. Note how the activity has shifted relative to the dots after the stimulation, indicating that the swim rhythm has been reset.

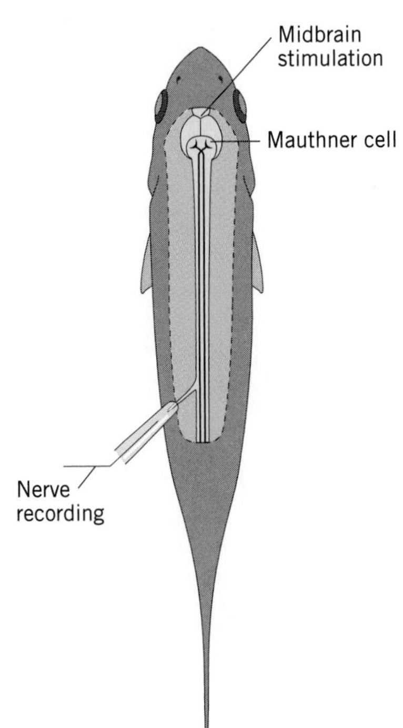

FIGURE 19-8. The setup for testing the effect of Mauthner neuron stimulation on fictive swimming in goldfish. The brain is exposed, and the cerebral hemispheres are removed. Electrodes are placed so that the mesencephalic locomotor region and a Mauthner neuron can be stimulated. The fish is paralyzed with curare, and recordings of swim motor patterns are taken from a spinal nerve.

The function of this interaction is probably to ensure a smooth transition from the violent initial escape flip to vigorous and repetitive swimming that will take the fish away from danger.

Escape movements like swimming are coordinated with ongoing behavior in at least two ways. The triggering stimulus inhibits ongoing muscle action and initiates new action appropriate to the desired escape, and it resets the rhythm of the ongoing behavior to produce a smooth transition from escape to rapid locomotion.

A Case Study of Escape

The emphasis in the previous sections of this chapter has been on two elements of escape behavior, the rapid initial component that moves the animal out of immediate danger, and the more variable and more modulated response that often follows the initial movement. However, escape behavior also often involves rather complex adjustments that might be necessary according to the physical circumstances in which an animal finds itself when a stimulus that evokes escape is detected. The nature of these adjustments is not always obvious but may be revealed by a thorough neuroethological analysis of the behavior and its underlying mechanisms.

Cockroach Escape

Consider, for example, the escape of a cockroach. As many an apartment dweller can attest, some cockroaches have a highly developed evasive system, scurrying to safety when the occupant tries to swat the interloper with a newspaper. Several decades before his work on the moth ear, Kenneth Roeder attempted to uncover the physiological basis of this escape system. Drawing on the work of others as well as his own, he suggested that cockroaches use the fine filiform hairs on the cerci to detect approaching danger. Cerci are the peglike appendages at the posterior end of the abdomen that can serve as primitive auditory receptors (Figure 19-10; see Chapter 12). The hairs on the cerci move when the air around them is disturbed by an approaching newspaper or other potentially lethal object. This movement stimulates sensory neurons connected to the hairs. The sensory neurons in turn make excitatory synaptic contact with a set of GIs that rapidly send their action potentials to the thorax, and there presumably excite the leg motor neurons that initiate running.

For decades after Roeder first described this escape system in 1948, the relationship between a known type of sense organ, a set of large and identifiable interneurons, and the escape behavior of the insect was considered one of the most elegant examples of the neural basis of a simple behavior. As time passed, however, it became apparent that the simple pathway from sensory fibers to motor neurons via GIs could not be the only basis for escape. It was not until thorough neuroethological studies of the interaction of a cockroach and a predator had been conducted that a satisfactory explanation for the behavior was finally developed.

There were two main difficulties with the original view. First, the delay between the arrival of a stimulus at the cerci and the first movements of the legs during an escape run was greater than researchers calculated it should be considering the duration of sensory transduction, the conduction time of action potentials in the GIs, and synaptic delay introduced by the synapses in the pathway. Considering that the circuit was thought to be arranged to minimize delay, this extra time, enough for several

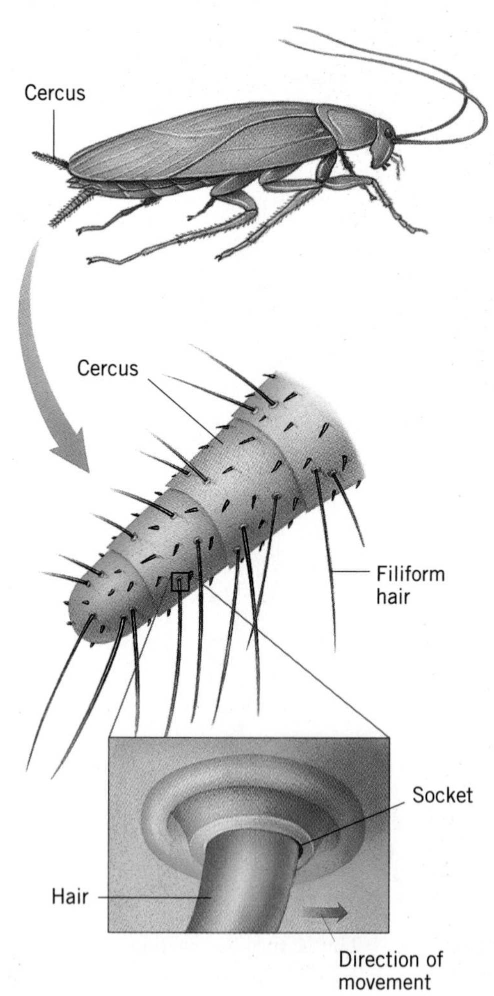

FIGURE 19-10. The common American cockroach, showing the cerci. The fine filiform hairs on each cercus sit in sockets that allow the base of the hair to move mainly in one plane.

more synapses to be present in the circuit, was paradoxical. Second, researchers were unable to find any monosynaptic connection between the GIs and motor neurons that control the leg extensor muscles, even though minimizing conduction time from the cerci to the motor neurons would seem to require just a single synapse between the GIs and the motor neurons. In addition, strong motor activity in the leg muscle motor neurons could be evoked by stimulating the cerci in animals in which the GIs had been destroyed, indicating the presence of parallel pathways for escape, a redundancy hard to explain in the accepted view.

An appreciation for the true complexity of the escape response of cockroaches did not emerge until the escape behavior itself was carefully reexamined by Jeffrey Camhi and his colleagues. These researchers showed that the insect did not merely run straight ahead to escape danger but instead turned away from the stimulus before it began to run. Toads, which are known to be natural predators of the cockroaches being studied, were put into an arena along with individual cockroaches. Films of encounters between the animals revealed that unless the cockroach is standing facing away from the toad, or walking directly away from it, the insect's first movement in reaction to a toad strike is a turn away from the toad's advancing tongue (Figure 19-11). Only after the cockroach has turned does it begin to run. Analysis of the relative positions of the toad's tongue and the insect as the latter turns showed that the turn usually has the effect of moving the animal out of the direct path of the tongue. Running straight ahead would not have the same effect, since the toad usually strikes at the middle of the cockroach's body and the cockroach will not have moved forward far enough to avoid the tongue entirely by the time the tongue reaches its position.

Experiments in which the hairs on the cerci were covered with petroleum jelly verified that the cerci are the "wind" detectors that trigger the turn. Animals whose cerci had been covered were always caught by the toads, whereas intact animals escaped more than 50% of the time. In one case, an insect with "covered" cerci escaped several times. The experimenters surmised that they had inadvertently left a small portion of one cercus uncovered. Unfortunately for science, the toad caught and ate the cockroach before the researchers could verify their hypothesis.

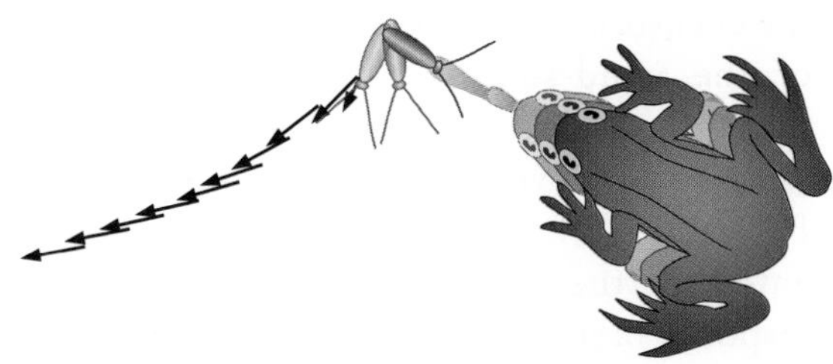

FIGURE 19-11. The escape turn and run of a cockroach in response to the strike of a toad. High-speed motion pictures were taken of an encounter between a cockroach and a toad. The frame in which the toad's tongue was first visible was designated zero, and the frames before and after this were assigned negative and positive numbers, respectively. The entire toad is shown in frame -1. In subsequent frames (0, 1), only the tongue and the front of the toad's head are shown. The first three positions of the body of the cockroach are shown corresponding to frames -1, 0, and 1. For subsequent frames, only arrows representing the position and orientation of the cockroach are shown. Note that the insect turns away from the approaching tongue before it begins to run forward.

Early study of escape in cockroaches suggested that wind-sensitive hairs on the cerci stimulated sensory cells, which in turn activated giant interneurons. These interneurons rapidly carried information to the thorax and directly triggered running. Later neuroethological work, however, showed that cockroaches turn before starting to run.

Physiological Basis of Cockroach Escape

If a cockroach is able to turn away from danger, it must have some way of determining the direction from which the threat is approaching. Camhi continued his neuroethological investigation of escape by determining how the cockroach can do this. Close examination of the wind-sensitive hairs on the cerci revealed that they are morphologically constrained to move only in a single plane (see Figure 19-10). Since not all the hairs are oriented in the same direction, some hairs respond best to wind from one direction, whereas other hairs respond best to wind from another direction. The directional selectivity of the cercal hairs is passed on to the GIs. Sensory afferents from hairs that respond to wind from just one direction tend to converge on just one or two of the GIs. (The cell bodies and dendritic fields of the GIs are shown in Figure 18-A.) Hence, different GIs respond most strongly when wind strikes the cerci from different directions. This obviously gives the insect information about the direction from which the air disturbance originates, and therefore a basis for determining which way to turn.

The directional selectivity of the cercal hairs and the consequent directional selectivity of the GIs, together can account for the insect's ability to recognize the direction from which an air stimulus comes and to turn away from it. However, these features do not explain the relatively long delay between the time that information about the stimulus arrives in the thorax and the time the legs start to move, nor the lack of direct synaptic connection between the GIs and the motor neurons controlling the leg muscles.

The extra delay is necessary for the insect to organize its escape movements according to the exact physical circumstances in which it finds itself. Since the cockroach must pivot in order to turn, it must take into account the starting positions of its legs as well as the direction of the air disturbance before it can direct the proper muscle action to cause a turn in the correct direction. This is because the consequence of any action of the muscles in a leg will depend on the starting position of the leg, as well as the direction from which the stimulus comes. For example, if the rear legs are in a neutral position, halfway between full flexion and full extension, they should both extend if the wind comes from the rear (Figure 19-12A). If the wind comes from one side, on the other hand, the leg closest to the stimulus should extend to pivot the insect away from the wind, while the other leg flexes (Figure 19-12B). Similarly for the other legs, the exact action that is appropriate will depend on the initial position of the leg and the direction from which the stimulus comes.

A cockroach "selects" the proper muscles to activate for the desired turn with the help of a complex network of more than 100 interneurons in the thoracic ganglia (the ganglia that contain the motor neurons controlling the leg musculature). The GIs synapse with these thoracic interneurons, which in turn synapse with other interneurons or with motor neurons controlling the leg muscles. The interneurons are strongly influenced by sensory input from the legs in such a way that the particular position of a leg influences the activation of the motor neurons innervating the muscles of that leg. For example, if a standing insect receives a wind stimulus from the right, it will turn to the left. If its right rear leg is already fully extended, it should flex the leg before extending it further. It will do so because the

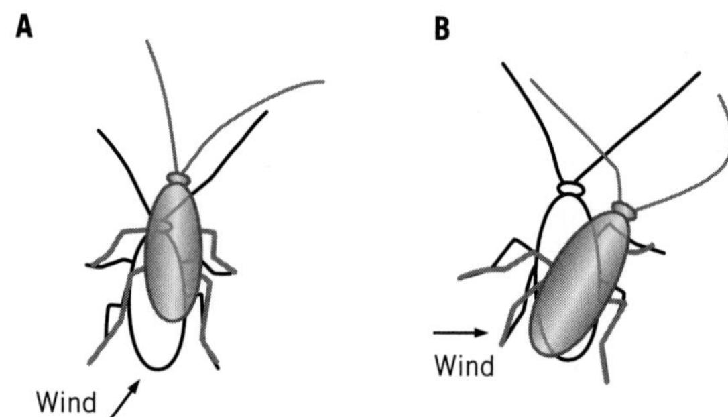

FIGURE 19-12. The actions of the middle and rear legs required to turn a cockroach in different directions, depending on the direction from which a wind stimulus comes. The front legs are omitted for clarity. The solid figure of the insect shows its position when the wind stimulus was first detected. The shadow figure shows the insect's position a few milliseconds later. Whether a leg flexes or extends depends on the direction of the turn and on the initial positions of the legs. (A) When the air disturbance comes mainly from behind the insect, both rear legs extend to move the insect forward. (B) When the disturbance comes from the insect's left, the left rear leg extends, but the right rear leg flexes in order to turn the insect to the right.

sensory feedback from the extended leg will bias the thoracic interneurons so as to inhibit the extensor and excite the flexor motor neurons. If the leg is initially flexed, the sensory input will bias the interneurons so as to excite the extensors instead. The synapses between the GIs and this network of interneurons are the source of the extra time required to initiate escape compared with the theoretical time.

The escape system of the cockroach has further complexities as well. For example, it has been shown that an escape response is still possible after the elimination of the GIs in the ventral nerve cord. Just as in the case of crayfish, cockroaches have a parallel escape system, one that is a little slower and more variable but does not depend on activity of any of the GIs. This system is strongly activated by touch, either to the antennae or to the body. It may be used to detect predators such as spiders, which do not have enough body mass to create an air disturbance that will activate the cercal escape system.

The discussion of the escape maneuvers of moths, cockroaches, and other animals has centered on the mechanisms by which an animal is able to orchestrate a proper response to a threatening stimulus. It may seem that the animal "decides" how to respond: Should the moth turn away from the bat or engage in unpredictable evasive action? Should the cockroach turn right or left? However, this kind of "decision" is quite different from your decision to skip a lecture. The neural mechanisms described here represent a kind of neural decision making that allows an animal to select a sequence of actions that are most likely to be effective in avoiding capture, but the "selection" is entirely automatic. The animal has no more choice in the matter than you do when you cough if a bit of pizza enters your trachea as you eat.

The turn and run of an escaping cockroach is orchestrated by complex neural pathways. The direction from which the stimulus comes is encoded in the pattern of action of the giant interneurons of the ventral nerve cord. The organization of the leg movements necessary to bring about the correct turn is done by a network of thoracic interneurons that takes into account the current positions of the legs.

Additional Reading

General

Camhi, J. M. 1984. *Neuroethology: Nerve Cells and the Natural Behavior of Animals.* Sunderland, Mass.: Sinauer.

Ewert, J.-P. 1980. *Neuroethology: An Introduction to the Neurophysiological Fundamentals of Behavior.* New York: Springer-Verlag.

Huber, F., and H. Markl. 1983. *Neuroethology and Behavioral Physiology: Roots and Growing Points.* Berlin: Springer-Verlag.

Roeder, K. D. 1963. *Nerve Cells and Insect Behavior.* Cambridge, Mass.: Harvard University Press.

Young, D. 1989. *Nerve Cells and Animal Behaviour.* Cambridge: Cambridge University Press.

Research Articles and Reviews

Heiligenberg, W. 1991. The neural basis of behavior: A neuroethological view. *Annual Review of Neuroscience* 14:247–67.

Hoyle, G. 1970. Cellular mechanisms underlying behavior: Neuroethology. *Advances in Insect Physiology* 7:349–444.

Miller, L. A., and J. Olesen. 1979. Avoidance behavior in green lacewings: I. Behavior of free flying green lacewings to hunting bats and ultrasound. *Journal of Comparative Physiology. A: Sensory, Neural, and Behavioral Physiology* 131:113–20.

Ritzmann, R. E., and A. J. Pollack. 1990. Parallel motor pathways from thoracic interneurons of the giant interneuron system of the cockroach, *Periplaneta americana. Journal of Neurobiology* 21:1219–35.

Roeder, K. D. 1970. Episodes in insect brains. *American Scientist* 58:378–89.

Svoboda, K. R., and J. R. Fetcho. 1996. Interactions between the neural networks for escape and swimming in goldfish. *Journal of Neuroscience* 15:843–52.

CHAPTER 20

Analysis of Simple Behavior

Understanding the neural basis of any behavior is a significant challenge. One way to meet this challenge is to analyze behavior in terms of functional components that are easier to understand than the whole behavior, and then reassemble the whole behavior from those components. In this chapter, you will read about several approaches to the study of behavior that have been useful in the analysis of the mechanisms that underlie it.

Because even a simple behavior involves many neural circuits that work together to make neural "decisions,"obtaining a complete description of the neural basis of any behavior is a formidable task. This task can be made more manageable by breaking the behavior into functional subunits that can be analyzed separately. Furthermore, in some cases, the neural circuits that underlie a behavior can be treated as control circuits operationally similar to man-made circuits that are widely used to regulate everything from the temperature in a house to a complex manufacturing process. Analyzing a neural control circuit in familiar terms usually makes both the circuit and its operation easier to understand, and hence helps researchers better understand the whole behavior of which it is a part.

Biological Servomechanisms

Many artificial control circuits use feedback to help control a process. Feedback is an engineering term for the transfer of a copy of an output signal to some controlling device. A similar use of sensory feedback is common in biological systems in cases where the nervous system is controlling directed movements or goal-oriented events (see Chapter 17). When sensory information is used for feedback, the sensory signals become an integral part of the mechanism for the behavior being controlled, and the mechanism can be analyzed as a **servomechanism**, a man-made device that uses feedback to regulate output. Such an analysis is often helpful for understanding how the neural circuit functions.

Principles of Servomechanisms

Servomechanisms can be simple devices, consisting of just three main components (Figure 20-1A). An effector produces a desired action, a sensor measures the action, and a comparator checks the output of the sensor against some desired and preestablished set point. The comparator "subtracts" the sensor signal from the set point, in a process known as **negative feedback**. When the output measured by the sensor is equal to or greater than the set point, the effector is turned off.

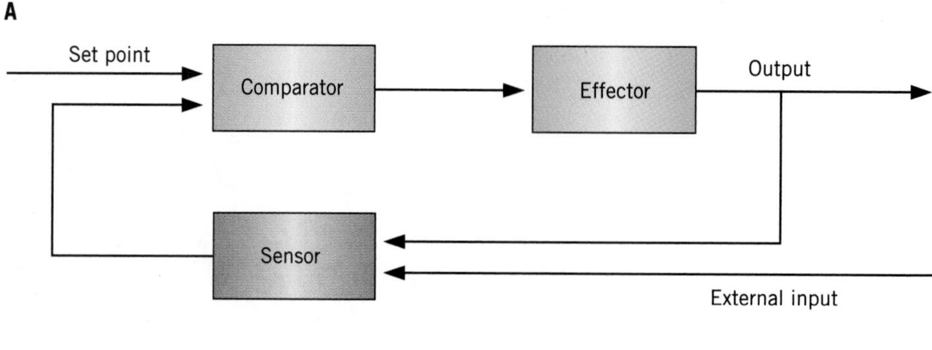

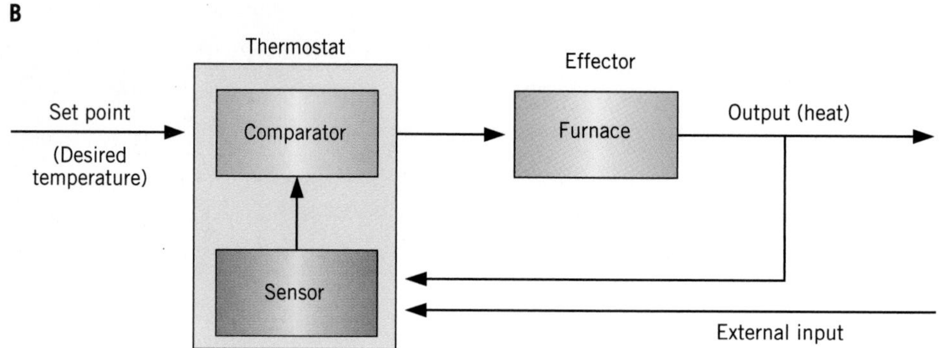

FIGURE 20-1. The organization of a servomechanism. (A) A generalized feedback loop. (B) Loop based on control of a household furnace. A set point (setting the thermostat) establishes the initial condition of the output produced by the effector (the furnace). The output (heat) is fed back to the comparator (the thermostat), which determines whether to turn on the furnace to generate more heat. The functional components of a feedback loop are indicated by distinct colors in this and subsequent diagrams.

External input can also affect the sensor and hence can influence the effector directly.

A familiar example of a servomechanism is the thermostat and the heating and cooling systems in your house (Figure 20-1B). Consider a furnace first. The furnace is the effector, and its output is heat. The sensor and the comparator are functionally combined in the thermostat, and the setting of the thermostat is the set point for the furnace. When the temperature of the air around the thermostat drops below the set point, the furnace turns on. This causes the temperature of the air to rise. When the temperature of the air reaches the set point of the thermostat, the furnace shuts off. As the air cools, the temperature will again drop below the set point, and the cycle will repeat. The external input represents a change in temperature that is not caused by the furnace, such as the slow leak of heat through windows or a sudden drop in temperature if someone holds the door open on a cold winter day. An air-conditioning unit works similarly, except here the effector turns on when the thermostat registers a temperature *above* the set point, and the effector cools rather than heats the air.

Figure 20-1 illustrates a general diagram of a servomechanism, as well as one specific example of such a device. These diagrams represent most, but not all, types of servomechanisms. In the case of the furnace, two functions, the sensor and the comparator, are incorporated into a single device, the thermostat. Even though there is only one device, both functions are nevertheless fulfilled. Furthermore, there is usually more than one way a particular function can be carried out. The furnace operates discontinuously, being either on or off. The generalized circuit in Figure 20-1A, however, implies that output is continuous, a common feature of homeostatic biological systems. You could set up a heating system to operate continuously by hooking up a variable voltage supply to a heater that was on all the time. Then some heat would always be produced, but the amount would depend on the set point and sensor feedback signals. Alternatively, you could combine the heating and cooling functions so that both were on all the time, but in different proportions, to produce a particular temperature.

There are quite a few examples of servomechanisms in the biological world. Regulation of your body temperature is one. When your body temperature falls too low, you shiver, which generates heat to warm you. When your body temperature rises too high, you sweat, which cools you through evaporation. Hence, your body has both heating and cooling mechanisms, with two "thermostats," one for heating and one for cooling, located in the hypothalamus. Temperature regulation also illustrates the point that the set point can be changed. Pyrogens, substances that induce fever, raise the set point for body temperature.

Some biological servomechanisms have features that are not reflected in the simple diagram shown in Figure 20-1A. Consider the stretch reflex mediated by the mammalian muscle spindle organ. This organ helps to regulate the length of a skeletal muscle (see Figure 10-9 and Chapter 16). Suppose you hold a cup as someone pours coffee into it. The increased weight of the cup as it fills with coffee will cause your arm to extend at the elbow because the initial force generated by your biceps muscle is not sufficient to hold up the cup with coffee in it. As your arm extends, the muscle spindles in the biceps stretch, sending a sensory signal back to the spinal cord that excites the biceps motor neurons to increase their rate of firing. When you think of this sequence in terms of a servomechanism (Figure 20-2), it is clear that the extrafusal muscle (the biceps) is the effector and the spindle is the sensor. The α motor neuron that innervates the extrafusal muscle is the comparator. It receives the set point signal from the brain, as well as the feedback signal from the spindle.

The muscle spindle system, however, has additional features beyond those discussed so far. One important feature is that the set point for the muscle spindle system can be set dynamically, meaning that it can be adjusted during an action that the spindle is helping to control. As you can see in Figure 20-2, the sensor (the muscle spindle) is stimulated not only by changes in the length of the extrafusal muscle due to contraction or passive movements ("external input" in the figure) but also by contraction of the intrafusal muscle fibers. This additional input, which is delivered via the γ motor neurons that control the intrafusal muscles, acts as an additional set point for the spindle. Because it is

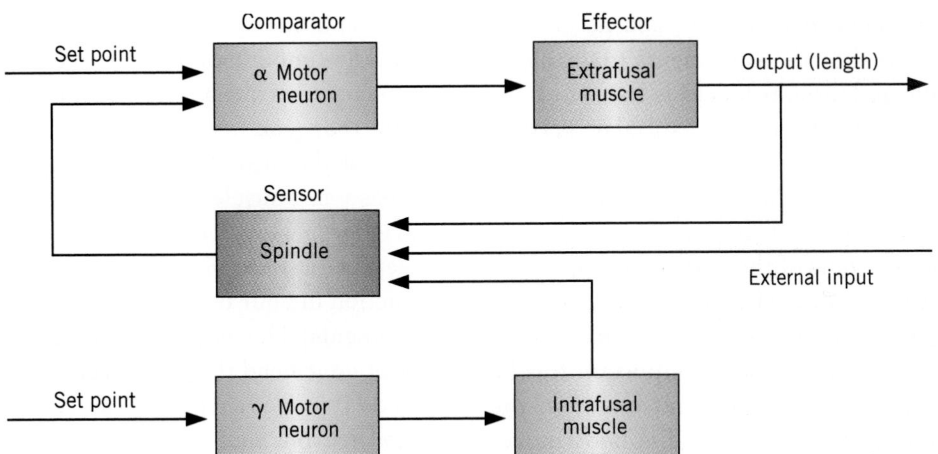

FIGURE 20-2. The mammalian muscle spindle system as a servomechanism, labeled to show the anatomical structures that serve as comparator, effector, and sensor. Compare this figure with Figure 10-7, which shows the anatomical structures. The γ motor neuron allows the output of the spindle to be adjusted continuously, so that the feedback loop will work under dynamic as well as static conditions.

independent of the α motor neuron-extrafusal muscle loop, this set point can be adjusted by changes in the firing rate of the γ motor neurons as the extrafusal muscle contracts, allowing a dynamic regulation of the desired muscle length even during an active contraction.

A servomechanism is a device in which an output is compared with a preestablished set point. Many biological systems, such as those for regulation of body temperature, act as servomechanisms. Reflexes like the stretch reflex activated by the muscle spindle are also based on servomechanisms.

Control of Movement and Balance

One of the main functions of biological servomechanisms is to allow an animal to make automatic adjustments to changing environmental circumstances without having to involve higher centers in the brain. Even those biological systems that involve complex adjustments of many different parts of the body at the same time, and hence typically require the action of many different and complex neural circuits in the nervous system, can operate without conscious involvement.

An excellent example of neural circuits that make bodywide adjustments to changing environmental circumstances is the set of circuits that constitute the vestibular system, which controls body orientation and balance in vertebrates (see Chapter 14). The vestibular system consists of the semicircular canals and otolith organs of the vestibular apparatus (see Figure 14-6). The semicircular canals provide information about rotational movements of the head. They are important for the control of eye movements that maintain the stability of a retinal image, and of body movements that maintain postural stability, during any movement. The otolith organs provide information about orientation with respect to gravity and about linear movements of the head and body. This information is used to control muscles of the limbs as well as those of the trunk. Because of the complexity of the vestibular reflexes driven by input from the vestibular apparatus, these reflexes can be difficult to understand. Analyzing their functions in terms of a servomechanism, however, makes it easier to grasp how they work.

One of the simplest functions of the vestibular system is to stabilize the head in space. To control head position, sensory signals from both the semicircular canals and the otoliths are directed via the medial vestibular nucleus that spans the medulla and pons to motor neurons that innervate the neck muscles. Deviations of the head from its intended position (the set point) are sensed by the vestibular organs and are corrected by action of appropriate neck muscles in a feedback loop. Imagine, for example, that you are sitting looking straight ahead. If there is a slight reduction of force in the muscles on the left side of your neck, your head will start tilting slightly to the right. The tilting movement will excite hair cells in the semicircular canals in your head, sending excitatory sensory signals back to the medial and other vestibular nuclei. These feedback signals in turn will cause increased excitation of the motor neurons that innervate the left side neck muscles and increased inhibition of the motor neurons that innervate the right side neck muscles, as shown in Figure 20-3A. These actions will pull your head back to an upright position.

These kinds of adjustments occur continually. The amount of head movement that is necessary to evoke a reflex action is usually so slight that you are not consciously aware either of the movement or of the subsequent increase in muscle force that counters the movement. However, as you start to fall asleep, you lose muscle tone and suffer a reduction in your unconscious control of head movements. This can cause the embarrassing jerk of your head that tells everyone you have dozed off in class. The excitatory input to the motor neurons that control your neck muscles will decline as you fall asleep. As a consequence, your head will begin to tilt. Because your brain centers are less active when you doze than when you are awake, your head

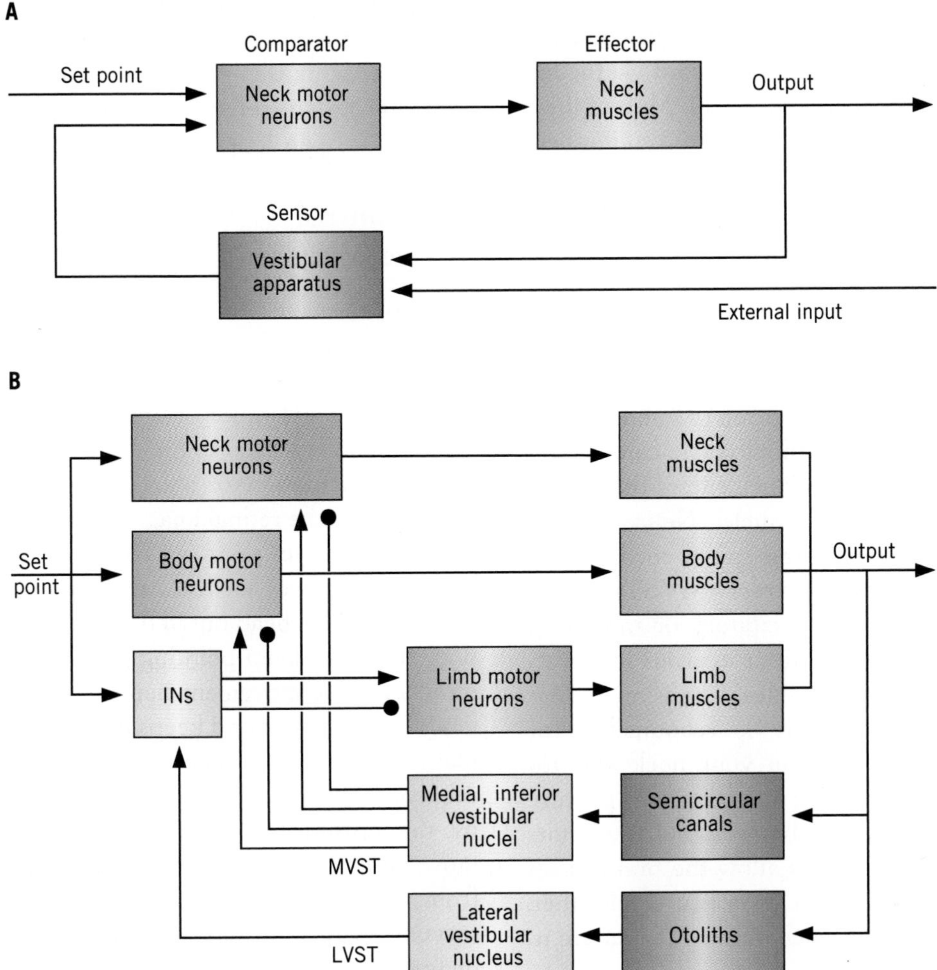

FIGURE 20-3. (A) Diagram of the servomechanism that controls neck muscles involved in maintaining a particular head position. (B) Simplified diagram of functional relations between the vestibular apparatus and neck, body, and limb muscles involved in balance. Omitted for the sake of clarity are minor tracts between the sense organs and the vestibular nuclei, minor tracts between the vestibular nuclei and motor neurons or interneurons, and connections to other brain stem nuclei and to the cerebellum. MVST, medial vestibulospinal tract; LVST, lateral vestibulospinal tract; INs, interneurons in the brain stem that control the motor neurons of the limb muscles.

will move farther than usual before the strength of the sensory input from your vestibular apparatus is sufficient to trigger an increase in motor activity. By that time a strong muscle contraction is required to keep your head from falling farther, so the movement of your head is abrupt and obvious.

The vestibular apparatus also helps to control the balance of the entire body. This action is more complex than stabilizing the head because the sensors (the vestibular organs) are in the head and the head can move independently of the body. As shown in simplified form in Figure 20-3B, separate pathways are involved for control of head and neck muscles and for control of limb muscles. Although several steps are interposed between the sensory input and the motor neurons that control the effectors, you should be able to pick out the feedback loops in Figure 20-3B by comparing the diagram with Figures 20-3A and 20-1A.

Information from the otoliths is segregated from information from the semicircular canals (Figure 20-3B). Input from the otoliths mainly enters the lateral vestibular nucleus, although there is some input from them to the inferior nucleus that is not illustrated in the figure. Neurons from the lateral vestibular nucleus send axons down to the spinal cord via the lateral vestibulospinal tract (LVST). Most of these neurons form excitatory synapses on interneurons in the spinal cord that in turn either excite or inhibit motor neurons that innervate limb muscles. The otolith/LVST pathway hence mediates actions of your arms and legs associated with balancing, such as the flailing movements your arms make as you learn to walk along a narrow rail and the subtle actions of your leg muscles that help you to stand upright.

Input from the semicircular canals that affects posture and balance enters the medial and inferior vestibular nuclei. Neurons there send axons down to the spinal cord via the medial vestibulospinal tract (MVST). These neurons form either excitatory or inhibitory synapses directly on motor neurons innervating neck or trunk muscles. The semicircular canals/MVST pathway is responsible for balancing your head on your neck and for maintaining your body in a particular posture relative to gravity. In both cases, the circuits of the servomechanism allow the brain to set some desired position or posture and then allow unconscious mechanisms to maintain it.

Biological servomechanisms allow an animal to make automatic adjustments to changing environmental circumstances. A typical example is the vestibular system, which maintains the balance of the body and the stability of the head in space. These functions are carried out by typical servo feedback loops in which a comparison is made between an intended or set position and the actual position of the head or body.

Open Loop Systems

Not all neural control circuits operate via a feedback loop, even when it seems at first glance that all the pieces are in place for such a loop to work. This is because in some cases feedback is too slow to allow precise control. Neural circuits that lack a feedback component are often referred to as **open loop circuits** or **open loop systems**.

The Control of Eye Movement

Some of the neural circuits that control your eye movements when you move your head are typical closed loop systems, whereas others are excellent examples of open loop systems. If you turn your head to the left while you are looking at a stationary object, neural circuits that stabilize the retinal image will activate eye muscles to direct your eyes to the right. As long as your eyes move at the same angular velocity as your head but in the opposite direction, they will keep pointing in their original direction. Eye movement hence stabilizes the image of the object and keeps it focused in one place on the retina. When you move your head relatively slowly, this stabilization is effected by the **optokinetic reflex**, a compensatory movement of the eyes in the direction opposite from the rotation of the head. It is a typical servomechanism in which the slippage of the image over the retina is the signal that is fed back to the extrinsic eye muscles to cause the eyes to move in compensation.

If you rotate your head quickly, however, the optokinetic reflex does not work well. When head movement is rapid, the signal to the eye muscles arrives too late to keep the image firmly stabilized on the retina, due to the time it takes to activate the reflex. This problem is overcome through the use of an **open loop reflex**, in which there is no sensory feedback to help direct the action that constitutes the reflex movement. Instead, information from a sense organ is routed directly to an effector organ for maximum speed of response (Figure 20-4A). The knee-jerk reflex familiar from a physician's examination is an example of an open loop reflex; the response is not influenced by any feedback. (The reflex is open loop when it is evoked in a physician's office. It is normally closed loop when it helps you keep your balance as you stand, however, since it is mediated by muscle spindle organs in the way described previously for a biceps muscle.) In the case of eye movements, when the head is moved rapidly, the open loop **vestibulo-ocular reflex (VOR)** takes over. In this reflex, the vestibular organs send information via the vestibular nuclei in the brain stem directly to the motor neurons that control the muscles of the eye.

It may seem that this type of system would allow no room for regulation or adjustment of the reflex, but this is not the case. Open loop reflexes often include **feed forward control.** This type of control is sometimes termed parametric control, and is considered synonymous with open loop systems. In this book, feed forward control will be considered a component that may be present in open loop systems (Figure 20-4B) but is not necessary (Figure 20-4A). When feed forward control is present, a copy of the sensory signal that drives the effector is fed to another neural center, which in turn sends information to the effector, as described in the following section. The purpose of such an arrangement is to allow the reflex to be calibrated accurately even in the absence of sensory feedback. In a feedback loop, a copy of sensory output generated by the effector is brought back to affect the control signal generated by the comparator, as shown in Figure 20-1. In a feed forward loop, the sensory signals generated by the effector are not monitored. Instead, the sensory signal that triggers the reflex is used to adjust the gain of the reflex, that is, how big a response is generated by a sensory input of a particular magnitude.

Not all neural control circuits operate via feedback. Some, such as the vertebrate circuit that keeps a stable image on the retina when the head is moved rapidly, are carried out by an open loop arrangement. In this case, the sensory signal not only activates the effector but also is fed forward to another neural center, which adjusts the strength of the reflex appropriately.

The Vestibulo-ocular Reflex

How does the vestibulo-ocular reflex actually work? Fast head rotations cause strong stimulation of the semicircular canals. Signals

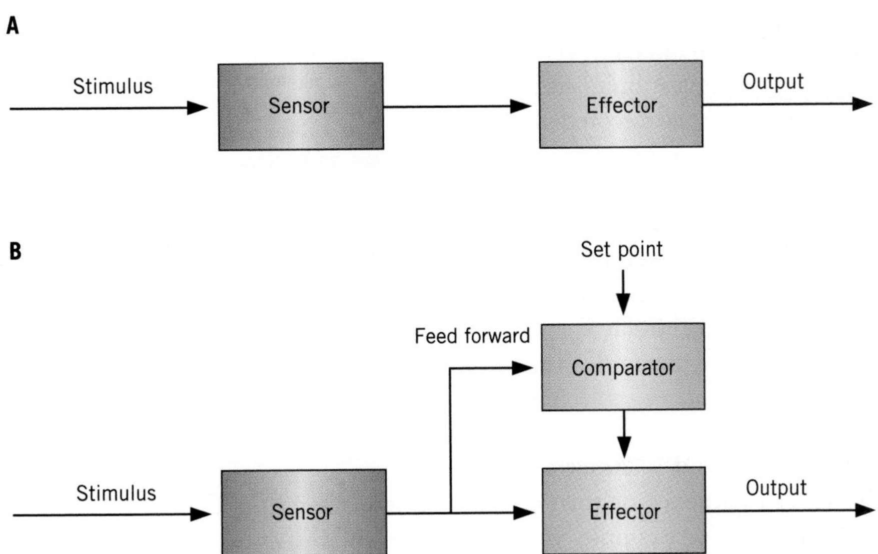

FIGURE 20-4. (A) Schematic diagram of an open loop reflex. (B) An open loop circuit with a sensory feed forward component. Feed forward allows for adjustment of the reflex to ongoing circumstances.

from these canals are sent to the vestibular nuclei in the brain stem. From there, they are distributed to various nuclei containing the motor neurons that innervate the ocular muscles and cause compensatory eye movements.

Consider, for example, a quick horizontal rotation of the head to the left (Figure 20-5). As a result of such a movement, there will be a strong excitatory input from the vestibular nuclei on both sides to the right abducens nucleus. In this nucleus are the motor neurons that innervate the right lateral eye muscle, as well as interneurons that project to the left oculomotor nucleus, which contains motor neurons that control the left medial eye muscle. At the same time, the motor neurons that innervate the left lateral and right medial eye muscles will be inhibited, to prevent them from opposing the required eye movement. Contraction of the right lateral and the left medial muscles will move the eyes to the right to compensate for the head movement, at a speed that is matched to the speed of head rotation.

In order for the VOR to work properly, it is obvious that the relationship between vestibular stimulation and eye movements must be properly adjusted. However, this relationship can change due to growth of the animal or as a result of some diseases. This is where the feed forward branch comes in. The semicircular canals, in addition to sending input into the vestibular nuclei, also send axons directly into the vestibulocerebellum (see Chapter 18), which is devoted to processing vestibular information. It, in turn, sends axons back to the vestibular nuclei, where these axons can influence the gain of the VOR, increasing or decreasing the amount of muscular activity associated with a particular amount of semicircular canal stimulation. The circuit requires an adjustment of synaptic strength and hence is a form of learning (see Chapter 24).

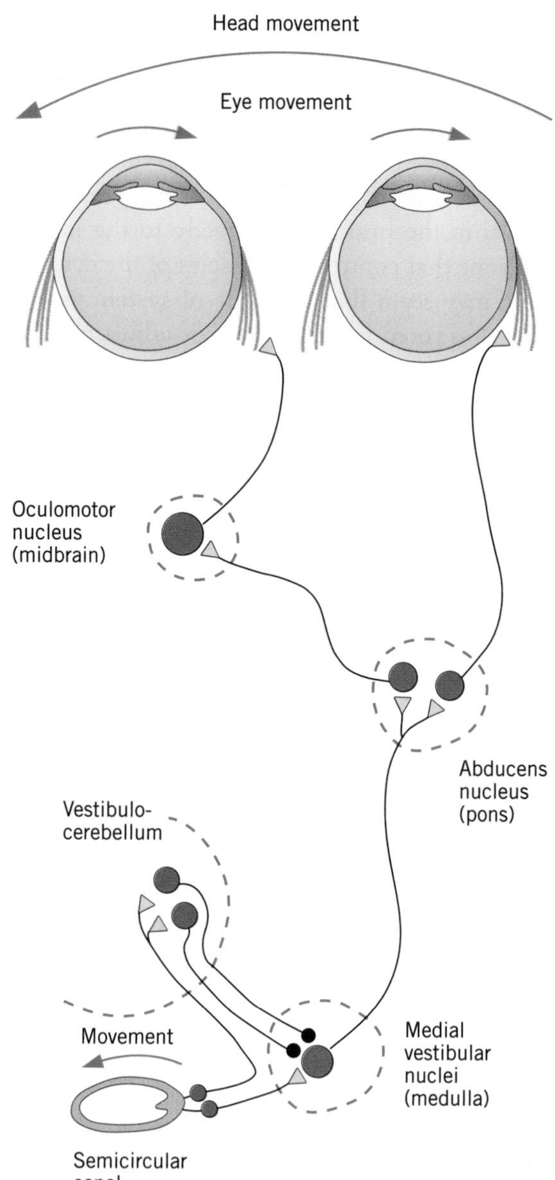

FIGURE 20-5. The main neural pathways through which the vestibulo-ocular reflex (VOR) is carried out for a horizontal rotation of the head. To keep the diagram relatively simple, pathways that inhibit the motor neurons of the left lateral and right medial eye muscles have been omitted, as have pathways from the other semicircular canal. Feed forward modulation of the inhibitory pathway from the vestibulocerebellum to the medial vestibular nuclei allows the gain of the reflex to be adjusted on the basis of experience.

As with other reflexes, it is important that an animal be able to influence or turn off the vestibulo-ocular reflex. For example, you may want to direct your eyes to look at a moving object while you yourself are moving. Exerting such control is an important function of the set point shown in Figure 20-4B and is a major role of the cerebellum. The vestibulocerebellum receives sensory input not only from the semicircular canals but also from the otolith organs. Furthermore, it receives signals from the lateral geniculate nuclei, the superior colliculi, and other places. All of these inputs are integrated in the cerebellum to help direct the eyes to the correct location according to the needs and actions of the body at any time.

The vestibulo-ocular reflex works by the influence of sensory input from the semicircular canals on the motor neurons that control the eye muscles, so that the eyes will move in a direction opposite to that in which the head is rotated. The gain of the vestibulo-ocular reflex is adjusted via a feed forward circuit involving the vestibulocerebellum, which also helps establish the set point of the reflex.

Positive Feedback Circuits

The engineering concept of negative feedback has been a great help to neurobiologists as they have attempted to understand neural systems. Another engineering concept, positive feedback, has also helped. **Positive feedback** in a circuit refers to feedback that is added to the comparator rather than subtracted from it as is negative feedback, causing an output that keeps getting larger until some limit is reached. A familiar example is the ear-splitting whine that can build up when a microphone is placed too close to the loudspeaker that broadcasts its signal. The output of the loudspeaker is itself picked up by the microphone, amplified, and broadcast back out the loudspeaker in a loop that results in louder and louder sound, until the microphone is moved or the amplifier adjusted.

Positive feedback is present in many biological systems as well; you have already read about several examples. At the cellular level, the influx of sodium ions during the rising phase of an action potential (see Chapter 5) is an instance of positive feedback. The more Na^+ that enters the cell, the more depolarized the membrane becomes and the more sodium channels open, leading to more influx of sodium, and so on, until sodium inactivation shuts off the process. At the level of an organism, the mate-finding mechanisms in many insects provide good examples. Some female moths release a pheromone to attract a mate (see Chapter 13). A mature male moth that detects this pheromone will fly upwind into the pheromone stream, always steering so as to increase the strength of the chemical signal it is detecting, until it finds the female. This is a positive feedback system because the response of the moth causes an increase rather than a decrease of the sensory input.

Positive feedback systems occur in many places in the central nervous system as well. They work because there is always some other input that prevents the circuit from generating an output that is out of control. In the case of the moth, the flying male responds to a signal that it itself does not produce; it can never detect more pheromone than the female has released. In the central nervous system, there may be inhibitory inputs to neurons that exert positive feedback effects on one another, as well as other mechanisms that are still poorly understood, to control the system.

One place in which positive feedback has been hypothesized to operate is in the vertebrate motor control system. James Houk and his collaborators have suggested that positive feedback loops exist between the motor cortex and the cerebellum, and that these loops are instrumental in generating the response of an animal to a particular sensory input (Figure 20-6).

When fibers from the motor cortex are excited by some sensory input, they send an excitatory signal to the cerebellum via nuclei in the brain stem (see Chapter 18). In the cerebellum, the excitatory input activates cells in the dentate nucleus that (via the thalamus) feed back to the motor cortex, still further exciting neurons there. This positive feedback loop can have several functions. It can amplify the strength of firing of the neurons in the motor cortex, and hence can produce a strong response in the spinal motor neurons that the cortical cells activate and a correspondingly strong behavioral response as well. It can also

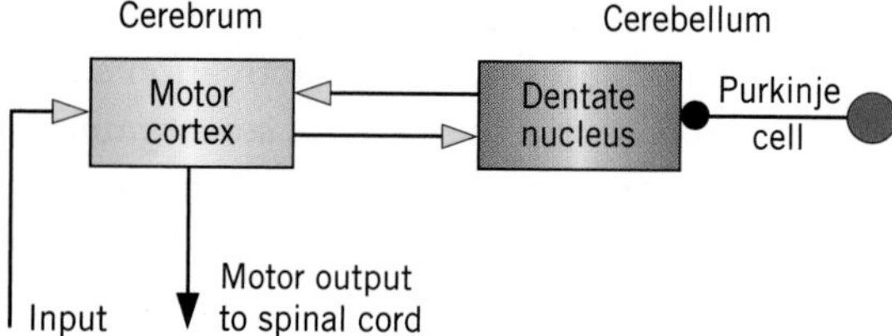

FIGURE 20-6. A hypothetical positive feedback arrangement between neurons in the motor cortex in the telencephalon and the dentate nucleus in the cerebellum. Each box represents groups of neurons, which send excitatory signals to each other to create positive feedback loops. The brain stem nuclei through which these signals are sent have been omitted from this diagram for clarity. Activity may be initiated by sensory input (left) and regulated by inhibitory input from Purkinje cells to the dentate nucleus (right). The motor output from the cortex is amplified and increased in duration by the positive feedback.

increase the duration of the motor response by keeping neurons active longer. If activated neurons excite other neurons, positive feedback can spread the excitation to more neurons in the cortex. The feedback loop can be controlled and shaped by the dampening effect of inhibitory input from Purkinje cells in the cortex onto the neurons of the deep cerebellar nuclei, as shown in Figure 20-6.

Positive feedback, a phenomenon in which feedback increases rather than decreases the output of a circuit or system, is common in some neural circuits. It may amplify the strength and duration of a neural response, as well as spread excitation to other neurons. The feedback circuit can be regulated by independent inhibitory input.

Computational Neural Circuits

Only a small fraction of an animal's behavioral repertoire is based on simple open or closed reflex loops. More commonly, an animal will engage in a specific behavior based on an evaluation by its nervous system of relevant sensory inputs. A flying bat, for instance, is able to distinguish between a piece of debris, such as a leaf fragment swirling in the air, and an edible moth. The bat analyzes the echo that it receives from the object and behaves according to the result of the analysis, ignoring the leaf and attacking the moth. From a neurobiological perspective, an important question is how the bat is able to identify the object it has detected and decide which behavior is most appropriate.

In behavior that involves identification of sensory stimuli, it is rarely obvious how the nervous system might make the proper identification. Nor is it obvious how the animal might subsequently select the appropriate behavior based on knowledge of the stimulus. One way to investigate this kind of behavioral selection is to treat the problem as one of information processing. For example, a flying bat must analyze a returning echo to determine not only how far away the object that produced it is but also how big it is. Furthermore, by analyzing the echo, some bats will be able to determine whether the object is changing its apparent shape, as a flying moth will when the wings beat.

The main issue is how the nervous system is actually able to carry out analysis of this kind. By considering how information about the stimulus is encoded in neural activity, and how that information is used and transformed as it moves through the central nervous system, neuroscientists can often gain a better understanding of how the nervous system carries out some specific function. This kind of investigation requires two steps. First, investigators must carry out careful behavioral experiments to determine exactly which aspects of a stimulus are critical in order for a specific behavior to be elicited. Detailed knowledge of the critical stimulus parameters allows researchers to go on to the second step, which is to determine the means by which neurons transform and combine information about the stimulus so that the animal can select an appropriate behavioral response.

You have already read about one example of this approach in the discussion of the barn owl's computational map of auditory space in Chapter 12. This approach, called computational neuroscience (Box 20-1), has yielded important new insights into the mechanisms by which the interactions of neurons can bring about a particular behavior. It has also helped to accelerate the integration of computers as useful tools in mainstream neuroscience.

As you consider specific examples of the computational approach to behavior, think of the work in a broad context. Your objective as you read about these studies should be to understand how the actions of individual neurons or groups of neurons can bring about a specific behavior. The details of how this is accomplished are sometimes complex. If you find yourself awash in details that only seem to confuse you, it may help to think about the purpose of the neural mechanism you are studying. You will probably find that putting the details in a broader context will make understanding those details easier.

Much behavior involves analysis of sensory input by an animal so that it can determine an appropriate response. The problem can be considered one of information processing. This approach, called computational neuroscience, has brought new insights into the neural basis of behavior.

The Jamming Avoidance Response

The computational approach was used with great success by Walter Heiligenberg and his colleagues to analyze a unique behavioral response of weakly electric fish. You may recall from Chapter 14 that some species of these fish can generate sinusoidal electrical signals by repetitively activating their electric organs. The resulting **electric organ discharge (EOD)** of these "wave-type" species has a frequency that is characteristic for each species of fish. For example, EOD frequencies of species of the genus *Eigenmannia*, a genus studied extensively by Heiligenberg, are typically several hundred hertz. The EOD is used for electrolocation (locating animate and inanimate objects) in the silt-laden rivers of South America in which the fish live. It is important that each fish be able to recognize its own EOD and not be confused by the EOD from a nearby fish because competing electrical signals of the same frequency severely interfere with the ability of the fish to electrolocate.

Individuals of *Eigenmannia* respond to competing EODs of similar frequency by temporarily adjusting the frequency of their own EODs to be sufficiently different from that of the EOD of the nearby fish so that they can still electrolocate. If a fish detects the EOD of a neighbor, the fish shifts the frequency of its own EOD away from the frequency of its neighbor's EOD. If the neighbor's EOD is at a frequency slightly lower than its own, the fish shifts its EOD up in frequency. If the neighbor's EOD is at a higher frequency, the fish shifts its EOD down (Figure 20-7). (Natural variation of EOD frequencies in different individual fish means that the frequencies of the two EODs are unlikely to be exactly the same.) This shift in EOD frequency in the presence of another weak oscillating electrical signal of slightly different frequency is known as the **jamming avoidance response (JAR)**.

The JAR is one of the few vertebrate examples of a behavior in which the neural mechanisms underlying a complete behavioral loop have been worked out, from sensory input (the EOD from itself and another fish) to motor output (the shift in frequency of the fish's own EOD). The mechanisms that underlie it have been worked out by careful and thorough neuroethological analysis of both the behavior and the neural connections that make the behavior possible. The JAR is described here specifically as an example of the computational approach. By following step-by-step how the nervous system analyzes and transforms a sensory signal into an appropriate behavior, you can gain a good understanding of how a series of simple neuronal interactions can yield an adaptive and complex behavioral result. A significant part of an animal's behavioral repertoire consists of

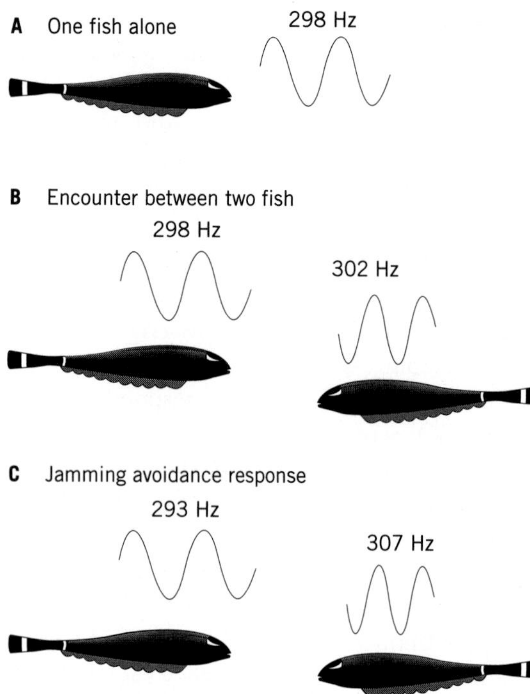

FIGURE 20-7. The jamming avoidance response (JAR) of weakly electric fish. Each fish generates an electric organ discharge (EOD) of a characteristic frequency (A). When another fish appears on the scene with an EOD of similar frequency (B), each fish shifts the frequency of its own EOD away from the frequency of its neighbor's EOD (C).

actions that are generated from an analysis of sensory input in this way, so understanding one specific example in depth will help you to understand how a great deal of behavior of this kind can be generated.

> The jamming avoidance response (JAR) of a "wave-type" weakly electric fish is a shift in frequency of its electric organ discharge (EOD) away from the frequency of the EOD of a neighboring fish.

Properties of the Sensory Signal

Heiligenberg's study, like the neuroethological investigations described in Chapter 19, began with an analysis of the electrical signal that an electric fish receives when another fish is nearby and the behavior the fish shows as a result of receiving this signal. From this work, it became clear that a fish "decided" to shift its EOD up or down in frequency based on an analysis of the input it received about its own EOD and the EOD of its neighbor. (Of course, as far as can be determined, the process is entirely automatic and does not involve any conscious act on the part of the fish.) In order to do this, the fish must be able to determine whether the frequency of the EOD of its neighbor is higher or lower than the frequency of its own EOD.

How the fish determines in which direction to shift its EOD can best be understood by knowing how the information it receives from its electroreceptors changes when a second EOD is also present. When an electric fish is alone, it senses only the electrical current, the signal, produced by its own EOD. Since the EOD is sinusoidal, the signal the fish senses also varies sinusoidally, from a peak positive value to a peak negative value and back again. When another fish is nearby, the EODs of the two fish interact to produce a combined signal that is an algebraic sum of the fish's own EOD and that of its neighbor.

The combined signal has two properties that allow the fish to determine in which direction to shift the frequency of its EOD. In order to understand how these properties arise and how a fish interprets them to make a decision, consider what happens when two sinusoidal waveforms interact. If two sinusoidal waves of the same frequency, S_1 and S_2, are summed, the resulting wave, $S_1 + S_2$, will be larger than either of the original waves but will have the same temporal properties as the originals (Figure 20-8A). This is because the peaks and troughs of the two waves line up exactly with one another, thereby producing a new, larger wave whose peaks and troughs are also lined up with the original waves, as illustrated in the figure.

When two sinusoidal waves of different frequencies interact, the resulting combined waveform is not just a larger version of the two original waves (Figure 20-8B). Instead, because the peaks of wave S_1 do not line up

BOX 20-1

Computing Brains: Computers and Neurobiology

Ever since the first computers were developed, researchers have been fascinated by comparisons between computers and brains, in terms of both their design and their computational capacity. Since the introduction of high-powered, inexpensive desktop computers in the late 1980s, the use of computers and computational approaches in neuroscience has become so widespread that a separate subdiscipline of neuroscience, called **computational neuroscience,** is now recognized. Computational neuroscience encompasses all neurobiological work in which computational methods or approaches play an integral part. These approaches fall into two main areas: the use of computers to model and simulate neural function, such as the models of central pattern generators described in Chapter 16, and the development of theoretical frameworks for describing the brain as an information-processing device, as described in this chapter.

A key component of computational neuroscience is the use of a model or a computational approach as an aid to understanding neural function. Some neural processes are so complex that it is difficult to imagine how the nervous system might accomplish them and also difficult to test ideas about how they function. Stereoscopic vision, the ability to perceive the world in three dimensions, falls into this category. One important issue is how the visual cortex is able to match corresponding elements of the right and left

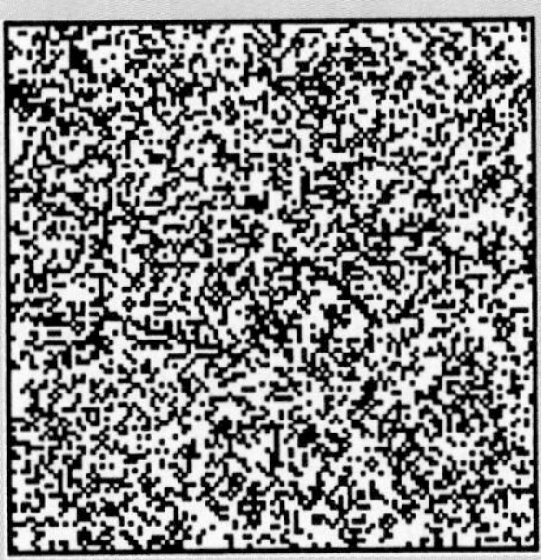

FIGURE 20-A. A random dot stereogram. The two boxes are computer-generated, but in such a way that one of the boxes has a set of black and white squares displaced to one side compared with their locations in the other box. When the two images are fused, you will see an image either floating above or sunk beneath the level of the page, depending on which method of fusion you use. You can fuse the images by holding the page up about a foot from your eyes and focusing on something across the room as if you could see through the book. With practice, many people are able to form a fused image of the two squares. Alternatively, you can cross your eyes to form the fused image. The central image will seem to be below the level of the page if you use the crossed-eye method.

(continued)

Computing Brains: Computers and Neurobiology *(continued)*

visual fields with one another to form an unambiguous image of a scene. The two images of black and white squares shown in Figure 20-A illustrate the point. Viewed stereoscopically, an image pops out, but only if corresponding squares are correctly matched to one another in the brain.

Computational models of stereoscopic vision have not solved the problem of how the visual cortex makes the proper matches, but they have provided researchers with important clues about how the process might work. By developing computer models that incorporate an idea of how stereopsis might work, and testing the models with actual images, researchers can rapidly explore the feasibility of many different ideas. Success of the model can then lead to physiological experiments to test its predictions, whereas failure of the model can provide an important clue about how reality might differ from the model. In the case of stereoscopic vision, computational models suggest that the nervous system keeps trying to find the aim and focus of the eyes that provide the strongest output from binocular neurons (neurons that receive input from both eyes) in specific cortical layers. Models based on this idea have been quite successful in generating an output that clearly represents a stereoscopic combination of the images given to it, so physiologists now search for neural mechanisms to account for such a process.

---

exactly with the peaks of wave S_2, the peaks of the combined wave, $S_1 + S_2$, will be displaced a bit relative to each of the original waves, causing a shift in the time at which the peak of the combined wave occurs relative to when the peaks of either of the original signals occur. The extent to which the peaks of S_1 and S_2 are out of alignment drifts over time, causing the peak amplitude of the combined wave to vary as well.

The EODs from two fish interact in the same way. Consider a case in which a nearby neighbor fish is producing an EOD, S_2, of a slightly higher frequency than the fish's own EOD, S_1. Under these conditions, the peak-to-peak amplitude of the combined signal, S_1 + S2, will vary, waxing and waning as the two EODs interact (Figure 20-9A). This variation occurs because sometimes the trough (or peak) of one wave will occur at the same time as the trough (or peak) of the other ("reinforcement" in Figure 20-9), producing an amplitude in the combined waveform larger than that of the larger of the original waves. At other times, the peak of one wave will occur at the same time as the trough of the other ("interference" in Figure 20-9), which will produce a combined amplitude smaller than that of either of the original waves. The drift of S_1 relative to S_2 produces a combined signal that waxes and wanes in its peak-to-peak amplitude. This modulation of peak-to-peak amplitude is called a beat pattern. You can experience a similar modulation or "beat" in sensory input yourself if you and a friend try to hum a tone at the same pitch at the same time. When the two tones are close to one another in frequency, you will hear an increase and decrease in the loudness of the sound as the sound waves of the two tones alternately reinforce and interfere with one another.

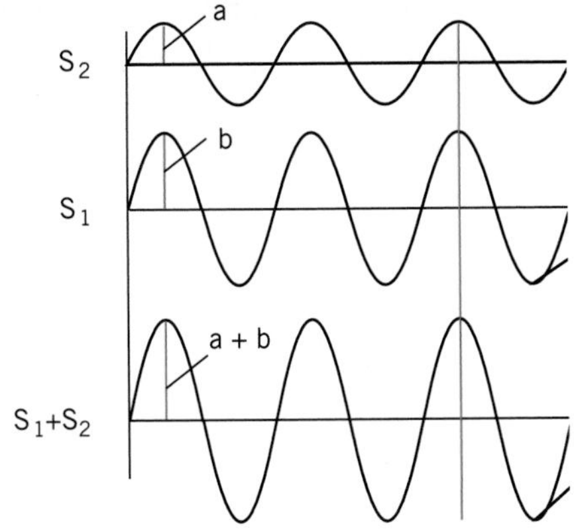

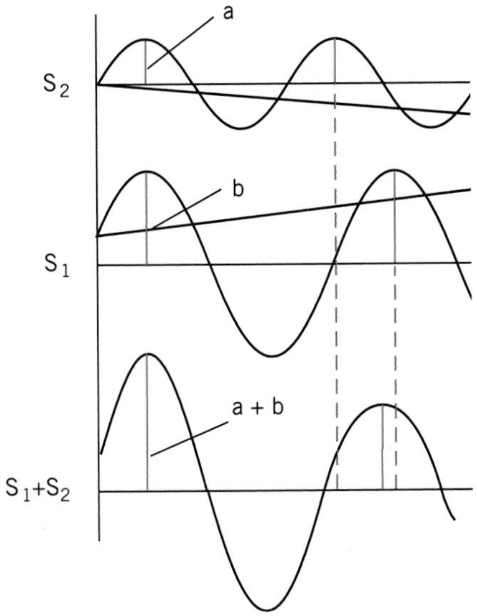

FIGURE 20-8. Summation of two sine waves. (A) Waves of the same frequency. When the positive peaks of the two waves coincide, the amplitude of the combined wave that results is augmented. Because the two waves have the same frequency, the peak of the combined wave will always be aligned with the peaks of the two original waves. (B) Waves of different frequency. Because the waves do not align exactly with each other, the combined wave will vary in amplitude from cycle to cycle and will show a temporal shift in the occurrence of its peak relative to the peaks of the original waves.

Interaction of two EOD signals also produces temporal variation like that shown in Figure 20-8A in the combined waveform. When waveforms are viewed together as in Figure 20-8, it is easy to compare prominent features like the peaks of each wave. Weakly electric fish, however, are able to detect the time at which the waveform crosses zero more easily than the peak of the wave, and they use this to determine temporal shift. For any EOD signal, the time at which the signal goes from negative to positive during each sinusoid is called the zero crossing. When two signals of different frequency are present together, the time at which the combined signal crosses zero will shift systematically relative to zero crossing for one of the original signals as the combined signal waxes and wanes.

Eigenmannia uses both the temporal shift and the variation in amplitude of the combined signal to determine whether the competing EOD is higher or lower in frequency than the signal it is generating itself, and hence whether it should raise or lower the frequency of its own EOD. It is able to do this because there is always a unique combination of change in amplitude and temporal shift that indicates whether a competing EOD is higher or lower in frequency than the fish's own EOD.

Figure 20-9A illustrates the unique combinations. Consider first the case in which the fish's own EOD, S_1, has a lower frequency than the competing EOD, S_2 (Figure 20-9A). As the amplitude of the combined wave, $S_1 + S_2$, waxes (increases), the zero crossing of $S_1 + S_2$ comes *later* than the zero crossing of S_1. This is called a phase delay. However, after the combined waveform has peaked and begins to wane (decline) in amplitude, the $S_1 + S_2$ time of zero crossing begins to come *before* zero crossing for S_1. This early crossing is called a phase advance. Hence, when the frequency of the fish's own EOD is lower than that of the neighbor's EOD, zero crossing of the combined signal is first delayed, then advanced as the amplitude of the combined signal waxes and wanes, respectively.

The converse relationship holds when S_1 has a frequency higher than the frequency of the second signal, S_2. In that case, the rise in amplitude of the combined signal will be associated with a phase advance (Figure 20-9B), and the fall in amplitude of the combined signal will be associated with a phase delay. Hence, if the fish can determine whether a

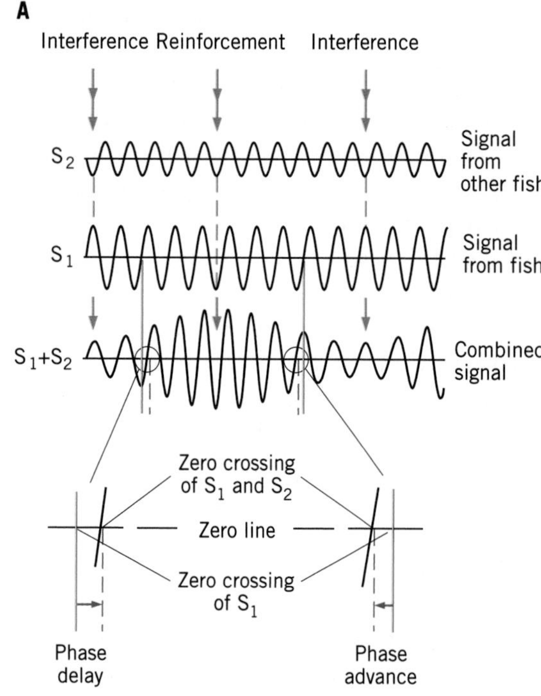

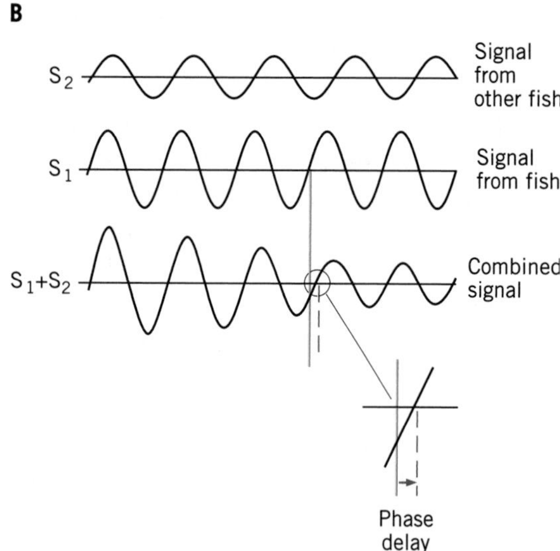

FIGURE 20-9. The electric organ discharge (EOD) of a weakly electric fish (S_1) and the signal detected by some of the receptors of that fish in the presence of a second EOD (S_2) from a neighboring fish ($S_1 + S_2$). (A) The case when S_2 has a higher frequency than S_1. Note that there are periodic variations of the resulting signal in both the amplitude and the time at which the waveform crosses zero. In this case, as the amplitude increases, the time of zero crossing is delayed (a phase delay); as the amplitude decreases, the time of zero crossing is advanced (a phase advance). (B) The case when S_2 has a lower frequency than S_1. Here a phase delay is associated with a decline in amplitude. An increase in amplitude is associated with a phase advance, but this is not shown in the figure.

phase delay or phase advance is associated with an increase in signal amplitude, it can determine whether the neighbor's EOD is at a higher or lower frequency than its own, and hence whether it should decrease or increase the frequency of its own EOD.

A fish is able to determine the direction in which to shift its EOD based on when the combined signal from the interaction of the two EODs crosses zero, and whether the amplitude of the resulting signal is rising or falling at that time.

Computing the Parameters

At the neuronal level, the ultimate question about how the JAR works is what neural mechanisms drive the frequency of the EOD in the appropriate direction given an external EOD signal of a particular frequency. To answer this question, it is necessary to uncover the transformations, the "neural computations," that make it possible for a fish to extract the requisite information about EOD frequency from the sensory input and then to transform it into an appropriate action.

The process starts with the excitation of specialized tuberous electroreceptors in the fish's skin (see Chapter 14). The signals generated by the receptors are sent to the brain, where the information is processed sequentially by four different brain centers, the electrosensory lateral line lobe, the torus semicircularis, the nucleus electrosensorius, and the prepacemaker nucleus, before being sent on to the pacemaker nucleus, which regulates the frequency at which the electric organ generates the EOD (Figure 20-10). Each brain region makes a "computation" on the input it receives, extracts some particular piece of information from it, and passes that information on to the next center. The end result is a specific neural signal that causes either an increase or decrease in the frequency of the fish's EOD.

The **electrosensory lateral line lobe,** (ELL), introduced in Chapter 14 as part of the medulla, receives input from two types of electroreceptors, P-type and T-type receptors. These receptors respond to different aspects of the combined EOD signal (Figure 20-11). The P-type receptor is stimulated by the amplitude of the local electrical field generated by the combined EOD. That is, the greater the amplitude of the stimulus signal, the more frequently the receptor fires. Hence, as the combined signal received by a fish waxes and wanes, the firing rates of the P receptors rise and fall. The other receptor, the T-type receptor, fires once at each zero crossing of the signal. Hence, the P and T receptors provide the two critical pieces of information that the fish needs to determine whether to raise or lower the frequency of its own EOD in the presence of the EOD from a neighboring fish.

The ELL processes the information it receives about the amplitude and timing of the electrical field in two ways. The T-type afferents converge on interneurons in the ELL, which in turn fire at a time that is determined by the average time of arrival of the input signals. Because they reduce the small amount of variability in the responses of the sensory afferents, the interneurons produce a more precise indication of the timing of zero crossing than any single T-type afferent can by itself. Put another way, these interneurons sharpen the timing information, making it more precise. The ELL carries out no other processing or computation on the information about the timing of the EOD.

On the other hand, information about amplitude, which is carried by the P-type afferents, is altered. The P receptors themselves fire more or less all the time, but with a frequency that waxes and wanes as the amplitude of the combined electrical field generated by the two EODs waxes and wanes. The fish, however, needs to know when the amplitude of the combined signal is increasing, as well as when it is decreasing, because its response depends on whether a phase advance or phase delay of zero crossing is coupled with an amplitude increase or decrease. For this reason, it would be useful to have separate neural signals for amplitude increases and for amplitude decreases.

Two separate signals are obtained in the following way. The P receptors send their information to two types of neurons in the ELL (Figure 20-12). In one case, several P receptors make direct excitatory synapses on neurons known as E-units (excitatory-units). The properties of each E-unit are such that it will be excited by the P receptors that make synaptic contact with it when the rate of firing of the P receptors is increasing but not when

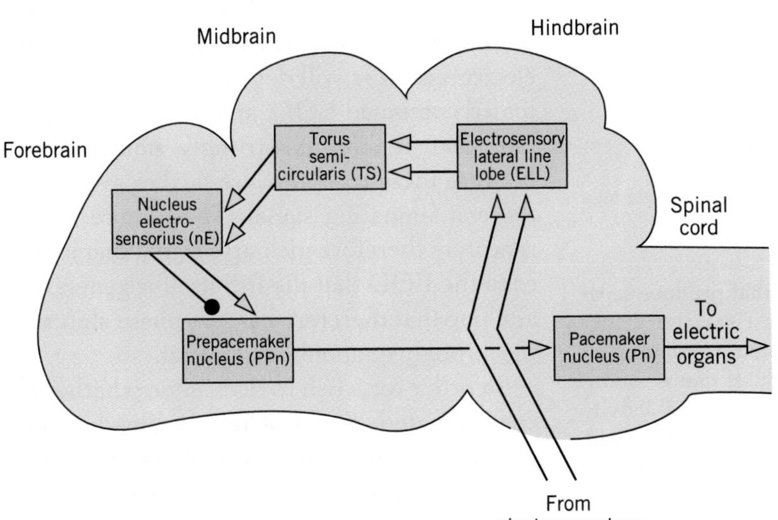

FIGURE 20-10. Summary diagram of the flow of information in the brain of a weakly electric fish that leads to the JAR. Information about the amplitude and timing of an EOD is processed in four regions of the brain. Each region makes a particular neural computation of the input it receives, so that in the end the fish can determine whether it should increase or decrease the frequency of its own EOD.

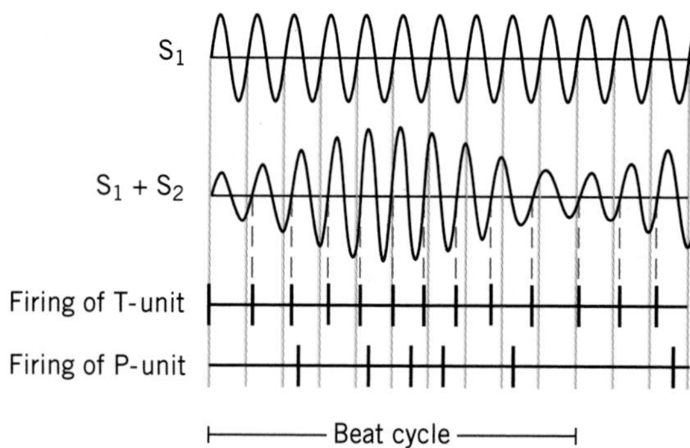

FIGURE 20-11. The firing of the two types of electroreceptors in the presence of an EOD signal from the fish itself (S_1) and a combined signal caused by the presence of a second EOD from a neighboring fish ($S_1 + S_2$). T-units fire at the time of zero crossing of the combined signal. P-units fire more or less continually but at a frequency that is modulated by the amplitude of the combined signal. Their rate of firing is high when the amplitude is high, and low when the amplitude is low.

the rate of firing is decreasing. Consequently, the E-units fire as the amplitude of the combined EOD signal is increasing.

The other type of ELL neuron that receives information from the electroreceptors is called an I-unit (inhibitory-unit). I-unit neurons normally fire spontaneously. However, they receive inhibitory input from cells that are

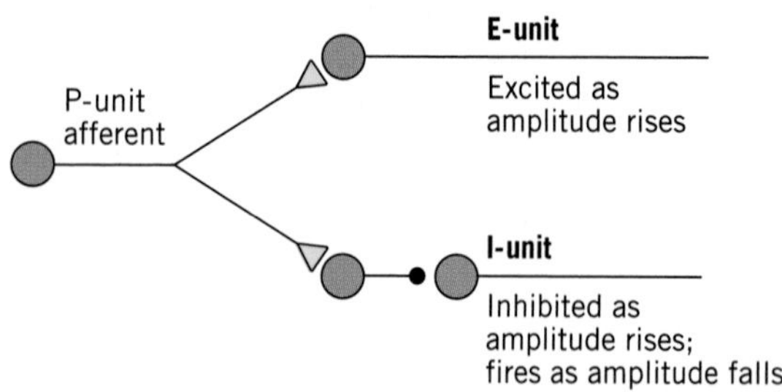

FIGURE 20-12. Neural mechanisms that produce separate information about amplitude rises and amplitude declines in the ELL. P receptors respond to an EOD with a train of action potentials that is modulated by amplitude, with high frequencies of activity being associated with peak amplitude. Input from the P receptors is sent directly to E-units, which are excited when amplitude is rising, and via inhibitory neurons to I-units, which fire when it is falling.

excited by the P receptors. Hence, the I-unit is inhibited when the amplitude of the EOD is increasing but is not inhibited when the amplitude is decreasing. As a consequence, the I-unit fires when the amplitude of the combined EOD is decreasing. The first stage of processing of the electrosensory signal, then, encodes increases in signal amplitude in the activity of E-units and decreases in signal amplitude in the activity of I-units.

The sharpened timing signals from cells in the ELL, together with the outputs of the E- and I-units, are sent to the torus semicircularis in the midbrain, which in lower vertebrates, as described in Chapter 3, processes auditory information. In electric fish, the torus is topographically organized, and neurons there carry out several computations that are necessary for a successful JAR. The first processing step in the torus is the computation of differential phase, which is accomplished by neurons that compare timing information from different parts of the body surface.

Electroreceptors are distributed all over the body of an electric fish, although they are somewhat concentrated at the rostral end. Hence, a fish picks up an EOD everywhere, from its head to its tail. However, the combined signal that a fish senses when another fish is nearby is not the same all over its body. If two fish face each other, as shown in Figure 20-7, the contamination of one fish's EOD by the EOD generated by the other fish will be greatest at the fish's head. No matter what the orientation of the two fish to one another, some electroreceptors will detect a strongly contaminated combined EOD, and others will receive one that is not as strongly contaminated because these receptors are farther away from the contaminating signal. These more distant receptors therefore pick up a signal that is closer to the EOD that the fish itself is generating, and one that therefore has less phase shift than the strongly contaminated signal.

In order for a fish to determine whether the contaminated signal (the combined signal) shows a phase advance or a phase delay associated with a rise in signal amplitude, the fish compares the times of zero crossing of the

EOD signal from receptors from all over its body. It does this in the torus. The time of zero crossing will vary from place to place along the body depending on how strongly the signal that a T receptor sees is contaminated by the second EOD. By combining information about phase differences with information about where on the body (that is, between which body points) there are large or small phase differences, neurons in the torus are able to separate phase advances from phase delays. This information is encoded in the activity of two types of toral neurons, called phase advance and phase delay units according to the condition they signal.

Elsewhere in the torus, the information about amplitude and phase converges on a group of cells called sign-selective neurons. Each sign-selective neuron receives input from one of the types of amplitude units (an E- or I-unit). This input is gated, in some way not yet established, by input from either a phase advance or a phase delay neuron. One possibility is that the amplitude units make direct excitatory synapses with the sign-selective units, and that the phase advance or phase delay neurons gate these synapses via facilitory presynaptic connections with the amplitude units (Figure 20-13). Under these conditions, the sign-selective units might fire only when an amplitude unit is active at the same time that the phase advance or phase delay unit that makes presynaptic contact with it is active as well.

There are four possible combinations of phase and amplitude input to sign-selective neurons, all found in the torus: phase advance coupled with amplitude decrease, phase advance with amplitude increase, phase delay with amplitude decrease, and phase delay with amplitude increase. Each combination produces a unique functional type of sign-selective neuron (Table 20-1). Each of these functional types will fire when and only when a particular combination of amplitude change and phase condition is present, and the presence of another EOD will always produce two of these conditions. A higher-frequency EOD from a neighboring fish will excite sign-selective units that signal an amplitude increase in association with phase delay, as well as those that signal an amplitude

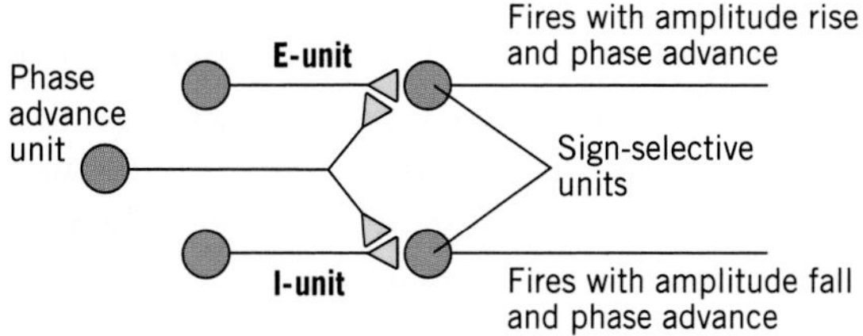

FIGURE 20-13. Neural interactions that produce sign-selective output, conveying information about the sign of a phase shift (i.e., whether it is a phase advance or a phase delay). Each sign-selective neuron receives input from E- or I-units, signaling increases or decreases in amplitude, respectively. A single phase discriminator neuron, a phase advance unit in this case, is hypothesized to make a facilitory presynaptic connection onto the presynaptic terminals of the E- or I-units. The result is to shape the activity of each sign-selective neuron so that it fires mainly when a particular combination of amplitude change and phase change occurs. The figure shows only the effects of a phase advance unit. Similar interactions are thought to occur for phase delay units.

TABLE 20-1. Conditions Signaled by Sign-selective Neurons

Condition	Sign-selective Neurons Firing
Higher-frequency EOD	Amplitude increase with phase delay
	Amplitude decrease with phase advance
Lower-frequency EOD	Amplitude increase with phase advance
	Amplitude decrease with phase delay

decrease in association with phase advance. A lower-frequency EOD from a neighbor will excite the other two types of sign-selective units—those that signal an amplitude increase in association with phase advance, and those that signal an amplitude decrease in association with phase delay.

Several regions of the brain help to compute the information necessary for a fish to shift the frequency of its own EOD in the presence of another fish. The ELL generates three signals: one timing signal from the T receptors and two from the P receptors. Of these, one signal indicates when the amplitude of the summed EOD is increasing, the other when it is decreasing. Neurons in one part of the torus generate signals that indicate whether there is a phase advance or a phase delay. Information from these neurons converges on sign-selective cells elsewhere in the torus. Each sign-selective neuron provides a unique indication of the presence of amplitude increases or decreases together with phase advances or delays.

Shifting the EOD

The computations carried out by the ELL and the torus generate a pattern of activity in sign-selective neurons that reflects whether a neighbor fish's EOD is higher or lower in frequency than the fish's own EOD. The sign-selective neurons send their information on to the **nucleus electrosensorius (nE)**, where this information is combined to produce activity in one of two classes of neurons. One class of neurons fires when an increase in amplitude is associated with a phase delay and a decrease in amplitude is associated with a phase advance. This combination of events occurs when the frequency of the EOD of the neighboring fish is higher than that of the fish's own EOD. The other class of neurons fires when an increase in amplitude is associated with a phase advance and a decrease in amplitude is associated with a phase delay. This combination occurs when the frequency of the EOD of the neighboring fish is lower than that of the fish's own EOD (Table 20-1). Hence, the output of the nE is neural activity that can be sent directly to brain structures that control EOD frequency.

The information from the nE is sent to the **prepacemaker nucleus (PPn)** and other related nuclei in the forebrain. Here, the final command for altering the fish's own EOD is generated. In the PPn, neurons that drive the **pacemaker nucleus (Pn)** in the medulla are inhibited if the fish should decrease the frequency of its EOD, or excited if it should increase the frequency of its EOD. These inhibitory or excitatory effects bring about a reduction or increase, respectively, in the frequency at which the pacemaker nucleus drives the electric organ.

A summary of the main demonstrated or hypothesized neural interactions that have been described here is shown in Figure 20-14. The final consequence of these interactions is to allow a fish to shift the frequency of its EOD in the proper direction in the presence of the EOD from another fish. A fish's behavioral response, shifting the frequency of its EOD away from the frequency of a neighbor's EOD, is essential for the fish to be able to electrolocate even while in the presence of other fish.

When viewed in a summary diagram such as Figure 20-14, the neural circuitry used for generation of a JAR may seem quite complex. However, it should be clear from this discussion of the mechanisms that the complexity can be reduced if the process is considered as a series of simple steps, each having a particular function. The E- and I-unit neurons indicate the amplitude changes (increases or decreases) that occur as a result of the interactions of the two EODs. The phase advance and phase delay units signal phase changes (advance or delay). In the torus, information from these four sets of neuron types is combined to produce two distinct signals: one for positive and one for negative differences in EOD frequency, as listed in

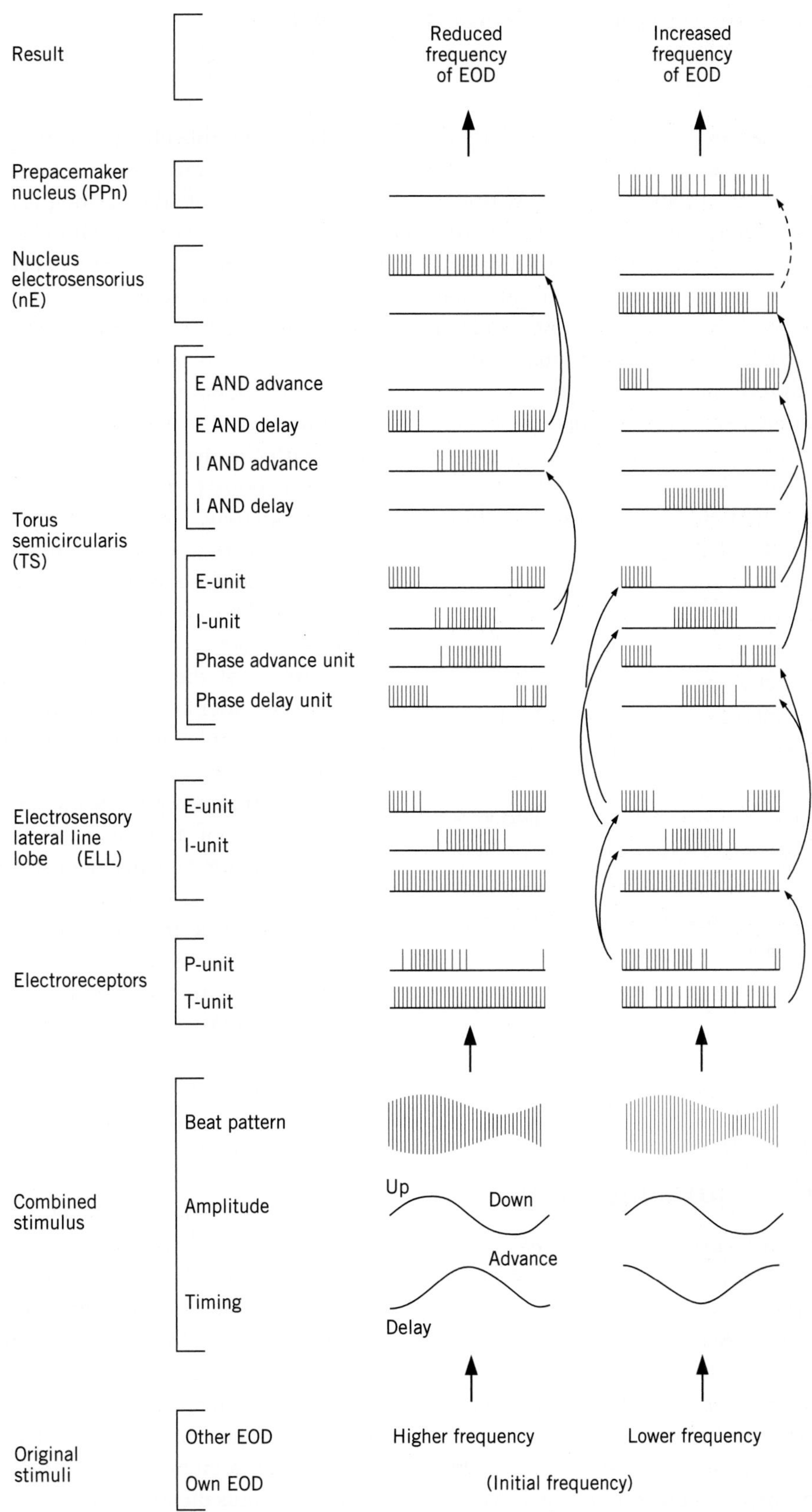

FIGURE 20-14. Summary of how the interaction of electrical signals from the EODs of two electric fish can lead to a JAR, in which each fish will adjust its own EOD away from the frequency of its neighbor. Information is processed from bottom to top. The arrows that connect stylized traces of neural activity show which elements are combined to produce the higher-level activity. By combining information from lower levels, a particular part of the brain makes a neural computation and produces a signal that is used at higher levels, until a clear and unambiguous signal is generated at the PPn to raise or not to raise the frequency of the fish's EOD.

Table 20-1. These signals interact directly with the pacemaker controller to adjust the frequency of the fish's EOD. Although Figure 20-14 may seem daunting at first glance, when it is examined step-by-step and looked at from the point of view of how specific decisions are made, it will show you from beginning to end how a specific sensory input can be analyzed and transformed to a motor command that drives an appropriate behavior, the jamming avoidance response, that allows the fish to locate objects in its environment even when that environment includes other fish of the same species.

By combining the information from the sign-selective neurons of the torus, the nucleus electrosensorius generates activity in neurons according to whether the EOD of the neighbor fish is at a higher or lower frequency than the fish's own EOD. In the prepacemaker nucleus, this activity is translated into a clear and unambiguous instruction to the pacemaker nucleus to increase or decrease the frequency of the EOD that is being generated.

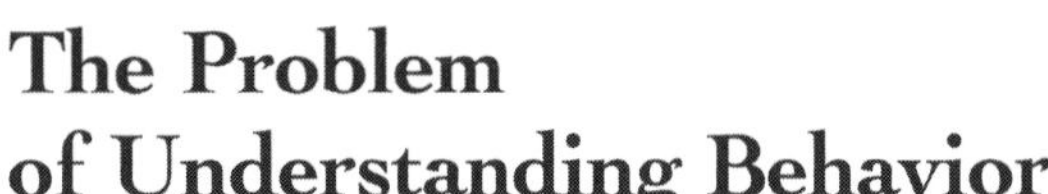

The Problem of Understanding Behavior

Explaining the neural basis of animal behavior is an obvious objective of much of neurobiological research, yet it is one of the most difficult to meet. Asking how the nervous system generates and coordinates a particular sequence of actions is akin to asking how a television set works. On the one hand, you can get a short answer that mentions electrons and phosphors but that doesn't provide much detail. On the other hand, you can spend considerable effort learning about electricity and electronics so you can gain a solid understanding of the principles involved. Similarly, when you ask how a particular behavior is generated by the nervous system, you might get a general answer about coordinating mechanisms and neural control of muscle action. If you really want to understand what happens inside the nervous system, however, you need to have some familiarity with the details of how neurons work, which this book provides.

Unfortunately, understanding behavior is more difficult than understanding a television set, partly because neural circuits are more complex and less well understood than a television set's electronic circuits. This can make explanations of the neural basis of behavior difficult to understand, because explanations may be too complex to grasp easily, or because they may be couched in vague terms when a full explanation is not yet available.

The theme of this chapter is that even though behavior and the neural circuits that underlie it can be complex, it is sometimes possible to break the behavior into simpler components and to study those. These components can be studied from the point of view of feedback loops in servomechanisms, open loop circuits, or neural circuits that carry out specific computational tasks, or in other ways. From such study, it is possible to make some headway in understanding how a complex sequence of actions such as the maintenance of balance by a cat (see Chapter 18) or the shift of electric organ discharge to a different frequency by an electric fish can be carried out. The examples given in this chapter provide a feel for the approach that neuroscientists can take to dissect the neural mechanisms that underlie behavior. If the details become overwhelming, it may help to step back for a moment and think about what point the details illustrate before you return to a consideration of the details themselves.

Explaining in detail the neural basis of animal behavior is one of the most difficult tasks of neurobiology. Sometimes it is possible to consider a whole behavior as consisting of simpler components. These components may then be studied from the point of view of feedback loops in servomechanisms, open loop circuits, or neural circuits that carry out specific computational tasks.

Additional Reading

General

Churchland, P. S., and T. J. Sejnowski. 1992. *The Computational Brain*. Cambridge, Mass.: MIT Press.

Heiligenberg, W. 1991. *Neural Nets in Electric Fish*. Cambridge, Mass.: MIT Press.

Leigh, R. J., and D. S. Zee. 1983. *The Neurology of Eye Movements*. Philadelphia: Davis.

Wilson, V. J., and G. M. Jones. 1979. *Mammalian Vestibular Physiology*. New York: Plenum Press.

Research Articles and Reviews

Bullock, T. H., R. H. Hamstra, and J. Scheich. 1972. The jamming avoidance response of high-frequency electric fish. *Journal of Comparative Physiology* 77:1–48

Heiligenberg, W., and G. Rose. 1985. Phase and amplitude computations in the midbrain of an electric fish: Intracellular studies of neurons participating in the jamming avoidance response of *Eigenmannia*. *Journal of Neuroscience* 5:515–31.

Hauk, J. C., J. Keifer, and A. G. Barto. 1993. Distributed motor commands in the limb premotor network. *Trends in Neurosciences* 16:27–33.

Robinson, D. A. 1981. The use of control systems analysis in the neurophysiology of eye movements. *Annual Review of Neuroscience* 4:463–503.

CHAPTER 21

Neural Basis of Complex Behavior

An important objective for many investigators is to account for behavior in terms of the actions of individual nerve cells. For much complex behavior, however, this goal is still beyond reach. Even the mechanisms by which an animal makes a simple decision are still not fully understood. Nevertheless, substantial progress is being made in understanding complex functions, including even those in the human brain. This understanding also allows researchers to deal with a variety of malfunctions of the brain.

Neural Decision Making

Not every behavior can be analyzed in terms of reflexes or by looking at specific computations carried out by the nervous system. Some behaviors are too complex for these simplifying approaches and may require that the researcher focus on some other aspect of the behavior to help unravel the neural mechanisms that underlie it.

Consider frogs and toads, for example. These amphibians both hunt and are hunted—they prey on small animals like flies and at the same time are prey for animals like herons and snakes. Hence, a frog or toad must be able not only to identify and catch its own prey but also to identify and avoid potential predators. A wrong choice can mean starvation on the one hand or being someone's meal on the other. Frogs and toads both tend to orient toward and approach visual images that may represent prey and to avoid images that may represent predators. Approach and avoidance behavior can each be studied in its own right, but another way to gain insight into how these behaviors are organized in the brain is to consider the mechanisms by which the animal is able to select the appropriate behavior in response to a particular stimulus.

A Behavioral Choice

How does a frog or toad choose between approach and avoidance? This question has been studied extensively by Jörg-Peter Ewert and his colleagues. Their first step toward answering this question was to characterize the behavior, to discover its features and identify its limits. This approach, as you may recall from Chapter 19, is that used in the field of neuroethology. Actually, Ewert's work is among the earliest neuroethological work carried out on vertebrate animals.

If you were in Ewert's laboratory with him, you would only have to watch these animals to discover that movement is important. In general, a toad such as *Bufo bufo* (or a frog—the following applies to both) will not respond at all to stimuli that are not moving. But factors besides motion must also be important in determining the animal's response. Some of these factors have to do with the internal state of the animal. Toads usually respond to small moving objects. However, if the toad has just eaten it will ignore such an object. Other factors have to do with the stimulus itself. If a toad is hungry, a moving object like a fly will evoke orientation (a turning movement) toward the fly and visual fixation. If the fly is within range, the toad will flick out its tongue to capture it. On the other hand, relatively large moving visual stimuli evoke variable behavior, depending in part on the exact nature of the stimulus. Toads may react to a large stimulus in a threatening manner—lifting themselves up from the ground or gulping in air to inflate their bodies. The toad may react to other stimuli by avoiding them, by either crouching or hopping away.

Experiments with toads by Ewert and others have shown that size and movement, not odor or other stimuli, are the main bases of this prey-predator discrimination. Showing a hungry toad a moving black square projected onto a white surface evokes either an avoidance response turning away from or an orienting response toward the stimulus by the toad, depending on the size of the square (Figure 21-1). If the stimulus is rectangular rather than square, the direction of movement relative to the long axis of the figure is the determining factor. Movements along the long axis evoke orienting responses, as if the toad interprets the stimulus as a worm,

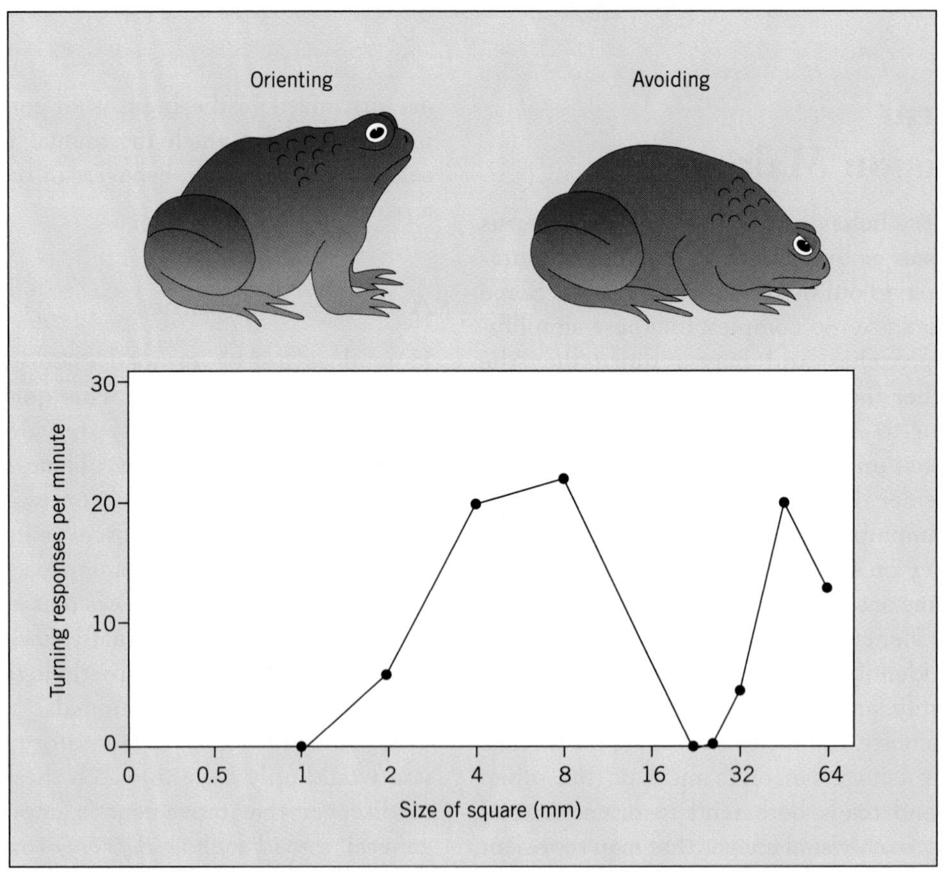

FIGURE 21-1. The relationship between the size of a square, moving stimulus and the number of orienting and avoiding responses per minute for a toad. Toads turn toward small stimuli that may be potential prey items; they turn away from large stimuli that may represent a predator or other threat. Note the clear separation of responses into orienting and avoiding behavior. When the stimulus is intermediate in size, the toad usually does not react at all.

whereas movements along the short axis evoke avoidance, as if the toad interprets the stimulus as a looming predator (Figure 21-2). Apparently, a rectangle that is set on end and moving from side to side "looks" large to a toad, whereas a rectangle that is moving up or down along its long axis "looks" small.

A hungry toad will turn toward and strike at a small moving target but will crouch to avoid a large one. In the case of rectangular targets, apparent size is determined by the direction in which the target is moving relative to its long axis. Movements along the long axis evoke an attack (as if the target is a worm), whereas movements along the short axis evoke avoidance (as if the target is a predator).

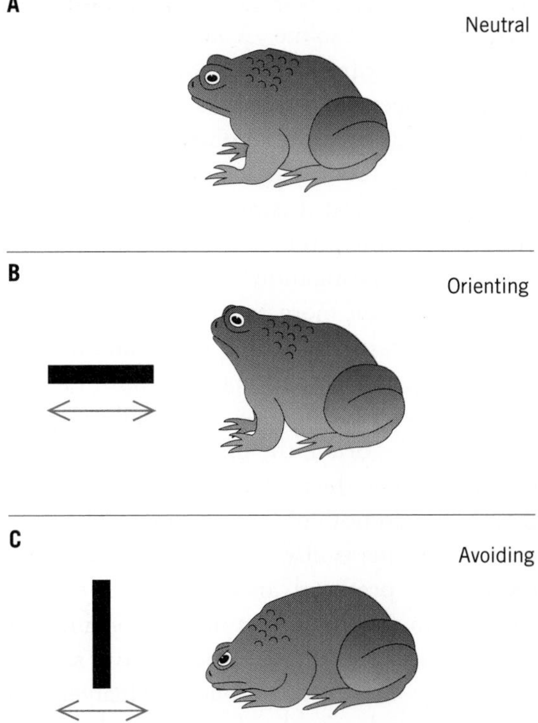

FIGURE 21-2. Typical responses of toads to various stimuli. (A) In the absence of a moving stimulus, the toad sits in a neutral position. (B) The toad orients toward and pays attention to a wormlike stimulus moved along its long axis. (C) If a rectangular stimulus is moved along its short axis, the toad usually shows avoidance behavior. The direction of movement relative to the long axis is the significant variable, not whether the rectangle is oriented upright or lengthwise.

Neural Basis of Size Discrimination

How does a toad make an accurate distinction between prey and predator? At this stage in your study of neurobiology, you might expect that no single mechanism by itself will be responsible, and you would be right. Several components of the nervous system work in concert to help the animal distinguish small from large visual stimuli, and to elicit the appropriate behavioral response to each, as described in the following sections. However, you probably also realize that a good researcher must test even the simplest hypotheses about how stimulus discrimination occurs and how an appropriate behavior is selected.

Since prey seems to be distinguished from predator largely on the basis of visual input, one possibility would be that the eyes contain specialized cells that allow the toad to distinguish between small and large stimuli without ambiguity. Experiments with the visual system of toads have shown that it has similarities to that of other vertebrate visual systems. Receptive fields of retinal ganglion cells are generally circular, with a central excitatory region and an inhibitory surround. Further, some ganglion cells respond phasically, being especially sensitive to movement, whereas others respond better to diffuse illumination. The axons of the ganglion cells project topographically from the retina to the optic tectum and the thalamus, where processing of visual signals takes place.

The most important finding from studies of toad or frog visual systems, however, is that there are no retinal ganglion cells that by themselves constitute a "prey detector" or a "predator detector." That is, no ganglion cells or other cells in the retina respond exclusively and unambiguously to small or large moving stimuli. In the ganglion cells that are specialized for detecting moving objects of various sizes, the distinction between large and small images is not absolute. A large stimulus can excite even a ganglion cell that responds best to small objects, and vice versa. Hence, determination of the size of an object is assisted by the response tendencies of different types of ganglion cells, but it cannot be done without ambiguity in the retina.

Ewert's investigation of neurons in the brain have been more successful in elucidating the neural basis of the discrimination of preylike (small) from predator-like (large) stimuli. Ewert has focused especially on two regions of the brain, the optic tectum and the thalamus, since the retinal ganglion cells send their axons to both of these regions. The posterior thalamus is also referred to as the **pretectum**, or pretectal area, since it lies just anterior to the optic tectum. Recordings of neural activity in these regions have shown that many neurons respond exclusively to moving visual stimuli of a particular size. Neurons in the tectum show different response properties than do neurons in the pretectum. Some neurons in the pretectum give strong responses when the toad is presented with any large stimulus but show no response to small or wormlike stimuli (Figure 21-3A). In the tectum, on the other hand, neurons show the opposite response. They are strongly excited when the animal is presented with small or wormlike stimuli that move either vertically or horizontally (along the long axis of a long and thin figure). This response is inhibited when large stimuli are presented, however, so that tectal neurons show no response to a large stimulus (Figure 21-3B).

Some retinal ganglion neurons respond preferentially but not exclusively to small or large images. In the brain, neurons in the pretectum respond strongly only to large stimuli. The optic tectum, on the other hand, contains neurons that respond exclusively to small moving stimuli and are inhibited by large stimuli.

Organizing the Behavior

Given that there are neurons in some brain regions that respond vigorously to moving images of different sizes, this information could be used by the brain to help organize the appropriate behavioral response. How can this be done? Stimulation and lesion experiments by Ewert's group have suggested that orientation and avoidance behaviors are organized in different regions of the brain, and that the two regions interact to select the actual response that the toad will exhibit. For example, electrical stimulation of certain well-defined areas of the optic tectum elicits orientation or strike behavior in unrestrained toads. Stimulation of regions of the thalamus, on the other hand, especially in the pretectal area, elicits avoidance behavior. Neither orientation nor avoidance can be elicited by electrical stimulation of other regions in the brain, suggesting that the decision to carry out avoidance, orientation, or strike behavior is made in or around these areas of the brain.

The means by which a toad decides whether to orient toward or avoid a stimulus is understood in broad outline. The main mechanism seems to be an inhibition of neurons in the tectum by neurons in the pretectum when large visual stimuli are presented. This hypothesis is supported by experiments in which electrical or chemical lesions are made in the brain. Electrical lesions are made by briefly passing strong electrical current into a restricted region of the brain. Chemical lesions are made by injecting a high concentration of kainic acid into a small region of the brain, which kills all the cells in that region. Focal lesions in the pretectum change the patterns of responses of tectal neurons. Before a lesion is made, these neurons respond to small moving stimuli but are inhibited by stimuli perceived to be large by the toad (Figure 21-4A). After a lesion, these same neurons respond vigorously to any moving stimulus, even a large one (Figure 21-4B). Behaviorally, toads lose all avoidance responses. An animal in which pretectal lesions have been made not only fails to show avoidance to large objects but also actively orients toward

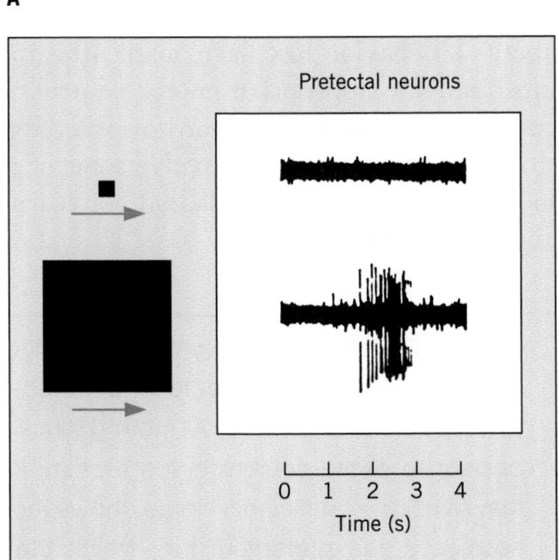

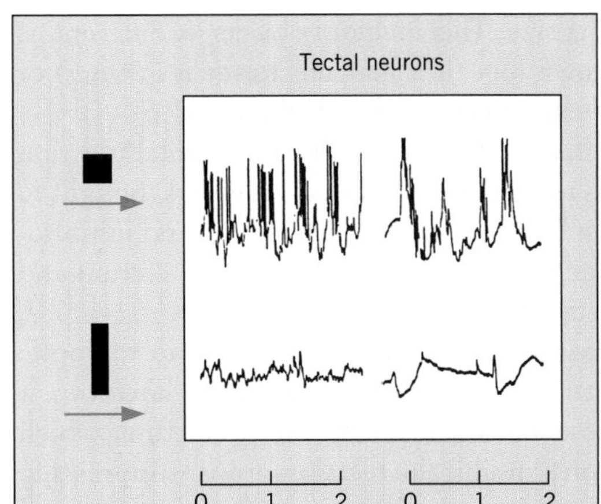

FIGURE 21-3. The activity of (A) pretectal and (B) tectal neurons in response to the presentation of small and large moving stimuli. The pretectal recording was taken with an extracellular electrode; the tectal recording was taken with an intracellular electrode. Pretectal (thalamic) neurons respond to large stimuli, and tectal neurons to small ones.

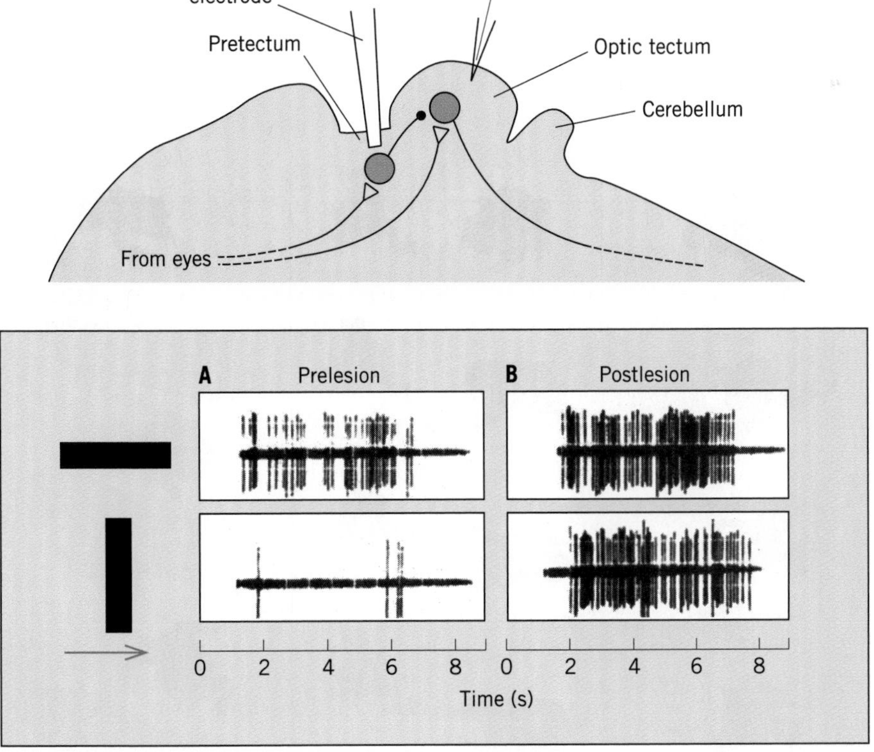

FIGURE 21-4. The effect of selective inactivation of pretectal neurons by injection of strong electrical current on the responses of tectal neurons to large or wormlike moving stimuli. (A) Responses of tectal neurons to small or large moving stimuli in an intact common toad. (B) After the electrical lesion is made, tectal neurons respond vigorously to all stimuli, including large ones.

and strikes at any object that it sees, regardless of its size. This includes objects as different as its own foot (if it happens to see it moving) or the experimenter's hand (Figure 21-5).

Based on these results, the normal reaction of a toad to moving visual stimuli is thought to be a balance between excitatory and inhibitory activity of neurons in the optic tectum and pretectum, respectively (Figure 21-6). A moving stimulus excites neurons in the optic tectum. If the stimulus is large, however, it also excites neurons in the pretectum, which in turn inhibit the tectal neurons, suppressing orientation. At the same time, the pretectal neurons will also activate avoidance behavior. If the stimulus is small, the activity of the excited tectal neurons converges on neurons that control orientation toward or a strike at the stimulus. The behavior of the toad is hence a balance between the excitatory and inhibitory signals indirectly generated by retinal ganglion cells that report on the size and movement of animals in the toad's field of view. Of course, once the initial decision about a behavior has been made, the toad must activate appropriate motor control circuits to move its body toward or away from the stimulus, circuits that involve some of the motor centers of the medulla that you learned about in Chapter 18.

Control of orienting and avoidance behavior seems to reside in the tectal and pretectal areas of the toad brain. All moving stimuli appear to excite neurons in the tectum. If the stimulus is sufficiently large, however, neurons in the pretectum that inhibit the approach response also are excited, suppressing approach and eliciting avoidance instead. Motor patterns involved in approach or avoidance are organized in part in the medulla.

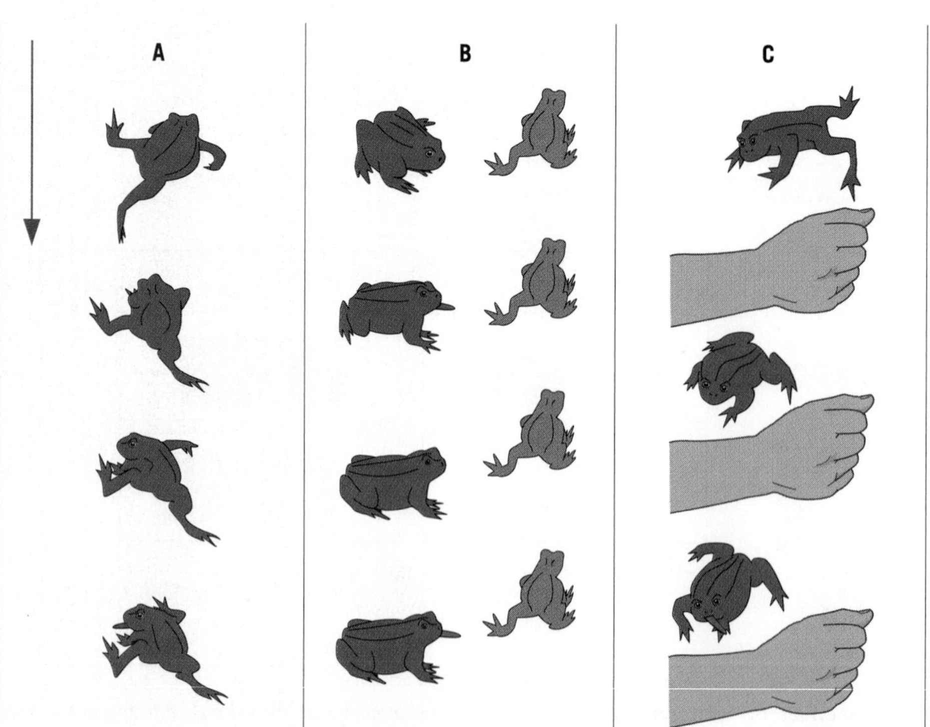

FIGURE 21-5. Behavioral responses of a toad with brain lesions in the pretectum to various large stimuli, including itself (A), another toad (B), and the experimenter's hand (C). Original images traced from filmed encounters.

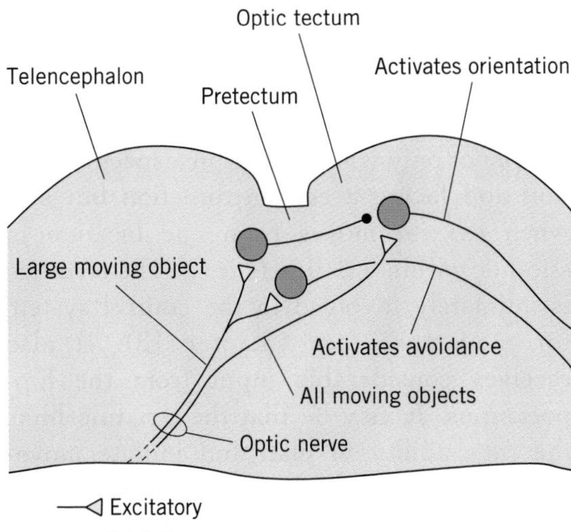

FIGURE 21-6. Lateral view of the brain of a toad, showing the types of excitatory and inhibitory interactions that are thought to underlie the elicitation of orientation or avoidance behavior. Fibers from the optic nerve are distributed to the pretectum and the tectum. Those to the pretectum carry mainly information about large moving objects, whereas the others carry information about all moving objects. Inhibition of fibers to the motor centers by neurons from the pretectum prevents orientation toward large objects.

Neural Activity and Complex Behavior

In principle, it is possible to study the neural basis of any behavior, no matter how complex, by recording the activity of individual neurons and relating this activity to the behavior of the animal. Sometimes the recorded neural activity can be analyzed in terms of feedback loops, feed forward effects, or the action of computational neural circuits, as discussed in Chapter 20. Sometimes the activity can best be understood in terms of some particular decision that an animal must make, such as distinguishing a predator from prey. Sometimes, however, there is no simple way to analyze neural activity. It may be that although individual neurons show activity during a behavior, there is no apparent relationship between the patterns of this activity and what the animal does. In other cases, it may simply be that no neurons that show significant activity related to the behavior of interest have been found.

A relatively new approach to this problem is to investigate the activity of populations of neurons rather than focusing on the activity of individual ones. Certainly it has been possible for over half a century to simultaneously record the summed electrical activity of many neurons from the brain, but this method does not allow researchers to determine the contributions of the individual neurons in the group whose summed activity is being recorded, nor does it allow researchers to identify the specific locations at which changes in activity might be occurring. Instead, several successful new approaches have been used over the past few decades. In one method, researchers use many electrodes to record from a large group of neurons all at once and use computers to help analyze the resulting data. Another method is to use noninvasive methods to record events that are measures of neural activity without actually recording from individual neurons themselves.

In some cases, it is not possible to relate behavior to the activity of individual neurons. In these instances, it may be possible to investigate the activity of populations of neurons, either by recording their electrical activity or by measuring other indicators of neural action.

Neural Correlates of Behavior

Studying populations of neurons has been effective in investigations of some of the more abstract aspects of behavior. One feature of behavior common to all mammals is an ability to become familiar with a place, enabling the animal easily to find its way around a territory. Neural activity correlated with an animal's location in an arena is especially strong in neurons in the dorsal region of the hippocampus (see Chapter 3). Neurons that show such correlated activity are called **place cells** because of their tendency to fire when the animal is in a particular place or is facing a specific site in an explored space. However, place cells are not merely passive indicators of location. In rats, they may change their patterns of firing when the animal receives visual or vestibular cues that its orientation within a space is changing. Their activity may also be influenced by the animal's purpose as it moves about an arena. For example, their activity shows a stronger correlation with place when the animal is following a specific route than when it is exploring or moving about apparently aimlessly.

Neurons in other regions of the brain also show spatially correlated activity. Neurons in the posterior parietal cortex, for example, code for a rat's movement in space, signaling whether the rat is at rest or moving, and if moving, whether the motion is to the right or left. In the striatum, the part of the basal ganglia consisting of the caudate nucleus and the putamen, neurons have been found that are active not only when a rat is in a specific location and facing a certain direction but also when the rat moves from one location to another within a defined arena. The striatum is intimately involved in the control system for movement (see Chapter 18). It also receives considerable input from the hippocampus. It may be that the striatum links the rat's ability to plan and initiate movements to the role of the hippocampus in giving place and movement a particular spatial context. Posterior parietal neurons may help

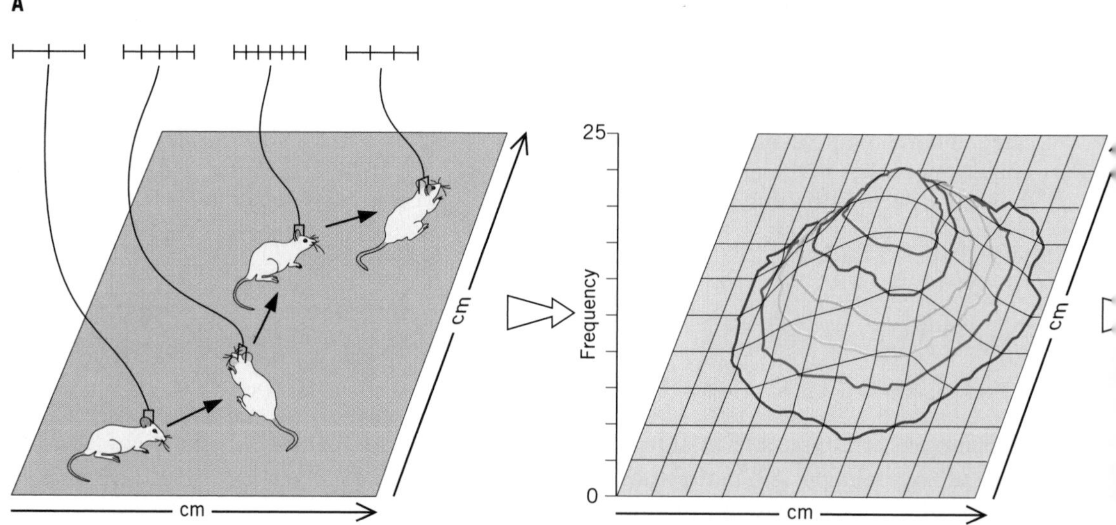

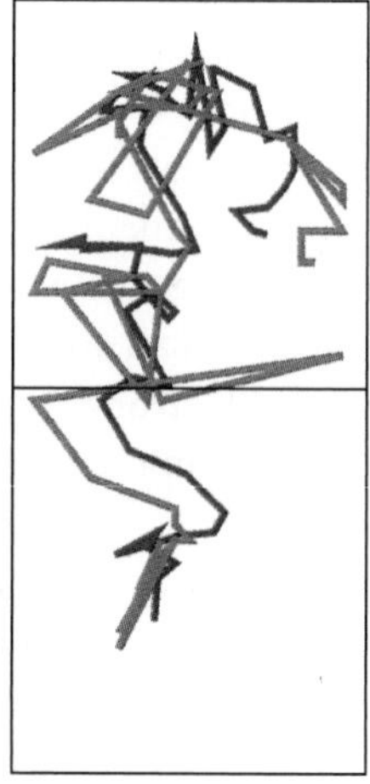

FIGURE 21-7. Collective neural activity in rat hippocampal neurons is a function of the spatial location of the rat in an arena with which it is familiar. (A) Generating a map of spatially related activity in hippocampal neurons. Neural activity is recorded from 80 neurons, by a multielectrode array implanted in the rat's hippocampus, as the rat wanders around an enclosed arena with which it is familiar. For the sake of clarity, activity in only one neuron is depicted here. The neural activity is stored in a computer, along with a record of the rat's location. After several minutes of recording,

the animal integrate sensory input with changes in place, since this brain region receives considerable convergent sensory and motor input. Whatever the role of hippocampal, parietal, or striatal neurons, it is clear that their activity is complex in relation to spatial orientation.

Simultaneous recordings from many individual place cell neurons have suggested a new way of looking at what these neurons do. In one set of experiments, Bruce McNaughton and colleagues recorded from between 70 and 140 individual hippocampal cells simultaneously while a rat wandered around an arena about the size of a small desktop (approximately 7600 cm^2). After the rat had become familiar with the arena, the researchers correlated the firing frequency of each neuron with the rat's location as it moved around the area.

Analysis of the recordings showed two significant features. First, the activity of any single cell usually has only a loose relationship to the location of the rat (Figure 21-7A). Many neurons certainly do change their level of activity (frequency of firing) when the rat is in one part of the arena as compared with some other part, but for most cells it was not possible to say very precisely where in the arena the rat was based only on information about the activity level of that individual cell. Second and in contrast, there is a strong correlation

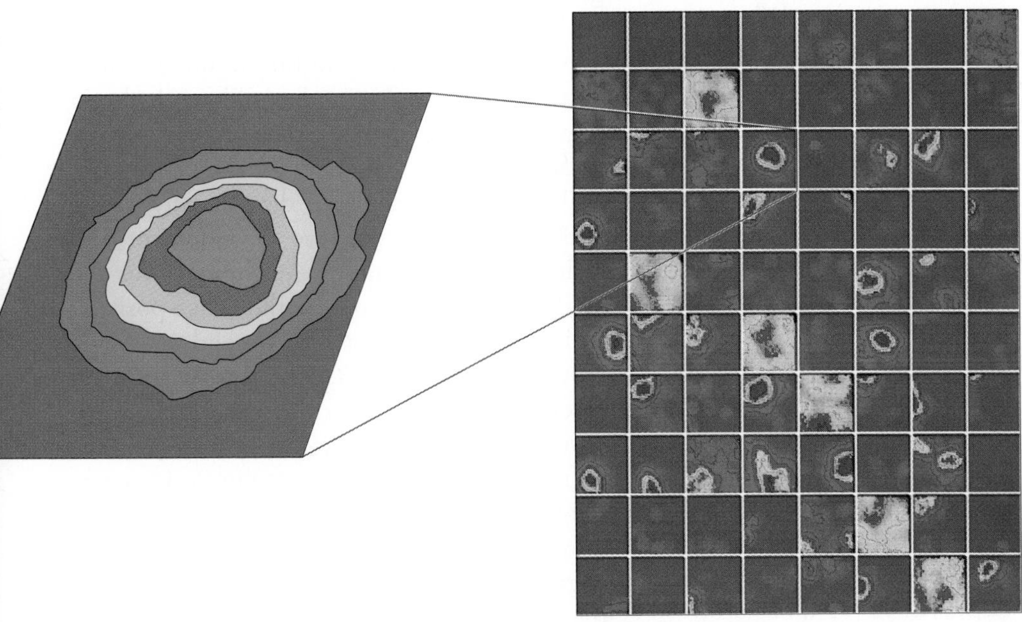

the data are plotted in a three-dimensional graph that relates the physical location of the rat to the firing frequency of the neuron. This three-dimensional frequency plot is converted into a color-coded panel that represents the spatial distribution of activity for that neuron. A similar procedure is used to process the data from every other neuron, generating 80 color-coded panels, which together depict the overall pattern of firing in relation to the rat's location in the arena. Note that some neurons hardly fire at all no matter where the rat is (top left panel), whereas others fire all the time no matter where the rat is (third from left, second row). (B) The predictive power of the collective spatial firing of hippocampal neurons. Researchers compute a single number representing the collective firing of all the neurons for any position of the rat in the arena. From this number, they determine where in the arena the rat most likely was at the time the recording was taken. Comparison of the actual location of the rat as tracked by a video system (black path) with the predicted location (red path) shows remarkably good agreement.

between the collective firing of all the neurons whose activity is being monitored and the location of the rat. That is, when the activity of all the neurons is considered together, this collective activity is a strong predictor of the rat's spatial location within the arena. Hence, each spot in the arena evokes a unique collective pattern of activity in the neurons, and as the rat moves from one place to another, the pattern of activity changes as well. This relationship between place and neural activity is so strong that McNaughton and his coworkers can tell to within 2 to 5 cm where the rat is in the arena just by knowing what the pattern of activity is in the neurons from which recordings are being taken (Figure 21-7B).

Not only are unique neural patterns associated with places familiar to the rat, but it appears that a rat is quickly able to build up unique patterns of response to new places. When a rat enters an arena it has not been in before, there is little or no correlation between the animal's location and the collective pattern of firing in the monitored neurons. However, after a period of 10 to 20 minutes, strong patterns associated with the new place develop. Development of these new patterns of neuronal activity does not seem to interfere with retention of the old neural patterns that are associated with the original arena. These results suggest that neurons in the hippocampus of the rat act together to encode spatial location relatively quickly and with a high degree of accuracy. This hypothesis would have been difficult to support based only on recordings from a single cell at a time.

The location of an animal in its surroundings may be signaled by neurons in several places in the brain. Neurons in the striatum of rats show activity correlated with the animal's position or movements within a known environment. Place cells in the hippocampus are also active in relation to the animal's spatial orientation. Recordings from groups of such neurons suggest that the cells collectively give a precise indication of where the animal is in a defined arena.

Studying Abstract Mental Function

In spite of the many successes of the single- and multiple-cell approaches, some researchers are not convinced that the analysis of such data can explain abstract mental functions like reasoning. Furthermore, since moral considerations quite properly preclude invasive experimentation on humans, studying single neurons or even populations of neurons will never help us to understand those mental functions that are especially human, such as abstract thinking or using language. Psychologists, of course, have studied these matters for more than a century, using a variety of tests to make inferences about how cognitive functions are organized in the brain. They and other scientists have also gained significant information from studies of brain malfunctions caused by disease or injury. Nevertheless, for most of the twentieth century an in-depth understanding of how the brain organizes abstract mental functions eluded researchers.

The situation changed dramatically in the 1980s, however, with the development and refinement of noninvasive techniques for the study of brain function. Methods like positron emission tomography (PET) and magnetic resonance imaging (MRI) have allowed researchers to peek into a living brain while it carries out specific tasks, and thereby to see which parts of the brain are active as it carries out these tasks (Box 21-1). The wealth of new information has even spawned a new field, **cognitive neuroscience**, devoted explicitly to the study of complex, so-called higher, mental functions. The main interest of cognitive neuroscientists is to understand these complex functions in terms of the operation of specific neural pathways within the brain, even in humans. The kind of analysis of function that is possible at this time is less detailed than the analysis of single-cell function you have already read about. However, the information that has become available from the application of modern imaging techniques to the human brain has provided new insight into how the brain is organized, as well as opening many tantalizing possibilities for future research.

BOX 21-1

Seeing Thinking: Noninvasive Techniques for Studying Function

Several noninvasive whole-body techniques have been developed since the 1980s that have allowed direct visualization of parts of the living brain, as well as a peek into its functional organization. These are based on the technique known as tomography, which originally was applied in clinical medicine for the reconstruction of a two- or three-dimensional image of the body through the use of multiple-angle X-ray pictures. With the advent of faster and more powerful computers to aid in the calculations necessary for the interpretation of the images, the method has been expanded and applied especially to the brain, and has become known as **CT** or **CAT scanning,** meaning **computed (axial) tomography** scanning. A computer allows researchers to obtain a reconstructed picture of a "slice" through the soft tissues of the body entirely without invasive procedures of any kind.

A newer technique, the **positron emission tomography** (**PET**) scan, works on similar principles. In this method, a substance that has been labeled with a radioactive atom such as ^{15}O is given to the subject, orally or by injection. ^{15}O gives off only a mild form of radiation, and hence presents no great danger. As the radioactive material decays, it releases positrons ("positive electrons"), each of which, when it collides with a normal electron, produces two gamma rays. These rays are picked up by an array of detectors around the body of the subject. Computers then construct an image of a slice through the brain from the detector data.

PET scans differ from CAT scans in that they can give information about brain activity rather than brain structure. Subjects can be given radioactively labeled 2-deoxy-D-glucose (2DG), an analog of glucose. Active neurons, which require more glucose than do resting neurons, will take up 2DG along with the glucose. However, the 2DG cannot be metabolized and will accumulate in active neurons, hence labeling an active brain region. Another technique is to inject a radiolabeled substance into the blood, which allows changes in blood flow through the brain to be registered. Since blood flow increases in active brain areas, changes in brain activity can be detected. Text figure 21-8 shows four PET scan images.

A third technique is known as **magnetic resonance imaging** (**MRI**), named for the use of a "resonant frequency" radio signal to excite protons of hydrogen atoms. The subject is placed in a strong magnetic field, then exposed to a high-frequency radio signal. When the frequency of this signal is adjusted to the resonant frequency of the protons, they jump from a low to a high energy state. Turning off the radio signal allows the protons to fall into their earlier state, releasing energy in the form of electromag-netic radiation. This harmless radiation is picked up by detectors, and its distribution is analyzed to produce an

(continued)

Seeing Thinking: Noninvasive Techniques for Studying Function (*continued*)

image, as in a PET scan. Most hydrogen in the brain by far is in fatty tissue like myelin and in water. Because different parts of the brain differ in the amount of myelin or water they contain, MRI scans yield excellent, high-contrast anatomical images that allow researchers to distinguish these different parts from one another.

When MRI was first introduced, it had to be combined with PET to provide a functional picture of brain activity to go with the anatomical image produced by MRI. However, researchers have recently been able to combine functional imaging with MRI by taking advantage of a transient local increase in the proportion of oxygenated to nonoxygenated hemoglobin that occurs in the blood in the region of metabolically active neural tissue. This changed ratio changes the magnetic properties of the region, allowing researchers to identify the active tissue at the same time as they obtain an image of brain structure. The combination of functional and anatomical images, known as **functional MRI,** or **fMRI,** has led to a proliferation of studies in which brain imaging is used (Figure 21-A).

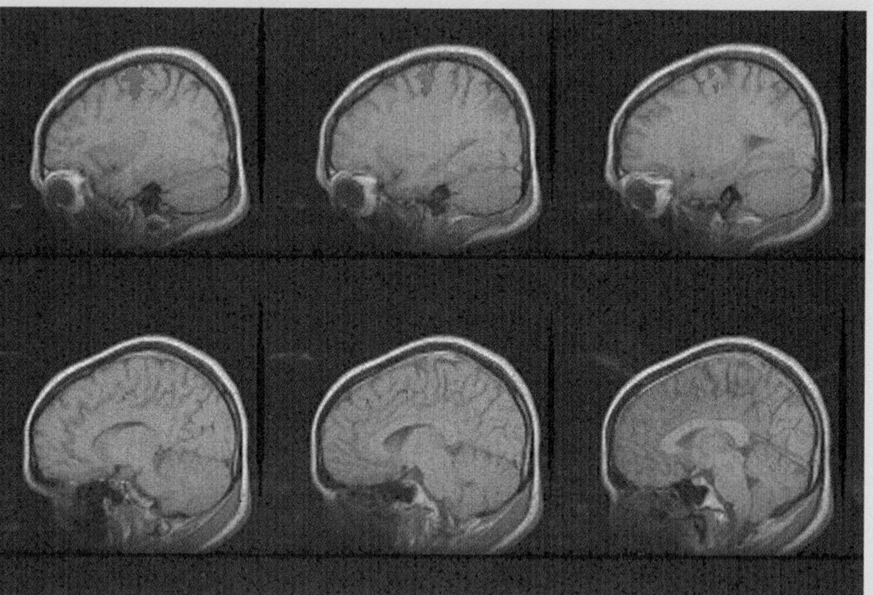

FIGURE 21-A. Functional MRI images of a human brain. The subject repeatedly touched the thumb of one hand to each of the other fingers of the same hand in sequence while the fMRI images were obtained. The six images shown here are part of a series taken through the entire head spaced 3.75 mm apart. An image was acquired about every 100 msec. The image of the head and brain is equivalent to an ordinary MRI image. The red areas are the brain regions that are activated by the task. Shown in the top three images is activation of the primary somatic and motor areas. Shown in the bottom images are activation of the supplementary motor area along the medial surface of each hemisphere just above the corpus callosum and activation of part of the cerebellum. (Courtesy of Dr. Venkata S. Mattay, Daniel R. Weinberger, and Joseph A. Frank.)

The introduction of sensitive, noninvasive techniques for imaging the human brain has made it possible to study so-called higher mental functions such as abstract thought, and has given rise to a newly designated field of research, cognitive neuroscience.

Localization of Function

Even in earliest historic times there were two conflicting ideas about the relationship between mental abilities or behavior and specific regions of the brain. On the one hand, the ability of people to recover from some types of head wounds suggested that function is distributed over much of the brain. Even today it is astonishing to consider Phineas Gage and his physical recovery from the complete destruction of a substantial part of his brain (see Chapter 1). On the other hand, the loss of function after damage to specific brain areas, such as the loss of sight that follows a crushing blow to the occipital lobes, suggested that function is localized in specific areas. Here, too, Phineas serves as an example. The dramatic change in Phineas's personality following his injury certainly suggests that the damaged regions of his brain had a specific role to play in his behavior, even if that role is not well understood.

One of the earliest well-known proponents of localization of function was Franz Gall, who worked at the end of the eighteenth century and the beginning of the nineteenth century. Gall's work became associated with phrenology, the idea that specific regions of the brain were the seats of faculties and feelings like memory and courage. By the middle of the nineteenth century, the proponents of phrenology popularized and carried it far beyond anything Gall condoned. As a consequence, for many years there was a strong reaction against any but the most rudimentary notion of localization of brain function.

Nevertheless, it has become increasingly clear over the last century of brain research that every part of the brain has some specific functional role. Even regions like the frontal cortex, about which nearly nothing was known 40 years ago, is now believed to consist of regions that have distinct functions. The surprise has been not that there are functional subdivisions but that parts of the brain that are physically distant from one another may work together in carrying out some particular mental task.

The main foundation for this new understanding has been provided by imaging techniques like PET, which have been used with particular success by Marcus Raichle, Michael Posner, and their colleagues for the study of the involvement of parts of the brain in human language. These techniques allow researchers to identify areas of the brain in which there is increased neural activity by pinpointing regions in which there is an increase in blood flow or an increase in metabolic activity. The procedure involves taking a scan of the brain when the subject is at rest, then repeating the scan when the subject is engaged in some particular task. By using a computer to subtract one image from the other, researchers can obtain a picture that highlights just the areas of the brain in which there is increased neural activity.

Consider, for example, what happens when someone reads, hears, or speaks a word. If a word is briefly flashed on a screen in front of a subject, two parts of the brain become active over background levels: the visual cortex and the **visual association area**, in the occipital lobes (Figure 21-8A). The visual association area, which lies anterior to the visual cortex, aids in the interpretation of visual images. If the subject lies with eyes closed and listens to a word, the **angular gyrus** in the posterior temporal lobe and **Wernicke's area** at the junction of the temporal and parietal lobes become active (Figure 21-8B). The angular gyrus is thought to integrate auditory input with input from other senses, and Wernicke's area helps decipher spoken words. If a subject is asked to say a word out loud, two other areas of the brain,

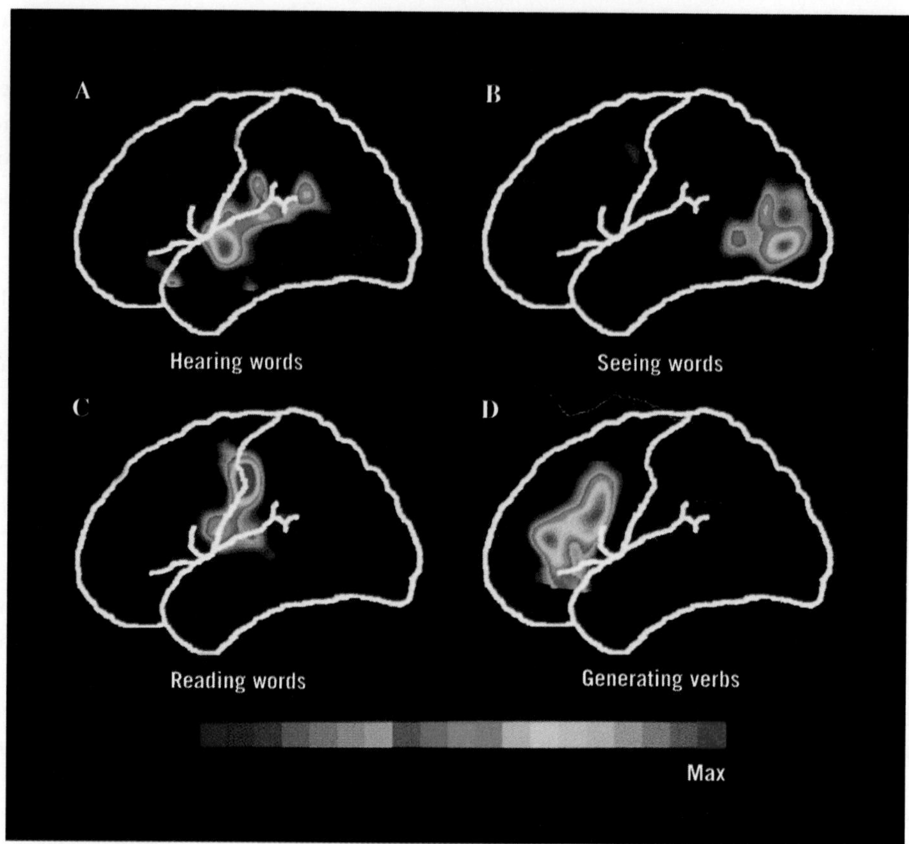

FIGURE 21-8. The areas of the human brain that become active during different activities. The brain scans are generated by subtracting the background activity of the brain from the activity that is present when the designated task is being carried out. See text.

Broca's area and the part of the primary motor cortex that controls movements of the lips and tongue, become active (Figure 21-8C). Broca's area is important in generating grammatical speech, and of course the motor area is responsible for producing the movements of the throat, lips, and tongue that allow the subject to pronounce a word.

Other regions of the brain are involved in mental manipulation of words. Suppose that a subject is presented with a noun and asked to say a verb that represents an appropriate action for the noun. If the word is "pencil," for example, the subject might say "write." Subtracting all the brain activity associated with reading, hearing, and saying a word still leaves a region of increased activity in the inferior frontal cortex, part of the lower surface of the frontal lobes, near the eyes (Figure 21-8D).

The significance of these studies is that they show the involvement of several distinct and separate areas of the brain in deciphering and generating language. Well-known regions such as Wernicke's and Broca's areas have long been known to be necessary for the understanding and generation of grammatical speech. But other, less well-known brain regions are also required for any type of mental manipulation of words or their meanings. These brain regions are not physically located near the traditional language areas, showing that the brain uses several specific, local regions that may be widely separated in order to carry out what to us seems like a single, seamless function.

Not only are several brain regions involved in what appears to be one task, but the conditions of the task can influence the way in which it is carried out. For example, directing attention to one part of a task rather than another can change the order in which parts of the brain are engaged. If a subject is asked to look at several words and to pick out one that has two particular features, all the brain areas involved in these tasks become active. Suppose a subject were shown two nouns like "train" and "rose," and asked to pick out the one that named a manufactured object. Particular regions of the brain would become active. Now suppose the subject were shown the same pair of nouns, except that one of them is printed with a letter in bold, like the letter **"i"** here in "tra**i**n." When the subject is asked to pick out the word with a letter in bold, a different set of brain areas becomes active. If the subject is then asked to pick out of a list a single word that named a manufactured object and contained a letter in bold, all of the brain areas involved in either task become active.

The real experiment comes when the subject is asked to pick out a word having both qualities but to concentrate on one quality more than the other. Brain PET scans alone cannot detect any differences between brain regions that become active under these conditions because the scans cannot be taken fast enough. However, recordings of brain electrical activity with focal scalp electrodes whose placement is guided by the PET experiments have shown that the brain regions involved in the task to which attention is directed become active a bit before those involved in the other task. If the task on which the subject is to concentrate is changed, so is the order in which brain regions become active.

The processing of verbal information by the brain can be affected by the subject's familiarity with the task to be carried out, as well as by where a subject directs his or her attention. For example, if a subject is allowed to practice the task of giving a verb upon presentation of a noun in a specific list of words, the brain regions that become active during the task change. A naive subject uses regions such as the anterior cingulate, left prefrontal, and left posterior temporal cortices. The neural activity in these regions is significantly reduced after practice, and other regions become active instead. This change in activation suggests that after practice, even a task involving semantic associations becomes routine and engages only brain regions that are also active in simple word repetition tasks. Presentation of a new list re-engages the original brain areas.

The main impact of studies of this type has been to make researchers aware that the so-called higher mental functions are not actually single tasks but are treated by the brain as a series of subtasks. Each subtask is then carried out by some particular part of the brain. For example, hearing or reading one word and generating some related word involves separate parts of the brain for receiving and interpreting the sensory input, interpreting the meaning of the word, and generating the related word. Since the areas of the brain in which these functions are carried out are quite distant from one another, there must be neural pathways that connect the different regions so they can act functionally as an integrated whole.

Careful study of patients with lesions caused by a small, localized stroke has suggested a similar picture for functions other than language. A particular mental function is carried out by several separate, discrete areas of the brain. Disruption of the pathways between these areas will impair the ability of a patient to perform the associated mental function. However, as the studies of the effects of practice demonstrate, the brain is extremely flexible regarding the regions it will use to carry out any one particular task, depending on specific circumstances and immediate past history. As further PET and other imaging studies are conducted, it seems likely that many new insights into the organization of the human brain will be forthcoming.

The human brain is organized so that different parts carry out quite specific functions. In consequence, carrying out any sort of abstract task, such as assigning meaning to a word, involves several different brain regions, regions that are usually quite distinct and even distant from one another. Directing attention to a task seems to enhance or give priority to the task to which attention is directed, and practicing a task changes the pattern of activation of brain regions.

Lateralization of Function

Study of brain activity during specific tasks has revealed not only that brain function is distributed in different, localized regions but that it is lateralized as well. That is, the functions and capabilities of the right and left hemispheres of the brain are not identical. For example, in humans it has been recognized for many decades that in most people the capacity to understand and use language is centered in the left cerebral hemisphere. If Broca's area or Wernicke's area in the left hemisphere is damaged, the victim usually shows severe deficits in articulating or understanding speech, respectively. Damage to the corresponding regions of the right hemisphere, however, rarely has these effects.

This is not to say that the right half of the cerebral cortex has no independent cognitive abilities. The right hemisphere's ability to interpret sensory input and reason from it was first demonstrated in humans by Roger Sperry, Michael Gazzaniga, and their coworkers in a series of elegant experiments with split-brain patients. **A split-brain patient** is a person in whom the corpus callosum, the broad band of nervous tissue that joins the right and left cerebral hemispheres, has been surgically severed. First performed in 1940, this operation, which results in functional as well as physical separation of the right and left cerebral hemispheres, has been used to treat otherwise intractable and severe cases of epilepsy. The procedure works because it separates two brain regions that, by exciting one another, are in some cases responsible for initiating epileptic attacks. The surgery is generally successful, leaving a patient who is apparently normal in all respects. However, detailed examination of such patients reveals that after the operation, the right and left cerebral hemispheres function independently of one another.

Investigation of the capabilities of the separated cerebral hemispheres of these patients is possible because of two features of human neuroanatomy. First, as in all vertebrates, the right cerebral hemisphere controls muscles on the left side of the body, and the left hemisphere controls those on the right side. This phenomenon occurs because the motor tracts leaving the motor cortex cross over to the other side of the brain before they enter the spinal cord. As described in Chapter 18, the crossover occurs in the medulla, in the region known as the **pyramidal decussation**. This decussation is below the level of the corpus callosum, and hence is unaffected by severing the callosum. Somatosensory input from touch receptors and proprioceptors to the cerebral hemispheres also crosses over in the brain stem (see Chapter 14).

Second, output from the optic nerve to the visual centers in the occipital lobes is divided in such a way that when the eyes are directed straight ahead, an image in the right side of the visual field is conveyed exclusively to the left hemisphere, and an image in the left field is sent to the right hemisphere (Figure 21-9). This condition is also unaffected by surgical section of the corpus callosum, since the optic chiasm, where the crossover takes place, is below and anterior to the callosum.

Sperry and his colleagues took advantage of these features of human anatomy to study the capabilities of the two cerebral hemispheres in human split-brain patients. Their objective was to test the idea that the cerebral hemispheres could function independently of one another after the corpus callosum had been cut. To do this, the investigators arranged a subject in such a way that he could see images flashed briefly on the right or left half of a screen in front of him, then asked him to indicate what he had seen. If an image is presented briefly in the right half of the visual field, the image will be sent to the left visual cortex, and the left cerebral hemisphere will be able to respond. For example, if the picture of a banana is flashed onto the right half of the screen, the patient could describe the banana, since the language centers are also in the hemisphere that receives the image of the banana. On the other hand, if the image is in the left half of the visual field, the patient will be unable to say what he has seen since the right hemisphere has no verbal ability.

In order to answer the question of whether the right hemisphere could also interpret an image, the researchers had to take advantage of the separate motor control of the two halves of the body by the two cerebral hemispheres. They did this by presenting the subject with a group of objects, one of which corresponded in some way to the image flashed on the screen, and asking him to choose the object that best matched or was related to the image. The key element of the experiment was that the subject was not allowed to look at the objects and had to select one using one hand only. When the image was flashed onto the right side of the screen, the subject was able to use his right hand to pick up the object because the right hand was controlled by the left cerebral hemisphere, which had seen the image (Figure 21-10). He could not choose the correct object (except by chance) with his left hand because the right cerebral hemisphere did not know what had been flashed on the screen. On the other hand, when an image was flashed onto the left half of the screen, the patient could easily select the corresponding object with his left hand. This result indicated that the right hemisphere was fully competent to interpret visual images and to respond to spoken instructions, even though it was mute.

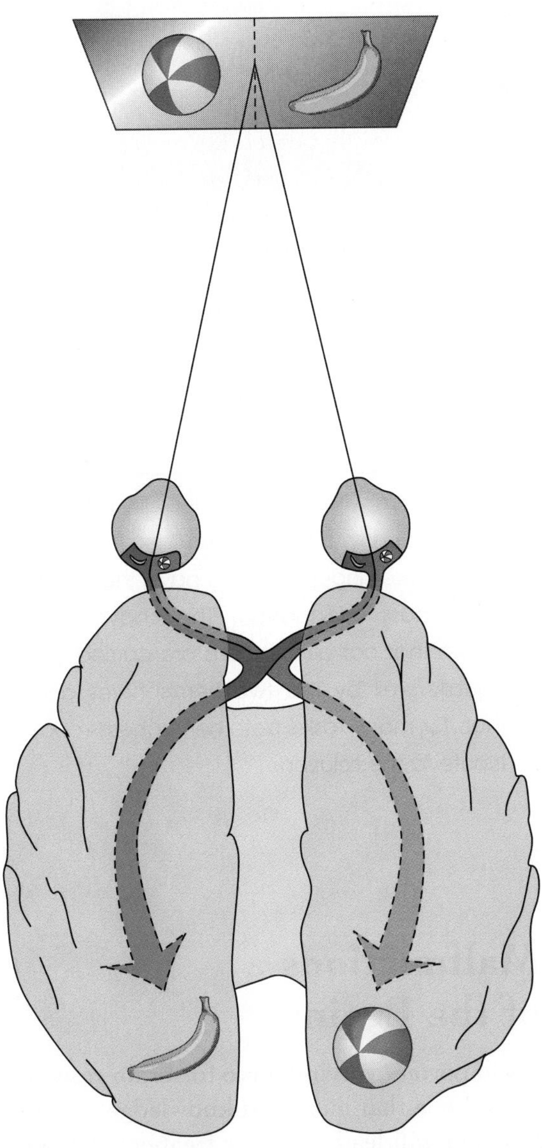

FIGURE 21-9. Diagram showing the distribution of visual field information to the right and left hemispheres of the human brain. Images in the right field are sent exclusively to the left hemisphere, and vice versa.

Many studies of the cognitive abilities of the right cerebral hemisphere have now been carried out using Sperry's techniques. The results have shown that except for linguistic abilities, the right hemisphere can carry out many of the same kinds of tasks as the left. However, the abilities of the two hemispheres are not identical for each type of task. In general, the right hemisphere performs better at cognitive tasks that require parallel processing of many small items of information, such as identifying faces. The left hemisphere, on the other hand, does better at tasks that must be broken into logical elements to be analyzed, such as solving logical puzzles.

Although it is clear that the independent cerebral hemispheres show differences in what functions they can best perform, it does not necessarily follow that in the intact brain these functions are carried out exclusively in one or the other of the hemispheres. In fact, there is extensive evidence that both cerebral hemispheres contribute to every higher cognitive function. The brain seems to be organized in functional modules and performs various functions by combining the actions of different modules, some in the right hemisphere, some in the left.

A striking demonstration of the necessity of this cooperation for some tasks can be seen in

FIGURE 21-10. Diagram of the experimental arrangement for the human split-brain patient experiment. The subject is seated in front of a screen, eyes fixed on the middle of the screen. Images that are flashed onto the left side of the screen go to the right hemisphere of the brain, and vice versa. In this illustration, the subject is asked to pick an object from the tabletop that corresponds to what he has seen on the screen. He can pick out the correct object with his right hand only if the image has been flashed onto the right half of the screen.

the performance of split-brain patients on the wire figure test. In this test, a subject is blindfolded and allowed to manipulate in one hand a geometrical figure such as a pyramid or a cube made of stiff wire. The wire figure is then placed with other figures. When the blindfold is removed, the subject must identify by sight the shape that he or she manipulated. Normal individuals can do this easily. Split-brain subjects, however, are unable to identify the shape when the blindfold is removed, no matter which hand they used to manipulate the shape. This inability indicates that each of the hemispheres makes an essential contribution to the solution of the problem. When the lines of communication through the corpus callosum are severed, the ability of the brain to solve the problem is lost, because neither hemisphere can do so by itself. This result also indicates that different brain areas must interact for successful completion of some tasks, as also indicated by the brain imaging studies of reading and word interpretation described in the previous section.

Experiments with human split-brain patients have shown that the right and left cerebral hemispheres independently have the ability to process information. They have also shown that not all functions are carried out equally well by the two hemispheres and that, for many tasks, both hemispheres contribute to the solution.

Malfunctions of the Brain

An important driving force for brain research is the hope that increased knowledge of brain function will lead to better treatment of various brain malfunctions that afflict humans. Because of the devastating human and social consequences of many of these malfunctions, the importance of this motive for neuroscience research can hardly be overestimated. However, by treating these malfunctions as experiments of nature, however tragic, researchers may also gain a better understanding of the so-called normal brain. Increased knowledge, in turn, may lead to better treatment, even though better treatment may not have been the original objective of the study. This is the underlying rationale for much basic research.

The symptoms of many brain malfunctions reveal the enormous complexity of the nervous system in general and the human brain in particular. In fact, the most important generalization that has arisen from study of brain malfunctions is that in many of the most serious diseases, symptoms are induced by failure of neurotransmission in specific sets or types of neurons. The result is a disease whose manifestations are often neurologically widespread and devastating to the individual.

Many examples illustrate this point. Huntington's disease, described in Chapter 18, is caused by degeneration of cholinergic and GABAergic neurons in the striatum, the part of each set of basal ganglia that lies mostly anterior to the thalamus. This degeneration, which ordinarily begins well into adulthood (past age 30), leads to uncontrollable movements, grotesque facial expressions, and loss of mental function, before death finally claims the victim. The course of the disease from minor motor disturbances to eventual death is a reflection of the progressive degeneration of neurons susceptible to the disease.

Alzheimer's disease is also caused at least in part by neural degeneration. The disease is characterized initially by loss of memory, followed eventually by complete mental disorientation and loss of control of bodily functions. It is associated with several morphological changes in the cells of the brain, mainly the formation of senile plaques and intracellular neurofibrillary tangles, followed by the eventual death of the afflicted neurons. Also associated with the disease are deficits in cholinergic transmitter systems in the cortex, manifested as a reduction in levels of choline acetyltransferase, the enzyme that catalyzes synthesis of acetylcholine (ACh) (see Chapter 7). The resulting loss of ACh activity is due mainly to degeneration of cholinergic neurons in the basal forebrain. Unfortunately, researchers do not yet know either what causes the disease or the nature of the cellular damage that causes degeneration and brings on the symptoms. Some researchers believe that several factors may induce the disease, complicating efforts to understand its roots.

Deficits in specific neurotransmitter systems in the brain have also been implicated in other brain malfunctions. For example, bipolar depressive illness (manic depression), a serious disorder of mood, is characterized by striking deficiencies in serotonergic and noradrenergic transmission, and many of the psychoactive drugs that are used to treat its symptoms act mainly by increasing catecholamine availability in the brain. The presence of schizophrenia, which is a disorder of thought processes rather than one of mood or emotion, is strongly correlated with an abnormally high level of activity of dopaminergic transmitter systems. Antipsychotic drugs like chlorpromazine that are used to treat schizophrenia powerfully block dopamine receptors.

The cause or causes of any of these diseases are not completely known, but it is clear that some have a significant genetic component. For example, Huntington's disease has been proved to be caused by an allele of a single gene. The allele is located on chromosome 4 and is transmitted as an autosomal dominant. It codes for a protein whose function is not known. Even in its normal allele, this gene often contains what are called repeats, sections of DNA that code over and over for the same amino acid. In the Huntington's allele, the number of repeats is excessive; instead of a dozen or so repeats, individuals with Huntington's disease may have anywhere from about 40 up to several hundred repeats, resulting in the synthesis of a protein containing long chains of a single amino acid. The presence of these amino acid chains somehow interferes with normal metabolism of the protein. In consequence, the protein seems to "clog" the metabolic machinery of the neuron, eventually causing the neuron to die.

It is not yet understood why the defective gene, which is present in every cell, affects only some of the neurons in the brain. Nor is it understood why the loss of the affected neurons should cause the widespread symptoms that it does. It is evident that many of the symptoms of disturbed motion stem from interference with the basal ganglia and their control of voluntary movement. In Huntington's disease, there is sharply decreased output from the basal ganglia to other parts of the brain, due to the loss of cholinergic and GABAergic neurons. The cause of other symptoms is not known.

Other brain malfunctions have also been shown to have a significant genetic component, although few appear to be caused almost exclusively by the action of a single autosomal dominant (or even recessive) gene. For example, some manifestations of Alzheimer's disease appear to have a genetic basis, the relevant gene being located on chromosome 21. The gene may be related to the gene for Down's syndrome, which is also located on this chromosome; an early-onset form of Alzheimer's occurs in nearly all individuals with Down's syndrome who live to be older than 35. Further, schizophrenia has been shown to run in families in such a way as to suggest that a genetic component is important in that disease as well. The component may contribute nothing more than a susceptibility to the disease when certain environmental conditions occur, but it is there nonetheless. In neither of these cases, however, do researchers know what specific gene or genes might be involved, nor what the product or function of these genes might be.

Even Parkinson's disease (see Chapter 18), for many decades considered to be caused entirely by some environmental agent, may have a genetic component. The cause of the degeneration of neurons in the substantia nigra, which is the root of the disease, is unknown, but a toxic environmental agent that builds up in the body over the course of years has been strongly implicated. The main evidence for this view is that the disease first appeared in the eighteenth century, at the time of the Industrial Revolution. Even though the symptoms of Parkinson's disease are striking, there are no descriptions of these symptoms by the Greeks, Egyptians, Chinese, or other ancient peoples who left medical records, suggesting that the disease was not present at that time. Nevertheless, research on five generations of a single Italian family, as well as research on other families, strongly suggests a genetic basis of the disease. It may be, however, that the penetrance of the allele (i.e., the extent to which the allele is expressed when it is present) may vary from population to population, being relatively low in most populations.

A full understanding of the root cause of and effective treatment for these diseases still eludes us. Nevertheless, it is clear that in every case an important if not unique component is a malfunction of synaptic transmission in one or more places in the brain. The recognition that changes in transmitter function can have such widespread and profound effects on brain function has opened brain research to many new possibilities. Application of noninvasive imaging techniques to the study of organic malfunctions of the brain, combined with the biochemical studies of drug action, is yielding new hypotheses about how different parts of the brain cooperate to carry out a particular function, and how disruption of this cooperation can bring about specific deficits in mental function.

Most brain malfunctions are caused by loss of neurotransmitter in specific neural circuits, either because of degeneration of the neurons that normally provide it or because of failure of the neurotransmitter to stimulate its target. A significant contributing factor to many brain malfunctions is one or more genes that either cause or make the person susceptible to a specific disease.

Additional Reading

General

Ewert, J.-P. 1980. *Neuroethology: An Introduction to the Neurophysiological Fundamentals of Behavior*. Berlin: Springer-Verlag.

Posner, M. I., and M. E. Raichle. 1994. *Images of Mind*. New York: W. H. Freeman.

Roland, P. E. 1993. *Brain Activation*. New York: Wiley-Liss.

Sacks, O. W. 1985. *The Man Who Mistook His Wife for a Hat and Other Clinical Tales*. New York: Summit Books.

Research Articles and Reviews

Doran, M., and D. G. Gadian. 1992. Magnetic resonance imaging and spectroscopy of the brain. In *Quantitative Methods in Neuroanatomy*, ed. M. G. Stewart, 163–79. New York: Wiley.

Ewert, J.-P. 1987. Neuroethology of releasing mechanism: Prey-catching in toads. *Behavioral and Brain Sciences* 10:337–405.

Gazzaniga, M. S. 1994. Consciousness and the cerebral hemispheres. In *The Cognitive Neurosciences*, ed. M. S. Gazzaniga, 1391–400. Cambridge, Mass.: MIT Press.

Raichle, M. E. 1994. Images of the mind: Studies with modern imaging techniques. *Annual Review of Psychology* 45:333–56.

Wilson, M. A., and B. L. McNaughton. 1993. Dynamics of the hippocampal ensemble code for space. *Science* 261:1055–58.

Zola-Morgan, S. 1995. Localization of brain function: The legacy of Franz Joseph Gall (1758–1828). *Annual Review of Neuroscience* 18:359–83.

PART 6

The Malleability of Neural Systems

One of the truly remarkable features of any nervous system is its ability to adapt to varying conditions. This changeability, or malleability, is manifest right from the beginning, during early growth. The experiences of a newly born animal and the environmental conditions to which it is exposed have a profound influence on the capabilities that its nervous system has when that animal reaches adulthood. It is therefore not surprising that these experiences often affect the animal's behavior. Experience can shape the nervous system and hence influence the behavior of adult animals, as well. In addition, hormones released by the endocrine system can have a dramatic effect on the malleability of the nervous system, shaping it both structurally and functionally.

Part 6 The Malleability of Neural Systems 508

CHAPTER 22

Development

Malleability appears first at about the time of birth, before the nervous system is fully mature. Hence, a full appreciation of how the environment can influence a growing nervous system can best be gained by first considering the principles by which nervous systems develop. In this chapter, you will learn about how neurons grow and how proper neural connections are formed. In the next chapter, you will see how the connections that are formed before birth can be modified.

Nervous systems develop just as do other organ systems. They also follow some of the same rules that have been found to apply to those other systems. However, the unique requirements of a nervous system also impose new requirements on its development.

Development of the Nervous System

Although developmental biology has been a distinct field of study for well over a century, only in the last few decades has **developmental neurobiology**, the study of the developmental processes that underlie the formation of the nervous system, been recognized as a discipline in its own right. The two fields obviously overlap. Questions regarding the origin of neurons as distinct cell types, or the rules that determine the final size, shape, and location of the major parts of the nervous system, are clearly similar to questions regarding the origin or developmental processes that shape any organ system. However, developmental neurobiology must also address questions that are unique to the nervous system, such as how each growing axon finds its destination, and how it identifies the target cells with which to synapse.

A full consideration of development is beyond the scope of this book. Nevertheless, developmental processes that influence the nervous system cannot be understood without some understanding of developmental processes in general. Every animal originates as a single fertilized egg (zygote). Hence, all the cells of the adult animal must arise from a series of divisions, starting from the first cell division of the zygote. Consequently, every cell has what is called a **lineage**, the sequence of divisions that produced it. Conversely, all the cells that are generated from divisions of a single cell are called the **progeny** of that cell. One of the fundamental questions of developmental biology is what factors determine the ultimate fate of a cell's progeny. On the one hand, genetic and cytoplasmic factors intrinsic to each cell may be the main elements that determine the fate of a cell's progeny. This is called **determinate development**, also known as **mosaic development**. On the other hand, influences from outside the cell may play a more significant role. This is called **indeterminate development**, also known as **regulated development**.

Developmental neurobiology is particularly concerned with developmental issues that are unique to the nervous system, such as the directional growth of axons and the formation of the proper synapses between a neuron and its target cell. These can best be understood in the context of developmental processes in general. There are two main types of development among animals: determinate and indeterminate.

Determinate Development

Except for echinoderms (starfish and their relatives), all invertebrates show determinate development. In this type of development, even the four cells that result from the first and second divisions of the fertilized egg are already committed to producing specific parts of the animal. This commitment of each cell's progeny to a particular fate comes about because of the uneven distribution of cytoplasmic factors in the egg, and the interaction of these factors with the cell's genetic instructions. As a result of this early commitment, the most striking feature of animals with determinate development is that the death of a single cell will result in the absence of all or nearly all the cells, tissues, or organs that would have developed from that cell because the remaining cells have only limited ability to fill in for the missing one. Therefore, if the death of the cell occurs early in development, the embryo is incomplete and soon dies.

An excellent example of determinate development is development of the nematode *Caenorhabditis elegans*. As an adult, this minute roundworm has 959 somatic cells, 302 of which are neurons. The lineage of each cell in the adult is known (Figure 22-1), and experiments have shown that there is no variability in this lineage during normal development. Furthermore, the animal shows exceptionally little ability to compensate for the experimental deletion of individual cells, even relatively late in development, when it would seem that only minor adjustments in a cell's genetic program would be required. With few exceptions, death of a single cell causes the loss of all the body cells that would have arisen from it; other cells have essentially no ability to generate replacements.

The complete lineage of every cell in the body of more complex invertebrates is not so easy to determine because in these animals it is difficult to identify and follow all the progeny of individual cells. Nevertheless, considerable effort has been devoted to finding the lineages of neurons and other cells in these more complex animals. There are several important reasons for this effort. From a developmental point of view, it is important to know how the diversity of cell types is generated during development. What are the precise rules by which cells give rise to neurons of one type or another? From a neurobiological point of view, it is important to know whether there is any relationship between the lineage of a cell and the functional roles the cell plays. Understanding how function relates to developmental history not only may help researchers understand how the nervous system operates but also may provide important insights into how structure or function changes over evolutionary time.

Most invertebrates exhibit determinate development, in which the fate of a cell's progeny is largely determined by the genetic and cytoplasmic factors that the cell inherits from its parents. There is little flexibility in this type of development.

Indeterminate Development

Indeterminate development is characteristic of the echinoderms and chordates, including vertebrates. In this type of development, cell lineage is strongly influenced by the interaction of the genome with epigenetic (nongenetic) factors. The genetic instructions of each cell for the course of its development are still important, but the commitment of a cell to a

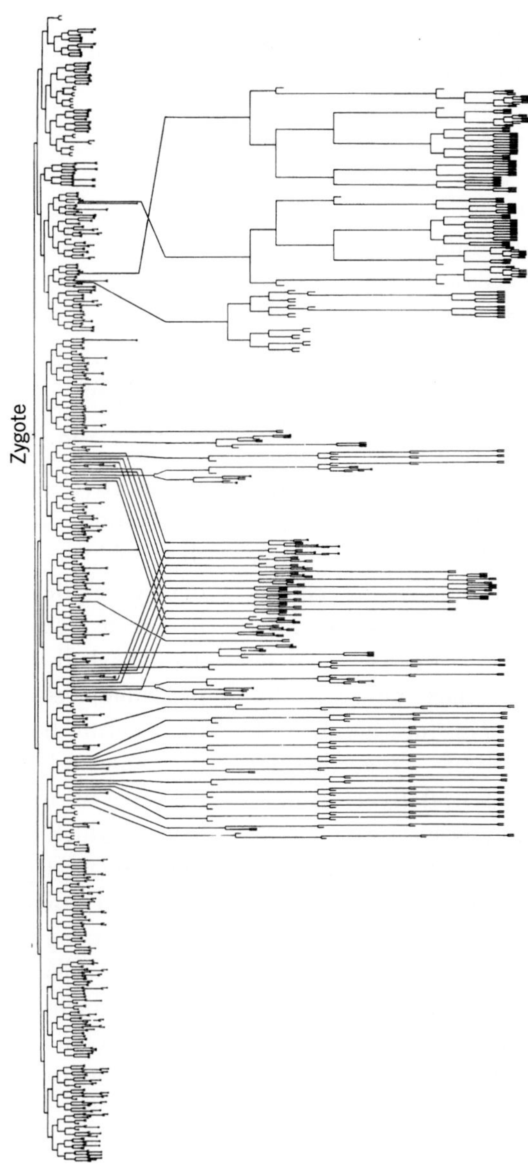

FIGURE 22-1. An extreme example of determinate development: the complete cell lineage of the roundworm *Caenorhabditis elegans*. The lines in the diagram connect the progeny of every cell, starting from the zygote. Each branch point represents a cell division, and the termination of each line represents one cell in the adult animal.

particular destiny, such as being a nerve cell or a liver cell, is also strongly influenced by extrinsic factors such as molecules released by neighboring cells. These external factors interact with the cell's genome to determine the cell's fate, a process known as induction.

Some compounds induce the formation of neural tissue by inhibiting factors that would normally keep cells from differentiating into neurons. Others, like retinoic acid, provide an anterior-posterior orientation to the developing neural tissue, promoting and accelerating the differentiation of the posterior end of the neural tube. Retinoic acid diffuses through cell membranes of adjacent cells and interacts with retinoic acid receptors in the nucleus of the target cell, where the receptors activate genes that then direct the subsequent fate of the cell. Because cell fate is a consequence of induction rather than a genetic program triggered automatically by a cell's parentage, removal of one or a few cells will not necessarily have any discernible effect on the development of the embryo; other cells will simply be induced to take over the function of the missing ones.

Cell lineages are difficult to determine in vertebrate embryos, in part because the embryo grows large and in part because of the indeterminacy of cell fate. Nevertheless, investigation of lineage in relatively simple vertebrates like amphibians has shown that the lineage of a cell can be determined in a probabilistic fashion. That is, at some early stage in development there is a certain probability that the progeny will include cells of a particular type. For example, in fish, the uniquely identifiable Mauthner neurons (see Box 18-1 in Chapter 18) may originate from any of several cells that are present at the 512-cell stage of development, even though there is only one unique pair of Mauthner cells.

Vertebrates and echinoderms exhibit indeterminate development, in which cells are strongly influenced by external factors present in their surroundings. Consequently, the lineage of a particular cell may be different in one individual of a species than it is in another.

Building a Nervous System

Irrespective of whether an animal exhibits determinate or indeterminate development, there will be a specific pattern to the formation of new cells that form the nervous system. In the nematode *Caenorhabditis elegans*, the nerve cord is established by the differentiation of 12 nerve progenitor cells distributed along the length of the ventral body wall (Figure 22-2). These progenitor cells eventually give rise to all of the neurons in the central nervous system. In insects, which have distinct body segmentation, the segmental ganglia develop from a type of cell called a **neuroblast**, 60 to 68 of which lie in each body segment. Each neuroblast gives rise to several cells known as **ganglion mother cells**, each of which in turn divides once to form a single pair of neurons or glia. These collectively form the ganglia associated with each segment.

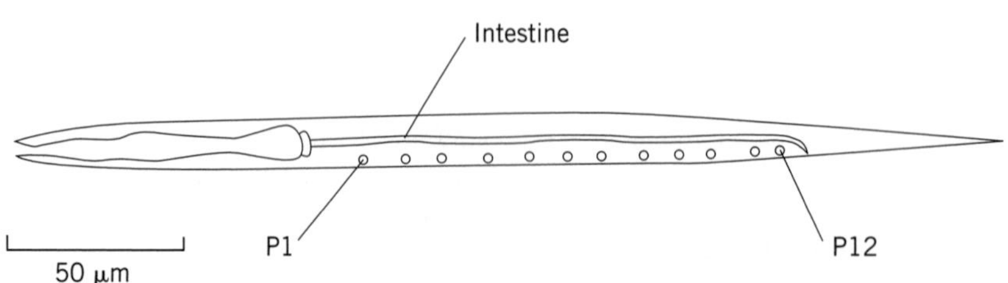

FIGURE 22-2. Origins of the nervous system in a roundworm. Neural development starts from a series of 12 progenitor cells that lie along the ventral body wall. Collectively, progenitor cells give rise to the neurons of the central nervous system.

In vertebrates, the nerve cord and brain are formed from a continuous sheet of tissue that develops at the end of the gastrula stage of embryonic development (the stage in which the hollow blastula has invaginated to form two layers of cells), as described in Chapter 3. A depression called the **neural plate** is formed from a groove that appears on the dorsal surface of the gastrula. The neural plate closes and, under the influence of inducer substances, forms the **neural tube,** the precursor to the brain and spinal cord. The inducers are released by a specialized group of cells under the neural plate. Once released, they block the chemical signals that keep the ectodermal tissue from differentiating into neural tissue. As the embryo grows, the neural tube expands and differentiates. The anterior part forms the brain, and the rest forms the spinal cord, as shown diagramatically in Figure 22-3 for a human embryo.

During embryogenesis, neural cells destined to form part of the brain generally proliferate from a germinal layer, a sheet of cells that divide repeatedly to generate new cells. With few exceptions, such as the cerebellum of the rat, this division ceases before birth, which means that adult vertebrates cannot replace neurons that die, and all the developmental changes that take place in the nervous system after birth are due to changes in connectivity between neurons. The newly formed neural cells migrate away from their place of origin to their final location. In so doing, they actually migrate past cells that had been born earlier (Figure 22-4). This means that the oldest (earlier-born) neurons in the cellular layers of the cortex are on the *inside,* and the youngest are on the *outside,* an order that is the opposite of what it would have been if the cells had stayed in birth order. Hence, the final shape of any vertebrate central nervous system is strongly influenced by the factors that bring about and guide the migration of cells. The mode of action of these factors, and hence the means by which the morphological organization and structure of the nervous system are determined, are still not well understood.

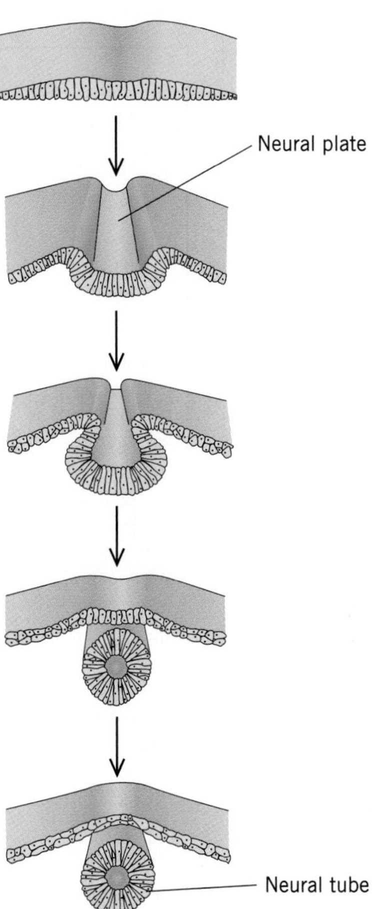

FIGURE 22-3. Formation of the neural tube, the precursor of the central nervous system, from the dorsal part of the developing gastrula, one of the early stages of development. A long depression, the neural plate, is formed first. The neural plate then collapses inward, forming a tube along the length of the dorsal surface of the embryo. Specific cells along the length of this tube release molecules that induce the differentiation of specific types of neurons.

Invertebrate nervous systems usually develop from specific precursor cells that give rise to the cells of the central nervous system. In vertebrates, neurons develop by differentiation of the neural tube and commonly migrate to locations distant from their origins.

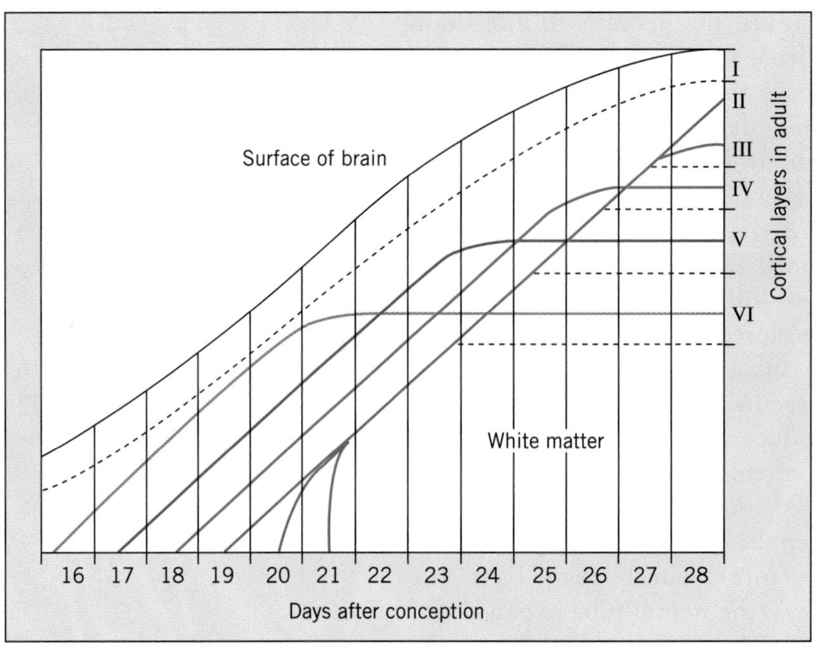

FIGURE 22-4. Inside-out development of the cortex of a rat fetus. Each line represents cells in one layer of cortex. The cells that are born first (leftmost on the horizontal axis) form the *inner* layer of the cortex, as the later-born cells migrate past them.

Forming Connections

How a nervous system attains its final size and shape is only one of the fundamental questions of developmental neurobiology. An even more critical question is how individual neurons make proper connections (synapses) with their target cells.

The Necessity for Specific Connections

It is obvious that in an adult animal neural connections are far from random. Consider the specific connections that are necessary between neurons at different functional levels for locomotion to be coordinated. First, there must be proper connections between the neurons of the network and the motor neurons so that the right movements are generated by activity of the network. Second, there must be proper connections between incoming sensory neurons and interneurons in the nerve cord or ganglion so that sensory feedback can have the desired effect on an ongoing motor performance. Third, there must be proper connections among the neurons that constitute a pattern-generating network (see Chapter 16) in order for the network to produce a rhythmic pattern of output. Finally, there must be proper connections between motor centers in the brain and the networks in the spinal cord or body ganglia so that activity in the rhythm-producing neural network can be regulated and the desired speed of progression can be generated.

Even in the vertebrate brain, every carefully studied part shows similar requirements for specific synaptic connections between neurons. Sensory systems, for example, exhibit extraordinarily precise arrays of specific synaptic connections between neurons. Consider the retinotopic projection of ganglion cells from the retina to the brain (see Chapters 9 and 11). In toads and frogs, ganglion cells send their axons into the optic tectum, where they form synapses with local neurons. Since the optic tectum is topographically organized, each ganglion cell must project to a particular part of the optic tectum, as illustrated in Figure 22-5. Similar relationships exist for other sensory systems that are topographically organized, and indeed for other brain systems as well.

An important step in the development of the nervous system is the establishment of proper connections between neurons. A good example of an array of precise connections is the retinotopic projection of retinal ganglion cells onto neurons in the optic tectum in lower vertebrates.

Neural Specificity

Although for much of the twentieth century most researchers accepted the view that connections between neurons could be highly specific in an adult animal, they did not agree on how this specificity was achieved. Just before the middle of the century, many developmental biologists believed that the nervous system grew in a rather random fashion. The specific neural pathways that could be identified in an adult animal, it was believed, were the result of a process of trial and error on the part of the animal during its early growth, a process that weeded out inappropriate neural circuits and strengthened appropriate ones.

This idea dominated thought at the time, in part because researchers could not understand how the alternative could be possible. It did not seem creditable that during development highly precise neural connections could form from the millions of possibilities. Much of the credit for changing this view, and indeed for laying the foundation for modern work in developmental neurobiology, belongs to Roger Sperry, who strongly advocated the view that quite specific connections are made between neurons during development, an idea he called the principle of **neural specificity.**

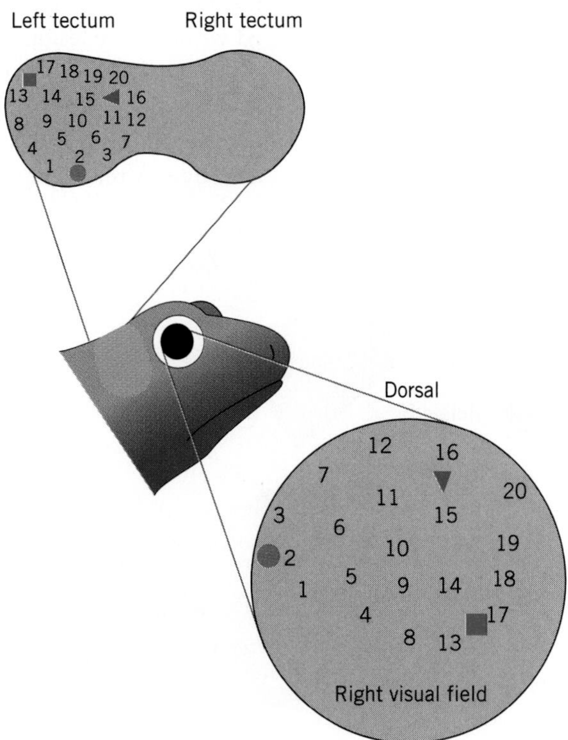

FIGURE 22-5. The retinotopic projection from the retina of a frog to its optic tectum. In the diagram, the visual field of the right eye is marked by numerals and symbols; the corresponding parts in the left optic tectum to which the retinal ganglion cells project are marked by corresponding numerals and symbols. Input from the other eye is not shown here.

Sperry used the visual system of frogs in his experiments. Not only do retinal ganglion cells in this animal map precisely to specific parts of the tectum (see Figure 22-5), but, in addition, frogs survive surgery well, making possible a variety of experiments. To begin, Sperry showed that once the topographical projection of the retina onto the tectum has been established in an adult animal, it cannot be changed. He demonstrated this by rotating the eyeball of a frog 180° in its socket without damaging the optic nerve. After the operation, the frog sees an inverted and right-left reversed visual world with that eye. For example, a fly in the upper left quadrant of the visual field of the right eye (which will be above and behind the frog as a result of the position of the eye on the side of the frog's head) will now stimulate retinal cells that previously were stimulated by images in the lower right quadrant. Since the ganglion cells in this region of the retina still connect with the lateral rostral region of the tectum, as they did before the eye was rotated, the frog interprets the location of the fly as below and in front of itself rather than above and behind (Figure 22-6). This change in the frog's visual perception is permanent; for the rest of its life, the frog behaves as if the visual field of the rotated eye is inverted.

Before Sperry conducted his eye rotation experiments, other researchers had already shown that cutting the optic nerve and letting it regenerate had no permanent effects on the frog's visual system. The frog temporarily becomes blind in the eye from which the severed optic nerve originated, but after some time, normal vision reappears. This result was interpreted to mean that the cut axons of the retinal ganglion cells regenerated and then reestablished their original connections in the optic tectum, allowing the frog to see once more.

Sperry wanted to know what would happen if he combined the two experiments, cutting the optic nerve and at the same time rotating the eyeball 180°. Two outcomes seemed possible. First, the regenerating ganglion cells could grow back to make new connections appropriate to the new, inverted, position of the retina, so the frog would be able to locate objects correctly in its visual field. Alternatively, the regenerating ganglion cells could grow back to the same regions of the tectum that they had originally occupied. In this case, the frog would behave just like those animals that had an eyeball rotated without any lesion of the optic nerve.

The results were unequivocal. Immediately after section of the optic nerve and rotation of the eye, the frog behaves as if blind in the operated eye. After regeneration of the optic nerve,

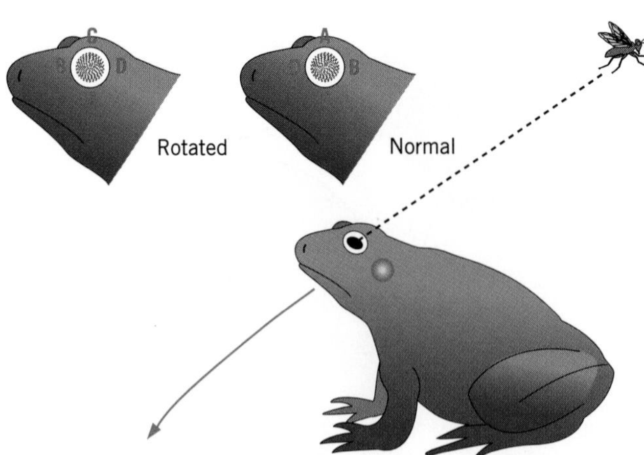

FIGURE 22-6. The effects of eye rotation on the reaction of a frog to objects in its visual field. After eye rotation, objects in the upper rear quadrant of the visual field stimulate optic tectal cells that before the rotation would have been stimulated by objects in the lower front quadrant. The frog therefore acts as if an object above and to its rear is actually below and in front of it.

however, the frog behaves as if it has a rotated eye without any lesion of the optic nerve. In other words, the operated eye of this frog possesses a right-left and up-down reversed visual field, just as shown in Figure 22-6. Sperry concluded that the axons of the retinal ganglion cells grow back to form synapses at about the same locations they had occupied originally.

These and similar experiments suggested that neurons could form specific synaptic connections with other neurons during development, and that these connections could be reestablished after being severed, even in an adult animal. They did not, however, give any clue about how the correct connections were actually formed. Building on the work of other researchers, Sperry suggested that each target neuron manufactured and released a specific chemical substance to which the neurons seeking that target would respond. He called this proposal the **chemoaffinity hypothesis** because it suggested an attraction, or "affinity," between each growing neuron and the neurons on which it synapses, based on chemical cues released by the target neurons. Although later experiments showed that the idea of specific chemical attractants between individual pairs of neurons was not correct, the idea that chemical cues guide neurons as they seek their targets was an important insight. The current view of how the process works is described in the next section.

During development, neurons grow and make synaptic connections in a precise and orderly fashion, a process Roger Sperry called neural specificity. Sperry proposed that the basis of neural specificity is a chemical affinity between growing neurons and their targets.

Finding the Target

How neurons develop properly is one of the central questions of developmental neurobiology. There are really three components to this question—what regulates the initial growth of an axon, how does a growing axon select the correct path, and how does an axon identify its specific target?

The Growth of Axons

When neurons first differentiate, they lack axons and dendrites. Neurites begin to grow from specific sites on the cell surface as determined intrinsically within each cell. External factors seem to be unimportant for this initial growth, as indicated by the consequences of occasional developmental mistakes. For example, during normal development in the rat brain, pyramidally shaped cells in the cerebral cortex invert. This inversion consists of a 180° rotation of the cell body, so that the cell ends with its apex pointing up rather than down (Figure 22-7A). This inversion takes place before the cell begins to sprout its single axon from the base of the cell and its single but highly branched dendrite from the apex of the cell.

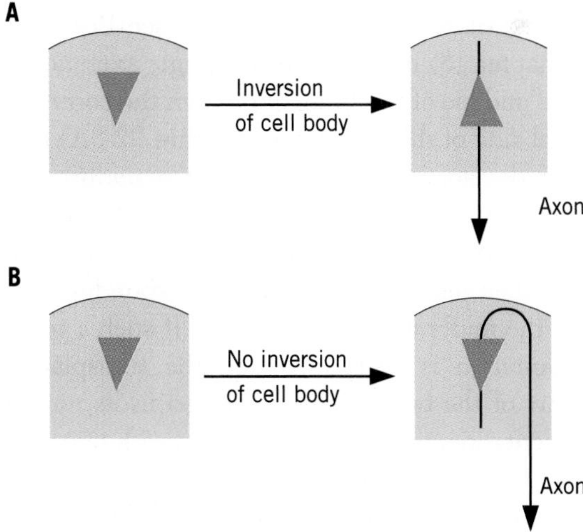

FIGURE 22-7. Pyramidal cells (triangles) in developing rat cortex normally invert shortly after they differentiate (A). Sometimes, a cell may fail to invert (B). Whether or not the cell inverts, the axon always starts growing out of the soma at one particular site; if the cell body is improperly oriented, however, the axon soon reorients itself and heads in the correct direction.

In some cases, however, inversion is incomplete or does not take place at all. In such instances, the axon and dendrite still start growing from the usual site on the cell membrane, even though those sites are not oriented properly relative to the rest of the cortex. This suggests that the sites at which axonal and dendritic growth is initiated are intrinsic to the neuron and not determined by the cell's surroundings.

Dendrites of noninverted neurons in the brain, which are always and normally quite short, never show any tendency to correct their position during subsequent development. They grow straight out from the cell body irrespective of the orientation of the soma. If the cell is not inverted, the dendrite grows in the wrong direction and remains improperly oriented. Axons of noninverted cells, on the other hand, correct their orientations. If the cell fails to invert, the axon will start growing toward the surface of the brain because that is the direction faced by the base of the cell (Figure 22-7B). After growing in this direction for a short distance, however, the axon will swing around and begin growing toward the interior, its normal direction of growth.

Experiments show that many neurons are capable of redirecting the growth of axons that start off in the wrong direction. For example, the giant Mauthner neurons of fish (see Box 18-1, Chapter 18) normally send a single axon across the midline of the brain and down the contralateral side of the spinal cord (Figure 22-8A) during development. The part of the midbrain in which the cell bodies of the Mauthner cells lie can be removed from one fish and implanted into another, just anterior to the corresponding part of the midbrain of the host fish. If such a transplantation is performed and the transplanted part of the brain stem is inserted in its normal orientation, the host fish develops with two pairs of Mauthner axons, each crossing the midline and descending to the spinal cord (Figure 22-8B). But if the transplanted brain tissue is inserted in a right-left inverted orientation, the transplanted Mauthner cells are oriented so that the normal direction of growth of their axons points anteriorly instead of posteriorly. These reoriented axons start to grow in an anterior direction, although they soon correct themselves, swing around, and grow posteriorly (Figure 22-8C).

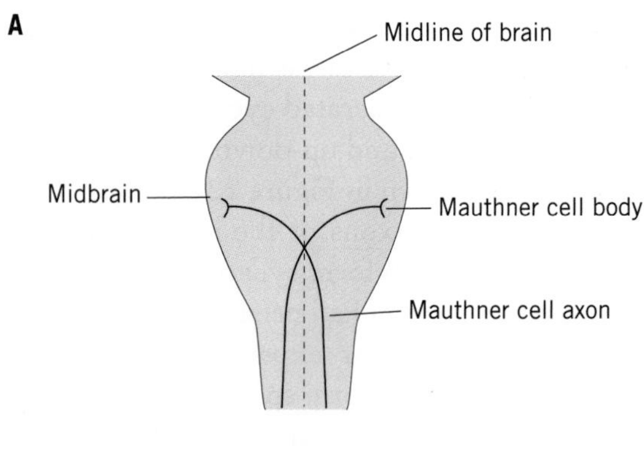

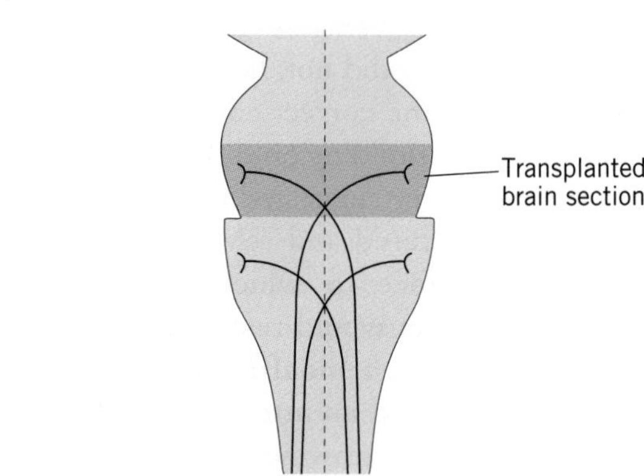

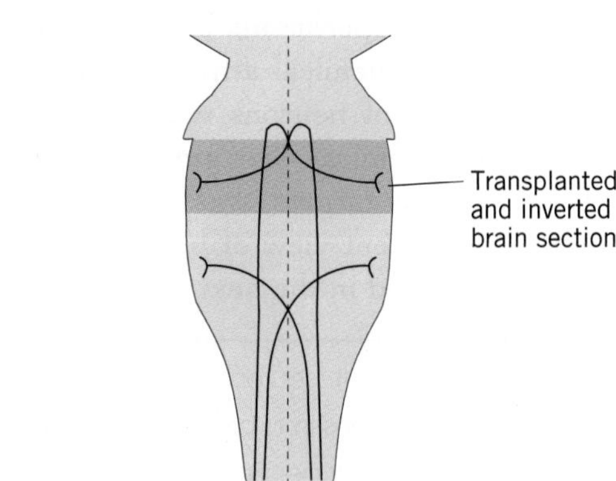

FIGURE 22-8. (A) Normal growth of Mauthner cells in the midbrain of a fish. Each cell sprouts an axon that crosses the midline and grows posteriorly into the spinal cord. (B) Growth of extra Mauthner cells from properly positioned transplanted tissue. The extra Mauthner cells grow axons in the proper direction. (C) Growth of inverted extra Mauthner cells. Axons from the extra Mauthner cells start to grow anteriorly but soon swing around and grow in the correct direction.

The foregoing examples clearly demonstrate that external factors are able to influence the direction in which a developing axon will grow. Which factors are necessary and the means by which they act, however, are not as clear. The most popular and thoroughly investigated hypothesis has been that directional growth is guided by specific chemical cues, or by gradients of chemicals in the developing tissues. Alternatively, electrical and physical cues may be used. For example, some embryos have a gradient of electrical current from one end to the other. Since growing axons have also been shown to orient to an imposed electrical field, it has been suggested that this electrical gradient could act to orient growing axons. It is also known that growing axons prefer to have physical contact with some edge or other border, which then might serve as an important cue for its growth. However, attempts to explain all axonal growth in terms of a single factor, be it chemical, electrical, or physical, have usually failed when sufficient care has been taken to separate the various factors. Hence, no single factor acting alone is likely to be responsible for guiding axonal growth. A variety of factors may act in concert, or each may act at different times during the growth of an axon.

Axons grow from the bodies of neurons at specific membrane sites. If the neuron is improperly oriented, the axon will soon correct its path. Several theories have been proposed to explain this phenomenon, but evidence suggests that no single factor is the sole cause of the directional growth of axons.

Pathfinding

Global processes such as those just described can send an axon off in the right direction, but in many instances axon growth involves such specific and stereotypic changes in direction that general directional cues cannot inadequately account for them. For this reason, investigators believe that axons also use specific local cues to guide their growth. By responding to a few strategically placed directional cues, an axon can find its way over remarkably long distances. Developmental neurobiologists use the term **pathfinding** to describe the process by which a growing axon is able to navigate to its target.

Axons grow in an ameboid fashion, sending branches out this way and that, keeping some and retracting others (Figure 22-9A). An axon gives all the appearance of actively seeking and responding to local cues or guidance factors as it wends its way to its destination. The key to axonal growth and pathfinding is the enlarged tip at the end of every branch of a growing axon. This tip, called a **growth cone**, sends out a large number of fine, threadlike projections called **filopodia** (Figure 22-9B). These filopodia probe the environment and respond to chemical guidance cues.

Guidance cues can act over short or long distances; they may also have various effects on growing axons. Some guidance molecules are short-range, being bound to the surfaces of neighboring cells or to an extracellular matrix of complex molecules. The growth cone must come into contact with the molecules in order to have its path affected by them. Other guidance molecules are long-range, being released into the extracellular fluid and diffusing to locations some distance from the point of release. Both short-range and long-range guidance cues may attract or repulse the filopodia of a growth cone. Membrane-bound proteins can cause the filopodia to adhere to the cells that express them, and hence to grow along a defined surface. They may also activate receptor proteins on the growth cone and trigger growth. Some molecules attract, whereas others repel the filopodia. Diffusible molecules working over a longer range can attract or repel filopodia by affecting the direction in which the growth cone elongates, effectively steering it toward its goal.

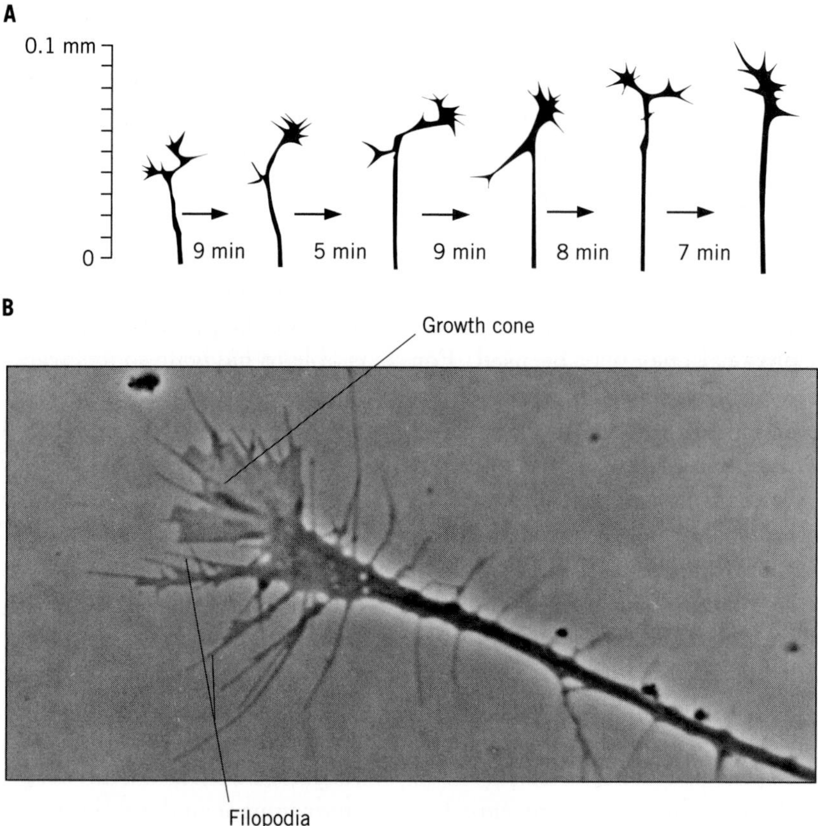

FIGURE 22-9. (A) Ameboid growth of an axon in tissue culture. Notice the growth cone at the tip of each axonal branch, and that the axon sends out branches that are later retracted. Numbers indicate the time that has passed from one sketch to the next. (B) Phase contrast micrograph of a growth cone. (Courtesy of Dennis Bray.)

The peripheral nervous systems of insects have provided especially clear instances of pathfinding. The cell bodies of insect sensory neurons lie in the periphery, near the sense organs from which they originate (Figure 22-10A). Hence, during development their axons must grow in toward the central nervous system in order to make synaptic contact with central neurons (Figure 22-10B). In a leg of a locust embryo, some sensory neurons begin to differentiate earlier than others. These neurons, called **pioneer neurons,** forge the path to the central nervous system that the remaining sensory neurons will also follow. Follower neurons do not need special pathfinding abilities, since they merely follow the pioneers. Follower neurons nevertheless can develop the ability to forge their own paths; if a pioneer neuron is eliminated (e.g., by blasting it with high-intensity laser light), the next sensory neuron to develop will differentiate into a pioneer and find its way to the central nervous system. This differentiation step is necessary. If the differentiation of all pioneer neurons is prevented (e.g., by briefly exposing the embryo to a high temperature), the surviving sensory neurons do not usually find the central nervous system.

Pioneer neurons find the correct route to the central nervous system by using several guidance cues. One critical cue is provided by surface molecules on a small group of neuron precursor cells called **guidepost cells** (Figure 22-10B). These cells differentiate from other cells at specific locations in the developing leg bud. They can be identified immunologically by their expression of unique surface molecules. As the pioneer neurons grow, their

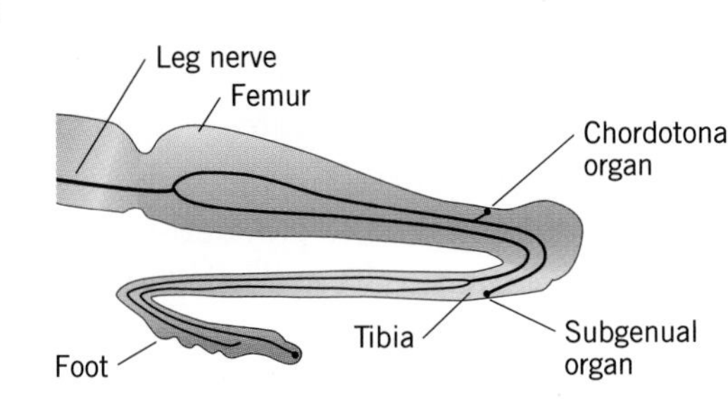

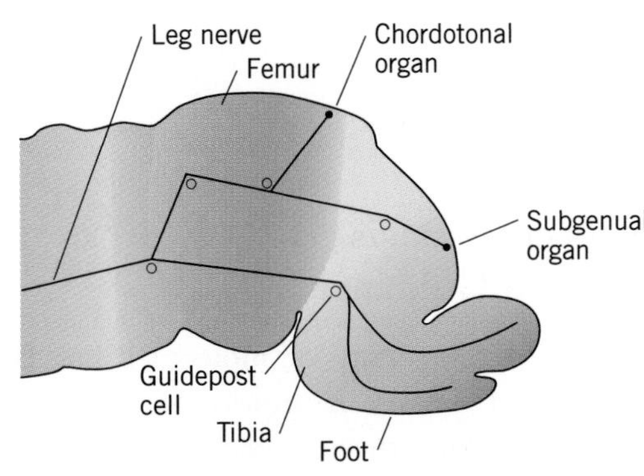

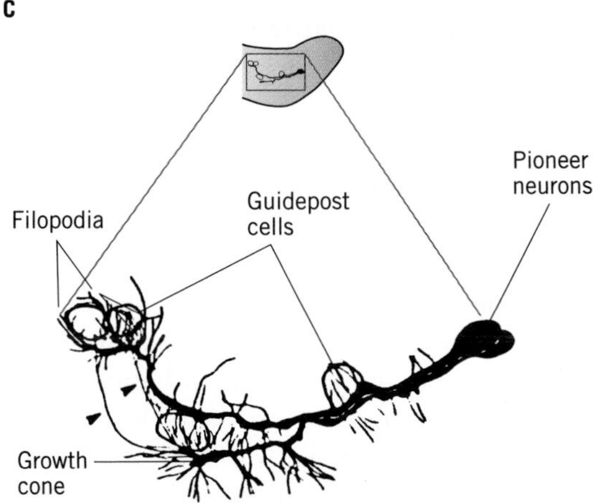

FIGURE 22-10. (A) Diagram of the main nerve trunks and some landmark sense organs in the rear leg of an adult locust. (B) Diagram of the locations of the guidepost cells that help the developing sensory neurons find their proper path in an embryonic locust rear leg. (C) Drawing of the growth of a pair of pioneer neurons and their interaction with guidepost cells.

filopodia make physical contact with the guidepost cells (Figure 22-10C), frequently wrapping tightly around them. When an appropriate guidepost cell has been located, the axon grows to that location, and the filopodia extend to search for further cues. The guidepost cells are critical to pathfinding. If they are destroyed, no other cells take their place. After such destruction, the growing axons wander around aimlessly in the limb bud, finding their way to the central nervous system only by chance.

In addition to guidepost cells, three other types of guidance cues are used by pioneer neurons in the developing limb bud. One type is a class of molecules known as **cell adhesion molecules (CAMs)**, which are expressed in the membranes of epithelial cells. Pioneer neurons show a strong affinity for epithelial cells, meaning that they preferentially grow along epithelial cells at the inner surface of a developing limb segment rather than in the center. This affinity is mediated by receptors on the filopodia of the growing axons for CAMs produced by the epithelial cells. Diffusible chemicals provide a second type of cue. In the limb bud, the pioneer neurons take a sharp turn just at the junction between the developing coxa and trochanter, two segments of the insect leg. This turn is produced by the repulsive effect of a diffusible molecule, semaphorin I, which is released by cells at the segmental boundary. A third type of cue is that provided by the polarity of the limb bud itself. This cue is not well understood but is presumed to be due to some diffusible material that provides a chemical gradient in the bud, with a high concentration at one end and a low concentration at the other. The gradient guides the growing axon toward the proximal end of the limb bud.

Axons in the insect central nervous system find their way in a similar fashion. Pathfinding is done by each individual developing neuron, which seeks out specific cues to use as markers for direction finding and turns. An important cue is the presence of other axons. Filopodia of the growing axons have a strong affinity for the axons of the neurons they use as markers; the filopodia make frequent and tight contact with these axons, as if to verify that they are indeed the markers they are looking for. If the target axon is missing (due to experimental deletion), the tip of the growing neuron will often wander around randomly in the nervous system.

Axons weave their way to their correct destinations by sending out fine filopodia from the growth cone. The filopodia search for guidance cues provided by marker cells such as guidepost cells, by cell adhesion molecules, or by diffusible attractive or repulsive molecules. In the central nervous system, chemical guidance cues may be expressed by other axons that then serve as signposts for the growth cones of later-developing axons.

Recognition of a specific target cell by a growing axon is mediated by the presence of chemical cues on the surface of the target. Once the correct target has been located, chemical cues provided by both the axon and its target initiate the formation of a synapse.

Target Recognition and Synapse Formation

Once an axon has grown to its destination, it must recognize and make contact with the specific neuron or muscle cell with which it normally synapses. Less is known about how this is done than is known about pathfinding, but evidence suggests that similar mechanisms are involved. That is, specific chemical cues on the surface of the target cell help the growth cone recognize the proper target. For example, a motor axon growing to a muscle will normally ignore muscle fibers that it does not typically innervate and will send branches to the muscle fibers that it does innervate. If muscle fibers of a specific type are experimentally duplicated, the motor axon will innervate all of the duplicated fibers.

Synapses between a developing axon and its target are formed after initial contact. The process requires two-way communication. The growing axon must stop growing and start to differentiate the proper presynaptic specialization. The postsynaptic cell must recognize the presence of the axon and start to build the proper postsynaptic structures. This communication between pre- and postsynaptic cells is mediated by molecules on the surfaces of the two cells. When these molecules are detected, they trigger appropriate changes in each of the cells to form the specialized patches of membrane.

Molecular Basis of Axon Guidance

A bewildering array of molecules has been implicated in axon guidance. Some researchers group these molecules according to their effect on a growing axon, referring to molecules as chemoattractors or chemorepulsers. Other researchers group molecules according to where they are located when they act: on the cell surface, embedded in the extracellular matrix that fills much of the space between cells during development, or over a wide area as diffusible substances. Unfortunately, no classification scheme for these molecules is entirely satisfactory because there is no relationship between the different ways of grouping them. For example, both chemoattractors and chemorepulsers may be surface-bound or freely diffusing molecules.

One way of grouping guidance molecules is according to structural similarity. Each molecule is a distinct protein, and hence is coded for by one gene. Proteins that have structural similarities are said to be members of single gene families; the implication is that the different genes are evolutionarily related and stem from a single ancient precursor. Proteins that have little or no similarity are members of different families. Even this scheme of classification is not without difficulties, since researchers may disagree over whether similarity between two proteins is sufficient to group them into a single family. Furthermore, in complex proteins, similarity to some particular protein may be confined to one domain (one specific arm of the protein), with another domain having similarity to another protein. In such a case, some researchers group all the proteins into a superfamily.

At present, more than a dozen families of guidance molecules have been identified. Some are currently represented by only a single example, so it is not clear which of these represent families that are widespread throughout the animal kingdom, and which are unique just to one or two species. For the present purpose, five representative and reasonably well-known families will be listed: the **cadherin, integrin, netrin,** and **semaphorin families,** and the **immunoglobulin superfamily** (abbreviated Ig), a large group that consists of molecules containing domains structurally similar to domains of vertebrate antibodies. Table 22-1 lists these five families, as well as a few individual named molecules that belong to them.

Guidance molecules work in several ways. Some, such as members of the cadherin and integrin families and the immunoglobulin superfamily, are cell adhesion molecules. Such molecules are synthesized in a growing axon (and in other cells) and inserted through the cell's membrane, where they bind with similar molecules present in the membranes of other cells or in the extracellular matrix. This binding promotes the adhesion of the growing axon to selected adjacent cells. By binding to molecules present in the extracellular matrix or on cells such as epithelial cells, guidance molecules on the growing axon constrain the axon to grow in certain directions only, because it is difficult for the axon to pull away from the cell or surface to which it adheres. The direction of growth can be changed if the neuron contacts molecules with a stronger binding affinity or if it encounters molecules that promote release from the adhesion surface.

Other guidance molecules influence the direction of growth by interacting with the production and placement of actin strands that are responsible for extension or retraction of growth cones or with other elements of the axon's cytoskeleton. They do this by binding with receptor molecules on the cell surface. Activation of a receptor by the guidance molecule initiates a biochemical cascade that often involves Ca^{2+} or a G protein or both, and that results in turning the growth cone toward or away from the filopodia that have contacted the guidance molecule. Growth can also be inhibited altogether. For example, as its name suggests, the molecule collapsin can cause the growth cone to shrink and disappear, halting further growth. This process or a functionally similar process must occur when an axon reaches its final destination. The distinction between adhesion molecules and molecules that work via receptors is not clear, however, since some adhesion molecules also act as receptors for the proteins to which they adhere, initiating responses within the cell, as well as serving as sites of adhesion. The precise molecular basis of these guidance events is still under intense investigation.

Nevertheless, it is abundantly clear that axon guidance is orchestrated by a complex array of site-specific and time-specific molecules to which a growing axon responds. Application of monoclonal antibodies (Box 22-1) to developing embryos has revealed that many specific guidance molecules appear at precise times during development (sometimes only for a matter of several hours) and are often localized to specific sites (Box 22-2). In the locust central nervous system, for example, two guidance molecules, **fasciclin I** and **fasciclin II**, are expressed on the surfaces of different parts of axons in a ganglion. Fasciclin I is expressed on horizontally oriented axons, and fasciclin II on vertically oriented ones. They may even be expressed on the horizontal and vertical portions of an axon of a single neuron, respectively. These fasciclins are guidance markers for other developing neurons as they wend their way through the neuropil, signaling to the growth cones of these neurons that they have located the right place to change direction. Further, the fasciclins are expressed on the axons only during a restricted time window during the development of the embryo; they appear only after some development has taken place, and they usually disappear again before development is complete.

BOX 22-1

Targeting Molecules: Monoclonal Antibodies and Immunocytochemistry

The 1975 discovery of the **monoclonal antibody technique,** a method for making antibodies specific for a particular antigen even when the antigen is not available in pure form, was a key event leading to greater understanding of the molecular basis of development. It had been known for decades that the mammalian immune system can make antibodies specific to individual proteins. Researchers had often taken advantage of this ability by inducing the immune system to make antibodies to tissues or tissue extracts in order to localize some of their important protein components. Until the monoclonal antibody technique was developed, however, it was not possible to produce a specific antibody in pure form and in sufficient quantity to be used in experiments. One of the main difficulties was that lymphocytes, the white blood cells that synthesize antibodies, do not divide, making it impossible to maintain a line of cells that would produce a particular antibody in quantity.

The discovery that lymphocytes could be induced to fuse with a type of mouse cancer called a *myeloma* made possible the large-scale production of a single type of antibody. Myeloma cells can divide indefinitely, so the result of fusion of a myeloma cell and a lymphocyte is a cell that can synthesize a single antibody and yet can grow and reproduce indefinitely.

Producing monoclonal antibodies is conceptually simple, but the process requires considerable effort (Figure 22-A). First, the tissue of interest is ground up (homogenized) and injected into a mouse. The immune system of the mouse responds to the foreign proteins by making antibodies to them. Lymphocytes are then extracted from the spleen of the mouse, placed with mutant myeloma cells, and treated so that some of the cells of the two types fuse. The mixture of myeloma cells, lymphocytes, and fused hybrid cells is transferred to a medium called HAT (for hypoxanthine, aminopterin, and thymidine, three main constituents), in which the mutant myeloma cells cannot grow. The normal lymphocytes, unable to reproduce, also die in a short time, leaving only the fused hybrid cells. The suspension of surviving hybrid cells is diluted so the cells can be separated and placed individually in small depressions (wells) on a slab of glass. Each cell is allowed to divide repeatedly, producing what is called a clone, a genetically identical population of cells that derive from a single precursor, in each well.

Because an individual lymphocyte synthesizes only a single type of antibody, the hybrid cell formed when a lymphocyte fuses with a myeloma cell also produces only a single type of antibody. This will be an antibody to

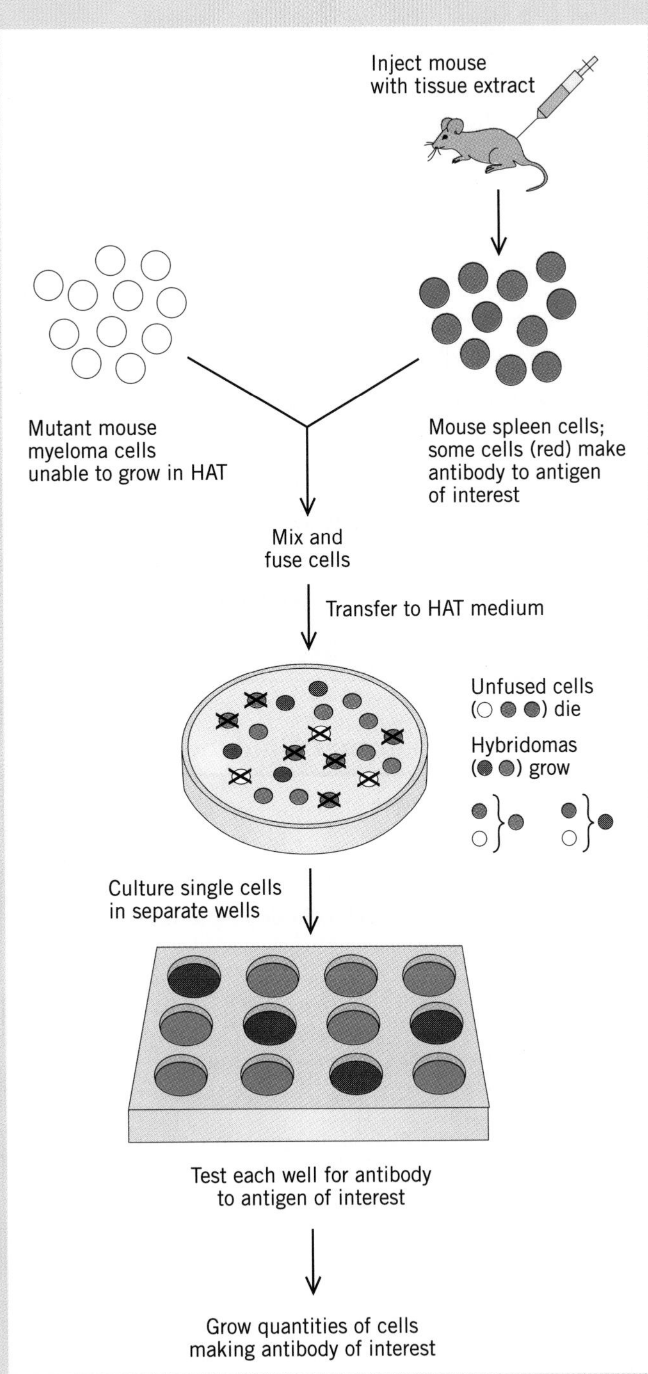

FIGURE 22-A. Schematic of the procedure for producing monoclonal antibodies. Lymphocytes from a mouse spleen and myeloma cells are mixed and transferred to a special medium (called HAT) in which unfused myeloma cells cannot grow. (The lymphocytes do not divide, and hence also die after some time.) The surviving cells are products of the fusion of spleen and myeloma cells. These are cultured in separate containers and screened for the production of antibodies of interest. Those that produce the desired antibody are then grown in large quantity and the antibody is harvested.

(continued)

Targeting Molecules: Monoclonal Antibodies and Immunocytochemistry (*continued*)

just one part of one of the many proteins originally injected into the mouse. The antibodies produced by the hybrid cells are tested one by one against the tissue of interest, and those hybrids that appear to be synthesizing appropriate antibodies are saved. The antibodies are called monoclonal antibodies because they are derived from the cloned cells.

Through the method of immunocytochemistry, monoclonal antibodies are used to determine the distribution of specific proteins in samples of the original tissue against which the antibodies were raised (see Box 7-3). This is done by conjugating (binding) the antibody to some stainable substance or marker. The tissue is exposed to the combined antibody-marker, which binds with the protein that the antibody recognizes, and the location of the protein-antibody-marker complex is pinpointed by staining the tissue with a compound that will produce a visible product when it reacts with the marker.

TABLE 22-1. Families of Neuronal Guidance Molecules

Cadherins

N-cadherin (vertebrates, insects)

Immunoglobulin superfamily (Ig) of cell adhesion molecules

Fasciclin II, III (insects)

L1 (vertebrates)

Neural cell adhesion molecule (N-CAM) (vertebrates)

Neuroglian (insects)

Integrins

Integrin (vertebrates)

PS integrin (insects)

Netrins

Netrin-1, -2 (vertebrates)

UNC-6 (nematodes)

D-netrin (insects)

Semaphorins

Collapsin I (vertebrates)

Semaphorin I, II (insects)

Semaphorin III (vertebrates)

Guidance molecules can be grouped into families of structurally similar molecules. Many of these families contain cell adhesion molecules, which constrain axons to grow in a specific direction. Other guidance molecules directly affect the direction of growth by influencing the cytoskeleton or actin production. Chemical cues may be expressed only during a narrow time window and at specific sites during development.

Extrinsic Factors Affecting Neuronal Growth

Molecular cues play a role in neural development beyond simply influencing the direction in which axons grow. Some molecules are also important **trophic factors**, substances that are necessary for the health and proper function of an individual cell. In some cases, these trophic factors may be critical not only for the initial growth and differentiation of a neuron but also for its regeneration if it is injured, or even for its day-to-day survival.

Nerve Growth Factor

That trophic factors might have an important role in the development of the nervous system was suggested by Ramón y Cajal early in the twentieth century, but the first real experimental verification of the concept did not come until 1951. In that year, the Italian researcher Rita Levi-Montalcini showed that sarcoma tissue from a mouse had a dramatic effect on the growth of sensory neurons and on neurons of sympathetic ganglia from chick embryos. Sarcoma tissue, being cancerous, exhibits vigorous, continuous growth. The original purpose of implanting such tissue into a chick embryo was to determine its effect on the growth and branching of peripheral nerves. Levi-Montalcini recognized that the mouse sarcoma not only led to nerve growth but did so in a highly selective fashion, affecting some types of neurons quite strongly and others not at all. She and coworker Stanley Cohen suggested that this effect was due to the release of a trophic substance, which was named **nerve growth factor (NGF)**. For the discovery of NGF and their pioneering work on it, Levi-Montalcini and Cohen shared the Nobel Prize in Physiology or Medicine in 1986.

The effect of nerve growth factor on tissue cultures of neurons is extraordinarily powerful and rapid. If a dorsal root ganglion from a chick embryo is placed in tissue culture, the cell bodies of sensory neurons in the ganglion will begin to grow new neurites. The growth is visible after 24 hours but is not particularly striking (Figure 22-11A). If NGF is added to the medium, however, there is an enormous proliferation of neurites from many of the neurons in the ganglion (Figure 22-11B). NGF has similar effects on sympathetic ganglia, although it has no apparent effects on motor neurons or on nerves of the parasympathetic system.

Nerve growth factor has been studied extensively since its discovery. It can be found in all vertebrate classes, from fish to mammals, and consists of two identical polypeptide chains of about 13 kdaltons each. There are different forms of NGF in different species, but NGF from any source is equally effective on any susceptible target cell, indicating that the receptors for the peptide are highly conserved across species. Nevertheless, differences in immunological reactivity shown by NGF from various animals indicate that the molecules from diverse species are slightly different in structure.

What exactly does NGF do, and how does it work? Experiments have shown that NGF is necessary for the normal development of sympathetic neurons and of certain types of sensory neurons. It is present in the tissues normally innervated by these neurons, and the neurons will not develop properly if NGF is absent at an early stage of development. For example, blocking the effects of NGF in a growing embryo by exposing the embryo to the antibody for NGF results in the death of

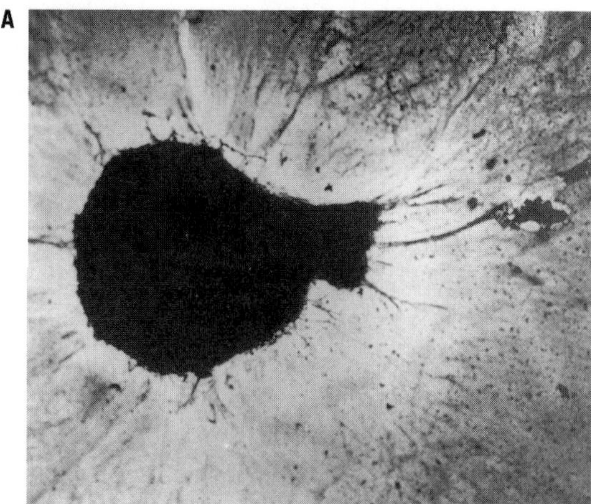
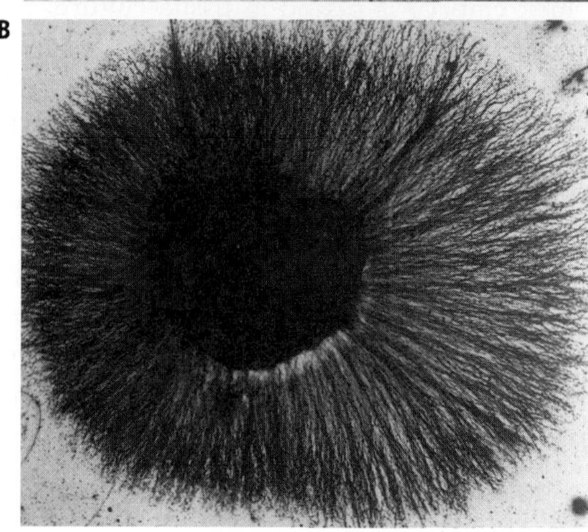

FIGURE 22-11. Excised dorsal root ganglia from chick embryos 24 hours after they have been placed in tissue culture. (A) A ganglion placed in normal tissue culture medium. (B) A ganglion placed in tissue culture medium to which about 10 ng nerve growth factor has been added.

virtually all sympathetic and many sensory neurons. Adding NGF antibody to ganglia in tissue culture also completely blocks the vigorous NGF-induced outgrowth of neurites normally seen in sympathetic or sensory nerve ganglia. In the sensory nervous system, NGF is critical for proper development of some small, unmyelinated sensory fibers, especially those that convey nociceptive information (see Chapter 14). The effects of NGF are specific to these neurons; fast-conducting, myelinated fibers develop normally in the absence of NGF.

In order to exert its effects, NGF must enter the distal tips of the neurites of susceptible neurons. It does this by first binding with appropriate receptor proteins in the membrane; these are called tyrosine kinase A (TrkA) receptors. TrkA receptors that have bound with molecules of NGF detach themselves from the membrane, bringing the NGF molecule with them through the membrane into the cell. Inside the neuron, the receptor-NGF complex in some cases may remain in the distal part of the neurite to stimulate neurite growth and elongation. In other cases, the entire complex may be moved via axonal transport back to the cell body, where it stimulates DNA transcription and promotes the growth of the neuron. The molecular mechanisms by which NGF exerts these effects are not yet understood.

Nerve growth factor is a trophic substance that stimulates the growth of certain types of vertebrate neurons. Its presence is necessary for the proper development of neurons of the sympathetic nervous system and of some small, unmyelinated sensory neurons, especially nociceptive ones.

Other Neurotrophic Factors

The discovery of nerve growth factor prompted a search for other trophic factors that could influence the development of neurons. Because NGF promoted neural outgrowth, other molecules that had similar effects on populations of neurons were also classified as trophic factors, although subsequent work has made clear that all neurotrophic factors do not have the same effects on neurons. At present, several families of neurotrophic factors have been identified based on their structural similarities. The three most important of these—the **neurotrophin family**, the **ciliary neurotrophic factor family**, and the **transforming growth factor superfamily**—and the main neurotrophic factors that belong to them are listed in Table 22-2.

Members of the neurotrophin family target specific populations of neurons, mainly different sensory neurons, and, for NGF, neurons of the sympathetic nervous system. Table 22-2 lists some of the main populations of sensory and other neurons affected by neurotrophins. Recent studies have revealed that the target population can be quite specific. For example, NT-3 is required for survival of many vertebrate sensory neurons. Detailed examination of the affected neurons showed that NT-3 is required specifically for survival of fast-adapting hair follicle receptor cells and slowly adapting Merkel cells (see Chapter 14). Other skin mechanoreceptors are not affected by absence of this neurotrophin.

Identification of neurons that are susceptible to the presence or absence of neurotrophins has been aided by the identification of the receptors that mediate their actions. The known receptors specific for individual members of the neurotrophin family of neurotrophic factors are listed in Table 22-3. The Trk receptors are all high-affinity, meaning that they bind readily and strongly to the specific neurotrophin with which they are associated. The neuron types on which specific Trk receptors have been located so far match the neuron types that the corresponding neurotrophins are known to affect. However, all neurotrophins will also bind with a low-affinity receptor known as p75, which has widespread distribution and also binds a number of other growth factors. At present, most researchers think that the cell-specific effects of the neurotrophins are mediated exclusively via the high-affinity Trk receptor molecules, but this leaves the role of the p75 receptor unclear. There is some evidence that the action of NGF is enhanced when it binds with both p75 and TrkA.

The original research on NGF led to the idea that neurotrophic factors are necessary to

TABLE 22-2. Neurotrophic Factors

Neurotrophic Factor	Target Neuron Population
Neurotrophin family	
Brain-derived neurotrophic factor (BDNF)	Proprioceptive mechanoreceptors; vestibular neurons
Nerve growth factor (NGF)	Sympathetic neurons; nociceptive and other small-fiber sensory neurons
Neurotrophin-3 (NT-3)	Proprioceptive and some tactile mechanoreceptors; auditory neurons
Neurotrophin-4/5 (NT-4/5)*	Visceral sensory neurons
Neurotrophin-6 (NT-6)†	Not known
Ciliary neurotrophic factor family	
Ciliary neurotrophic factor (CNTF)	Sympathetic and sensory neurons
Transforming growth factor family	
Transforming growth factor (TGF-β1, -β2, -β3)	Dopaminergic central neurons
Glial cell line–derived neurotrophic factor (GDNF)	Dopaminergic and other central neurons

*Neurotrophin 4 and neurotrophin 5 were described separately. They are now recognized to be the same molecule.
†Present in fish. Not yet found in mammals.

BOX 22-2

Scanning Through Tissue: Using Confocal Microscopy to See Cells

Fluorescent markers, whether injected into single cells, used as labels on monoclonal antibodies or other cell markers, or as part of voltage-sensitive dyes, have allowed neurobiologists to see the fine branching of neurons, the location of biologically significant proteins within cells, or the time course of a cell's functional response to chemical or electrical activation. However, the value of conventional fluorescence microscopy is limited because all of the fluorescence in the prepared tissue will be visible at once. Because only some of this fluorescence will come from parts of the cell or tissue that are in focus, the image will inevitably be blurred.

A microscope that can reduce this problem, the **confocal microscope**, was developed in the late 1960s, but it was not until the 1980s, with the advent of powerful desktop computers capable of storing digitized images, that the full power of the instrument began to be realized. Both conventional fluorescence microscopy and confocal microscopy involve illuminating a specimen with light that will cause a target molecule to fluoresce. The difference is that in confocal microscopy the light is passed through a pinhole aperture before it is reflected off a mirror to the specimen, and the light of the fluorescent molecules is passed through a second pinhole aperture before reaching a detector. The optics of the apparatus are such that only the light that comes from the single narrow plane of focus of the microscope at the level of the specimen can pass through the second pinhole, thereby providing a sharp image of objects at that level of focus.

Modern instruments have taken the principle a step farther by using a laser beam as the source of illumination (Figure 22-B). This improves performance considerably because a laser is a coherent beam of photons at a single wavelength; it thereby stimulates little except the molecules of interest, reducing the background fluorescence that normally obscures the tagged parts of the tissue and thus generating a considerably cleaner image. Furthermore, the beam can be "scanned" over the specimen by moving the beam; hence the full name of the technique as it is used today, **laser scanning confocal microscopy**. On modern instruments, fine controls on the focusing devices automatically focus the laser beam at precisely determined intervals through the thickness of the specimen, producing optical sections at each focal plane. Computer image analysis can then be used to "stack" these sections, producing a reconstructed image with optimal resolution of both the fluorescently marked cell inclusions or molecules and the light microscopic detail that gives the anatomical context (Figure 22-C). The resulting images are often remarkably beautiful.

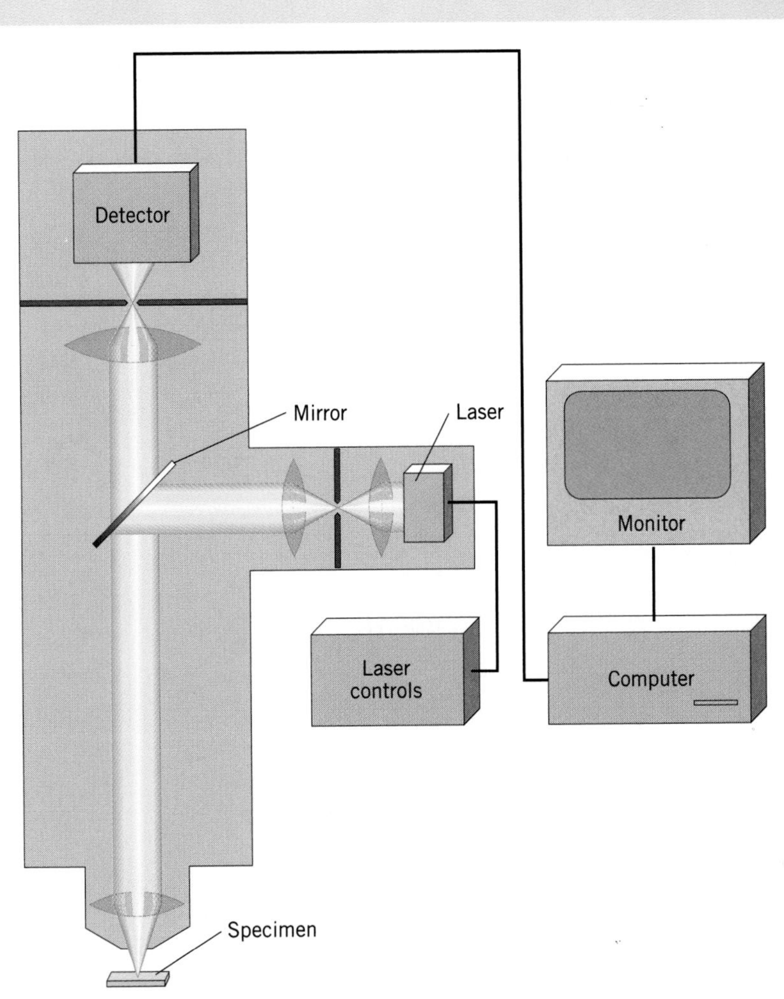

FIGURE 22-B. The principle of operation of a laser scanning confocal microscope. A laser that produces a beam of light at a wavelength appropriate to stimulate fluorescence in the specimen is aimed at a mirror that will reflect the beam down to the specimen. The light from the resulting fluorescence is at a different wavelength (one that the mirror will not reflect) and therefore passes through the mirror to the detector. By being passed through a pinhole, the light from the fluorescent molecules at a single plane only will be detected. The detected light is digitized and sent to a computer for display, storage, and later manipulation. Light filters and the controls for scanning the laser over the specimen have been omitted from this diagram.

A refinement of the method was introduced in the early 1990s: **two-photon laser scanning confocal microscopy**. As the name suggests, two separate beams of light are aimed at the specimen. Each beam by itself has a wavelength that is too long to excite fluorescence in the target molecules. However, if the two photons arrive at the molecule simultaneously, the

(continued)

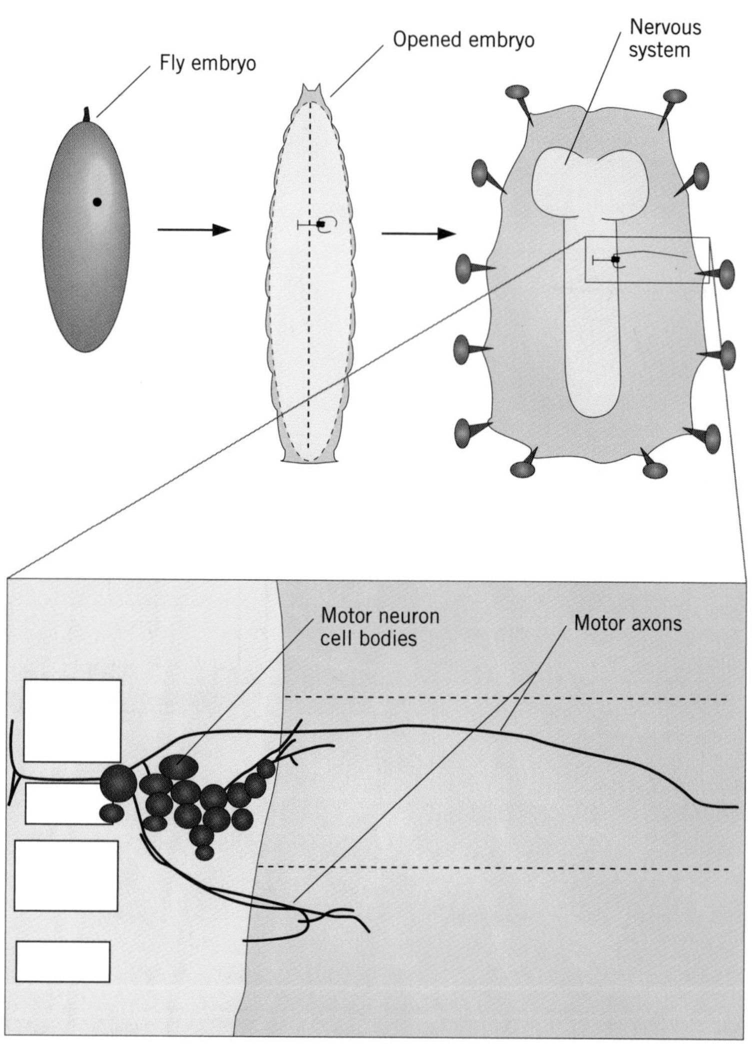
Fly embryo
Opened embryo
Nervous system
Motor neuron cell bodies
Motor axons

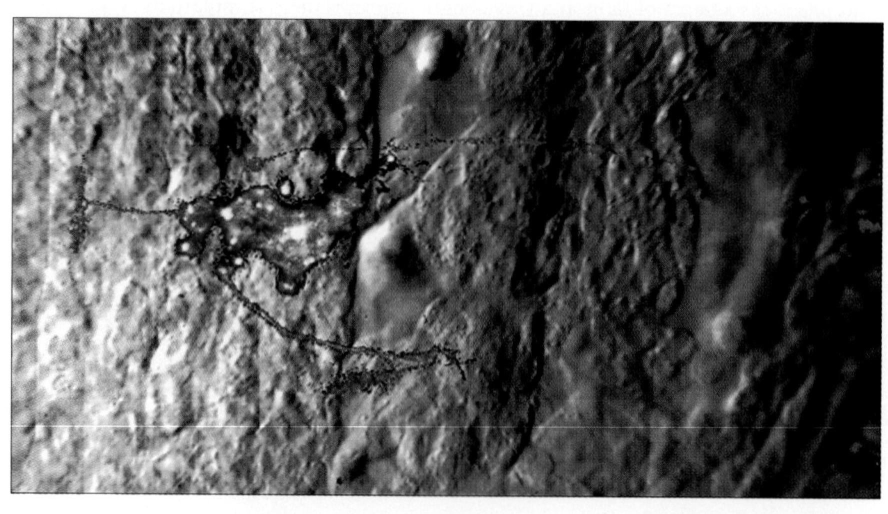

energy of the two combine and interact with the molecule as if a single photon at a shorter wavelength with the equivalent energy content had struck it. The main advantage of this method is that the two beams can be precisely focused on a single optical plane in the specimen, hence confining the bleaching of fluorescence that accompanies illumination to that plane. This provides much clearer images even in cases where relatively few fluorescent molecules are present because, when the focus is stepped to a new plane, none of the molecules will have been bleached by prior exposure to the fluorescence-inducing light.

The laser scanning confocal microscope has opened new vistas of cell structure and function for study, allowing researchers to see the precise distribution of molecules in a cell or tissue. Combined with other techniques that allow unstained cells or cellular elements to be seen, it offers an unparalleled glimpse into the way in which specific molecules shape and guide the growth of axons during development, as well as many other insights into the molecular basis of cell functions.

FIGURE 22-C (opposite). An image obtained with a laser scanning confocal microscope. The image shows tissue in a developing *Drosophila* embryo, with a cluster of motor neurons and their axons (shown extending into the peripheral tissue) marked with a fluorescent marker. The relationship of the motor neurons to the whole embryo is shown in the sketches above the image. The image was prepared by combining all the in-focus parts of the neurons and background in the stacked images that are obtained by scanning the entire thickness of the tissue, then discarding all the out-of-focus parts. (Courtesy of Aloisia T. Schmid and Christopher Q. Doe.)

TABLE 22-3. Receptors for Molecules of the Neurotrophin Family

Receptor	Trophic Molecule
TrkA	Nerve growth factor
TrkB	Brain-derived neurotrophic factor; neurotrophin-4/5
TrkC	Neurotrophin-3

promote neuron growth and ensure survival of neurons. However, the discovery of other neurotrophic factors has led to more detailed study of the actions and roles of these factors, and consequently to the recognition that neurotrophic factors may play quite different roles. Several circumstances make identification of these roles difficult. First, the effects of a neurotrophic factor in vivo (literally "in life," meaning in the body) may not be the same as its effects in vitro (literally "in glass," meaning under artificial conditions outside the body), when the cells with which it interacts are isolated from their normal chemical environment. Second, if a factor acts during development, it may act earlier or later than the time at which some other factor acts. Third, one factor may interact with another, so that eliminating one by itself does not necessarily have a strong effect on a target neuron.

In spite of these difficulties, several effects of neurotrophic factors have been documented. The classical effect of NGF is on the growth and survival of neurons. That is, NGF stimulates the growth of neurites in the proper direction and promotes the survival of neurons that are able to pick up NGF from their environment. Other neurotrophic factors also play important roles in survival (see next section), some during embryogenesis and others after an animal has reached maturity. Survival may be of the whole neuron, or of the synapses formed by it. However, neurotrophic factors can also promote the differentiation of neural precursors into specific neuronal types. For example, reducing NGF levels in neonatal rats leads to a reduction in the numbers of nociceptive neurons and an increase in the number of hair cell receptor neurons.

Even farther removed from the originally hypothesized function of neurotrophic factors, some molecules may function in recovery from injury. **Ciliary neurotrophic factor (CNTF),** for example, becomes abundant at the site of a wound. It may act as a trophic factor to promote repair directly, or it may stimulate astrocytes to help repair damage. The idea that neurotrophic factors may repair damage has also led to research on their effects on neurodegenerative diseases like Alzheimer's disease and Huntington's disease. There are few positive results from this research to date, but the enormous medical benefits to be gained should any neurotrophic factor be able to ameliorate the neural degeneration at the root of any of these diseases keep this an active area of research.

Neurotrophic substances may be grouped into molecular families on the basis of structural similarities. Neurotrophic substances may promote neuron growth and survival, the differentiation of neurons into specific types, or recovery from injury.

Cell Death

Death is the fate of every neuron. However, the death of a neuron is not always the consequence of accident, old age, or the death of the organism of which it is a part. Instead, death of selected neurons may occur during the course of normal development, in which case it is called **programmed cell death**. Such death is not specific to the nervous system; examples can also be found in other developing tissues.

Programmed cell death can manifest itself in one of two ways during development: as a consequence of overproliferation or of changing conditions in the developing animal. Overproliferation refers to the production of far more neurons than ever survive to adulthood. For example, in vertebrates, only about half of all the neurons that differentiate survive. This overproliferation of neurons leads to competition among the cells for necessary neurotrophic factors like NGF, **brain-derived neurotrophic factor (BDNF),** or ciliary neurotrophic factor (CNTF). Only those neurons that are successful in acquiring sufficient amounts of these factors during development will prosper. The remainder die. As described in the previous section, some of these factors are released extracellularly and picked up by neurons as they grow toward the target. Other neurotrophic factors are released only when proper synaptic contact is made with a target neuron or muscle. Hence, pruning of excess neurons occurs throughout development, from the time neurons first differentiate until they establish their correct synaptic connections.

Programmed cell death can also occur when conditions change inside an animal. It is as if some neurons, having served their function, are pruned away from the nervous system much like unwanted branches are pruned from a tree. This type of cell death is common in insects during metamorphosis, when an immature animal changes to a pupa and then to an adult. It also occurs in metamorphosing amphibians such as tadpoles. In insects, the process is closely regulated by the level of specific hormones in the insect's blood; you will learn more about this process in Chapter 25.

Much progress has been made in understanding the molecular and biochemical mechanisms that cause programmed cell death. In most cases, neurons (and other cells in the developing organism) appear to die by a form of cell suicide called **apoptosis**, an event that involves the activation of specific "cell death genes." This means that death is not simply due to the lack of some essential material like oxygen or a nutrient but is the result of a dedicated biochemical process that fragments the cell's nuclear DNA, ruptures the nuclear membrane, and hence causes the death of the cell. What is not yet clear is whether some cells survive because of an active process that continually rescues them, or whether survival is a kind of default state that must be overcome through triggering of suicide machinery. In either case, programmed cell death is not only a fundamental aspect of neural development; by pruning away surplus or unwanted cells, it also shapes the connections of the surviving neurons. Apoptosis and the influence of hormones on it in insects will be considered in more detail in Chapter 25.

Programmed cell death shapes development by competition between neurons for neurotrophic factors and proper synaptic contacts, and via planned elimination. Apoptosis, a specific mechanism of programmed cell death, is an active process in that specific biochemical machinery that terminates the cell's life is activated.

Additional Reading

General

Jacobson, M. 1991. *Developmental Neurobiology*. 3d ed. New York: Plenum Press.

Purvis, D., and J. W. Lichtman. 1985. *Principles of Neural Development*. Sunderland, Mass.: Sinauer.

Research Articles and Reviews

Bentley, D., and H. Keshishian. 1982. Pioneer neurons and pathways in insect appendages. *Trends in Neurosciences* 5:354–58.

Fraser, A., N. McCarthy, and G. I. Evan. 1996. Biochemistry of cell death. *Current Opinion in Neurobiology* 6:71–80.Goodman, C. S. 1996. Mechanisms and molecules that control growth cone guidance. *Annual Review of Neuroscience* 19:341–77.

Karlstrom, R. O., T. Trowe, and F. Bonhoeffer. 1997. Genetic analysis of axon guidance and mapping in the zebrafish. *Trends in Neurosciences* 20:3–8.

Oppenheim, R. W. 1991. Cell death during development of the nervous system. *Annual Review of Neuroscience* 14:453–501.Snider, W. D., and D. E. Wright. 1996. Neurotrophins cause a new sensation. *Neuron* 16:229–32.

Tanabe, Y., and T. M. Jessell. 1996. Diversity and pattern in the developing spinal cord. *Science* 274:1115–23.

CHAPTER 23

Developmental Plasticity

Although neural connections are specified in the genetic code, it is nevertheless abundantly clear that a variety of factors, both intrinsic and extrinsic to an animal, can modulate, alter, or shape these connections. Such changes can occur not only in the developing nervous system but even in adult animals. In this chapter, you will learn about the effects of the environment on the development and function of neural systems.

As you learned in Chapter 22, early development of the nervous system is intrinsic (also termed **activity-independent**), having no requirement for neural activity or external input. It is instead orchestrated by the properly timed activation of genes that direct growth or code for guidance molecules and receptors that determine the path taken by the growing axon. However, the nervous system has a great capacity to be shaped by neural activity, a capacity often called malleability or **plasticity**. This plasticity is especially obvious during the later stages of development, when the connections between neurons in the nervous system are being made.

Researchers especially became aware of the developmental plasticity of the nervous system through experiments in the 1960s which showed that a fully functional nervous system cannot be built solely through activity-independent processes. If the eyes of a mammal like a cat or a monkey are occluded (sewn shut so the animal cannot see) from birth to an age of several months, the animal will be functionally blind after the eyes are opened. Physiological tests show the retinas to be perfectly normal, so this result cannot be due to damage to the photoreceptors. The effect is caused by the failure of the proper connections in the visual centers in the brain to develop because of the lack of normal visual input. As described later in this chapter, this and many similar experiments starkly demonstrate the extraordinary effect on the nervous system of withholding or manipulating a sensory system's normal input in the late stages of development.

Subsequent experiments have both revealed the basis of the phenomenon and allowed researchers to generalize the result. It is not just external input that is critical to normal development; the electrical activity in neurons that the input elicits is important as well. It is now clear that whereas intrinsic factors can direct the growth of a nervous system according to a general plan, the wiring requires neural activity for fine-tuning. Hence, this aspect of development is referred to as being **activity-dependent**. In motor systems, all the electrical activity that is necessary for proper connections to form between spinal neurons and the motor neurons they innervate occurs spontaneously during development. In sensory systems, too, some of the necessary activity may be generated spontaneously, but some is generated by exposure of the system to the animal's environment.

Much development is intrinsic to the nervous system, but some development is more malleable and requires that neural activity be present for proper functional connections to be made. In sensory systems, this activity may be generated internally by the developing system, or it may be evoked by external stimuli.

The Role of Experience in the Development of the Visual System

Most studies of developmental plasticity have dealt with sensory systems, in part because manipulation of sensory stimuli is relatively easy, and in part because plasticity is especially striking there. The visual system has been a special favorite for study of developmental plasticity because the physiology of the visual system is well known, visual stimuli are easy to manipulate, and development of the system is influenced by both activity-independent and activity-dependent factors.

As you learned in Chapter 11, the axons of retinal ganglion cells form a topographical map of the retina in the optic tectum in lower vertebrates (see Figure 22-5) and in the lateral geniculate nucleus (LGN) in mammals. From the LGN, other neurons project to the primary visual cortex, forming a topographical map there as well. The pathfinding required of the ganglion cells for them to reach not only the appropriate part of the brain but also the right sets of cells within that part to form a topographical map is intrinsic. The growing axons are guided by local chemical and other cues to the appropriate destination in the brain, and are distributed throughout the target area according to their reactions to neurotrophic factors expressed by the target neurons. Develop-ment of the connections between layer 4 of the primary visual cortex (V1) and other layers is also activity-independent.

The map first formed in the cortex is rough-grained because the axons of the cells from the ganglion cells (in lower vertebrates) or from the LGN (in mammals) initially branch extensively in the brain to form synapses over several millimeters, hence reducing the map's precision (Figure 23-1A). After birth, however, stimulation of the eye by light causes activity of the ganglion and LGN cells, which causes the axonal arbors (branches) of these neurons to narrow as more distant branches are pared away (Figure 23-1B), leading to development of a fine-grained retinotopic map that confers greater visual acuity to the animal. Neural activity is necessary for

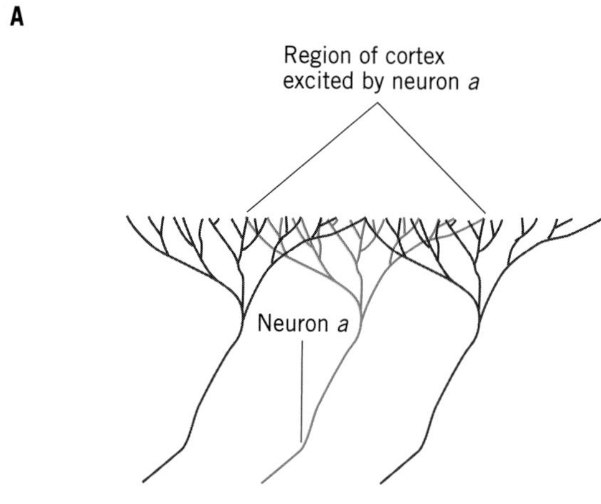

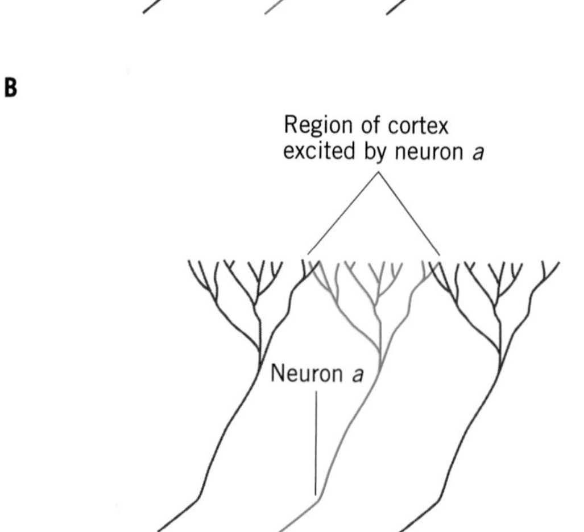

FIGURE 23-1. Schematic representation of the distribution of axon branches in the visual area of a vertebrate. (A) The condition at birth; note the substantial overlap between cells. (B) The condition at 6 weeks; note that there is essentially no overlap of branches at this time.

the fine-grained map to develop. Crushing the optic nerve from one eye of a fish and treating the eye with tetrodotoxin (TTX, as described in Chapter 5) to prevent activity in the regenerating nerve prevents the formation of the fine-grained map, but not the rough projection.

In addition to being necessary for the establishment of the fine-grained topographical map, neural activity is also necessary for the proper formation of other connections. In the mammalian LGN, the segregation of input to alternating, eye-specific layers of cells within the nuclei is activity-dependent, as is segregation by eye in layer 4 of the visual cortex. Further, establishment of ocular dominance columns in the visual cortex also depends on neural activity. These activity-dependent processes are described in greater detail in subsequent sections.

Development of the visual system occurs by a combination of intrinsic and activity-dependent events. Pathfinding, formation of a rough topographical map in the LGN or the cortex, and intracortical connections are activity-independent. The refinement of the topographical map and the formation of eye-specific layers and ocular dominance columns are activity-dependent.

Activity Dependence of Development in the Thalamus

As shown in Figure 11-13, the adult mammalian LGN is a layered nucleus, in which adjacent layers receive input from different eyes. The axons of ganglion cells make synaptic contact with neurons in the LGN even before the photoreceptors have developed. Nevertheless, experiments show that the formation of the eye-specific layers is activity-dependent. For example, it is possible to eliminate all neural activity in the optic nerves by the long-term infusion of TTX during the time when ganglion cell axons are growing to the LGN and forming synapses on LGN cells. This procedure completely abolishes the formation of eye-specific layers in the LGN. The TTX does not seem to inhibit normal growth of axonal branches; TTX-treated axons have even more branches spread over a larger area than normal.

Neural activity per se is not sufficient for formation of eye-specific layers; a specific pattern or timing of activity is necessary. The pattern of electrical activity in ganglion cells is produced by spontaneous waves of electrical activity in the retina (Figure 23-2). Loading retinal cells with the Ca^{2+} indicator fura-2 (see Box 7-2) allows the increase in free intracellular Ca^{2+} associated with cell electrical activity to be recorded. Such recordings reveal that waves of depolarization regularly sweep over small areas of the retina. Over a period of time, the entire retina experiences these waves. The waves of depolarization cause the generation of action potentials in the ganglion cells, which in turn helps to organize the layers of the LGN.

Apparently, the synapses that initially form between ganglion cells and their LGN targets must be strengthened in order to persist. The mechanism of this strengthening is the convergence of input onto specific LGN cells. The near-synchronous barrage of input to the LGN from the ganglion cells in the excited regions of the retina strengthens the synapses between

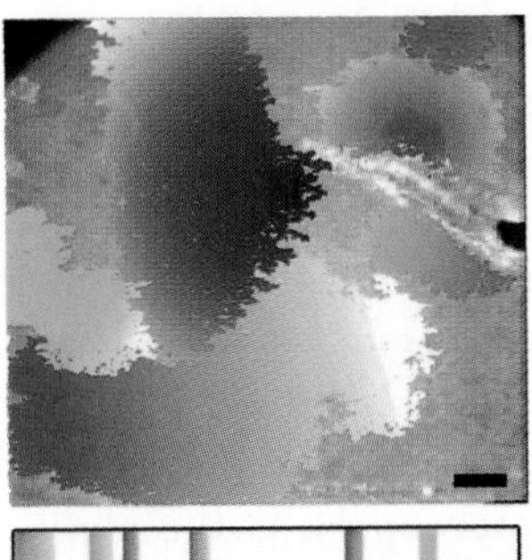

FIGURE 23-2. False color image showing spontaneous waves of neural activity in a region of a ferret retina. Each color represents one wave propagating from the densest to the lightest shade. Colors were assigned to waves arbitrarily. The waves appeared at the times shown at the scale bar under the main image. The scale bar represents 4 minutes of recording.

these neurons and the LGN cells on which these synapses converge. Since the neurons that are firing synchronously are from the same eye, the consequence is the strengthening of synapses from ganglion cells of one eye onto specific LGN neurons. Meanwhile, similar waves of electrical activity are being generated in the other eye, causing synapses from different ganglion cells that converge on other LGN cells to be strengthened. The LGN develops eye-specific layers because although initially synapses from different eyes do converge on the same LGN cells, these synapses are not strengthened. A wave of electrical activity that sweep over a particular region of one retina will only rarely be synchronous with a wave of electrical activity that sweeps over the corresponding region of the other retina. Consequently, all synapses that convey information from the two eyes to any one cell in the LGN will eventually be eliminated, leaving just the eye-specific layers.

In mammals, electrical activity in ganglion cells produced by spontaneous waves of depolarization in the retina is necessary for the segregation of eye-specific layers in the lateral geniculate nuclei.

Binocular Vision and Ocular Dominance

Activity-dependent visual development has been studied most intensively in the cortex, using stereoscopic vision in mammals as a model system (see Box 20-1). **Stereoscopic vision,** also variously known as **binocular vision** or **stereopsis,** is the ability to see in three dimensions; it arises from an overlap of the visual fields of the two eyes. In primates, in predators like cats and hawks, and in other animals for which good depth perception is important, the eyes are located on or near the front of the head, which allows most of the visual fields of the two eyes to overlap. Convergence of input from the eyes onto neurons in the primary visual cortex allows the image provided by each eye to be compared. The results of this comparison are sent to other visual cortical areas (see Chapter 11) so that the relative depth of objects in the scene can be determined.

The organization of the visual cortex into ocular dominance columns, columns of neurons that respond primarily or exclusively to input from one eye, appears to be critical for the ability of an animal to perceive depth of field. As described in Chapter 11, ocular dominance columns are arrayed in rows, an arrangement that can best be demonstrated by marking the parts of layer 4 of the visual cortex that receive input from one eye (see Figure 11-19B). One well-established technique for such marking is autoradiography (Box 23-1). Application of autoradiography to the visual cortex from a normal mammal such as a cat or a monkey shows that layer 4 is about evenly divided between the parts that receive input from one eye and the parts that receive input from the other. The labeled rows are about the same width as the unlabeled ones, and about the same total area is devoted to input from each eye.

Investigation of the physiology of cortical cells in the ocular dominance columns reveals that most cells can at least be somewhat driven by (excited by) a stimulus in the appropriate region of the retina of either eye, but that most are driven more strongly by input from one eye than by input from the other. The degree to which a particular cell is driven by input from one of the eyes is the **ocular dominance** of that cell. Think of ocular dominance as a scale that indicates the extent to which the cell responds to input from one eye, just as temperature is a scale that indicates the heat content of a fluid or solid.

The degree to which cells in the visual cortex are dominated by input from one eye can be assessed by systematically recording from a large number of cells and generating a histogram of ocular dominance (Figure 23-3). An electrode is inserted into the visual cortex of the experimental animal while an appropriate visual stimulus is presented to each eye in turn.

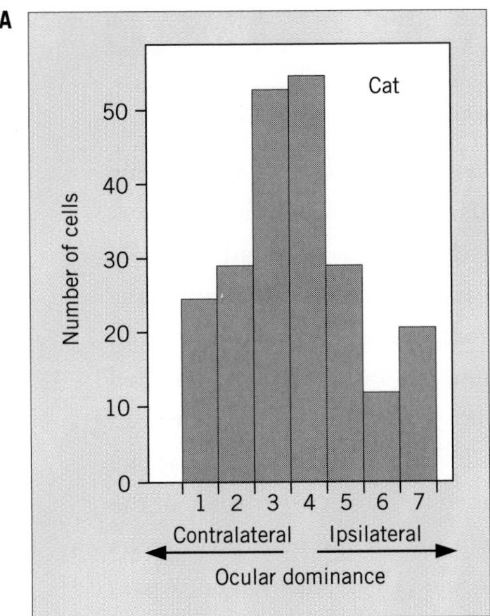

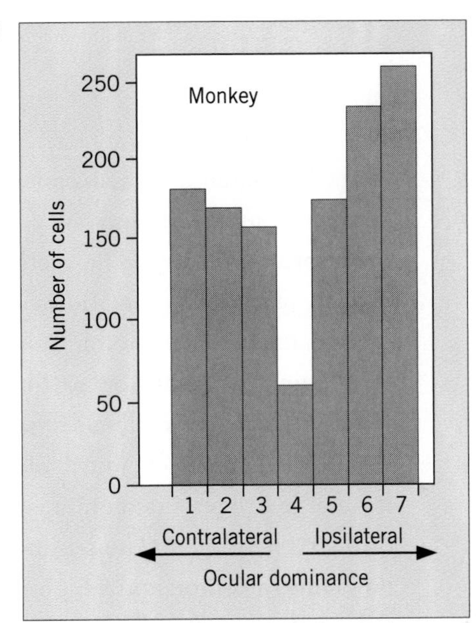

FIGURE 23-3. Histograms showing ocular dominance of cortical visual neurons in mammals. (A) Ocular dominance in a normal cat. In this animal, most neurons are strongly binocularly driven. (B) Ocular dominance in a normal monkey. In monkeys, most cells are strongly biased to respond to input from one eye or the other, and relatively few are truly binocular.

The graphs are constructed by assigning each cell to an ocular dominance group according to which eye drives it the most strongly and then counting the total number of cells that fall into each group. Cells in group 1 show contralateral ocular dominance; they respond vigorously to stimuli applied to the eye on the side *opposite* to the side of the brain from which the recording is being taken, and little, if at all, to stimuli applied to the ipsilateral (same side) eye. Cells in group 7 show ipsilateral ocular dominance, meaning that they respond vigorously to stimuli applied to the eye on the *same* side as the side of the brain from which the recording is being taken, and little, if at all, to stimuli applied to the contralateral (opposite) eye. Cells in group 6 respond somewhat less vigorously to ipsilateral input and a bit more strongly to contralateral input, and so on. Cells that respond equally strongly to input from either eye are placed in group 4.

Figure 23-3 contains two histograms, one showing ocular dominance in a cat and one showing ocular dominance in a monkey. The differences between the shapes of these histograms are not important here, but the histograms do reinforce the fact that animals may differ dramatically in their physiology. In this case, the visual cortex of cats contains mostly neurons that are driven about the same by input from either eye (ocular dominance groups 3, 4, and 5), whereas the visual cortex of monkeys contains many more cells that are driven quite strongly or exclusively by one eye or the other (ocular dominance groups 1, 2, 6, and 7).

Stereoscopic vision depends on overlap of the visual fields of the two eyes and the organization of the visual cortex into ocular dominance columns. In addition to showing eye specificity, neurons in the visual cortex also show ocular dominance.

Activity Dependence in the Cortex

All adult mammals with binocular vision exhibit ocular dominance columns. However,

BOX 23-1

Counting Grains: Autoradiographic Localization of Neurons

A major challenge in neuroanatomical studies is how to make visible only a selected subset of neurons. One solution is to use the technique known as **autoradiography**. The method involves the uptake of a radioactively labeled marker by specific neurons in the tissue of interest, coating the labeled tissue with photographic emulsion or placing it next to film, and developing the emulsion or film to determine the location of the labeled marker. Labeled molecules such as amino acids can be synthesized through ordinary chemical methods, substituting tritium, a radioactive form of hydrogen that contains two neutrons in its nucleus, for some of the ordinary hydrogen. The tritium decays when one of the neutrons breaks down into a proton and a high-energy β^- particle (an electron). The β^- particles substitute for light to expose the photographic emulsion.

Labeled tissue is placed on a microscope slide and covered with liquid emulsion or placed next to a piece of unexposed film, then stored in the dark for a period of some days. When the emulsion or film is developed, the radioactive atoms will have caused exposure of the emulsion or film just over them, producing a constellation of minute reduced silver grains on the film wherever the radioactive atoms were concentrated. The tissue on the slide is then stained to make its structure visible; the location of the grains on the film relative to the tissue indicates the location of the labeled marker.

The trick, of course, is to introduce the marker into only the specific neurons of interest. For this purpose, researchers can take advantage of the uptake of simple amino acids by neurons. Like other cells, neurons absorb the amino acids that they need for metabolism from the fluids that bathe them. If an amino acid like proline is labeled with tritium and injected into one eye of an anesthetized mammal, retinal ganglion cells will absorb the proline, incorporate it into proteins, and transport those proteins along their axons into the brain. For reasons that are not entirely clear, some of the labeled protein is subsequently released at the synapses formed by these ganglion cells in the lateral geniculate nucleus (LGN) and picked up by the neurons with which they synapse. These neurons, in turn, transport the label to the visual cortex, where it will accumulate in the terminals of the LGN neurons. Hence, only the regions of the cortex that receive synaptic input from the injected eye will take up the labeled proline. By removing the visual cortex from the animal at just the right time, flattening it out on a microscope slide, and storing it next to unexposed photographic emulsion for a period of time, a researcher can obtain a picture of the distribution of the label in the cortex (see Figure 23-5).

In addition to its application as a neural tracer as described here, autoradiography has also been used to study the distribution of single molecules in the brain. For example, a hormone or neurotransmitter can be labeled with a radioactive tag and introduced into the brain. Only those neurons that have the proper receptors will bind with or absorb the radioactively labeled compound. After some time, the brain can be fixed for histological preparation, and selected pieces of brain tissue set aside with photographic emulsion. Development of the emulsion and staining of the tissue will subsequently reveal the location of neurons that are sensitive to (i.e., can absorb or bind with) the labeled compound. Such information can identify neurons that have receptors for particular neuromodulators or hormones (see Chapter 25).

development and maintenance of the phenomenon during the late stages of development of the visual system are strongly dependent on an animal's visual experience during the period shortly after birth, and hence on the presence of specific patterns of neural activity in neurons that project to the visual cortex from the LGN. This activity dependence has been demonstrated by a number of experiments in which visual experience is altered in some way.

The Effects of Closing One Eye

Early experiments deprived one eye of normal input. Kittens are born with their eyes still shut. Normally, the eyes open within a few weeks of birth. If one eye is sutured shut at birth (under anesthesia), so that the eyelid cannot open as usual, that eye is deprived of its normal input. Because just one eye is closed, this is called a monocular deprivation experiment. If the sutures are removed when the kitten reaches about 3 months of age, allowing the eye to open, several dramatic effects on the animal's visual system are apparent. Most strikingly, the animal acts as if the previously sutured eye is blind. This behavioral blindness persists throughout the life of the cat. However, recordings taken from cells in the retina of the eye that had been closed, as well from cells of the LGN, while light stimuli are projected onto the retina, show that the retinal and LGN neurons exhibit normal physiology.

The visual cortex, however, is not normal. Both its physiology, as revealed by recordings from simple cortical cells, and its morphology, as revealed by autoradiography of cortical layer 4, are affected. Recordings from cells in the cortex after the experimental procedure show that nearly all the cells that respond to visual input do so only to input from the intact eye. Just a few out of several hundred neurons show any response at all to input from the sutured eye (Figure 23-4). The morphology of the visual cortex is also strongly affected. Instead of the normal, equally spaced distribution of ocular dominance bands seen in autoradiographs of the cortex of intact animals, layer 4 of the visual cortex of an experimental animal shows a strongly asymmetrical distribution of bands. After a suture experiment, nearly all the space in the cortex is devoted to input from the intact (unsutured) eye, and little is devoted to input from the closed eye. Figure 23-5 shows an example from a monkey. These cortical abnormalities explain the animal's behavioral blindness. The primary visual cortex is the part of the brain where the subjective sensation of vision is developed. Injury to or improper development of all or part of this

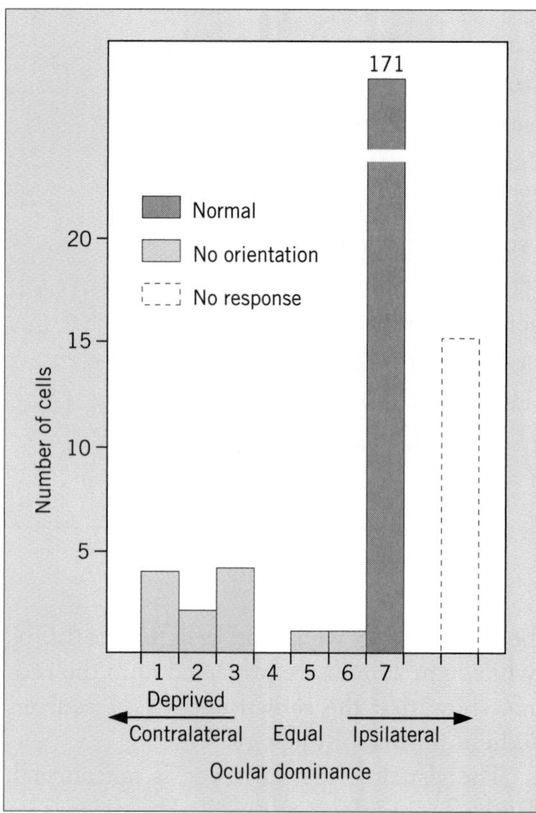

FIGURE 23-4. The physiology of visual cortical cells in a cat in which one eye (labeled "contralateral" in this diagram) has been sutured shut from birth to the age of 3 months. Note that nearly all cells respond to input from the good eye only.

cortex will therefore render the animal partly or completely blind.

These experiments clearly demonstrate that depriving one eye of its normal input has a critical effect on the normal physiological and morphological development of the visual cortex. Further experiments have shown that visual input must come during a particular period in the early life of the animal. If one eye of a kitten is sutured shut at birth, then opened again after two weeks, there is no effect on ocular dominance in the adult cat. The visual system of the cat is essentially normal, both physiologically and morphologically. If one eye is sutured shut at age 2 months, left closed for several months, and then opened again, there is also no effect on the cortex. However, if the closure includes the time from about age 23 to 40 days, there is a strong effect on the visual system. Comparable results have been obtained with monkeys, although the time during which an eye must be closed is several months rather than a few weeks.

From these experiments, researchers have concluded that there is a **critical period** for the development not just of binocular vision but of the entire cortical visual system, during which visual input must be present if the system is to develop normally. Depriving a mammal of visual input for a short time before this period, or for any amount of time after this period, has minimal effects. But depriving it of input during the critical period results in profound and permanent disturbances in the visual system.

Development of the visual cortex has a critical period, a time during which visual input must be present for normal development to proceed. If a mammal that has binocular vision is deprived of vision in one eye during this period, both the physiology and the morphology of the visual cortex are seriously disrupted.

The Effects of Closing Both Eyes

What mechanisms might underlie the profound effects of monocular deprivation described in the previous section? One possibility is that neurons carrying signals from the two eyes compete for synaptic sites. Many neurons compete to form synapses. Motor neurons, for example, form synapses when they are successful in sequestering an appropriate neurotrophic factor, as described in Chapter 22. In the visual cortex, it may be that a neuron must be electrically active in order to compete successfully. Depriving one eye of visual input prevents the neurons that normally carry signals from that eye from exhibiting any significant neural activity. (Spontaneous waves of electrical activity no

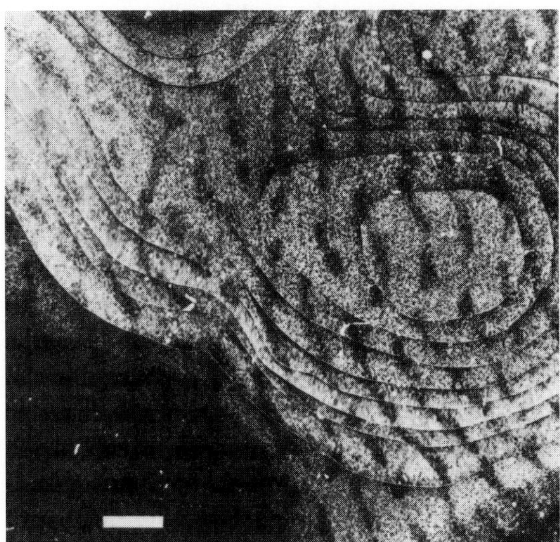

FIGURE 23-5. Bands of ocular dominance columns in the visual cortex of a monkey about 1½ years old in which one eye was sutured shut at age 2 weeks. The figure shows the distribution of radioactivity in the visual cortex after injection of radioactive proline into the intact eye 18 months later. Notice the nearly complete dominance of the cortex by synaptic sites from the normal eye (the light areas), compared with the area that receives synaptic input from the closed eye. Compare this figure to Figure 11-19B. Scale bar 1 mm.

longer occur in the retina by the time of birth.) If activity is necessary to make synapses, these inactive neurons cannot compete successfully for synaptic sites in the visual cortex, so they make few contacts, while the active neurons make widespread synapses. One way to test this hypothesis is to even the competition for synaptic sites by depriving *both* eyes of input, that is, by keeping both eyes shut during the critical period. If the effects of unilateral eye closure are due mainly to an imbalance in the input from the two eyes, then suturing both eyes shut might produce less disturbance than does unilateral suturing.

The behavioral effects of suturing shut both eyes of a monkey from birth to about 3 or 4 months of age are just as dramatic as those that result from closing just one eye: the animal is blind, even though each retina and LGN responds normally to appropriate stimuli. However, the physiological and morphological effects on the visual cortex are quite different. Physiologically, the neurons of the visual cortex do not show the same skewed distribution of sensitivity to inputs from the right and left eyes as do those in monkeys with unilaterally sutured eyes. The physiological responses are still not normal, but they are symmetrical. Most neurons in binocularly deprived animals respond to input from only one eye or the other; only a few respond to input from both eyes (Figure 23-6). This finding suggests that there is little overlap of input from one eye with input from the other at the level of the cortex. Autoradiography of the visual cortex using radioactively labeled proline injected into one eye reveals no significant difference from normal animals in the distribution of inputs from the two eyes. This means that input from one eye does not dominate most of layer 4 in the visual cortex in these animals as it does in those that have been monocularly deprived.

From these results, researchers have concluded that during development the neurons that carry input from the two LGNs compete for synaptic sites in the visual cortex. When both eyes are open, or when both eyes are sutured shut and hence deprived of light, all

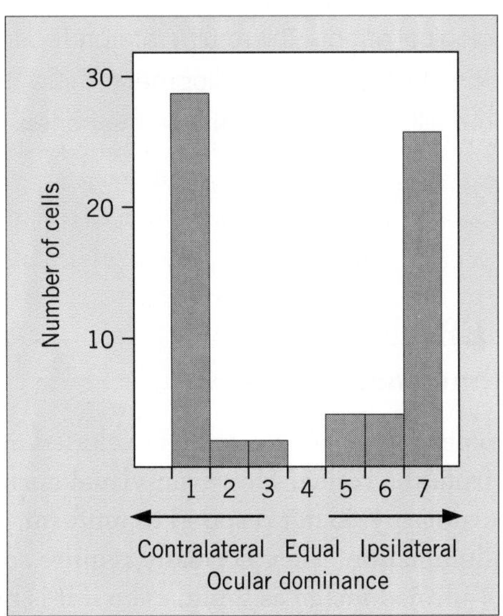

FIGURE 23-6. The physiology of visual cortical cells in a monkey in which both eyes have been sutured shut from birth to the age of 3 months. Note that most cells respond to input from one eye or the other, but almost none respond to input from both eyes.

LGN neurons are more or less equally excited. If light is available, they all carry many action potentials; if light is not available, they carry few. In either case, no set of LGN neurons has a competitive advantage. On the other hand, if one eye is open and the other shut, the LGN neurons carrying information from the intact eye will obviously be much more active than those carrying information from the closed one. This imbalance in input results in an imbalance in the distribution of synapses on visual cortical cells. The imbalance in synapses is thought to occur because the active cells are able to compete successfully for synaptic sites on neurons of the visual cortex, whereas the inactive neurons are not. The results of the suturing experiments manifest themselves only at the cortical level, because it is only here that competition for synaptic sites takes place. Neither the retina nor the LGN requires such competition for normal development.

Depriving a mammal with binocular vision of input to both eyes during the critical period disrupts the physiology of neurons in the visual cortex but has little effect on the distribution of bands of ocular dominance in the cortex. In contrast, monocular deprivation produces stronger physiological and morphological effects. This result is interpreted to mean that during development there is competition for synaptic sites in the cortex.

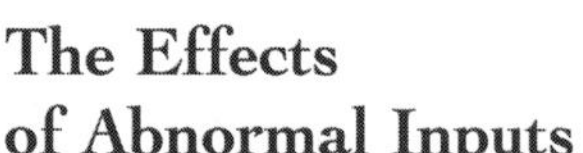

The Effects of Abnormal Inputs

The experiments just described excluded most light from the retina. However, visual cortical cells ordinarily do not respond to uniform, diffuse illumination. They normally require some sort of formed image as input, even if it is only a simple line. The question then arises whether the physiological effect of suturing the eye is due to the exclusion of light alone or to the loss of the particular type of input that normally stimulates cortical neurons. This question can be addressed by allowing diffuse light to enter the retina, but preventing the formation of images. Such an experiment can be performed by fitting an eye with a translucent cap that will allow only diffuse light to enter. The translucent cap functions in much the same way as the frosted glass of a lightbulb, which allows light to pass but not a clear image of the filament inside the bulb.

The effect of fitting such a cap over one eye during the critical period is the same as the effect of excluding light altogether. That is, the loss of form input has just as severe an effect on the physiology of visual cortical neurons and on the morphology of the ocular dominance columns as does the complete exclusion of light. If only one eye is fitted with a cap, the results are the same as if that eye had been sutured shut. If both eyes are fitted with caps, the results are the same as if both eyes had been sutured. In other words, it is the loss of specific formed images that is crucial, not the loss of light per se.

Further refinements of occlusion experiments have revealed other important parameters of the input required by the visual system for normal development. One group of experimenters, for example, reasoned that since most cortical neurons in normal animals respond to input from both eyes (although not with the same vigor), whereas cortical neurons in either monocular or binocular deprived animals do not, perhaps overlap of the visual fields of the eyes was essential. These researchers caused the eyeballs of the eyes in cats to diverge, a condition called strabismus, by severing the ocular muscles that move the eyeballs medially. This has the effect of moving the visual fields of the two eyes out of register with one another. The result of this operation is similar to that of binocular occlusion experiments. Most cortical cells respond to stimuli presented to one eye or the other, whereas few respond to stimuli presented to both eyes. As in binocular occlusion experiments, there is no significant change in the morphology of ocular dominance columns in the visual cortex as revealed by autoradiography.

So many variations of visual deprivation experiments such as these have been done that it is easy to lose sight of the significance of their results. The important point to keep in mind is that the neurons carrying information to the visual cortex are successful or not successful in forming effective synapses with cortical cells depending on the kind of information they carry. If light but no form is available, the adult animal will be unresponsive to form input. If form input is present but is not presented to the same cells at the same time, as in the strabismus experiments, the adult will have no effective binocular vision. During normal development, an animal experiences a wide range of visual experience, which provides continually varying input to the eyes. This input forges the appropriate connections between neurons in the cortex so that the adult will be able to interpret the types of stimuli that it will see during its lifetime. Because the same image will stimulate corresponding parts of the retinas of the two eyes in the binocular fields of view, input also forges appropriate connections between the two eyes to allow binocular vision.

The effects of visual deprivation during the critical period are of more than just theoretical interest to some people. An extremely small number of human infants are born with congenital cataracts. Until about the middle of this century, it was not possible to remove cataracts surgically and correct vision with contact lenses. Babies born with cataracts were therefore destined to remain blind their entire lives, even though they could distinguish light from dark. When surgical procedures for the removal of cataracts were first developed, the operation was performed on a number of adults who had been blind with cataracts since birth. The expectation naturally was that these individuals would then be able to see. To the surprise of physicians and the acute dismay of the patients, this was not the result. Instead, these individuals never attained good functional use of their eyes. Subsequent studies of infants in which cataract surgery was performed at different ages has suggested that surgery before the age of 8 weeks allows fully functional vision to develop, but surgery after 12 weeks essentially has no effect on vision. Hence, adults who undergo such surgery can never recover vision because by adulthood no operation is ever going to repair the malformed neural connections in their visual cortexes.

The main conclusion that has been drawn from these studies is that the vertebrate visual system has a critical period in development during which some synapses are made or preserved. In consequence, the environment shapes the kind of visual processing that will be possible in the adult, in part by a process of competition among neurons for synaptic sites in the visual cortex.

Depriving the eyes of formed images during the critical period without depriving them of light has the same effect on the visual system as does keeping out light.

Molecular Mechanisms of Developmental Plasticity

The experiments described in the preceding sections reveal the dramatic effects of depriving one or both eyes of normal input during development. What is still needed is a sound molecular explanation for these effects. Most molecular studies of the development in the visual system have concentrated on two particular aspects: the formation of ocular dominance columns and the formation of proper synaptic connections for the establishment of binocularly sensitive neurons. Although these phenomena are related, they can conveniently be treated separately, especially since different molecular factors have been implicated in their development.

Neurotrophic Factors

Closure of an eye or elimination of form vision from an eye causes a striking growth of axon branches of neurons carrying information from the intact eye, accompanied by lack of growth of neurons carrying information from the experimental eye. The demonstrated effects of neurotrophins on the growth of axons (described in Chapter 22) has naturally directed considerable attention toward the possible role of these molecules in visual cortical development, with the result that the neurotrophins have been strongly implicated in the process.

Although the mechanism by which the neurotrophins exert their effects is not known, the experimental evidence for their involvement is dramatic. For example, infusion of nerve growth factor (NGF) into the visual cortex of a monocularly occluded rat during the critical period can prevent the shift in ocular dominance toward the open eye that normally occurs. In other experiments, infusion of antibodies to NGF has the effect of lengthening the critical period, suggesting that NGF helps to stabilize and fix synapses in the visual cortex. Supplying excess NGF prevents the shift in synaptic connections that normally accompanies monocular occlusion, whereas preventing NGF from exerting its normal effects lengthens the period during which synaptic connections can be modulated.

Other neurotrophins also seem to be involved. For example, infusion of brain-derived neurotrophic factor (BDNF) after monocular occlusion in kittens prevents the formation of any kind of ocular dominance columns, presumably by overloading the system with the molecule so that local competition for it by active neurons is unnecessary. In addition, neurotrophin-3 (NT-3) and NT-4/5, other members of the neurotrophin family, have also been implicated in the process. Only NT-6 has not been shown to have an effect on development in the visual cortex. The involvement of more than one neurotrophic factor has impeded understanding of the role they all play because eliminating only one of them does not usually have especially dramatic effects. Furthermore, study of the distribution of tyrosine kinase (Trk) receptors shows that the locations of these receptors shift during development from axons to dendrites and cell bodies, suggesting that the neurotrophins themselves may shift roles as development proceeds.

Nearly all members of the neurotrophin family of neurotrophic factors seem to be necessary for activity-dependent development of the visual cortex. Infusion of neurotrophins can counter the effects of monocular deprivation experiments.

The Role of NMDA Receptors

Monocular occlusion experiments also have an effect on the formation of proper synaptic connections for the establishment of binocularly sensitive neurons. Here, experiments have focused most strongly on the role played by ionotropic glutamate receptors of the type known as N-methyl-D-aspartate (NMDA). Glutamate is the neurotransmitter at many synapses in the LGN and in the visual cortex. As described in Chapter 8, NMDA receptors underlie the phenomenon of long-term potentiation (LTP), in which a synaptic response is augmented through the coincidence of synaptic inputs to NMDA-bearing neurons. Hence, the strengthening of synaptic associations via NMDA receptors seems like a natural means by which the convergence of synaptic input from the two eyes can be forged and maintained.

The experimental evidence for the involvement of NMDA receptors in the process is striking. NMDA receptors can be blocked by the NMDA blocking agent, amino phosphonovaleric acid (APV). If APV is infused into the visual cortex during the critical period in cats, monocular occlusion fails to elicit the usual shift in cell physiology from cells mostly driven binocularly to cells mostly driven monocularly. Further, APV infused during this period also prevents formation of or destroys previously formed orientation selectivity in visual cortical cells. Since it has been shown that neurons in the visual cortex can exhibit LTP, and LTP is thought to be functionally dependent on NMDA receptors, these experiments suggest that the development of experience-dependent (or, more precisely, activity-dependent) connections in the cortex is due to the presence of LTP. Further evidence is the finding that the number of functional NMDA receptors in the visual cortex falls sharply after the critical period, when neural activity no longer has the capacity to alter synaptic connections.

The notion that NMDA receptors in the visual cortex mediate the formation of binocularly driven neurons, orientation selectivity, and perhaps other activity-dependent developmental processes is an appealing one, but the evidence, although suggestive, is not all firm. For example, studies show that APV may interfere with all neural activity in the cortex, not just activity that links input from the two eyes. If the effects of APV are general, then infusing it is functionally the same as applying TTX, which also disrupts normal development. Resolution of these issues awaits further research.

Formation of proper synaptic connections for neurons to be sensitive to input from both eyes may depend on a form of long-term potentiation mediated by NMDA receptors. Infusion of blockers of NMDA receptors prevents the normal shift in ocular dominance due to monocular occlusion during the critical period in cats.

Plasticity in Other Systems

The visual system is not the only sensory system that is acutely responsive to events during early development. Experiments with the somatosensory cortex of mammals have shown that it, too, exhibits a great deal of developmental plasticity. For example, in the somatosensory cortex of the rat, the region devoted to input from a forelimb is adjacent to the region devoted to input from a hindlimb. If a forelimb is amputated prenatally, the region in the somatosensory cortex that normally would have received input from it is instead innervated by input from the hindlimb, so that the hindlimb area is twice as large as normal. Presumably, as in the visual system, neurons are competing for synaptic targets, and following amputation, more cortical area is available to neurons from the hindlimb because of the lack of competition from neurons from the forelimb.

Furthermore, mammals are not the only animals in which development is affected by experience. Even in insects, whose development is much more deterministic than that of vertebrates, deprivation from sensory input in the larval stage produces dramatic effects. You learned in Chapter 12 that some insects have cerci, the peglike appendages at the end of the abdomen that are covered with fine, filiform hairs. These hairs are sensitive to both sound and air movements. In crickets, cercal afferents synapse with giant interneurons similar to those found in cockroaches (see Chapter 19). Each of the giant interneurons shows its own characteristic, complex dendritic arbor that helps to identify it. The size and shape of dendritic branching can be influenced by input from the cerci.

If a cercus is amputated from an immature cricket, the cell bodies and dendrites of all the sensory neurons that normally convey information about air movements or sound from the cercus to the giant interneurons are eliminated. If the cercus is removed early enough in the period between molts (see Chapter 25), a new cercus will regenerate. However, after amputation of the cercus, even after the full-sized and fully functional cercus has regenerated, there is a significant reduction in the size of the dendritic arbor of the giant interneurons that receive input from the sensory neurons on the regenerated cercus (Figure 23-7). If the amputation is done late in the life of the insect, the reduced dendritic arbor of the giant interneurons persists throughout the insect's life. If the amputation is done early enough so that the insect has time to go through several molts, there may be complete morphological recovery of the dendritic field in the adult. However, there is still some loss of the normal functional connection between the cercal sensory neurons

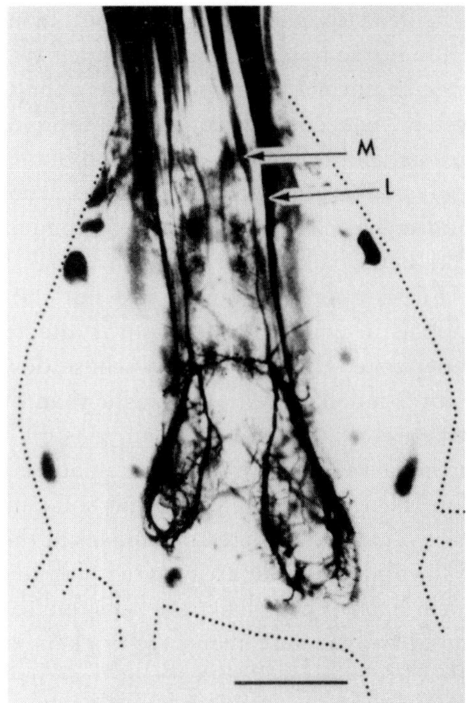

FIGURE 23-7. The effect of denervation of a cercus on the dendritic arbor of the giant interneurons of a cricket that normally receive input from the cercal afferents. The cercus on the left side was amputated in each larval stage (it regenerates after each molt). Note the reduced size of the dendritic branches of the giant interneurons that normally receive input from cercal afferents.

and the giant interneurons. These experiments show that loss of sensory input to the giant interneurons for even a relatively short period has a significant deleterious effect on the structure and function of these neurons.

Lack of input is not the only experimental condition that can modify the structure or function of a sensory system. In crickets, the cercal afferent-to-giant synapse normally loses responsiveness when the cerci are exposed to repeated pulses of sound. However, immature crickets raised from hatching to adulthood in an environment of continuous sound pulses respond differently. As adults, such insects can be exposed to a series of sound pulses without losing their ability to respond to subsequent sounds as nomally raised crickets do. Hence, the sensory system has been modified as a result of early experience.

Work on sensory systems other than vision, and on animals other than vertebrates, shows that the results of mammalian vision deprivation experiments can be generalized to other sensory systems and other animals. The sensory experience of an animal strongly shapes the morphology of sensory connections in the nervous system, as well as the sensory capabilities of the animal as an adult.

Many sensory systems in both vertebrate and invertebrate animals show a sensitivity to input during an early period of development. For example, the somatosensory region of a rat's brain will reorganize itself after the early loss of a limb, and the structure and functional responses of giant interneurons in crickets will be disturbed if sensory input from the cerci is missing.

Developmental Plasticity in Adult Animals

Development is ordinarily thought of as confined to embryos or immature animals. Yet in many ways, developmental processes continue well into adulthood. If you think of the formation of new synapses as being a manifestation of development, then developmental processes continue in any adult animal in which new synapses can be formed. In vertebrates, new synapses form in the hippocampus, the cerebellum, and other brain regions in which learning takes place, as you will learn in Chapter 24. Learning and the formation of new synapses occur in invertebrates as well. Hence, from this point of view, developmental processes continue throughout the lives of all animals.

The Reorganization of Sensory Cortex

In vertebrates, somatosensory information is processed in the somatosensory cortex. Since the discovery of the topographical organization of this cortex, it has served as the quintessential example of topographical mapping in the vertebrate brain. As you have just learned, studies with neonatal mammals have made it clear that the organization of many cortical sensory areas is subject to environmental influences during development. It had been assumed, however, that this topographical map, once fixed during development, was immutable thereafter.

In the 1980s it became apparent that the somatosensory cortex and by extension other sensory cortical areas are capable of reorganization to a remarkable extent even in full adulthood. For example, in squirrel monkeys as in other vertebrates, the main areas of the primary somatosensory cortex are arranged topographically such that the surface of the animal's body is represented on the surface of the cortex (Figure 23-8A). If you examine the region of the cortex devoted to the hand, you will find that most of that part receives input from the palm side, with a few spots that receive input from the back of the hand interspersed among them (Figure 23-8B). In the hand, the median nerve innervates mainly the palm side of the thumb, the forefinger, and a part of the middle finger. If this nerve is cut

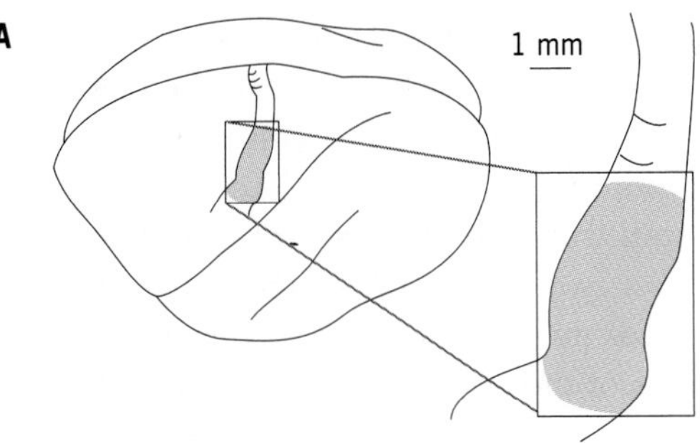

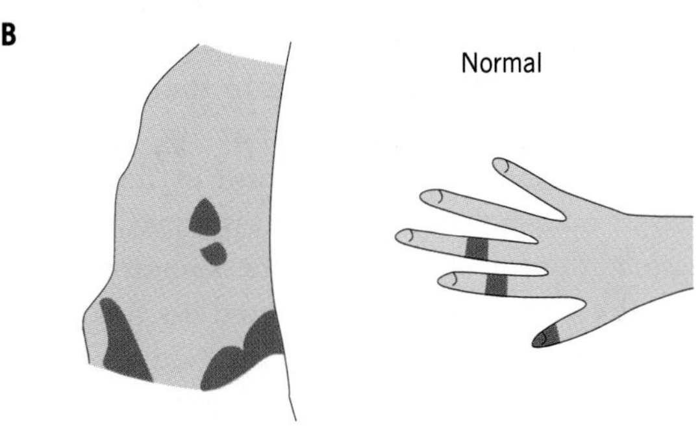

FIGURE 23-8. (A) The brain of a squirrel monkey viewed from the left side, showing in outline the regions of the somatosensory cortex devoted to different parts of the body. The expanded view on the right shows the region devoted to the hand, including the fingers. (B) Detailed view of the representation of the skin surface of the hand. Red sections represent the parts of the hand that are red in the accompanying drawing of a hand. (C) The reorganized representation of skin surface 144 days after section of the median nerve at the wrist. The nerve is tied so it cannot regenerate. The median nerve innervates the palm side of the first three fingers. After section of the nerve, the cortical regions devoted to these parts of the hand become reorganized to receive input from the back of the fingers instead, as shown by the regions in red. The small, solid black region remains unresponsive to input from any part of the hand.

and tied off (to prevent regeneration) at the wrist of a forelimb, the part of the somatosensory cortex that was devoted to the denervated part of the hand is reorganized. After several months, the areas that had been devoted to input from the palm side of the thumb and the first two fingers of the hand now become devoted mainly to input from the dorsal regions of the fingers instead (Figure 23-8C). Similar cortical reorganization has been found in owl monkeys and the type of bat known as a flying fox, following amputation of one or more digits of a hand.

There are certainly limits to the amount of cortical reorganization that is possible. Denervation of an entire hand or limb in an adult mammal will not result in massive reorganization of the parts of the somatosensory cortex that had been devoted to analyzing input from the lost part of the body. For the rest of the life of the animal, most of those parts of the cortex that had received input from the denervated region will simply receive little sensory input.

Careful and detailed study of the representation of skin area in the somatosensory cortex has revealed that even the organization of input within a cortical area is susceptible to change. Ordinarily, the cortical areas representing the fingers on a hand are sharply delineated from one another, whereas those that represent adjacent areas of skin on a single finger are integrated, being less distinct and slightly overlapping. In one set of experiments, monkeys were trained to respond to stimuli that touched several fingers of one hand. After 4 to 6 weeks of intensive training, the areas of the somatosensory cortex devoted to the stimulated fingers became integrated rather than being distinct. Conversely, cortical patches that represented adjacent areas of skin on a single finger, which were not stimulated in sequence, became sharply delineated. This reorganization shows that sensory areas of the brain have considerable capacity for organizing themselves to handle in the most efficient way possible the kind of sensory input that experience shows is common.

Modifications or modulation of brain regions devoted to sensory processing similar to those encountered during embryonic development can be seen even in adult animals. The somatosensory cortex can reorganize itself to some extent following loss of input from a part of the skin due to amputation or denervation of part of an appendage. Somatosensory areas may also be reorganized to handle a specific pattern of input that the animal experiences.

The Effects of Experience on the Brain

Perhaps the most striking demonstration of the capacity of an adult mammalian brain to respond to experience arises from experiments by William Greenough and colleagues in which rats are raised in a "simple" or a "complex" environment. A simple environment is a plain cage in which rats are raised individually. The rats are provided with all the food and water they need, but they cannot interact with other rats, and there is nothing in the cage except a little bedding, a water spout, and the food bin (Figure 23-9A). Complex environments are more like the animal's natural habitat. In such environments, animals are usually raised in groups in an arena that contains many objects such as tubes and toys for climbing on or hiding in, balls to roll around, and plenty of materials for making nests (Figure 23-9B). To add variety, the items in the arena may also be changed daily.

Greenough and his coworkers found that rats raised in simple and complex environments differ in behavior, in gross brain morphology, and in brain chemical content. Behaviorally, rats raised in a complex environment are able to solve simple and complex mazes more easily. Morphologically, their cerebral cortexes weigh proportionately more relative to the rest of the brain. Chemically,

A

B

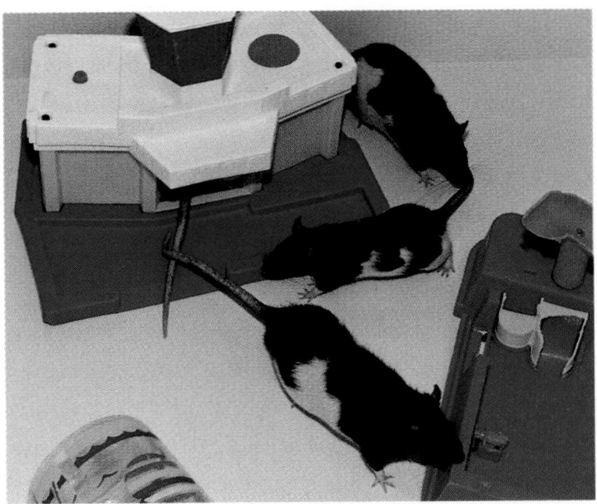

FIGURE 23-9. (A) A simple environment for raising a rat. The rat is caged with no artifacts and without social contact with other rats. (B) A complex environment. Rats are raised in groups in large arenas in which there are many artifacts for the rats to interact with. (Courtesy of Rodney A. Bates and William T. Greenough.)

their cortexes contain significantly more total protein, as well as more of certain chemicals such as acetylcholine, than do the cortexes of rats raised in a simple environment. There are also increases in glial density and in vascularization in the cortexes of rats living in complex environments.

The implication of these results is that physical changes are induced in the brain by the more varied experience of the rats living in a complex environment, and that these changes cause the behavioral differences that can be demonstrated between rats raised in the two types of environments. This implication has been fully supported by direct study of neural structure and synapses in the cortex. For example, studies of the visual cortex, as well as other brain regions, have shown that the number and complexity of dendritic branching there are significantly increased in the brains of rats exposed to a complex environment compared with dendrites of animals that are raised in a simple environment. Furthermore, the total number of synapses and the number of synapses per neuron that are formed between neurons in these brain regions are also significantly greater in animals that have been raised in a complex environment than in those raised alone in simple cages. These effects have also been shown to be related to learning induced by the environment, not just to the greater levels of activity shown by the rats living in an arena stuffed with toys.

The early experiments of this type were carried out on young rats, in which experience-dependent plasticity is not entirely unexpected. However, the most startling result has been that these effects also occur in adult animals. If rats are raised together until adulthood, then separated into two groups, one group living in a complex environment and one in a simple environment, similar enhancements of cortical thickness, dendritic branching and complexity, total number of synapses, and the number of synapses per neuron can be demonstrated in the complex-environment group. No special long periods of exposure are required in order for complex environments to exert a measurable effect in adult animals; effects can be measured after a period as short as a week, although longer exposure causes stronger effects. These effects are therefore not confined to a critical period that is present only during the early life of the animal; they can occur at any time. Not every effect of a complex environment in a young rat is duplicated in an old one, however. The changes in vascularization that occur in young animals do not appear in old ones.

The most general conclusion from these experiments is that the effects of the environment on synapses or synapse formation can extend beyond the simple processing of sensory information. The nature of the environment can affect brain structure, function, and composition strongly enough to be manifested in the overt behavior of the animal. Furthermore, environmental effects on the cortex can occur not just in a young animal but also in one that has already reached adulthood. The implications of these results for human life and social policy have not escaped notice.

Rats that live in so-called complex environments show better performances in mazes; an increase in the relative mass of the cerebral cortex; and an increase in the number of dendritic branches, the total number of synapses, and the number of synapses per neuron in many regions of the brain when compared with rats living in simple environments. These environmental effects on the nervous system appear in adult animals as well, and hence are not confined to early stages of development.

Additional Reading

General

Jacobson, M. 1993. *Developmental Neurobiology*, 2d ed. New York: Plenum Press.

Purvis, D., and J. W. Lichtman. 1985. *Principles of Neural Development*. Sunderland, Mass.: Sinauer.

Research Articles and Reviews

Greenough, W. T., and A. M. Sirevaag. 1991. A neuroanatomical approach to substrates of behavioral plasticity. In *Developmental Psychobiology: New Methods and Changing Concepts*, ed. H. N. Shair, G. A. Barr, and M. A. Hofer, 448–59. New York: Oxford University Press.

Kaas, J. H., S. L. Florence, and N. Jain. 1997. Reorganization of sensory systems of primates after injury. *Neuroscientist* 3:123–130.

Katz, L. C., and C. J. Shatz. 1996. Synaptic activity and the construction of cortical circuits. *Science* 274:1133–38.

LeVay, S., T. N. Wiesel, and D. H. Hubel. 1980. The development of ocular dominance columns in normal and visually deprived monkeys. *Journal of Comparative Neurology* 191:1–51.

Murphey, R. K. 1986. The myth of the inflexible invertebrate: Competition and synaptic remodeling in the development of invertebrate nervous systems. *Journal of Neurobiology* 17:585–91.

Thoenen, H. 1995. Neurotrophins and neuronal plasticity. *Science* 270:593–98.

Wiesel, T. N., and D. H. Hubel. 1965. Comparison of the effects of unilateral and bilateral eye closure on cortical unit responses in kittens. *Journal of Neurophysiology* 28:1029–40.

CHAPTER 24

Behavioral Plasticity: Learning

Perhaps the most remarkable manifestation of neural malleability is the change in neural function that can occur as a consequence of an animal's experience. This malleability, called learning, manifests itself at both the cellular and the molecular level, the former by changes in the structure and function of neural circuits, the latter by gene expression and the activation or inhibition of biochemical pathways. In this chapter, you will read about some of the mechanisms that underlie learning in simple and complex animals.

The malleability of the nervous system is an extraordinarily widespread phenomenon that has attracted a considerable amount of attention quite apart from its role in neural development. This is because it is at the root of the ubiquitous phenomenon of **learning**. Learning has been defined in many ways; here it will be considered a change in the behavior of an animal as a consequence of the animal's experience. This is a useful definition because it ties learning to the behavior of the animal and because it encompasses both simple and complex forms of learning, since the change may last from just a few minutes to more than a century. Closely tied to learning is **memory**. Memory is more difficult to define because it can refer both to stored information about an experience and to the process by which storage occurs. Memory is necessary for any experience to have an effect beyond the time at which it occurs.

The behavioral manifestations of learning and memory in animals have been studied for well over a hundred years, but only with the advent of intracellular recording and molecular techniques has it become possible to identify some of the specific cellular and molecular mechanisms that underlie the two phenomena. This chapter will introduce you to these mechanisms.

Learning is a manifestation of the malleability of the nervous system because it is a change in the behavior of an animal based on experience. Memory refers to the stored experience and to the process by which it is stored. Memory is a requirement for learning.

Simple Learning

Nearly all investigators will agree that the type of learning that we ourselves engage in every day has greater intrinsic interest than other forms of learning. However, some investigators have suggested that even the simplest forms of learning may reveal something of the molecular mechanisms that underlie more complex forms; therefore, they are useful subjects for study not only in their own right but also for what they may reveal about more complex learning. The similarity of the molecular mechanisms that are believed to underlie simple and complex learning has certainly validated this approach.

Habituation

Habituation, among the simplest of all forms of learning, is the cessation of a response to a stimulus after repeated presentation of the stimulus. You are probably familiar with many examples of habituation. You may have seen a dog respond to an unusual sound by pricking up its ears; if the sound is repeated several times, the dog's response declines and eventually disappears altogether. The cessation of the dog's response to the sound is an example of habituation. Even many reflexes, normally thought of as unchangeable, can habituate.

One reflex that can habituate and that has been studied in detail is the gill withdrawal reflex of the sea hare *Aplysia*. Like other mollusks, this sluglike marine animal has a siphon, mantle, and gill (Figure 24-1). A gentle tactile stimulus to the siphon or mantle causes a slight withdrawal of the siphon and gill. Repetitions of the stimulus will result in a gradual cessation of the withdrawal response; it habituates. Typically, gentle touch applied every 3 minutes over a period of about 4 hours will bring about habituation. After the reflex has habituated and the animal is left undisturbed for a period of about 2 hours, the withdrawal reflex will reappear. Longer- or shorter-term habituation can be produced by longer or shorter stimulus regimens.

Eric Kandel and his coworkers, as well as other researchers, have made considerable progress in uncovering the physiological basis of habituation and related phenomena in *Aplysia*. One reason for Kandel's choice of *Aplysia* as his experimental animal was that a considerable body of knowledge about the animal and its nervous system was already available when he began his work. In particular, the neural circuit underlying the gill withdrawal reflex was known in detail. The circuit consists of the classic three neural elements of a reflex (as discussed in Chapter 16): sensory neurons from the siphon or mantle, motor neurons to the gill muscles, and interneurons interposed between the two (Figure 24-2). Just as in many other reflexes, there are two pathways from the sensory neuron to the motor neuron—one a direct monosynaptic pathway, and the other via the interneuron. However, the direct monosynaptic pathway contributes most to the reflex.

Kandel and his colleagues first showed that depression of the synapse between the sensory and motor neurons was the main underlying cause of habituation. During habituation, there is a reduction in the effectiveness of a presynaptic action potential in eliciting an action potential in the postsynaptic neuron. Such a loss of synaptic effectiveness is called

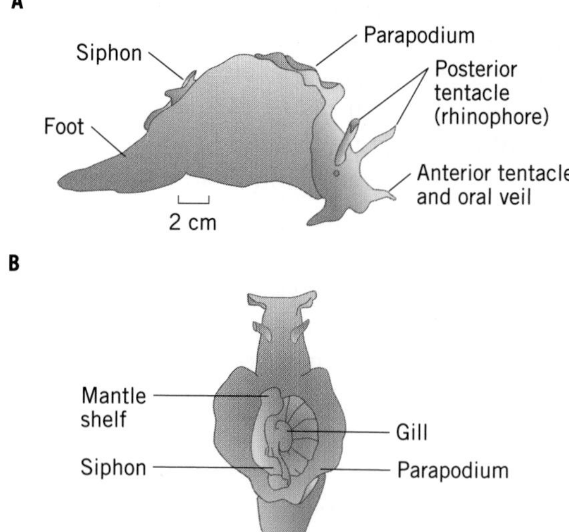

FIGURE 24-1. (A) The sea hare *Aplysia californica*. This mollusk "breathes" by means of oxygen exchange through its gill. Water flows in between the parapodia, under the mantle shelf, over the gill, and out through the siphon. Since the gill must be thin to allow for the exchange of oxygen, it is susceptible to mechanical damage. For this reason, *Aplysia* exhibits a number of reflexes to retract the gill when the possibility of damage arises. (B) Dorsal view of *Aplysia*, with the parapodia opened and the mantle shelf displaced to expose the gill.

synaptic depression. In the context of the gill withdrawal reflex, this means that when sensory neurons in the siphon or mantle are stimulated repeatedly, there is a gradual reduction in the probability that their action potentials will generate action potentials in the motor neurons with which they synapse. As a consequence, the motor response in the gill becomes weaker and weaker and eventually disappears altogether.

Additional experiments showed that synaptic depression is caused by a reduction in the amount of neurotransmitter released from the sensory neuron. Two events contribute to this. First, there is inactivation of certain calcium channels, so that less Ca^{2+} is available to trigger the release of neurotransmitter and less neurotransmitter is released at the synapse, as described in Chapter 7. Second, there is also a reduction in the number of presynaptic vesicles that are available for mobilization. Therefore, there are fewer vesicles available to respond to the presence of Ca^{2+} by releasing their neurotransmitter.

Synaptic depression is a general neuronal phenomenon. However, like facilitation, described in Chapter 8, synaptic depression is not present at all synapses. Many synapses show no significant depression whatsoever when they are activated repeatedly. Examples are the motor neurons innervating the diaphragm muscles in a mammal or bird, which fire reliably throughout the life of the animal.

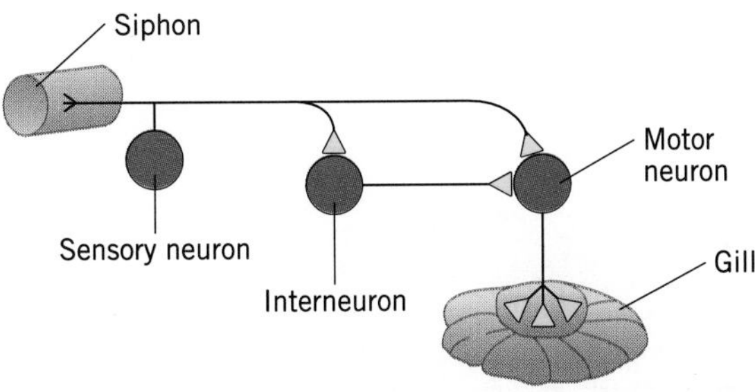

FIGURE 24-2. Simplified diagram of the neural circuit underlying the gill withdrawal reflex of *Aplysia*. Stimulation of the siphon skin excites sensory neurons that directly or indirectly synapse with gill motor neurons. Each neuron in the diagram actually represents many individual cells.

Habituation is a simple form of learning that has been studied at the cellular level. Habituation of the gill withdrawal reflex of the sea hare *Aplysia* is due to a reduction in the amount of neurotransmitter released at the synapse between sensory and motor neurons. This reduction in release is due to a reduction in the amount of internal Ca^{2+} available at the synapse, and to a reduction in the number of vesicles available for release.

Dishabituation and Sensitization

Two other behavioral phenomena, dishabituation and sensitization, are related to habituation and have also been studied as simple forms of learning. **Dishabituation** is the full recovery of the original strength of a habituated response after the presentation of some strong, novel stimulus. **Sensitization** is an increase in the strength of *any* reflex that is caused by one or more strong stimuli other than the stimulus that usually evokes that reflex.

Aplysia shows both dishabituation and sensitization. After an *Aplysia* has been touched on the siphon every few minutes for several hours, it no longer responds to such a touch by withdrawing its gill. If the animal then receives a strong tap on the tail, the next touch on the siphon once more causes the gill to withdraw. The gill withdrawal reflex has been dishabituated. *Aplysia* can be sensitized by a strong tap on the tail even in the absence of previous habituation. Ordinarily, a very light touch to the siphon, weaker than that used to induce habituation, will cause no reaction in the gill. After sensitization, *Aplysia* responds to such a light touch by withdrawing the gill. Dishabituation and sensitization are similar in that both involve an increase in responsiveness of the animal to a stimulus that previously had been too weak to elicit a response.

Dishabituation and sensitization also have a similar physiological basis, a type of heterosynaptic facilitation. You will recall from

Chapter 8 that heterosynaptic facilitation is an increase in responsiveness of a postsynaptic neuron after a stimulus is applied presynaptically at a synapse. The neural circuit and synaptic interactions proposed to account for dishabituation and sensitization are depicted in Figure 24-3A. The key element is that one or more interneurons make presynaptic connections with the terminals of the sensory neurons. These interneurons receive input from other parts of the body, such as the tail. When these interneurons are activated with sufficient strength, they heterosynaptically cause an increase in the efficacy of the sensory-to-motor

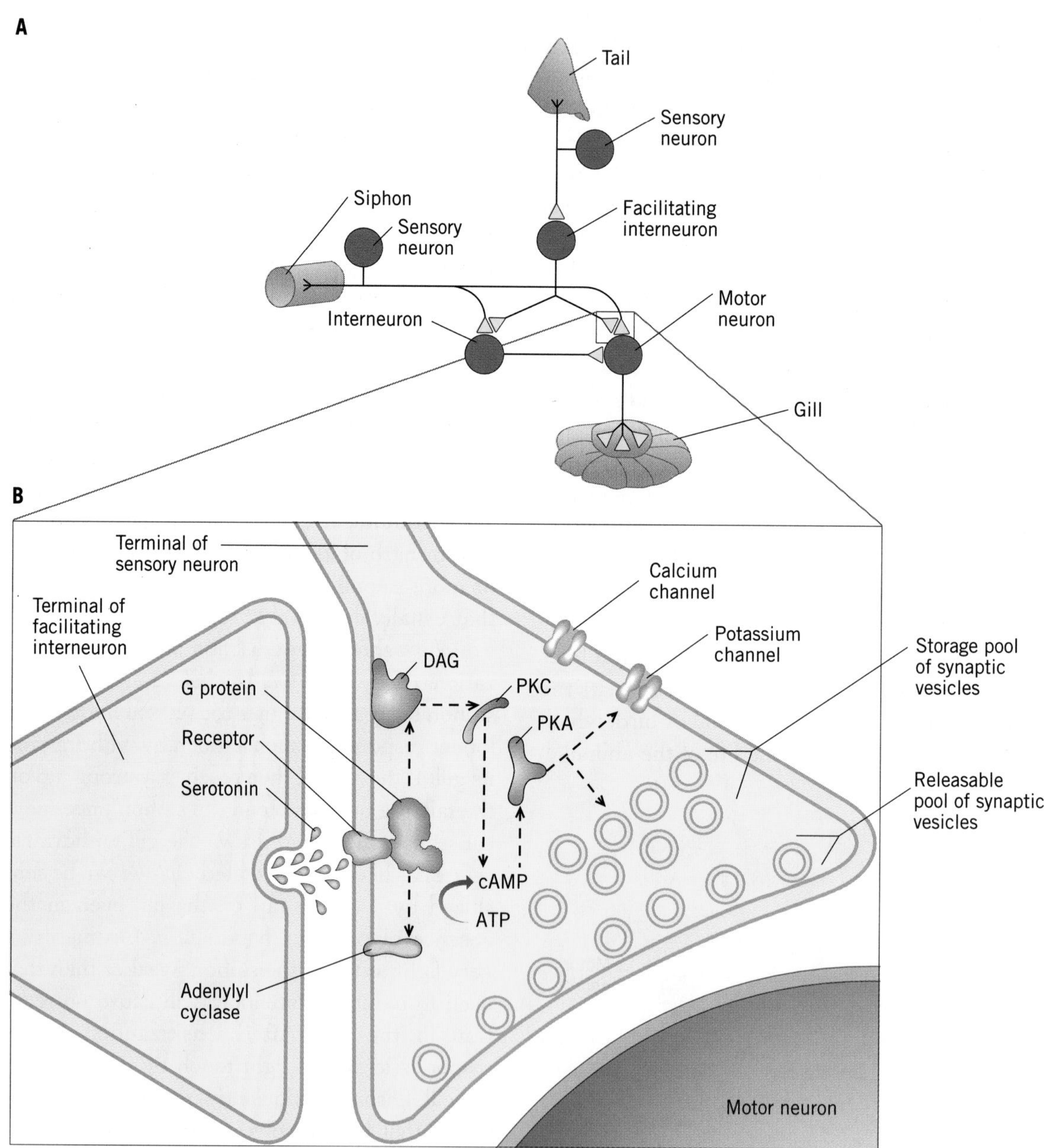

FIGURE 24-3. The neural basis of dishabituation and sensitization in *Aplysia*. (A) Simplified diagram showing the relationship of facilitating interneurons to the sensory synapses that show synaptic depression. The input to the facilitating interneurons may come via other interneurons, not necessarily directly from tail sensory neurons. (B) Summary of mechanisms that are thought to contribute to dishabituation and sensitization. Facilitation can change the cell's electrical characteristics, as well as the number of vesicles available in the releasable pool.

neuron synapses, hence augmenting (or restoring) the gill withdrawal reflex in response to touch on the siphon or mantle.

At the molecular level, dishabituation and sensitization are thought to be due to serotonin, the neurotransmitter released by the facilitating interneurons. Serotonin has two effects on the presynaptic terminals of the sensory neurons on which these interneurons synapse (Figure 24-3B). First, it brings about an increase in the amount of neurotransmitter released by the sensory neuron by increasing the amount of Ca^{2+} that enters the terminal upon depolarization. It does this by activating the cyclic adenosine monophosphate-dependent (cAMP-dependent) second messenger mechanism described in Chapter 6. Binding of serotonin with its receptors on the membrane of the sensory neuron activates a G protein. This G protein then activates adenylyl cyclase, which catalyzes the conversion of adenosine triphosphate (ATP) to cAMP. The increase in cAMP concentration turns on cAMP-dependent protein kinase A (PKA) in the cell; this kinase promotes phosphorylation of a K^+ channel, either directly or via a regulatory protein associated with it. The result is blockage of the potassium channel, which tends to prolong the duration of any action potential that invades the terminal. A longer action potential increases the time during which voltage-gated calcium channels are open, allowing more Ca^{2+} to enter the terminal and therefore causing an increase in the amount of neurotransmitter that it releases.

A second action of serotonin that counters habituation or promotes sensitization is to increase the amount of neurotransmitter that is available for release at the synapse. This is brought about by cooperative action of PKA and protein kinase C (PKC) in mobilizing vesicles from a storage pool to a releasable pool. PKC is turned on by membrane-bound diacylglycerol, which is itself activated via a different G protein than the one that activates adenylyl cyclase. The reduction in available free Ca^{2+} and in available synaptic vesicles that are brought about by habituation is hence countered by the action of serotonin in promoting the entry of Ca^{2+} into the presynaptic terminal of the sensory neuron and in increasing the number of synaptic vesicles available there for release. As a consequence, an action potential in the sensory neuron leads to a stronger motor response, and hence a more vigorous withdrawal of the gill, than previously.

Sensitization ordinarily lasts for only a few minutes, but under some circumstances it can last for hours. As you will learn later, long-term changes in synaptic efficacy are associated with protein synthesis. With persistent facilitation of a sensory-to-motor synapse, the cascade of biochemical events initiated by serotonin has other, longer-term effects in addition to the short-term ones already described. The PKA that is activated by G proteins has a direct effect not only on channels and on vesicle availability but also on the genome of the cell, an effect that requires the kinase to translocate to the nucleus. It is thought that in the nucleus the protein kinase initiates a chain of events that produces two types of protein. One type brings about a long-term increase in the amount of neurotransmitter released at the synapse. The protein does this by increasing the availability of PKA, hence causing a long-term closure of sensitive potassium channels and thereby broadening action potentials. The other type of protein promotes the formation of new synaptic connections.

Dishabituation and sensitization in *Aplysia* result from the action of serotonin on sensory-to-motor synapses. Serotonin exerts its effects via a second messenger system that has short-term effects on the availability of free Ca^{2+} and of synaptic vesicles, and long-term effects on the number of synaptic vesicles released and on the formation of new synapses.

Conditioning

Studying simple types of learning such as sensitization and habituation can provide insights into cellular mechanisms of synaptic plasticity. However, some researchers have questioned the relevance of these mechanisms for the more complex types of learning usually associated with vertebrates. After all, habituation seems at first glance to be quite a different process from learning that your class meets at 2:00 P.M. three days a week. Nevertheless, work begun in the 1980s and described in the next section has revealed a remarkable similarity of mechanisms between simple and complex types of learning in *Aplysia* and other animals.

Perhaps the most exhaustively studied type of complex learning is that known as **associative conditioning**, or just **conditioning.** In this type of learning, an animal forms an association between two different stimuli. One stimulus, called the **unconditioned stimulus (US),** evokes a particular reflex response. The other, called the **conditioned stimulus (CS),** normally evokes no such response. When the conditioned stimulus is presented shortly before the unconditioned stimulus in a series of trials, the animal eventually responds to the presentation of the conditioned stimulus alone by exhibiting the reflexive behavior. Conditioning was first studied extensively by the Russian psychologist Ivan Pavlov in the early part of the twentieth century. In his most famous series of experiments, he conditioned a dog to salivate to the sound of a bell by ringing the bell each time just before he gave meat to the hungry dog. After several trials, Pavlov rang the bell but did not give the dog any meat. When the dog began salivating anyway, Pavlov had demonstrated that the dog had formed an association between the sound and the meat.

Conditioning is a complex form of behavior in which an animal forms an association between an unconditioned stimulus, which normally evokes a reflex, and a conditioned stimulus, which normally evokes no such reaction. After conditioning, the conditioned stimulus alone will evoke the reflex.

Invertebrate Mechanisms

Conditioning in invertebrates has been studied most extensively in *Aplysia* by Kandel and his collaborators. Remarkably, these researchers have been able to demonstrate that the mechanisms by which conditioning works in *Aplysia* have elements in common with those that underlie simpler forms of learning. Kandel and his coworkers used a variation of the classical approach to conditioning, using the gill withdrawal reflex. Instead of training *Aplysia* to withdraw its gill when a novel stimulus was presented, they trained it to withdraw its gill quite strongly to a stimulus that ordinarily would produce only a partial or weak withdrawal. They used electrical shock to the tail as the unconditioned stimulus. This stimulus causes a vigorous withdrawal of the gill. The conditioned stimulus was a gentle touch applied to the skin of the siphon or the mantle. Such stimuli usually yield only partial or weak withdrawal of the gill. After presenting a series of paired CSs and USs, however, the researchers found that the animal responded to the CS alone with a vigorous withdrawal of the gill.

Conditioning a change in the strength of a response rather than an entirely new response has several advantages. First, because the conditioned stimulus is a gentle touch to the mantle or siphon and the response is a withdrawal of the gill, researchers are likely dealing with the same sensory-to-motor synapse that was studied during investigations of habituation and sensitization (Figure 24-4). Hence, at least some of the mechanisms of conditioning might have similarities to the mechanisms of simple learning studied previously.

A second advantage is that researchers can demonstrate specificity of the conditioned response. One of the criticisms of studies of sensitization is that the phenomenon represents a generalized response to any stimulus, and hence is not a specific behavioral change linked to a specific stimulus. To show that a conditioned response is a learned association between the US and the CS, it is necessary to show that the animal has formed a specific association between the two stimuli and is not merely making a generalized response to a wide range of stimuli. In *Aplysia*, an animal trained to give a vigorous withdrawal of the gill when it is gently touched on the mantle will move its gill weakly, if at all, when its

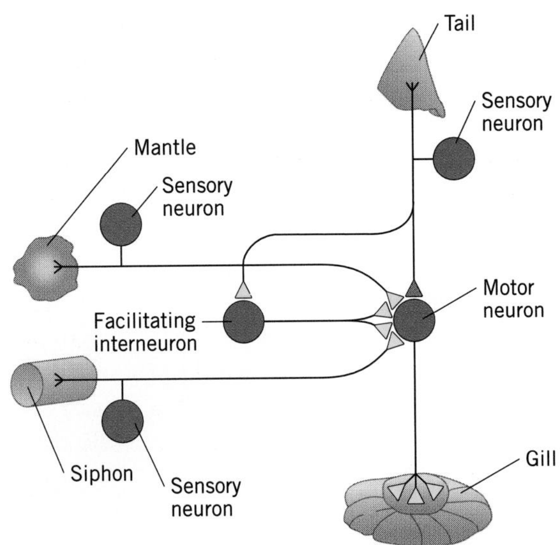

FIGURE 24-4. Schematic diagram showing the neural circuit thought to underlie conditioning of the gill withdrawal reflex in *Aplysia*. The shaded synapses and gill represent the unconditioned pathway. Strong stimulation of the tail causes excitation of the motor neurons that move the gill. Gentle touch applied to another region, such as the skin of the mantle, will ordinarily have little effect on the gill, because the sensory-to-motor synapses are not strongly activated. Pairing gentle touch with tail stimulation results in facilitation of the appropriate sensory-to-motor neuron synapse. After several pairing trials, gentle touch by itself will also produce a vigorous withdrawal of the gill.

siphon is touched gently. If it is trained to respond to touch on the siphon, it will not respond to touch on the mantle. Hence, the association forged in the nervous system between the conditioned and the unconditioned stimuli is specific, not general, as in the case of sensitization.

The association between CS and US is made at the synaptic terminals of sensory neurons from the siphon or mantle when conditioned and unconditioned stimuli are paired over a series of trials. Both the CS and the US have effects on these terminals. The CS, gentle touch on the siphon or mantle, excites the sensory terminals directly, causing weak gill withdrawal (Figure 24-5A). When these sensory terminals depolarize, calcium channels open, allowing Ca^{2+} to enter the terminals. The terminals of the sensory neurons contain calmodulin, a molecule that when activated by Ca^{2+} can interact with adenylyl cyclase to promote the conversion of ATP to cAMP. Hence, one effect of depolarization is a transient activation of adenylyl cyclase, which causes a small and transient rise in cAMP level as well. Because the rise in cAMP level is small, there are no significant long-term effects on potassium channels or on the physiology of synaptic transmission.

Application of the US, electrical stimulation of the tail, causes two main effects. First, there is a strong synaptic input from tail sensory neurons to motor neurons controlling the gill musculature, causing the strong withdrawal of the gill. In addition, there is a heterosynaptic facilitation of the terminals of the sensory neurons that originate in the skin of the siphon and mantle, neurons that also synapse on the gill motor neurons (Figure 24-5B). This facilitation is the basis of dishabituation and sensitization, and works via the second messenger system described in a previous section. Serotonin is released by facilitating interneurons, and it activates G proteins in the terminals of the sensory neurons. The G proteins in turn activate adenylyl cyclase, which produces cAMP. The cAMP activates PKA, causing the broadening of subsequent spikes and the mobilization of synaptic vesicles that cause dishabituation and sensitization. However, these effects last for only a few minutes.

When a CS such as weak mechanical stimulation of the skin is followed within a few seconds by an US such as an electrical shock to the tail, these two processes interact (Figure 24-5C). Adenylyl cyclase is the molecule at which the interaction takes place. When adenylyl cyclase is activated by calmodulin just before it is also stimulated by the G protein pathway, it produces substantially more cAMP than the mere sum of the amount that would have been produced via the calcium channel/calmodulin pathway and the G protein pathways alone. The elevated levels of cAMP produce the two effects already described: spike broadening and an increase in the mobilization of vesicles when the sensory neuron is subsequently excited. After a series of stimulus pairings, the effects

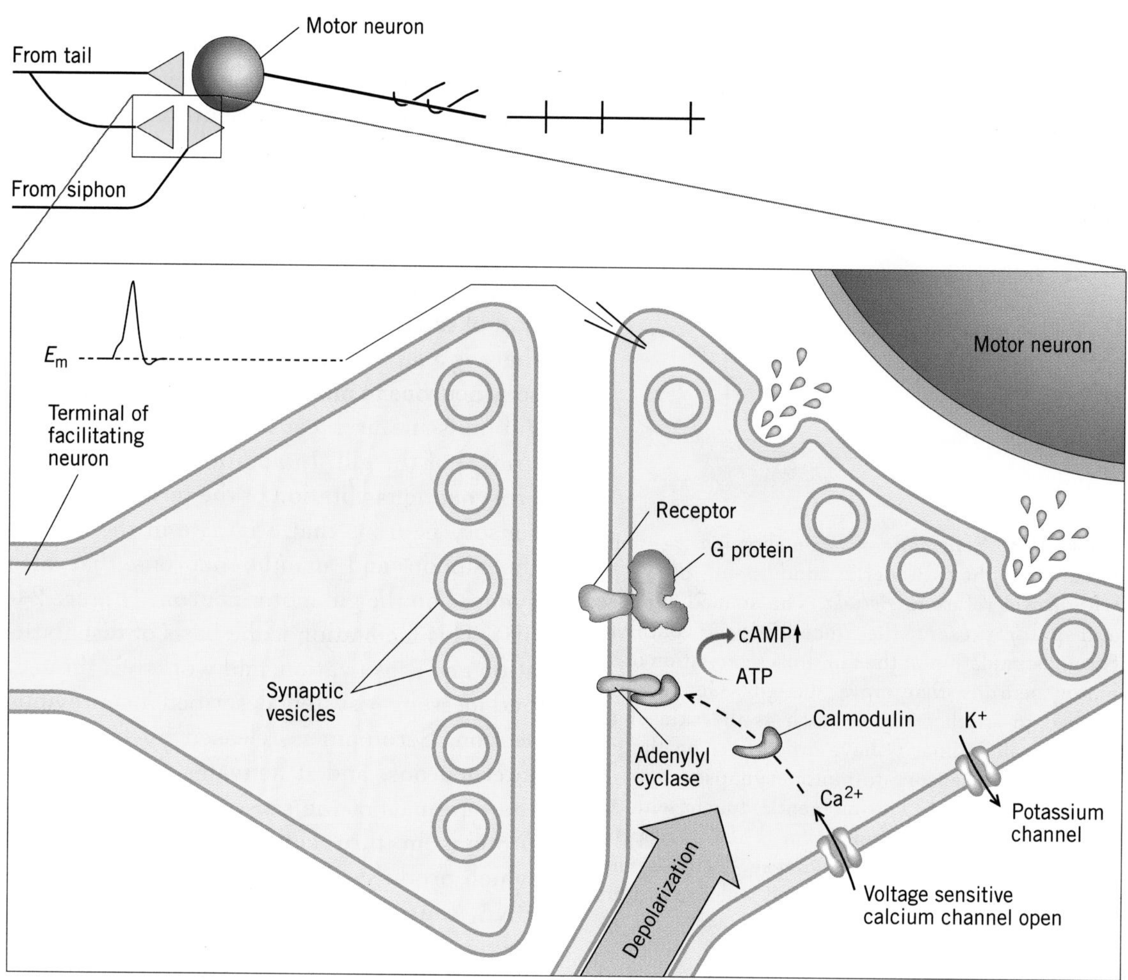

FIGURE 24-5 (A). Summary of the molecular processes in a sensory-to-motor neuron synapse that lead to a conditioned response. Applying the conditioned stimulus, a light touch to the siphon, before training causes a weak response in the motor neurons innervating the gill and a small increase in the level of cAMP in the presynaptic terminal due to the activation of calmodulin by the influx of Ca^{2+} and the subsequent activation of adenylyl cyclase.

of the elevated cAMP become significant enough that stimulation of the skin alone will cause a vigorous gill withdrawal (Figure 24-5D). The pairings must be in the specific order CS followed by US. If stimulation of the tail precedes stimulation of the skin, there is no enhancement of adenylyl cyclase action, presumably because the augmentation of adenylyl cyclase action requires that the enzyme be activated by calmodulin before being further stimulated by G protein; hence, there is no conditioning when the sequence is reversed.

The mechanism of conditioning described in *Aplysia* appears to be of general significance. This is suggested by studies of certain mutants of the fruit fly *Drosophila melanogaster*. Normal wild-type fruit flies can be conditioned to avoid or respond to various odors by pairing an odor with electrical shock or with the presentation of a sugar solution. However, a number of *Drosophila* mutants cannot be conditioned. Some mutations are structural, affecting brain structures known as mushroom bodies, or other sites of learning in the brain, but other mutations affect specific

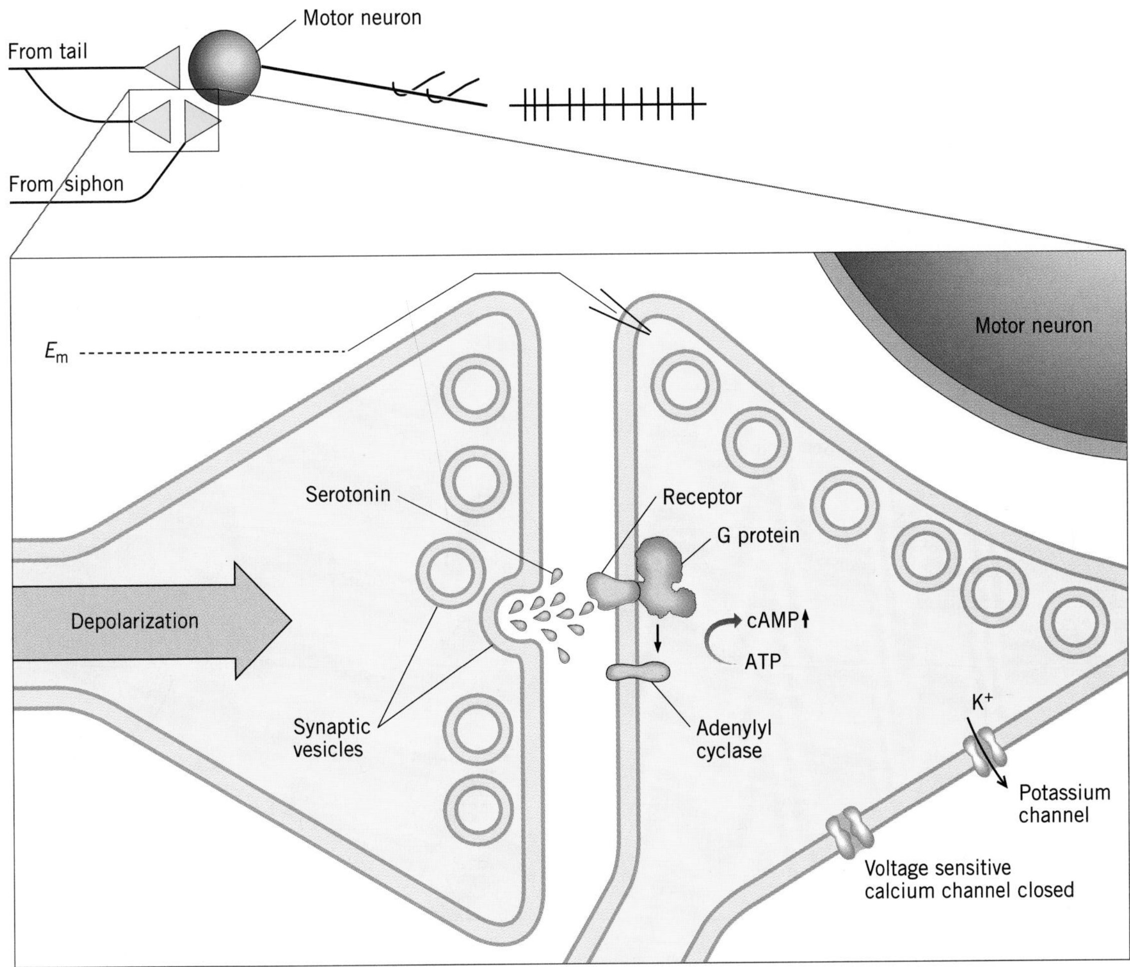

FIGURE 24-5 (B). Applying the unconditioned stimulus, a strong touch to the tail, causes a strong response in the gill motor neurons, as well as a small facilitation of the siphon-to-gill-motor-neuron synapse. This facilitation is caused by a small increase in the level of cAMP in the presynaptic terminal due to the activation of adenylyl cyclase by a G protein.

biochemical pathways. The genes known as *dnc* (dunce) and *rut* (rutabaga), for example, encode proteins that interact with the biochemical sequence described previously. The *dnc* gene encodes the enzyme cAMP PDE, a phosphodiesterase (PDE) specific for the degradation of cAMP. (A cGMP PDE that breaks down cGMP is found in vertebrate rods and cones, as described in Chapter 11.) Hence, the level of cAMP in neurons in normal flies is always less than what might be expected considering only the synthetic capability of adenylyl cyclase. During conditioning, the activity of adenylyl cyclase is elevated, hence raising the level of cAMP. Mutation of the *dnc* gene results in abnormally high cAMP levels in the neurons in which it is expressed. Hence the fly cannot be conditioned because cAMP levels cannot be elevated further by the pairing of CS and US. The *rut* gene encodes an adenylyl cyclase; its mutation causes low levels of cAMP, and also an inability of the fly to be conditioned, because cAMP levels cannot be elevated at all.

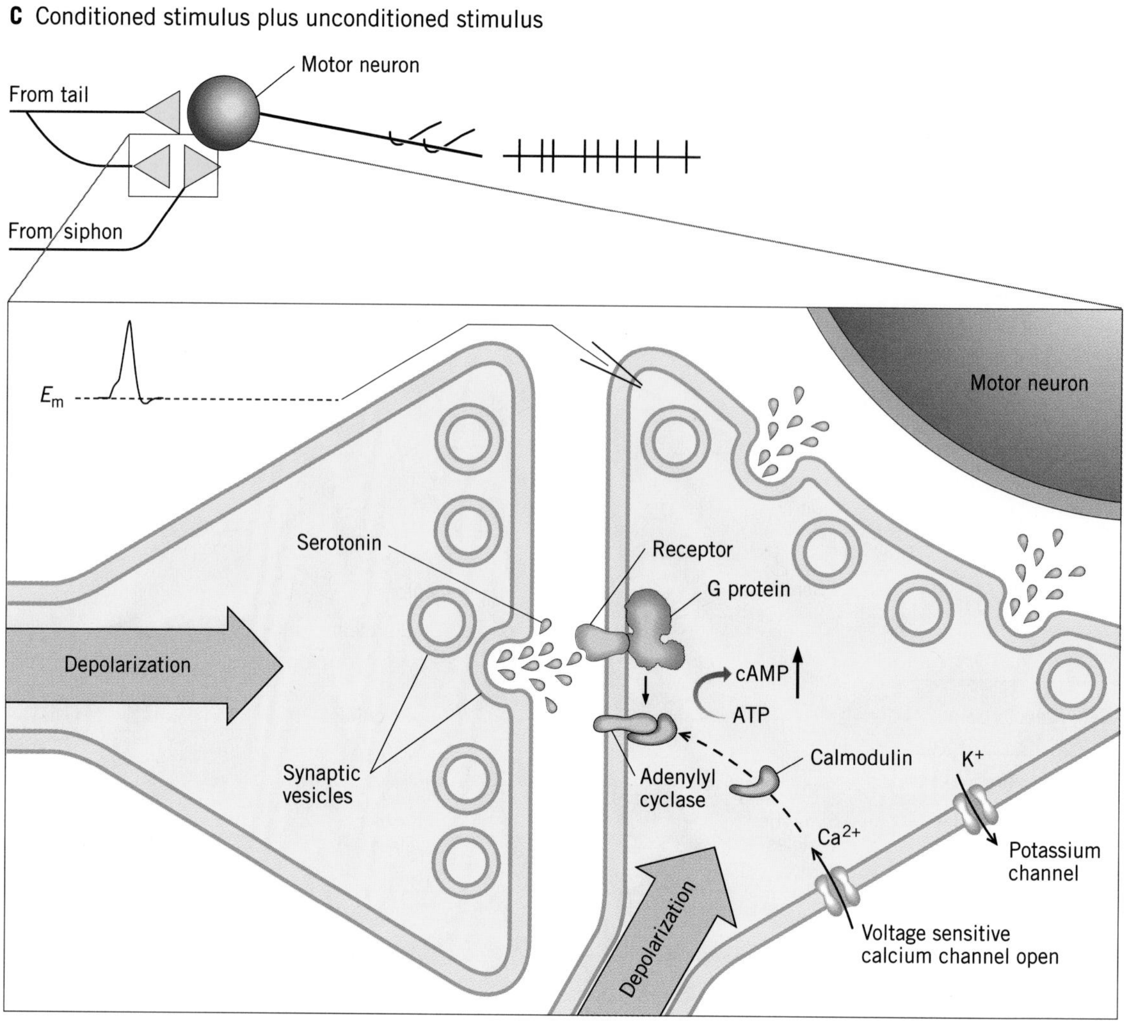

C Conditioned stimulus plus unconditioned stimulus

FIGURE 24-5 (C). Repeatedly applying the CS to the siphon just before applying the US to the tail causes a large increase in production of cAMP in the presynaptic terminal. This increase in cAMP levels closes potassium channels and mobilizes more synaptic vesicles (not shown).

Pairing gentle touch to the siphon or mantle with electrical shock to the tail in *Aplysia* causes a strong conditioned gill withdrawal when gentle touch alone is applied. Conditioning is believed to occur because of an enhancement of adenylyl cyclase action when it is activated sequentially by elevated free Ca^{2+} (caused by gentle touch) and via G proteins (activated by tail shock). Increased adenylyl cyclase levels may underlie conditioning in invertebrates like insects as well.

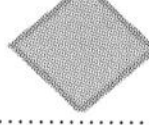

Conditioning in Vertebrates and Long-term Potentiation

The cellular basis of conditioning has been studied extensively in vertebrate animals as well as in *Aplysia*. Although not identical to the mechanisms for conditioning in invertebrates, vertebrate mechanisms show important functional similarities. The most important of these is a convergence of inputs originating from two different stimuli onto a single cell. This convergence results in the augmentation of a postsynaptic response that allows a new or stronger behavioral response to one of the stimuli than was present before the two stimuli were paired.

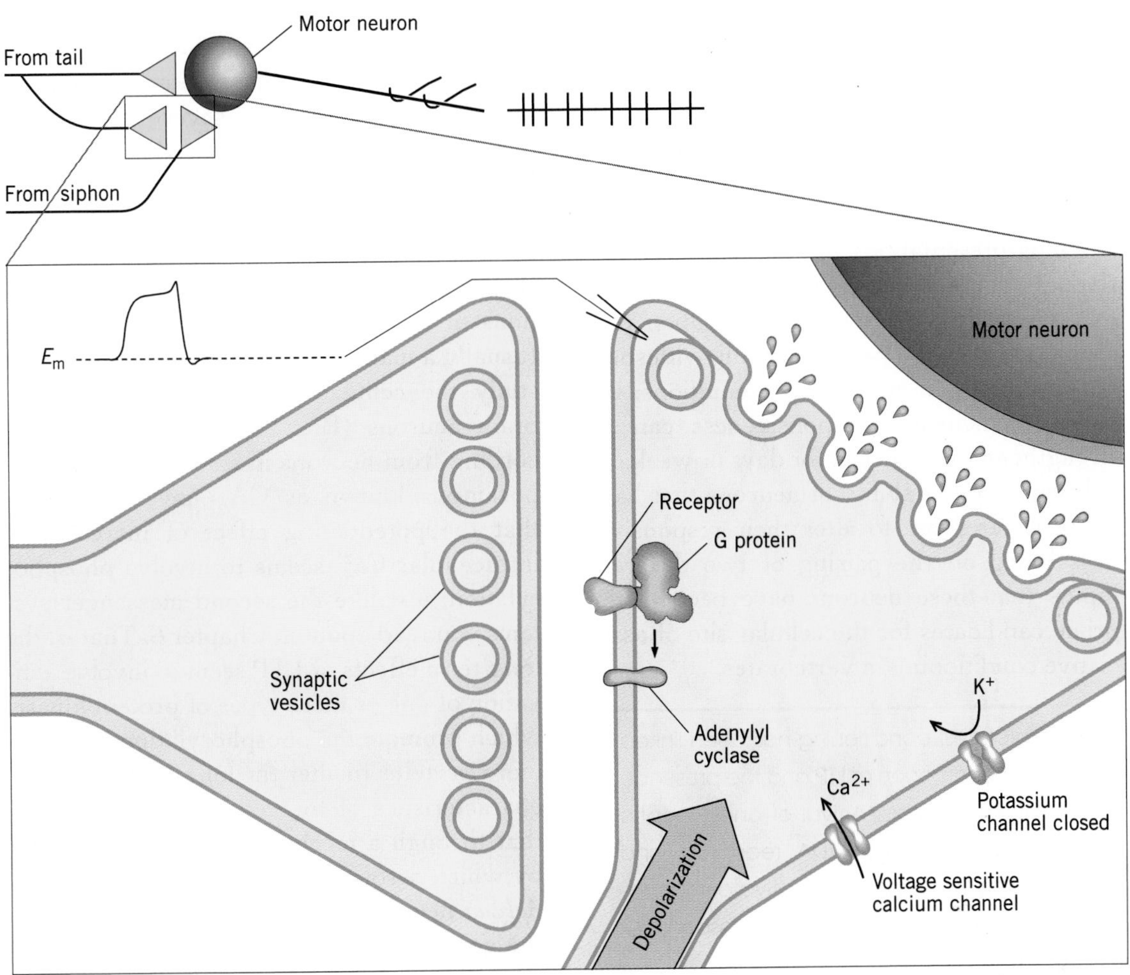

FIGURE 24-5 (D). After a number of pairings of the CS and US, presentation of the CS alone produces a significantly stronger response than before the training because of the combined effects of a broader action potential and more vesicles in the releasable pool. For simplicity, only one of the hundreds of receptor proteins and enzymes is shown.

Strong evidence suggests that vertebrate conditioning is based on the phenomenon known as long-term potentiation (LTP). LTP refers to an increase in the amplitude of a postsynaptic response that may last for many days (see Chapter 8). It is associated with the type of glutamate receptor known as an NMDA receptor, described in Chapter 6. In one form of LTP, rapid, repetitive activation of NMDA receptors by glutamate causes movement of Mg^{2+} away from the mouth of the channel associated with the receptor, unblocking it. The open channel allows Ca^{2+} to enter the cell, raising the concentration of internal free Ca^{2+} and activating various Ca^{2+}-dependent kinases that make non-NMDA receptors more responsive to neurotransmitter (see Chapter 8).

However, such rapid, repetitive input is not the only way in which LTP can be induced. NMDA receptors can also be unblocked by any depolarization of the synaptic knob in which they are located. Hence, if inputs from two different sources converge via separate synapses on a single neuron that contains NMDA receptors, activation of one presynaptic neuron just before the other can unblock the NMDA receptors sufficiently for the second input to cause a significant rise in internal free Ca^{2+}. This rise in Ca^{2+} then causes an increase in the responsiveness of the neuron to subsequent inputs. It can also promote the synthesis of NO, a retrograde messenger (see Figure 8-11, Chapter 8) that diffuses back to the presynaptic neuron to increase the amount of neurotransmitter it releases. One set of paired stimulus presentations increases the responsiveness of the postsynaptic neuron to one of the stimuli, but not a great deal. Upon presentation of several pairs of stimuli, with some minutes between stimulus presentations, the increased neuronal responsiveness can be strengthened so as to last for days or weeks. It is because of the ability of neurons that have NMDA receptors to alter their responsiveness based on the pairing of two different inputs that these neurons have become the prime candidates for the cellular site of associative conditioning in vertebrates.

In vertebrates, conditioning has been linked to the phenomenon of LTP and the presence of NMDA receptors. Arrival of one synaptic input can unblock NMDA receptors and allow a second input to activate mechanisms for enhancing neuronal responsiveness.

Long-term Potentiation and the Hippocampus

The hippocampus, part of the limbic system (see Chapter 3), has been shown to be necessary for several types of learning and memory formation in rats and other mammals. In humans, damage to the hippocampus causes deficits in learning about people, places, or things. In rats, lesions to the hippocampus cause deficits in spatial learning, meaning the ability of the animal to locate a particular place by using spatial cues (see Chapter 21). The role of the hippocampus in learning and memory is thought to be mediated by NMDA receptors and the phenomenon of LTP that they underlie. Many cells in the hippocampus show the phenomenon of LTP. Furthermore, application of drugs that are known to interfere with LTP causes severe behavioral effects similar to those caused by physical lesions of the hippocampus.

Study of LTP in the hippocampus has frequently been conducted in hippocampal brain slice preparations. Brain slices are thin sections of a part of the brain of a vertebrate (usually a mammal) in which researchers can study the cellular physiology of individual brain neurons (Box 24-1). Researchers recording from neurons in a section of the hippocampus known as CA1 have discovered that the potentiating effect of increases of intracellular Ca^{2+} seems to involve phosphorylation, just like the second messenger systems you read about in Chapter 6. That is, the long-term effects of LTP seem to involve activation of one or more types of protein kinase, which promote the phosphorylation of one or more proteins to alter the long-term electrical characteristics of the cell. It is also possible that through a mechanism analogous to that by which serotonin works on the genome of *Aplysia* neurons there is a long-term effect on the genome and hence potentially the structure of these hippocampal neurons as well.

In 1992, Susumu Tonegawa and his associates demonstrated links between LTP, learning, and the formation of memory in the hippocampus in a particularly dramatic fashion. They investigated the role of one specific protein kinase, known as alpha-calcium-calmodulin-dependent kinase II (α-CaMKII), in LTP. The precise role of this kinase in LTP is still not known, but based on several lines of evidence it is believed to be important in the induction of greater responsiveness in postsynaptic neurons that contain NMDA channels. For his experiments, Tonegawa genetically engineered a line of mice that were normal in all respects except that the gene responsible for synthesis of α-CaMKII was rendered inoperable. A mouse in which both copies of a single gene are rendered inoperable is called a **knock-out mouse;** the technique for producing such animals, **targeted gene replacement,** has become an important tool for the study of

BOX 24-1

Slicing the Brain: Recording from Brain Slabs

One of the drawbacks of working with a vertebrate brain is that there is just too much of it. Not only are there billions of neurons, but their connections are so intricate that unraveling the morphology and function of the circuits they form is a task to daunt the brashest of researchers. Fortunately, most researchers love a challenge, and the challenge of how to study the brain has been met by some groups in a rather unexpected way—they simply cut it into smaller pieces.

The **brain slice technique** is almost that simple. The brain of a mammal, often a rat, is removed and carefully sliced into slabs about 0.5 mm thick. Of course the cells at the edge of the cut are damaged, either from mechanical shock or from being cut apart, but the cells even 25 μm from the surface are usually surprisingly healthy after this procedure. The brain slice is placed in a chamber that is flooded with a solution containing the proper proportion of inorganic ions, glucose (for nutrition), and oxygen, to allow the neurons to survive for many hours.

There are several compelling advantages of a brain slice over a whole brain studied in situ. First, it is much easier to make intracellular recordings from brain neurons if the tissue is not subject to the periodic pulsing of blood caused by a beating heart. Second, it is also easier to study neurons from a particular region of the brain if an electrode does not have to penetrate several millimeters of cells before it can get to the part that a researcher wishes to study. Third, it is possible to study the pharmacology of known synapses because specific drugs or other pharmacological agents can easily be applied to the brain slice. In addition to these advantages, in some cases the specific role of individual neurons in circuits formed in restricted local regions of the brain can be studied because it is possible to record from both pre- and postsynaptic neurons across a known synapse.

Brain slices have been used to study many phenomena of the central nervous system, from the properties of synapses to the mechanisms of circadian rhythms. The technique has been especially popular in the study of LTP because LTP has been shown to be prominent in one specific region of the brain, the hippocampus, and because this region is easy to slice and remove. Study of the hippocampal brain slice has provided many insights into the mechanisms by which LTP works.

behavior at the molecular level (Box 24-2). In Tonegawa's experiments, these mice could not synthesize α-CaMKII. Investigation of a hippocampal brain slice from a mutant mouse revealed little or no LTP in cells from the CA1 or other regions of the hippocampus. This finding suggests that α-CaMKII is necessary for normal expression of LTP in the hippocampus. The neurons that were investigated seemed normal in all other respects, so the deficit was quite specific to LTP.

Tonegawa next investigated the behavior of the knockout mice. In most respects, the mice behaved normally. They had a tendency to respond more vigorously to stimuli than did wild-type mice, but otherwise there was nothing to distinguish them. However, they had great difficulty in performing certain types of spatial learning tasks. If normal mice are placed in a pool of murky water and left to swim until they find a submerged (and hence invisible) platform, they soon learn to find the platform even without being able to see it. However, the mutant mice take much longer to master the task; they also take longer to swim to the platform once they have mastered it. From a variety of other tests and controls, Tonegawa and his colleagues concluded that the mutant mice use cues other than distant spatial ones (such as the appearance of the wall of the swim chamber rather than the positions of objects in the room relative to the platform) to locate the platform. Since the hippocampus, and specifically the process of LTP in the neurons there, is thought to underlie spatial learning, this result supports the suggestion that α-CaMKII is necessary for LTP.

These experiments suggest that LTP and memory are related, and that the hippocampus is the brain structure that mediates memory formation and learning. However, they did not address the possible role of NMDA receptors, which are thought to underlie LTP, in the process. To investigate more precisely the importance of NMDA receptors for learning, Tonegawa and his group developed a more sophisticated type of knockout mouse. In this mouse, a specific type of NMDA receptor is eliminated from area CA1 of the hippocampus only, and only after the mouse reached an age of 2 to 3 weeks. No other cells in the brain (or elsewhere) were affected, and even the CA1 region developed normally up to about the third postnatal week.

Study of the CA1 cells of these knockout mice revealed several differences from normal mice. First, the knockout mice were unable to develop LTP. No stimulus paradigm was found that could elevate the amplitude of an excitatory postsynaptic potential after appropriate stimulation. In contrast, cells in other regions of the hippocampus, regions that were unaffected by the mutation and hence had normal NMDA receptors, showed normal LTP. A second difference was that place cells showed different characteristics. Recall that the hippocampus of rats and mice contains cells, known as place cells, that fire preferentially when the animal enters a specific place within an explored arena, as described in Chapter 21. The fields of place cells in these mutant knockout mice were up to 50% larger than normal; further, collectively they proved of virtually no predictive value in allowing researchers to determine where they were in a test arena. Behaviorally, the mutant mice performed poorly in the water maze test, being unable to learn to find the underwater platform. Furthermore, the mice seemed unable to learn or become familiar with a new arena.

LTP in mammals can be demonstrated in neurons in the hippocampus, a structure that seems to be necessary for certain types of learning. Studies with brain slices and with knockout mice have strongly suggested that LTP is the basis of spatial learning in the hippocampus, and further, that NMDA receptors and certain types of protein kinase are necessary for the process.

Memory

If learning is to change behavior, there must be a mechanism for storing the experience that brought about the new behavior, otherwise the animal would simply revert back to its original behavior after the experience is over. For even the simplest forms of learning, this storage is an essential part of the phenomenon. The stored information is what we refer to as memory. Investigation of memory

BOX 24-2

Eliminating Specific Genes: The Knockout Mouse

The techniques of molecular biology have had a dramatic impact on all areas of biology. Few are as amazing, however, as the ability to produce animals deficient in just a single, known gene. Developed independently in the mid-1980s by Mario Capecchi and Oliver Smithies, the technique known as targeted gene replacement has now become a standard tool in neurobiology for the study of the role of specific proteins in adult mammals. The technique is most usually applied to mice; the animals that are produced are called knockout mice because in them the function of a single gene is eliminated, or knocked out.

The procedure for producing a knockout mouse consists of two main steps. The first is to introduce into mouse cells a mutated copy of the gene under study (Figure 24-A). To begin, the gene is altered in some fashion,

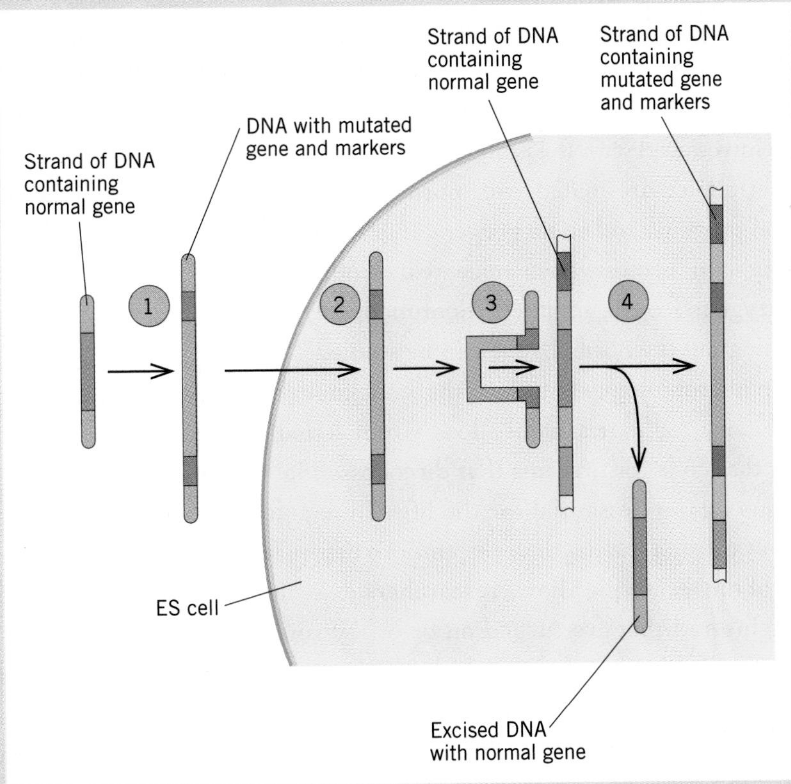

FIGURE 24-A. Introduction of a mutated gene into mouse cells. The altered gene, along with appropriate markers, is cloned (1) and placed in a medium with mouse embryo stem (ES) cells. The gene with its markers enters the ES cells (2) and through homologous replacement (3) substitutes for the normal gene (4). By searching for the markers that were inserted along with the mutated gene, ES cells are screened to select those in which the insertion has been successful.

(continued)

combined with genetic markers that will allow its presence in cells and in the mouse to be detected, and then cloned. The altered (mutated) gene is then introduced into embryo-derived stem cells (ES cells) from a mouse, that is, cells taken from the blastocyst stage of a mouse embryo. These cells are used because they can be grown in culture, manipulated, and later reintroduced into another early mouse embryo. The mutated gene becomes a part of the genome of the ES cells by the spontaneous process of homologous recombination, by which the mutated gene and markers replace the original gene in the mouse DNA. The ES cells containing the mutated gene are then grown in culture, and through the use of the markers that entered the ES cells' genome along with the mutated gene, ES cells that contain the mutated gene in the proper position are selected.

The second main step is to grow an adult mouse that is homozygous for the mutated gene (Figure 24-B). To begin, ES cells carrying the targeted mutation are injected into embryos at the blastocyst stage. The blastocysts are implanted into a pseudopregnant mouse (one hormonally prepared to carry embryos to term, but not pregnant) and are allowed to develop. The genetically altered cells will contribute to each mouse that develops from a blastocyst, producing mice called chimeras, which have some body cells (and germ cells) that are normal and some that contain the targeted mutation. Chimeric mice can be identified by expression of one of the marker genes introduced with the mutated gene, such as having a multicolored coat. Chimeric mice are mated with normal (wild-type) mice to produce some normal progeny, and some progeny that are heterozygous for the mutation. Mating two heterozygous mice will produce about 25% mice that are homozygous for the mutated (nonfunctional) gene, in which the effect of knocking out the normal gene can be studied.

For neurobiological studies, the knockout technique described here is satisfactory for genes whose loss is not lethal to the animal. However, genes that code for proteins that direct essential stages of development, or proteins that are essential for the life of the animal, cannot be eliminated without causing the death of the embryo before it is ever born. A new variation of the technique allows researchers to avoid this difficulty. Genes can be engineered that are turned on or off by some controllable event, such as the application of a brief pulse of high temperature, called heat shock. With the introduction of appropriately mutated genes into the genome of an insect like a fruit fly, it is possible to produce a fly in which the normal gene is active until the fly matures, and then to turn it off to study the effect of loss of the gene in the adult insect. This technique has been used successfully to study the role of cAMP response element binding protein in the formation of long-term memory, as described in this chapter.

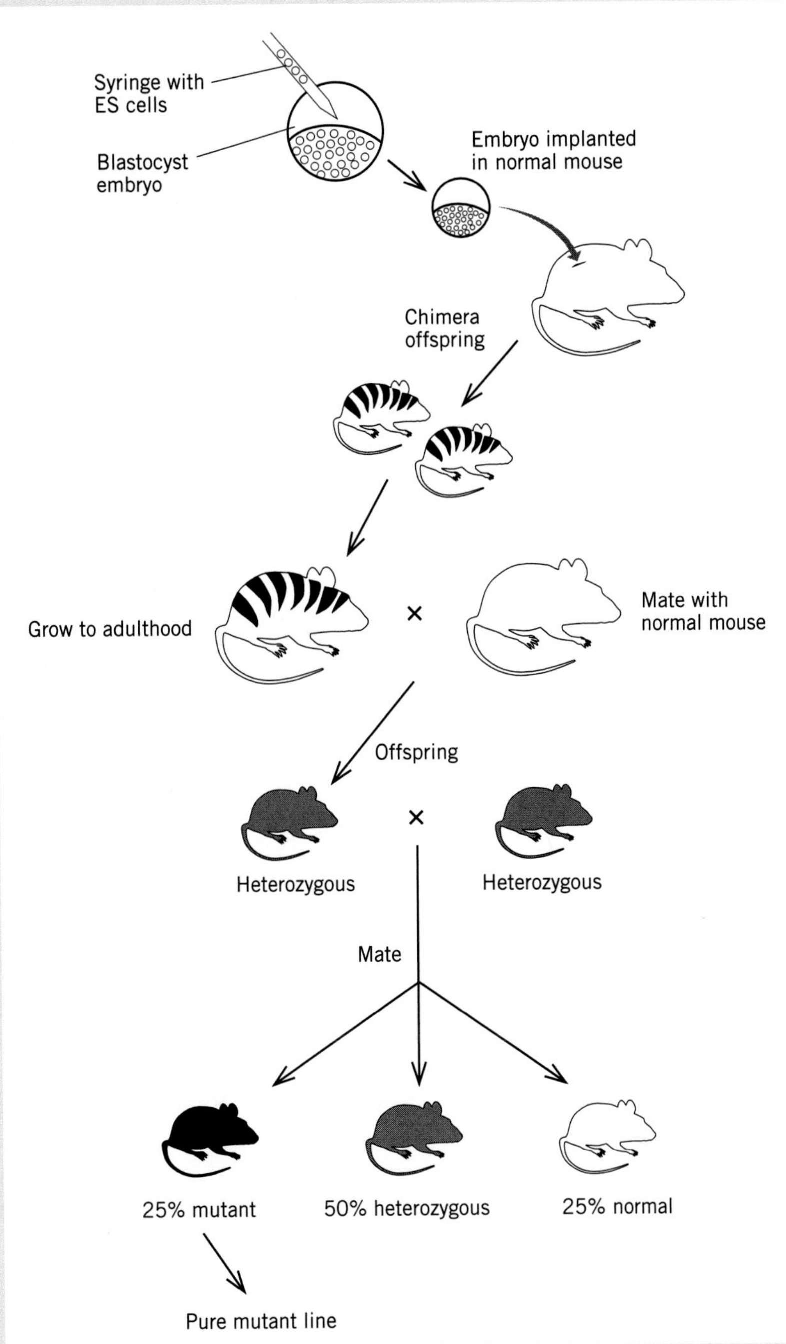

FIGURE 24-B. Growing mice homozygous for the mutated gene (knockout mice). ES cells are injected into embryos at the blastocyst stage, and the embryos are implanted into female mice. The young mice are called chimeras because they contain a mixture of normal cells and cells in which the mutated gene (and markers) has replaced the normal gene. Mating chimeras with normal wild-type mice produces some normal mice and some mice that have one copy of the normal gene and one copy of the mutant gene in each cell. These heterozygous mice can be identified through expression of the marker, here indicated by black coat color. Mating heterozygous mice to one another produces litters, about 25% of which are mice that contain two copies of the mutated gene in each cell. These homozygous mutant mice can then be studied.

includes questions of what memory is, how it is formed, where in the nervous system it resides, and how it is retrieved. After a brief consideration of the types of memory, the discussion in this chapter will concentrate on the molecular basis of memory formation, and where different kinds of memory may be located.

Types of Memory

Even 30 years ago, memory was generally considered to be a single, unified phenomenon. However, research has now made it clear that there are several types of memory. For example, not only is there a clear difference between remembering someone's name and remembering how to ride a bicycle, there are also differences in how long a memory lasts; the memory of an experience may last from just a few minutes up to more than a century. Unfortunately, recognition by researchers that there are different kinds of memory has not been accompanied by a well-accepted scheme for its classification. The main difficulty is that the type of memory is defined by the conditions under which it is evoked, and these conditions do not always lend themselves to being grouped into mutually exclusive categories.

In spite of this uncertainty, it will be useful here to introduce some of the main kinds of memory. One important distinction is between **declarative memory** and **nondeclarative memory.** Declarative memory, also called **explicit memory,** refers to the memory of facts and events, and allows for the establishment of relationships between objects and events with which the animal has had experience. There are different kinds of declarative memory. One type, called **episodic memory,** allows you to recall what you have experienced. For example, remembering that you sat in the library the first time you read this paragraph requires episodic memory. A different kind of memory, **semantic memory,** allows you to remember the names and definitions of the different types of memory that you read about. You may have experienced the frustration of recalling the circumstances in which you read something, but not the content. This may happen because the two kinds of memory are separate. Declarative memory seems to be present only in the vertebrates, and is most obvious in birds and mammals. For example, a rat's ability to remember the route to take through a maze is a manifestation of declarative memory. The best evidence suggests that declarative memory is stored either in or by the hippocampus.

Nondeclarative memory, in contrast, refers to the memory formed from skill acquisition of any kind, as well as from simple conditioning and nonassociative learning like habituation. All animals have nondeclarative memory. The memory of motor or cognitive skills such as riding a bicycle or doing long division (without a calculator!) is called **procedural memory,** as is the memory of any conditioned response. Another type of nondeclarative memory, called **priming,** involves the process that lets you remember having seen a word before without any knowledge of what the word means or the context in which you saw it. Nondeclarative memory is formed in brain regions other than the hippocampus (see later discussion).

Memories may have different durations. **Short-term memory,** or **working memory,** refers to memory formed and retained for a relatively short time, from a few minutes up to several hours. Remembering definitions you crammed in just before you took an exam is an example of short-term memory. You will not remember the definitions for the next exam. In contrast is **long-term memory,** which can be retained indefinitely. Being able to recall five years from now that you took a course in neurobiology is an example of long-term memory. Unfortunately, the classification of memory based on how long it lasts is independent of and unrelated to the type of memory that is under investigation. Research on short- and long-term memory usually involves declarative memory. However, simple forms of nondeclarative learning such as sensitization and habituation involve both short- and long-term forms of memory as well. On the other hand, not all declarative memory has a short-term component. For example, you may be able to recall a whole scene from your past just from smelling a distinctive odor that you encountered only once before when you were in the place you first smelled it.

Memory is information stored in the nervous system about an animal's experience. Several main types are recognized: declarative, including episodic and semantic memory, and nondeclarative, including procedural and priming memory. Declarative memory is memory of facts and events, whereas nondeclarative memory refers to skill acquisition and conditioning. Memory can also be subdivided into short-term (working) and long-term memory.

Memory Formation

Long-term learning requires the formation of some kind of long-lasting change in the central nervous system, which requires the synthesis of protein. Many experiments have been conducted to demonstrate that protein synthesis is necessary for long-term learning. These experiments usually involve exposing an animal to an inhibitor of protein synthesis before it is trained in some avoidance procedure, and demonstrating that after this treatment, the animal will not recall the procedure for more than a few hours. For example, goldfish can be trained to avoid an electrical shock by swimming to one end of an aquarium when a light is turned on. Normally, fish can remember such training for about a month. If an inhibitor of protein synthesis is injected into the fish just before training is started, the fish learns to avoid the shock but the next day shows no sign of remembering the training.

Experiments at the molecular level, in both invertebrates and vertebrates, have been carried out to elucidate the mechanisms by which long-term memory is formed. Among the invertebrates, the most detailed work has been conducted on the gill withdrawal reflex in *Aplysia* and on olfactory avoidance conditioning in *Drosophila*. In both animals, any procedure that blocks the synthesis of new proteins blocks the formation of long-term memory, whether the procedure is a biochemical block in the whole animal or in tissue culture, or genetic disruption of the pathway by which protein synthesis is induced. The mechanism by which new synthesis is induced is remarkably similar in the two animals. An appropriate stimulus induces activation of cAMP and hence PKA. Activation of PKA by cAMP causes separation of a catalytic subunit of PKA from the original tetrameric molecule, a subunit that translocates to the nucleus of the cell.

In the nucleus, the PKA subunit phosphorylates a regulatory protein known as cAMP response element binding protein (CREB). CREB is a transcription factor (a molecule that promotes the synthesis of RNA from a strand of DNA) that binds with a segment of DNA called the cAMP response element (CRE) (Figure 24-6). The binding of CREB to CRE promotes transcription of the associated gene, and hence the synthesis of new proteins. Some of the protein gene products that are produced when CREB is phosphorylated apparently serve to enhance synaptic transmission by influencing the amount of neurotransmitter that is released. Others are regulatory proteins that control the expression of yet other genes, in a complex chain that may lead to the synthesis of structural proteins for the formation of new synapses.

Molecular genetic studies of the CREB pathway have convincingly shown its specificity for long-term learning. Genetic disruption of CREB through use of an inducible negative CREB transgene (a form of the CREB gene that produces a defective product, and that can be turned on in an adult animal through some specific treatment such as exposure to high temperature) completely eliminates long-term conditioning in *Drosophila* without having any measurable impact on short-term learning, just as does inhibition of protein synthesis. Conversely, activation of an inducible activator CREB

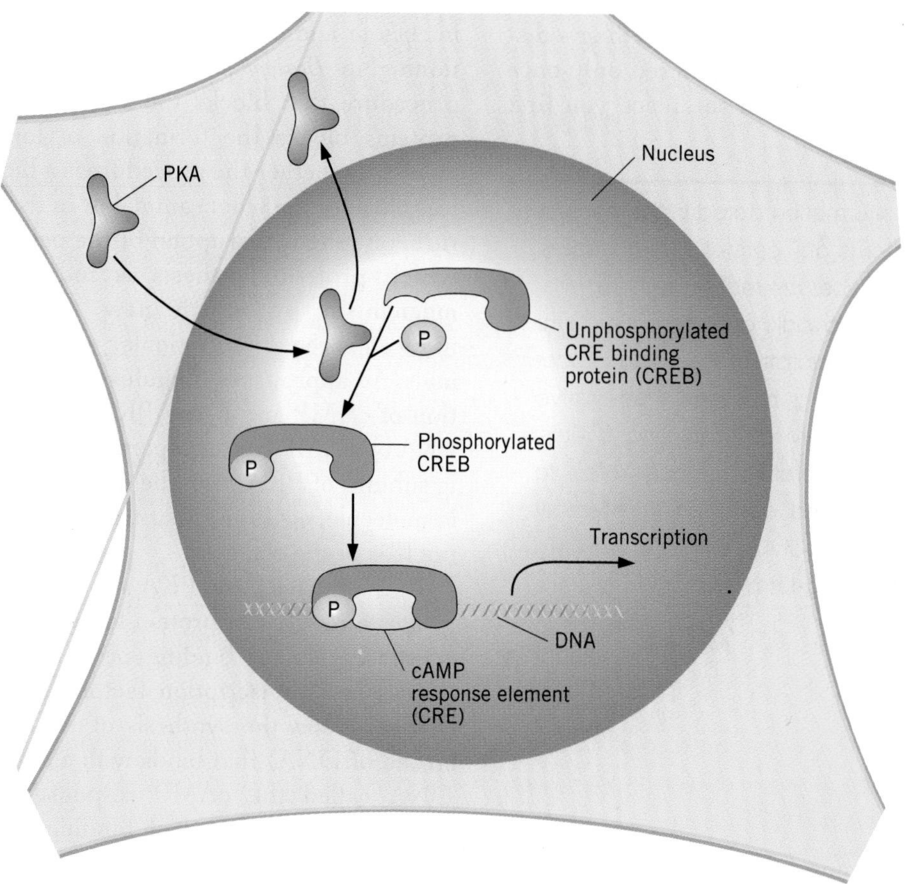

FIGURE 24-6. Activation of genes and the synthesis of new protein by the action of PKA in *Drosophila*. One subunit of PKA translocates to the nucleus, where it phosphorylates cAMP response element binding (CREB) protein. The phosphorylated protein binds with a cAMP response element (CRE) on a strand of DNA, which initiates transcription of the associated gene.

transgene (a gene that will produce active CREB even in the absence of PKA phosphorylation) will produce long-term memory even after training that in normal flies produces only a short-term effect.

The same cAMP-to-PKA-to-CREB pathway underlies the formation of long-term memory in *Aplysia*, and perhaps even in mammals. Because the neurons involved in learning are known, it is possible to inject enhancers or inhibitors of CREB into the cell bodies of the sensory neurons at whose terminals the long-term facilitation that represents long-term learning takes place. These experiments produce the predicted results: inhibition of CREB suppresses long-term learning and memory even if long-term training is used, whereas activation of CREB induces it, even if short-term training is applied. In mammals as well, a CREB analog has been described that appears to have the same general function, and disruption of cAMP and the PKA pathway interferes with the formation of long-term memory. Hence, research indicates a remarkable similarity of molecular mechanisms for memory formation in a remarkable diversity of animals.

The dependence of long-term memory on the synthesis of new protein is not just of theoretical interest. Physicians take advantage of this dependence in certain circumstances. Inhibitors of protein synthesis, called amnesiacs for their action, may be administered to children in conjunction with especially painful medical procedures, such as bone marrow biopsies for leukemia. By temporarily halting protein synthesis, the compounds block the formation of any long-term memory concerning anything the child experienced during the few hours after the amnesiacs were administered. As a consequence, the patient does not remember the procedure from one occasion to the next, and therefore does not form the extreme fear of and aversion to it that might otherwise develop.

New protein must be synthesized in order for long-term memory to form. In both invertebrates and vertebrates, protein synthesis is stimulated via increases in cAMP, activation of PKA, and phosphorylation in the nucleus of the neuron of CREB, a transcription factor. Activation of CREB, which initiates protein synthesis, is necessary for long-term memory to form.

Memory Storage

Although it would seem logical that memory would be stored in or around the neurons in which protein synthesis takes place, this is not necessarily the case. This is shown by the loss of an ability to form new long-term memory in individuals with hippocampal lesions, without impairment of recall of already formed memory. Hence, it is important to investigate the problem of where in the nervous system memory is stored. Research on this question in the first half of the twentieth century seemed to point clearly to a single answer with regard to the mammalian brain: everywhere. Over a period of 30 years, the psychologist Karl Lashley carried out hundreds of experiments in which he tried to disrupt recall in rats trained to run mazes. Whether he tried brain lesions, tissue ablation, electrical interference, or other methods, he always obtained the same result: the amount of "memory loss" exhibited by his animals was proportional to the amount of brain tissue he removed or damaged; it was not affected by where the tissue was located. He concluded that memory was distributed throughout the brain rather than localized in one place.

However, in many cases it is obvious that local storage of memory must occur. In *Aplysia*, memory of both simple and complex learning experiences, like habituation and conditioning, resides in a particular set of synapses between sensory and motor neurons. In the fruit fly *Drosophila*, the memory of olfactory avoidance conditioning is stored in the mushroom bodies, a particular part of the brain. In vertebrates, recognition that there are different types of memory has led to more precise tests of memory localization, and, in spite of Lashley's seemingly conclusive results, has produced evidence that memory is stored in specific brain sites. One especially successful study of procedural memory has been conducted by Richard Thompson and his collaborators, using a classical conditioning paradigm.

Thompson applied a paradigm similar to that developed by Pavlov, but he used rabbits as his experimental animals. The US, a gentle puff of air applied to the cornea of the eye, causes the rabbit to blink its eye. The CS, a brief tone, has no effect on the eye when it is presented alone. After the CS and US have been presented together for several trials, the rabbit will respond by blinking when the tone is presented alone. The reflex eye blink is mediated by pathways from the eye to the nucleus of the fifth cranial nerve, and from there via the sixth and seventh cranial nuclei to the motor neurons controlling the eyelid (Figure 24-7). At the same time, information about both the CS (the tone) and the US (the air puff) is transmitted via various brain stem nuclei to the cerebellum. Information about the CS enters the cerebellum via mossy fibers, and information about the US enters via climbing fibers from the inferior olivary nucleus. These signals interact at two levels in the cerebellum where the association between them is forged, at the Purkinje cells in the cerebellar cortex, and in the interpositus nucleus.

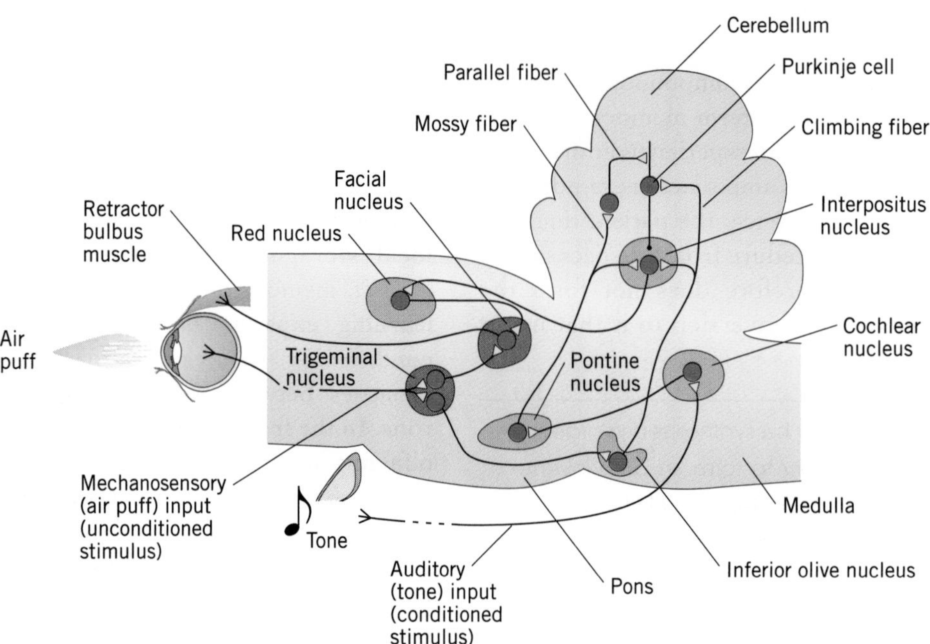

FIGURE 24-7. Hypothetical neural circuits in a rabbit for unconditioned and conditioned reflexes. The US is a gentle puff of air on the cornea of the eye, to which the rabbit responds by blinking. The CS is a tone, which after being paired with the air puff also produces a blink. The puff evokes an unconditioned reflex response via the trigeminal and facial nuclei in the brain stem. The response, a blink, is produced by contraction of the retractor bulbi muscles of the eyelid. At the same time, the puff activates fibers in the inferior olive that send information to the cerebellum. There, information about the air puff is combined with information about the tone, which has entered via mossy fibers from the pontine nucleus. Once the association has been made, the tone alone can cause an eye blink via the interpositus nucleus in the cerebellum and the red nucleus in the pons.

The pathways thought to underlie the response, shown in Figure 24-7, are deduced from a number of studies. First, electrical stimulation of parts of the proposed pathways can induce or mimic the conditioning. This is a critical test. If an air puff is thought to evoke a sensory signal that is sent to the cerebellum via the inferior olivary nucleus, then an electrical stimulus in the appropriate spot of the inferior olive ought to mimic the effects of the air puff. It does. When electrical stimulation of the inferior olive is paired with a tone in the same way that an air puff is, the rabbit will be conditioned to the tone. Electrical stimulation of other parts of the proposed pathway works as well. For example, sequences of direct electrical stimulation of mossy fiber tracts paired with stimulation of climbing fiber tracts will yield the normal learned response when the mossy fiber tracts are stimulated alone. In other words, stimulation of the climbing fiber tracts can take the place of the actual US, the air puff, and stimulation of the mossy fiber tracts can take the place of the CS, the tone. These results indicate that learning takes place in the cerebellar cortex, where these two tracts converge.

Physical and functional lesion studies also support the pathways for conditioning that are shown in Figure 24-7. Lesions of specific areas will abolish memory of the task, or permanently prevent learning of the task. For example, mice with the mutation *Purkinje cell degeneration* (*pcd*), in which the Purkinje cells all degenerate at about the age of 3 to 4 weeks, cannot be trained to eye blink to a tone, although normal mice can. Lesions of the interpositus nucleus also prevent conditioning. These results suggest that the memory of the association between the tone and the air puff is localized at the Purkinje cells and in the interpositus nucleus.

Furthermore, temporary functional lesions pinpoint the location of the memory as well.

Injection of muscimol, an agonist of the neurotransmitter GABA, into specific sites in the brain will temporarily block normal nerve transmission in the region of the injection. Injection of muscimol into the red nucleus, an output station for the conditioned response, during conditioning blocks expression of the conditioned response. However, after the effects of muscimol wear off, the animal shows the conditioned response perfectly well. Injection of muscimol into the cerebellum, however, blocks expression of the conditioned response not only during the period of disruption of transmission but also afterward.

Not all learning in vertebrates involves such localized and specific brain regions, as work by Lashley and others has demonstrated. In rats, Lashley was unable to find a specific locus for the memory of how correctly to navigate a maze. One reason may be that an animal's brain contains multiple representations of any event. Hence, a rat learns many different aspects of a maze as it masters the maze path. Destruction of a part of the brain containing information about any one part, such as olfactory cues, would not necessarily cause a collapse of the animal's ability to navigate the maze because visual, tactile, and auditory cues would still be available.

This idea of multiple brain representations of events is supported by work with newly hatched chicks that have been trained to avoid pecking at a seed coated with a strongly aversive chemical. Two regions of the forebrain seem to be intimately involved in the learning. Yet a variety of lesion experiments of these regions show that not every lesion that is expected to cause amnesia of the training actually does so. These and other results suggest that memory formation is complex, and may involve both local and distributed storage. Specific aspects of an experience may be stored in a discrete locus, but there may be more than one storage site, perhaps because events can be related to more than one experience. It may also be that different types of memory are stored separately. Hence, declarative memories, such as those formed by running a maze, might not be stored in the same place as procedural memories, such as those resulting from conditioning.

In vertebrates, some types of memory are stored in specific, localized regions of the brain, as certain conditioned responses are stored in the cerebellum. Other types of memory, however, may be stored in a distributed form in several places.

Additional Reading

General

Cohen, N. J., and H. Eichenbaum. 1993. *Memory, Amnesia, and the Hippocampal System*. Cambridge, Mass.: MIT Press.

Dudai, Y. 1989. *The Neurobiology of Memory: Concepts, Findings, Trends*. Oxford: Oxford University Press.

Squire, L. R. 1987. *Memory and Brain*. Oxford: Oxford University Press.

Research Articles and Reviews

Abrams, T. W., and E. R. Kandel. 1988. Is contiguity detection in classical conditioning a system or a cellular property? Learning in *Aplysia* suggests a possible molecular site. *Trends in Neurosciences* 11:128–35.

Byrne, J. H., and E. R. Kandel. 1996. Presynaptic facilitation revisited: State and time dependence. *Journal of Neuroscience* 16:425–35.

Cohen, T. E., S. W. Kaplan, E. R. Kandel, and R. D. Hawkins. 1997. A simplified preparation for relating cellular events to behavior: Mechanisms contributing to habituation, dishabituation, and sensitization of the *Aplysia* gill-withdrawal reflex. *Journal of Neuroscience* 17:2886–99.

Massicotte, G., and M. Baudry. 1991. Triggers and substrates of hippocampal synaptic plasticity. *Neuroscience and Biobehavioral Reviews* 15:415–23.

Thompson, R. F. 1991. Are memory traces localized or distributed? *Neuropsychologia* 29:571–82.

Tully, T., T. Preat, S. C. Boynton, and M. del Vecchio. 1994. Genetic dissection of consolidated memory in *Drosophila*. *Cell* 79:35–47.

Wilson, M. A., and S. Tonegawa. 1997. Synaptic plasticity, place cells and spatial memory: Study with second generation knockouts. *Trends in Neurosciences* 20:102–6.

CHAPTER 25

Hormones and the Nervous System

The nervous system can be shaped and molded by factors inside the body, as well as by those external to it. Hormones are the most important of such internal factors. Hormones not only modulate neural circuits but also can turn on or off whole sequences of behavior. In addition, they can guide the development or change the very structure of the nervous system.

Sensory input or experiences from which the animal learns are not the only factors or events that can shape the structure of the nervous system and alter the way it functions. Hormones and neuromodulators, whether circulating throughout the body or restricted to some local region of the brain, can also have strong effects on neural structure and function. Furthermore, just as external events may have temporary or permanent effects on the nervous system, so too may the effects of internal chemicals be temporary or permanent.

The Neuroendocrine System

Fifty years ago, the endocrine and nervous systems were treated as separate and distinct functional entities. Students were taught that the nervous system controlled highly specific reactions that needed to be carried out quickly and that the endocrine system controlled more diffuse, slower events. The nervous system communicated at specific sites (synapses), and the endocrine system communicated through chemicals released into the blood. It was clear that there were interactions between the two systems, but there was no question as to which organs and functions belonged to which system. Today, the distinction between endocrine and neural function is so blurred that researchers can easily justify treating them as a single neuroendocrine system, with the traditional endocrine functions lying at one end of a continuum and the traditional neural functions lying at the other. Between the two extremes lie functions that cannot readily be identified as belonging to one system or the other.

There are at least three reasons for considering the endocrine and nervous systems together in this way. The first is that they have functional similarities. Both systems respond to environmental stimuli. The nervous system reacts to sensory input such as visual images and sound, and the endocrine system reacts to factors such as the day-night cycle and the rhythms of the seasons. Furthermore, both integrate and control bodywide events, such as a specific behavior or growth of the body. They are also functionally similar in that they both use chemical messengers for communication.

A second reason for considering the endocrine and nervous systems together is that increased knowledge has blurred the traditional distinctions between them. Substances like adrenaline, oxytocin, and vasopressin, all first described as hormones, are also neurotransmitters, being released by some neurons and affecting other neurons. Conversely, substances like dopamine, first considered typical neurotransmitters, are now known to have an endocrine function as well.

A third reason is that the endocrine and nervous systems interact intimately. When you watch a horror movie, it is your autonomic nervous system that releases epinephrine (adrenaline) in your body. The hormone then triggers the sweaty palms and racing heart you associate with being frightened. Conversely, as you will learn in this chapter, hormones have direct effects on neural function, leading to specific behavior, as well as on the growth and subsequent structure of the nervous system.

Because of the intimate relationship and interaction between the endocrine and nervous systems, many neurobiologists and endocrinologists interested in this interaction now consider themselves as representatives of a new discipline known as **neuroendocrinology.** Researchers in this discipline focus on all aspects of neuroendocrine function, from questions of how hormones affect neurons at the molecular level to how release of a hormone can orchestrate a complex behavioral sequence.

The endocrine and nervous systems have so many functions in common and interact so intimately that they can be treated as a single neuroendocrine system, in which traditional neural and endocrine action lie at the two extremes of a continuum of function. One manifestation of the interactions between the neural and endocrine systems is the action of hormones on the structure and function of the nervous system. The discipline of neuroendocrinology deals with these issues.

Insect Metamorphosis

Insect metamorphosis provides a dramatic illustration of the actions of hormones on the nervous system. Metamorphosis is the transformation of an immature insect, called a larva, into an adult. The larvae of some insects, such as grasshoppers, are physically similar to the adults. Most insects are more like butterflies, however, and change body form dramatically during development, a process called complete matamorphosis. Following a succession of larval stages, the insects change from larva to pupa and later from pupa to adult (Figure 25-1A). Metamorphosis transforms the inside of an insect just as dramatically as it does the outside. Muscles, the gut, and even the nervous system undergo morphological and functional reorganization. This reorganization has profound consequences, for afterward the insect exhibits new behavior appropriate to the adult body form and adult life.

The key event in metamorphosis is molting, the process by which an insect sheds its old external skeleton (the cuticle) and exposes the new one. An insect must molt several times during its lifetime because the cuticle is inelastic and must be shed periodically in order for the insect to become larger. During the stages between molts, the insect eats and lays down a new, larger cuticle under the old one. The molts from larva to pupa and from pupa to adult are the most interesting from a neurobiological point of view because they are accompanied by dramatic changes in neural structure and function, as well as by changes in the shape and appearance of the animal.

Hormones regulate the cycle of molting, as illustrated in Figure 25-1B. Several hormones are involved in this cycle, but two of these, **ecdysone** and **eclosion hormone (EH),** are especially important. Ecdysone is a **steroid hormone,** structurally related to cholesterol and to the vertebrate sex hormones (Figure 25-2A, B). All steroid hormones have a common structure of three six-carbon rings and one five-carbon ring. You will learn more about the

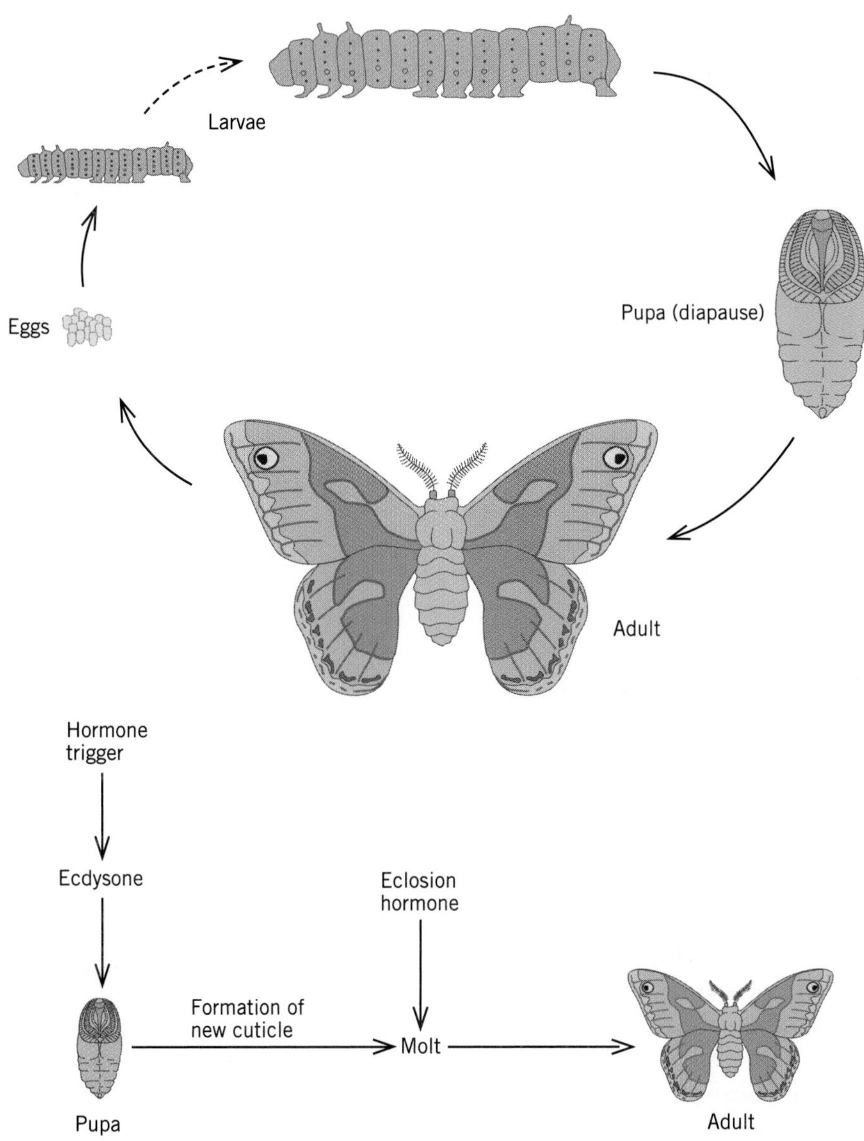

FIGURE 25-1. The life cycle of a typical insect that undergoes metamorphosis. Most insects undergo a dramatic morphological transformation during the course of development from a larva, a caterpillar-like stage, to a free-flying adult. The pupa is an intermediate transitional stage between the larva and the adult. The transformation is accompanied by appropriate changes in behavior. The changes in morphology and behavior are coordinated and controlled by hormonal actions, as shown in the lower part of the figure.

mode of action of these hormones in subsequent sections of this chapter. EH is a peptide consisting of 62 amino acids, with a molecular weight of about 8.8 kdaltons (Figure 25-2C). Steroid hormones, since they are lipid-soluble, pass directly into cells to bind with intracellular receptor molecules. Peptide hormones interact with receptor molecules on the cell surface.

When an insect has eaten enough to move from one stage to another, an environmental cue, often light, stimulates the release of a trigger hormone from the brain. This hormone causes the release of ecdysone from endocrine glands in the thorax of the insect. The sharp increase in ecdysone titer initiates the developmental changes necessary for the molt (Figure 25-3). During molts from one larval stage to another, the main effect of ecdysone is to stimulate the formation of new cuticle by epidermal cells. Ecdysone acts on

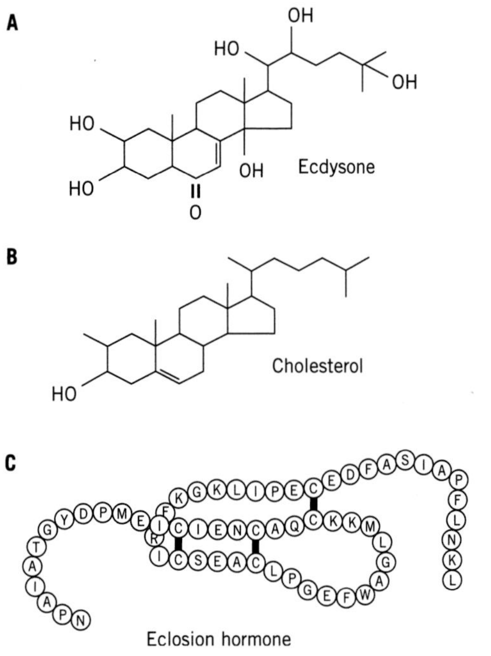

FIGURE 25-2. (A) The chemical structure of ecdysone, an insect steroid hormone. Insects synthesize ecdysone from cholesterol (B), to which it is chemically related. (C) The amino acid sequence of eclosion hormone, a 62-amino acid peptide. The hormone is cross-linked in three places by sulfur-sulfur bonds at cysteine. Each letter stands for a single amino acid.

all parts of the body, including the nervous system, to direct the transformation of a larva to a pupa and of a pupa to an adult. In all cases, ecdysone also primes the nervous system to make it responsive to EH.

When the changes appropriate to the insect's stage of growth have been completed, the animal waits for another environmental cue, usually the light of dawn detected by the ocelli (see Chapter 11), to stimulate the release of EH, which stimulates eclosion (molting). The old cuticle is shed at this time. EH is synthesized and stored in two pairs of specialized **neurosecretory cells** whose cell bodies are in the brain. Neurosecretory cells do not form typical synapses with other neurons; instead they synthesize and release hormones. The axons of the four neurosecretory cells travel the length of the ventral nerve cord, then leave the cord and swing up to lie along the hindgut. They release EH from varicosities along the length of each axon into the central nervous system, and from varicosities along the gut into the blood. Hence, the transition from one growth stage to another involves two hormonally controlled steps. Ecdysone controls the physical development of the insect and primes the nervous system to be sensitive to EH, then eclosion hormone stimulates the nervous system to initiate molting.

Metamorphosis, the physical transformation of a larval insect into a mature adult, often involves a dramatic reorganization of both the physical structure and the behavior of the insect. This transformation is mainly regulated by the coordinated action of ecdysone, a steroid, and eclosion hormone, a peptide.

Hormonal Effects on Neural Structure

A favorite subject for insect neuroendocrinological studies is the tobacco hornworm, *Manduca sexta*. This insect's common name comes from the characteristic curved spine at the tail end of the caterpillar. (The tomato hornworm that can decimate homegrown tomato plants in the summer is a close relative.) The adults of the entire group are swift and agile flyers, from which they gain their adult name, hawk moths.

Three mechanisms of neural reorganization can be identified during metamorphosis: structural change in existing neurons, death of larval neurons, and growth of new neurons. Structural changes occur in many neurons. For example, many motor neurons in the abdomen of the insect innervate different muscles in the adult hawk moth than they do in the caterpillar stage, and also exhibit a different pattern of dendritic branching in the two life stages. Motor neuron MN-1, for instance, which lies in the fourth abdominal ganglion of the larva, has a single main dendritic arbor contralateral to the cell body (Figure 25-4A). In the adult, however, this same motor neuron

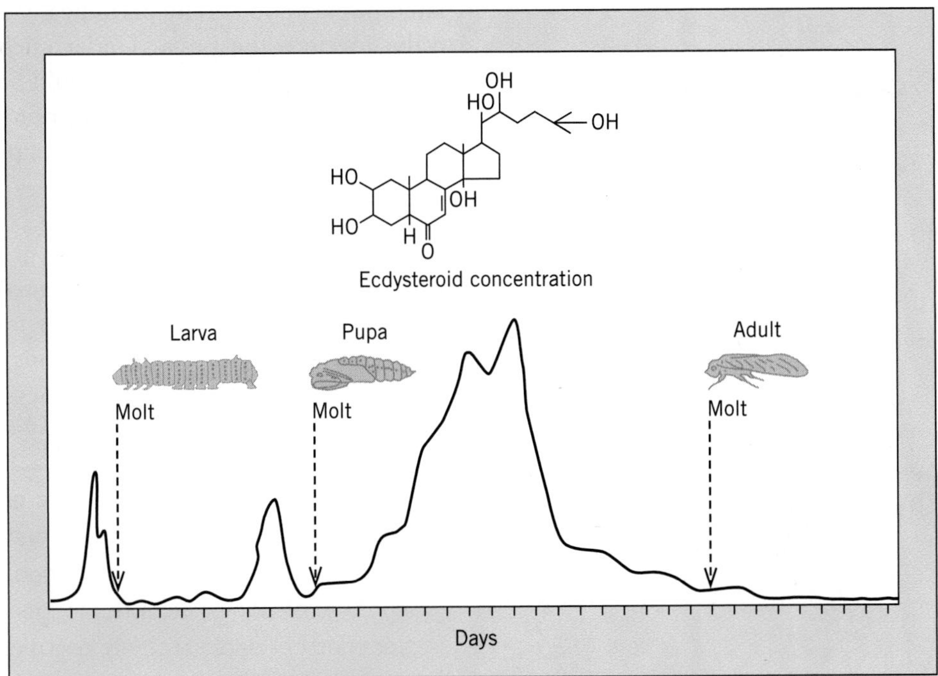

FIGURE 25-3. The change in blood titers of ecdysone in the hawk moth *Manduca sexta* just before the last larval molt, and before the molts to the pupal and adult stages. Ecdysis (molting) occurs at the dotted lines, induced by an increased concentration of eclosion hormone in the blood.

has a new ipsilateral arbor, as well as a larger contralateral arbor (Figure 25-4C). The pupa shows a different, somewhat intermediate, pattern of dendritic branching (Figure 25-4B).

In the caterpillar, MN-1 innervates a lateral muscle that, when contracted, flexes the abdomen to one side. A similar muscle lies on the other side and flexes the abdomen in the opposite direction. These muscles are normally used individually or alternately to move the abdomen laterally. The abdomen of the adult is shorter and stiffer than the abdomen of the caterpillar. Furthermore, being dorsoventrally flattened, it can bend only up and down. In the adult, the muscle innervated by MN-1 flexes the abdomen dorsoventrally, a movement that requires simultaneous contraction of the muscles on the right and left sides. The new bilateral dendritic arbor of MN-1 ensures that the neuron will respond equally well to input from either side of the body. Manipulation of ecdysone titers in isolated abdomens has shown that the presence of ecdysone stimulates the growth of the new dendritic branching.

Whereas neurons like MN-1 undergo reorganization during metamorphosis, other neurons die. Experiments in which the fate of individual, identified neurons in the hornworm is followed throughout metamorphosis show that certain motor neurons and interneurons always die at certain times, according to a fixed schedule adhered to in all animals. For example, interneurons that are scheduled to die do so before motor neurons. Furthermore, within each type of neuron an order of death is followed in most cases: small neurons usually die before large ones. Cell death that is not the result of old age or accident is referred to as programmed cell death, as described in Chapter 22 in the context of development. In the case of metamorphosis, the process is part of the extended development of the insect that culminates in its metamorphosis into an adult.

Neuronal death during metamorphosis is triggered at least in part by a change in the cells' hormonal environment. The surge of ecdysone that initiates the developmental

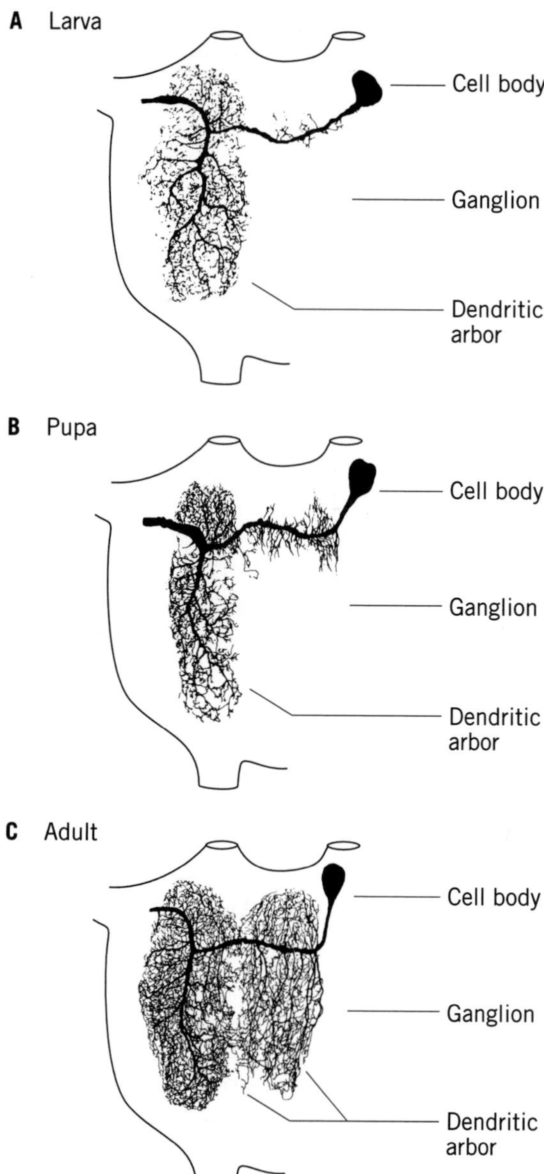

FIGURE 25-4. Dorsal view of an identified motor neuron in the fourth abdominal ganglion of the tobacco hornworm, showing the differences in morphology of this neuron at different life stages of the insect. (A) Larva (caterpillar). (B) Pupa. (C) Adult. Notice the elaborate pattern of dendritic branching on the right side of the adult neuron that is absent in the larval and pupal stages.

changes in epidermal and other body cells is a transient one. That is, ecdysone concentrations rise to very high levels for a short time, then fall again. This drop in ecdysone titer after the initial surge is an important component of the trigger for cell death. High ecdysone levels can be maintained in an isolated pupal abdomen by periodically injecting ecdysone into the pupa. Under such conditions, neurons that normally would have died with a drop in ecdysone levels remain alive and healthy as long as high ecdysone levels are maintained.

New neurons also develop. Just before the birth of the larval insect, a number of neurons stop developing, remaining arrested in an immature, inactive stage. During metamorphosis into the adult, these neurons become fully functional, providing the neural substrate for many adult-specific behaviors.

Ecdysone has strong effects on the nervous system of an insect undergoing metamorphosis. It induces a dramatic change in the structure of some neurons, triggers programmed cell death in others, and promotes growth and differentiation in undeveloped neurons.

Hormonal Effects on Neural Function

The behavior of insects that undergo complete metamorphosis depends on the stage of life they are in. For example, caterpillars crawl, pupae do little but wriggle, and adult moths walk and fly. Hormones orchestrate the changes in behavior from one life stage to the next. Experiments by James Truman and his colleagues on the silkworm moth have shown that in the molt from the pupa to the adult, EH has two distinct functions: to activate eclosion behavior and to trigger the switch from pupal to adult behavior by turning off the former while turning on the latter.

Before any insect can molt, it must be developmentally complete under the old cuticle. For example, when a pupa is developmentally ready to molt into an adult, there must be a morphologically complete adult insect inside the shell of pupal cuticle, waiting to emerge. In nature, a pupa may wait for many hours after it is developmentally an adult before it begins to emerge. Its emergence is triggered by an internal clock, which is normally cued by specific environmental stimuli such as the occurrence of dawn or dusk. Hence, if the physical development of a silkworm moth that normally emerges just after dawn is complete by midafternoon, the pupa must wait until the next day to molt. An adult that has not emerged from its pupal shell, but is ready to do so, is referred to as a pharate adult; it can be used in many experiments to study the hormonal regulation of behavior.

The first function of EH is to activate the behavior (eclosion behavior) that frees the insect from the old cuticle. Imagine yourself stuffed into a large, closed paper sack, unable to use your arms and legs. To get out you might push against the sack with your head to tear it, then wriggle yourself out by squirming around. This is essentially what the pupa does. Eclosion starts when the pupa pushes against the old cuticle until it splits along several lines of weakness at the head. The insect then engages in two main movements, abdominal rotation and a kind of shrugging, which together inch the insect out of its old cuticle.

Eclosion behavior is a centrally generated rhythmic motor activity that does not require sensory feedback for its execution (see Chapter 16). If the pupal cuticle is manually peeled off a pharate adult, the moth does not need to undertake eclosion behavior (Figure 25-5). Nevertheless, it does so anyway, showing that the presence of the old cuticle is not a necessary trigger to evoke the behavior.

Results from several kinds of experiments indicate that EH must be present in order for eclosion to take place. If a brain from a pupa on one developmental schedule is placed into a headless, isolated abdomen of a pupa that was on a different developmental schedule, the abdomen will molt according to the schedule of the implanted brain (which sends the trigger to release EH from the neurosecretory neurons) rather than the schedule of the abdomen. Furthermore, if EH is injected into brainless pupae, they will molt. In addition, if an isolated nervous system from a pharate adult is exposed to EH, fictive molting, the expression of the motor pattern appropriate to eclosion without any actual movement (Chapter 16), can be recorded

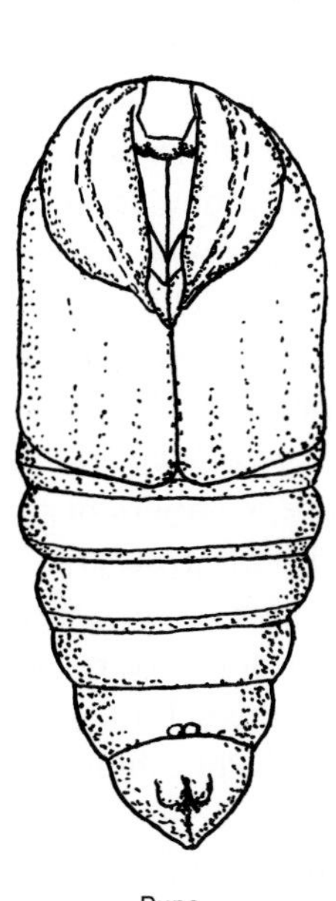

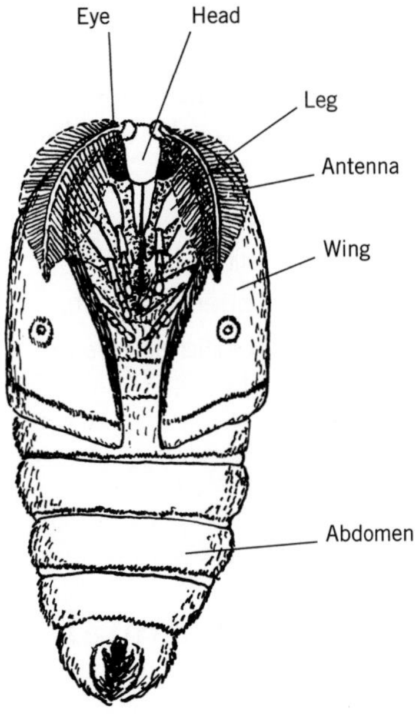

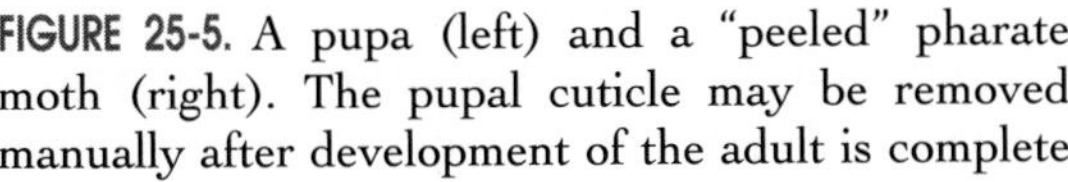
FIGURE 25-5. A pupa (left) and a "peeled" pharate moth (right). The pupal cuticle may be removed manually after development of the adult is complete but before molting occurs. Nevertheless, the insect exhibits only pupal behavior until it has undergone eclosion behavior.

from the stumps of the motor nerves that innervate abdominal muscles. Note, however, that the central nervous system will not respond to EH until after the surge in blood ecdysone level that signals the start of development into the next stage. Injection of EH before that time has no effect.

The second function of EH is to turn off pupal behavior and turn on adult behavior, as demonstrated by experiments with pharate adults from which the pupal cuticle has been removed many hours before eclosion would normally occur. Even after their cuticle has been removed, these insects exhibit only pupal behavior, engaging in abdominal wriggling in response to any stimulus to which they will react; they do not exhibit adult behavior. They do not try to right themselves if they are turned over, they do not walk, nor, even if they are males, do they show any responsiveness to the sex pheromone of the female. However, these insects will still undergo "eclosion" at the normal time, going through all the motions of eclosion even in the absence of the pupal cuticle. After this "eclosion," the insect exhibits only the behavior of a normal adult, and never again the behavior of a pupa. Furthermore, if eclosion is triggered early by injection of EH into a pupa containing a pharate adult, before the time at which the pupa would normally have molted, the silkworm moth undergoes early eclosion and then behaves as an adult.

Eclosion hormone has two distinct effects. It activates molting itself and, after the molt, turns off pupal behavior and initiates adult behavior.

Mechanisms of Hormonal Action on the Insect Nervous System

As described in the previous sections, ecdysone and EH are the two main hormones that regulate eclosion behavior. A third important hormone, juvenile hormone, influences the kind of cuticle that is made before each molt and in adulthood regulates the insect's reproductive behavior. Ecdysone and juvenile hormone are both lipid-soluble, and hence exert their effects via intracellular receptors. EH, a peptide, interacts with EH receptors on the cell membrane.

Ecdysone

There are two main reasons that the molecular mode of action of ecdysone has been the subject of considerable interest during the past decade. First, evidence suggests that all steroid hormones work via essentially the same mechanism, meaning that members of the same superfamily of proteins are responsible for binding all steroid hormones, whether insect or mammalian. Second, the geneticist's magic wand, the fruit fly *Drosophila*, exhibits the same hormonally controlled metamorphosis as do the moths described in the previous sections. Hence, the powerful tools of modern molecular genetics can be applied to the problem of how these hormones exert their effects at the molecular level.

Ecdysone exerts its effects directly on the genome of each cell that is sensitive to it. Being a steroid, it can easily pass through the membrane of any cell. Hence, the tissue specificity of ecdysone is accomplished through the presence or absence of appropriate receptor proteins. Cells that are responsive to ecdysone contain two proteins, ecdysone receptor protein (EcR) and a protein product of the *ultraspiracle* gene, known as USP. Both proteins are members of the nuclear receptor superfamily of proteins, to which proteins that bind vertebrate steroid hormones also belong, and both are required for a vigorous response of the cell to ecdysone.

When ecdysone binds with both proteins, the hormone-receptor complex binds with short segments of DNA called ecdysteroid response elements and thereby promotes gene transcription (Figure 25-6). The full response of the cell is orchestrated in two stages. Ecdysone first activates early-response genes.

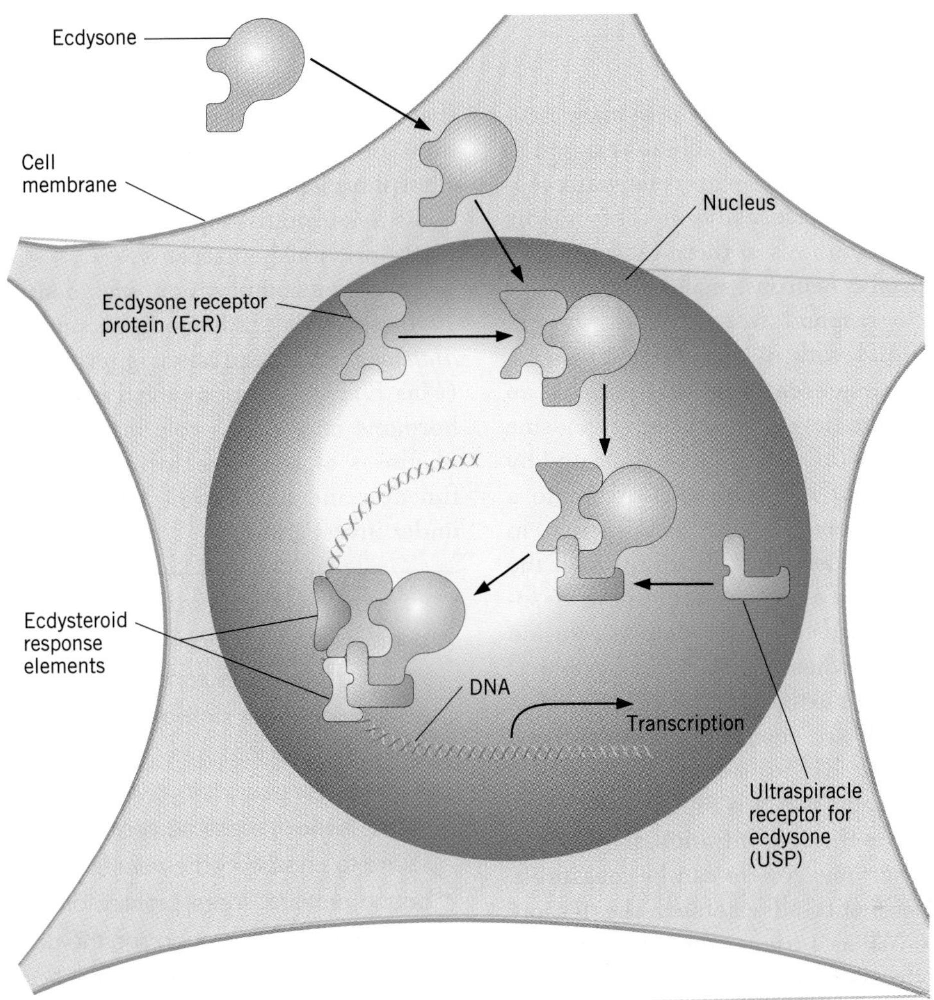

FIGURE 25-6. The mechanism of action of ecdysone on cells in insects. Ecdysone binds with two receptors, EcR and USP, to form a hormone-receptor complex. The complex binds with ecdysteroid response element sequences of DNA to promote the expression of genes.

Many of the protein products of these genes are themselves gene regulatory products. Either alone or in combination with more hormone-receptor complex, they activate late-response genes that are responsible for orchestrating the response of the cell to the hormone, whether it is the growth of new neurites, initiating apoptosis (see Chapter 22), or making the cell responsive to still other hormones.

The full sequence of molecular events that mediate a cell's response to ecdysone has not yet been determined. Not all the proteins involved in the regulation of this response have been identified, and the functions of the proteins that have been identified have not been firmly established.

Ecdysone, in common with vertebrate steroid hormones, controls the expression of genes by binding with a pair of protein receptors that are members of a nuclear receptor superfamily of proteins. Binding of the hormone-receptor complex initiates gene transcription.

Eclosion Hormone

One of the effects of ecdysone is to make neurons in the nervous system able to respond to EH. Since EH interacts with cells via receptors on the cell surface, ecdysone presumably promotes the synthesis of these receptor proteins in specific neurons, making it possible for them to respond when EH is present. Binding of EH with its receptor activates a second messenger cascade that causes an increase in the levels of cyclic guanosine monophosphate (cGMP). This is indicated by the evidence that injection of cGMP into a pharate adult moth is just as effective in inducing molting as is EH. In addition, if the nervous system of a pharate adult moth is isolated and exposed to cGMP, fictive eclosion occurs, just as when the nervous system is exposed to EH. Furthermore, if enzymes that degrade cGMP are chemically inhibited, the effect of adding EH to isolated abdomens is enhanced. Finally, if EH is injected, a significant increase in the concentration of cGMP in the central nervous system can be measured. These experiments all establish the importance of cGMP as a trigger for eclosion.

These effects of EH are more complex than was first thought. One complication is that in situ hybridization (Box 25-1) shows that only four neurons express the gene for EH. These neurons release the hormone into the nervous system, where immunocytological studies show that a larger population of neurons exhibits a mild elevation of cGMP levels within 5 to 10 minutes after application of EH. After about 30 to 40 minutes, however, a subset of about 50 neurons show substantial increases in cGMP levels. This secondary increase of cGMP level is synaptically driven, since it can be prevented by section of the nerve cord. The neurons that show this elevation in cGMP may be responsible for triggering the activity of the neural networks that drive eclosion behavior.

The specific role of cGMP is to promote the synthesis and subsequent phosphorylation of at least two proteins in the nervous system. There is some evidence that these proteins are membrane-bound. The exact location, identity, and function of these proteins are currently under investigation, but they may be receptor or channel molecules, or may be associated with such molecules. Phosphorylation of these proteins may then allow a neuron to respond to signals to which it was previously insensitive.

A further complication was identified in a 1996 report that indicated a second hormone, *Manduca sexta* ecdysis-triggering hormone (Mas-ETH), is also involved in molting. This hormone may play a role in determining the readiness of the larva to molt, but its exact function and its relationship to EH are still under investigation.

Eclosion hormone exerts its effects on behavior by directly affecting the activity of selected sets of neurons. These neurons, in turn, synaptically activate a subset of 50 other neurons that may activate eclosion. Activation of neurons involves an increase of cGMP levels in these neurons, which in turn seems to phosphorylate several membrane-bound proteins. These proteins are thought to play a key role in changing the activity of the selected neurons. A second hormone, Mas-ETH, may also play a role in eclosion.

Juvenile Hormone

Insect nervous systems are also responsive to hormones other than those that regulate metamorphosis. In crickets, for example, sexually mature females show a strong phonotactic response to the calling song of a male, turning and walking toward a calling male. Females that have just molted to the adult stage do not show phonotaxis because the phonotactic response takes about three to four days to develop. During these first few days after the molt to adulthood, the level of a hormone called **juvenile hormone (JH)**

BOX 25-1

Targeting Peptide Synthesis: In Situ Hybridization for Localization of mRNA for Specific Peptides

Just as is the case in studies of neurotransmitters, an important question to be answered in studies of hormones is where the hormone is synthesized. Knowing the site of synthesis will yield important clues about how release of the hormone is controlled and, in some cases, clues about the possible targets of the hormone as well. One recent technique localizes peptides or small proteins by using **in situ hybridization.**

This technique allows a researcher to determine the location of mRNA in a piece of tissue. Since each strand of mRNA codes for a specific peptide or protein, locating the mRNA that codes for the molecule of interest allows a researcher to find the cells that synthesize that molecule. A radioactive label or some cytochemical marker is added to cRNA (or cDNA, a strand complementary to the strand of mRNA of interest). The tagged cRNA is then placed on the target tissue so it can hybridize (bind to) the target mRNA. After unbound cRNA is washed away, suitable treatment of the tissue will reveal the precise location of the labeled cRNA in the tissue. By selecting cRNA that encodes for a peptide, or for enzymes that catalyze synthesis of a peptide, it is possible to study the distribution of the cells that contain the message for that peptide or enzyme (Figure 25-A).

FIGURE 25-A. Cellular localization of mRNA that codes for eclosion hormone in the central nervous system of a tobacco hornworm caterpillar by in situ hybridization. (Courtesy of Lynn M. Riddeford and James W. Truman.)

increases in the female's blood. JH is neither a protein nor a steroid, but a lipid-soluble molecule known as a sesquiterpenoid. Its name indicates the role it plays in young insects in preventing premature molting into an adult, but in an adult the hormone is necessary for the insect to become sexually mature and to express behavior associated with reproduction. The importance of JH is suggested by the precocious appearance of phonotactic responsiveness in 1-day-old female crickets within a few hours after they have been injected with JH.

In the mature female, phonotaxis depends on the action of a single pair of auditory interneurons known as L1, which lie in the first thoracic ganglion of the cricket central nervous system. The L1 cells receive input from afferents from the ears and respond by firing action potentials when they are stimulated sufficiently strongly. The threshold of L1 is determined in part by the density of receptors for acetylcholine (ACh), the neurotransmitter released by the ear afferents. In 1-day-old adult females, the threshold of responsiveness of the pair of L1 neurons is so high that normal intensities of calling song do not excite the cells.

The rising levels of JH are responsible for the onset of phonotactic responsiveness by acting directly on the threshold of the L1 cells by promoting the expression of genes that direct the synthesis of additional nicotinic AChR channels, the receptor protein with which the ACh binds. The normal effect of JH on 3- to 4-day-old crickets can be prevented by applying a blocking agent for transcription of DNA, rendering the cricket unresponsive to the male's calling song. This effect is specific to the particular neuron of the bilateral pair of L1s that is treated. Applying JH to only one of the bilateral pair of L1 neurons produces a cricket that will respond to a male's calling only on the side that sends input to the treated neuron. Treating the entire ganglion with JH and then treating just one side with a blocking agent produces a cricket that will respond to a male's call only on the side that sends input to the unblocked neuron. Hence, the increase in sensitivity of a female cricket to a male's calling song during the first few days of adult life is due to the action of JH in promoting the synthesis of greater numbers of AChR molecules; the presence of more AChR molecules causes a reduction of L1's response threshold to acoustic input and makes the female more responsive to the call of the male.

Although the molecular basis of JH action is not well understood, a few points are clear. First, several JH receptor proteins have been identified. They are large, about 30 kdaltons each, and appear to be members of a novel nuclear binding protein family. It has also been shown in some cells (but not yet in neurons) that the hormone-receptor complex directly activates gene transcription. Hence, the overall mode of action of JH may be similar to that of ecdysone.

In insects, juvenile hormone regulates the onset of behavior associated with sexual maturity. It turns on phonotaxis in female crickets by increasing the number of receptors for ACh on a pair of interneurons that direct locomotion toward a calling male. This probably works via the activation of genes by JH following its binding to a member of a family of unique nuclear binding receptors.

Steroid Hormones and the Vertebrate Brain

All vertebrates, indeed most animals, have two sexes, often distinguishable by external physical differences. The behavioral and structural difference between the sexes is referred to as sexual dimorphism because there are two varieties, or "morphs," male and female. Behavior that is different in males and females is called sexually dimorphic behavior. Both morphological and behavioral differences are due to the effects of steroid sex hormones on the body of the animal, including the nervous system. In mammals, the default morphology of an individual is female; that is, an individual develops as a female in the absence of sex hormones. Male physical attributes (and behavior) develop due to the presence of testosterone, the male sex hormone. The behavioral effects of testosterone can be shown by manipulating the hormonal environment of a developing animal. For example, injecting testosterone into a female rat in utero produces an animal that is malelike in its behavior.

Hormones influence behavior by affecting neural structure and neural function. Hormones affect neural structure by influencing the development and growth of the nervous system and directing the formation of specific neural circuits that will allow sex-specific behavior to be carried out. Hormones affect neural function by activating specific neural circuits so that appropriate behavior may be expressed at appropriate times. Hence, the sexually dimorphic behavior common to all vertebrates is a manifestation of differences in neural structure and function caused by specific hormones. Testosterone induces male structure as well as malelike behavior, and the female sex hormones (estrogens) induce femalelike behavior. One rather confusing aspect of the effects of the steroid hormones on behavior in mammals is that testosterone is readily metabolized. It may even be chemically converted to an estrogen. Hence, in some instances the active agent that is directly responsible for a masculinizing effect on neural structure and behavior is estrogen, although the agent started life as testosterone.

The focus of much work in neuroendocrinology is on sexual dimorphism, how hormones cause it, and how they operate on it to produce male and female behavior.

Behavioral differences between male and female animals are manifestations of structural and functional differences in their nervous systems. This sexual dimorphism is caused in part by the different actions of male and female steroid sex hormones on developing and adult nervous systems.

Steroids and Mammalian Sexual Dimorphism

Structural sexual dimorphism is manifested in specific regions of the central nervous system, especially those that control reproductive behavior. In mammals, certain parts of the **preoptic area** of the hypothalamus, as well as the **spinal nucleus of the bulbocavernosus (SNB)** in the spinal cord, show dimorphism. In rats, for example, the **sexually dimorphic nucleus of the preoptic area (SDN-POA)** appears to be necessary for the normal expression of mounting, a characteristic male mating behavior. The SDN-POA is more than twice as large in males as in females, due to the presence of more neurons in this area of male brains. The SNB in male rats is also significantly larger than the corresponding area of the female spinal cord, mainly because this region contains motor neurons that innervate penile muscles that are absent or very small in adult female rats. Conversely, the **anteroventral periventricular nucleus** in the preoptic area of female rats is larger than the corresponding region of the male brain. Cells in this nucleus secrete oxytocin, a hormone important in stimulating maternal behavior. Many other mammals show a similar pattern of sexual dimorphism in specific brain and spinal regions.

The sexual dimorphism of males appears during development as a result of the presence of testosterone, the male sex hormone. Castrating a newborn male rat will result in an adult whose SDN-POA is significantly smaller than that of a normal male. Conversely, treating a normal female rat with doses of testosterone just after birth results in an SDN-POA that is larger than that usually found in female rats. This effect is dose-dependent: the greater the amount of testosterone that is injected, the larger the SDN-POA. It is also dependent on just when the hormone is injected. Later injections have smaller effects. With the injection of enough testosterone at the right time during development, it is possible to produce a female rat whose SDN-POA is just as large as that normally found in a male rat.

The influence of sex hormones on neural structure is usually readily apparent, but the influence of sex hormone on neural function is not easy to understand. Neural function can be assessed by noting the behavior expressed by an animal after a particular experimental manipulation. Treatment usually consists of removing the gonads (testes or ovaries) of the animal to eliminate uncontrolled production of sex hormones, then injecting known doses of a hormone and noting the effects. Mating behavior is most commonly studied, although related activities such as maternal behavior can also be investigated. It has been difficult to make generalizations about steroid effects on mammalian reproductive behavior because results vary significantly depending on the species of animal studied (for example, rats may give different results than hamsters) and on the age of the animal when treatment is applied.

In spite of these difficulties, decades of research have shown that there is a critical period during development for sensitivity to male sex hormones. In the rat, this period is from a few days before birth to the end of the first week after birth. During this time, exposure to testosterone will have a strong and permanent effect on the type of mating behavior subsequently exhibited by the animal. Even large doses of estrogen administered to an adult rat cannot completely overcome this effect. Similarly, exposure of a rat to estrogen during the critical period strongly disposes the rat to femalelike behavior, a disposition that is not entirely countered by administration of testosterone later in life. Hence, the early effects of a hormone can be modified but not completely overcome by later administration of another hormone.

It is also important for the interpretation of these results to realize that although for simplicity researchers often speak of sexual dimorphism as if animals were either completely male or completely female, it is understood that sex differences lie on a continuum. Behaviorally and physiologically, no animal—human or otherwise—is exclusively "male" or "female."

The presence of sex hormones increases the size of certain regions of the brain while decreasing the size of others compared to the size of those same regions of brains that are not exposed to these hormones. Sex hormones also influence reproductive behavior. Testosterone tends to have a masculinizing effect, whereas estrogens have a feminizing effect. The effects of testosterone are most pronounced when the animal is exposed to the hormone during a brief critical period, which in rats is from a few days before birth to the end of the first week after birth.

Steroid Effects on the Brain and Behavior

Steroid hormones exert their effects by binding with receptors that are present in specific neurons in the brain. Neurons that accumulate steroids can be identified and located in histological preparations of brain tissue through autoradiography (described in Box 23-1), using radioactively labeled steroid hormone. Working with this and similar techniques, it has been determined in rats, for example, that estrogens accumulate in a number of regions along the ventral surface of the brain, principally in parts of the preoptic area, the hypothalamus, the median eminence (just above the pituitary), and the pituitary gland, as well as in the septum, the amygdala, and parts of the dorsal midbrain (Figure 25-7).

Many studies have been conducted to investigate the behavioral relevance of these and other sexually dimorphic areas of the nervous system to sexual behavior. In general, the results show that specific behaviors are controlled by specific regions of the brain. For example, destruction of the medial preoptic area in adult male rats by microinjection of a chemical that causes neural lesions effectively eliminates copulation. Only about 10%

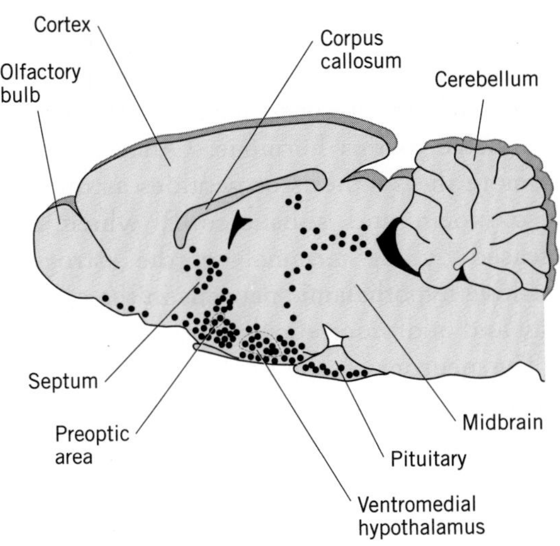

FIGURE 25-7. The distribution of radioactively labeled estrogen in the brain of a female rat. The dots represent localized radioactivity (refer to Box 23-1), and hence cells that selectively take up this hormone when it is available. Note that there is a concentration of estrogen uptake in the preoptic area and several other regions of the brain. The amygdala, which also takes up estrogen, cannot be seen in this view of the brain.

of treated rats will copulate when given the opportunity, compared with 100% of control rats. The effect is specific to the behavior itself and is not due to changes in motivation. The interest of a male rat in copulation can be assessed experimentally by training it to press a bar for access to a female ready to mate, and measuring how many times or for how long the rat will press the bar. Such tests show that rats in which the medial preoptic area has been destroyed press the bar just as much as control animals. Similar experiments have shown that certain brain regions control specific sexual behavior in females as well.

The most thoroughly studied example of the influence of hormones on regions of the brain that control a sexually dimorphic behavior is lordosis, the characteristic posture assumed by many mammalian females as they prepare to receive a male during copulation (Figure 25-8). It consists of a concavely bowed back and elevated hindquarters, and is driven mainly by activity of motor neurons that innervate muscles of the back and the hind legs. Lordosis is a reflex, initiated by stimulation of Ruffini endings (see Chapter 14) in the animal's skin over the back and flank. The afferents from these sensory receptors synapse with a number of interneurons in the spinal cord, which in turn synapse with motor neurons controlling the muscles whose contractions produce lordosis. However, the presence of estrogen is essential for expression of the reflex. Progesterone also has an effect on the strength of the reflex. A female that is not physiologically ready to mate, that is, that does not have high levels of estrogen and progesterone in the blood, will not exhibit lordosis. Conversely, injection of hormones can make an unresponsive female responsive.

Estrogen and progesterone exert their effects via a complex neural circuit consisting of neurons in several regions of the brain, as well as the motor centers in the spinal cord (Figure 25-9). The steroids act on neurons in the ventromedial hypothalamus and other hypothalamic regions, which send axons to the central gray and reticular formation of the midbrain. The **central gray** is the part of the mesencephalon that surrounds the hollow core of the brain stem, lying medial and ventral to the colliculi. Both the central gray and the midbrain reticular formation project axons to the medullary reticular formation. This region, along with the lateral vestibular nucleus of the pons, sends axons down the spinal cord to synapse on interneurons in the motor centers of the lumbar region that are responsible for organizing lordosis behavior.

FIGURE 25-8. The lordosis response of a female rat. This posture, showing elevated hindquarters, is typical of many mammals during mating.

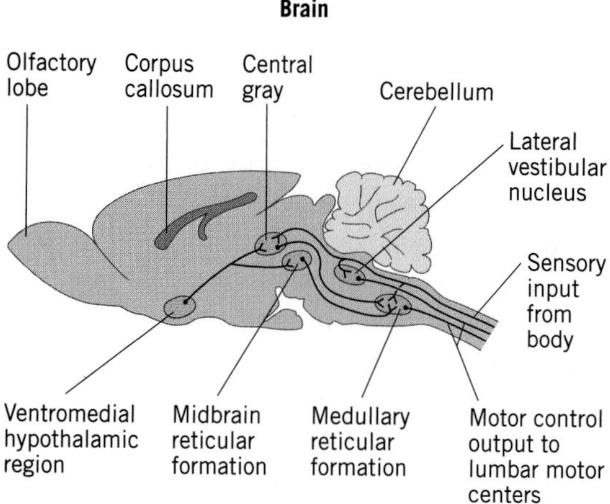

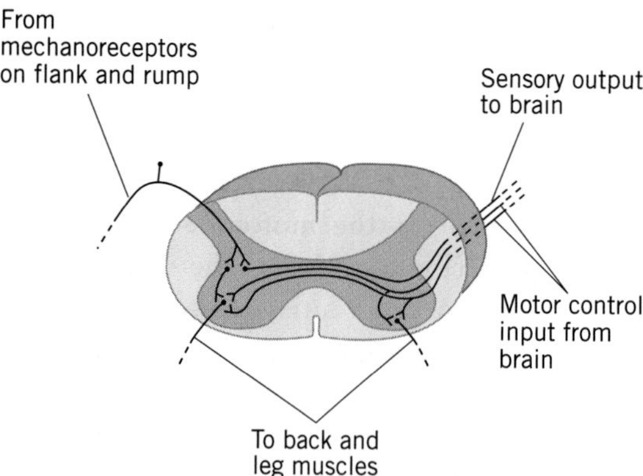

FIGURE 25-9. The neural circuit underlying hormonal regulation of the lordosis response of female rats. Estrogen activates estrogen-sensitive neurons in the hypothalamus, which in turn excite neurons in the midbrain central gray and midbrain reticular formation. Input from the midbrain regions acts synergistically with sensory input from skin receptors to produce signals to the spinal cord. These signals allow skin input to elicit a motor response. Without the estrogen-induced excitation of neurons in the medullary reticular formation, input from skin receptors will not induce a reflex lordosis response.

Simplified, the roles of hormone-sensitive neurons in the hypothalamus and the brain regions that they influence are as follows. Estrogen activates neurons in the hypothalamus, promoting protein synthesis and an increase in the spontaneous firing rate in these neurons. The protein synthesis is an essential mediator of the effect of estrogen, since blocking it completely blocks lordosis. Several kinds of proteins are synthesized, in several surges of synthesis. Among the synthesized proteins are receptors for progesterone, whose presence makes the neurons responsive to this hormone. Other proteins promote the synthesis of peptides as diverse as oxytocin and substance P, which are released by the terminals of the estrogen-sensitive hypothalamic neurons in the central gray and midbrain reticular formation.

The neurons in the central gray and the midbrain reticular formation that are stimulated by hypothalamic neurons in turn excite neurons in the medullary reticular formation. Here (and to some extent in the central gray area as well) the excitatory synaptic input interacts with input from the spinal cord that carries information from the Ruffini endings in the skin to excite neurons that effectively gate the reflex in the spinal cord. Meanwhile, signals from the lateral vestibular nucleus, also primed by input from skin receptors, help to coordinate the behavior. Information from the vestibular nucleus and the reticular formation converges on motor centers in the spinal cord and evokes the reflex.

Estrogen and progesterone (which augments the strength of the lordosis response) are key to the reflex. In the absence of sufficient estrogen, there is no activation of hypothalamic neurons, no activation of midbrain neurons, and hence insufficient input to excite neurons in the medullary reticular formation. Consequently, the excitatory input from skin receptors will be insufficient to excite medullary neurons above threshold. Hence, although the necessary activation of muscles is organized in the vestibular nucleus, the sensory input at the spinal level is not strong enough to allow it to be expressed. Only input from medullary neurons to the spinal motor centers, combined with mechanosensory input from the skin, will evoke the lordosis reflex.

Although all the examples of hormone action on neural activity discussed in this and previous sections have emphasized an effect on neurons of the central nervous system, hormones can and do exert effects on peripheral structures as well. For example, in the frog *Xenopus laevis*, males produce a calling song of trills rapidly modulated in amplitude; females produce no courtship song but can generate a slower, unmodulated release call. The physiological basis of this difference is that males have a greater number of fast-twitch muscle fibers in laryngeal muscles, more motor axons innervating these muscles, and weaker neuromuscular synapses. Motor nerve impulses in females always produce contraction of laryngeal muscles, yielding an unmodulated call. In males, motor impulses must facilitate in order to produce muscle contraction. (Facilitation is discussed in Chapter 8.) This feature, combined with the contractile characteristics of the muscle fibers and the pattern of impulses, generates a modulated call that waxes and wanes in amplitude.

The differences in structure and function of the laryngeal neuromuscular system in male and female *Xenopus* are under hormonal control. The musculature of the larynx, that is, the types and numbers of muscle fibers, is masculinized by testosterone during development, as is the number of axons that innervate these fibers. Masculinization of the larynx of female frogs can be induced by injection of testosterone during development, and can be prevented in males by drugs that block the binding of testosterone with its receptor. Conversely, the increase in synaptic efficiency that occurs in females, which is due to a higher quantal content in motor axons of females (see Chapter 7), is induced by estrogen. A similar peripheral effect of testosterone is seen in male rats, in which the hormone induces hypertrophy of muscles associated with penile erection. In this case, the hypertrophied muscles release a neurotrophic factor that induces the motor neurons that innervate them to develop larger dendritic fields. These studies demonstrate that not all sexually dimorphic behavioral effects of steroid hormones are due just to hormonal effects on the central nervous system. The effector organs (muscles in this case) can also be strongly influenced by hormones in ways that enable males to engage in different behavior than females.

Steroid hormones accumulate in specific regions of the brain of male and female vertebrates. These regions are responsible for much of the sexual behavior of these animals. Lordosis, the stereotypical posture adopted by a female rat during copulation, has been studied extensively. The components of the neural circuit have been worked out, and the role of estrogen in inducing the behavior elucidated. In some cases, hormones also affect peripheral structures to bring about sexually dimorphic behavior.

Steroid Regulation of Cyclic Behavior

Although the preceding discussion has treated the effects of a hormone as if they were permanent, the concentrations of steroid hormones in the blood typically wax and wane throughout an animal's reproductive cycle. High concentrations of estrogens or testosterone activate female or male sexual behavior, respectively, and low concentrations leave the animal uninterested in such activities. In some songbirds that sing seasonally, this cyclic activation of mating behavior is especially prominent and is correlated with a remarkable structural dimorphism.

The brains of many songbirds have specialized regions that control singing (Figure 25-10). Two of these, the **high vocal center (HVC)** and the **robust nucleus of the archistriatum (RA)**, are lateralized, just as the speech center is in humans, meaning that the area is somewhat larger on one side of the brain, usually the left, than it is on the other. In species in which both males and females sing, the song centers are approximately similar in size and neural complexity in male and female birds. However, in birds like zebra finches and canaries, in which only the males sing, the HVC, the RA, and an area called the X region are much larger and more complex

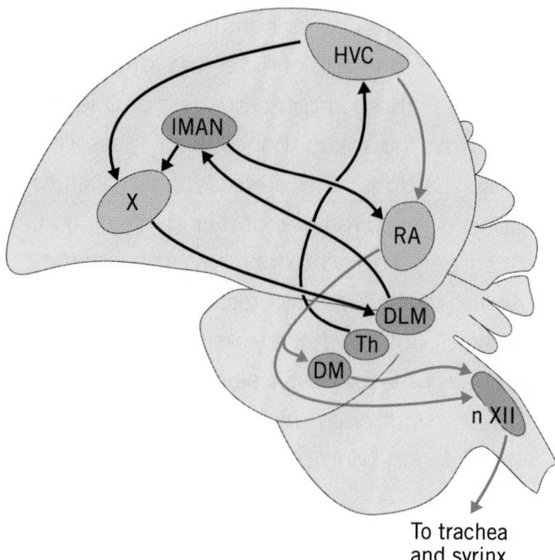

FIGURE 25-10. The main brain centers involved in singing in an adult songbird. The HVC is the so-called high vocal center, one of the main control areas. The other regions are also involved to varying degrees. The syrinx is the voice box of birds.

in male birds than in females. For some of these brain regions, the difference in size is obvious in prepared tissue even without the use of any quantitative measurements (Figure 25-11). In most cases, however, differences in morphology can be demonstrated only by careful, quantitative study.

The larger size of the song regions of the male brain is induced mainly by testosterone. In finches, for example, treatment with testosterone of birds whose gonads have been removed at birth produces enlargement of the song regions of the bird's brain, whether the bird is a genetic male or female. Such treatment also always leads to singing. Birds from which the testes are removed and to which no testosterone treatment is given do not sing, nor do birds treated with estrogen, whether the subjects are genetically male or female.

The males of some songbirds, like canaries, sing seasonally. That is, they sing at one time of the year (spring) to attract mates and do not sing at other times. The HVCs in these birds show striking morphological changes correlated with this seasonal change in behavior, enlarging and regressing according to whether the bird is entering its singing or silent phase. These changes, like those that occur during maturational development in finches, can be induced by appropriate hormonal treatments.

Studies from canaries have particularly interested neurobiologists because the basis for the brain growth seems to be an increase in neurogenesis, as well as an increase in dendritic branching and density of synapses. Detailed analysis of the enlarging vocal control regions has shown that there is a significant increase in the size of existing neurons and proliferation of new dendritic branching there. However, there is evidence that a proliferation of new neurons occurs as well. This brain region is unusual in that new neurons are generated throughout life, even into adulthood. However, neurogenesis increases markedly at the time of maximum growth of song centers. This region of the brain then stands as the only known, documented example of the growth of new neurons in an adult brain. As such, it has attracted a good deal of attention; if it can be determined how hormones stimulate new cellular growth, it may be possible for scientists to learn to induce new growth in other (particularly human) brains as well. The work has also attracted attention because it is an amazingly clear example of the dependence of a single behavior, singing, on just a few well-defined brain regions. To some extent, these regions seem to be assembled and disassembled every year as the bird goes through its seasonal cycle of reproductive behavior.

In some songbirds, song control regions in the brain are larger in male birds that sing than in females that do not. In seasonal singers like canaries, the size and complexity of the song control regions of the brain wax and wane under the influence of changing concentrations of testosterone. The increase in size of these regions is due in part to the birth of neurons and in part to an increase in dendritic branching.

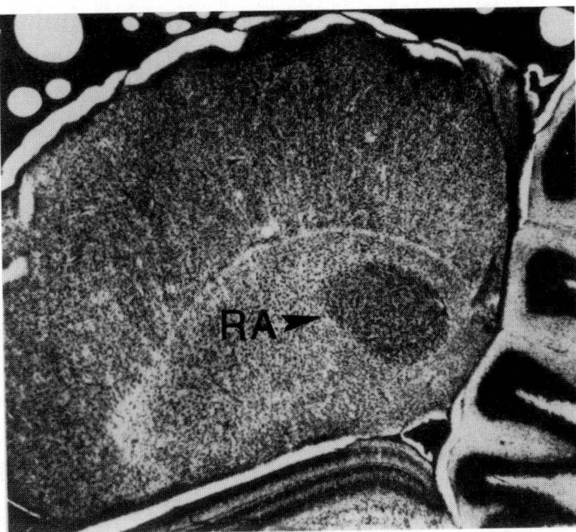

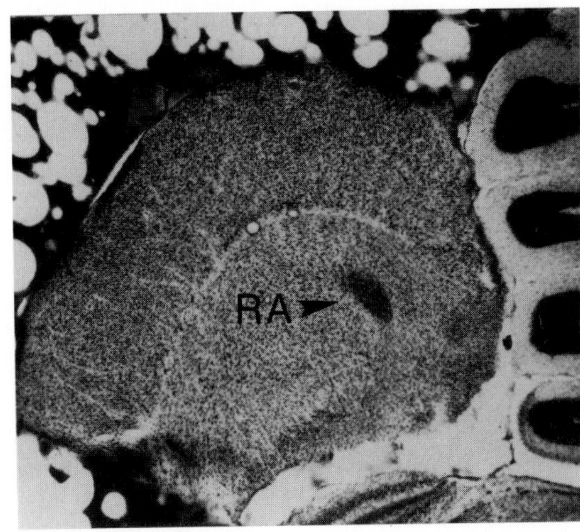

A Male

B Female

FIGURE 25-11. Sexual dimorphism in the brain of a songbird. Many of the regions of the brain that are used in the generation of song, such as the robust nucleus of the archistriatum (RA) shown here, are considerably larger in the brains of males during the months they are singing than they are in females.

Vertebrate Peptide Hormones

Neurobiology textbooks that deal with neuroendocrinology often emphasize the steroid hormones because of their obvious and dramatic effects on sexual behavior. However, animals actually synthesize and use many more glycoprotein and peptide hormones than they do steroids. For example, all the hormones secreted by the pituitary gland are either glycoproteins or peptides. Furthermore, many of the peptide hormones are now known to play several roles in the body. Not only do they have their traditional roles as hormones, regulating such aspects of homeostasis as water balance, but many are also synthesized and released as neuromodulators in the brain.

Two peptide hormones can serve as examples. **Vasopressin**, known as antidiuretic hormone (ADH) in the older literature, and **oxytocin** are secreted by the posterior pituitary. Vasopressin, also called **arginine vasopressin (AVP)**, acts on the kidney as a typical hormone to control the amount of water that is resorbed from urine. At high concentrations, it can also increase blood pressure by causing constriction of the smooth muscle in arterioles. Oxytocin stimulates uterine contraction after birth and elicits milk ejection in lactating females. The hormones are synthesized by neurosecretory cells whose cell bodies lie in the hypothalamus, and they are released from the nerve terminals in the pituitary into the bloodstream. That these specific effects are exerted in the body rather than in the central nervous system is indicated by experiments in which the stalk of the posterior pituitary is severed, preventing the release of AVP and oxytocin into the bloodstream without interfering with release of the hormones into the brain. The most immediate and obvious effects are loss of control of water resorption in the kidney and loss of milk ejection in response to suckling.

Both hormones also have behavioral effects mediated by their action on the brain, where they may influence sexually dimorphic behavior. For example, oxytocin helps to elicit maternal behavior via oxytocin-secreting neurons present in the medial preoptic area of the hypothalamus and elsewhere. Vasopressin, on the other hand, decreases female sexual receptivity, apparently by inhibiting the lordosis response in mating behavior. Both hormones also influence learning and memory, via pathways that have not yet been fully identified. Oxytocin, for example, can interfere with memory formation, and vasopressin facilitates conditioning. The variety of effects of vasopressin are mediated in part by specific receptors in different cells. Only one type of receptor has so far been identified for oxytocin.

Many hormones are peptides secreted by the pituitary gland. These hormones may play a traditional role in the control of bodily processes. They may also be released in the brain, where they may help control sexually dimorphic behavior, or may play roles in other aspects of an animal's behavior.

Molecular Mechanisms of Vertebrate Hormone Action

A hormone can exert an effect on a nerve cell only if its presence initiates some change in the cell's activity, be it metabolic, electrical, or both. Any change must start with the binding of the hormone with an appropriate receptor.

Mode of Action of Steroid Hormones

Traditionally, vertebrate steroids like testosterone and estrogen have been thought to interact with neurons exclusively via nuclear receptor proteins, in a way remarkably similar to that described in a previous section for ecdysone. After entering a neuron, a steroid hormone binds with one or more members of a family of nuclear binding proteins. The hormone-receptor complex acts as a transcription factor in the nucleus, initiating transcription of DNA and leading to protein synthesis, as shown for ecdysone in Figure 25-6. The proteins that are synthesized may play any of several roles in the cell, from changing the cell's metabolic activity to regulating the expression of other genes, either activating or suppressing further transcription.

In some cases, however, the response of an individual cell to a steroid is so fast (on the order of a few minutes) that it is unlikely to require the synthesis of proteins. For example, estrogen increases the electrical excitability of specific neurons in the amygdala within 2 minutes of its application to cortical brain slices from this region. Furthermore, it does so even in the presence of inhibitors of protein synthesis and after elimination of all synaptic input, indicating that this effect is mediated by nongenomic mechanisms, that is, by mechanisms that do not involve activation of genes and the synthesis of new proteins.

Considerable research effort has been expended in the past several decades to identify the mechanisms of nongenomic steroid effects. One possibility is that the hormone binds with one or more receptors within the cell, and the receptor-steroid complex then initiates some biochemical response. An alternative possibility that has gained considerable support is that some receptors for steroid hormones may be bound to the cell membrane, just as are the receptors for neurotransmitters, neuromodulators, neurotrophic factors, and peptide hormones. Not all researchers are yet convinced of the presence of such receptors, but evidence is beginning to persuade many.

Traditionally, steroid hormones have been thought to exert their effects on neurons by binding with intracellular hormone receptors in the nucleus and activating one or more genes to change the metabolic activity or electrical responsiveness of the cell. However, evidence is accumulating that some steroid effects are mediated by membrane-bound receptors.

Mode of Action of Peptide Hormones

Peptide hormones, being insoluble in lipid, cannot readily cross the cell membrane. Consequently, they must exert their effects by binding with receptors in the cell membrane. All receptors for peptide hormones are coupled to G proteins and belong to a single receptor superfamily characterized by seven transmembrane domains. Activation of a receptor by the presence of the hormone for which it is specific activates the associated G protein and initiates a biochemical cascade that causes an effect in the neuron. The effect on the neuron may range from a local change of its electrical or transmission properties to a long-term effect mediated by activation of specific genes in the cell nucleus. You have probably noticed that this mechanism is the same as the mechanism of action of neuromodulatory neurotransmitters; this similarity serves to emphasize that in many cases there is no distinction between a hormone and a neuromodulator; a single molecule may serve both functions.

The precise effects of a particular hormone on a neuron will vary from hormone to hormone, depending on the type of G protein to which the receptor for that hormone is coupled and the biochemical pathway that is consequently engaged. All of the pathways described in Chapter 6 for neuromodulatory neurotransmitters have been described for various types of hormones. These include activation of adenylyl cyclase and an increase in cAMP activity, inhibition of adenylyl cyclase and a decrease in cAMP activity, activation of the inositol triphosphate (IP_3) pathway, and others. Some of these pathways operate via cAMP-dependent protein kinase A, some via phospholipid-dependent protein kinase C, and yet others via changes in intracellular Ca^{2+} levels.

Although the bewildering array of hormonal receptor types, G proteins, and second messenger systems that can be found in the nervous system can be confusing to researchers as well as to students, scientists are beginning to understand how these elements are used to control physiological processes in neurons and other cells. Take vasopressin as an example. At present, three types of receptors for AVP have been identified: V_{1a}, V_{1b}, and V_2. Receptors of the V_1 type are coupled with G proteins that activate adenylyl cyclase and the protein kinase A system. V_{1a} receptors are found in blood vessels and mediate effects on blood pressure. V_{1b} receptors are found in the brain and mediate release of other hormones. Receptors of the V_2 type are coupled with G proteins that activate phospholipase C and the protein kinase C system; they are found in the kidney and mediate the antidiuretic action of AVP.

It remains to be seen which specific receptor type mediates some of the behavioral effects of the hormone, and whether there are additional receptors or subtypes. It is already clear, however, that just as is the case for neurotransmitters, it is the nature and properties of the receptor for a hormone that will determine a cell's response to that hormone. Hence, a single hormone may have entirely unrelated effects on different cells. The effects of a hormone (or neuromodulator) on a cell will be determined by the types of receptors that are present in its membrane. This makes it possible for cells to turn sensitivity to a hormone on and off by expressing or destroying the receptors for that hormone. There is already molecular evidence that cells do this; for example, the distribution of receptors for some hormones varies at different stages of an animal's reproductive cycle.

Peptide hormones act via membrane-bound receptors and a biochemical cascade initiated by a specific type of G protein, mediated by second messenger molecules to initiate cell changes. Any given hormone may bind with several different receptors. Each receptor may initiate specific second messenger pathways, and hence result in different cellular responses. The response of a cell to a hormone is regulated in part by the types of receptor it expresses and when it expresses them.

Hormones and the Nervous System: A Paradigm for Neurobiology

It is entirely fitting that a chapter describing the effects of hormones on the nervous system should close this book. In many ways, neuroendocrinology is like neurobiology in miniature, reflecting every important theme found in the field as a whole, from the level of molecules to that of behavior. Furthermore, knowledge of neuroendocrinological mechanisms also has enormous practical implications for human medicine and for the human condition itself, just as is true for many other aspects of the field of neurobiology. For these reasons, it seems useful to mention the main points of similarity here as a way to draw together some of the common themes that run through neurobiology, and hence through this book.

You have learned that, at the molecular level, neurons regulate their responses to hormones through receptor proteins that interact directly with the cell's genome, or that are coupled to a second messenger system that triggers a complex biochemical cascade. The effects range from short-term changes in a neuron's electrical activity to the synthesis of new proteins that can more or less permanently alter a neuron's function. The connection to the broader field of neurobiology is that the second messenger cascades activated by hormones are the very same cascades used by neurons to respond to neuromodulatory neurotransmitters and neurotrophic molecules, to transduce sensory signals, and to modulate their own properties on the basis of the animal's experience. These biochemical signaling cascades hence constitute a fundamental property of neurons that allow them to process information.

You have also learned that hormones act at the level of neural circuits by turning on or off specific behavior, or by sensitizing or suppressing an animal's response to some stimulus. The connection to neurobiology generally is that neural circuits are the basis of all behavior, from a simple reflex withdrawal away from a noxious stimulus to a complex mating dance. Animals are able to perform integrated actions because their nervous systems are organized into neural circuits that coordinate these actions. As exemplified by the effects of hormones, these circuits can be turned on or off, and can even be blended to achieve new combinations of movements, making behavior adaptable to circumstances and enhancing an animal's chances for survival.

The effects of hormones on neurons at the molecular level are translated into an effect on a neural circuit, and from there into an integrated behavior that involves the entire animal. Similarly, any aspect of neural function starts at the molecular level and works through neural circuits to affect the whole animal. For example, biologically significant sensory input such as the sound of a predator will influence the behavior of the animal by a series of events that start with transduction at the molecular level in a sensory neuron, producing activation of a neural circuit whose action is modulated by the conditions in which the animal finds itself, and ending with an adaptive behavior involving the whole animal.

Even from the standpoint of the impact of neuroscience on human affairs, neuroendocrinology stands as an example for neurobiology as a whole. The practical ability to diagnose and treat medical conditions associated with hormonal imbalance, the enormous potential of hormone treatments for memory failure associated with old age, to say nothing of the social impact of research on the biological basis of human sexuality, are all of great significance. But better medical diagnosis and treatment, improvements in the human condition, and the social impact of neurobiological research are significant consequences of virtually all neurobiological research, from dealing with spinal injury or mental disease to the implications of human split-brain procedures, as you have learned in chapters throughout this book.

Any subject must be introduced one part at a time. Having now completed your introduction to neurobiology, think about the story of the blind wise men and the elephant. Each man, touching a different part of the animal, thought that an elephant was something different. The man touching its side claimed it was large and solid, the man handling its ear said it was thin and pliable, the man holding its trunk said it was long and flexible like a snake, and so on. Neurobiology can be like that. Some researchers emphasize molecular biochemistry, some emphasize cellular electrical properties, and some emphasize behavior. But in the end, neurobiology is an integrated whole that can describe how any animal lives and how it adapts to its environment. As you continue your study of neurobiology, you most likely will forget some details, so concentrate on the principles of neural organization that you have learned. It is these principles that will enable you better to understand the breakthroughs of the future as you read about them in years to come.

The subfield of neuroendocrinology reflects the important themes of neurobiology as a whole, from the molecular mechanisms of signaling in neurons and the modulation of neural circuits, to the integration of whole behavior. It also exemplifies the impact of neurobiology on human affairs, from its medical importance to its significance for human thought and social policy.

Additional Reading

General

Becker, J. B., S. M. Breedlove, and D. Crews, eds. 1992. *Behavioral Endocrinology*. Cambridge, Mass.: MIT Press.

Brown, R. E. 1994. *An Introduction to Neuroendocrinology*. Cambridge: Cambridge University Press.

Griffin, J. E., and S. R. Ojeda. 1992. *Textbook of Endocrine Physiology*. 2d ed. New York: Oxford University Press.

Research Articles and Reviews

Breedlove, S. M. 1992. Sexual dimorphism in the vertebrate nervous system. *Journal of Neuroscience* 12:4133–42.

Emson, P.C. 1993. *In-situ* hybridization as a methodological tool for the neuroscientist. *Trends in Neurosciences* 16:9–16.

Nottebohm, F. 1991. Reassessing the mechanisms and origins of vocal learning in birds. *Trends in Neurosciences* 14:206–11.

Pfaff, D. W., S. Schwartz-Giblin, M. M. McCarthy, and L.-M. Kow. 1994. Cellular and molecular mechanisms of female reproductive behaviors. In *The Physiology of Reproduction*, vol. 2, ed. E. Knobil and J. D. Neill, 107–220. New York: Raven Press.

Stout, J., V. Hayes, D. Zacharias, J. Henley, A. Stumpner, J. Hao, and G. Atkins. 1992. Juvenile hormone controls phonotactic responsiveness of female crickets by genetic regulation of the response properties of identified auditory interneurons. In *Insect Juvenile Hormone Research: Fundamentals and Applied Approaches*, ed. B. Mauchamp, F. Couillaud, and J. C. Baehr, 265–83. Paris: Institut National de la Recherche Agonomique.

Truman, J. W. 1992. The eclosion system of insects. *Progress in Brain Research* 92:361–74.

APPENDIX

Terms of Orientation

Over the years, biologists have developed a number of terms that describe the locations of parts of the body relative to other parts. These terms are based on the planes of three-dimensional space, as illustrated in Figure A-1. Each plane has a specific name, as shown. Furthermore, each plane defines one or more directional terms. To understand what this means, consider a globe of the world. Just as the plane established by the equator defines the directions of north and south, so too do each of the imaginary planes intersecting an animal's body define one or more sets of directions. The terms are shown in Figure A-1 and are listed and defined in Table A-1.

Note especially that just as the terms north and south are relative (you can head south even from near the South Pole), so too are the directional terms used on the body. For example, it is quite correct to say that the nerve cord of a dog is ventral to the skin of its back, even though the nerve cord does not lie in the lower half of the dog's body. Since these terms are relative, they cannot be used to describe an absolute location. A statement to

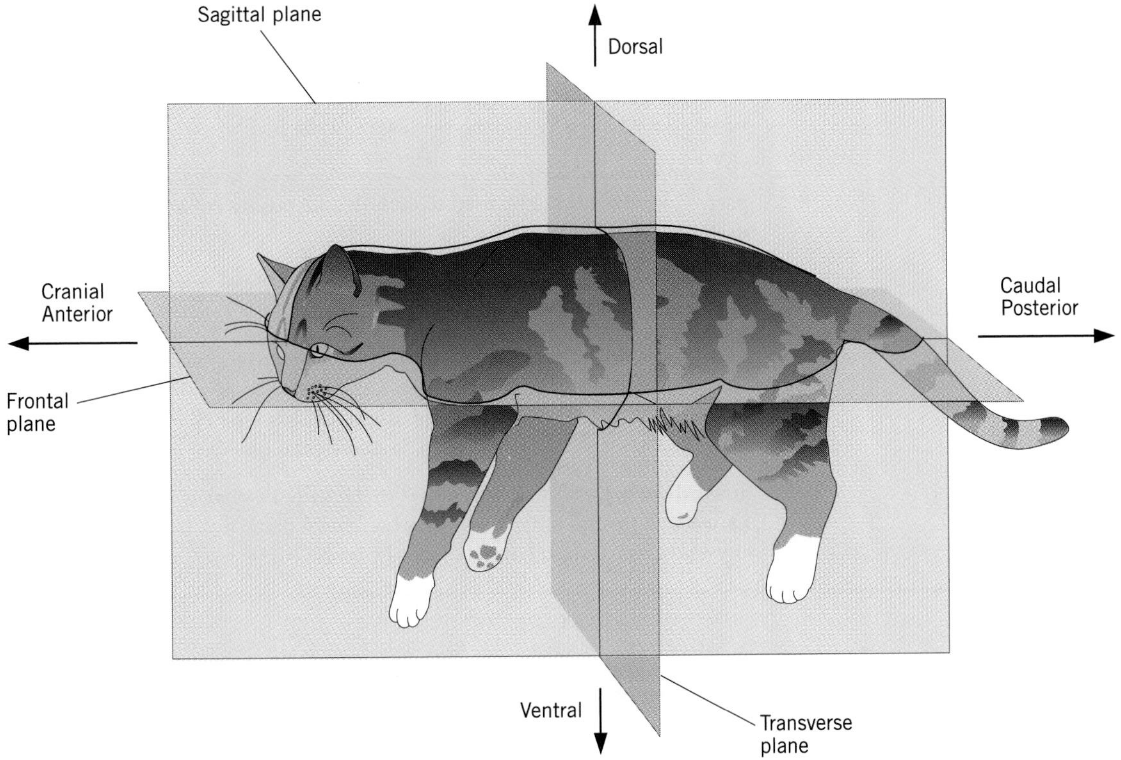

FIGURE A-1. The planes of three-dimensional space and the directions along the body that they define. See also the definitions listed in Table A-1.

TABLE A-1. Terms of Relative Anatomical Position

Term (Antonym)	Meaning
Anterior (posterior)	Toward the front of the body; closer to the front of the body than some other part. Anterior and posterior are defined by a transverse plane that may be positioned anywhere along the length of the body.
Caudal (rostral)	Toward the tail of the animal; closer to the tail than some other part. This term is effectively synonymous with posterior.
Distal (proximal)	In an appendage, farther along the appendage away from the body than some other part. This term cannot be used to describe the position of a body part by itself.
Dorsal (ventral)	Toward the back of the body; closer to the back than some other part. Dorsal and ventral are defined by a horizontal plane that may be positioned anywhere from top to bottom along the body.
Lateral (medial)	In the body (but not in appendages), farther from the midline of the body than some other part. This term cannot be used to describe the position of a body part by itself.
Left (right)	Toward the side of the body that is oriented toward the west if the animal is facing north. The term is not used in a comparative way; the terms lateral and medial are used comparatively. Right and left are defined by a sagittal plane that bisects the body exactly (a midsagittal plane).
Medial (lateral)	In the body (but not in appendages), closer to the midline of the body than some other part. This term cannot be used to describe the position of a body part by itself.
Posterior (anterior)	Toward the tail end of the body; closer to the tail end of the body than some other part. Anterior and posterior are defined by a transverse plane that may be positioned anywhere along the length of the body.
Proximal (distal)	In an appendage, along the appendage closer to the body than some other part. This term cannot be used to describe the position of a body part by itself.
Rostral (caudal)	Toward the head of the animal; closer to the head than some other part. This term is effectively synonymous with anterior.
Right (left)	Toward the side of the body that is oriented toward the east if the animal is facing north. The term is not used in a comparative way; the terms lateral and medial are used comparatively. Right and left are defined by a sagittal plane that bisects the body exactly (a midsagittal plane).
Ventral (dorsal)	Toward the belly of the body; closer to the belly than some other part. Dorsal and ventral are defined by a horizontal plane that may be positioned anywhere from top to bottom along the body.

the effect that the nerve cord in a vertebrate is dorsal and that in an invertebrate is ventral is, strictly speaking, not correct. What is missing is the unwritten or unspoken corollary that these structures lie dorsal or ventral to most other structures of the body.

Two pairs of terms in Table A-1 have somewhat more restricted meanings than do the rest. These are the terms *medial* and *lateral*, and *proximal* and *distal*. Medial and lateral refer to the position of parts of the body relative to one another, but in reference to the

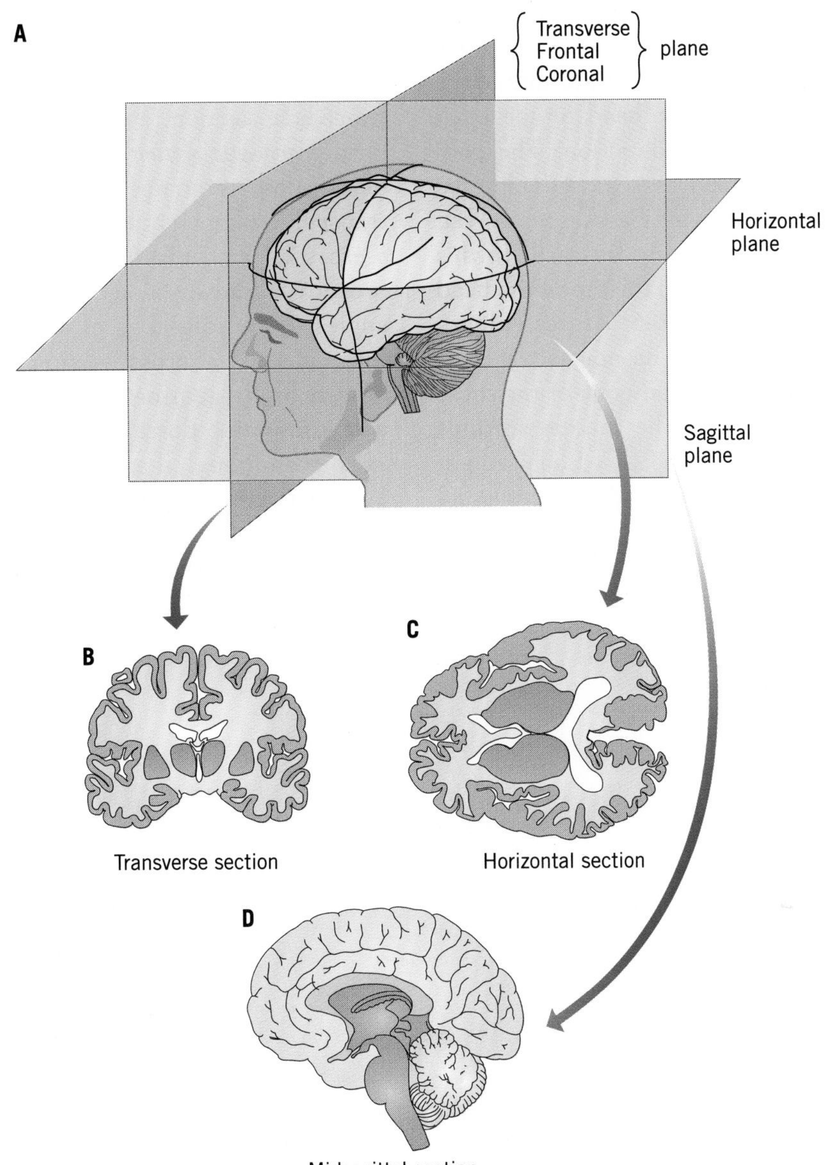

FIGURE A-2. (A) The planes of three-dimensional space as applied to the human head. Note that the transverse plane can also be called the frontal plane or the coronal plane. (B) A midsagittal section through a human brain. (C) A transverse (= frontal, = coronal) section through a human brain. (D) A horizontal section through a human brain.

midline of the body. Medial means closer to the midline of the body, and lateral means farther away from the midline. For example, you can say that the temporal lobe of the brain lies lateral to the thalamus. Unlike terms such as anterior and posterior, each of which refers to opposite directions away from the defining plane, the terms medial and lateral are symmetrical. Your left shoulder is lateral to your heart, but so is your right shoulder.

The terms proximal and distal also have a special use: they are applied to position along an appendage. Structures that are farther away from the body are said to be distal to those closer in, and those that are closer to the body are proximal to those farther out. Like other directional terms, proximal and distal are relative. For example, your elbow lies distal to your shoulder but proximal to your wrist.

Terms of orientation as applied to the human body can be confusing because humans stand upright. Not all authors agree on whether the top of the head or the back should define the dorsal direction. The convention in this book will be to treat the human body as if it were bent over at the waist, but with the head up and facing forward. In this position, both the back and the top of the head are dorsal relative to other structures (Figure A-2A). With the head in its normal orientation, slicing a brain along any one of the three planes will yield one of the unique sections illustrated in Figure A-2B, C, D. Each section is identified by the name of the plane along which the brain was sliced to prepare it. Note that sections along the transverse plane can also be called frontal or coronal sections.

You should be familiar with one additional common term for a section, the *cross section*. A cross section is usually synonymous with a transverse section, but not always so. The term *transverse section* is usually a technical description, referring to a section along the transverse plane as defined above. However, a cross section is frequently defined less rigorously to mean a section taken at right angles to the long axis of an elongate structure. Hence, a cross section of either the human brain stem or the human spinal cord will show the dorsal/ventral and right/left axes, even though if a person is standing bent over at the waist but looking straight ahead, as described earlier, the brain stem will be oriented nearly vertically and the spinal cord will be horizontal.

GLOSSARY

Note: Numbers in parentheses after each entry indicate the chapter in which the term or phrase is first introduced. Acronyms are defined under the full term or phrase given for them. Synonyms are defined under the most commonly used term.

– A –

Absolute refractory period (5) A brief period immediately following an action potential during which no new action potential can be generated

Acetylcholinesterase (AChE) (6) The enzyme that inactivates acetylcholine at synapses

AChE (6) See acetylcholinesterase

Acoustic interneuron (19) A neuron in the central nervous system that carries acoustic (sound) information

Across-fiber code (10) A type of coding for sensory information in which the collective pattern of response of many sensory neurons in a sense organ is used to convey information about stimulus quality; also called a population code

Action potential (5) A transient, all-or-none reversal of membrane potential that sweeps along the membrane of a nerve or muscle cell

Activation gate (5) One of the two gates of a sodium channel; the gate opens when the membrane depolarizes, allowing sodium ions to pass through the open channel

Active electric sense (14) The electric sense in animals that are electrogenic, meaning they can generate electrical fields of their own

Active transport (2) The movement of ions or other molecules across a cell membrane against a concentration gradient of the molecule, a process that requires energy in the form of ATP

Active zone (7) The region in a presynaptic neuron in which synaptic vesicles are stored

Activity-dependent development (23) A type of neural development that requires neural activity or external input for its proper expression

Activity-independent development (23) A type of neural development that is intrinsic to the nervous system, and thus does not require neural activity or external input

Adaptation (9) The decline in the amplitude of the response of a sensory neuron in the presence of a constant stimulus

Adrenergic neuron (7) A neuron that releases epinephrine (adrenaline) as a neurotransmitter

Afferent fiber (3) Any neurite (axon or dendrite) that carries information toward the central nervous system

Afferent nerve (3) A nerve (a bundle of neurites) consisting entirely of afferent fibers

All or none (5) The characterization of an action potential; the potential either reaches its full amplitude or does not occur at all

Alzheimer's disease (21) A degenerative disease of old age characterized initially by loss of memory, followed by mental disorientation and loss of control of bodily functions; associated with death of cholinergic neurons in many regions of the brain

Amacrine cell (11) A type of cell found in the vertebrate retina. Amacrine cells receive input from bipolar cells and connect with ganglion cells and other amacrine cells

AMPA receptor (8) A type of ionotropic glutamate receptor that also responds to α-amino-3-hydroxy-5-methyl-4-isoxazole-proprionic acid. The associated channel allows passage of K^+ and Na^+, but no significant amount of Ca^{2+}

Ampulla (14) In vertebrates, the enlargement near the base of each semicircular canal in the inner ear

Ampullary organ (14) One of the two types of sensory receptors of vertebrates that have an electric sense. Ampullary organs are embedded in the skin but open to the surface through a canal; they are specialized to detect weak electrical signals from other animals

Anaxonal neuron (2) A neuron that has no neurites or extremely short ones. Neurons of this type seem to occur exclusively in sensory systems, such as the vertebrate visual and auditory systems

Angular gyrus (21) In vertebrates, a posterior region of the parietal lobe of the brain that integrates auditory input with input from other senses

Antagonistic muscles (15) Muscles acting in opposition to one another around a single joint, or otherwise having opposite actions when they contract

Anterior lobe (18) In mammals, one of the three physical subdivisions of the cerebellum

Anterograde transport (2) Axonal transport of materials from the cell body toward the periphery in neurites. There are several speeds of anterograde transport

Anteroventral periventricular nucleus (25) In mammals, a nucleus of the preoptic area of the hypothalamus that in rats is larger in females than in males

Apoptosis (22) A form of programmed cell death in which the cell commits suicide by activating genes that orchestrate the fragmentation of the cell's DNA and rupture of the nuclear membrane

Association neuron (16) In a simple reflex arc, the interneuron interposed between the sensory neuron and the motor neuron

Associative conditioning (24) A form of complex learning in which an animal forms an association between a stimulus that normally evokes little or no response (the conditioned stimulus) and one that normally evokes a reflex response (the unconditioned stimulus). The association is formed as a consequence of the paired presentation of the two stimuli

Associative LTP (8) See heterosynaptic LTP

Astrocyte (2) An astroglial cell; see astroglia

Astroglia (2) A type of glia somewhat starlike in shape

Arginine vasopressin (AVP) (25) The full name for vasopressin; also known as antidiuretic hormone (ADH)

Autonomic nervous system (3) In vertebrates, strictly defined, the peripheral nerves and ganglia that help to regulate autonomic (involuntary) functions. Often used to include centers in the brain such as the hypothalamus that also regulate autonomic functions

Autoradiography (23) A technique in which a radioactively labeled substance is given to an animal, and the distribution of the substance is subsequently determined by exposing photographic emulsion or film to tissue from the animal

AVP (25) See arginine vasopressin

Axon (2) In a neuron, the neurite that conveys information away from the cell body

Axonal transport (2) The process by which materials in neurons are moved along neurites

Axoplasmic transport (2) See axonal transport

— B —

Basal ganglia (3) In vertebrates, a group of nuclei mostly in the forebrain that are involved in motor control, consisting of the caudate, putamen, globus pallidus, subthalamic nucleus, and the substantia nigra

Basilar membrane (12) A membrane in the cochlea on which the hair cells are located; sound transmitted to the cochlea causes movement of the basilar membrane

BDNF (22) See brain derived neurotrophic factor

Best excitatory frequency (12) The frequency to which an auditory cell will respond at the lowest stimulus strength

Binding protein (2) A type of membrane-associated protein that helps one cell adhere to another or to an extracellular matrix

Binocular vision (23) See stereoscopic vision

Biofeedback (15) Learning to control a bodily process by observing some external effect that it has

Bipolar cell (11) A type of cell found in the vertebrate retina. Bipolar cells connect the outer plexiform layer with the inner plexiform layer

Bipolar neuron (2) A neuron with two neurites, one extending from each end of the cell body. This structure is typical of many invertebrate sensory neurons

Blob (11) In the mammalian visual cortex, a column of cells that extends through layers 1 to 3 and 5 to 6 that stains strongly for the presence of cytochrome oxidase; these columns appear as spots ("blobs") when cut across

Blood-brain barrier (3) The barrier that surrounds the brain and spinal cord, keeping them chemically isolated from the rest of the body

Brain derived neurotrophic factor (BDNF) (22) A member of the neurotrophin family of trophic growth factors that influences the growth and development of specific types of neurons

Brain slice technique (24) A method for removing and recording from thin slabs of tissue from a vertebrate brain

Brain stem (3) The collective term for the medulla, pons, and midbrain. The brain stem controls the most vital body functions

Broad generalist (13) A type of olfactory receptor in insects that responds to a wide variety of odorants

Broca's area (21) In humans, a part of the frontal lobe of the brain that is important in generating grammatical speech

–C–

Cadherin family (22) A family of guidance molecules that help determine the direction of growth of developing axons. Cadherin family members are cell adhesion molecules

CAM (22) See cell adhesion molecule

Capacitance (5) An electrical term that refers to the ability of a cell membrane to store electrical charge

CAT scan (21) A brain scan taken by computed axial tomography

Catecholaminergic neuron (7) A neuron that releases a catecholamine as neurotransmitter

Caudate nucleus (3) In vertebrates, one of the basal ganglia

Celiac ganglion (3) In vertebrates, a sympathetic autonomic ganglion that lies in the body cavity and innervates the upper intestines

Cell adhesion molecule (CAM) (22) A type of membrane-bound molecule on developing neurons that binds with receptors or other cell adhesion molecules and thereby promotes growth of the axon in particular directions in a developing embryo

Cell theory (2) The idea that cells are the smallest living units of all life, proposed in 1838–39 by Schleiden and Schwann

Center (11) In the vertebrate retina, the middle part of a circular receptive field of a bipolar cell or ganglion cell

Central canal (3) In vertebrates, the small fluid-filled tube in the center of the spinal cord

Central gray (25) In vertebrates, a region of the mesencephalon that surrounds the cerebral aqueduct and lies medial and ventral to the colliculi

Central nervous system (CNS) (3) The main mass of neural tissue of an animal. In vertebrates, the brain and spinal cord

Central pattern generator (CPG) (16) A group of neurons in the central nervous system that, due to the properties of the neurons and the connections between them, can produce a specific pattern (usually rhythmic) of output in the absence of sensory feedback

Central sulcus (3) The major sulcus of the mammalian brain. In humans, it runs from about the top of the head out and down toward the temporal lobe on both the right and left sides

Centrifugal control (10) Functionally the same as efferent control; sometimes used to mean modulation of sensory signals within the central nervous system

Cerci (12) Hair-covered, peg-like structures located at the end of the abdomen in some insects

Cerebellum (3) In vertebrates, the part of the hindbrain that sits atop the brain stem and helps to coordinate movement

Cerebral aqueduct (3) In vertebrates, the narrow channel in the midbrain that connects the third and fourth ventricles

Cerebral cortex (3) In vertebrates, the cortex of the cerebral hemispheres

Cerebral hemispheres (3) In vertebrates, the paired lobes that sit on top of the rest of the brain, collectively called the cerebrum. They are responsible for higher mental functions, sensory processing, and motor control

Cerebral peduncle (3) In vertebrates, one of a pair of thick bundles of axons that connect the cerebellum with the rest of the brain

Cerebrocerebellum (18) In vertebrates, the functional subdivision of the cerebellum that makes extensive connections with the cerebral hemispheres

Cerebrospinal fluid (CSF) (3) In vertebrates, the fluid that fills the ventricles and the central canal

Cerebrum (3) In vertebrates, the collective term for the cerebral hemispheres

Characteristic frequency (12) The best excitatory frequency of an auditory neuron

Chemical synapse (2) The type of synapse in which communication is by chemical synaptic transmission

Chemical (synaptic) transmission (2) The transfer of information from one neuron to another across a synapse via a chemical intermediary called a neurotransmitter

Chemoaffinity hypothesis (22) In neural development, the idea that developing neurons synthesize specific chemicals that serve as attractants for developing neurons and hence allow proper synaptic connections to be formed

Chemoreceptor (9) A sensory receptor for any chemical substance

Cholinergic neuron (7) A neuron that releases acetylcholine as a neurotransmitter

Ciliary neurotrophic factor (CNTF) (22) A trophic factor produced by Schwann cells and some other peripheral tissues that acts as a growth factor on selected neurons

Ciliary neurotrophic factor family (22) A family of trophic factors that help promote neural growth, differentiation, and in some cases, repair after injury

Cingulate gyrus (3) In vertebrates, a part of the cerebral hemispheres that lies above the corpus callosum, along the medial face of the parietal and frontal lobes; part of the limbic system

Climbing fiber (3) In vertebrates, one of two types of fibers (axons) that carry information into the cerebellum from the brain

CNS (3) See central nervous system

CNTF (22) See ciliary neurotrophic factor

Cochlea (12) In vertebrates, the spiral-shaped part of the inner ear within which the transduction of sound energy into neural activity takes place

Cochlear nerve (12) In vertebrates, a part of cranial nerve VIII that carries auditory information and terminates in the dorsal and ventral cochlear nuclei in the medulla

Coding (10) The means by which information about sensory quality or strength is conveyed to the central nervous system

Cognitive neuroscience (21) The study of complex, so-called higher mental functions in vertebrates, especially mammals

Columnar organization (9) In vertebrates, the organization of neurons in a sensory cortex into functionally distinct columns of cells

Command neuron (18) One of a pair of neurons that are both necessary and sufficient to control the onset, duration, and vigor of a complete behavior

Complex cell (11) In the visual cortex of mammals, a type of neuron that is excited by a properly oriented edge or corner, often requiring movement

Compound eye (11) A type of eye that has multiple lenses, each lens directing light to one or a few photoreceptive elements. Common in arthropods, but also found in a few annelids

Computational map (12) A topographical map in a sensory region of the brain that is formed from the precise and systematic convergence of two inputs that represent separate parameters, projected onto an array of neurons that forms the map

Computational neuroscience (20) An area of neuroscience in which computational methods are used to model neural function or computational approaches are used to guide thinking about neural function

Computed (axial) tomography (21) A technique for generating an image of a part of a living brain. The method involves taking x-ray pictures from many different angles through a portion of the brain and reconstructing the densities of the tissue by computer analysis of the pictures

Conditioned stimulus (CS) (24) In associative conditioning training, a stimulus that evokes a simple reflex response after it has been paired with an unconditioned stimulus

Conditioning (24) See associative conditioning

Conductance (5) The reciprocal of electrical resistance, and hence a measure of ion mobility across a cell membrane. The higher the conductance of the membrane to an ion, the more easily the ion can cross the membrane

Cone cell (11) A type of receptor cell in the vertebrate retina. The outer segments of cone cells are cone shaped; cones subserve color vision

Confocal microscope (22) A type of fluorescence microscope in which the light used to excite the fluorescence and the light from the image are passed through pinholes to allow sharp focus on just one plane of the specimen

Conformational change (2) A change in the structure of a protein. In gated proteins, a conformational change will typically widen or narrow the channel or move an obstructing part of the protein away from or into the channel, hence opening or closing the channel for the passage of ions

Connective (3) In invertebrates, a relatively large nerve trunk consisting of axons that connect one ganglion of the central nervous system to another

Connexin (2) The structural protein six copies of which makes up one connexon

Connexon (2) The type of ion channel that forms gap junctions (electrical synapses)

Constant field equation (4) See the Goldman equation

Corpus callosum (3) In vertebrates, the broad band of neural tissue that physically and functionally binds together the two cerebral hemispheres

Corpus striatum (3) In vertebrates, the caudate nucleus, putamen, and globus pallidus (paleostriatum); part of the basal ganglia

Cortex (3) In vertebrates, the outer, gray layer of any part of the brain, containing synapses, dendrites, and the cell bodies of neurons; often used to mean just the cortex of the cerebral hemispheres

Corticospinal tract (3) In vertebrates, a bundle of axons that run from the cortex to a destination in the spinal cord

CPG (16) See central pattern generator

Cranial nerve (3) In vertebrates, one of the 12 peripheral nerves that connect the brain with other parts of the body

Critical period (23) During development, a time during which normal sensory input must be present in order for normal development to proceed

Crossed extension reflex (16) The excitation of extensor muscles in a limb on the other side of the body from a limb undergoing reflex flexion

Crystalline cone (11) In insects, the part of the eye that serves as a lens to focus light onto the sensory neurons

CS (24) See conditioned stimulus

CSF (3) See cerebrospinal fluid

CT scan (21) A brain scan taken by computed (axial) tomography

Cuneate nucleus (14) In vertebrates, a nucleus of the medulla that is part of the lemniscal pathway, which carries somatosensory information such as touch, vibration, and kinesthetic information

Cupula (10) A gelatinous matrix in which the hair-like projections of receptor cells in neuromast organs of fish are embedded

Cupula (14) In the vertebrate semicircular canal, the gelatinous mass that sits over the hair cells, and whose movement causes activity in the vestibular neurons

— D —

Deafferentation (16) Elimination of sensory input to all of or a restricted region of the central nervous system, usually to determine the role such input plays in behavior

Decerebrate rigidity (18) A characteristic posture of a vertebrate after the brain stem has been severed between the red nucleus and the other motor nuclei. It is marked by unusually strong activation of the main extensor muscles in the parts of the limbs closest to the body

Declarative memory (24) The memory of facts, events, and the relationships between objects and events with which an animal has had experience

Dendrite (2) In a neuron, the neurite that conveys information toward the cell body; most neurons have several dendrites

Dense-core vesicle (2) A type of synaptic vesicle that appears solid in electron micrographs; characteristic of neurons that secrete hormones or that release some peptides

Depolarization (5) A change of the membrane potential of a cell such that the interior of the cell becomes less negative. Usually a shift of the potential toward zero

Desensitization (6) A slow structural change that renders a receptor protein insensitive to its neurotransmitter

Determinate development (22) Development in which genetic and cytoplasmic factors predominate in determining the fate of each cell; also called mosaic development

Developmental neurobiology (22) The subfield of neurobiology that deals with the development of the nervous system, including the mechanisms by which growing axons find their targets and the formation of proper synapses between neurons

Diencephalon (3) In vertebrates, the part of the forebrain that is adjacent to the midbrain. It contains the thalamus and the hypothalamus

Directional tuning curve (18) A graph showing the response of a single motor cortical neuron during voluntary movement of an arm in different directions

Dishabituation (24) The full recovery of the original strength of a habituated response after the presentation of some strong, novel stimulus

Dorsal cochlear nucleus (12) In vertebrates, a nucleus in the medulla in which some of the axons of the cochlear nerve terminate

Dorsal column nuclei (14) In vertebrates, the collective name for the cuneate and gracile nuclei in the medulla

Dorsal root ganglion (3) In vertebrates, one of a series of ganglia that lie just outside the spinal column and contain the cell bodies of sensory neurons that send their axons into the dorsal horn of the spinal cord

Dynein (2) A protein involved in retrograde axonal transport. Dynein transiently binds the material being transported to neurotubules in such a way that the material moves back toward the cell body

— E —

Ecdysone (25) In insects, a steroid hormone that controls the growth of new cuticle and other developmental changes necessary before an insect can molt during the molting cycle

Eclosion hormone (EH) (25) In insects, a peptide hormone that induces the molt of the insect to its next developmental stage during the molting cycle

Efferent control (10) The influence over or regulation of a sense organ or sensory signals by the central nervous system

Efferent fiber (3) Any neurite (axon or dendrite) that carries information away from the central nervous system

Efferent nerve (3) A nerve (a bundle of neurites) consisting entirely of efferent fibers

Efflux (4) The flow of ions out of a cell

EH (25) See eclosion hormone

Electric organ discharge (20) The electrical signal generated by weakly electric fish

Electrical synapse (2) The type of synapse in which communication is by electrical synaptic transmission

Electrical (synaptic) transmission (2) The transfer of information between neurons via direct flow of electrical current from one neuron to another

Electrical tuning frequency (12) For the hair cells of the mammalian cochlea, the frequency at which the membrane potential oscillates when the cell is depolarized

Electroantennogram (13) The electrical potential that can be recorded from the antenna of an insect when the antenna is exposed to various odorants

Electrocommunication (14) The use of a weak electrical field generated by one animal for communication with another animal of the same species

Electrode (4) A small probe capable of conducting electricity; used to detect an electrical potential in living tissue

Electrogenic fish (14) Fish that are capable of generating an electrical field of their own as well as detecting fields from outside themselves

Electrolocation (14) The use by an animal of a weak electrical field for navigation or orientation

Electromyographic recording (EMG) (15) The technique of electromyography

Electromyography (15) Recording the electrical activity of muscle by placing extracellular electrodes on or near the muscle

Electroreception (14) The direct detection of electrical fields by animals

Electroreceptor (9) A sensory receptor for electrical current typically found in the skin of electric fish

Electrosensory lateral line lobe (ELL) (14) In electric fish, a specialized region of the medulla in which the sensory afferents from electroreceptors typically terminate

ELL (14) See electrosensory lateral line lobe

Emergent property (11) A feature of one or a group of neurons that arises from the synaptic interactions of the neurons that drive that neuron or group of neurons

EMG (15) See electromyographic recording

End plate potential (epp) (6) The electrical potential generated in vertebrate muscle at the neuromuscular junction as a result of stimulation by a motor neuron

Enteric nervous system (3) In vertebrates, a division of the autonomic nervous system that consists of the ganglia of the myenteric and submucosal plexuses along the alimentary canal

Entorhinal cortex (13) In vertebrates, a part of the telencephalon located on the ventral surface of the temporal lobe; it receives olfactory information

EOD (20) See electric organ discharge

Episodic memory (24) A kind of declarative memory representing an animal's experience

Epp (6) See end plate potential

Epsp (6) See excitatory postsynaptic potential

Equilibrium potential (4) For an individual ion, the electrical potential at which the movement of the ion in one direction across a cell membrane due to the electrical potential difference across the membrane equals the movement of the ion in the other direction due to the difference in concentration

Excitation (6) Excitatory synaptic transmission

Excitatory postsynaptic potential (epsp) (6) A potential that is generated in a postsynaptic neuron by activation of an excitatory synapse, usually depolarizing

Excitatory (synaptic) transmission (6) Synaptic transmission that has a depolarizing effect on the postsynaptic neuron and in which the probability that the postsynaptic neuron will fire an action potential is increased. In neurons that cannot generate action potentials, excitation increases the amount of neurotransmitter that is released

Execution level (of motor performance) (18) The level in the motor hierarchy at which the muscles to be used in a behavior and the timing of their use is selected

Executive level (of motor performance) (18) The level in the motor hierarchy at which the behavior itself is decided upon, the vigor or speed of its performance is selected, its frequency (if it is repetitive) is set, and its duration is determined

Explicit memory (24) See declarative memory; memory of facts and events

Extensor muscle (15) A muscle that functionally increases the angle of a joint

Extracellular recording (5) Recording in which the electrodes are placed outside the cell or cells from which the recording is being made

Extrafusal muscle (10) The part of a vertebrate muscle that is responsible for generating the force of contraction

Extrapyramidal motor system (3) In vertebrates, the basal ganglia and associated regions important in motor control

— F —

Facilitation (8) An increase in the amplitude of postsynaptic responses as a result of repetitive activation of the synapse

Fasciclin I, II (22) Immunoglobulin-like glycoproteins present in some developing neurons, and which serve as markers for other developing axons

Fast axonal transport (2) Anterograde axonal transport at the rate of about 410 mm/day

Fast-twitch glycolytic muscle fiber (FG) (15) A type of fast-contracting muscle fiber that fatigues readily and has low oxidative capacity and high glycolytic capacity

Fast-twitch oxidative glycolytic muscle fiber (FOG) (15) A type of fast-contracting muscle fiber that is fatigue resistant and has high oxidative capacity and intermediate glycolytic capacity

Feedback (16) Neural signals, often sensory in origin, that go back to affect or influence the neuron or neurons that produced them or caused them to be produced

Feed forward control (20) A type of neural circuit in which a copy of the sensory signal that drives a reflex is fed to another neural center that in turn sends information on to the reflex arc

FG (15) See fast-twitch glycolytic muscle fibers

Fictive (motor) pattern (16) A motor pattern recorded in the absence of any movement of the animal, as for example from an isolated nervous system or from a paralyzed animal

Filopodia (22) In neural development, the fine, finger-like projections of a growth cone that probe the environment in search of the correct path for the growing axon to take

Flexor muscle (15) A muscle that functionally decreases the angle of a joint

Flocculonodular lobe (18) In vertebrates, the posteriormost part of the cerebellum

fMRI (21) See functional MRI

FOG (15) See fast-twitch oxidative glycolytic muscle fibers

Forebrain (3) In vertebrates, the anteriormost part of the brain, consisting of the diencephalon and the telencephalon

Fornix (3) In vertebrates, a nerve tract that lies just above the third ventricle and connects the mammillary bodies and hippocampus; part of the limbic system

Fovea (11) In the vertebrate retina, the region specialized for high visual acuity

Free nerve ending (14) A type of sensory nerve ending that may respond to mechanical, thermal, or noxious stimuli. As a mechanoreceptor it responds primarily to light touch or tickle

Frequency control (of muscle force) (15) Increasing muscle force by increasing the firing rate of the motor neurons that are active at any instant

Frontal lobe (3) In vertebrates, the anterior portion of the cerebral hemispheres

Functional MRI (fMRI) (21) Functional magnetic resonance imaging; the combination of functional imaging of brain activity with conventional MRI imaging for clear anatomical pictures

Fusion pore (7) A transient channel formed between the interior of a synaptic vesicle and the outside of the cell in which that vesicle resides during release of neurotransmitter

– G –

G protein (2) A membrane protein that when activated by a receptor initiates a biochemical cascade that has a long-term effect on the properties of a neuron, or that opens or closes an ion channel and thereby changes the membrane potential of the neuron

Gain control (17) The regulation by the nervous system of the magnitude of a reflex response relative to the strength of the stimulus that evokes it

Ganglion (3) Usually, a discrete mass of cell bodies and neurites surrounded by a connective tissue sheath. In invertebrates, the central nervous system frequently consists of a chain of ganglia joined by large trunks of nerves. In vertebrates, ganglia are always located outside the central nervous system. Also an obsolete term for a brain nucleus, now used only for the basal ganglia

Ganglion cell (11) In the vertebrate retina, the innermost type of nerve cell. Ganglion cells constitute the output cells of the retina; their axons form the optic nerve

Ganglion cell layer (11) In the vertebrate retina, the innermost layer, consisting of ganglion cells

Ganglion mother cell (22) In an insect embryo, a type of cell derived from a single neuroblast that gives rise to a pair of individual neurons or glial cells within each ganglion

Gap junction (2) A channel that provides a pathway for ions and other materials to pass directly from one neuron to the other

Gate (2) A part of an ion channel that can move so as to obstruct or open the channel to the passage of ions

Gate control theory of pain (14) The hypothesis that the reduction in the intensity of pain that occurs when the skin around an injured area is rubbed is due to inhibition of pain interneurons by inhibitory input from surrounding mechanoreceptors

Gated channel (2) An ion channel that can open or close and hence can allow or prevent the flow of ions through it

Generator potential (9) The electrical potential generated by a sufficiently strong stimulus in sensory neurons capable of generating spikes

GI (18) See giant interneuron

Giant interneuron (GI) (18) A neuron in an invertebrate in which the axon diameter is especially large, often more than 10 times larger than other axons in the nervous system

Glia (2) Non-neural cells in the nervous system

Globus pallidus (3) In vertebrates, one of the basal ganglia

Glomerulus (13) A discrete, globular tangle of densely packed dendrites and axons found in the vertebrate olfactory bulb and in invertebrate olfactory processing centers

Goldman equation (4) An equation that can be used to calculate the membrane potential of a cell given the internal and external concentrations of relevant ions and the relative permeabilities of the membrane to these ions

Golgi stain (2) A stain for making individual neurons visible in histological preparations of neural tissue

Golgi tendon organ (14) In vertebrates, a type of mechanoreceptor located at the junction of a muscle and its tendon that provides information about muscle force

Gracile nucleus (14) In vertebrates, a nucleus of the medulla that is part of the lemniscal pathway, carrying somatosensory information such as touch, vibration, and kinesthetic information

Graded (synaptic) transmission (7) Chemical synaptic transmission in which the amount of neurotransmitter released is proportional to the depolarization of the presynaptic terminal. Usually applied to synapses or neuromuscular junctions at which action potentials are not generated

Granule cell (13) In the vertebrate olfactory bulb, a class of neurons that makes synaptic contact between mitral and tufted cells outside the glomeruli

Gray communicating ramus (3) A small nerve branch through which a sympathetic postganglionic fiber from a sympathetic ganglion of the paravertebral chain re-enters the spinal nerve to travel to the innervated body organ

Gray matter (2) The regions of the vertebrate central nervous system occupied by cell bodies and unmyelinated dendrites

Growth cone (22) The enlarged, growing end of a neurite

Guanine nucleotide binding protein (2) A G protein

Guidepost cell (22) In insect development, a type of cell that differentiates from other cells at specific locations in the developing leg bud, and acts as a marker for growing axons

Gustation (13) The sense of taste

Gustducin (13) A type of G protein similar to transducin that mediates transduction in vertebrate gustatory receptors

Gyrus (3) In the mammalian brain, a band of tissue defined by infoldings (sulci) on each side

– H –

Habituation (24) The decline in a behavioral response to a stimulus after the stimulus has been presented many times

Hair cell (10) The primary sensory receptor in the lateral line of fish and amphibians

Hair cell (12) The primary sensory receptor in the inner ear of vertebrates

Hair follicle receptor (14) In mammals, a type of fast-adapting mechanoreceptor wrapped around hair follicles in hairy skin; it responds primarily to slight movement of the hair

Heterosynaptic facilitation (8) The facilitation of a different synapse than the one activated. Compare homosynaptic facilitation

Heterosynaptic LTP (8) A form of long-term potentiation that requires depolarization of the neuron in which it occurs simultaneously with activity of a neuron that synapses on that neuron. Because of its presumed involvement in learning, it is also called associative LTP

High vocal center (HVC) (25) In songbirds, a region of the brain that is involved in the control of singing. It is larger in males of bird species in which only the males sing

Hindbrain (3) In vertebrates, the posterior part of the brain adjacent to the spinal cord, consisting of the medulla, pons, and cerebellum

Hippocampal formation (3) In vertebrates, a banana-shaped region in the forebrain that lies adjacent to the hypothalamus; part of the limbic system

Hippocampus (3) In vertebrates, a nucleus in the forebrain that is the principal component of the hippocampal formation; part of the limbic system. The hippocampus is necessary for several types of learning and memory

Histofluorescence (7) A method for localizing serotonin or catecholamine neurotransmitters in neural tissue. The method involves exposing a tissue slice to formaldehyde vapors, which react chemically with the neurotransmitter to form a fluorescent compound that is visible under a fluorescence microscope

Homosynaptic facilitation (8) The facilitation of the activated synapse. Compare heterosynaptic facilitation

Homosynaptic LTP (8) A form of long term potentiation that requires a tetanizing stimulus across the synapse at which long term potentiation is observed

Horizontal cell (11) A type of cell found in the vertebrate retina. Horizontal cells connect to photoreceptors and bipolar cells

Horseradish peroxidase (HRP) (3) An enzyme used as a marking agent to allow individual neurons or tracts of neurons to be studied

HRP (3) See horseradish peroxidase

Huntington's chorea (18) See Huntington's disease

Huntington's disease (18) A late-onset, inherited brain disease characterized by uncontrollable movements, grotesque facial expressions, and eventual loss of mental function. It is caused by the degeneration of cholinergic and GABAergic neurons in the striatum

HVC (25) See high vocal center

Hygroreceptor (9) A sense organ that is sensitive to the water content of air; found in insects

Hyperpolarization (5) An increase in membrane polarization and hence a change of the membrane potential of a cell from its resting level away from zero

Hypothalamus (3) In vertebrates, a part of the diencephalon that acts as a homeostatic regulatory center, regulating such parameters as body temperature and body water level via the pituitary gland and the autonomic system; also important in sexual behavior

–I–

Identifiable neuron (18) A neuron that is recognizable by having similar morphological and physiological characteristics in all individuals of a species

Immunocytochemistry (7) See immunohistochemistry

Immunoglobulin superfamily (22) A superfamily of guidance molecules with structural similarity to vertebrate antibodies. The molecules help determine the direction of growth of developing axons. Members of this group are cell adhesion molecules

Immunohistochemistry (7) A method for making visible specific neurotransmitters or other molecules in neural tissue. The method involves binding the material of interest to an antibody that has been joined to a molecule that can itself be made visible by chemical techniques

Immunostaining (7) See immunohistochemistry

Inactivation (6) The process by which a neurotransmitter is functionally removed from the vicinity of a synapse

Inactivation gate (5) One of the two gates of a sodium channel. This gate shuts shortly after the activation gate opens, thereby closing the channel

Indeterminate development (22) A form of development in which epigenetic factors (factors external to the cell) predominate in determining each cell's fate; also called regulated development

Inferior colliculus (3) In mammals, a part of the midbrain important in processing auditory input. Corresponds in part to the torus semicircularis in lower vertebrates

Inferior mesenteric ganglion (3) In vertebrates, a sympathetic autonomic ganglion that lies in the body cavity and innervates the bladder, rectum, and genitals

Inferior olivary nucleus (18) In vertebrates, a nucleus in the medulla that is the source of all climbing fiber input to the cerebellum

Influx (4) The flow of ions into a cell

Inhibition (6) Inhibitory synaptic transmission

Inhibitory (synaptic) transmission (6) A form of synaptic transmission that usually has a hyperpolarizing effect on a postsynaptic neuron, and in which the probability that the postsynaptic neuron will fire an action potential is decreased. In neurons that cannot generate action potentials, inhibition decreases the amount of neurotransmitter that is released

Inhibitory postsynaptic potential (ipsp) (6) A potential generated in a postsynaptic neuron by activation of an inhibitory synapse; usually hyperpolarizing

Inner hair cell (12) In the mammalian cochlea, the auditory receptor cell

Inner nuclear layer (11) In the vertebrate retina, the layer of cell bodies with nuclei that lies between the inner and outer plexiform layers. It consists of the cell bodies and nuclei of horizontal, amacrine, and bipolar cells

Inner plexiform layer (11) In the vertebrate retina, the innermost layer of synapses, containing synapses among bipolar, amacrine, and ganglion cells

Inner segment (11) In vertebrates, the proximal part of a photoreceptor cell, including the cell body

Innervate (15) The process of making or having a synaptic contact between a neuron and a target. The term usually refers to synaptic contact between a motor neuron and a muscle fiber

In situ hybridization (25) A method for making the distribution of specific molecules visible in neurons through the use of molecular genetic techniques

Insula (13) In mammals, the part of the cerebral hemispheres of the brain that lies behind and underneath the temporal and parietal lobes

Integrating system (1) The parts of the nervous system that make decisions about information and organize action. The integrating system works with the sensory and motor systems to carry out neural functions

Integration (8) The process by which signals in the nervous system interact; the decision-making capability of neurons

Integrin family (22) A family of guidance molecules that help determine the direction of growth of developing axons. Integrin family members are cell adhesion molecules

Intensity (10) In a sensory modality, the subjective sensation corresponding to stimulus strength, as for example the brightness of light or the loudness of sound

Intermediate filament (2) The name by which neurofilaments are known in non-neural cells

Interneuron (2) Neurons that are interposed between sensory neurons and motor neurons; generally confined to the central nervous system

Intracellular recording (4) Electrical recording in which the electrode is inserted into the interior of a cell

Intrafusal muscle (10) In vertebrate muscle, the fibers of the muscle spindle organ; the fibers are contractile except in the middle, and at the ends attach to the extrafusal muscle

Ion channel (2) A hollow-cored protein that spans the membrane of a cell and that allows ions to pass from one side of the membrane to the other through the hollow core

Ionophoresis (7) See iontophoresis

Ionotropic receptor (2) A receptor that causes the rapid opening or closing of an ion channel and a quick, strong electrical response in the postsynaptic neuron

Ion pump (2) In membranes, proteins that actively transfer one or more ions from one side of the membrane to the other

Iontophoresis (7) The controlled release of ions from a microelectrode, done by passing current through the electrode

Ion-selective channel (2) An ion channel that allows passage of some types of ions only

Ipsp (6) See inhibitory postsynaptic potential

— J —

Jamming avoidance response (JAR) (20) A shift in the frequency of the electric organ discharge of a weakly electric fish in the presence of the discharge from another nearby fish

JAR (20) See jamming avoidance response

JH (25) See juvenile hormone

Juvenile hormone (JH) (25) In insects, a sesquiterpenoid hormone that prevents premature molting into an adult in young insects, and that in the adult is necessary for sexual maturity

— K —

Kinesin (2) A protein involved in anterograde axonal transport. Kinesin transiently binds the material being transported to microtubules in such a way that the material moves away from the cell body

Kinesthesis (14) The sense of limb position

Knockout mouse (24) A genetically engineered mouse in which a single gene has been inactivated through the method of targeted gene replacement

— L —

Labeled line code (10) A type of coding for sensory information in which each neuron carries information about only one specific stimulus quality

Lamina (11) In insects, the part of the optic lobe just beneath the compound eye

Laser scanning confocal microscopy (22) A technique in which a laser beam is used to scan a specimen at one specific focal plane in a confocal microscope, a procedure that produces a sharper image by reducing background fluorescence

Lateral geniculate nucleus (LGN) (3) In higher vertebrates, a thalamic nucleus that processes visual input

Lateral inhibition (9) An arrangement of neurons common in sensory systems in which adjacent neurons in the array make reciprocal inhibitory connections with each other, resulting in contrast enhancement

Lateral vestibulospinal tract (LVST) (14) In vertebrates, a spinal cord tract that carries vestibular information from brain stem vestibular nuclei to motor centers in the spinal cord. Information carried in this tract originates mainly from the maculae of the inner ears

Law of specific nerve energies (9) The statement that each sensory nerve normally carries information about only one type of stimulus

Leakage channel (2) An ion channel that is open to the flow of ions all the time; an ungated channel

Learning (24) A change in the behavior of an animal as a consequence of the animal's experience

Lemniscal pathway (14) Tracts in the spinal cord and brain stem that convey information from touch, vibration, and kinesthesis receptors to nuclei in the thalamus and then to the somatosensory cortex

Length constant (8) The distance along a membrane at which a constant voltage will decay to about 37% of its original value; functionally, the distance that a postsynaptic potential can spread. Quantitatively expressed as $\lambda = \sqrt{r_m / r_i}$

Lentiform nucleus (3) The putamen and globus pallidus together; part of the basal ganglia

LGN (11) See lateral geniculate nucleus

Ligand-gated channel (2) A gated ion channel that opens or closes in the presence of a particular type or class of signaling molecule, the ligand

Ligand-sensitive channel (2) See ligand-gated channel

Limbic system (3) In vertebrates, a group of nuclei and brain regions that form a kind of ring around the brain stem. It is devoted mostly to the control of emotional reactions

Lineage (22) In development, the sequence of divisions that produces any one cell in an adult animal

Lipid bilayer (2) The composition of cell membranes, a double layer of lipid molecules in which the (hydrophobic) tails of the molecules in each of the two layers face each other

Lobula (11) In insects, part of the lobula complex of the optic lobes, in which some of the processing of visual information takes place

Lobula complex (11) In insects, the part of the optic lobes closest to the rest of the brain

Lobula plate (11) In insects, part of the lobula complex of the optic lobes in which some of the processing of visual information takes place

Long term memory (24) A type of memory that can be retained indefinitely

Long term potentiation (8) A type of facilitation induced by a train of stimuli in which the facilitory effect may last several days

LTP (8) See long term potentiation

Lucifer yellow (3) A fluorescent compound used to stain individual neurons by intracellular injection

LVST (14) See lateral vestibulospinal tract

– M –

M channel (11) See magnocellular pathway

Macroglomerulus (13) An enlarged glomerulus in the antennal lobe in the brains of some male insects, in which processing of information about the species specific sex pheromone is carried out

Macula (14) In vertebrates, the part of the utricle or saccule of the inner ear in which the receptor cells are located

Magnetic resonance imaging (MRI) (21) A technique for generating an image of a part of a living brain. The method involves placing the subject in a strong magnetic field, perturbing the field, and recording the location of the resulting electromagnetic radiation. Structures in the brain can be distinguished from one another because they contain different materials, and different materials respond differently to the perturbation of the magnetic field

Magnetic sense (14) The ability to detect the presence and orientation of a magnetic field

Magnetite (14) Crystals of ferrous ferrite, an oxide of iron, found in some animals thought to possess a magnetic sense

Magnetoreceptor (9) A sensory receptor for a magnetic sense

Magnocellular layer (11) In the mammalian lateral geniculate nuclei, a layer of neurons with relatively large cell bodies that receive input from M-type ganglion cells

Magnocellular pathway (11) In the mammalian visual system, the pathway from M-type ganglion cells via the magnocellular layers of the lateral geniculate nuclei to the visual cortex. This pathway processes information especially concerning motion

Mammillary body (3) In birds and mammals, one of a pair of nuclei that are visible on the ventral surface of the thalamus

Mauthner cell (18) In fish, one of a pair of identifiable neurons that originate in the brain and are involved in escape swimming

Mechanoreceptor (9) A sensory receptor for mechanical (movement) energy

Medial geniculate nucleus (3) In higher mammals, a thalamic nucleus that processes auditory input

Medial lemniscus tract (14) A somatosensory tract in the brain stem that is part of the lemniscal pathway, carrying information about touch, vibration, and kinesthesis

Medial vestibulospinal tract (MVST) (14) A spinal cord tract that carries information from brain stem vestibular nuclei to motor centers in the spinal cord. Information carried in this tract originates mainly in the semicircular canals of the inner ears

Middle temporal region (MT) (11) In higher mammals, a visual processing area, also known as V5

Medulla (3) See medulla oblongata and also next entry

Medulla (11) In insects, the part of the optic lobe between the lamina and the lobula complex, in which some processing of visual information takes place

Medulla oblongata (3) In vertebrates, the most posterior part of the brain; the connecting link between the spinal cord and the rest of the brain

Meissner's corpuscle (14) In mammals, a type of fast-adapting mechanoreceptor found near the surface in hairless skin, that responds primarily to light touch on the skin

Membrane potential (4) The difference in electrical potential across the membrane of a cell

Memory (24) The stored representation of experience in the central nervous system; also refers to the process by which storage occurs

Meninges (3) In vertebrates, the tissue layers that surround the brain

Mepp (7) See miniature end plate potential

Merkel's nerve complex (14) In mammals, a type of slowly adapting mechanoreceptor found near the surface in hairy and hairless skin that primarily responds to pressure on the skin

Mesencephalicus lateralis dorsalis (MLD) (12) The bird homologue of the mammalian inferior colliculus; used to process auditory information

Mesencephalic locomotor region (18) In vertebrates, a region of the pontine tegmentum, stimulation of which will elicit locomotion

Mesencephalon (3) The midbrain

Metabotropic receptor (2) A receptor protein that initiates a series of biochemical reactions in the postsynaptic neuron, and has only a slow, weak electrical effect, if any, on the postsynaptic neuron

Microelectrode (4) Typically a fine, hollow tube of glass, pulled out to form a minute tip and filled with a conducting solution. Also a fine-tipped metal wire. Glass microelectrodes are used for intracellular recording, and wire microelectrodes are used for extracellular recording

Microfilament (2) Short, ubiquitous filaments in neurons, about 5 nm in diameter

Microglia (2) A type of glia

Microtubule (2) The name by which neurotubules are known in non-neural cells

Microneurography (14) A method for recording neural activity in single axons from nerves in awake human subjects

Midbrain (3) In vertebrates, the middle part of the brain between the hindbrain and the forebrain

Miniature end plate potential (Mepp) (7) A small end plate potential that can appear spontaneously at a neuromuscular junction and is thought to be caused by the postsynaptic effects of only one or a few quanta of neurotransmitter

Mitral cell (13) A type of neuron in the vertebrate olfactory bulb that in the glomeruli receives input from olfactory neurons

Mixed nerve (3) A cranial or spinal nerve that contains both afferent and efferent fibers

MLD (12) See mesencephalicus lateralis dorsalis

Modality (9) The subjective sensation of a particular sense, such as sight or hearing

Model (16) A representation of a neural network by a hypothetical arrangement of neurons that will generate a particular type of output given certain characteristics and input conditions. More generally, any representation, usually mathematical, of a neural (or other) process

Monoclonal antibody technique (22) A method for labeling a single type of protein with its antibody; the antibodies are produced by a single line of hybridoma cells

Monopolar neurons (2) Neurons with a single neurite extending from the cell body. Usually, this neurite branches to form a true axon and dendrites. This structure is typical of invertebrate interneurons and motor neurons, and of many vertebrate sensory neurons

Monosynaptic reflex (16) A reflex arc containing only a single synapse, and hence only two neurons

Mosaic development (22) See determinate development

Mossy fiber (3) In vertebrates, one of two types of neurons that bring information into the cerebellum

Motor cortex (3) The region of the cortex that is devoted to control of the motor neurons innervating body muscles; it includes the primary motor cortex, the premotor cortex, and the supplementary motor area

Motor nerve (3) A cranial or spinal nerve consisting entirely of neurons that carry information from the central nervous system toward a muscle or gland; motor nerves are efferent nerves

Motor neuron (2) A neuron that sends neural signals to muscles or glands

Motor protein (2) A type of protein that is able to transform energy into physical movement through the hydrolysis of ATP

Motor system (1) The part of the nervous system that controls motor output

Motor unit (15) A motor neuron plus all the muscle fibers innervated by that motor neuron

MRI (21) See magnetic resonance imaging

MT (11) See middle temporal region

Multimodal receptor (14) A receptor cell that is sensitive to several different types of stimuli, such as mechanical and chemical inputs. Ordinary receptors are sensitive primarily to one type of stimulus

Multipolar neuron (2) A neuron with a single long axon and many dendrites extending from the cell body. This structure is typical of some invertebrate sensory neurons and of vertebrate interneurons and motor neurons

Multiunit smooth muscle (15) In vertebrates, smooth muscle typically innervated by several autonomic motor neurons, and with little electrical coupling between muscle cells

Muscarinic acetylcholine receptor (6) A type of ion channel that binds the neurotransmitter acetylcholine and that when activated produces a slow electrical response in the postsynaptic neuron

Muscle spindle organ (10) A type of stretch-sensitive sense organ present in the muscles of most vertebrates

MVST (14) See medial vestibulospinal tract

Myelin (2) The layers of glial membrane tightly wrapped around axons or dendrites of some vertebrate neurons; myelin speeds up conduction of nerve impulses

Myenteric plexus (3) In vertebrates, a network of interconnected ganglia located between the longitudinal and circular muscles along the length of the alimentary canal

— N —

NA (12) See nucleus angularis

nAChR (6) See nicotinic acetylcholine receptor

Narrow specialist (13) A type of olfactory receptor in insects that responds to only a single type of chemical substance

nE (20) See nucleus electrosensorius

Negative feedback (20) In a neural circuit, a negative signal that loops back to decrease the input to a component in the circuit, producing an output that gets smaller

Neostriatum (3) In vertebrates, the caudate nucleus and putamen together; part of the basal ganglia

Nernst equation (4) An equation used to calculate the equilibrium potential for any ion species in a cell given the concentrations of the ion inside and outside that cell

Nerve (3) A bundle of axons or dendrites of nerve cells that enter or leave the central nervous system

Nerve growth factor (NGF) (22) A member of the neurotrophin family of trophic growth factors that promotes vigorous sprouting and growth of neurites of selected neurons

Nerve tract (3) A bundle of axons that sends information from one region of the central nervous system to another

Netrin family (22) A family of guidance molecules that helps determine the direction of growth of developing axons

Neural plate (22) In vertebrate development, a depression formed on the dorsal surface of the gastrula stage of the embryo that develops into the nervous system

Neural specificity (22) The idea that specific connections are made between individual neurons during the development of the nervous system

Neural tube (22) In vertebrate development, the long, hollow tube formed by the infolding of the sides of the neural plate. The neural tube is the rudimentary beginning of the brain and spinal cord

Neurite (2) Any fine, branch-like projection from the cell body of a neuron

Neurobiology (1) The study of the nervous system, how it is organized and how it functions

Neuroblast (22) In a developing insect embryo, a type of cell from which the ganglia of the body segments develop

Neuroendocrinology (25) The subfield of neurobiology devoted to the study of hormone interactions with the nervous system

Neuroethology (19) The subfield of neurobiology devoted to the explicit study of the neural basis of behavior; some neuroethologists concentrate especially on biologically important behavior such as communication, escape, or prey capture

Neurofilament (2) A twisted coil of rodlike strands of protein, about 10 nm in diameter

Neuroglia (2) See glia

Neuromast organ (10) The sense organ that is part of the lateral line system of fish and amphibians

Neuromodulation (6) Neuromodulatory (synaptic) transmission

Neuromodulator (6) A neurotransmitter that is used in neuromodulatory transmission

Neuromodulatory transmission (6) A form of synaptic transmission in which the main effect on the postsynaptic neuron is a change in the neuron's electrical or metabolic characteristics. Neuromodulatory effects are mediated by metabotropic receptors, may last from tens of milliseconds to weeks, and typically have little effect on the membrane potential

Neuron (2) In nervous systems, the cell type that conveys information; a nerve cell

Neuron doctrine (2) The idea that the nervous system, like other bodily systems, is composed of discrete cellular elements, each entirely bounded by its own membrane

Neuron theory (2) See neuron doctrine

Neuropil (3) In invertebrates, the central region of a ganglion, containing dendrites and synapses

Neuroscience (1) The study of the nervous system, how it is organized and how it functions; sometimes used to include also medical disciplines such as neurology

Neurosecretory cell (25) A neuron that synthesizes and releases hormones

Neurotransmitter (2) Any small molecule that is used as a chemical intermediary in synaptic transmission

Neurotrophin family (22) A group of molecules structurally and functionally related to nerve growth factor; its members influence the growth and development of neurons

Neurotubule (2) A hollow protein tubule consisting of tubulin and with an outer diameter of about 23–25 nm; used in axonal transport

NGF (22) See nerve growth factor

Nicotinic acetylcholine receptor (nAChR) (6) A type of a ion channel that binds the neurotransmitter acetylcholine and produces a quick electrical response in the postsynaptic neuron

Nissl substance (2) An especially prominent rough endoplasmic reticulum of neurons that synthesizes the proteins and phospholipids destined for insertion into membranes

NL (12) See nucleus laminaris

NM (12) See nucleus magnocellularis

NMDA receptor (8) A type of ionotropic glutamate receptor that also responds to N-methyl-D-aspartate. The associated channel allows passage of K^+, Na^+, and, under certain circumstances, Ca^{2+}. This type of receptor is thought to underlie the phenomenon of LTP

Nociceptor (9) A receptor cell for noxious stimuli. In mammals, probably in all vertebrates, pain is the subjective sensation felt from stimulation of these receptors

Node of Ranvier (2) A gap between two adjacent sections of myelin sheathing around vertebrate neurites

Nondeclarative memory (24) Memory formed from skill acquisition of any kind, as well as from simple conditioning and non-associative learning like habituation

Nonspiking neuron (7) A neuron that is incapable of generating action potentials due to lack of voltage sensitive ion channels

Noradrenergic neuron (7) A neuron that releases norepinephrine as a neurotransmitter

NSF (7) N-ethylmaleimide sensitive factor, a protein necessary for docking of synaptic vesicles with the presynaptic membrane, formation of a fusion pore, and subsequent release of neurotransmitter

Nucleus (3) In the central nervous systems of vertebrates, a distinct group of neuron cell bodies, dendrites, and synapses

Nucleus angularis (NA) (12) A nucleus in the brain stem of birds. In owls, it aids in the determination of intensity differences in sounds that impinge on the two ears

Nucleus electrosensorius (NE) (20) A nucleus in the brain of weakly electric fish that generates the information as to which direction the electric organ discharge should be shifted in the jamming avoidance response

Nucleus laminaris (NL) (12) A nucleus in the brain stem of birds. In owls, it aids in the determination of temporal differences in sounds that impinge on the two ears

Nucleus magnocellularis (NM) (12) A nucleus in the brain stem of birds. In owls, it provides precise information about when a sound has occurred

Nucleus of the lateral lemniscus (12) In vertebrates, a nucleus in the pons that is a relay station for auditory information

Null direction (11) In the retina, the direction of movement of a visual stimulus that produces no response in a directionally selective neuron

–O–

Occipital lobe (3) In vertebrates, the posteriormost region of each cerebral hemisphere of the brain; site of the primary visual cortex

Ocellus (11) In insects, simple, single-lens eyes used to detect light and shadow but not for the formation of images

Ocular dominance (23) In vertebrates, the degree to which any one neuron in the visual cortex responds to input from one eye rather than the other

Ocular dominance column (11) In vertebrates, a column consisting of cells that respond primarily to input from one eye, and which is located adjacent to a column consisting of cells that respond mainly to input from the other eye

Odorant binding protein (13) A class of protein found in the fluid that surrounds the receptive areas of olfactory receptor cells. These proteins bind with an odorant and move it to the receptor cell membrane that contains the receptor molecules

Off-center bipolar cell (11) In the vertebrate retina, a bipolar cell that hyperpolarizes when the center of its receptive field is stimulated by light

Off-center ganglion cell (11) In the vertebrate retina, a ganglion cell that gives an "off" response, that is, decreases its rate of firing, when the center of its receptive field is stimulated by light

Olfaction (13) The sense of smell

Olfactory bulb (13) A projection of the brain that in humans lies directly over the nasal cavity and in other animals lies posterior to the nose in the anterior part of the brain

Olfactory cortex (13) In vertebrates, a portion of cortex devoted to processing olfactory information, located on the ventral surface of the temporal lobe of the telencephalon. Also called the pyriform cortex

Olfactory tract (13) In vertebrates, the bundle of axons leading from the olfactory bulb to the primary olfactory cortex on the anterior ventral surface of the telencephalon

Oligodendrocyte (2) See oligodendroglia

Oligodendroglia (2) A type of glia that forms myelin around neurites in the central nervous systems of vertebrates

Ommatidium (11) In arthropods, the name for the photoreceptive unit of compound eyes, consisting of several photoreceptive and other cells

On-center bipolar cell (11) In the vertebrate retina, a bipolar cell that depolarizes when the center of its receptive field is stimulated by light

On-center ganglion cell (11) In the vertebrate retina, a ganglion cell that gives an "on" response, that is, increases its rate of firing, when the center of its receptive field is stimulated by light

Open loop circuit (20) Any type of circuit, electrical or neural, in which there is no feedback to help regulate the performance of the circuit

Open loop reflex (20) A reflex based on a neural circuit in which there is no sensory feedback to help regulate the performance of the circuit

Open loop system (20) An open loop circuit

Optic lobe (11) In insects, the part of the brain that lies just under the compound eyes, and in which most of the processing of visual information takes place

Optic tectum (11) Also called the tectum. In lower vertebrates, it is the part of the midbrain (the dorsal roof) devoted to the processing of visual and other information

Optokinetic reflex (20) A compensatory movement of the eyes in the direction opposite to a (slow) rotation of the head, driven by slip of an image on the retina

Orbitofrontal cortex (13) In vertebrates, a region on the ventral surface of the frontal lobe of the brain where conscious recognition of odors is thought to occur

Organ of Corti (12) The portion of the cochlea in which sound energy is transduced into neural activity; it consists of the basilar and tectorial membranes and the hair cells

Orientation column (11) In vertebrates, a column of cells in the visual cortex, all of the neurons in which respond preferentially to a line or edge at some particular angle

Oscillator (16) See central pattern generator

Oscilloscope (5) A device for the measurement and display of changes in voltage over time

Otolith (14) In vertebrates, a stone-like particle embedded within a gelatinous mass in the vestibular organ. The same as a statolith in invertebrates

Otolith organ (14) In vertebrates, the term for a statocyst organ, the sense organ that detects the direction of gravity and changes in orientation of the animal

Outer hair cell (12) In the mammalian cochlea, a type of cell that helps to determine the mechanical properties of the cochlear membranes, and hence influences hearing

Outer nuclear layer (11) In the vertebrate retina, the outermost layer containing cell nuclei, consisting of the nuclei of the photoreceptor cells

Outer plexiform layer (11) In the vertebrate retina, the outermost layer of synapses, containing synapses between the photoreceptors, horizontal cells, and bipolar cells

Outer segment (11) In vertebrates, the distal-most part of a photoreceptor cell, where transduction of light energy takes place

Oxytocin (25) In vertebrates, a peptide hormone released by the posterior pituitary gland. It stimulates uterine contraction, elicits milk ejection, and promotes maternal behavior in female mammals

— P —

P channel (11) The parvocellular pathway

Pacemaker neuron (5) A neuron that in the absence of any input spontaneously generates individual spikes or bursts of spikes

Pacemaker nucleus (Pn) (20) A nucleus in the medulla of weakly electric fish that drives the electric organ for the production of the electric organ discharge

Pacinian corpuscle (14) In vertebrates, a type of fast-adapting mechanoreceptor found deep in hairy and hairless skin, that primarily responds to pressure on the skin

Paleostriatum (3) The globus pallidus, one of the basal ganglia

Parabrachial nucleus (13) In vertebrates, a nucleus in the dorsal part of the pons near the attachment of the cerebellum through which gustatory information passes (except in primates)

Parahippocampal gyrus (3) In vertebrates, a part of the cerebral hemispheres that lies along the dorsal medial face of the temporal lobe; part of the limbic system

Parallel processing (11) The idea that the various features of a sensory signal are encoded and interpreted by different sets of neurons with distinct features that are intrinsic to the neurons

Parasympathetic autonomic system (3) In vertebrates, the division of the autonomic nervous system that tends to relax the body, inhibiting action and bringing about a state of rest

Paravertebral chain (3) In vertebrates, one of a pair of chains of sympathetic autonomic ganglia that lie next to the vertebral column

Parietal lobe (3) In vertebrates, the middle portion of the cerebral hemispheres of the brain, between the frontal lobe and the occipital lobe

Parkinson's disease (7) A disease of the motor system characterized by deficits in voluntary movement, and, in its most severe stage, by a frozen inability to make almost any voluntary movements. It is brought on by a degeneration of dopaminergic neurons in the substantia nigra

Parvocellular layer (11) In mammalian lateral geniculate nuclei, a layer of neurons with relatively small cell bodies that receive input from P-type ganglion cells

Parvocellular pathway (11) In the mammalian visual system, the pathway from P-type ganglion cells via the parvocellular layers of the lateral geniculate nuclei to the visual cortex. This pathway processes information especially concerning objects and their colors

Passive electric sense (14) The electric sense in animals that can detect externally generated electrical fields but that are not capable of generating electrical fields of their own

Patch clamp (5) A method of recording the currents generated by ions flowing through just one or a few ion channels

Pathfinding (22) In neural development, the process by which a growing axon is able to navigate to its target, especially the mechanisms involved at the molecular level

Pattern generator (16) See central pattern generator

Periglomerular cell (13) In the vertebrate olfactory bulb, a type of neuron that makes synaptic contact with the mitral cells and olfactory neurons in one glomerulus and sends an axon to another glomerulus where it contacts other mitral cells

Peripheral nervous system (3) All the nerves that run to and from the central nervous system as well as any ganglia that may lie outside it

PET scan (21) A scan of the body using positron emission tomography

Phasic receptor (9) A type of receptor cell in which adaptation is so rapid that the response of the cell to a stimulus declines to zero very quickly

Phasi-tonic receptor (9) A type of receptor cell that has a component of both rapid and slow adaptation

Photoreceptor (9) A receptor cell specialized to respond to light energy

Pioneer neuron (22) In insect development, a neuron that forges a path to the central nervous system that is followed by sensory neurons that develop in the periphery

Pitch (17) During the flight of an insect or bird, the angle above the horizontal of a line through the long axis of the animal's body

Place cell (21) In vertebrates, a neuron in the dorsal region of the hippocampus that shows activity correlated with an animal's orientation or location in a familiar environment

Plasticity (23) The ability of some part of the nervous system to change its functional characteristics. Neural plasticity is the basis of learning

Pn (20) See pacemaker nucleus

Polysynaptic reflex (16) A reflex arc in which there are two or more synapses, and hence at least three neurons

Pons (3) In vertebrates, the part of the hindbrain that lies between the medulla and the midbrain. Nuclei in the pons are especially concerned with balance and eye reflexes

Population code (10) See across fiber code

Positive feedback (20) In a neural circuit, a positive signal that loops back to increase the input to a component in the circuit, producing an output that increases in size

Positron emission tomography (21) A technique for generating an image of a part of a living brain. The method involves giving the subject a radioactively labeled substance and measuring the resultant radiation. Active brain regions take up the substance, allowing their locations to be determined

Postcentral gyrus (3) In vertebrates, the prominent ridge of brain tissue that lies just posterior to the central sulcus, containing the somatosensory cortex

Posterior lobe (18) The middle of three anatomical subdivisions of the mammalian cerebellum

Postganglionic fiber (3) The axon of a postganglionic neuron

Postganglionic neuron (3) A neuron of the autonomic system whose cell body lies in a ganglion outside the central nervous system and whose axon terminates in or on a target organ

Post-inhibitory rebound (16) The tendency of a neuron to fire a series of action potentials upon being released from inhibition

Postsynaptic inhibition (6) Inhibitory synaptic transmission

Postsynaptic neuron (2) At a synapse, the neuron that receives a signal from another neuron

Posttetanic potentiation (8) A form of facilitation in which a train of impulses, called a tetanus, is required to produce the facilitation

PPn (20) See prepacemaker nucleus

Precentral gyrus (3) The prominent ridge of brain tissue that lies just anterior to the central sulcus; it contains the primary motor cortex

Preferred direction (11) In an eye, the direction of movement of a visual stimulus that elicits a maximal response in a directionally selective visual neuron

Preganglionic fiber (3) The axon of a preganglionic neuron

Preganglionic neuron (3) A neuron of the autonomic system whose cell body lies in the central nervous system and whose axon terminates in a peripheral autonomic ganglion

Premotor cortex (18) In mammals, the part of the motor cortex that lies anterior to the primary motor cortex. This region sends axons to the brain stem motor nuclei

Preoptic area (25) In the mammalian brain, a region of the hypothalamus involved in mating behavior. One part of this region, the sexually dimorphic nucleus of the preoptic area, is larger in males than in females

Prepacemaker nucleus (PPN) (20) In brains of weakly electric fish, a nucleus in the forebrain that sends commands to the pacemaker nucleus to increase or decrease the frequency of the electric organ discharge

Presynaptic inhibition (6) A form of inhibition in which the inhibitory effect is brought about by a reduction in the amount of neurotransmitter released by the target neuron, hence reducing the probability that excitation of the target neuron will produce an action potential in a postsynaptic neuron

Presynaptic neuron (2) At a synapse, the neuron that passes on a signal to another neuron

Pretectum (21) In the amphibian brain, the posterior area of the thalamus, just anterior to the optic tectum

Primary motor cortex (18) In mammals, the part of the motor cortex that lies just anterior to the central sulcus. This area sends axons directly to the motor areas of the spinal cord

Primary visual cortex (11) In vertebrates, visual area V1 in the occipital lobe

Priming memory (24) A type of nondeclarative memory, such as remembering that you have seen a word before without any knowledge of what the word means or the context in which you saw it

Procedural memory (24) A type of nondeclarative memory, the memory of motor or cognitive skills such as riding a bicycle or doing long division by hand; also the memory of any conditioned response

Progeny (22) In development, all the cells that are generated from divisions of a single cell

Programmed cell death (22) The directed death of a neuron due to circumstances such as lack of proper synapse formation or an altered hormonal environment

Projection neurons (2) Especially in vertebrates, neurons that send neurites from one place to some distant location in the central nervous system

Proprioceptor (9) A sensory receptor that provides an animal with information about the relative position or movement of its body parts

Prosencephalon (3) See forebrain

Protein kinase (6) An enzyme that promotes the phosphorylation of another protein

PTP (8) See posttetanic potentiation

Purkinje cell (3) In vertebrates, a type of neuron found in the cerebellum. Purkinje cells send their axons out of the cortex to the deep cerebellar nuclei

Putamen (3) One of the main nuclei comprising the basal ganglia

Pyramidal decussation (21) In vertebrates, the region of the medulla at which the motor tracts from each half of the cortex cross over to the other side of the body

Pyramidal (motor) system (18) In vertebrates, collectively, the neurons from the motor cortical areas that pass through the pyramids in the medulla

Pyramidal tract (3) In vertebrates, one of two large tracts in the medulla that contain the axons carrying motor information from the motor cortex to the spinal motor centers

Pyramids (18) In mammals, the region of the medulla at the junction with the spinal cord at which the motor tracts from the cortex cross over to the other side of the body

Pyriform cortex (13) The (primary) olfactory cortex of the brain, located on the ventral surface of the temporal lobe of the telencephalon

– Q –

Quality (9) The "type" of a stimulus, such as the tone of sound or the color of light

Quantal transmission (7) The idea that neurotransmitter is released in discrete packets, each representing the contents of a single vesicle

Quantum (7) An individual packet of neurotransmitter. In the vesicular hypothesis, one quantum represents the contents of one synaptic vesicle

—R—

RA (25) See robust nucleus of the archistriatum

Radioimmunolabeling (7) A method for localization of specific neurotransmitters in neural tissue. The method involves labeling the material of interest by exposing it to a radioactively labeled antibody, then placing the material next to radioactively sensitive emulsion or film. Radioactive decay exposes the emulsion or film at the sites of the label

Range fractionation (9) A division of labor among sensory neurons in a sense organ such that each neuron responds to only a part of the range of stimuli to which the organ as a whole is sensitive

Raphe nucleus (3) In vertebrates, one of a group of mid-line nuclei of the reticular formation in the medulla, pons, or midbrain, whose destruction may lead to permanent coma

Reafference (10) Sensory signals that are generated as a consequence of an animal's own movements

Receptive field (9) In sensory systems, the specific region of a sensory surface that when stimulated causes a change in activity of a neuron

Receptor (2) See receptor protein and also the next entry

Receptor (9) See receptor cell and also sensory receptor

Receptor cell (9) A peripherally located neuron, often anaxonal, that can detect and respond to a sensory stimulus. May also be used generically to mean any neuron that can respond to a sensory stimulus

Receptor layer (11) In the vertebrate retina, the outermost layer, consisting of the photoreceptor cells

Receptor potential (9) The electrical potential that appears in sensory neurons when a stimulus is applied

Receptor protein (2) A protein with which signaling molecules like serotonin or GABA can bind to start the chain of events that brings about a neuron's response to the signaling molecule

Reciprocal inhibition (16) An arrangement of two neurons so that excitation of one causes inhibition of the other and vice versa. The two neurons may inhibit each other directly or via other neurons interposed between the reciprocally inhibited pair

Reciprocal synapses (2) Two chemical synapses adjacent to each other, one transmitting in one direction, the other in the opposite direction

Recording (4) The process of detecting and displaying an electrical potential from a living cell; carried out with electrodes

Recruitment (of motor neurons) (15) The addition of one or more motor neurons to the pool of neurons that is active at any one instant in a muscle

Red nucleus (3) In vertebrates, a nucleus located in the tegmentum of the midbrain; devoted to motor control

Reflex (16) A simple, relatively stereotyped behavior caused by a specific stimulus

Reflex arc (16) The neural circuit that underlies a reflex, consisting at minimum of a sensory neuron, a motor neuron, and, usually, an interneuron

Reflex gating (17) The process by which a reflex is produced only when both the stimulus that evokes it and some other behavior are present at the same time

Reflex modulation (17) A continuous change in the strength or the sign of a reflex response as an animal moves through one cycle of an ongoing repetitive behavior

Reflex reversal (17) The reversal of the effect of a particular sensory stimulus under certain circumstances

Refractory period (5) A brief period following an action potential during which it is more difficult than normal or impossible to excite a neuron to generate another action potential. See also absolute and relative refractory periods

Regulated development (22) See indeterminate development

Reissner's membrane (12) A membrane of the cochlea that separates the upper channel of the cochlea into two chambers, the scala vestibuli and the scala media

Relative refractory period (5) A brief period following an action potential during which it is more difficult to generate another action potential because the threshold for elicitation of a new action potential is raised

Releasable pool (7) Collectively, synaptic vesicles that are attached to the presynaptic membrane at the active zone and are ready to release their contents into the synaptic cleft

Resistance reflex (17) A reflex in which muscles are activated that oppose a movement that is imposed on some part of the body

Resting potential (4) The membrane potential of a cell when it is at "rest," that is, not electrically active

Reticular activating system (3) See reticular formation

Reticular formation (3) A network of neurons in the medulla, the pons, and the midbrain that helps to regulate brain arousal as well as having other functions

Reticular nuclei (18) A pair of motor nuclei in the brain stem

Reticular theory (2) The idea, no longer believed, that the nervous system is a network of living material with multiple nuclei that lacks any boundary between individual elements. This implies cytoplasmic continuity from one location in the network to another

Retina (11) In vertebrate eyes, the layer of photoreceptor cells and other neurons that lines most of the inside of an eyeball

Retinal cell (11) In insects, a sensory neuron of the compound eye

Retinotopic organization (9) In vertebrates, the topographical relationship between the retina and the tectum or the visual cortex in the brain, in which there is a point-to-point map of the retina projected onto the brain

Retinotopy (9) See retinotopic organization

Retinula (11) In insects, the light-sensing structure of the eye formed by parts of the retinal cells

Retrograde messenger (8) A molecule that diffuses from a postsynaptic cell back to the presynaptic one, where it influences subsequent synaptic transmission

Retrograde transport (2) The transport of materials from the periphery toward the cell body in a neurite

Re-uptake (6) See uptake

Reversal potential (6) The membrane potential at which a postsynaptic potential changes from hyperpolarizing to depolarizing or vice versa. Usually determined experimentally by applying successively greater current pulses to the postsynaptic neuron while stimulating the presynaptic neuron, until the postsynaptic potential is reduced to zero and then changes polarity

Rhabdome (11) In insects, the light-sensing structure of the eye, formed by eight rhabdomeres

Rhabdomere (11) In insects, the specialized photoreceptive area of each retinal cell. Eight rhabdomeres together form a rhabdome

Rhombencephalon (3) See hindbrain

Robust nucleus of the archistriatum (RA) (25) In songbirds, a region of the brain involved in the control of singing. It is sexually dimorphic in birds in which only the males sing, being larger in males

Rod cell (11) In the vertebrate retina, a type of receptor cell characterized by a cylindrical outer segment and great sensitivity to light

Roll (17) During the flight of an insect or bird, a rotation of the body around a line through its long axis

Ruffini end organ (14) In vertebrates, a type of slowly adapting mechanoreceptor found deep in hairy skin that primarily responds to pressure on the skin

– S –

Saccule (14) In the vertebrate inner ear, the part of the vestibular system that mainly helps to detect linear acceleration of the head and body

Saltatory conduction (5) The method of conduction of action potentials along myelinated axons, in which the action potential jumps along the axon from one node of Ranvier to another

Schwann cell (2) In vertebrates, a type of glia found only in the peripheral nervous system, where it forms myelin around the neurites of peripheral nerves

SDN-POA (25) See sexually dimorphic nucleus of the preoptic area

Second messenger (6) A chemical intermediary released inside a cell that activates biochemical reactions, which in turn cause some longer-term change in the cell's activity

Selective permeability (4) The property of a nerve cell membrane that allows some ions to cross the membrane relatively easily, but other ions to cross only with great difficulty, if at all

Semantic memory (24) A kind of declarative memory that represents items like faces and names of things

Semaphorin family (22) A family of guidance molecules that help determine the direction of growth of developing axons

Semicircular canal (14) In the vertebrate inner ear, a fluid-filled canal that is part of the vestibular system, and that helps to detect rotational acceleration of the head and body

Semi-permeability (4) See selective permeability

Sense organ (9) A complex, multicellular structure specialized to detect one particular type of sensory stimulus

Sensitization (24) An increase in the strength of any reflex caused by one or more strong stimuli different from the stimulus that usually evokes that reflex

Sensory nerve (3) A cranial or spinal nerve consisting entirely of neurons that carry information from a sensory structure such as a sense organ to the central nervous system. Sensory nerves are afferent nerves

Sensory neuron (2) A neuron that carries sensory information into the central nervous system

Sensory receptor (9) A single celled or simple multicellular structure specialized to detect a particular type of stimulus; sometimes referred to just as a receptor

Sensory specificity (9) The separation of different sensory modalities and different sensations associated with them in the central nervous system

Sensory system (1) The part of the nervous system that deals with sensory input. This usually includes all the sensory structures of the body, the sensory nerves, and the areas of the nervous system that are responsible for interpreting sensory information

Septal nucleus (3) A nucleus in the forebrain that lies at the anterior end of the cingulate gyrus; part of the limbic system

Serial processing (11) The idea that the various features of a sensory signal are encoded and interpreted by a hierarchy of cells in which the response properties of each cell in the hierarchy is shaped by the input it receives from cells in the layer above it

Serotonergic (7) A term applied to neurons that release serotonin as a neurotransmitter

Servomechanism (20) A device in which feedback is used to adjust the output of the device based on how it is performing

Sexually dimorphic nucleus of the preoptic area (SDN-POA) (25) In mammals, a nucleus of the preoptic area of the hypothalamus that controls mounting during mating. It is significantly larger in mature males than in females

Short-term memory (24) A type of memory formed and retained for a relatively short period of time, from a few minutes to several hours

Signaling protein (2) A membrane-bound protein that receives or responds to chemical messages originating from outside the neuron

Sign-conserving synapse (11) In the vertebrate retina, a synapse formed by a nonspiking neuron at which presynaptic depolarization leads to postsynaptic depolarization and presynaptic hyperpolarization leads to postsynaptic hyperpolarization

Sign-inverting synapse (11) In the vertebrate retina, a synapse formed by a nonspiking neuron at which presynaptic depolarization leads to postsynaptic hyperpolarization and presynaptic hyperpolarization leads to postsynaptic depolarization

Simple cell (11) In the visual cortex of mammals, a type of neuron that is excited by properly positioned and oriented straight borders or bars of light

Size principle (15) The observation that recruitment of muscle fibers occurs in an orderly fashion, from those innervated by the smallest motor neurons to those innervated by the largest ones

Slow axonal transport (2) Orthograde axonal transport at the rate of between 0.5 and 6 mm/day

Slow-twitch oxidative muscle fibers (SO) (15) Slow-contracting and relatively weak muscle fibers that are fatigue resistant and have high oxidative capacity and low glycolytic capacity

SMA (18) See supplementary motor area

SNAP (7) Soluble NSF attachment protein, a type of protein that helps bind a synaptic vesicle to the presynaptic membrane at a synapse during docking

SNAP-25 (7) 25 kilodalton synaptosomal-associated protein, a t-SNAP associated with the membrane of a presynaptic neuron at a synapse, and which is thought to aid in docking a vesicle with the membrane of the presynaptic neuron prior to the release of neurotransmitter

SNARE hypothesis (7) The SNAP receptor hypothesis; the idea that docking of synaptic vesicles to the membrane of a presynaptic neuron at the synapse occurs by mutual binding among v-SNAREs, t-SNAREs, and cytoplasmic proteins such as NSF and the SNAPs

SNB (25) See spinal nucleus of the bulbocavernosus

SO (15) See slow-twitch oxidative muscle fibers

Sodium inactivation (5) Closure of the inactivation gate of the sodium channel after the channel has been open a short while, which prevents further flow of sodium ions through the channel

Sodium-potassium exchange pump (2) A type of ion pump capable of exchanging three atoms of sodium for two atoms of potassium against their concentration gradients across a membrane, powered by the hydrolysis of ATP

Sodium pump (2) See sodium-potassium exchange pump

Solitary nuclear complex (13) In vertebrates, a nucleus in the medulla that receives afferent input via cranial nerves VII, IX, and X from gustatory receptor cells on the tongue and in the mouth and throat

Soma (2) The part of a neuron that contains the cell nucleus; the cell body

Somatic nervous system (3) The nerves and ganglia involved in sensory input and voluntary actions

Somatosensory cortex (9) In vertebrates, the part of the cortex posterior to the central sulcus that is devoted to processing input from the somatosensory system

Somatosensory system (9) In vertebrates, the sensory system that deals with information from the body, such as touch, vibration, pain, heat, cold, and kinesthesis

Somatotopic organization (9) The topographical organization of the somatosensory system, in which there is a point-to-point mapping of the surface of the body onto the somatosensory cortex

Somatotopy (9) See somatotopic organization

Space constant (8) See length constant

Spatial summation (8) In a neuron, summation of two or more inputs of the same type from different locations that occur at about the same time

Specific nerve energies, law of (9) See law of specific nerve energies

Spike (5) An action potential, so called because it usually has a sharp, pointed appearance on an oscilloscope

Spike initiation zone (6) The region of an axon at which an action potential is normally initiated when the neuron is excited

Spinal cord (3) In vertebrates, the part of the central nervous system that is encased in the bones of the vertebral column

Spinal nucleus of the bulbocavernosus (SNB) (25) In the mammalian spinal cord, a region that controls pelvic, and, in the male, penile muscles used during mating. The nucleus is sexually dimorphic, being larger in males

Spinocerebellum (18) In vertebrates, the functional subdivision of the cerebellum that makes extensive connections with the spinal cord

Spinocortical tract (3) In vertebrates, any tract that carries information from the spinal cord to the cortex

Spinothalamic pathway (14) In vertebrates, tracts in the spinal cord and brain stem that convey information from nociceptive, temperature, and certain touch receptors to nuclei in the thalamus and to the somatosensory cortex

Spiral ganglion (12) In vertebrates, the ganglion located in the cochlea, within which lie the cell bodies of the auditory neurons

Split-brain patient (21) A person in whom the corpus callosum has been severed surgically; the procedure was formerly done for the control of severe epilepsy

Statocyst organ (10) A sense organ that detects the direction of gravity and changes in the orientation of an animal

Statolith (10) A stone-like concretion resting on or attached to sensory hairs in a statocyst organ

Steady state (4) A condition in which the number of ions of any type that enter a neuron or other cell equals the number of ions of that same type that leaves it

Stemmata (11) The single-lens eyes of immature insects

Stereocilia (12) The hair-like projections of the hair cells of the vertebrate inner ear and the neuromast cells of the lateral line of fishes and amphibians

Stereopsis (23) See stereoscopic vision

Stereoscopic vision (23) The three-dimensional vision that is achieved in vertebrates whose visual fields overlap significantly

Steroid hormone (25) A type of hormone that is structurally related to cholesterol, is lipid soluble, and can pass directly through cell membranes

Stimulus intensity (10) The objective strength of a stimulus, variation of which gives rise to the subjective sensation of intensity, as for example the brightness of light or the loudness of sound

Stimulus strength (10) See stimulus intensity

Storage pool (7) Collectively, synaptic vesicles that are anchored to the cytoskeleton and are not ready to release their contents

Stretch-activated channel (14) A type of ion channel that opens when the cell is stimulated mechanically

Stretch-inhibited channel (14) A type of ion channel that closes when the cell is stimulated mechanically

Stretch-sensitive channel (2) A type of ion channel that opens or closes when a mechanical stimulus is applied to the cell

Striate cortex (11) See primary visual cortex

Striatum (3) The neostriatum, part of the basal ganglia

Submucosal plexus (3) In the vertebrate intestine, a network of interconnected ganglia located within the submucosa

Substantia nigra (3) In vertebrates, a part of the tegmentum of the midbrain, functionally part of the basal ganglia and important in motor control; it is the structure that is damaged in Parkinson's disease

Subthalamic nucleus (3) In vertebrates, a nucleus in the diencephalon at the junction with the mesencephalon, functionally part of the basal ganglia and important in motor control

Sulcus (3) In vertebrates, an infolding of tissue on the surface that define an edge of a gyrus

Summation (8) In a neuron, the combined postsynaptic effect of two or more excitatory presynaptic inputs

Superior cervical ganglion (3) In vertebrates, a sympathetic autonomic ganglion that lies in the body cavity and innervates the head and neck

Superior colliculus (3) In mammals, a part of the midbrain important in oculomotor reflexes; it corresponds in part to the tectum in lower vertebrates

Superior mesenteric ganglion (3) In vertebrates, a sympathetic autonomic ganglion that lies in the body cavity and innervates the large intestine

Superior olivary nucleus (12) In vertebrates, a nucleus in the medulla that acts as a relay station for auditory information

Supplementary motor area (SMA) (18) In mammals, the part of the motor cortex that lies in front of the premotor cortex on the top of the brain. This region sends axons to the basal ganglia and the thalamus as well as to some of the motor nuclei of the brain stem

Surround (11) In the vertebrate retina, the outer, doughnut-shaped part of the circular receptive field of a bipolar cell or ganglion cell

Sympathetic autonomic system (3) In vertebrates, the division of the autonomic nervous system that tends to arouse the body for action

Synapse (2) The specialized site at which communication between one neuron and another takes place

Synapsin-I (7) A protein that binds vesicles to the cytoskeleton of a presynaptic neuron

Synaptic cleft (2) At a chemical synapse, the small space between the pre- and postsynaptic neurons

Synaptic depression (24) A reduction in the effectiveness of a synapse in eliciting an action potential in a postsynaptic neuron

Synaptic transmission (2) The process by which one neuron communicates with another. Transmission may occur electrically or via a chemical intermediary

Synaptic vesicle (2) The roughly spherical intracellular organelle in which chemical neurotransmitter is stored in presynaptic neurons

Synaptobrevin (7) A v-SNARE associated with the membrane of a vesicle that helps bind the vesicle to the membrane of the presynaptic terminal during vesicle docking; also known as VAMP

Synaptotagmin (7) A specialized protein located in the membrane of a synaptic vesicle that is thought to aid in docking a vesicle with the membrane of the presynaptic neuron prior to the release of neurotransmitter

Synergistic muscles (15) Muscles that have a similar action around a joint or other part of the body when they contract

Syntaxin (7) A type of t-SNARE associated with the membrane of a presynaptic neuron at a synapse that is thought to aid in docking a vesicle with the membrane of the presynaptic neuron prior to the release of neurotransmitter

—T—

Targeted gene replacement (24) A method for replacing one known gene in an animal embryo with a mutant copy of it for the purpose of studying the effect of the loss of function of the normal gene. The animal so produced is called a knockout animal

Tectorial membrane (12) In the vertebrate ear, a membrane of the cochlea that lies over the basilar membrane, and which by its movement relative to the basilar membrane in the presence of a sound stimulus causes the stereocilia of the hair cells to vibrate

Tectum (3) In lower vertebrates, the part of the midbrain that lies over the cerebral aqueduct. It is important in processing visual input

Tegmentum (3) In vertebrates, the part of the midbrain that lies under the cerebral aqueduct. It is important in processing auditory input and in orientation reflexes and posture

Telencephalon (3) In vertebrates, the anteriormost part of the forebrain, comprising mainly the cerebral hemispheres

Temporal code (10) A code for sensory information that conveys information by variations of a response over time (besides the variation due to adaptation)

Temporal lobe (3) In vertebrates, the sausage-shaped lobe of each cerebral hemisphere that lies ventral and lateral to the parietal lobe

Temporal summation (8) In neurons, summation of two or more inputs of the same type that occur at different times

Tetrodotoxin (TTX) (6) A neurotoxin derived from puffer fish that blocks voltage-sensitive sodium channels

Thalamus (3) In vertebrates, a part of the diencephalon that serves as a relay station for sensory input

Thermoreceptor (9) A receptor for heat or cold

Threshold (5) The minimum depolarization of the membrane potential necessary to evoke an action potential in a neuron

Time constant (8) The time it takes a transient change of the membrane potential to decline to 37% of the peak value of the change; functionally, the time a postsynaptic potential will last at a particular site. Quantitatively expressed as $\tau = r_m \cdot c_m$

Tonic muscle fiber (15) A muscle fiber that gives a slow and weak mechanical response to stimulation by its motor neuron, but that can contract over relatively long periods without significant fatigue

Tonic receptor (9) A receptor cell in which adaptation is very slow; the response of the cell to a continuously applied stimulus never declines to zero

Tonotopic organization (9) In vertebrates, the topographical organization of the vertebrate cochlea, in which there is a point-to-point mapping of the cochlear surface onto the auditory cortex; more generally, the topographical mapping of input from an auditory sense organ in any animal onto some part of the central nervous system

Tonotopy (9) See tonotopic organization

Topographical organization (9) The point-to-point representation of a sensory surface in the central nervous system

Torus semicircularus (3) In lower vertebrates, a part of the midbrain important in processing auditory information; see also next entry

Torus semicircularus (14) In electric fish, the midbrain structure that receives axons from the ELL carrying electrosensory information

Tract (3) See nerve tract

Transducin (11) In vertebrate eyes, a type of G protein that aids transduction in rod and cone cells

Transduction (9) In the context of sensory systems, the transformation of the energy of a stimulus to a change in membrane potential of a sensory cell

Transforming growth factor superfamily (22) A superfamily of neurotrophic factors that affect the growth and development of dopaminergic and other neurons

Transport protein (2) A specialized type of membrane protein that facilitates the transfer of ions or molecules across the membrane

Trophic factor (22) A chemical substance that is necessary for the continued health and proper function of an individual cell

t-SNARE (7) A target snap receptor, a type of protein associated with the membrane of a presynaptic neuron at a synapse, and which is thought to aid in docking a vesicle with the membrane of the presynaptic neuron prior to the release of neurotransmitter

TTX (6) See tetrodotoxin

Tuberous organ (14) One of the two types of sense receptors of animals that have an electric sense. Tuberous organs are embedded in the skin, are not usually open to the surface, and are present only in animals with an active electric sense

Tubulin (2) A 120 kilodalton globular protein consisting of α and β dimers and which can be assembled into a hollow-cored neurotubule

Tufted cell (13) In vertebrates, a type of neuron in the olfactory bulb that in a glomerulus receives input from olfactory neurons

Tuning (12) In auditory systems, the feature that a cell may respond only to a relatively narrow band of sound frequency

Tuning curve (12) A graph showing the range of frequency and intensity to which any cell in the auditory pathway will respond

Twitch muscle fiber (15) A muscle fiber that gives a short but relatively strong and fast contraction to a single nerve impulse

Two-photon laser scanning confocal microscopy (22) A technique similar to scanning confocal microscopy, but in which two laser beams, individually containing too little energy to excite fluorescence, are scanned together over a single plane in a specimen; this confines fluorescent excitation to a single plane and produces a clear image by sharply reducing fluorescence from out-of-focus planes in the specimen

Two-tone suppression (12) The reduction of the response of a hair cell or auditory neuron to one tone by the presence of a second tone of a different frequency

Tympanic organ (12) The true ear present in some insects, consisting of a membrane over an air sac. The sensory neurons are attached directly to the membrane in such a way that they are excited when the membrane vibrates

—U—

Unconditioned stimulus (24) In associative conditioning learning, a stimulus that causes a simple reflex response in the absence of any learning by the animal

Uncus (3) A part of the cerebral hemispheres that lies along the medial face of the temporal lobe; part of the limbic system

Unitary smooth muscle (15) The smooth muscle found along the gut, containing extensive electrical coupling via gap junctions among the muscle cells

Uptake (6) The removal of neurotransmitter from a synapse by transport into a neuron. Functionally, one of the means of transmitter inactivation

US (24) See unconditioned stimulus

Utricle (14) In the vertebrate inner ear, the part of the vestibular system that mainly helps to detect the direction of the force of gravity

—V—

VAMP (7) Vesicle-associated membrane protein, also known as synaptobrevin, a v-SNARE associated with the membrane of a vesicle that helps bind the vesicle to the membrane of the presynaptic terminal during vesicle docking

Vasopressin (25) A peptide hormone that controls water resorption and blood pressure peripherally, and influences sexual receptivity and learning and memory in the central nervous system; also known as antidiuretic hormone (ADH)

Ventral cochlear nucleus (12) In vertebrates, a nucleus in the medulla in which some of the axons of the cochlear nerve terminate

Ventral posteriomedial nucleus (13) In primates, a nucleus in the thalamus that processes gustatory information

Ventricle (3) One part of the system of interconnected cavities inside the vertebrate brain

Vesicular hypothesis (7) The idea that in chemical synaptic transmission vesicles are the storage organelles for molecules of neurotransmitter in presynaptic terminals, and that consequently neurotransmitter is released in packets

Vestibular ganglion (14) In vertebrates, the ganglion containing the cell bodies of afferent neurons running from the vestibular apparatus of the inner ear to the brain

Vestibular nerve (14) In vertebrates, part of cranial nerve VIII, which carries axons of vestibular afferents from the vestibular apparatus of the inner ear to the vestibular nuclei in the medulla

Vestibular nuclei (14) In vertebrates, a group of four nuclei in the medulla that serve as motor nuclei and as relay stations for information from the vestibular apparatus

Vestibular organ (14) In vertebrates, the organ of gravity, orientation, and motion detection

Vestibular system (14) In vertebrates, structures such as the semicircular canals and otolith organs that together detect or respond to the force of gravity, the orientation of the head, and accelerations

Vestibular tract (14) In vertebrates, one of several tracts in the spinal cord that carry vestibular information from the medulla to the motor centers of the spinal cord

Vestibulocerebellum (18) The functional subdivision of the cerebellum that makes extensive connections with the semicircular canals. This part can set the gain of the vestibulo-ocular reflex by a feed forward signal

Vestibulo-ocular reflex (VOR) (20) A compensatory movement of the eyes in the direction opposite to a (fast) rotation of the head, driven by input from the vestibular organs

Visual association area (21) A portion of the occipital lobe of the brain that lies anterior to the visual cortex

Visual cortex (11) The region of the occipital lobe of the cerebral hemispheres of the brains of birds and mammals in which visual information is processed

Voltage clamp (5) A technique for recording the currents generated by ions crossing the cell membrane during controlled changes in the membrane potential

Voltage-gated channel (2) An ion channel that opens or closes according to the level of the voltage difference across the neuronal membrane

Voltage-sensitive channel (2) See voltage gated channel

VOR (20) See vestibulo-ocular reflex

v-SNARE (7) Vesicular SNAP receptor, a type of protein associated with the membrane of a synaptic vesicle that helps bind the vesicle to the membrane of the presynaptic terminal during docking

– W –

W cells (11) In the mammalian retina, ganglion cells that have relatively large diameter axons, are located in the fovea, and help control the pupils and eye movement

Wernicke's area (21) In humans, a part of the temporal lobe that helps to interpret spoken words

White communicating ramus (3) In vertebrates, a small nerve branch through which a sympathetic preganglionic fiber from the spinal cord enters a sympathetic ganglion of the paravertebral chain

White matter (2) In vertebrates, the regions of the central nervous system occupied by bundles of myelinated axons

Working memory (24) Short-term memory

– X –

X cells (11) In the mammalian retina, ganglion cells that have relatively small diameter axons, are located all over the retina, and help in the resolution of an image

– Y –

Y cells (11) In the mammalian retina, ganglion cells that have relatively large diameter axons, are located in the periphery of the retina, and help to detect movement

Yaw (17) During the flight of an insect or bird, a deviation from the flight path in which the body is turned away from the direction in which the animal is flying. During a normal turn, the direction in which the animal is flying changes along with the direction in which its body is pointing

CREDITS

CHAPTER 1

1-1(A) Harlow, J. M. 1868. Recovery from the passage of an iron bar through the head. *Bulletin of the Massachusetts Medical Society* 2, 3–20.

1-1(B) Reprinted with permission from Damasio, H., T. Grabowski, R. Frank, A. M. Galaburda, and A. R. Damasio. 1994. The return of Phineas Gage: Clues about the brain from the skull of a famous patient. *Science* 264:1102–1105. Copyright 1994 American Association for the Advancement of Science.

1-4 Photo courtesy of Tove Heller, University of Kaiserslautern.

1-5 Photo courtesy of Deborah Allen.

CHAPTER 2

2-A Micrograph courtesy of Mario Saltarelli and William T. Greenough, University of Illinois at Urbana-Champaign.

2-C B. J. Schnapp, R. D. Valle, M. P. Sheetz, and T. S. Reese, from Lodish, H., D. Baltimore, A. Berk, S. L. Zipursky, P. Matsudaira, and J. Darnell. 1995. *Molecular Cell Biology,* 3d ed. New York: W. H. Freeman & Co.

2-3(A) From A. Peters, S. L. Palay, and H. deF. Webster. 1991. *The Fine Structure of the Nervous System,* 3d ed. New York: Oxford University Press. Copyright (c) 1991 Alan Peters.

2-3(B) From Sotelo, C., R. Llinas, and R. Baker. 1974. Structural study of inferior olivary nucleus of the cat: morphological correlates of electrotonic coupling. *Journal of Neurophysiology* 37:541–559.

2-4(A) From A. Peters, S. L. Palay, and H. deF. Webster. 1991. *The Fine Structure of the Nervous System,* 3d ed. New York: Oxford University Press. Copyright (c) 1991 Alan Peters.

2-5(A) From Toner, P. G., and K. E. Carr. 1971. *Cell Structure,* 2d ed. Baltimore: Williams and Wilkins.

2-5(B) From A. Peters, S. L. Palay, and H. deF. Webster. 1991. *The Fine Structure of the Nervous System,* 3d ed. New York: Oxford University Press. Copyright (c) 1991 Alan Peters.

2-6 From Raine, C. S. 1994. Neurocellular anatomy. In *Basic Neurochemistry,* 5th ed., ed. G. J. Siegel, B. W. Agranoff, R. W. Albers, and P. B. Molinoff. New York: Raven Press; and from A. Peters, S. L. Palay, and H. deF. Webster. 1991. *The Fine Structure of the Nervous System,* 3d ed. New York: Oxford University Press. Copyright (c) 1991 Alan Peters.

2-9(B) Adapted from Catterall, W. A. 1991. Structure and function of voltage-gated sodium and calcium channels. *Current Opinion in Neurobiology* 1:5–13.

2-12(A) From Maunsbach, A. B., E. Skriver, M. Soederholm, and H. Hebert. 1986. Electron microscopy of the Na,K-ion pump. *Proceedings of the XIth International Congress on Electron Microscopy,* Kyoto, pp. 1801–1806.

2-12(B) Adapted from Hilgemann, D. W. 1994. Channel-like function of the Na,K pump probed at microsecond resolution in giant membrane patches. *Science* 263:1429–1432.

CHAPTER 3

3-A From Warr, W. B., J. S. de Olmos, and L. Heimer. 1981. Horseradish peroxidase. The basic procedure. In *Neuroanatomical Tract-Tracing Methods.* Lennart, H. and M. J. Robards, eds., pp. 207–262. New York: Plenum Press. Reprinted by permission of Plenum Press.

3-B F. Delcomyn.

3-4(A) Micrograph courtesy of Chris Comer, University of Illinois at Chicago.

3-5 F. Delcomyn.

3-6(A) Diane Major and Henry Hall, MIT, from Nauta, W. J. H., and M. Feirtag. 1986. *Fundamental Neuroanatomy.* New York: W. H. Freeman & Co.

3-15 Diane Major and Henry Hall, MIT, from Nauta, W. J. H., and M. Feirtag. 1986. *Fundamental Neuroanatomy.* New York: W. H. Freeman & Co.

3-18 Diane Major and Henry Hall, MIT, from Nauta, W. J. H., and M. Feirtag. 1986. *Fundamental Neuroanatomy.* New York: W. H. Freeman & Co.

3-28(B) Micrograph by G. Gabella, from Lees, G. M. 1986. Anatomy, histology, and electron microscopy of sympathetic, parasympathetic, and enteric neurons. In *Autonomic and Enteric Ganglia. Transmission and its Pharmacology,* ed. A. G. Karczmar, K. Koketsu, and S. Nishi. 27–60. New York: Plenum Press. Reprinted by permission of Plenum Press.

CHAPTER 4

4-A(a) Reprinted with permission from *Nature:* Hodgkin, A. L., and A. F. Huxley. 1939. Action potentials recorded from inside a nerve fibre. *Nature (London)* 144:710–711. Copyright (c) (1939) Macmillan Magazines Limited.

4-B(a) Micrograph courtesy of Robert T. Doyle and Philip G. Haydon, Iowa State University.

CHAPTER 5

5-5(A, B) From Hodgkin, A. L., and Huxley, A. F. 1967. *The Conduction of the Nervous Impulse.* Liverpool: Liverpool University Press. Copyright (c) 1967 by H. L. Hodgkin.

5-6 From Hodgkin, A. L., and Huxley, A. F. 1967. *The Conduction of the Nervous Impulse.* Liverpool: Liverpool University Press. Copyright (c) 1967 by H. L. Hodgkin.

5-7 Reprinted with permission from *Nature:* Sigworth, F. J., and F. Neher. 1980. Single Na^+ channel currents observed in cultured rat muscle cells. *Nature (London)* 287:447–449. Copyright (c) (1980) Macmillan Magazines Limited.

CHAPTER 6

6-A Reproduced from Palay, S. 1956. Synapses in the central nervous system. *Journal of Biophysical and Biochemical Cytology Supplement 2,* pp. 193–202 by copyright permission of The Rockefeller University Press.

6-C Omikron/Photo Researchers.

6-2(A, B) Electron micrograph courtesy of Peter Brink, State University of New York.

6-6 From Nishi, S., and K. Koketsu. 1960. Electrical properties and activities of single sympathetic neurons in frogs. *Journal of Cellular and Comparative Physiology* 55:15–30. Copyright (c) (1960) Wiley-Liss, Inc. Reprinted by permission of Wiley-Liss, Inc., a subsidiary of John Wiley & Sons, Inc.

CHAPTER 7

7-B(a, b) Reprinted with permission from Conner, J. A., W. J. Wadman, P. E. Hockberger, and R. K. S. Wong. 1988. Sustained dendritic gradients of Ca^{2+} induced by excitatory amino acids in CA1 hippocampal neurons. *Science* 240:649–653. Copyright 1988 American Association for the Advancement of Science.

7-1(B) From Katz, B., and R. Miledi. 1967. A study of synaptic transmission in the absence of nerve impulses. *Journal of Physiology* 192:407–436. Copyright (c) 1967 The Physiological Society.

7-2 From Fatt, P., and B. Katz. 1952. Spontaneous subthreshold potentials at motor nerve endings. *Journal of Physiology* 117:109–128. Copyright (c) 1952 The Physiological Society.

7-3 From Boyd, I. A., and Martin, A. R. 1956. The endplate potential in mammalian muscle. *Journal of Physiology* 132:74–91. Copyright (c) 1956 The Physiological Society.

7-4 Reproduced from Hirokawa, K. Sobue, K. Kanda, A. Harada, and H. Yorifuji. 1989. The cytoskeletal architecture of the presynaptic terminal and molecular structure of synapsin 1. *Journal of Cell Biology* 108:111–126 by copyright permission of The Rockefeller University Press.

7-5(A) From Couteaux, R., and M. Pecot-Dechavassine. 1974. Les zones specialisees des membranes presynaptiques. *Comptes Rendus des Séances de l'Academie des Sciences* (Paris) 278:291–293. Reprinted by permission of the French Academy of Sciences.

7-5(B) Reproduced from Heuser, J. E., Reese, T. S., Dennis, M. J., Jan, Y., Jan, L., and Evans, L. 1979. Synaptic vesicle exocytosis captured by quick freezing and correlated with quantal transmitter release. *Journal of Cell Biology* 81:275–300 by copyright permission of The Rockefeller University Press.

7-7 From Heuser, J. E. 1976. Synaptic vesicle exocytosis revealed in quick-frozen frog neuromuscular junctions treated with 4-amino-pyridine and given a single electrical shock. In *Approaches to the Cell Biology of Neurons,* ed. W. M. Cowan and J. A. Ferrendelli, 215–239. Bethesda, MD: Society for Neuroscience.

7-8 From Raine, C. S. 1994. Neurocellular anatomy. In *Basic Neurochemistry,* 5th ed., ed. G. J. Siegel, B. W. Agranoff, R. W. Albers, and P. B. Molinoff, pp. 3–32. New York: Raven Press.

CHAPTER 9

9-1(A, B) Spoken phrase adapted from the authorized version of the revised SPIN (Speech Perception in Noise) test. Courtesy of Chen Liu and Robert C. Bilger, University of Illinois at Urbana-Champaign.

9-2 From Randall, D., W. Burggren, and K. French, *Eckert Animal Physiology. Mechanisms and Adaptations,* 4th ed. New York: W. H. Freeman & Co.

9-3(A, B) N.A.S./M.W.F. Tweedie/Photo Researchers.

9-4 From Lacher, V. 1964. Elektrophysiologische Untersuchungen an einzelnen Rezeptoren für Geruch, Kohlendioxyd, Luftfeuchtigkeit und Temperatur auf den Antennen der Arbeitsbiene und der Drohne (*Apis mellifica* L.). *Zeitschrift für vergleichende Physiologie* 48:587–623, figure 17, p. 606. Copyright (c) 1983 Springer-Verlag. Reprinted by permission.

9-6 Adapted from Terzuolo, C. A., and Y. Washizu. 1962. Relation between stimulus strength, generator potential and impulse frequency in stretch receptor of Crustacea. *Journal of Neurophysiology* 25:59–66.

9-7 Adapted from Loewenstein, W. R. 1961. Excitation and inactivation in a receptor membrane. *Annals of the New York Academy of Sciences* 94:510–534.

9-8(A) From Boeckh, J., K. E. Kaissling, and D. Schneider. 1965. Insect olfactory receptors. *Cold Spring Harbor Symposia on Quantitative Biology* 30, 263–280.

9-10 Adapted from Ottoson, D. and G. M. Shepherd. 1965. Receptor potentials and impulse generation in the isolated spindle during controlled extension. *Cold Spring Harbor Symposia on Quantitative Biology* 30, 105–114.

9-11 Adapted from Ottoson, D., and G. M. Shepherd. 1965. Receptor potentials and impulse generation in the isolated spindle during controlled extension. *Cold Spring Harbor Symposia on Quantitative Biology* 30, 105–114.

9-16 Adapted from Thompson, R. F. 1993. *The Brain.* New York: W. H. Freeman & Co., Fig. 8-13b.

CHAPTER 10

10-5(B) Adapted from Flock, A. 1965. Transducing mechanisms in the lateral line canal organ receptors. *Cold Spring Harbor Symposia on Quantitative Biology* 30, 133–145.

CHAPTER 11

11-A(a, b) From Baylor, D. A., Lamb, T. D., and Yau, K.-W. 1979. The membrane current of single rod outer segments. *Journal of Physiology* 288:589–611. Copyright (c) 1979 The Physiological Society.

11-B Photomicrograph by Akimichi Kaneko, from J. Dowling. 1987. *The Retina.* Cambridge, MA: Belknap Press of Harvard University Press. Copyright (c) 1987 Akimichi Kaneko.

11-4(A) From Farber, D., and E. Adler. 1986. Issues and questions in cell biology of the retina. In *The Retina: A Model for Cell Biology Studies, Part 1,* ed. R. Adler and D. Farber, 2–16. New York: Academic Press. Reprinted by permission of Academic Press.

11-13 Photomicrograph courtesy of Jonathan W. Marks and Joseph G. Malpeli, University of Illinois at Urbana-Champaign.

11-19(B) From LeVay, S., T. N. Wiesel, and D. H. Hubel. 1980. The development of ocular dominance columns in normal and visually deprived monkeys. *Journal of Comparative Neurology* 191:1–51. Copyright (c) 1980 John Wiley & Sons. Reprinted by permission of Wiley-Liss, a subsidiary of John Wiley & Sons, Inc.

11-20(B) From Bosking, W. H., Y. Zhang, B. Schofield, and D. Fitzpatrick. 1997. Orientation selectivity and the arrangement of horizontal connections in tree shrew striate cortex. *Journal of Neuroscience* 17:2112–2127.

Copyright (c) 1997 Society for Neuroscience; reprinted by permission.

11-21(A) From Hubel, D., and M. Livingstone. 1987. Segregation of form, color, and stereopsis in primate area 18. *Journal of Neuroscience* 7:3378–3415. Copyright (c) 1987 Society for Neuroscience; reprinted by permission.

11-23(B) From Sherk, T. E. 1978. Development of the compound eyes of dragonflies (Odonata). I. Adult compound eyes. *Journal of Experimental Zoology* 203:61–80. Copyright (c) 1978 John Wiley & Sons. Reprinted by permission of Wiley-Liss, a subsidiary of John Wiley & Sons, Inc.

11-24 From Gribakin, F. G. 1975. Functional morphology of the compound eye of the bee. In *The Compound Eye and Vision of Insects*, ed. G. A. Horridge, 154–176. Oxford: Clarendon Press.

11-27 From Riehle, A., and N. Franceschini. 1984. Motion detection in flies: parametric control over ON-OFF pathways. *Experimental Brain Research* 54:390–394, Figure 1C, p. 391. Copyright (c) 1984 Springer-Verlag; reprinted by permission.

CHAPTER 12

12-A Adapted from Arthur, R. M., R. R. Pfeiffer, and N. Suga. 1971. Properties of 'two-tone inhibition' in primary auditory neurones. *Journal of Physiology* 212:593–609.

12-4(A, B) From Harrison, R. V., and I. M. Hunter-Duvar. 1988. An anatomical tour of the cochlea. *In Physiology of the Ear*, ed. A. F. Jahn and J. Santos-Sacchi, 159–171. New York: Raven Press. Reprinted by permission of Lippincott-Raven Publishers.

12-5(B) From Pickles, J. O., J. Brix, and O. Gleich. 1990. The search for the morphological basis of mechanotransduction in cochlear hair cells. In *Information Processing in Mammalian Auditory and Tactile Systems*, ed. L. A. Rowe, pp. 29–43. New York: Wiley-Liss. Copyright (c) 1990 John Wiley & Sons. Reprinted by permission of Wiley-Liss, a subsidiary of John Wiley & Sons, Inc.

12-8(A) From Gerhardt, H. C. 1974. The significance of some spectral features in mating call recognition in the green treefrog (*Hyla cinerea*). *Journal of Experimental Biology* 61:229–241. Copyright 1974 The Company of Biologists, Limited.

12-8(B) Reprinted with permission from Capranica, R. R. 1966. Vocal response of the bullfrog to natural and synthetic mating calls. *Journal of the Acoustical Society of America* 40:1131–1139. Copyright 1966 Acoustical Society of America.

12-9 Adapted from Knudsen, E. I., and M. Konishi. 1978. A neural map of auditory space in the owl. *Science* 200:795–797.

12-10 Adapted from Knudsen, E. I., and M. Konishi. 1978. A neural map of auditory space in the owl. *Science* 200:795–797.

12-11 Adapted from Suga, N. 1984. The extent to which biosonar information is represented in the bat auditory cortex. In *Dynamic Aspects of Neocortical Function*, ed. G. M. Edelman, E. W. Gall, and W. M. Cowan, 315–373. New York: John Wiley & Sons.

12-13 From Randall, D., W. Burggren, and K. French, *Eckert Animal Physiology. Mechanisms and Adaptations*, 4th ed. New York: W. H. Freeman & Co.

12-15(A, B) From Rheinlaender, J., and H. Romer. 1990. Acoustic cues for sound localisation and spacing in orthopteran insects. In *The Tettigoniidae. Biology, Systematics and Evolution*, ed. W. J. Bailey and D. C. F. Rentz, 248–264. London: Crawford House Press. Copyright (c) 1990 J. Rheinlaender.

CHAPTER 13

13-A(a, b) From Schneider, D., Kafka, W. A., Beorza, M. and Bierl, B. A. 1977. Odor receptor responses of male gypsy and nun moths (Lepidoptera,Lymantriidae) to disparlure and its analogues. *Journal of Comparative Physiology A* 113:1–15, figure 1, p. 3, and figure 2, p. 4. Copyright (c) 1977 Springer-Verlag. Reprinted by permission.

13-C(a) Adapted from Kauer, J. S. 1988. Real-time imaging of evoked activity in local circuits of the salamander olfactory bulb. *Nature (London)* 331:166–168

13-C(b) From Kauer, J. S. 1991. Contributions of topography and parallel processing to odor coding in the vertebrate olfactory pathway. *Trends in Neurosciences* 14:79–85.

13-9 Reprinted with permission from *Physiology and Behavior*, vol. 56, by T. Yamamoto, T. Shimura, N. Sakai, and N. Ozaki. Representation of hedonics and quality of taste stimuli in the parabrachial nucleus of the rat, pp. 1197–1202, 1994. Elsevier Science Inc., 655 Avenue of the Americas, New York, NY 10010-5107.

13-14 From Seeliger, G., and T. R. Tobin. 1982. Sense organs. In *The American Cockroach*, ed. W. J. Bell and K. G. Adiyodi, 217–245. London: Chapman and Hall, Figure 9.9a and 9.9b, p. 236.

13-15 Adapted from Boeckh, J., and Ernst, K.-D. 1987. Contribution of single unit analysis in insects to an understanding of olfactory function. *Journal of Comparative Physiology A* 161:549–565.

CHAPTER 14

14-6(B, C, D) Johnson, K. O. 1988. Spatial pattern representation and transformation in monkey somatosensory cortex. *Proceedings of the National Academy of Sciences* 85:1217–1321.

14-12 Adapted from J. M. Goldberg and C. Fernandez. 1971. Physiology of peripheral neurons innervating semicircular canals of the squirrel monkey. I. Resting discharge and response to constant angular accelerations. *Journal of Neurophysiology* 34:635–660.

14-15 From Randall, D., W. Burggren, and K. French, *Eckert Animal Physiology. Mechanisms and Adaptations*, 4th ed. New York: W. H. Freeman & Co.

14-16 Adapted from Szabo, T. 1974. Anatomy of the specialized lateral line organs of electroreception. In *Handbook of Sensory Physiology*, Vol. 3, part 3, ed. A. Fessard, 13–58. New York: Springer-Verlag.

14-17 From Blakemore, R. P., and R. B. Frankel. 1981. Magnetic navigation in bacteria. *Scientific American* 245(6): 58–65. Copyright (c) David L. Balkwill.

CHAPTER 15

15-1(A) NCRI/Science Photo Library/Photo Researchers.

15-1(B) M. I. Walker/Science Source/Photo Researchers.

15-1(C) Michael Abbey/Science Source/Photo Researchers.

15-4 G. W. Willis, M.D./Biological Photo Service.

15-8 Adapted from Atwood, H. L. 1973. An attempt to account for the diversity of crustacean muscles. *American Zoologist* 13:357–378.

CHAPTER 16

16-7(A, B) From Selverston, A. I., Miller, J., and M. Wadepuhl. 1983. Cooperative mechanisms for the production of rhythmic movements. In *Neural Origin of Rhythmic Movements* (Symposia of the SEB, vol. 37), 55–88. Cambridge: Cambridge University Press. Reprinted by permission of the Society of Experimental Biology.

16-8 Adapted from Grillner, S., Wallen, P., Viana Di Prisco, G. 1990. Cellular network underlying locomotion as revealed in a lower vertebrate model: transmitters, membrane properties, circuitry, and simulation. *Cold Spring Harbor Symposia on Quantitative Biology* 55:779–789.

16-10(A, B) Reprinted with permission from *Nature:* Meyrand, P., J. Simmers, and M. Moulins. 1991. Construction of a pattern-genera ing circuit with neurons of different networks. *Nature (London)* 351:60–63. Copyright (c) (1991) Macmillan Magazines Limited.

CHAPTER 17

17-4(A) From Foth, E., and Graham, D. 1983. Influence of loading parallel to the body axis on the walking coordination of an insect. I. Ipsilateral effects. *Biological Cybernetics* 47:17–23, figure 1, p. 18, and figure 3, p. 19. Copyright (c) 1983 Springer-Verlag. Reprinted by permission.

17-4(B) From Foth, E., and D. Graham. 1983. Influence of loading parallel to the body axis on the walking coordination of an insect. II. Contralateral effects. *Biological Cybernetics* 47:149–157, figure 4C, p. 152. Copyright (c) 1983 Springer-Verlag. Reprinted by permission.

17-5 From Pearson, K. G., D. N. Reye, and R. M. Robertson. 1983. Phase-dependent influences on wing stretch receptors on flight rhythm in the locust. *Journal of Neurophysiology* 49:1168–1181.

17-6 Intracellular recording adapted from data supplied by William Heitler, University of St. Andrews.

CHAPTER 18

18-A From Daley, D. L., N. Vardi, B. Appignani, and J. M. Camhi. 1981. Morphology of the giant interneurons and cercal nerve projections of the American cockroach. *Journal of Comparative Neurology* 196:41–52. Copyright (c) 1981 John Wiley & Sons. Reprinted by permission of Wiley-Liss, a subsidiary of John Wiley & Sons, Inc.

18-B(a) Reprinted with permission from *Nature:* Canfield, J. G., and R. C. Eaton. 1990. Swimbladder acoustic pressure transduction initiates Mauthner-mediated escape. *Nature (London)* 347:760–762. Copyright (c) (1990) Macmillan Magazines Limited.

18-B(b) Photomicrograph courtesy of Steve Zotolli, Williams College.

18-D (a,b,c) From Evarts, E. V. 1968. Relation of pyramidal tract activity to force exerted during voluntary movement. *Journal of Neurophysiology* 31:14–27.

18-2 Adapted from Dumont, J. P. C., and J. J. Wine. 1987. The telson flexor neuromuscular system of the crayfish. II. Segment-specific differences in connectivity between premotor neurones and the motor giants. *Journal of Experimental Biology* 127:279–294.

18-8B From Georgopoulos, A., R. Caminiti, J. F. Kalaska, and J. T. Massey. 1983. Spatial coding of movement: a hypothesis concerning the coding of movement direction by motor cortical populations. In *Neural Coding of Motor Performance,* ed. J. Massion, J. Paillard, W. Shultz, and M. Wiesendanger, *Experimental Brain Research Supplement 7*: 327–336, Figure 4, p. 332. Copyright (c) 1983 Springer-Verlag. Reprinted by permission.

CHAPTER 19

19-A(a,b) From Roeder, K. D. 1971. Acoustic alerting mechanisms in insects. *Annals of the New York Academy of Sciences* 188:63–79.

19-B Reprinted with permission from Roeder, K. D. 1966. Auditory system of noctuid moths. *Science* 154:1515–1521. Copyright 1966 American Association for the Advancement of Science.

19-2(A, B) Photographs by Fredrick A. Webster, from Roeder, K. D. 1963. *Nerve Cells and Insect Behavior.* Harvard University Press, Cambridge, MA.

19-3(A) Reprinted from Surlykke, A., and L. A. Miller. 1982. Central branchings of three sensory axons from a moth ear (*Agrotis segetum,* Noctuidae). *Journal of Insect Physiology* 28:357–364 with kind permission from Elsevier Science Ltd, The Boulevard, Langford Lane, Kidlington 0X5 1GB, UK.

19-3(B) Left: From Boyan, G. S. and J. H. Fullard. 1986. Interneurones responding to sound in the tobacco budworm moth *Heliothis virescens* (Noctuidae): morphological and physiological characteristics. *Journal of Comparative Physiology A* 158:391–404, figure 1A, p. 393. Copyright (c) 1986 Springer-Verlag. Reprinted by permission. **Right:** From Boyan, G. S., L. Williams, and J. H. Fullard. 1990. Organization of the auditory pathway in the thoracic ganglia of noctuid moths. *Journal of Comparative Neurology* 295:248–267. Copyright (c) 1990 John Wiley & Sons. Reprinted by permission of Wiley-Liss, a subsidiary of John Wiley & Sons, Inc.

19-4(A, B) From Boyan, G. S., L. Williams, and J. H. Fullard. 1990. Organization of the auditory pathway in the thoracic ganglia of noctuid moths. *Journal of Comparative Neurology* 295:248–267. Copyright (c) 1990 John Wiley & Sons. Reprinted by permission of Wiley-Liss, a subsidiary of John Wiley & Sons, Inc.

19-5 From King, J. R., and C. M. Comer. 1996. Visually elicited turning behavior in Rana pipiens: comparative organization and neural control of escape and prey capture. *Journal of Comparative Physiology A* 178:293–305, figures 7A and 8, p. 301. Copyright (c) 1996 Springer-Verlag. Reprinted by permission.

19-6 From Pfluger, H.-J., and M. Burrows. 1978. Locusts use the same basic motor pattern in swimming as in jumping and kicking. *Journal of Experimental Biology* 75:81–93. Copyright 1978 The Company of Biologists, Limited.

19-7 Adapted from Pearson, K. G. 1983. Neural circuits for jumping in the locust. *Journal of Physiology (Paris)* 78:765–771.

19-9 From Svoboda, K. R., and J. R. Fetcho. 1996. Interactions between the neural networks for escape and swimming in goldfish. *Journal of Neuroscience* 16:843–852. Copyright (c) 1996 Society for Neuroscience; reprinted by permission.

19-11 Adapted from Camhi, J. M., W. Tom, and S. Volman. 1978. The escape behavior of the cockroach *Periplaneta americana.* II. Detection of natural predators by air displacement. *Journal of Comparative Physiology A* 128:203–212.

19-12(A, B) Adapted from Ney, S. W., and R. E. Ritzmann. 1992. Motion analysis of leg joints associated with escape turns of the cockroach, *Periplaneta americana. Journal of Comparative Physiology A* 171:183–194.

CHAPTER 20

20-A Generated and printed with the assistance of Nancy Hyland, University of Illinois at Urbana-Champaign.

20-11 Adapted from Heiligenberg, W. 1991. *Neural Nets in Electric Fish.* Cambridge, Mass.: MIT Press.

20-14 Adapted from Heiligenberg, W. 1991. *Neural Nets in Electric Fish.* Cambridge, Mass.: MIT Press.

CHAPTER 21

21-A Image courtesy of Venkata S. Mattay, M.D., and Daniel R. Weinberger, M.D., Clinical Brain Disorders Branch, and Joseph A. Frank, M.D., Laboratory of Diagnostic Radiology Research, NIMH, NIH, Bethesda, MD 20892.

21-1 Adapted from Ewert, J.-P. 1980. *Neuroethology. An Introduction to the Neurophysiological Fundamentals of Behavior.* Berlin: Springer-Verlag.

21-3(A) From Ewert, J.-P. 1971. Single unit response of the toad's (*Bufo americanus*) caudal thalamus to visual objects. *Zeitschrift für vergleichende Physiologie* 74:81–102, figure 2A, 2E, p. 87. Copyright (c) 1971 Springer-Verlag. Reprinted by permission.

21-3(B) From Matsumoto, N., W. W. Schwippert, and J.-P. Ewert,. 1986. Intracellular activity of morphologically identified neurons of the grass frog's optic tectum in response to moving configurational visual stimuli. *Journal of Comparative Physiology A* 159:721–740, figure 6, p. 727. Copyright (c) 1986 Springer-Verlag. Reprinted by permission.

21-4(A, B) From Ewert, J.-P. 1989. The release of visual behavior in toads: stages of parallel/hierarchical information processing. In *Visuomotor Coordination. Amphibians, Comparisons, Models, and Robots,* ed. J.-P. Ewert and M. A. Arbib, 39–120. New York: Plenum Press. Reprinted by permission of Plenum Press.

21-5 Adapted from Ewert, J.-P. 1980. *Neuroethology. An Introduction to the Neurophysiological Fundamentals of Behavior.* Berlin: Springer-Verlag.

21-6 Adapted from Ewert, J.-P. 1987. Neuroethology of releasing mechanisms: prey-catching in toads. *Behavioral and Brain Sciences* 10:337–405.

21-7(A, B) Reprinted with permission from Wilson, M. A., and B. L. McNaughton. 1993. Dynamics of the hippocampal ensemble code for space. *Science* 261:1055–1058. Copyright 1993 American Association for the Advancement of Science.

21-8 From Posner, M. I., and M. E. Raichle. 1997. *Images of Mind.* New York: Scientific American Library. Courtesy of Marcus E. Raichle, Washington University.

CHAPTER 22

22-C Photomicrograph courtesy of Aloisia Schmid and Christopher Q. Doe, University of Illinois at Urbana-Champaign.

22-1 From Purves, D., and J. W. Lichtman. 1985. *Principles of Neural Development.* Sunderland, Mass: Sinauer Associates, adapted from Sulston, J. E., E. Schierenberg, J. G. White and J. N. Thomson. 1983. The embryonic cell lineage of the nematode *Caenorhabditis elegans. Developmental Biology* 100:64–119.

22-8(A, B, C) Adapted from Hibbard, E. 1965. Orientation and directed growth of Mauthner's cell axons from duplicated vestibular nerve roots. *Experimental Neurology* 13:289–301.

22-9(B) Photomicrograph courtesy of Dennis Bray, University of Cambridge.

22-10(A, B) Adapted from Bentley, D. R., and H. Keshishian. 1982. Pioneer neurons and pathways in insect appendages. *Trends in Neurosciences* 5:354–358

22-10(C) From Taghert, P. H., M. J. Bastiani, R. K. Ho, and C. S. Goodman. 1982. Guidance of pioneer growth cones: filopodial contacts and coupling revealed with an antibody to lucifer yellow. *Developmental Biology* 94:391–399. Reprinted by permission of Academic Press.

22-11(A, B) Reprinted with permission from Levi-Montalcini, R. 1964. Growth control of nerve cells by a protein factor and its antiserum. *Science* 143:105–110. Copyright 1964 American Association for the Advancement of Science.

CHAPTER 23

23-2 Reprinted with permission from Feller, M. B., D. P. Wellis, D. Stellwagen, F. S. Werblin, and C. J. Shatz. 1996. Requirement for cholinergic synaptic transmission in the propagation of spontaneous retinal waves. *Science* 272:1182–1187. Copyright 1996 American Association for the Advancement of Science.

23-3 **Cat:** Adapted from Hubel, D. H., and Wiesel, T. N. 1962. Receptive fields, binocular interaction and functional architecture in the cat's visual cortex. *Journal of Physiology* 160:106–154. **Monkey:** Adapted from Hubel, D. H., 1978. Effects of deprivation on the visual cortex of cat and monkey. *Harvey Lectures,* Series 72, 1976–1977, 1–51. New York: Academic Press.

23-4 Adapted from Wiesel, T. N., and D. H. Hubel. 1965. Comparison of the effects of unilateral and bilateral eye closure on cortical unit responses in kittens. *Journal of Neurophysiology* 28:1029–1040.

23-5 From LeVay, S., T. N. Wiesel, and D. H. Hubel. 1980. The development of ocular dominance columns in normal and visually deprived monkeys. *Journal of Comparative Neurology* 191:1–51. Copyright (c) 1980 John Wiley & Sons. Reprinted by permission of Wiley-Liss, a subsidiary of John Wiley & Sons, Inc.

23-6 Adapted from Wiesel, T. N., and D. H. Hubel. 1974. Ordered arrangement of orientation columns in monkeys lacking visual experience. *Journal of Comparative Neurology* 158:307–318.

23-7 From Murphey, R. K., S. G. Matsumoto, and B. Mendenhall. 1976. Recovery from deafferentation by cricket interneurons after reinnervation by their peripheral field. *Journal of Comparative Neurology* 169:335–346. Copyright (c) 1976 John Wiley & Sons. Reprinted by permission of Wiley-Liss, a subsidiary of John Wiley & Sons, Inc.

23-8(A) Adapted from Merzenich, M. M., J. H. Kaas, M. Sur, and C.-S. Lin. 1978. Double representation of the body surface within cytoarchitectonic areas 3b and 1 in "S1" in the owl monkey (*Aotus trivirgatus*). *Journal of Comparative Neurology* 181:41–74.

23-8(B, C) Adapted from Merzenich, M. M., J. H. Kaas, J. T. Wall, M. Sur, R. J. Nelson, and D. J. Felleman. 1983. Progression of change following median nerve section in the cortical representation of the hand in areas 3b and 1 in adult owl and squirrel monkeys. *Neuroscience* 10:639–665.

23-9(A, B) Photographs courtesy of Rodman A. Bates and William T. Greenough, University of Illinois at Urbana-Champaign.

CHAPTER 25

25-A Photomicrograph courtesy of Lynn Riddiford and James W. Truman, University of Washington.

25-3 Reprinted from Truman, J. W. 1992. The eclosion system of insects. *Progress in Brain Research* 92:361–374 with kind permission from Elsevier Science—NL, Sara Burgerhartstraat 25, 1055 KV Amsterdam, The Netherlands.
25-4 From Truman, J. W., and S. E. Reiss. 1988. Hormonal regulation of the shape of identified motoneurons in the moth Manduca sexta. *Journal of Neuroscience* 8:765–775. Copyright (c) 1988 Society for Neuroscience; reprinted by permission.
25-5 From Truman, J. W. 1973. How moths "turn on": a study of the action of hormones on the nervous system. *American Scientist* 61:700–706. Reprinted by permission from Sigma Xi, the Scientific Research Society.
25-7 Adapted from Carter, C. S. 1992. Neuroendocrinology of sexual behavior in the female. In *Behavioral Endocrinology*, ed. J. B. Becker, S. M. Breedlove, and D. Crews, 71–95. Cambridge, Mass: MIT Press
25-9 Adapted from Pfaff, D. W. 1989. Features of a hormone-driven defined neural circuit for a mammalian behavior. *Annals of the New York Academy of Sciences* 563:131–147.
25-10 Adapted from Nottebohm, F. 1991. Reassessing the mechanisms and origins of vocal learning in birds. *Trends in Neurosciences* 14:206–211.
25-11 From Arnold, A. P. 1980. Sexual differences in the brain. *American Scientist* 68:165–173. Copyright (c) 1980 Arthur P. Arnold.

INDEX